Sulfur Metabolism in Plant Growth: Mechanisms and Applications

Sulfur Metabolism in Plant Growth: Mechanisms and Applications

Editor: Sasha Hendrickson

STATES
ACADEMIC PRESS
www.statesacademicpress.com

States Academic Press,
109 South 5th Street,
Brooklyn, NY 11249, USA

Visit us on the World Wide Web at:
www.statesacademicpress.com

ISBN: 978-1-63989-500-7 (Hardback)

Cataloging-in-Publication Data

Sulfur metabolism in plant growth : mechanisms and applications / edited by Sasha Hendrickson.
p. cm.
Includes bibliographical references and index.
ISBN 978-1-63989-500-7
1. Plants--Metabolism. 2. Plant physiology. 3. Growth (Plants). 4. Plants--Development.
5. Plants--Effect of anaerobiosis on. I. Hendrickson, Sasha.
QK881 .S85 2022
572.42--dc23

Table of Contents

Preface

In my initial years as a student, I used to run to the library at every possible instance to grab a book and learn something new. Books were my primary source of knowledge and I would not have come such a long way without all that I learnt from them. Thus, when I was approached to edit this book; I became understandably nostalgic. It was an absolute honor to be considered worthy of guiding the current generation as well as those to come. I put all my knowledge and hard work into making this book most beneficial for its readers.

Sulfur is a nutrient essential for plant growth. Sulfates are the primary source of sulfur for plants, which they take up from the rhizosphere. Gaseous sulfur dioxide is also absorbed by plants through the stomata and is converted into sulfite. All the inorganic sulfur absorbed by plants is assimilated by them by converting it into cysteine. This is again processed and changed into methionine. The secondary products containing sulfur have a characteristic smell and act as a defense mechanism for plants against herbivores, insects and other pathogens. These compounds also act as signalling molecules for fundamental cellular functions. From theories to research to practical applications, case studies related to all contemporary topics of relevance to this field have been included in this book. Also included herein is a detailed explanation of the various mechanisms and applications of sulfur metabolism in plant growth. This book is a vital tool for all researching or studying sulfur metabolism in plants as it gives incredible insights into emerging trends and concepts.

I wish to thank my publisher for supporting me at every step. I would also like to thank all the authors who have contributed their researches in this book. I hope this book will be a valuable contribution to the progress of the field.

Editor

1

The multi-protein family of sulfotransferases in plants: composition, occurrence, substrate specificity, and functions

Felix Hirschmann, Florian Krause and Jutta Papenbrock *

Institute of Botany, Leibniz University Hannover, Hannover, Germany

Edited by:
Stanislav Kopriva, University of Cologne, Germany

Reviewed by:
Masami Yokota Hirai, RIKEN Plant Science Center, Japan
Tamara Gigolashvili, University of Cologne – Biocenter, Germany
Frédéric Marsolais, Agriculture and Agri-Food Canada, Canada

****Correspondence:***
Jutta Papenbrock, Institute of Botany, Leibniz University Hannover, Herrenhäuser Straße 2, D-30419 Hannover, Germany
e-mail: jutta.papenbrock@botanik.uni-hannover.de

All members of the sulfotransferase (SOT, EC 2.8.2.-) protein family transfer a sulfuryl group from the donor 3′-phosphoadenosine 5′-phosphosulfate (PAPS) to an appropriate hydroxyl group of several classes of substrates. The primary structure of these enzymes is characterized by a histidine residue in the active site, defined PAPS binding sites and a longer SOT domain. Proteins with this SOT domain occur in all organisms from all three domains, usually as a multi-protein family. *Arabidopsis thaliana* SOTs, the best characterized SOT multi-protein family, contains 21 members. The substrates for several plant enzymes have already been identified, such as glucosinolates, brassinosteroids, jasmonates, flavonoids, and salicylic acid. Much information has been gathered on desulfo-glucosinolate (dsGl) SOTs in *A. thaliana*. The three cytosolic dsGl SOTs show slightly different expression patterns. The recombinant proteins reveal differences in their affinity to indolic and aliphatic dsGls. Also the respective recombinant dsGl SOTs from different *A. thaliana* ecotypes differ in their kinetic properties. However, determinants of substrate specificity and the exact reaction mechanism still need to be clarified. Probably, the three-dimensional structures of more plant proteins need to be solved to analyze the mode of action and the responsible amino acids for substrate binding. In addition to *A. thaliana,* more plant species from several families need to be investigated to fully elucidate the diversity of sulfated molecules and the way of biosynthesis catalyzed by SOT enzymes.

Keywords: *Arabidopsis thaliana*, glucosinolate, histidine residue, phosphoadenosine 5′-phosphosulfate, sulfotransferase

INTRODUCTION

Members of the sulfotransferase (SOT) family have been found in all organisms investigated to date. All of these enzymes catalyze the transfer of a sulfuryl group from 3′-phosphoadenosine 5′-phosphosulfate (PAPS) to an appropriate hydroxyl group (**Figure 1**), hydroxyl amine or unprotonated amine of various substrates with the parallel formation of PAP.

The SOTs catalyze the sulfation of a wide range of compounds and produce sulfate esters, sulfamates, and sulfate conjugates (Klaassen and Boles, 1997). A sulfate conjugate is more water soluble than a non-sulfated molecule (Weinshilboum and Otterness, 1994), thus facilitating excretion and bioactivation.

Due to the unifying use of the co-substrate PAPS, all SOT proteins are characterized by a histidine residue in the active site, defined PAPS binding sites and a defined SOT domain (Pfam: PF00685; Finn et al., 2014). Proteins with this SOT domain occur in all organisms from all three domains investigated so far, usually as a multi-protein family. Originally, the SOT proteins in mammals were classified on the basis of their affinity for different classes of substrates. One group of SOT proteins, mainly membrane-associated, accepts as substrates macromolecules, such as proteins and peptides, and glycosaminoglycans (Niehrs et al., 1994). The second group, usually soluble proteins, accepts as substrates small organic molecules, such as flavonoids, steroids, and xenobiotics, with diverse chemical structures. In plants, the best criteria for forming subgroups within the multi-protein family is still a matter of debate, because either sequence identity/similarity or their substrate specificity could be chosen. Several compounds have been found in different plant species, such as: brassinosteroids, coumarins, flavonoids, gibberellic acids, glucosinolates (Gls), phenolic acids, sulfate esters such as choline-*O*-sulfate, and terpenoids that might be sulfated by SOT proteins. However, only for some of these substrates has the catalyzing SOT protein been identified. Not all sulfated compounds are necessarily sulfated by SOTs. Sulfolipids contain a 6-deoxy-6-sulfoglucose sugar head group, referred to as sulfoquinovose. The sulfoquinovose precursor UDP-sulfoquinovose is biosynthesized from UDP-glucose by a UDP-sulfoquinovose synthase associated with a ferredoxin-dependent glutamate synthase using sulfite as cosubstrate (Shimojima et al., 2005). Much information has been gathered on desulfo-glucosinolate (dsGl) SOTs in *Arabidopsis*, differing in their affinity to indolic and aliphatic dsGls. However, determinants of substrate specificity and the exact reaction mechanism still need to be clarified. Probably, the three-dimensional structures of more plant proteins have to be solved to analyze the mode of action and the responsible amino acids for substrate binding.

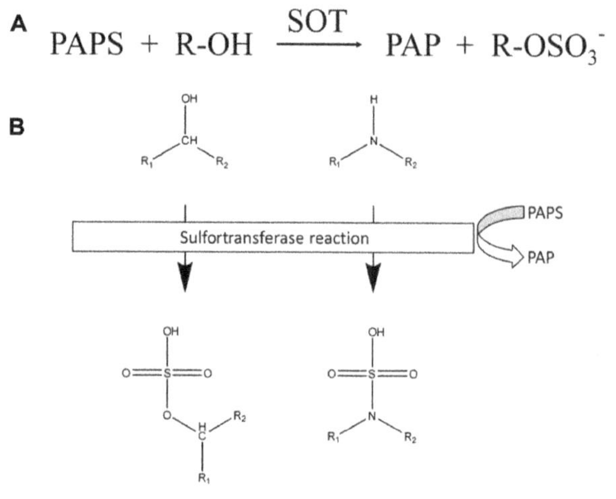

FIGURE 1 | Reactions catalyzed by SOTs. (A) Chemical equation of reactions catalyzed by SOTs. **(B)** Schematic overview of SOT targets. Chemical structure of targeted hydroxyl and amide groups and their sulfated products, sulfate ester and the sulfamate group.

PRIMARY STRUCTURE OF SOTS, PAPS BINDING REGIONS, AND ALIGNMENT OF THE HIGHLY CONSERVED REGIONS

Generally, SOTs can be divided into membrane-bound proteins and soluble cytosolic proteins. So far, only a few membrane-bound SOTs have been characterized in plants. They are either bound to the plasma membrane, as shown for the gallic acid glucoside SOT from *Mimosa pudica* L. (Varin et al., 1997a), or localized in the Golgi apparatus, as shown for the tyrosylprotein SOTs (TPSTs) from *Asparagus officinalis* L. (Hanai et al., 2000) and *Arabidopsis thaliana* (L.) Heynh. (Komori et al., 2009). The term cytosolic SOT might indicate a localization in the cytoplasm, yet the name implies that the proteins can be purified from plant cells and kept in solution (Hernàndez-Sebastiá et al., 2008). The exact localization of most cytosolic plant SOTs still remain unknown.

Sequence alignments of eleven cytosolic SOTs from plants, animals, and bacteria resulted in the identification of four highly conserved regions I to IV (Marsolais and Varin, 1995; **Figure 2**). Further analyses showed that especially the regions I and IV are highly conserved, for example throughout the SOT family of *A. thaliana* (Klein and Papenbrock, 2004). The regions I, II, and IV are responsible for the binding of the co-substrate PAPS (Varin et al., 1997b). The first structural approach to clarify the relevance of the regions for PAPS binding was determined by X-ray crystallography analyses of a mouse estrogen SOT (Kakuta et al., 1997). Region I is localized close to the N-terminus and includes the PAPS binding domain (PSB domain) that interacts with the 5′-phosphate of PAPS. Region II starts with a characteristic highly conserved histidine, responsible for proton acceptance during the sulfuryl transfer (Kakuta et al., 1998). In the C-terminal part of region II the two amino acids Arg130 and Ser138 are responsible for the binding of the 3′-phosphate of PAP and form a 3′ P-motif (Kakuta et al., 1997). This motif can be found in 18 SOTs out of 22 from *A. thaliana* and from almost all other plant SOTs. In many plant SOTs, a conserved hydrophilic site containing poly-glutamic acid (Poly-Glu) of unknown function can be found between region III and IV. Similar motifs have been found in a human chondroitin 6-SOT, but at a different position (Fukuta et al., 1998). Region IV is localized at the C-terminus and contains a P-loop related GxxGxxK motif (**Figure 2**).

In *A. thaliana*, 18 protein sequences with high similarity to known SOTs have been originally identified by BLAST approaches (Klein and Papenbrock, 2004). Later, another three SOTs were added (Klein and Papenbrock, 2008). These were formerly annotated in NCBI as "nodulation-related protein" and are now annotated as "P-loop containing triphosphate hydrolase family protein." In addition to these SOTs, a TPST has been identified (Komori et al., 2009). Furthermore, a not yet literarily mentioned protein, Q9SCR3, with a *Sulfotransfer_1* domain (PF00685), is available in the Pfam database [http://pfam.xfam.org/protein/Q9SCR3 (accessed 23.06.2014)]. About 75% of the amino acid sequence is identical to *A. thaliana* SOT (AtSOT19). Therefore, it might be a redundant entry or a product of a different splicing process. Interestingly, AtTPST is exceptional in its structure compared to the remaining AtSOTs. With 500 amino acids, it is not only bigger, but is also the solely identified transmembrane, *cis*-Golgi localized AtSOT. Furthermore, it shows no sequence similarity to human TPSTs; and no other typical features like the regions I to IV and the characteristic highly conserved histidine were identified (Komori et al., 2009). It is also the only *Arabidopsis* SOT that contains a *Sulfotransfer_2* domain (PF03567), instead of a *Sulfotranfer_1* domain (PF00685). Hence, it is only associated by function and not by sequence. Excluding the pseudogenic sequence *AtSOT2* and TPST, the SOT protein lengths range between 273 and 403 amino acids with an average length of 321 amino acids. Only seven out of 21 *AtSOTs* contain introns.

There are several nomenclatures for *A. thaliana* SOTs used in the literature. The most common ones are listed in **Table 1**, including information about the preferred SOT substrates. In this review, the nomenclature first introduced by Klein and Papenbrock (2004) is used.

So far, only one SOT from plants (AtSOT12 from *A. thaliana*) was structurally solved (Smith et al., 2004). Therefore, most SOT proteins lack structural analyses and detailed enzymological characterizations. The identified motifs only give a hint on the proteins' general function as a SOT, but no information about their specificity and affinity toward certain substrates.

Most proteins identified as putative SOTs contain at least one out of seven related Pfam motifs that are based on Hidden Markov Models (HMM). The most important HMMs referring to SOTs are the SOT domains *Sulfotransfer_1* (PF00685), *Sulfotransfer_2* (PF03567), and *Sulfotransfer_3* (PF13469), which have an average length of 230.1, 218.3, and 224 amino acids, respectively. According to the model information in the Pfam database, the SOT domain 1 shows an average coverage of its contributing protein sequences of 64%. An average of 16% of all amino acid residues that are covered by the HMM are identical to it. The average coverage of the SOT domains 2 and 3 in their respective sequences are 67 and 47%, with average sequence identities of 15 and 14%. In addition, HMMs for more specific SOT subfamilies have been

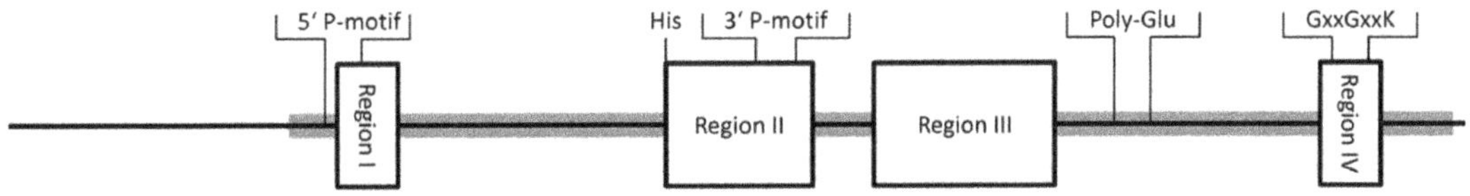

FIGURE 2 | Conserved regions I to IV of plant SOTs. Regions are shown as boxes, with size and position in the protein relative to the average size of SOTs from *Arabidopsis thaliana*. Functional amino acids were obtained from structural analyses of mouse SOTs as described above. The PAPS binding regions (5′P-motif, 3′P-motif, GxxGxxK), the proton acceptor histidine (His), and a poly-glutamic acid (Poly-Glu), that can be found in many plant SOTs, are labeled. The position of the Pfam domain PF00685 is shaded in gray.

Table 1 | Summary of the members of the SOT family in *Arabidopsis* and their putative substrates.

NCBI accession	*Arabidopsis* gene ID	Nomenclature I	Nomenclature II	Nomenclature III	Amino acids	Preferred substrate	Reference
NP_199182	AT5G43690	AtSOT1		AtSULT202B4	331		
NP_190689	AT3G51210	AtSOT2		Pseudogene	67		
NP_194358	AT4G26280	AtSOT3		AtSULT202C1	314		
NP_180325	AT2G27570	AtSOT4		AtSULT202B3	273		
NP_190093	AT3G45070	AtSOT5	AtST3a	AtSULT202B1	323	Flavonol	Gidda and Varin (2006), Hashiguchi et al. (2013)
NP_190094	AT3G45080	AtSOT6	AtST3b	AtSULT202B2	329		
NP_174139	AT1G28170	AtSOT7		AtSULT202B8	326		
NP_172799	AT1G13420	AtSOT8	AtST4b	AtSULT202B7	331	Flavonol glycosides	Hashiguchi et al. (2014)
NP_172800	AT1G13430	AtSOT9	AtST4c	AtSULT202B5	351		
NP_179098	AT2G14920	AtSOT10	AtST4a	AtSULT202B6	333	Brassinosteroids	Marsolais et al. (2007)
NP_565305	AT2G03750	AtSOT11		AtSULT202D1	351		
NP_178471	AT2G03760	AtSOT12	AtST1	AtSULT202A1	326	Flavonone, brassinosteroids, salicylic acid	Lacomme and Roby (1996), Marsolais et al. (2007), Baek et al. (2010), Hashiguchi et al. (2013)
NP_178472	AT2G03770	AtSOT13		AtSULT202E1	324	Flavonol	Hashiguchi et al. (2013)
NP_196317	AT5G07000	AtSOT14	AtST2b	AtSULT203A2	347		
NP_568177	AT5G07010	AtSOT15	AtST2a	AtSULT203A1	359	Hydroxyjasmonate	Gidda et al. (2003)
NP_177550	AT1G74100	AtSOT16	AtST5a	AtSULT201B3	338	Phenylalanine and tryptophan derived dsGls	Piotrowski et al. (2004), Klein et al. (2006)
NP_173294	AT1G18590	AtSOT17	AtST5c	AtSULT201B2	346	Benzyl and methionine derived dsGls	Piotrowski et al. (2004), Klein et al. (2006)
NP_177549	AT1G74090	AtSOT18	AtSTb	AtSULT201B1	350	Phenylalanine and methionine derived dsGls	Piotrowski et al. (2004), Klein et al. (2006), Luczak et al. (2013)
NP_190631	AT3G50620	AtSOT19			340		
NP_179175	AT2G15730	AtSOT20			344		
NP_195168	AT4G34420	AtSOT21			403		
NP_563804	AT1G08030	TPST			500	Tyrosylprotein	Komori et al. (2009)

I: Nomenclature used in this review introduced by Klein and Papenbrock (2004). II: Nomenclature introduced by Piotrowski et al. (2004). III: Nomenclature introduced by Hashiguchi et al. (2013). The term TPST was introduced by Komori et al. (2009).

deposited in Pfam. Generally, they show a lower number of hits in the database, with mostly increased sequence coverage and a higher average sequence identity as compared to the more general SOT domains 1–3. The aryl SOT domains *Arylsulfotrans_1* and *Arylsulfotrans_2* (PF05935 and PF14269) and the Stf0 SOT domain *Sulphotransf* (PF09037) represent groups of sequences with more specific occurrence, especially in prokaryota. The two aryl SOT domains show an average coverage of 83 and 57%, with an identity of 30 and 28%, respectively. For the Stf0 SOT domain the average sequence coverage is 81%, with an average identity of 33%. The galactose-3-*O*-SOT domain (PF06990) shows coverage of 78% with 24% identity.

SULFOTRANSFERASE FAMILIES IN DIFFERENT PLANT GENOMES

Sulfotransferases have a broad range of substrates and therefore many functions. In previous literature it has been stated that SOTs are present in all kingdoms except in Archaea (Klein and Papenbrock, 2008; Chen et al., 2012). Nevertheless, according to the protein family database Pfam [http://pfam.xfam.org/ (accessed 23.06.2014)] there are Archaea sequences with a characteristic conserved SOT domain.

There are only few studies that aim to identify all SOTs of a plant species. A requirement to do this is a fully sequenced genome, but due to their eclectic functions, we assume that SOTs are present in almost every plant species. The Pfam database already stores 538 putative plant SOT sequences, 459 of which have a *Sulfotransfer_1* domain (PF00685), 49 a *Sulfotransfer_2* domain (PF03567), 16 a *Sulfotransfer_3* domain (PF13469), and 10 a *Sulphotransf* domain (PF09037). The actual number might be less, because of redundant entries. The *Sulfotransfer_3* (PF13469) and the *Gal-3-O_sulfotr* domain (PF06990) are only present in algae. The *Arylsulfotrans* domain (PF05935) is not present in plants, while *Arylsulfotran_2* (PF14269) is only found in a single *Ricinus communis* L. sequence.

While so far 22 putative *A. thaliana* SOTs were identified, 35 genes coding proteins with a SOT domain were reported in *Oryza sativa* L., including six genes likely to be pseudogenes (Chen et al., 2012). In phylogenetic analyses, they are clustered into seven subfamilies. However, microarray data revealed that the genes within subfamilies are expressed in a different manner, indicating individual functions. When 17 AtSOTs were added to the distance trees, they did not group together with any of the *O. sativa* genes. This was taken as a hint for independent evolution of *O. sativa* and *A. thaliana* SOTs by gene duplication or loss. This was supported by the finding that half of the *O. sativa SOTs* contain introns, which is hardly the case for *AtSOTs* (Klein and Papenbrock, 2004).

Comparative genomics studies were conducted in *Brassica rapa* L. with *A. thaliana*, in order to identify all Gl biosynthesis genes (Zang et al., 2009; Wang et al., 2011). Thirteen putative *desulfo-glucosinolate SOT* (*dsGl SOT*) genes were identified. Two genes are paralogs of *AtSOT16*, 1 of *AtSOT17*, and 10 of *AtSOT18*. One *AtSOT18* paralog appears to be nonfunctional, because of transposon insertion, and one carries a frame shift. None of the genes contains introns, as it is the case for *AtSOT* genes. All paralogs share at least 70% sequence identity with their *AtSOT* counterparts, with the exception of one SOT from *B. rapa* (*BrSOT18*, 68%; Wang et al., 2011). The higher number of *BrSOTs* is explained by the triplication of the *B. rapa* genome and later duplication, transposition, or tandem duplication of the genes.

In *Brassica napus* L., so far only twelve putative genes encoding SOTs were identified (Rouleau et al., 1999; Marsolais et al., 2000). Additionally, there are at least five isoforms of dsGl SOTs in *B. napus*, which have similar substrate affinities as their *A. thaliana* homologs (own unpublished results). Regarding that *B. napus* is an allotetraploid species formed by the hybridization of *B. rapa* and *B. oleracea*, a much higher number of *SOT* genes can be expected.

To group these diverse enzymes into families and subfamilies remains a difficult task. Klein and Papenbrock (2004, 2008) ordered 21 *A. thaliana* SOTs in eight groups, according to their amino acid sequence identity. However, already characterized SOTs with the same substrate specificity did not group together, and even high sequence identity of more than 85% among two SOTs did not reveal equal enzymological characteristics. Neither sequence identity, nor generated trees ordered already characterized SOTs in groups according to their substrate specificities. Three dsGl SOTs were on one separate branch, but flavonoid and brassinosteroid SOTs could not be distinguished.

Hernàndez-Sebastiá et al. (2008) generated a phylogenetic tree including 78 SOTs from 13 different plant species. This approach faced the same problems as the one by Klein and Papenbrock (2004, 2008) and it was again concluded that the prediction of SOT substrates by high primary sequence identities is limited. For example, it was speculated that AtSOT13 was a brassinosteroid SOT, because of its close distance to AtSOT12, but Hashiguchi et al. (2013) showed that AtSOT13 uses flavonoids as preferred substrates.

Another attempt included, besides 17 *A. thaliana*, also *B. napus* and *Flaveria* spp. sequences (Hashiguchi et al., 2013). *AtSOT2* was excluded, because it is most likely a pseudogene. According to a dendrogram, three families were defined with two, three and five subfamilies, respectively. The families had an amino acid sequence identity of at least 45% and the subfamilies of at least 60%. But again, except for the dsGl SOTs, the SOTs did not group together according to their substrate specificities.

Labonne et al. (2009) tried to identify a putative *SOT* of *Turnera krapovickasii* Arbo (Passifloraceae) by phylogenetic analysis. The sequence was aligned with 28 SOTs from *A. thaliana*, *B. napus*, *Vitis vinifera* L., *O. sativa*, *Hordeum vulgaris* L., *Populus trichocarpa* Hook. The SOT from *T. krapovickasii* was on a branch by itself and alignments with characterized SOTs revealed low sequence identity. Therefore, it was not possible to identify the function of the respective SOT.

Overall, past attempts indicate that it is difficult to order plant SOTs according to their amino acid sequence. Only for dsGl SOTs does it seem to be possible, because they are clustered together on a separate branch in all approaches. Therefore, only enzymatic assays with additional mutational studies can give reliable information about substrate specificity and function.

SUBSTRATES FOR SULFOTRANSFERASES

BIOSYNTHESIS OF THE CO-SUBSTRATE PAPS

3′-phosphoadenosine 5′-phosphosulfate is an obligate co-substrate for sulfation reactions catalyzed by SOTs. In plants, PAPS does not represent an intermediate of reductive sulfate assimilation as in fungi and some bacteria, but it seems to play an exclusive role as a sulfuryl donor for sulfation reactions. PAPS is synthesized from ATP and sulfate in a two-step reaction (**Figure 3**). In the first step, ATP sulfurylase (EC 2.7.7.4) catalyzes sulfate activation. The enzyme hydrolyses the bond between the β- and the γ-phosphates of ATP and then adds sulfate to the γ-phosphate. The activation step is necessary, because sulfate is metabolically inert. The energy is stored in the phosphoric acid-sulfuric anhydride bond of the reaction product, adenosine 5′-phosphosulfate (APS), allowing sulfate to undergo further reactions. The energetic balance of the sulfate adenylylation reaction favors ATP formation. Therefore, the reaction products, APS, and pyrophosphate (PP_i), need to be maintained at a low concentration by the enzymes inorganic pyrophosphatase that hydrolyses PP_i, APS reductase (EC 1.8.4.9) and APS kinase (EC 2.7.1.25; AKN) that metabolize APS. APS reductase catalyzes the first step of sulfate reduction. APS kinase catalyzes the ATP-dependent phosphorylation on the 3′-position of APS. *In vitro* tests have shown that excess APS inhibits APS kinase. The product PAPS is the substrate for the SOT proteins.

In general, the availability of PAPS for sulfation *in vivo* depends on its synthesis, transport, degradation, and utilization as investigated in mammals (Klaassen and Boles, 1997). Recently, it was shown that the transporter PAPST1 in the chloroplast envelope membrane is not only involved in the provision of PAPS for the extraplastidic sulfation reactions, but is also capable to transport PAP in an antiport manner. The loss of *PAPST1* leads to a decreased production of sulfated compounds like Gl, increased production of dsGl, and the modulation of primary sulfate assimilation, another indication for the strong interconnectedness of primary and secondary sulfur metabolism (Gigolashvili et al., 2012). The by-product of the sulfation reaction, PAP, has gene regulatory attributes. In turn, PAP is regulated by the adenosine bisphosphate phosphatase SAL1 that dephosphorylates PAP to adenosine monophosphate. 3′-Phosphoadenosine 5′-phosphate accumulates at drought stress and high light conditions. Mutational studies indicated that PAP inhibits 5′–3′ exoribonucleases in the cytosol and nucleus, which causes changes in expression of stress-responsive genes. It was suggested that a PAP-SAL1 retrograde pathway alters gene expression as part of the stress response

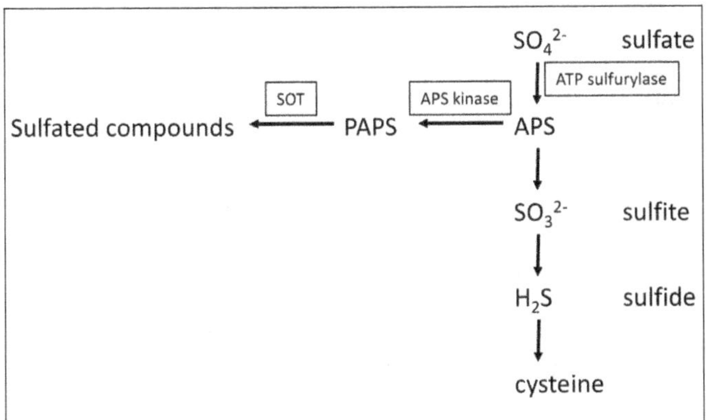

FIGURE 3 | Biosynthesis of APS and PAPS and the connection of primary and secondary sulfur metabolism.

(Estavillo et al., 2011). Additionally, a correlation between the increases of PAP with changes in the sulfur metabolism was reported. Further analysis of *sal1* knock out mutants led to the conclusion that changes of gene expression due to sulfur limitation is triggered by internal sulfur deficiency and not by low external sulfur levels. PAP accumulation also resulted in an increase of enzymatic oxygenation of fatty acids, an increase of jasmonic acid synthesis and a decrease of Gls. Possible explanations for the Gl decrease could be inhibition of dsGl SOTs or the disruption of PAPS transport from plastids to the cytosol (Gigolashvili et al., 2012; Lee et al., 2012).

SUBSTRATES FOR PLANT SOTs

The first isolated and characterized plant SOTs were flavonol 3′- and flavonol 4′-SOT of *Flaveria chloraefolia* (Varin et al., 1992) and later of *F. bidentis* (L.) Kuntze (Varin et al., 1997b). These SOTs sequentially sulfate specific hydroxyl groups of the flavonol quercetin to quercetin tetrasulfate. Flavonol biosynthesis was demonstrated to be regulated by auxin and ethylene. In turn, the flavonol quercetin and quercetin sulfates affect root development processes, such as the basipetal root auxin transport, elongation growth, and gravitropism (Faulkner and Rubery, 1992; Lewis et al., 2011).

So far, four flavonoid SOTs have been characterized in *A. thaliana*: AtSOT5, AtSOT8, AtSOT12, and AtSOT13 (see **Table 1** for details). Hashiguchi et al. (2013) compared characteristics and substrate specificities of AtSOT5, AtSOT12, and AtSOT13.

AtSOT13 and AtSOT5 showed the highest activity with the flavonol galangin (3,5,7-trihydroxy-2-phenyl-4H-chromen-4-one), while AtSOT12 showed the highest activity for the flavanone naringenin [(2S)-5,7-dihydroxy-2-(4-hydroxyphenyl)-2,3-dihydro-4H-chromen-4-one] and was the only SOT that sulfates anthocyanidin. Interestingly, the AtSOTs showed no or comparably low activity with quercetin. It was speculated that the position-3 hydroxyl group of quercetin inhibits the catalytic activity. It was also shown that only AtSOT12 is able to use 3-hydroxyflavone as substrate, while 7-hydroxyflavone is used by all three AtSOTs. After comparisons of K_m-values using kaempferol as substrate, it was concluded that particular hydroxyl groups of kaempferol are specifically sulfated by the AtSOTs.

AtSOT8 was also characterized by Hashiguchi et al. (2014). The pH optimum at 5.5 was lower than for previously characterized flavonoid SOTs. Thus it was speculated that AtSOT8 might be located in the vacuole. Comparison of V_{max}/K_m-values showed that AtSOT8 prefers flavonol glycosides instead of their aglycone counterparts as substrates. Also, there was only activity to flavonoids with a hydroxyl group at position 7. Hence, it was suggested that AtSOT8 might be a flavonol glucoside-7 SOT. Surprisingly, neither sulfated glucoside flavonoids could be detected *in vivo* in *A. thaliana* by LC/MS, nor were there any accordant database entries. Possible explanations for the non-detected sulfated glucoside flavonoids could be low or condition-dependent occurrence.

AtSOT10 showed activity with brassinosteroids (Marsolais et al., 2007), specifically brassinosteroid biosynthetic end products. In summary, it was speculated that it inactivates brassinosteroids and therefore is involved in plant development processes. In numerous studies, overexpression of brassinosteroid catabolic genes led to brassinosteroid-deficient phenotypes. However, overexpression and T-DNA insertion in null mutants of AtSOT10 did not show brassinosteroid-deficient phenotypes, emphasizing difficulties of transferring *in vitro* results to *in vivo* insights (Sandhu and Neff, 2013).

Of all investigated *A. thaliana* SOTs so far, AtSOT12 has the broadest substrate specificity. Besides using flavonoids as substrates, it was also shown to be active with brassinosteroids and salicylic acid. Within the brassinosteroids, it showed preference for 24-epibrassinosteroids (Marsolais et al., 2007). It was stereospecific for 24-epibrassinosteroids and accepted mammalian hydroxysteroids and estrogens, too. The most preferred substrate was the metabolic precursor 24-epicathasterone ($K_m = 6.9$ μM), which showed inhibitory effects above 5 μM. The K_m-value for salicylic acid is comparably high (440 μM; Baek et al., 2010). Salicylic acid is a signal molecule in plant defense, and cellular concentrations increased up to 40 μM after pathogen infection indicating that sulfation of salicylic acid is a response to pathogen attack. This theory was supported by the fact that *atsot12* knock out mutants were less resistant to the pathogen *Pseudomonas syringae*, while AtSOT12 overexpressing lines showed a higher resistance.

Two *B. napus* brassinosteroid isoforms, BNST3 and BNST4, were enzymatically characterized. Recombinant BNST3 stereospecifically sulfated 24-epibrassinosteroids and preferred 24-epicathasterone ($K_m = 1.4$ μM), which is a biosynthetic intermediate of 24-epibrassinolide. Because of the biological inactivity of 24-epibrassinolide sulfate, it was hypothesized that BNST3 is involved in brassinosteroid inactivation (Rouleau et al., 1999). BNST4 also preferred 24-epibrassinosteroids ($K_m = 4.9$ μM), but also showed a broad substrate specificity with other steroids, also indicating a role in detoxification (Marsolais et al., 2004). Overall, they showed similar substrate specificities toward brassinosteroids as AtSOT12 (Marsolais et al., 2007).

AtSOT15 specifically sulfates 11- and 12-hydroxyjasmonate, which is a signaling molecule in plant defense and development. K_m-values indicate a higher affinity to 12- than to 11-hydroxyjasmonate (10 μM and 50 μM, respectively). 12-hydroxyjasmonate naturally occurs in *A. thaliana* and it was suggested that sulfation might function in inactivation of 12-hydroxyjasmonic acid (Gidda et al., 2003).

Komori et al. (2009) identified a 62 kDa, Golgi-localized, transmembrane protein, that sulfates tyrosylproteins in *A. thaliana*. The recombinantly expressed TPST sulfated tyrosine residues of precursor polypeptides of the "plant peptide containing sulfated tyrosine 1" (PSY1) and phytosulfokine (PSK). PSY1 and PSK are peptide hormones, which promote growth and cell proliferation (Matsubayashi and Sakagami, 1996; Amano et al., 2007). The activity with both substrates indicates broad substrate specificity. TPST showed a higher activity with PSY1, which was explained by a closer proximity of an acidic region to the sulfated tyrosine residue. *TPST* loss-of-function mutants showed numerous abnormal attributes, which led to the conclusion that sulfated peptides or proteins are involved in plant growth and development. Previously, in microsomal membrane preparations from carrot cells, rice, and asparagus TPST activity was shown (Hanai et al., 2000). In rice, the K_m-value was 71 μM at a pH of 7.0–8.5 in the presence of manganese ions. The enzyme kinetic values such as K_m and V_{max} *A. thaliana* TPST remain to be determined. TPST also sulfates peptide root meristem growth factors (RGFs), which are involved in postembryonic root development. Loss-of-function *tpst-1* mutants showed reduction in root meristem size and loss of coordination between cell elongation and expansion in the elongation–differentiation zone. Addition of RGF restored the meristem activity to ~70% and addition of RGF, PSK, and PSY1 restored the activity comparable to the wild-type. Sulfation of RGFs was found to be critical for its function. Further experiments showed that RGFs positively regulate the expression of PLETHORA transcription factors that mediate the pattern of the root stem niche (Matsuzaki et al., 2010).

Other examples of already characterized SOTs like a choline-*O*-sulfate SOT of the halophytic *Limonium* species (Rivoal and Hanson, 1994) and a plasma membrane-associated gallic acid SOT of *M. pudica* L. (Varin et al., 1997a) undercut the diversity of substrates and functions of these enzymes. Choline sulfate is an osmolyte that accumulates under saline conditions. The respective choline-*O*-SOT showed a fourfold higher activity under high salinity. The choline-*O*-SOT had its pH optimum at 9.0 and the K_m-value for choline was 25 μM. The 42 kDa membrane bound gallic acid SOT might be involved in the regulation of the seismonastic response. It showed strict substrate specificity and a K_m-value of only 3.0 μM.

Further studies indicate the existence of more SOTs, even though they were not especially isolated or characterized. It was shown *in vivo* that poplar trees convert hydroxylated metabolites of polychlorinated biphenyls (PCBs) into sulfated PCB. It was suggested that SOTs catalyze this reaction (Zhai et al., 2013). Sulfated polysaccharides occur in marine angiosperms, mangroves (Aquino et al., 2005, 2011), freshwater plants (Dantas-Santos et al., 2012), and algae (Ngo and Kim, 2013), which are likely to be sulfated by not yet identified SOTs.

GLUCOSINOLATES: PRODUCTS OF THE SOT REACTION

Glucosinolates are a group of over 200 nitrogen- and sulfur-containing natural products found in vegetative and reproductive tissues of 16 plant families within the Capparales (Clarke, 2010). They are well-known as the major secondary metabolites in agriculturally important crop plants of the Brassicaceae family, such as oilseed rape (*B. napus*), fodder and vegetables (e.g,. broccoli and cabbage). The model plant *A. thaliana.* Gl share a core structure containing a β-D-glucopyranose residue linked via a sulfur atom to a (*Z*)-*N*-hydroximino sulfate ester. They are distinguished by a variable R group derived from one of several amino acids, mainly tryptophan, phenylalanine and methionine (Mithen, 2001). The Gl pattern varies among the plant species and among *A. thaliana* ecotypes. In 39 *A. thaliana* ecotypes, 34 different Gls have been identified. Quantity and composition of Gls depend on the developmental stage of the plants and on the plant organ (Kliebenstein et al., 2001a).

Intact Gls are not toxic to cells. However, after cell damage Gls are hydrolyzed, catalyzed by thioglucosidase enzymes ("myrosinase"), to produce a variety of volatile hydrolysis products, such as thiocyanates, isothiocyanates, and nitriles. Only these breakdown products have a wide range of biological activities including both negative and positive effects (Fenwick and Heaney, 1983). In several studies these breakdown products were shown to be involved in plant defense against pathogens and herbivores. Thus, Gls are the best-characterized preformed defense compounds in the Brassicaceae and contribute to the protection against pathogens of the generalist type (Rausch and Wachter, 2005).

The last step in the Gl core structure biosynthesis of the different aliphatic, aromatic, and indole desulfo (ds) Gls is catalyzed by members of the SOT family. Glendening and Poulton (1990) partially purified a protein from *Lepidium sativum* L. that had PAPS-dependent dsGl SOT activity; however, at that time no molecular data was available. Later it was shown that three SOT proteins from *A. thaliana* are involved in Gl biosynthesis catalyzing the sulfation of dsGls to the intact Gls (Varin and Spertini, 2003; Piotrowski et al., 2004; Hirai et al., 2005).

Sulfotransferases involved in sulfation of desulfo-glucosinolates

The three dsGl AtSOT proteins (AtSOT16, AtSOT17, and AtSOT18) were predicted and then verified by different means (screening of many sulfated compounds, combining of knowledge, and integration of metabolomics and transcriptomics) for being responsible for the sulfation of dsGl (Varin and Spertini, 2003; Piotrowski et al., 2004; Hirai et al., 2005). Up till now, it was not unambiguously demonstrated why multiple *dsGl SOT* genes have been conserved during evolution in *A. thaliana* and in other Brassicaceae species.

The glucosylation and the sulfation reactions were assumed to be non-specific with respect to the side chain (Halkier, 1999). It is also hypothesized that first the side chains are elongated to synthesize so-called parent Gls, then the glycone moiety is developed and finally, several side chain modifications take place to produce the respective daughter Gl (Wittstock and Halkier, 2002). However, it is not clarified when the dsGls are sulfated by SOT proteins and whether there is a specificity for certain parent or daughter Gls. As the Gl pattern differs among *A. thaliana* ecotypes (Kliebenstein et al., 2001a), the investigation of the three dsGl SOTs from ecotype C24, which shows the broadest variety of Gls in comparison to other ecotypes, was most rational. In addition, one exemplary SOT from the fully sequenced ecotype Col-0 was investigated. To determine if and how these three dsGl SOT proteins might influence the Gl pattern, different *in vitro* enzyme assays were performed. Substrate specificity varies among the three proteins in the same ecotype (C24) and between ecotypes (C24 versus Col-0). AtSOT16 (C24) has the broadest substrate specificities. Tryptophan and phenylalanine-derived dsGl are the most preferred substrates, but it also accepts methionine-derived dsGl of chain length C3, C4, C5, C6, C7, and C8, although at much lower activities. AtSOT17 (C24) has narrow substrate specificities and does not act upon tryptophan-derived dsGl. Phenylalanine-derived benzyl dsGl is the most preferred substrate, but it also accepts methionine-derived dsGls, but has a strong preference for longer side chains, C6, C7, and C8. AtSOT18 (C24) also has narrow substrate specificities, does not act upon tryptophan-derived dsGl. It accepts phenylalanine-derived dsGl and methionine-derived dsGl, but has a strong preference for longer side chains, C6, C7, and C8. In summary, these three enzymes differ significantly in their affinity for the investigated substrates and the co-substrate PAPS (Klein et al., 2006; Klein and Papenbrock, 2009). It was speculated that the differences between *AtSOT16-18* could be an explanation for the different Gl patterns between organs, developmentally stages and growth environments reported by Brown et al. (2003). Anyhow, Møldrup et al. (2011) transformed genes involved in Gl biosynthesis into tobacco, thus successfully enabling it to synthesize Gls. In this approach, they could show that SOTs are not the bottleneck of Gl synthesis, but the supply of the co-substrate PAPS. Therefore, regulation of AtSOT16–18 could be, only taken together with other Gl and PAPS biosynthesis genes, partly responsible for Gl variation.

Up to now, the knowledge on secondary modifications of parent Gls is limited (Graser et al., 2001). In future work it could be interesting to verify the general acceptance that parent dsGls are sulfated before secondary modifications of Gls take place (Kliebenstein et al., 2001b). However, assuming the general acceptance is right, no secondarily modified Gls would exist in a ds form to interact with the SOTs. Therefore, it is possible, that artificially de-sulfated Gls with secondary modifications are sulfated *in vitro*, but with no *in vivo* relevance.

In A. thaliana ecotypes SOT18 proteins differ in their sequence and substrate specificity

It was shown that AtSOT18 proteins from two different *A. thaliana* ecotypes differ in their kinetic parameters as well as their substrate specificities. The primary structure of AtSTO18 proteins from the ecotypes Col-0 and C24 differ in two amino acids (Klein et al., 2006; Klein and Papenbrock, 2009). One could assume that there could be a correlation of AtSOT18 enzyme activities and differences in Gl profiles between these ecotypes. Therefore, AtSOT18 sequences from eight *A. thaliana* ecotypes with highly diverse Gl patterns were investigated: The AtSOT18 sequence from Col-0 showed the highest similarity to the largest number of other sequences in the alignment. The AtSOT18 proteins showed sequence deviations of maximal two amino acids in comparison to the AtSOT18 sequence from Col-0. The positions of the amino acid replacements were different in each sequence. The small differences in the primary sequence lead to important structural changes in secondary and tertiary structure that might be the key for different kinetic activities toward a broad range of substrates (Luczak et al., 2013). All recombinant AtSOT18 proteins showed low substrate specificity with an indolic Gl, while the specificity for aliphatic substrates varied. There was no correlation in the kinetic behavior with the major dsGl contents or with the ratio of C_3/C_4 dsGl in the respective ecotype. Therefore, it is unlikely that dsGl AtSOT18 enzymes play a major role in shaping the Gl profile in *A. thaliana* (Luczak et al., 2013). Interestingly, in humans, inter-individual variation in sulfation capacity may be important in determining an individual's response to xenobiotics, and recent studies have begun to suggest roles for SOT polymorphism in disease susceptibility (Gamage et al., 2006). Variations in concentration and composition of Gls in *A. thaliana* ecotypes and different environmental conditions can still neither be fully explained, nor predicted.

EXPRESSION OF SULFOTRANSFERASES

Sulfated compounds are mainly linked to biotic and abiotic stress response. This is supported by several expression studies of *SOTs*. So far, the mRNA levels of most characterized *A. thaliana SOTs* were rather low under normal growth conditions (Lacomme and Roby, 1996; Gidda et al., 2003; Piotrowski et al., 2004). This is supported by the fact that there is a relatively low number of *SOTs* in EST databases with the exception of *AtSOT15* and *AtSOT16* (Klein and Papenbrock, 2008). However, the expression of *AtSOT12*, *AtSOT15*, *AtSOT16*, and *AtSOT17* was significantly increased by treatment with jasmonate (Lacomme and Roby, 1996; Gidda et al., 2003; Piotrowski et al., 2004).

Only 17 *AtSOTs* were found to be present on 24 k Affymetrix chips and for many of those the absolute signal was quite low (Klein and Papenbrock, 2008). Hashiguchi et al. (2013, 2014) reported that microarray database research suggested that *AtSOT8* is mainly expressed in roots, while At*SOT13* is expressed in the early stages of the embryonic development. Interestingly, Hashiguchi et al. (2013) cloned *AtSOT13* from 2-week old seedlings.

Transcripts of *AtSOT10* were mainly detected in roots. Transcript levels were repressed 4 h after *trans*-zeatin treatment. After 8 h no transcripts were detectable anymore by qRT-PCR (Marsolais et al., 2007).

Northern Blot analysis revealed that *AtSOT12*, the encoded protein uses brassinosteroids, flavonoids, and salicylic acid as substrates, is moderately expressed in roots and leaves, and highly in flowers, while no expression was detected in stems and siliques. Furthermore, it was strongly induced by salt and sorbitol and slightly by cold, ABA, auxins, cytokinins, methyl jasmonate, salicylic acid, and interactions with bacterial pathogens (Lacomme and Roby, 1996; Baek et al., 2010). These results strongly indicate a function of *AtSOT12* in stress and hormone response. Similar results were obtained for the respective homologous genes in *B. napus*. *BNST3* and *BNST4* mRNA levels were quite low, but increased after treatment with salicylic acid, ethanol, xenobiotics, low oxygen stres, and the herbicide safener naphtalic anhydride (Rouleau et al., 1999; Marsolais et al., 2004). *BNST3* and *BNST4* induction also indicates a function in stress response and detoxification.

The protein encoded by *AtSOT15* uses hydroxyl jasmonate as substrate. Expression was induced upon methyljasmonate and 12-hydroxyjasmonate treatment. Probably, it inactivates the function of jasmonic acids and therefore enhances the hypocotyl growth (Gidda et al., 2003). Yamashino et al. (2013) showed that *AtSOT15* transcription is also regulated by an external coincidence mechanism. Database research indicated that *AtSOT15* might be a target of the phytochrome interacting transcriptional factors PIF4 and PIF5. PIF4 and PIF5 are controlled by the circadian clock, but also independently influenced by light and temperature. Further qRT-PCR analysis showed that *AtSOT15* was diurnally regulated by PIF4 and PIF5 at the end of a short day dark phase and/or high temperatures. Accordingly, *AtSOT15* is induced under conditions when hypocotyl growth takes places.

At first, the dsGl SOTs *AtSOT16-18* were reported to be constitutively expressed in all leaves, flowers, and siliques (Varin and Spertini, 2003). Later Northern Blot analysis revealed that *AtSOT16* mRNA level increased after treatment with coronatine (an analog of octadecanoid signaling molecules), jasmonic acid precursor 12-oxophytodienonic, ethylene precursor ACC and after treatment with jasmonic acid. UV-C illumination and wounding also induced *AtSOT16* expression. *AtSOT17* mRNA increased 2.4 fold and 1.2 fold, respectively, while *AtSOT17* expression only slightly increased (1.3 fold) after coronatine treatment (Piotrowski et al., 2004). Regarding the developmental stages, *AtSOT16* and *AtSOT17* mRNA levels were highest in two week old seedlings and lowest in flowering plants. In contrary, *AtSOT18* levels were quite low in young plants and slightly increased after 5–6 weeks. Only *AtSOT17* expression was influenced by a 12 hour dark / 12 hour light cycle. It was the highest at the end of the light phase and the lowest at the end of the dark phase. No differences in any of the three mRNA levels were detected, when *A. thaliana* was grown in media with tenfold sulfate concentration (Klein et al., 2006).

Huseby et al. (2013) investigated how the Gl biosynthesis is controlled by light and the diurnal rhythm. By qRT-PCR analyses, it was shown that *AtSOT16*, *AtSOT17*, and *AtSOT18* are up regulated in light and down regulated in darkness. Further experiments indicated that the three *dsGl AtSOTs* are controlled by different transcriptional factors. In *A. thaliana* mutants, lacking the transcription regulator HY5, *AtSOT18* was less up-regulated than in the wild-type, indicating the HY5 is in control of *AtSOT18*. Interestingly, HY5 not only promotes numerous genes, but also seemed to repress MYBs. MYBs are a group of transcription factors, which are also involved in the control of Gl biosynthetic genes (Gigolashvili et al., 2007a,b, 2008; Hirai et al., 2007; Sønderby et al., 2007; Malitsky et al., 2008; Sønderby et al., 2010; Li et al., 2013; Jensen et al., 2014). *AtSOT16* was significantly down regulated in *myb34 myb51 myb122-2* triple mutant, revealing the specific control of these transcription factors (Frerigmann and Gigolashvili, 2014). Furthermore, *MYB51*, an indolic Gl metabolism specific transcription factor, was found to be down regulated in the dark, resulting in repression of indolic dsGl specific *AtSOT16*. Another indolic Gl transcription factor, *MYB34*, was up regulated after re-illumination (Celenza et al., 2005). This was not the case for *MYBs* controlling aliphatic Gl biosynthesis. It was concluded that *MYB* factors controlling biosynthesis of indolic Gl have a specific function in light regulation of their target gene, unlike the aliphatic group of *MYB* (Huseby et al., 2013).

Even so, the interaction and hierarchy of HY5 and MYBs still remains unclear. *AtSOT16*, *AtSOT17*, and *AtSOT18* were also up regulated in *apk1 apk2* double mutants. Hence, a reduction in PAPS supply and therefore reduction in Gl concentration leads to an up regulation *dsGl SOTs* (Mugford et al., 2009).

The expression of the twelve putative *dsGl SOTs* in *B. rapa* was investigated by qRT-PCR (Zang et al., 2009). Two genes are paralogs of *AtSOT16*, one of *AtSOT17*, and ten of *AtSOT18*. Generally, *BrSOT16s* were most strongly expressed, followed by *BrSOT18s* and then the *BrSOT17s*. With the exception of one *BrSOT18*, all of them were expressed in all examined tissue types. One *BrSOT16* was expressed in all tissue types, except in the stamen, while the other one was strongly expressed in the stamen, but weakly in the floral bud and carpel. Some *BrSOT18s* were strongly expressed in the carpel and others in the stamen. Hence, the expression was not tissue-specific, but there was great variation in between tissue types. The expression of some *BrSOT18s* was developmentally regulated, but not of *BrSOT16s*. Again it was concluded that the expression could influence the Gl content, since SOTs play a crucial role in Gl biosynthesis.

The *TPST* gene is expressed in the whole plant, which was shown by analyzing *A. thaliana TPST-GUS* transformants, but especially strong in the root apical meristem and in the lateral root primordial and vascular tissues (Komori et al., 2009).

Expression of the 35 *O. sativa SOTs* was investigated by microarray database analysis (Chen et al., 2012). The overall expression was reported to be considerably low. Low expression levels were in the apical meristem and young leaves. Higher expression was found in the stigma, ovary and roots. Treatment with IAA and BAP led to up and down regulation of several *SOTs* also with differences in respect to tissue types and seedlings age. Furthermore, expression of eleven *SOTs* reacted to abiotic stress, such as high and low temperatures and dehydration. It was concluded that the individual responses of *SOTs* indicate functions in stress response and plant development.

Overall, *SOT* expressions suggest functions in plant defense, stress response, signaling and developmental regulation. Sulfation can either lead to activation or deactivation of the according substrate. *SOT* expression takes place basically in all organs and many stages in plant development. Interestingly, all *SOTs* studied so far, were induced by several conditions or stress signaling compounds, indicating a general stress response. Additionally, in the case of *dsGl AtSOTs* and *AtSOT15*, a diurnal and circadian control was detected. It seems plausible that this could be the case for other *SOTs*, too.

WHAT IS KNOWN ABOUT THE REACTION MECHANISM OF SULFOTRANSFERASES

So far, the reaction mechanism of plant SOTs remains largely unknown. Kinetic and inhibition studies of a flavonol 3′-SOT from *F. chloraefolia* A. Gray led to the hypothesis of an ordered Bi-Bi mechanism (Varin and Ibrahim, 1992). However, the few conducted experiments are not sufficient enough for a definite conclusion.

More information is available about human SOTs. By pre-steady state binding studies, isotopic trapping, quenched-flow, and classic inhibition studies, Wang et al. (2014) completely solved the kinetic mechanism of the human SOT SULT2A1. SULT2A1 sulfates dehydroepiandrosterone and regulates binding of steroids to their receptors and detoxifies steroid-like xenobiotics. The according mechanism was found to be rapid equilibrium random. In this mechanism, substrates are bound and products are released in a random order. The ligands are bound in separate binding sites and released independently of the presence of its partner, hence without contribution of sulfuryl-group interactions. Ligand-binding rate constants also indicated that ligand-protein interactions, which enable the chemical reaction, are either established prior to addition of the second substrate and/or they are engaged as the system moves toward the transition state. Furthermore, it was shown that the release of the PAP nucleotide is the rate-determining step of the reaction. Substrate inhibition was explained by trapping of PAP in a dead end complex (enzyme with bound PAP and substrate), which decreases the release of PAP. Since closely related enzymes often share the same mechanism, it was speculated that this could also be the case for other human SOTs. Anyhow, this cannot be done for plant SOTs without further experimental analysis. This is already illustrated, when regarding that SULT2A1 is a half-site reactive dimer, while yet investigated plant SOTs are monomers.

The mechanism of a monomeric SOT Stf0 from *Mycobacterium tuberculosis* was analyzed by electrospray ionization mass spectrometry and Fourier transform ion cyclotron resonance mass spectrometry (Pi et al., 2005). Stf0 forms trehalose sulfate, which is the core disaccharide of the potential virulence factor sulfolipid-1. Interestingly, the results also indicated a rapid equilibrium random mechanism, at which the sulfuryl group is transferred in the ternary complex. Again, there is one binding site for products and one independent binding site for substrates. Results also indicated that PAPS binding was competitively inhibited by PAP.

Further studies of non-herbal SOTs, human estrogen SOT (Zhang et al., 1998) and insect retinol dehydratase (Vakiani et al., 1998), also indicated random Bi-Bi mechanisms. So far, only for a *Rhizobium meliloti* NodH SOT a hybrid random ping-pong mechanism was suggested (Pi et al., 2004). Therefore, the investigation of the complete kinetic mechanisms of plant SOTs remains an interesting task, which could also give new insights of the overall evolution of SOTs.

HOW TO IDENTIFY THE SUBSTRATE SPECIFICITY?

CHANCES AND RESTRICTION OF MODELING

Simple online tools like SWISS-MODEL do not lead to satisfying Z-scores and therefore unreliable models. However, Cook et al. (2013) generated significant models of human SOTs by using more advanced programs such as MODELLER, GOLD, GROMACS, and AMBER. Models of human SOTs SULT1A1 and SULT1A2, which are Phase II detoxifying enzymes, were used for *in silico* docking studies. As substrates, 1455 small molecule drugs were tested. For SULT1A1, 76 substrates were predicted, of which 53 were already known substrates. Of the remaining 23 putative substrates, 21 were tested in enzyme assays and all of them were accepted as substrates. Of 22 predicted substrates for SULT2A1, eight were not previously mentioned in literature. Enzyme assays were carried out with four of the eight newly identified substrates, and all of them were accepted as substrates. For both SOTs neither a single false positive nor a false negative prediction occurred. Furthermore, 136 SULT1A1 and 35 SULT2A1 inhibitors were predicted. Two of those were exemplary tested in classical inhibition studies and both showed inhibitory effects.

But can these techniques be transferred to plant SOTs? Principally they could be transferred to plant SOTs, but it has to be kept in mind that SULT1A1 and SULT1A2 are extensively studied SOTs. Building reliable models for *in silico* docking studies requires knowledge about structure and mechanism of the protein. For example, SULT1A1 and SULT1A2 have a site cap, which regulates substrate specificity (Cook et al., 2012). This also had to be considered when generating the *in silico* models. Furthermore, it was shown that human SOTs have a high plasticity (Allali-Hassani et al., 2007; Berger et al., 2011) and that PAP binding leads to dramatic conformational changes, such as pre-formation of the acceptor binding pocket (Bidwell et al., 1999; Dajani et al., 1999; Berger et al., 2011). Hence, without prior structural knowledge about at least some of the plant SOTs, *in silico* modeling is still restricted. Nevertheless, *in silico* modeling of SOTs is a promising approach, especially because of the limitations in substrate identification based on phylogenetic analyses.

A. THALIANA AS A MODEL PLANT – SUITED FOR THE ELUCIDATION OF ALL SOT FUNCTIONS?

Elucidation of all SOT functions in *A. thaliana* as a model plant is difficult for several reasons. Until now, ten out of 22 identified putative *A. thaliana* SOTs have been enzymatically characterized *in vitro*. The identified substrates were peptides, flavonoids, brassinosteroids, GIs, hydroxyjasmonate, and salicylic acid. As discussed before, prospects of phylogenetic analyses are very limited for SOTs. Already small changes in the sequence can lead to wide variations in substrate specificity. Even reliable predictions of yet uncharacterized SOTs in the organism *A. thaliana* are not possible. Therefore, reliable function prediction of SOTs in other plant species on the basis of *A. thaliana* sequences seems very unlikely.

Even when comparing SOTs that use the same class of substrates from different plant species, not only differences in kinetic values, but also variation of specificity toward different substrates and specific hydroxyl groups are noticed. For example, flavonol SOTs (AtSOT5, AtSOT8, AtSOT12, AtSOT13) from *A. thaliana* prefer kaempferol or flavonol glycosides as substrate and sulfate the hydroxyl groups at 3- and 7-position (Hashiguchi et al., 2013, 2014). But flavonol SOTs from *F. chloraefolia* and *F. bidentis* (L.) Kuntze prefer quercetin as substrates and sulfate at 3′- and 4′-position (Varin et al., 1992, 1997b). Furthermore, all so far characterized *A. thaliana* SOTs sulfate a broad range of substrates. Most are functionally and biochemically related, but for example in case of AtSOT12, substrates with a wide range of biological functions are accepted as substrates.

Another difficulty is that SOTs are part of secondary metabolism and therefore fulfill species-specific functions. Hence, it is unlikely that all types of SOTs occur in *A. thaliana*. This is supported by the Pfam database research. In *A. thaliana*, only the TPST contains a *Sulfotransfer_2* domain (PF03567). The remaining AtSOTs all contain a *Sulfo-transfer_1* domain (PF00685), while the *Sulfotransfer_3* (PF13469) and *Gal-3-O_sulfotr* domain (PF06990) are only present in algae. *Arylsulfotran_2* domain (PF14269) is only found in a single *Ricinus communis* sequence. In addition, *SOT* homologues in different plant species differ in their number of paralogs. For example, there were nine homologues of *AtSOT18* found in *B. rapa* (Zang et al., 2009), which could differ in their characteristics.

All in all, it remains an important future task to clarify the biological functions and characteristics of the remaining *A. thaliana* SOTs, not only by *in vitro* enzymatic assays, but in consideration of mutation, expression and localization studies, as well as metabolomics. Findings could at least be partly transferred and give valuable hints about specific SOTs in other species. Since *A. thaliana* is the most studied plant, complete characterization of all AtSOTs could also give more information about the connection of primary and secondary metabolisms in plants in general.

FUTURE CHALLENGES

Plant SOT research still remains a biological field with many open questions, especially in comparison with mammalian SOTs. In the model organism *A. thaliana*, only ten out of 22 SOTs have been enzymatically characterized *in vitro* so far. In many of these cases, the *in vivo* function is not elucidated yet. Some compounds, which were found to be sulfated by SOTs *in vitro*, could not be detected *in vivo*, as it was the case for sulfated glucoside flavonoids, sulfated by AtSOT8 (Hashiguchi et al., 2014). Furthermore, the function of sulfation or the sulfated compound is often not completely understood. Hence, for a deeper understanding it is advisable to follow *in vitro* enzymatic characterization with mutation, expression and localization studies.

For the remaining twelve putative *A. thaliana* SOTs, disregarding the pseudogene *AtSOT2*, no accepted substrates have been identified so far. Due to the enormous number of putative substrates and the restricted reliability of phylogenetic analyses, complete functional elucidation of all AtSOTs is an ambitious goal. Hence, recombinant expression and offering randomly chosen substrates seems like looking for a needle in a haystack. A more promising approach could be to feed wild-type and mutant plants with ^{35}S, followed by mass spectrometry analysis. This can facilitate the identification of newly sulfated compounds *in vivo*. Next steps could be the isolation of these compounds, depending on its availability and chemical properties. If possible, the compounds could be bound to a column and used for affinity chromatography of total protein preparations. This would be a very systematic approach and was already partly used for the successful identification of TPST (Komori et al., 2009).

Especially in pharmaceutical research, *in silico* analysis has become a powerful tool (Song et al., 2009). With the help of three dimensional structures of the target molecules, computational drug design becomes more and more promising. The three dimensional structure of one *A. thaliana* SOT (AtSOT12) has already been solved, but without including substrates into the crystals (Smith et al., 2004). A deeper understanding could be reached with the help of more solved structures with and without substrates and with additional knowledge about the enzymatic mechanism. Definitely identified binding sites, combined with protein modeling could give more specific hints about putative substrates of SOTs.

Another interesting field would be the elucidation of SOTs from more plant species, especially highly specialized ones. In order to cope with additional stress, plants growing in challenging environments often biosynthesize specific compounds. Its properties are often promising from a biological point of view, for a better understanding of stress response, but also interesting for medical or biotechnological applications. Zosteric acid [*p*-(sulfo-oxy) cinnamic acid] from the seagrass *Zostera marina*, for example, has anti-fouling properties (Newby et al., 2006). In a patent, SOT involvement in biosynthesis was suggested, but not proven yet (Alexandratos, 1999). Additionally, in *Zostera*, *Halophila*, and *Thalassia* seagrass, the existence of sulfated flavones was indicated (Harborne and Williams, 1976). Furthermore sulfated polysaccharides were detected in seagrass (Aquino et al., 2005, 2011), freshwater plants (Dantas-Santos et al., 2012) and algae (Ngo and Kim, 2013). While the sulfation of polysaccharides is well-studied in humans (Kusche-Gullberg and Kjellén, 2003), no polysaccharide SOTs have been studied in plants yet. It is hypothesized that sulfated polysaccharides modify the cell wall in halophytes in order to increase salt tolerance (Aquino et al., 2005). They are also interesting for human nutrition and pharmaceutical products, because of their antioxidant, anti-allergic, anti-human immunodeficiency virus, anti-cancer and anticoagulant properties (Ngo and Kim, 2013).

Overall, substrate specificities, regulations, and catalytic mechanisms of plant SOTs are still poorly understood. Considering the large number of possible functions, further research on these enzymes remains a challenging field.

ACKNOWLEDGMENTS

We acknowledge support by Deutsche Forschungsgemeinschaft and Open Access Publishing Fund of Leibniz Universität Hannover. Experimental work in our laboratory is funded by the Deutsche Forschungsgemeinschaft (PA 764/10-1).

REFERENCES

Alexandratos, S. D. (1999). Synthesis and purification of zosteric acid. US Patent 5990336.

Allali-Hassani, A., Pan, P. W., Dombrovski, L., Najmanovich, R., Tempel, W., Dong, A., et al. (2007). Structural and chemical profiling of the human cytosolic sulfotransferases. *PLoS Biol.* 5:e97. doi: 10.1371/journal.pbio.0050097

Amano, Y., Tsubouchi, H., Shinohara, H., Ogawa, M., and Matsubayashi, Y. (2007). Tyrosine-sulfated glycopeptide involved in cellular proliferation and expansion in *Arabidopsis*. *Proc. Natl. Acad. Sci. U.S.A.* 104, 18333–18338. doi: 10.1073/pnas.0706403104

Aquino, R. S., Grativol, C., Mourão, P. S., and Meyer, P. (2011). Rising from the sea: correlations between sulfated polysaccharides and salinity in plants. *PLoS ONE* 6:e18862. doi: 10.1371/journal.pone.0018862

Aquino, R. S., Landeira-Fernandez, A. M., Valente, A. P., Andrade, L. R., and Mourão, P. A. (2005). Occurrence of sulfated galactans in marine angiosperms: evolutionary implications. *Glycobiology* 15, 11–20. doi: 10.1093/glycob/cwh138

Baek, D., Pathange, P., Chung, J.-S., Jiang, J., Gao, L., Oikawa, A., et al. (2010). A stress-inducible sulphotransferase sulphonates salicylic acid and confers pathogen resistance in *Arabidopsis*. *Plant Cell Environ.* 33, 1383–1392. doi: 10.1111/j.1365-3040.2010.02156.x

Berger, I., Guttman, C., Amar, D., Zarivach, R., and Aharoni, A. (2011). The molecular basis for the broad substrate specificity of human sulfotransferase 1A1. *PLoS ONE* 6:e26794. doi: 10.1371/journal.pone.0026794

Bidwell, L. M., Mcmanus, M. E., Gaedigk, A., Kakuta, Y., Negishi, M., Pedersen, L., et al. (1999). Crystal structure of human catecholamine sulfotransferase. *J. Mol. Biol.* 293, 521–530. doi: 10.1006/jmbi.1999.3153

Brown, P. D., Tokuhisa, J. G., Reichelt, M., and Gershenzon, J. (2003). Variation of glucosinolate accumulation among different organs and developmental stages of *Arabidopsis thaliana*. *Phytochemistry* 62, 471–481. doi: 10.1016/S0031-9422(02)00549-6

Celenza J. L., Quiel, J. A., Smolen, G. A., Merrikh, H., Silvestro, A. R., Normanly, J., et al. (2005). The *Arabidopsis* ATR1 Myb transcription factor controls indolic glucosinolate homeostasis. *Plant Physiol.* 137, 253–262. doi: 10.1104/pp.104.054395

Chen, R., Jiang, Y., Dong, J., Zhang, X., Xiao, H., Xu, Z., et al. (2012). Genome-wide analysis and environmental response profiling of SOT family genes in rice (*Oryza sativa*). *Genes Genom.* 34, 549–560. doi: 10.1007/s13258-012-0053-5

Clarke, D. B. (2010). Glucosinolates, structures and analysis in food. *Anal. Methods* 2, 310–325. doi: 10.1039/B9AY00280D

Cook, I., Wang, T., Falany, C. N., and Leyh, T. S. (2012). A nucleotide-gated molecular pore selects sulfotransferase substrates. *Biochemistry* 51, 5674–5683. doi: 10.1021/bi300631g

Cook, I., Wang, T., Falany, C. N., and Leyh, T. S. (2013). High accuracy in silico sulfotransferase models. *J. Biol. Chem.* 288, 34494–34501. doi: 10.1074/jbc.M113.510974

Dajani, R., Cleasby, A., Neu, M., Wonacott, A. J., Jhoti, H., Hood, A. M., et al. (1999). X-ray crystal structure of human dopamine sulfotransferase, SULT1A3: molecular modeling and quantitative structure-activity relationship analysis demonstrate a molecular basis for sulfotransferase substrate specificity. *J. Biol. Chem.* 274, 37862–37868. doi: 10.1074/jbc.274.53.37862

Dantas-Santos, N., Gomes, D. L., Costa, L. S., Cordeiro, S. L., Costa, M. S. S. P., Trindade, E. S., et al. (2012). Freshwater plants synthesize sulfated polysaccharides: heterogalactans from water hyacinth (*Eicchornia crassipes*). *Int. J. Mol. Sci.* 13, 961–976. doi: 10.3390/ijms13010961

Estavillo, G. M., Crisp, P. A., Pornsiriwong, W., Wirtz, M., Collinge, D., Carrie, C., et al. (2011). Evidence for a SAL1-PAP chloroplast retrograde pathway that functions in drought and high light signaling in *Arabidopsis*. *Plant Cell* 23, 3992–4012. doi: 10.1105/tpc.111.091033

Faulkner, I. J., and Rubery, P. H. (1992). Flavonoids and flavonoid sulphates as probes of auxin-transport regulation in *Cucurbita pepo* hypocotyl segments and vesicles. *Planta* 186, 618–625. doi: 10.1007/BF00198044

Fenwick, G. R., and Heaney, R. K. (1983). Glucosinolates and their breakdown products in cruciferous crops, foods and feedingstuffs. *Food Chem.* 11, 249–271. doi: 10.1016/0308-8146(83)90074-90072

Finn, R. D., Bateman, A., Clements, J., Coggill, P., Eberhardt, R. Y., Eddy, S. R., et al. (2014). Pfam: the protein families database. *Nucl. Acids Res.* 42, D222–D230. doi: 10.1093/nar/gkt1223

Frerigmann, H., and Gigolashvili, T. (2014). MYB34, MYB51, and MYB122 distinctly regulate indolic glucosinolate biosynthesis in *Arabidopsis thaliana*. *Mol. Plant* 7, 814–828. doi: 10.1093/mp/ssu004

Fukuta, M., Kobayashi, Y., and Uchimura, K. (1998). Molecular cloning and expression of human chondroitin 6-sulfotransferase. *Biochim. Biophys. Acta* 1399, 57–61.

Gamage, N., Barnett, A., Hempel, N., Duggleby, R. G., Windmill, K. F., Martin, J. L., et al. (2006). Human sulfotransferases and their role in chemical metabolism. *Toxicol. Sci.* 90, 5–22. doi: 10.1093/toxsci/kfj061

Gidda, S. K., Miersch, O., Levitin, A., Schmidt, J., Wasternack, C., and Varin, L. (2003). Biochemical and molecular characterization of a hydroxyjasmonate sulfotransferase from *Arabidopsis thaliana*. *J. Biol. Chem.* 278, 17895–17900. doi: 10.1074/jbc.M211943200

Gidda, S. K., and Varin, L. (2006). Biochemical and molecular characterization og flavonoid 7-sulfotransferase from *Arabidopsis thaliana*. *Plant Physiol. Biochem.* 44, 628–636 doi: 10.1074/jbc.M211943200

Gigolashvili, T., Berger, B., Mock, H. -P., Muller, C., Weisshaar, B., and Flügge, U.-I. (2007a). The transcription factor HIG1/MYB51 regulates indolic glucosinolate biosynthesis in *Arabidopsis thaliana*. *Plant J.* 50, 886–901. doi: 10.1111/j.1365-313X.2007.03099.x

Gigolashvili, T., Yatusevich, R., Berger, B., Müller, C., and Flügge, U.-I. (2007b). The R2R3-MYB transcription factor HAG1/MYB28 is a regulator of methionine-derived glucosinolate biosynthesis in *Arabidopsis thaliana*. *Plant J.* 51, 247–261. doi: 10.1111/j.1365-313X.2007.03133.x

Gigolashvili, T., Engqvist, M., Yatusevich, R., Müller, C., and Flügge, U.-I. (2008). HAG2/MYB76 and HAG3/MYB29 exert a specific and coordinated control on the regulation of aliphatic glucosinolate biosynthesis in *Arabidopsis thaliana*. *New Phytol.* 177, 627–642. doi: 10.1111/j.1469-8137.2007.02295.x

Gigolashvili, T., Geier, M., Ashykhmina, N., Frerigmann, H., Wulfert, S., Krueger, S., et al. (2012). The *Arabidopsis* thylakoid ADP/ATP carrier TAAC has an additional role in supplying plastidic phosphoadenosine 5-phosphosulfate to the cytosol. *Plant Cell* 24, 4187–4204. doi: 10.1105/tpc.112.101964

Glendening, T. M., and Poulton, J. E. (1990). Partial purification and characterization of a 3′-phosphoadenosine 5′-phosphosulfate: desulfoglucosinolate sulfotransferase from cress (*Lepidium sativum*). *Plant Physiol.* 94, 811–818. doi: 10.1104/pp.94.2.811

Graser, G., Oldham, N. J., Brown, P. D., Temp, U., and Gershenzon, J. (2001). The biosynthesis of benzoic acid glucosinolate esters in *Arabidopsis thaliana*. *Phytochemistry* 57, 23–32. doi: 10.1016/S0031-9422(00)00501-X

Halkier, B. A. (1999). "Glucosinolates," in *Naturally Occurring Glycosides: Chemistry, Distribution and Biological Properties*, ed. R. Ikan (New York: John Wiley & Sons Ltd), 193–223.

Hanai, H., Nakayama, D., Yang, H., Matsubayashi, Y., Hirota, Y., and Sakagami, Y. (2000). Existence of a plant tyrosylprotein sulfotransferase: novel plant enzyme catalyzing tyrosine O-sulfation of preprophytosulfokine variants in vitro. *FEBS Lett.* 470, 97–101. doi: 10.1016/S0014-5793(00)01299-0

Harborne, J. B., and Williams, C. A. (1976). Occurrence of sulphated flavones and caffeic acid esters in members of the fluviales. *Biochem. Syst. Ecol.* 4, 37–41. doi: 10.1016/0305-1978(76)90007-7

Hashiguchi, T., Sakakibara, Y., Hara, Y., Shimohira, T., Kurogi, K., Akashi, R., et al. (2013). Identification and characterization of a novel kaempferol sulfotransferase from *Arabidopsis thaliana*. *Biochem. Biophys. Res. Commun.* 434, 829–835. doi: 10.1016/j.bbrc.2013.04.022

Hashiguchi, T., Sakakibara, Y., Shimohira, T., Kurogi, K., Yamasaki, M., Nishiyama, K., et al. (2014). Identification of a novel flavonoid glycoside sulfotransferase in *Arabidopsis thaliana*. *J. Biochem.* 155, 91–97. doi: 10.1093/jb/mvt102

Hernàndez-Sebastiá, C., Varin, L., and Marsolais, F. (2008). "Sulfotransferases from plants, algae and phototrophic bacteria," in *Sulfur Metabolism in Phototrophic Organisms*, eds R. Hell, C. Dahl, D. Knaff, and T. Leustek (Amsterdam: Springer), 111–130.

Hirai, M. Y., Klein, M., Fujikawa, Y., Yano M., Goodenowe, D. B., Yamazaki, Y., et al. (2005). Elucidation of gene-to-gene networks in *Arabidopsis* by integration of metabolomics and transcriptomics. *J. Biol. Chem.* 280, 25590–25595. doi: 10.1074/jbc.M502332200

Hirai, M. Y., Sugiyama, K., Sawada, Y., Tohge, T., Obayashi, T., Suzuki, A., et al. (2007). Omics-based identification of *Arabidopsis* Myb transcription factors regulating aliphatic glucosinolate biosynthesis. *Proc. Natl. Acad. Sci. U.S.A.* 104, 6478–6483. doi: 10.1073/pnas.0611629104

Huseby, S., Koprivova, A., Lee, B.-R., Saha, S., Mithen, R., Wold, A.-B., et al. (2013). Diurnal and light regulation of sulphur assimilation and glucosinolate biosynthesis in *Arabidopsis*. *J. Exp. Bot.* 64, 1039–1048. doi: 10.1093/jxb/ers378

Jensen, L. M., Halkier, B. A., and Burow, M. (2014). How to discover a metabolic pathway? An update on gene identification in aliphatic glucosinolate biosynthesis, regulation and transport. *Biol. Chem.* 395, 529-543. doi: 10.1515/hsz-2013-0286

Kakuta, Y., Pedersen, L. G., Carter, C. W., Negishi, M., and Pedersen, L. C. (1997). Crystal structure of estrogen sulphotransferase. *Nature Struct. Biol.* 4, 904–908. doi: 10.1038/nsb1197-904

Kakuta, Y., Petrotchenko, E. V., Pedersen, L. C., and Negishi, M. (1998). The sulfuryl transfer mechanism. *J. Biol. Chem.* 273, 27325–27330. doi: 10.1074/jbc.273.42.27325

Klaassen, C. D., and Boles, J. W. (1997). The importance of 3′-phosphoadenosine 5′-phosphosulfate (PAPS) in the regulation of sulfation. *FASEB J.* 11, 404–418.

Klein, M., and Papenbrock, J. (2004). The multi-protein family of *Arabidopsis* sulphotransferases and their relatives in other plant species. *J. Exp. Bot.* 55, 1809–1820. doi: 10.1093/jxb/erh183

Klein, M., and Papenbrock, J. (2008). "Sulfotransferases and their tole in glucosinolate biosynthesis," in *Sulfur Assimilation and Abiotic Stress in Plants*, eds. N. Khan, S. Singh, and S. Umar (Berlin: Springer), 149–166. doi: 10.1007/978-3-540-7 6326-0_7

Klein, M., and Papenbrock, J. (2009). Kinetic parameters of desulfoglucosinolate sulfotransferases. *Physiol. Plant.* 135, 140–149. doi: 10.1111/j.1399-3054.2008.01182.x

Klein, M., Reichelt, M., Gershenzon, J., and Papenbrock, J. (2006). The three desulfo-glucosinolate sulfotransferase proteins in *Arabidopsis* have different substrate specificities and are differentially expressed. *FEBS J.* 273, 122–136. doi: 10.1111/j.1742-4658.2005.05048.x

Kliebenstein, D. J., Kroymann, J., Brown, P., Figuth, A., Pedersen, D., Gershenzon, J., et al. (2001a). Genetic control of natural variation in *Arabidopsis* glucosinolate accumulation. *Plant Physiol.* 126, 811–825. doi: 10.1104/pp.126.2.811

Kliebenstein, D. J., Lambrix, V. M., Reichelt, M., and Gershenzon, J. (2001b). Gene duplication in the diversification of secondary metabolism: Tandem 2-oxoglutarate-dependent dioxygenases control glucosinolate biosynthesis in *Arabidopsis*. *Plant Cell* 13, 681–693. doi: 10.1105/tpc.13.3.681

Komori, R., Amano, Y., Ogawa-Ohnishi, M., and Matsubayashi, Y. (2009). Identification of tyrosylprotein sulfotransferase in *Arabidopsis*. *Proc. Natl. Acad. Sci. U.S.A.* 106, 15067–15072. doi: 10.1073/pnas.0902801106

Kusche-Gullberg, M., and Kjellén, L. (2003). Sulfotransferases in glycosaminoglycan biosynthesis. *Curr. Opin. Struct. Biol.* 13, 605–611. doi: 10.1016/j.sbi.2003.08.002

Labonne, J. J. D., Goultiaeva, A., and Shore, J. S. (2009). High-resolution mapping of the S-locus in Turnera leads to the discovery of three genes tightly associated with the S-alleles. *Mol. Genet. Genomics* 281, 673–685. doi: 10.1007/s00438-009-0439-5

Lacomme, C., and Roby, D. (1996). Molecular cloning of a sulfotransferase in *Arabidopsis thaliana* and regulation during development and in response to infection with pathogenic bacteria. *Plant Mol. Biol.* 30, 995–1008. doi: 10.1007/BF00020810

Lee, B. R., Huseby, S., Koprivova, A., Chételat, A., Wirtz, M., Mugford, S. T., et al. (2012). Effects of fou8/fry1 mutation on sulfur metabolism: is decreased internal sulfate the trigger of sulfate starvation response? *PLoS ONE* 7:e39425. doi: 10.1371/journal.pone.0039425

Lewis, D. R., Ramirez, M. V., Miller, N. D., Vallabhaneni, P., Ray, W. K., Helm, R. F., et al. (2011). Auxin and ethylene induce flavonol accumulation through distinct transcriptional networks. *Plant Physiol.* 156, 144–164. doi: 10.1104/pp.111.172502

Li, Y., Sawada, Y., Hirai, A., Sato, M., Kuwahara, A., Yan, X., et al. (2013). Novel insights into the function of Arabidopsis R2R3-MYB transcription factors regulating aliphatic glucosinolate biosynthesis. *Plant Cell Physiol.* 54, 1335–1344. doi: 10.1093/pcp/pct085

Luczak, S., Forlani, F., and Papenbrock, J. (2013). Desulfo-glucosinolate sulfotransferases isolated from several *Arabidopsis thaliana* ecotypes differ in their sequence and enzyme activities. *Plant Physiol. Biochem.* 63, 15–23. doi: 10.1016/j.plaphy.2012.11.005

Malitsky, S., Blum, E., Less, H., Venger, I., Elbaz, M., Morin, S., et al. (2008). The transcript and metabolite networks affected by the two clades of *Arabidopsis* glucosinolate biosynthesis regulators. *Plant Physiol.* 148, 2021–2049. doi: 10.1104/pp.108.124784

Marsolais, F., Boyd, J., Paredes, Y., Schinas, A.-M., Garcia, M., Elzein, S., et al. (2007). Molecular and biochemical characterization of two brassinosteroid sulfotransferases from *Arabidopsis*, AtST4a (At2g14920) and AtST1 (At2g03760). *Planta* 225, 1233–1244. doi: 10.1007/s00425-006-0413-y

Marsolais, F., Gidda, S. K., Boyd, J., and Varin, L. (2000). Structural and functional similarity with mammalian enzymes. *Rec. Adv. Phytochem.* 34, 433–456. doi: 10.1371/journal.pone.0104594

Marsolais, F., Sebastià, C. H., Rousseau, A., and Varin, L. (2004). Molecular and biochemical characterization of BNST4, an ethanol-inducible steroid sulfotransferase from *Brassica napus*, and regulation of BNST genes by chemical stress and during development. *Plant Sci.* 166, 1359–1370. doi: 10.1016/j.plantsci.2004.01.019

Marsolais, F., and Varin, L. (1995). Identification of amino acid residues critical for catalysis and cosubstrate binding in the flavonol 3-sulfotransferase. *J. Biol. Chem.* 270, 30458–30463. doi: 10.1074/jbc.270.51.30458

Matsubayashi, Y., and Sakagami, Y. (1996). Phytosulfokine, sulfated peptides that induce the proliferation of single mesophyll cells of *Asparagus officinalis* L. *Proc. Natl. Acad. Sci. U.S.A.* 93, 7623–7627. doi: 10.1073/pnas.93.15.7623

Matsuzaki, Y., Ogawa-Ohnishi, M., Mori, A., and Matsubayashi, Y. (2010). Secreted peptide signals required for maintenance of root stem cell niche in *Arabidopsis*. *Science* 329, 1065–1067. doi: 10.1126/science.1191132

Mithen, R. (2001). Glucosinolates - biochemistry, genetics and biological activity. *Plant Growth Regul.* 34, 91–103. doi: 10.1023/A:1013330819778

Møldrup, M. E., Geu-Flores, F., Olsen, C. E., and Halkier, B. A. (2011). Modulation of sulfur metabolism enables efficient glucosinolate engineering. *BMC Biotechnol.* 11:12. doi: 10.1186/1472-6750-11-12.

Mugford, S. G., Yoshimoto, N., Reichelt, M., Wirtz, M., Hill, L., Mugford, S. T., et al. (2009). Disruption of adenosine-5′-phosphosulfate kinase in *Arabidopsis* reduces levels of sulfated secondary metabolites. *Plant Cell* 21, 910–927. doi: 10.1105/tpc.109.065581

Newby, B.-M. Z., Cutright, T., Barrios, C., and Xu, Q. (2006). Zosteric acid—An effective antifoulant for reducing fresh water bacterial attachment on coatings. *JCT Res.* 3, 69–76. doi: 10.1007/s11998-006-0007-4

Ngo, D.-H., and Kim, S.-K. (2013). Sulfated polysaccharides as bioactive agents from marine algae. *Int. J. Biol. Macromol.* 62, 70–75. doi: 10.1016/j.ijbiomac.2013.08.036

Niehrs, C., Beisswanger, R., and Huttner, W. B. (1994). Protein tyrosine sulfation, 1993 – an update. *Chem. Biol. Interact.* 92, 257–271. doi: 10.1016/0009-2797(94)90068-X

Pi, N., Hoang, M. B., Gao, H., Mougous, J. D., Bertozzi, C. R., and Leary, J. A. (2005). Kinetic measurements and mechanism determination of Stf0 sulfotransferase using mass spectrometry. *Anal. Biochem.* 341, 94–104. doi: 10.1016/j.ab.2005.02.004

Pi, N., Yu, Y., Mougous, J. D., and Leary, J. A. (2004). Observation of a hybrid random ping-pong mechanism of catalysis for NodST: a mass spectrometry approach. *Protein Sci.* 13, 903–912. doi: 10.1110/ps.03581904

Piotrowski, M., Schemenewitz, A., Lopukhina, A., Muller, A., Janowitz, T., Weiler, E. W., et al. (2004). Desulfoglucosinolate sulfotransferases from *Arabidopsis thaliana* catalyze the final step in the biosynthesis of the glucosinolate core structure. *J. Biol. Chem.* 279, 50717–50725. doi: 10.1074/jbc.M407681200

Rausch, T., and Wachter, A. (2005). Sulfur metabolism: a versatile platform for launching defence operations. *Trends Plant Sci.* 10, 503–509. doi: 10.1016/j.tplants.2005.08.006

Rivoal, J., and Hanson, A. D. (1994). Choline-O-sulfate biosynthesis in plants (Identification and partial characterization of a salinity-inducible choline sulfotransferase from species of Limonium (Plumbaginaceae). *Plant Physiol.* 106, 1187–1193.

Rouleau, M., Marsolias, F., Richard, M., Nicolle, L., Voigt, B., Adam, G., et al. (1999). Inactivation of brassinosteroid biological activity by a salicylate-inducible steroid sulfotransferase from *Brassica napus. J. Biol. Chem.* 274, 20925–20930. doi: 10.1074/jbc.274.30.20925

Sandhu, K. S., and Neff, M. M. (2013). The Arabidopsis gene ATST4a in not a typical brassinosteroid catabolic gene. *Plant Signal. Behav.* 8, 10. doi: 10.4161/psb.26847

Shimojima, M., Hoffmann-Benning, S., Garavito, R. M., and Benning, C. (2005). Ferredoxin-dependent glutamate synthase moonlights in plant sulfolipid biosynthesis by forming a complex with SQD1. *Arch. Biochem. Biophys.* 436, 206–214. doi: 10.1016/j.abb.2005.02.005

Smith, D. W., Johnson, K. A., Bingman, C. A., Aceti, D. J., Blommel, P. G., Wrobel, R. L., et al. (2004). Crystal structure of At2g03760, a putative steroid sulfotransferase from *Arabidopsis thaliana. Proteins* 57, 854–857. doi: 10.1002/prot.20258

Sønderby, I. E., Burow, M., Rowe, H. C., Kliebenstein, D. J., and Halkier, B. A. (2010). A complex interplay of three R2R3 MYB transcription factors determines the profile of aliphatic glucosinolates in *Arabidopsis. Plant Physiol.* 153, 348–363. doi: 10.1104/pp.109.149286

Sønderby, I. E., Hansen, B. G., Bjarnholt, N., Ticconi, C., Halkier, B. A., and Kliebenstein, D. J. (2007). A system biology approach identifies a R2R3 MYB gene subfamily with distinct and overlapping functions in regulation of aliphatic glucosinolates. *PLoS ONE* 2:e1322. doi: 10.1371/journal.pone.0001322

Song, C. M., Lim, S. J., and Tong, J. C. (2009). Recent advances in computer-aided drug design. *Brief. Bioinf.* 10, 579–591. doi: 10.1093/bib/bbp023

Vakiani, E., Luz, J. G., and Buck, J. (1998). Substrate specificity and kinetic mechanism of the insect sulfotransferase, retinol dehydratase. *J. Biol. Chem.* 273, 35381–35387. doi: 10.1074/jbc.273.52.35381.

Varin, L., Chamberland, H., Lafontaine, J. G., and Richard, M. (1997a). The enzyme involved in sulfation of the turgorin, gallic acid 4-O-(beta-D-glucopyranosyl-6′-sulfate) is pulvini-localized in *Mimosa pudica. Plant J.* 12, 831–837. doi: 10.1046/j.1365-313X.1997.12040831.x

Varin, L., Marsolais, F., Richard, M., and Rouleau, M. (1997b). Biochemistry and molecular sulfotransferases biology of plant sulfotransferases. *FASEB J.* 11, 517–525.

Varin, L., Deluca, V., Ibrahim, R. K., and Brisson, N. (1992). Molecular characterization of two plant flavonol sulfotransferases. *Proc. Natl. Acad. Sci. U.S.A.* 89, 1286–1290. doi: 10.1073/pnas.89.4.1286

Varin, L., and Ibrahim, R. K. (1992). Novel flavonol 3-sulfotransferase. Purification, kinetic properties, and partial amino acid sequence. *J. Biol. Chem.* 267, 1858–1863.

Varin, L., and Spertini, D. (2003). Desulfoglucosinolate sulfotransferases, sequences coding the same and uses thereof for modulating glucosinolate biosynthesis in plants. Patent WO 2003010318.

Wang, H., Wu, J., Sun, S., Liu, B., Cheng, F., Sun, R., et al. (2011). Glucosinolate biosynthetic genes in *Brassica rapa. Gene* 487, 135–142. doi: 10.1016/j.gene.2011.07.021

Wang, T., Cook, I., Falany, C. N., and Leyh, T. S. (2014). Paradigms of sulfotransferase catalysis – the mechanism of SULT2A1. *J. Biol. Chem.* 289, 26474–2680. doi: 10.1074/jbc.M114.573501

Weinshilboum, R. M., and Otterness, D. M. (1994). "Sulfotransferase enzymes," in *Handbook of Experimental Pharmacology*, ed. Kauffman (Berlin: Springer-Verlag), 45–78.

Wittstock, U., and Halkier, B. A. (2002). Glucosinolate research in the *Arabidopsis* era. *Trends Plant Sci.* 7, 263–270. doi: 10.1016/S1360-1385(02)02273-2

Yamashino, T., Kitayama, M., and Mizuno, T. (2013). Transcription of ST2A encoding a sulfotransferase family protein that is involved in jasmonic acid metabolism is controlled according to the circadian clock- and PIF4/PIF5-mediated external coincidence mechanism in *Arabidopsis thaliana. Biosci. Biotechnol. Biochem.* 77, 2454–2460. doi: 10.1271/bbb.130559

Zang, Y.-X., Kim, H. U., Kim, J. A., Lim, M.-H., Jin, M., Lee, S. C., et al. (2009). Genome-wide identification of glucosinolate synthesis genes in *Brassica rapa. FEBS J.* 276, 3559–3574. doi: 10.1111/j.1742-4658.2009.07076.x

Zhai, G., Lehmler, H.-J., and Schnoor, J. L. (2013). Sulfate metabolites of 4-monochlorobiphenyl in whole poplar plants. *Environ. Sci. Technol.* 47, 557–562. doi: 10.1021/es303807f

Zhang, H., Varmalova, O., Vargas, F. M., Falany, C. N., and Leyh, T. S. (1998). Sulfuryl transfer: the catalytic mechanism of human estrogen sulfotransferase. *J. Biol. Chem.* 273, 10888–10892.doi: 10.1074/jbc.273.18.10888

Conflict of Interest Statement: The authors declare that the research was conducted in the absence of any commercial or financial relationships that could be construed as a potential conflict of interest.

2

The significance of glucosinolates for sulfur storage in Brassicaceae seedlings

Tahereh Aghajanzadeh[1], Malcolm J. Hawkesford[2] and Luit J. De Kok[1]*

[1] *Laboratory of Plant Physiology, University of Groningen, Groningen, Netherlands*
[2] *Plant Biology and Crop Science Department, Rothamsted Research, Harpenden, UK*

Edited by:
Agnieszka Sirko, Institute of Biochemistry and Biophysics – Polish Academy of Sciences, Poland

Reviewed by:
Doug Van Hoewyk, Coastal Carolina University, USA
Elke Bloem, Julius Kühn-Institute – Institute for Crop and Soil Science, Germany

****Correspondence:***
Luit J. De Kok, Laboratory of Plant Physiology, University of Groningen, P.O. Box 11103, Groningen 9700 CC, Netherlands
e-mail: l.j.de.kok@rug.nl

Brassica juncea seedlings contained a twofold higher glucosinolate content than *B. rapa* and these secondary sulfur compounds accounted for up to 30% of the organic sulfur fraction. The glucosinolate content was not affected by H_2S and SO_2 exposure, demonstrating that these sulfur compounds did not form a sink for excessive atmospheric supplied sulfur. Upon sulfate deprivation, the foliarly absorbed H_2S and SO_2 replaced sulfate as the sulfur source for growth of *B. juncea* and *B. rapa* seedlings. The glucosinolate content was decreased in sulfate-deprived plants, though its proportion of organic sulfur fraction was higher than that of sulfate-sufficient plants, both in absence and presence of H_2S and SO_2. The significance of myrosinase in the *in situ* turnover in these secondary sulfur compounds needs to be questioned, since there was no direct co-regulation between the content of glucosinolates and the transcript level and activity of myrosinase. Evidently, glucosinolates cannot be considered as sulfur storage compounds upon exposure to excessive atmospheric sulfur and are unlikely to be involved in the re-distribution of sulfur in *B. juncea* and *B. rapa* seedlings upon sulfate deprivation.

Keywords: Brassicaceae, glucosinolate, hydrogen sulfide, myrosinase activity and expression, sulfur deficiency, sulfur dioxide, sulfur storage

INTRODUCTION

Glucosinolates are secondary sulfur compounds commonly found in relatively high levels in shoots, roots, and seeds of Brassicaceae and may account for up to 20% of the organic sulfur fraction (Schnug, 1990, 1993; Fahey et al., 2001; Castro et al., 2004; Halkier and Gershenzon, 2006; Clay et al., 2009; Del Carmen Martinez-Ballesta et al., 2013). These secondary sulfur compounds are responsible for the spicy flavor of many species of the Brassicaceae, e.g., mustard and radish (Schnug, 1990, 1993). Glucosinolates are derived from amino acids and have a core structure consisting of a β-D-glucopyranose residue linked to a (*Z*)-*N*-hydroximino sulfate ester via a sulfur atom and a variable side chain (Halkier and Gershenzon, 2006). On the basis of their precursor amino acids, specific aliphatic, indolyl, and aromatic glucosinolates can be distinguished, and furthermore *Brassica* species differ strongly in content and composition of these glucosinolates (Castro et al., 2004; Halkier and Gershenzon, 2006).

The glucosinolate content may be affected by the sulfur nutritional status of the plant; supplemental sulfur fertilization of Brassica in greenhouse and field experiments resulted in an up to a 20-fold increase in glucosinolate content in foliar tissues (Falk et al., 2007). However, the impact of sulfur fertilization on glucosinolate content varied strongly between plant species, growth stage and organs, and the rate of sulfur supplied (Kirkegaard and Sarwar, 1998; Castro et al., 2004; Falk et al., 2007; Antonious et al., 2009). For instance, in some cultivars of broccoli, sulfur fertilization only resulted in an increase in glucosinolate content in the heads, whereas in other cultivars it had no effect or even resulted a lower glucosinolate content (Falk et al., 2007).

Upon cellular injury, glucosinolates are enzymatically degraded by myrosinase (a thioglucosidase), resulting in a variety of breakdown products, including glucose, sulfate, and depending on specific chemical structure, isothiocyanates, nitriles, epithionitriles, oxazolidinethions, indolyl alcohols, thiocyanate, and amines (Fenwick et al., 1983; Halkier and Du, 1997; Bones and Rossiter, 2006; Halkier and Gershenzon, 2006; Kissen and Bones, 2009; Kissen et al., 2009; Ahuja et al., 2010). It is presumed that both glucosinolates and their breakdown products may play a role in the defense of plants against microorganisms, fungi, and insects (Ernst, 1993; Bones and Rossiter, 1996; Ahuja et al., 2010). Moreover, glucosinolates are presumed to have a sulfur storage role in plants and their degradation by myrosinase might have significance in the re-distribution of sulfur in plants under sulfur-deprived conditions (Schnug, 1990; Hirai et al., 2004, 2005; Bloem et al., 2007; Falk et al., 2007).

In addition to sulfate taken up by the roots, plants are able to utilize foliarly absorbed sulfur gases as a supplemental sulfur source for growth (De Kok et al., 2000a, 2007, 2009). For instance, continuous exposure of Brassica to atmospheric H_2S or SO_2 levels of $\geq 0.2\ \mu l\ l^{-1}$ was sufficient to cover the organic sulfur requirement to maintain growth in the absence of sulfate in the root environment (De Kok et al., 2000b, 2002; Yang et al., 2006a,b; Koralewska et al., 2008). In the current study, the impacts of sulfur nutrition (atmospheric and pedospheric) on the glucosinolate content and the transcript levels and activity of myrosinase were studied in seedlings of two Brassica species, which are characterized by a high (*Brassica juncea*, mustard greens) and low (*B. rapa*, mustard spinach) glucosinolate content. The aim of the study was to gain insight into the significance of glucosinolates in sulfur storage and the role of myrosinase in the re-distribution of sulfur in sulfate-deprived plants.

MATERIALS AND METHODS

PLANT MATERIAL AND H_2S AND SO_2 EXPOSURE

Seeds of *B. juncea* cv. Rugosa and *B. rapa* cv. Komatsuna; Van der Wal, Hoogeveen, The Netherlands) were germinated in vermiculite in a climate-controlled room. Day and night temperatures were 22 and 18°C (±1°C), respectively, relative humidity was 60–70%. The photoperiod was 14 h at a photon fluence rate of 300 ± 20 μmol m^{-2} s^{-1} (400–700 nm) at plant height, supplied by Philips GreenPower LED (deep red/white 120) production modules. Ten day-old seedlings were transferred to an aerated 25% Hoagland nutrient solution at 0.5 mM sulfate for 3 days and subsequently transferred to fresh Hoagland nutrient solution at 0 mM sulfate (−S, sulfate-deprived) or 0.5 mM sulfate (+S, sulfate-sufficient) in 13 l stainless steel containers (10 sets of plants per container; three plants per set). Plants were exposed to 0.25 μl l^{-1} H_2S or SO_2 for 7 days in 150 l cylindrical stainless steel cabinets (0.6 m diameter) with a polymethyl methacrylate top. Sealing of the lid of the container and plant sets prevented absorption of atmospheric H_2S or SO_2 by the solution. Day and night temperatures were 24 and 20°C (±2°C), respectively, and relative humidity was 40–50%. The photoperiod was 14 h at a photon fluence rate of 300 ± 20 μmol m^{-2} s^{-1} (400–700 nm) at plant height, supplied by Philips GreenPower LED (deep red/white 120) production modules. The temperature inside the cabinets was controlled by adjusting the cabinet wall temperature. The air exchange was 40 l min^{-1} and the air inside the cabinets was stirred continuously by a ventilator. Pressurized H_2S and SO_2 diluted with N_2 (1 ml l^{-1}) was injected into the incoming air stream and their concentration in the cabinet was adjusted to the desired level using electronic mass flow controllers (ASM, Bilthoven, The Netherlands). The sulfur gas level in the cabinets was monitored by an SO_2 analyzer (model 9850) equipped with a H_2S converter (model 8770; Monitor Labs, Measurement Controls Corporation, Englewood, CO, USA). Plants were harvested 3 h after the onset of the light period and the roots were rinsed in ice-cold demineralized water (for 3 × 20 s). Roots were separated from the shoots, weighed, and for glucosinolate, myrosinase activity and RNA isolation, plant material was frozen immediately in liquid N_2 and stored at −80°C. For analysis of dry matter, sulfate, and total sulfur, plant tissue was dried at 80°C for 24 h.

TOTAL SULFUR AND SULFATE CONTENT

The total sulfur content was analyzed using a modification of the method as described by Jones (1995). Dried shoots and roots were pulverized in a Retsch Mixer-Mill (Retsch type MM2; Haan, Germany) and 50–150 mg of the samples was weighed into porcelain ashing trays. A 50% $Mg(NO_3)_2 \cdot 6H_2O$ (w/v) solution was added until saturation of the material, and was dried overnight in an oven at 100°C. Subsequently, the samples were ashed in an oven at 650°C for 12 h. The residues were dissolved in 5 or 10 ml of 20% aqua regia (50 ml conc. HNO_3 and 150 ml conc. HCl in 1 l demineralized water) and quantitatively transferred to a volumetric flask and made up to 50 or 100 ml with demineralized water. One SulphaVer® 4 Reagent Powder Pillow (HACH, Permachem® reagents, Loveland, USA) containing $BaCl_2$ was added to 10 or 25 ml of extract, and the turbidity was measured with a spectrophotometer (HACH DR/400V, Loveland, CO, USA) at 450 nm. For measurement of the sulfate content, pulverized dried plant material was incubated for 3–4 h in demineralized water (10 mg/ml) at 50°C (Tausz et al., 1996; Yang et al., 2006a,b) and centrifuged at 30,000 *g* for 15 min. Anions were separated by HPLC on an Agilent IonoSpher 5A anion exchange column (250 × 4.6 mm; Agilent Technologies, Amstelveen, The Netherlands) and sulfate content was determined refractometrically according to Maas et al. (1986). The HPLC system consisted of a Knauer HPLC pump model 100 and a Knauer differential refractometer model 98.00 (Knauer, Berlin, Germany). The mobile phase contained 25 mM potassium biphthalate (pH 4.3) with 0.02% NaN_3 (w/v). There were no significant differences in sulfate content determined in dried and fresh plant material of *B. juncea* and *B. rapa*. This indicated that there was no increase in sulfate content in dried plant material caused by the degradation of glucosinolates. The organic sulfur content was calculated by subtracting the sulfate content from the total sulfur content determined in the same tissue sample.

GLUCOSINOLATE CONTENT

The glucosinolates were extracted and determined according to a modified method of Heaney and Fenwick (1980) and O'Callaghan et al. (2000). Frozen plant material was freeze-dried in a LyoLAB 3000 freeze drier (Heto-Holten A/S, Allerød, Denmark) for 3 days. Freeze-dried plant samples were pulverized in a Retsch Mixer-Mill (Retsch type MM2; Haan, Germany). The glucosinolates were extracted in boiling 90% methanol (50 mg in 3 ml) for 2 min. The extract was centrifuged for 2 min at 2,500 *g* and the residue was re-extracted twice with 3 ml boiling 70% methanol. Total glucosinolate content was determined based on its reaction with sodium tetrachloropalladate II (Na_2PdCl_4; Gupta et al., 2012; Ishida et al., 2012). The reaction mixture containing 60 μl extract and 1800 μl 2 mM Na_2PdCl_4 was incubated at 20°C for 30 min and the absorbance of the developed color measured colorimetrically at 450 nm (Thies, 1982). Sinigrin (Sigma–Aldrich, S1647) was used as an internal standard for all samples (13 μmol per extracts) and data were corrected for recovery rate (always higher than 80%).

MYROSINASE ACTIVITY

The myrosinase activity was determined by the photometric quantification of released glucose with sinigrin (2-propenyl glucosinolate, S1647, Sigma–Aldrich) as substrate, as described by Travers-Martin et al. (2008). Frozen plant material was homogenized in 200 mM Tris-HCl, 10 mM EDTA, pH 7.0 (1 g fresh weight per 5 ml) at 0°C with an Ultra Turrax (T25, IKA Werke, Staufen, Germany) and filtered through one layer of Miracloth. The filtered extract was centrifuged at 16,000 *g* at 4°C for 15 min. The reaction mixture contained a final volume one ml, 150 μl supernatant and 1 mM sinigrin and was incubated at 25°C for 30 min. The reaction was stopped by incubating the reaction mixtures at 100°C for 10 min. Subsequently the reaction mixtures were centrifuged at 10,000 *g* at room temperature for 10 min and the content of glucose in the supernatant was determined by using an enzymatic glucose assay kit (Sigma–Aldrich). The soluble protein content was determined by the method of Bradford (1976) using bovine serum albumin as a standard.

RNA EXTRACTION AND REAL-TIME QUANTITATIVE PCR OF MYROSINASE

Total RNA was isolated by a modified hot phenol method (Verwoerd et al., 1989). Frozen ground plant material was extracted in hot (80°C) phenol/extraction buffer (1:1, v/v), 1 g ml^{-1}. The extraction buffer contained 0.1 M Tris-HCl, 0.1 M LiCl, 1% SDS (w/v), 10 mM EDTA, pH 8.0). After mixing, 0.5 ml of chloroform–isoamyl alcohol (24:1, v/v) was added. After centrifugation (13,400 *g*) for 5 min at 4°C, the aqueous phases were transferred to new tube. After adding an equal volume of chloroform and isoamyl alcohol, the total RNA was precipitated by 4 M LiCl overnight at 4°C. Total RNA was collected and washed with 70% ethanol. Possible genomic DNA contamination was removed with a DNase treatment step (Promega, USA). Phenol–chloroform–isoamyl alcohol and chloroform–isoamyl alcohol were used for further purification and total RNA was precipitated by ethanol and dissolved in diethylpyrocarbonate-treated water. The quantity and quality of RNA was checked using ThermoNanoDrop 2000 and RNA in each was adjusted to the same concentration. The integrity of RNA was checked by electrophoresis by loading 1 μg RNA on a 1% TAE-agarose gel.

DNA-free intact RNA (1 μg) was reverse transcribed into cDNA with oligo-dT primers using a first strand cDNA synthesis kit (Promega, USA) according to the manufacture-supplied instructions. Subsequently, the cDNA was used as a template in real-time PCR experiments with gene-specific primers. To design primers, the full length complementary DNA of the *Arabidopsis* genes, actin (reference gene) and myrosinase (TGG1 and TGG2), which are mostly expressed in the shoot (Andersson et al., 2009), were used to query homologous *B. juncea* and *B. rapa* sequences. Expressed sequence tags (ESTs) were obtained from the publically available platform at NCBI. The primer sequences used for myrosinase and actin were (F 5′-CCGGTCGATGTTCTCCTAT-3′, R 5′-GAAGAAT TTCCACCGTAACAC-3′) and (F 5′-AGCAGCATGAAGATCAAG GT-3′, R 5′-GCTGAGGGATGCAAGGATAG-3′), respectively. RT-PCR was performed on Applied Bio Systems' 7300 real-time PCR system using the SYBR Green master mix kit (Thermo Scientific) based on manufacturer's instructions. The transcript level of the target gene and actin was measured using the comparative Ct method. Analysis of qPCR data was performed using three independent RNA preparations from separate plant shoots.

STATISTICAL ANALYSIS

Data from different experimental sets ware analyzed for statistical significance using an unpaired two-tailed Student's *t*-test ($P < 0.01$).

RESULTS

IMPACT OF ATMOSPHERIC AND PEDOSPHERIC SULFUR NUTRITION ON GROWTH AND SULFUR CONTENT

Exposure of *B. juncea* and *B. rapa* to 0.25 μl l^{-1} H_2S and 0.25 μl l^{-1} SO_2 for 7 days did not significantly affect plant biomass production, shoot to root ratio, and dry matter content (DMC) of shoots and roots at sulfate-sufficient conditions (**Table 1**). In addition, H_2S exposure did not affect the total sulfur, sulfate, and organic sulfur content of the shoots and roots of both species (**Figure 1**). Moreover SO_2 exposure also

Table 1 | Impact of H_2S, SO_2 and sulfate deprivation on biomass production and dry matter content (DMC) of shoots and roots of *B. juncea* and *B. rapa*.

	+S	+S + H_2S	+S + SO_2	−S	−S + H_2S	−S + SO_2
B. juncea						
Shoot biomass production	0.71 ± 0.18a	0.75 ± 0.27a	0.65 ± 0.19a	0.34 ± 0.08b	0.66 ± 0.10a	0.71 ± 0.17a
Root biomass production	0.14 ± 0.05c	0.13 ± 0.04c	0.13 ± 0.05c	0.16 ± 0.06bc	0.20 ± 0.05ab	0.23 ± 0.09a
Shoot DMC	10.4 ± 1.1a	9.8 ± 0.8a	10.4 ± 1.1a	11.3 ± 1.1a	9.7 ± 0.6a	10.0 ± 0.5a
Root DMC	8.9 ± 0.7a	7.9 ± 1.7ab	8.1 ± 1.3a	6.1 ± 1.0b	6.8 ± 0.7b	5.8 ± 0.8b
Shoot/root ratio	5.3 ± 0.7a	5.7 ± 1.2a	5.2 ± 0.9a	2.2 ± 0.5c	3.9 ± 0.6b	3.0 ± 0.6bc
B. rapa						
Shoot biomass production	1.14 ± 0.28a	1.07 ± 0.27a	0.90 ± 0.26ab	0.37 ± 0.17c	0.74 ± 0.30b	0.78 ± 0.22b
Root biomass production	0.19 ± 0.06ab	0.16 ± 0.05b	0.15 ± 0.04b	0.13 ± 0.06c	0.19 ± 0.07ab	0.23 ± 0.08a
Shoot DMC	9.5 ± 0.5b	9.3 ± 0.6b	9.8 ± 0.4b	11.7 ± 1.2a	9.6 ± 0.8b	9.8 ± 0.7ab
Root DMC	7.2 ± 1.4a	7.8 ± 1.3a	7.3 ± 0.8a	8.2 ± 2.3a	7.5 ± 1.1a	7.2 ± 0.6a
Shoot/root ratio	6.0 ± 1.0a	7.0 ± 1.2a	6.2 ± 0.5a	2.9 ± 0.8c	4.0 ± 0.8bc	3.5 ± 0.6bc

Ten day-old seedlings were grown on a 25% Hoagland solution containing 0.5 mM sulfate for 3 days and subsequently transferred to fresh 25% Hoagland solution at 0 (−S) or 0.5 mM sulfate (+S) and exposed to 0.25 μl l^{-1} H_2S or SO_2 for 7 days. The initial fresh weight of shoots and roots in B. juncea were 0.042 ± 0.008 g and 0.022 ± 0.003 g, respectively. The initial fresh weight of shoots and roots in B. rapa were 0.066 ± 0.025 g and 0.021 ± 0.003 g, respectively. Data on biomass production (g FW), shoot/root ratio and DMC (%) represent the mean of four independent experiments with three measurements with six to nine plants in each (±SD). Different letters indicate significant differences between treatments (P < 0.01, Student's t-test).

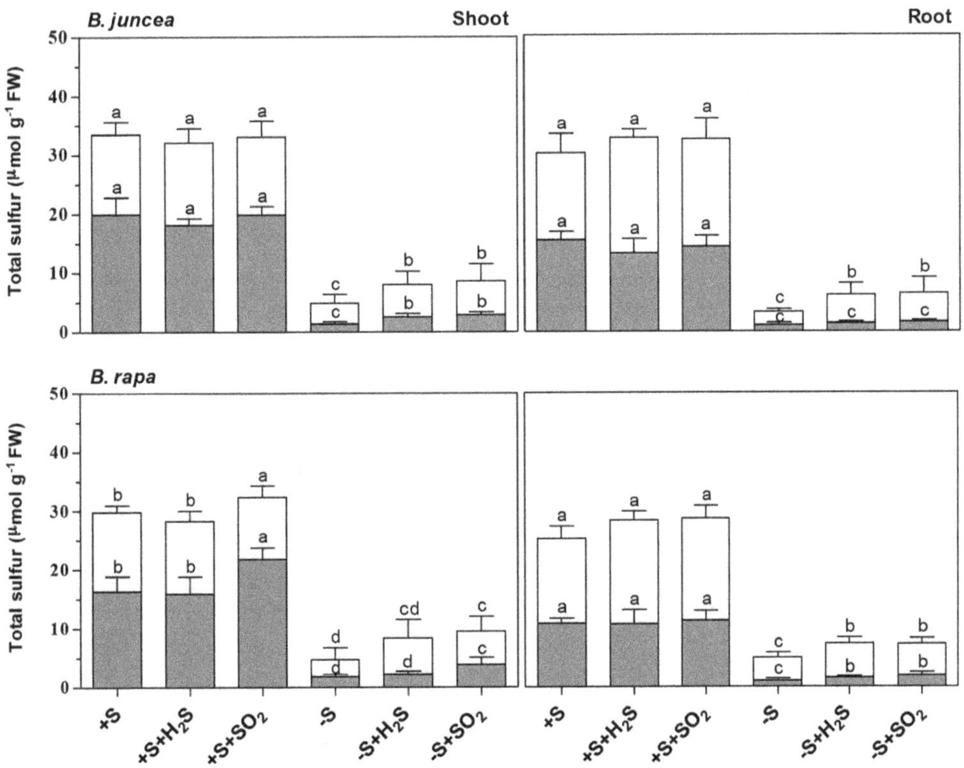

FIGURE 1 | Impact of H_2S, SO_2 and sulfate deprivation on total sulfur, sulfate and organic sulfur content of shoots and roots of *B. juncea* and *B. rapa*. For experimental details, see legends of **Table 1**. The sulfate and organic sulfur fraction is presented in gray and white, respectively. Data represent the mean of two experiments with three measurements with six to nine plants in each (±SD). Different letters indicate significant differences between treatments ($P < 0.01$, Student's *t*-test).

did not affect the sulfur, sulfate, and organic sulfur content of both shoots and roots of *B. juncea*. In *B. rapa*, however, SO_2 exposure resulted in a slight but significant increase in total sulfur content of the shoot, which could be attributed to an increase in the sulfate content, whereas the sulfur content of the root remained unaffected. Furthermore the organic sulfur content was hardly affected by the exposure of *B. rapa* to SO_2 in sulfate-sufficient conditions in both shoots and roots (**Figure 1**).

Sulfate deprivation resulted in a decreased biomass production of *B. juncea* and *B. rapa* (**Table 1**). However, shoot growth was more affected than root growth, resulting in a decrease on the shoot to root ratio upon sulfate deprivation. A 7-day sulfate deprivation resulted in 53 and 68% decreases of shoot biomass production of *B. juncea* and *B. rapa*, respectively. The root biomass production of *B. juncea* was not affected and that of *B. rapa* was reduced by 32%. The DMC of the shoot of *B. juncea* was not affected by sulfate deprivation, but that of the root was decreased (**Table 1**). However, the DMC of the shoot of *B. rapa* was significantly increased, whereas that of the root was hardly affected (**Table 1**). Sulfate deprivation resulted in strongly decreased total sulfur, sulfate, and organic sulfur contents of shoots and roots of both species (**Figure 1**). In particular, the proportion of sulfate was diminished and utilized for the synthesis for organic sulfur compounds (**Figure 1**).

Exposure of plants to 0.25 $\mu l\ l^{-1}$ H_2S and SO_2 alleviated either fully (*B. juncea*) or largely (*B. rapa*) the decrease in shoot biomass production upon sulfate-deprivation, demonstrating that the foliarly absorbed sulfur gases replaced sulfate taken up by the root as a sulfur source for growth. However, the root biomass production of H_2S and SO_2 exposed sulfate-deprived plants was even higher than that of plants at sulfate-sufficient conditions (**Table 1**). As a consequence, the shoot to root ratio was lower for sulfate-deprived H_2S and SO_2 exposed plants than that of sulfate-sufficient plants. The shoot DMC of *B. juncea* for sulfate-deprived plants was hardly affected by H_2S and SO_2 exposure and was quite similar to that of sulfate-sufficient plants (**Table 1**). The DMC of the root sulfate-deprived H_2S and SO_2 exposed plants remained lower than that of sulfate-sufficient plants. The DMC of the shoots and roots of sulfate-deprived *B. rapa* was only slightly affected by H_2S and SO_2 exposure and was quite similar to that of sulfate-sufficient plants (**Table 1**).

Both H_2S and SO_2 exposure of sulfate-deprived *B. juncea* and *B. rapa* resulted in an increase in total sulfur and organic sulfur content of both shoots and roots (**Figure 1**) however the overall total sulfur content was almost threefold and 2.5-fold lower than that observed for sulfate-sufficient *B. juncea* and *B. rapa*, respectively. The low total sulfur content could only in part be ascribed to a low apparent sulfate content upon sulfate-deprivation.

IMPACT OF ATMOSPHERIC AND PEDOSPHERIC SULFUR NUTRITION ON GLUCOSINOLATE CONTENT

There were considerable differences in the content of glucosinolates of shoots of *B. juncea* and *B. rapa* seedlings. The glucosinolate content of the shoot of *B. juncea* was twofold higher than that in the shoot of *B. rapa*, expressed either on a fresh weight basis and or relative to organic sulfur (**Figure 2**). However, the glucosinolate content of the roots of both *Brassica* species was quite similar. Neither H_2S and SO_2 exposure affected the glucosinolate content of shoots and roots of *B. juncea* and *B. rapa*, expressed either on a fresh weight or organic sulfur basis (**Figure 2**).

A 7-day sulfate deprivation resulted in 50% decrease in the glucosinolate content of both shoots and roots of *B. juncea* and a 20 and 45% decrease in content of the shoots and roots

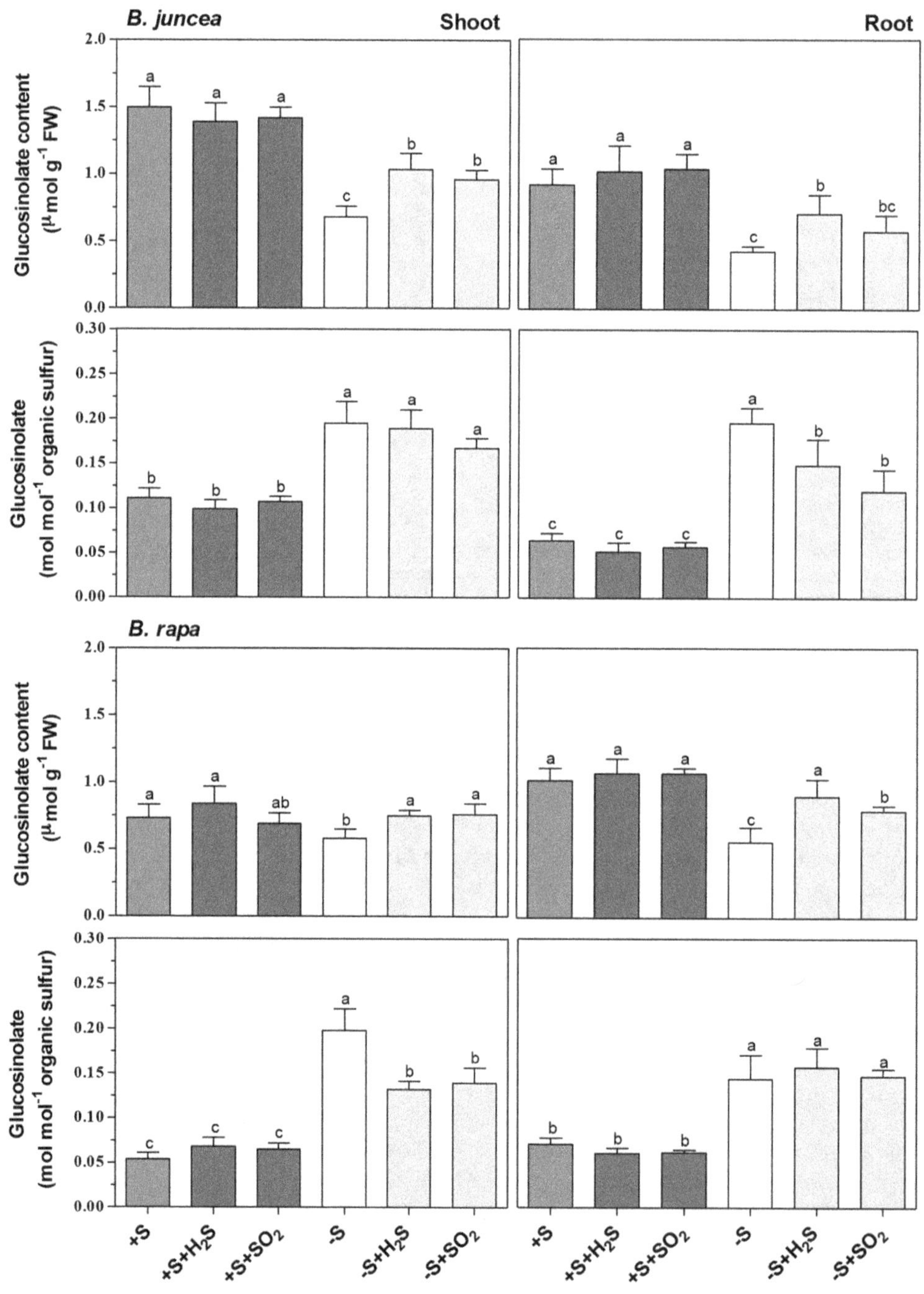

FIGURE 2 | Impact of H_2S, SO_2 and sulfate deprivation on glucosinolate content of shoots and roots of *B. juncea* and *B. rapa*. For experimental details, see legends of **Table 1**. Data represent the mean of two experiments with three measurements with nine plants in each (±SD). Different letters indicate significant differences between treatments ($P < 0.01$, Student's *t*-test).

of *B. rapa*, respectively (**Figure 2**). H_2S and SO_2 exposure of sulfate-deprived plants alleviated almost fully the decrease in glucosinolate contents of shoot and roots of *B. rapa*, and their contents were quite similar (except that of the roots of sulfate-deprived SO_2-exposed plants) to that of sulfate-sufficient plants (**Figure 2**). H_2S and SO_2 exposure of sulfate-deprived *B. juncea* also resulted in a higher glucosinolate content of both shoots and roots. However, their contents were significantly lower than that observed in sulfate-sufficient plants (**Figure 2**). On an organic sulfur basis, however, the glucosinolate content of shoots and roots of sulfate-deprived of *B. juncea* and *B. rapa*, both in absence and presence of H_2S or SO_2, was always higher than that observed in sulfate-sufficient plants (**Figure 2**). This indicated that proportionally, the content of other organic sulfur compounds (e.g., proteins), was more affected by sulfate deprivation than that of the glucosinolates, even in presence of foliarly absorbed H_2S or SO_2 as alternative sulfur sources for growth.

IMPACT OF ATMOSPHERIC AND PEDOSPHERIC SULFUR NUTRITION ON THE ACTIVITY AND EXPRESSION OF MYROSINASE

There were considerable differences in the activity of myrosinase in shoots and roots of *B. juncea* and *B. rapa* seedlings (**Table 2**). Both shoots and roots of *B. juncea* were characterized by a high myrosinase activity. *B. rapa*, however, was characterized by very low and hardly detectible myrosinase activity in the shoot and high activity in the root, the latter being quite similar to that observed in *B. juncea*. If glucoberin was used as substrate instead of sinigrin, comparable myrosinase activities were observed (data not presented). This demonstrated that the differences in activity of this enzyme in the shoots of *B. juncea* and *B. rapa* were unlikely to be explained by differences in substrate selectivity. Additionally the transcript level of myrosinase was substantially lower in shoots of *B. rapa* than that of *B. juncea* (**Figure 3**). In sulfate-sufficient conditions, H_2S and SO_2 exposure did not affect the activities of myrosinase enzyme of shoots and roots of *B. juncea* and *B. rapa* (**Table 2**) or the expression of the gene in the shoots (**Figure 3**).

A 7-day sulfate deprivation resulted in a decrease in myrosinase activity in roots of *B. juncea* and *B. rapa* (expressed on a fresh weight basis), while activity in the shoot of *B. juncea* was not affected (**Table 2**). However, the transcript levels of myrosinase in shoots of both *B. juncea* and *B. rapa* were decreased upon sulfate deprivation (**Figure 3**). Sulfate deprivation resulted in a decrease in soluble protein content of shoots and roots of both species (**Table 2**). If the myrosinase activity in the roots of both species was expressed on a protein basis, then specific activity remained unaffected and was around 90 nmol mg^{-1} protein min^{-1}, whereas in the shoot of *B. juncea* its activity increased from 23 in sulfate-sufficient to 39 nmol mg^{-1} protein min^{-1} in sulfate-deprived plants. Exposure of sulfate-deprived plants to H_2S or SO_2 resulted in an increase in soluble protein content, although it hardly affected the myrosinase activity in shoots and roots of *B. juncea* and *B. rapa* (expressed on a fresh weight basis; **Table 2**). H_2S or SO_2 exposure of sulfate-deprived plants did not affect the myrosinase transcript level in the shoot of *B. juncea*, whereas it was increased in the shoot of *B. rapa* (**Figure 3**).

DISCUSSION

Brassicaceae are fast growing species characterized by a relatively high sulfur requirement (Westerman et al., 2001a; Castro et al., 2003; Yang et al., 2006a,b). Under the experimental conditions used here, *B. juncea* and *B. rapa* seedlings had a relative growth

Table 2 | Impact of H_2S, SO_2 and sulfate deprivation on myrosinase activity and water-soluble protein content of shoots and roots of *B. juncea* and *B. rapa*. For experimental details see legends of Table 1.

	+S	+S + H_2S	+S + SO_2	−S	−S + H_2S	−S + SO_2
B. juncea						
Shoot						
Myrosinase activity	0.31 ± 0.03ab	0.28 ± 0.01b	0.34 ± 0.03ab	0.37 ± 0.02a	0.36 ± 0.02a	0.38 ± 0.02a
Soluble proteins content	13.5 ± 1.7a	12.9 ± 1.8a	13.7 ± 1.2a	9.5 ± 1.0b	12.1 ± 1.3a	12.5 ± 1.6a
Root						
Myrosinase activity	0.34 ± 0.07a	0.27 ± 0.04a	0.26 ± 0.06a	0.10 ± 0.01b	0.07 ± 0.01c	0.11 ± 0.03bc
Soluble proteins content	4.1 ± 0.5a	4.2 ± 1.0a	3.7 ± 0.9a	1.1 ± 0.1c	2.3 ± 0.4b	2.7 ± 0.48ab
B. rapa						
Shoot						
Myrosinase activity	<0.01a	<.0.01a	<0.01a	<0.01a	<0.01a	<0.01a
Soluble proteins content	11.4 ± 1.3a	12.6 ± 2.1a	11.8 ± 1.0a	8.6 ± 0.5b	10.5 ± 0.6a	11.6 ± 1.4a
Root						
Myrosinase activity	0.42 ± 0.03a	0.48 ± 0.07a	0.40 ± 0.03a	0.14 ± 0.01b	0.12 ± 0.02b	0.13 ± 0.02b
Soluble proteins content	4.2 ± 0.3a	4.9 ± 1.1ab	4.3 ± 0.7ac	1.2 ± 0.1d	2.5 ± 0.2bc	3.0 ± 0.1b

*For experimental details see legends of **Table 1**. Data on myrosinase activity (nmol mg^{-1} fresh weight min^{-1}) and soluble protein content (mg g^{-1} fresh weight) represent the mean of three measurements with three plants in each (±SD). Different letters indicate significant differences between treatments ($P < 0.01$, Student's t-test).*

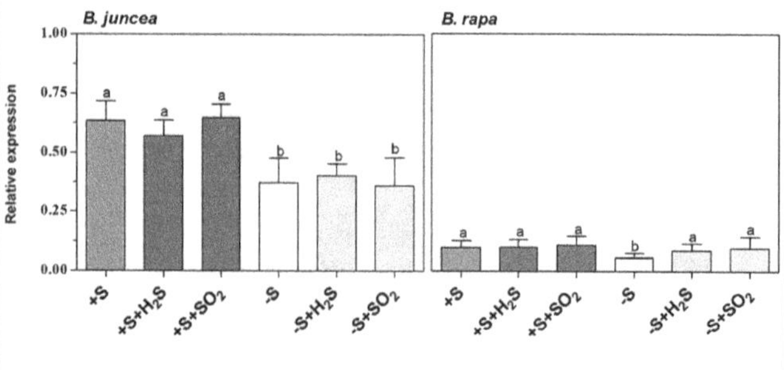

FIGURE 3 | Impact of H_2S, SO_2 and sulfate deprivation on transcript levels of myrosinase in shoots of *B. juncea* and *B. rapa*. For experimental details, see legends of **Table 1**. Relative gene expression of myrosinase was determined by qRT-PCR and the mRNA levels were compared to actin. Data on relative expression in each treatment represent the mean of three measurements with three shoots in each (±SD). Different letters indicate significant differences between treatments ($P < 0.01$, Student's *t*-test).

rate of 36.9 and 38.6% day^{-1} and a plant sulfur content of 33.4 and 29.4 μmol g^{-1} fresh weight (340 and 272 μmol g^{-1} dry weight), respectively (data derived from **Table 1**; **Figure 1**). Seedlings of Brassicaceae are often characterized by a relatively high ratio of sulfate to organic sulfur. For instance, in *B. oleracea* the sulfate content may account for more than 80% of the total sulfur in the shoot (Westerman et al., 2001b; Castro et al., 2003; Yang et al., 2006a,b). The sulfate content accounted for 60 and 50% of total sulfur in the shoots and roots of *B. juncea*, and for 55 and 43% in shoots and roots of *B. rapa*, respectively. *Brassica* species are also characterized by a relatively high content of secondary sulfur compounds, viz. glucosinolates, which strongly varies between species, and may in seedlings account for 10 to 23% of the organic sulfur fraction (Castro et al., 2004). The glucosinolates content of the shoot of *B. juncea* was twofold higher than that in the shoot of *B. rapa* and accounted for up to 30% of the organic sulfur fraction (on the basis of 3 S groups per molecule), presuming that the aliphatic glucosinolates are the major secondary sulfur compounds present in *Brassica* (Kushad et al., 1999; Van Dam et al., 2003; Cartea et al., 2008). In the root, content appeared to be quite similar in both species, where it accounted for 15% of the organic sulfur fraction.

Atmospheric sulfur gases, viz. SO_2 and H_2S, may be taken up by the plant shoot and used as a sulfur source for growth (De Kok et al., 2007). The foliar uptake of SO_2 is determined by its chemical/physical properties, viz. rapid dissociation in the water of the mesophyll apoplast, which is beyond regulatory control (De Kok et al., 2007). The foliar uptake of H_2S, however, is largely determined by the rate of metabolism in the shoot (De Kok et al., 2007). Exposure of *B. juncea* and *B. rapa* to 0.25 μl l^{-1} SO_2 and H_2S hardly affected the total sulfur content of both species. There was only a slight increase in the shoot total sulfur content of SO_2-exposed *B. rapa*, which could be attributed to an enhanced sulfate content. However, it was evident that at this atmospheric sulfur concentration of SO_2 and H_2S, plants were able to take up sufficient sulfur by the shoot to fully cover their organic sulfur requirement for plant growth. Since, the decrease in biomass production upon sulfate-deprivation was completely alleviated by SO_2 and H_2S exposure. Similarly to previous observations, sulfate-deprived plants invested relatively more biomass in their roots than those grown under sulfate-sufficient conditions, even upon SO_2 and H_2S exposure (Buchner et al., 2004; Yang et al., 2006a; Koralewska et al., 2008; Shahbaz et al., 2014).

In greenhouse and field experiments involving soil, hydroponic, and tissue culture media, sulfur fertilization generally resulted in an increased glucosinolate content of *Brassica* (Falk et al., 2007). However, SO_2 and H_2S exposure of *B. juncea* and *B. rapa* seedlings did not affect the glucosinolate content of the shoots and roots, either expressed on fresh weight or organic sulfur basis. This demonstrated that these sulfur compounds did not form a sink for excessive supplied atmospheric sulfur. SO_2 and H_2S exposure of *Brassica* generally resulted in enhanced levels of water-soluble, non-protein thiol compounds in the shoot (De Kok and Tausz, 2001; Westerman et al., 2001a; Buchner et al., 2004; Yang et al., 2006a; Koralewska et al., 2008; Shahbaz et al., 2014), which could be ascribed to an accumulation of cysteine and glutathione (De Kok and Tausz, 2001; De Kok et al., 2007). Apparently, an enhanced availability of these thiol compounds, which also function as reduced sulfur donors in the synthesis of glucosinolates (Schnug, 1990, 1993; Halkier and Gershenzon, 2006; Falk et al., 2007) did not affect the rate of synthesis of these secondary sulfur compounds in the shoot. *Allium* species (e.g., onion, garlic, leek) also contain secondary sulfur compounds viz. γ-glutamyl peptides and allins, which are synthesized from cysteine, via γ-glutamylcysteine or glutathione. In contrast to observations with glucosinolates, in *Brassica* the content of these secondary sulfur compounds, their precursors and/or degradation products were strongly enhanced in shoots of H_2S-exposed *Allium* (Durenkamp and De Kok, 2002, 2004; Durenkamp et al., 2005). SO_2 exposure, however, hardly affected the levels of these secondary sulfur compounds in *Allium* (Durenkamp et al., 2005).

Exposure of *Brassica* to H_2S resulted in a strong decrease in expression, in protein level and in enzyme activity of APS reductase in the shoot (Westerman et al., 2000a,b, 2001a,b; Durenkamp et al., 2007; Koralewska et al., 2008; Shahbaz et al., 2014). Evidently, a down-regulation of APS reductase, the key enzyme controlling the flux through the sulfate reduction pathway, did not affect the synthesis of glucosinolates via the channeling of the APS through the APS kinase/sulfotransferase pathway. The latter is essential for the synthesis of the sulfated moiety of the glucosinolates (Kopriva et al., 2012). This demonstrated that the synthesis of glucosinolates in the shoot of *Brassica* seedlings was under strict regulatory control and was not affected by an excess supply of foliarly absorbed sulfur, irrespective of the differences in content of these secondary sulfur compounds between *B. juncea* and *B. rapa*.

Sulfate deprivation resulted in a strong decrease in the total sulfur content of shoots and roots of *B. juncea* and *B. rapa*, which was for the greater part due to a decrease in sulfate content. It has been observed that in *Arabidopsis* sulfur deficiency resulted in a repression of the glucosinolate biosynthesis genes (Hirai et al., 2005). There was also a strong decrease in the glucosinolate content of both shoots and roots of *B. juncea* and *B. rapa* upon sulfate deprivation. However, the decrease in glucosinolate content was lower than that of the other organic sulfur compounds, resulting in an increase in its content expressed on organic sulfur basis. Sulfur in proteins generally accounts for more than 80% of the organic sulfur content (Stulen and De Kok, 1993). Apparently, sulfate deprivation had a higher impact on the proportion of sulfur in proteins in *Brassica* than the other organic sulfur compounds viz. glucosinolates. It has been observed that in shoots, sulfate deprivation resulted in a degradation of Rubisco, which proportion may account for 25–60% of the soluble proteins in photosynthetic tissue (Ferreira and Teixeira, 1992; Gilbert et al., 1997). The proportion of sulfur in the glucosinolates in both shoots and roots of *B. juncea* and *B. rapa* in sulfate-deprived plants exceeded 50% of the total organic sulfur fraction (on the basis of 3 S groups per molecule). These results indicated that glucosinolates cannot be considered as sulfur storage compounds and they were not utilized in the re-distribution of sulfur in *B. juncea* and *B. rapa* seedlings upon sulfate deprivation.

SO_2 and H_2S exposure resulted in an increase in total sulfur content of shoots and roots of the two *Brassica* species upon sulfate deprivation, which was mainly due to an increase in the organic sulfur fraction. Likewise, there was an increase in glucosinolate content, though the levels were always lower than that of sulfate-sufficient plants. However, on an organic sulfur basis, the glucosinolate content in both shoots and roots was substantially higher than that observed in sulfate-sufficient plants upon SO_2 and H_2S exposure. Apparently in sulfate-deprived conditions, a relatively greater proportion of atmospheric sulfur taken up by the shoot was used for the synthesis of glucosinolates than that of other organic sulfur compounds, e.g., proteins. Roots of sulfate-deprived plants depended on sulfur transported form shoot to root upon exposure to SO_2 and H_2S, although in which form the sulfur was transported from shoot to root under these conditions remains unknown. Organic sulfur may be transported in the phloem from source to sink, e.g., from shoot to root in different forms, viz. glutathione, *S*-methylmethionine or as glucosinolates (Brunold and Rennenberg, 1997; Bourgis et al., 1999; Grubb and Abel, 2006; Rennenberg and Herschbach, 2014).

Glucosinolates may be involved in plant defense against pathogens and herbivory (Bones et al., 1994; Brader et al., 2006; Grubb and Abel, 2006; Bednarek et al., 2011; Frerigmann et al., 2012; Schiestl, 2014) and is dependent upon breakdown activated by tissue damage and catalyzed by myrosinase (Bones et al., 1994; Grubb and Abel, 2006; Clay et al., 2009; Kissen and Bones, 2009; Kissen et al., 2009; Ahuja et al., 2010). Moreover, it has been suggested that myrosinase might have significance in the redistribution of sulfur in plants under sulfur-deprived conditions (Schnug, 1990; Hirai et al., 2004, 2005; Bloem et al., 2007; Falk et al., 2007).

There was a direct relation between the content of glucosinolates and the transcript level and activity of myrosinase in the shoot of *B. juncea* and *B. rapa*. Both low glucosinolate content and a low transcript level of myrosinase and activity of this enzyme characterized the shoot of the latter species. Even though myrosinase activity was decreased in both roots of sulfate-deprived *B. juncea* and *B. rapa*, SO_2 and H_2S exposure hardly affected the activity in shoots and roots under either sulfate-sufficient or sulfate-deprived conditions, despite an increase in glucosinolate content upon exposure in sulfate-deprived plants. There was apparently no direct co-regulation between the content of glucosinolates and the activity of myrosinase. From the current observation that the sulfur of glucosinolates was hardly re-distributed upon sulfate deprivation, the significance of myrosinase in the *in situ* turnover of these secondary sulfur compounds needs to be questioned.

AUTHOR CONTRIBUTIONS

Tahereh Aghajanzadeh, Malcolm J. Hawkesford, and Luit J. De Kok designed the research. Tahereh Aghajanzadeh carried out the experiments and analyzed the data. All authors contributed to writing the manuscript. Luit J. De Kok and Malcolm J. Hawkesford supervised the project.

ACKNOWLEDGMENTS

Rothamsted Research is supported via the 20:20 Wheat® Programme by the UK Biotechnology and Biological Sciences Research Council. We wish to thank Helen Jenkins and C. Elisabeth E. Stuiver for critically reading of the manuscript.

REFERENCES

Ahuja, I., Rohloff, J., and Bones, A. M. (2010). Defence Mechanisms of Brassicaceae: implications for plant-insect interactions and potential for integrated pest management. *Agron. Sustain. Dev.* 30, 311–348. doi: 10.1051/Agro/2009025

Andersson, D., Chakrabarty, R., Bejai, S., Zhang, J., Rask, L., and Meijer, J. (2009). Myrosinases from root and leaves of *Arabidopsis thaliana* have different catalytic properties. *Phytochemistry* 70, 1345–1354. doi: 10.1016/j.phytochem.2009.07.036

Antonious, G., Bomford, M., and Vincelli, P. (2009). Screening *Brassica* species for glucosinolate content. *J. Environ. Sci. Health B* 44, 311–316. doi: 10.1080/03601230902728476

Bednarek, P., Piślewska-Bednarek, M., van Themaat, E. V. L., Maddula, R. K., Svatoš, A., and Schulze-Lefert, P. (2011). Conservation and clade-specific diversification of pathogen-inducible tryptophan and indole glucosinolate metabolism in *Arabidopsis thaliana* relatives. *New Phytol.* 192, 713–726. doi: 10.1111/j.1469-8137.2011.03824.x

Bloem, E., Haneklaus, S., and Schnug, E. (2007). Changes in the sulphur and glucosinolate content of developing pods and seeds of oilseed rape (*Brassica napus* L.) in relation to different cultivars. *Landbauforsch. Völk.* 57, 297–306.

Bones, A. M., and Rossiter, J. T. (1996). The myrosinase-glucosinolate system, its organisation and biochemistry. *Physiol. plant.* 96, 194–208. doi: 10.1111/j.1399-3054.1996.tb00497.x

Bones, A. M., and Rossiter, J. T. (2006). The enzymic and chemically induced decomposition of glucosinolates. *Phytochemistry* 67, 1053–1067. doi: 10.1016/j.phytochem.2006.02.024

Bones, A., Visvalingam, S., and Thangstad, O. (1994). Sulphate can induce differential expression of thioglucoside glucohydrolases (myrosinases). *Planta* 193, 558–566. doi: 10.1007/BF02411562

Bourgis, F., Roje, S., Nuccio, M. L., Fisher, D. B., Tarczynski, M. C., Li, C., et al. (1999). S-methylmethionine plays a major role in phloem sulfur transport and is synthesized by a novel type of methyltransferase. *Plant Cell* 11, 1485–1497. doi: 10.1105/tpc.11.8.1485

Brader, G., Mikkelsen, M. D., Halkier, B. A., and Palva, E. T. (2006). Altering glucosinolate profiles modulates disease resistance in plants. *Plant J.* 46, 758–767. doi: 10.1111/j.1365-313X.2006.02743.x

Bradford, M. M. (1976). A rapid and sensitive method for the quantitation of microgram quantities of protein utilizing the principle of protein-dye binding. *Anal. Biochem.* 72, 248–254. doi: 10.1016/0003-2697(76)90527-3

Brunold, C., and Rennenberg, H. (1997). "Regulation of sulfur metabolism in plants: first molecular approaches," in *Progress in Botany*, eds H. D. Behnke, U. Lüttge, H. C. M. Karl Esser, J. W. Kadereit, and M. Runge (Berlin: Springer) 58, 164–186.

Buchner, P., Stuiver, C. E. E., Westerman, S., Wirtz, M., Hell, R., Hawkesford, M. J., et al. (2004). Regulation of sulfate uptake and expression of sulfate transporter genes in *Brassica oleracea* as affected by atmospheric H_2S and pedospheric sulfate nutrition. *Plant Physiol.* 136, 3396–3408. doi: 10.1104/pp.104.046441

Cartea, M. E., Velasco, P., Obregón, S., Padilla, G., and De Haro, A. (2008). Seasonal variation in glucosinolate content in *Brassica oleracea* crops grown in northwestern Spain. *Phytochemistry* 69, 403–410. doi: 10.1016/j.phytochem.2007.08.014

Castro, A., Aires, A., Rosa, E., Bloem, E., and Stulen, I. (2004). Distribution of glucosinolates in *Brassica oleracea* cultivars. *Phyton (B Aires)* 44, 133–143.

Castro, A., Stulen, I., and De Kok, L. J. (2003). "Nitrogen and sulfur requirement of *Brassica oleracea* L. cultivars," in *Sulfur Transport and Assimilation in Plants*, eds J. C. Davidian, D. Grill, L. J. De Kok, I. Stulen, M. J. Hawkesford, E. Schnug, et al. (Leiden: Backhuys Publishers), 181–183.

Clay, N. K., Adio, A. M., Denoux, C., Jander, G., and Ausubel, F. M. (2009). Glucosinolate metabolites required for an *Arabidopsis* innate immune response. *Science* 323, 95–101. doi: 10.1126/science.1164627

De Kok, L. J., Durenkamp, M., Yang, L., and Stulen, I. (2007). "Atmospheric sulfur," in *Sulfur in Plants: An Ecological Perspective*, eds M. J. Hawkesford and L. J. De Kok (Berlin: Springer), 91–106. doi: 10.1007/978-1-4020-5887-5_5

De Kok, L. J., Stuiver, C. E. E., Westerman, S., and Stulen, I. (2000a). "Atmospheric H_2S pollution – deposition and impact on sulphur metabolism in plants," in *Environmental Stress: Indication, Mitigation and Eco-Conservation*, eds M. Yunus, N. Singh, and L. J. De Kok (Berlin: Springer), 135–141.

De Kok, L. J., Westerman, S., Stuiver, C. E. E., and Stulen, I. (2000b). "Atmospheric H_2S as plant sulphur source: interaction with pedospheric sulfur nutrition - a case study with *Brassica oleracea* L.," in *Sulphur Nutrition and Sulphur Assimilation in Higher Plants: Molecular, Biochemical and Physiological Aspects*, eds C. Brunold, H. Rennenberg, L. J. De Kok, I. Stulen, and J.-C. Davidian (Bern: Paul Haupt Verlag), 41–55.

De Kok, L. J., Stuiver, C. E. E., Westerman, S., and Stulen, I. (2002). "Elevated level of hydrogen sulphide in the plant environment: nutrient or toxin,"in *Air Pollution and Plant Biotechnology-Prospects for Phytomonitoring and Phytoremediation*, eds K. Omasa, H. Saji, S. Youssefian, and N. Kondo (Tokyo: Springer-Verlag), 201–219.

De Kok, L. J., and Tausz, M. (2001). "The role of glutathione in plant reaction and adaptation to air pollutants," in *Significance of Glutathione to Plant Adaptation to the Environment*, eds D. Grill, M. Tausz, and L. J. De Kok (Dordrecht: Kluwer Academic Publishers), 185-201.

De Kok, L. J., Yang, L., Stuiver, C. E. E., and Stulen, I. (2009). "Negative versus positive functional plant responses to air pollution: a study establishing cause-effect relationship of SO_2 and H_2S," in *Air Quality and Ecological Impacts: Relating Sources to Effects*, ed. A. K. Legge (Amsterdam: Elsevier), 121–134.

Del Carmen Martinez-Ballesta, M., Moreno, D., and Carvajal, M. (2013). The physiological importance of glucosinolates on plant response to abiotic stress in *Brassica*. *Int. J. Mol. Sci.* 14, 11607–11625. doi: 10.3390/ijms140611607

Durenkamp, M., and De Kok, L. J. (2002). The impact of atmospheric H_2S on growth and sulfur metabolism of *Allium* cepa L. *Phyton (B Aires)* 42, 55–63.

Durenkamp, M., and De Kok, L. J. (2004). Impact of pedospsheric and atmospheric sulphur nutrition on sulphur metabolism of *Allium* cepa L. a species with a potential sink capacity for secondary sulphur compounds. *J. Exp. Bot.* 55, 1821–1830. doi: 10.1093/jxb/erh187

Durenkamp, M., De Kok, L. J., and Kopriva, S. (2007). Adenosine 5'-phosphosulphate reductase is regulated differently in *Allium cepa* L. and *Brassica oleracea* L. upon exposure to H_2S. *J. Exp. Bot.* 58, 1571–1579. doi: 10.1093/jxb/erm031

Durenkamp, M., Posthumus, F. S., Stuiver, C. E. E., and De Kok, L. J. (2005). "Metabolism of atmospheric sulfur gases in onion," in *Plant Responses to Air Pollution and Global Change*, eds K. Omasa, I. Nouchi, and L. J. De Kok (Tokyo: Springer-Verlag), 3–11.

Ernst, W. (1993). "Ecological aspects of sulfur in higher plants: The impact of SO_2 and the evolution of the biosynthesis of organic sulfur compounds on populations and ecosystems," in *Sulfur Nutrition and Sulfur Assimilation in Higher Plants; Regulatory, Agricultural and Environmental Aspects*, eds L. J. De Kok, I. Stulen, H. Rennenberg, C. Brunold, and W. Rauser (The Hague: SPB Academic Publishing), 295–313.

Fahey, J. W., Zalcmann, A. T., and Talalay, P. (2001). The chemical diversity and distribution of glucosinolates and isothiocyanates among plants. *Phytochemistry* 56, 5–51. doi: 10.1016/S0031-9422(00)00316-2

Falk, K. L., Tokuhisa, J. G., and Gershenzon, J. (2007). The effect of sulfur nutrition on plant glucosinolate content: physiology and molecular mechanisms. *Plant Biol.* 9, 573–581. doi: 10.1055/s-2007-965431

Fenwick, G. R., Heaney, R. K., and Mullin, W. J. (1983). Glucosinolates and their breakdown products in food and food plants. *Crit. Rev. Food Sci. Nutr.* 18, 123–201. doi: 10.1080/10408398209527361

Ferreira, R. M., and Teixeira, A. R. (1992). Sulfur starvation in Lemna leads to degradation of ribulose-bisphosphate carboxylase without plant death. *J. Biol. Chem.* 267, 7253–7257.

Frerigmann, H., Böttcher, C., Baatout, D., and Gigolashvili, T. (2012). Glucosinolates are produced in trichomes of *Arabidopsis thaliana*. *Front. Plant Sci.* 3:242. doi: 10.3389/fpls.2012.00242

Gilbert, S., Clarkson, D. T., Cambridge, M., Lambers, H., and Hawkesford, M. J. (1997). Sulfate-deprivation has an early effect on the content of ribulose 1,5-bisphosphate carboxylase/oxygenase and photosynthesis in young leaves of wheat. *Plant Physiol.* 115, 1231–1239. doi: 10.1104/pp.115.3.1231

Grubb, C. D., and Abel, S. (2006). Glucosinolate metabolism and its control. *Trends Plant Sci.* 11, 89–100. doi: 10.1016/j.tplants.2005.12.006

Gupta, S., Sangha, M. K., Kaur, G., Atwal, A. K., Kaur, P., Kumar, H., et al. (2012). Biochem. *Anal. Biochem.* 1, 121. doi: 10.4172/2161-1009.1000121

Halkier, B. A., and Du, L. (1997). The biosynthesis of glucosinolates. *Trends Plant Sci.* 2, 425–431. doi: 10.1016/S1360-1385(97)90026-1

Halkier, B. A., and Gershenzon, J. (2006). Biology and biochemistry of glucosinolates. *Annu. Rev. Plant Biol.* 57, 303–333. doi: 10.1146/annurev.arplant.57.032905.105228

Heaney, R. K., and Fenwick, G. R. (1980). Glucosinolates in *Brassica* vegetables. analysis of 22 varieties of brussels sprouts (*Brassica oleracea* var. gemmifera). *J. Sci. Food Agr.* 31, 785–793. doi: 10.1002/jsfa.2740310808

Hirai, M. Y., Klein, M., Fujikawa, Y., Yano, M., Goodenowe, D. B., Yamazaki, Y., et al. (2005). Elucidation of gene-to-gene and metabolite-to-gene networks in *Arabidopsis* by integration of metabolomics and transcriptomics. *J. Biol. Chem.* 280, 25590–25595. doi: 10.1074/jbc.M502332200

Hirai, M. Y., Yano, M., Goodenowe, D. B., Kanaya, S., Kimura, T., Awazuhara, M., et al. (2004). Integration of transcriptomics and metabolomics for understanding of global responses to nutritional stresses in *Arabidopsis thaliana*. *Proc. Natl. Acad. Sci. U.S.A.* 101, 10205–10210. doi: 10.1073/pnas.0403218101

Ishida, M., Nagata, M., Ohara, T., Kakizaki, T., Hatekeyama, K., and Nishio, T. (2012). Small variation of glucosinolat composition in Japanese cultivars of radish (*Raphanus sativus* L.) requires simple quantitative analysis for breeding of glucosinolate component. *Breed. Sci.* 62, 63–70. doi: 10.1270/jsbbs62.63

Jones, J. B. (1995). Determining total sulphur in plant tissue using the HACH kit spectrophotometer technique. *Sulphur Agric.* 19, 58–62.

Kirkegaard, J. A., and Sarwar, M. (1998). Biofumigation potential of Brassicas. *Plant Soil* 201, 71–89. doi: 10.1023/A:1004364713152

Kissen, R., and Bones, A. M. (2009). Nitrile-specifier proteins involved in glucosinolate hydrolysis in *Arabidopsis thaliana*. *J. Biol. Chem.* 284, 12057–12070. doi: 10.1074/jbc.M807500200

Kissen, R., Rossiter, J., and Bones, A. (2009). The 'mustard oil bomb': not so easy to assemble?! Localization, expression and distribution of the components of the myrosinase enzyme system. *Photochem. Rev.* 8, 69–86. doi: 10.1007/s11101-008-9109-1

Kopriva, S., Mugford, S. J., Baraniecka, P., Lee, B., Matthewman, C. A., and Koprivova, A. (2012). Control of sulfur partitioning between primary and secondary metabolism in *Arabidopsis*. *Front. Plant. Sci.* 3:163. doi: 10.3389/fpls.2012.00163

Koralewska, A., Stuiver, C. E. E., Posthumus, F. S., Kopriva, S., Hawkesford, M. J., and De Kok, L. J. (2008). Regulation of sulfate uptake, expression of the sulfate transporters Sultr1;1 and Sultr1;2, and APS reductase in Chinese cabbage (*Brassica pekinensis*) as affected by atmospheric H_2S nutrition and sulfate deprivation. *Funct. Plant Biol.* 35, 318–327. doi: 10.1071/FP07283

Kushad, M. M., Brown, A. F., Kurilich, A. C., Juvik, J. A., Klein, B. P., Wallig, M. A., et al. (1999). Variation of glucosinolates in vegetable crops of *Brassica oleracea*. *J. Agric. Food Chem.* 47,1541–1548. doi: 10.1021/jf980985s

Maas, F. M., Hoffmann, I., Van Harmelen, M. J., and De Kok, L. J. (1986). Refractometric determination of sulphate and other anions in plants separated by high-performance liquid chromatography. *Plant Soil* 91, 129–132. doi: 10.1007/BF02181825

O'Callaghan, K. J., Stone, P. J., Hu, X., Griffiths, D. W., Michael, R., Davey, M. R., et al. (2000). Effects of glucosinolates and flavonoids on colonization of the roots of *Brassica napus* by *Azorhizobium caulinodans* ORS571. *Appl. Environ. Microbiol.* 66, 2185–2191. doi: 10.1128/AEM.66.5.2185-2191.2000

Rennenberg, H., and Herschbach, C. (2014). A detailed view on sulphur metabolism at the cellular and whole-plant level illustrates challenges in metabolite flux analyses. *J. Exp. Biol.* 65, 5711–5724. doi: 10.1093/jxb/eru315

Schiestl, F. P. (2014). Correlation analyses between volatiles and glucosinolates show no evidence for chemical defense signaling in *Brassica rapa*. *Front. Ecol. Evol.* 2:10. doi: 10.3389/fevo.2014.00010

Schnug, E. (1990). "Glucosinolates – fundamental, environmental and agricultural aspects," in *Sulfur Nutrition and Sulfur Assimilation in Higher Plant; Fundamental, Environmental and Agricultural Aspects*, eds H. Rennenberg, C. Brunold, L. J. De Kok, and I. Stulen (The Hague: SPB Academic Publishing), 97–106.

Schnug, E. (1993). "Physiological functions and environmental relevance of sulfur-containing secondary metabolites," in *Sulfur Nutrition and Sulfur Assimilation in Higher Plant; Regulatory Agricultural and Environmental Aspects*, eds L. J. De Kok, I. Stulen, H. Rennenberg, C. Brunold, and W. E. Rauser (The Hague: SPB Academic Publishing), 179–190.

Shahbaz, M., Stuiver, C. E. E., Posthumus, F. S., Parmar, S., Hawkesford, M. J., and De Kok, L. G. (2014). Copper toxicity in Chinese cabbage is not influenced by plant sulphur status, but affects sulphur metabolism-related gene expression and the suggested regulatory metabolites. *Plant Biol.* 16, 68–78. doi: 10.1111/plb.12019

Stulen, I., and De Kok, L. J. (1993). "Whole plant regulation of sulfate uptake and metabolism – a theoretical approach and comparison with current ideas on regulation of nitrogen metabolism," in *Sulfur Nutrition and Assimilation in Higher Plants; Regulatory, Agricultural and Environmental Aspects*, eds L. J. De Kok, I. Stulen, H. Rennenberg, C. Brunold, and W. E. Rauser (The Hague: SPB Academic Publishing), 77–91.

Tausz, M., De Kok, L. J., Stulen, I., and Grill, D. (1996). Physiological responses of Norway spruce trees to elevated CO_2 and SO_2. *J. Plant Physiol.* 148, 362–367. doi: 10.1016/S0176-1617(96)80266-5

Thies, W. (1982). Complex-formation between glucosinolates and tetrachloropalladate (II) and its utilization in plant breeding. *Fette Seifen Anstrichmittel.* 84, 388–342. doi: 10.1002/lipi.19820840903

Travers-Martin, N., Kuhlmann, F., and Müller, C. (2008). Revised determination of free and complexed myrosinase activities in plant extracts. *Plant Physiol. Biochem.* 46, 506–516. doi: 10.1016/j.plaphy.2008.02.008

Van Dam, N. M., Witjes, L., and Svatoš, A. (2003). Interactions between aboveground and belowground induction of glucosinolates in two wild *Brassica* species. *New Phytol.* 161, 801–810. doi: 10.1111/j.1469 8137.2004.00984.x

Verwoerd, T. C., Dekker, B. M., and Hoekema, A. (1989). A small-scale procedure for the rapid isolation of plant RNAs. *Nucleic Acids Res.* 17:2362. doi: 10.1093/nar/17.6.2362

Westerman, S., Blake-Kalff, M. A. A., De Kok, L. J., Stuiver, C. E. E., and Stulen, I. (2001a). Sulfate uptake and utilization by two varieties of *Brassica oleracea* with different sulfur need as affected by atmospheric H_2S. *Phyton (B Aires)* 41, 49–62.

Westerman, S., Stulen, I., Suter, M., Brunold, C., and De Kok, L. J. (2001b). Atmospheric H_2S as sulfur source for *Brassica oleracea*: consequences for the activity of the enzymes of the assimilatory sulfate reduction pathway. *Plant Physiol. Biochem.* 39, 425–432. doi: 10.1016/S0981-9428(01)01258-X

Westerman, S., De Kok, L. J., and Stulen, I. (2000a). Interaction between metabolism of atmospheric H_2S in the shoot and sulfate uptake by the roots of curly kale (*Brassica oleracea* L). *Physiol. Plant.* 109, 443–449. doi: 10.1034/j.1399-3054.2000.100411.x

Westerman, S., Weidner, W., De Kok, L. J., and Stulen, I. (2000b). Effect of H_2S exposure on 35S-sulfate uptake, transport and utilization in curly kale. *Phyton (B Aires)* 40, 293–302.

Yang, L., Stulen, I., and De Kok, L. J. (2006a). Sulfur dioxide: relevance of toxic and nutritional effects for Chinese cabbage. *Environ. Exp. Bot.* 57, 236–245. doi: 10.1016/j.envexpbot.2005.06.002

Yang, L., Stulen, I., and De Kok, L. J. (2006b). Impact of sulfate nutrition on the utilization of atmospheric SO_2 as sulfur source for Chinese cabbage. *J. Plant Nutr. Soil Sci.* 169, 529–534. doi: 10.1002/jpln.200520574

Conflict of Interest Statement: The authors declare that the research was conducted in the absence of any commercial or financial relationships that could be construed as a potential conflict of interest.

3

To control and to be controlled: understanding the *Arabidopsis* SLIM1 function in sulfur deficiency through comprehensive investigation of the EIL protein family

Anna Wawrzyńska* and Agnieszka Sirko

Institute of Biochemistry and Biophysics, Polish Academy of Sciences, Warsaw, Poland

Edited by:
Stanislav Kopriva, University of Cologne, Germany

Reviewed by:
Ruediger Hell, University of Heidelberg, Germany
Margret M. Sauter, University of Kiel, Germany
Bok-Rye Lee, Chonnam National University, South Korea

****Correspondence:***
Anna Wawrzyńska, Institute of Biochemistry and Biophysics, Polish Academy of Sciences, Pawińskiego 5A Street, 02-106 Warsaw, Poland
e-mail: blaszczyk@ibb.waw.pl

Sulfur limitation 1 (SLIM1), a member of the EIN3-like (EIL) family of transcription factors in *Arabidopsis*, is the regulator of many sulfur deficiency responsive genes. Among the five other proteins of the family, three regulate ethylene (ET) responses and two have unassigned functions. Contrary to the well-defined ET signaling, the pathway leading from sensing sulfate status to the activation of its acquisition *via* SLIM1 is completely unknown. SLIM1 binds to the 20 nt-long specific UPE-box sequence; however, it also recognizes the shorter TEIL sequence, unique for the whole EIL family. SLIM1 takes part in the upregulation and downregulation of various sulfur metabolism genes, but also it controls the degradation of glucosinolates under sulfur deficient conditions. Besides facilitating the increased flux through the sulfate assimilation pathway, SLIM1 induces microRNA395, specifically targeting ATP sulfurylases and a low-affinity sulfate transporter, SULTR2;1, thus affecting sulfate translocation to the shoot. Here, we briefly review the identification, structural characteristics, and molecular function of SLIM1 from the perspective of the whole EIL protein family.

Keywords: transcription factor, EIL protein family, EIN3, SLIM1, *Arabidopsis*, sulfur, ethylene

INTRODUCTION

Sulfur is present in various compounds due to its ability to readily change the oxidation state. The majority of sulfur in living organisms is in the reduced form of organic sulfur and thiols, while the environment offers predominantly oxidized inorganic sulfate. Only plants (and algae) together with fungi and bacteria are capable of sulfate assimilation and its reduction, therefore playing a pivotal role in the biogeochemical sulfur cycle. The availability of sulfur in the soil fluctuates, therefore plants constantly have to adapt to the changing environment by reprogramming their metabolism. Modulation of gene expression at the level of transcription is a major control point in multiple biological processes, thus the main interest of many researchers was to identify transcriptional regulators specific for sulfur deficiency signaling. Sulfur limitation 1 (SLIM1) from *Arabidopsis thaliana*, so far the only described transcription factor strictly assigned to this pathway, was found in an elegant genetic approach exploiting the fluorescent sulfur deficiency responsive reporter (Maruyama-Nakashita et al., 2006). It has to be stressed out, however, that SLIM1 regulates only a set of genes of sulfur metabolism and also other factors are controlling the gene expression during sulfur limitation.

THE EIL FAMILY OF TRANSCRIPTIONAL REGULATORS

Sulfur limitation 1 was previously identified as the gene *ETHYLENE-INSENSITIVE-LIKE 3* (*EIL3*) coding for a putative transcription factor of unknown function (Guo and Ecker, 2004).

It belongs to a small family of proteins found exclusively in plants of which several members have been cloned and characterized across various species, including *Arabidopsis* (Chao et al., 1997), tobacco (Kosugi and Ohashi, 2000; Rieu et al., 2003), tomato (Tieman et al., 2001), maize (Gallie and Young, 2004), carnation (Iordachescu and Verlinden, 2005), rice (Mao et al., 2006), kiwi (Yin et al., 2010), and cucumber (Bie et al., 2013). In the *Arabidopsis* genome, there are six genes annotated to encode the EIL family proteins [ethylene-insensitive3 (EIN3) and EIL1–EIL5; Guo and Ecker, 2004]. EIN3 together with its functional homologues EIL1 and EIL2 are transcription factors controlling the expression of ethylene (ET)-responsive genes (Chao et al., 1997; Solano et al., 1998). EIL3/SLIM1 seems to be a specific regulator of sulfur deficiency response since only SLIM1 from the EIL family complemented the phenotype of the *slim1* mutants (Maruyama-Nakashita et al., 2006). Additional proof that SLIM1 mediated regulation is separated from the ET response pathway is that the set of SLIM1-dependent genes are not regulated by the ET precursor 1-aminocyclopropane 1-carboxylic acid (Maruyama-Nakashita et al., 2006). The roles of EIL4 and EIL5 in plant metabolism so far have not been defined (Guo and Ecker, 2004). The first cloned gene of the family, *EIN3*, was identified through positional cloning in the collection of ET-insensitive *Arabidopsis* mutants (Chao et al., 1997). The family is characterized by highly acidic N-terminal amino acids, five small clusters of basic amino acids scattered mostly in the first half of the protein and a proline-rich domain (Chao et al., 1997). The EIL family proteins are highly homologous to one another mainly in their N-terminal half of around 300 amino acid residues. Sequence-specific DNA-binding activities of EIN3, EIL1, EIL2 proteins have been demonstrated using electro-mobility shift assay (Solano et al., 1998; Kosugi and Ohashi, 2000). The location of the unique DNA-binding domain in the primary structure of an EIL protein was identified based on the SLIM1 sequence using the surface plasmon resonance technique (Yamasaki et al., 2005). The structure consists of five alpha-helices, packing together into a globular shape as a whole, possessing a novel fold dissimilar to known DNA-binding domain structures.

ETHYLENE SIGNALING PATHWAY

The best characterized protein of the *Arabidopsis* EIL family is EIN3, which together with EIL1, mediates most, if not all, plant responses to ET. The gaseous phytohormone ET regulates many aspects of the plant life cycle, including seed germination, root hair development, root nodulation, flower senescence, leaf abscission, and fruit ripening (Johnson and Ecker, 1998). The emission of ET is tightly controlled by internal signals during development as well as environmental stimuli, including nutritional deficiencies. An initially linear pathway of ET signaling was drawn using a number of molecular genetic studies (Guo and Ecker, 2004). However, latest research presents a much more complex pathway with multiple feedback loops and control levels (see Merchante et al., 2013 for review). A family of five endoplasmic reticulum-associated receptors perceives ET. There are two types of ET receptors in *Arabidopsis.* ETR1 and ERS1 contain three transmembrane domains and a conserved histidine kinase domain, and have been shown to form homodimers. ETR2, EIN4, and ERS2 have four membrane-spanning regions and a degenerate histidine kinase domain that lacks one or more elements necessary for catalytic activity. A copper cofactor, which is delivered by the copper transporter responsive to antagonist-1 (RAN1), is required for ET binding (Wang et al., 2002). In the absence of an ET signal, receptors activate a Ser/Thr kinase, CTR1, that dimerizes and suppresses the ET response (**Figure 1**). ET binding leads to the functional inactivation of receptors and the disability of CTR1 to phosphorylate a positive component of the pathway – the membrane located EIN2. The non-phosphorylated C-terminal end of EIN2 is cleaved off by an unknown mechanism and is transferred to the nucleus (Merchante et al., 2013). The level of EIN2 is regulated by the F-box proteins ETP1 and ETP2, and its degradation via the 26S proteasome. Two other F-box proteins, EBF1 and EBF2, control the level of transcription factors EIN3/EIL1 in the nucleus, thus shutting off the transcription of the ET response genes in the absence of the signal (**Figure 1**). Upon perception of ET, the C-terminal end of EIN2 stabilizes EIN3/EIL1 and induces degradation of EBF1 and EBF2. Additionally, the levels of mRNAs encoding *EBF1* and *EBF2* are negatively regulated by the exoribonuclease EIN5 in the presence of ET. The transcription factor EIN3 dimerizes and then activates the expression of target genes, including the transcription factor gene ethylene-response-factor1 (*ERF1*). ERF1, in turn, starts a transcriptional cascade of 100s of ET-regulated genes. The mechanism of ET signaling in plants is probably universal as all the elements identified in *Arabidopsis* are conserved in evolutionary distant plant species (Merchante et al., 2013).

EIN3 PROTEIN CONTROL

Recent studies have highlighted the role of ubiquitin/proteasome pathway in various aspects of plant growth and development as the paradigm for plant hormone signaling. A ubiquitin/proteasome pathway has been demonstrated in auxin, gibberellin, abscisic acid and jasmonate signaling, and implicated in the salicylic acid, cytokinin, and brassinosteroid responses (Frugis and Chua, 2002; Vierstra, 2003; Smalle and Vierstra, 2004; Dreher and Callis, 2007). Three groups independently discovered that EIN3 is degraded by the 26S proteasome-dependent pathway, and that EBF1 and EBF2 are two proteins mediating EIN3 degradation (Guo and Ecker, 2003; Potuschak et al., 2003; Gagne et al., 2004). EBF1 functions constitutively by keeping EIN3 below a critical threshold, thereby repressing the ET response at low hormone concentrations. EBF2 acts mainly in silencing the signal by removing activated EIN3 so the plants can more rapidly resume normal growth (Binder et al., 2007). It has been shown, that ET can induce *EBF2* expression forming a negative feedback loop to desensitize ET signaling (Potuschak et al., 2003).

Moreover, EIN3 protein seems also to be quantitatively controlled by other signals. It has been found that glucose can promote EIN3 degradation by an unknown mechanism (Yanagisawa et al., 2003), whereas light can positively regulate EIN3 and EIL1 stability (Lee et al., 2006). Furthermore, two different phosphorylation sites, oppositely affecting the level of EIN3, have been identified, pointing out the involvement of the MAPK-dependent pathway in ET signaling (Yoo et al., 2008). In this model, MKK9 cascade phosphorylates EIN3 to promote its stability, whereas phosphorylation by an MAPK pathway mediated by kinase CTR1 promotes EIN3 degradation. An additional phosphorylation site, highly conserved in all members of the EIL family, was recently proved to be of fundamental importance for the dimerization of tomato EIL1 and crucial for its transcriptional activity (Li et al., 2012). As such, EIN3 may represent a central regulator of plant growth, capable of integrating various external, and internal signals. This is understandable, since most phytohormones are involved in multiple processes and influence each other through complex crosstalk strategies (Santner and Estelle, 2010). For many years the synergy or antagonism between ET and jasmonic acid (JA) signaling has been observed in many developmental and defense-related processes (Pauwels and Goossens, 2011). At least part of this crosstalk is mediated by the interaction of EIN3/EIL1 with JAZ proteins, which are repressors in jasmonate signaling (Kazan and Manners, 2012). JAZ proteins bind to EIN3/EIL1, thus suppressing the DNA-binding ability of EIN3. The emerging model, providing a plausible explanation for the synergy in many processes regulated by both hormones, emphasizes the role of ET in EIN3/EIL1 stabilization and jasmonate in EIN3/EIL1 release from the JAZ protein repression (Zhu et al., 2011). Another layer of the crosstalk between those two pathways is the interaction between the jasmonate-activated transcription factor MYC2 and EIN3 (Song et al., 2014). MYC2 interacts with EIN3 to attenuate ET-enhanced apical hook curvature, while EIN3 represses MYC2 to downregulate jasmonate-regulated plant defense against generalist herbivores (Zhang et al., 2014; **Figure 1**).

There is strong evidence regarding the involvement of ET in plant responses to nutritional stresses. Changed levels of ET production were reported as a result of nitrogen, phosphorus, potassium, calcium, and iron deficiency (Lynch and Brown, 1997; Benlloch-Gonzalez et al., 2010). A direct molecular link between ET signaling and iron metabolism was found (Lingam et al., 2011).

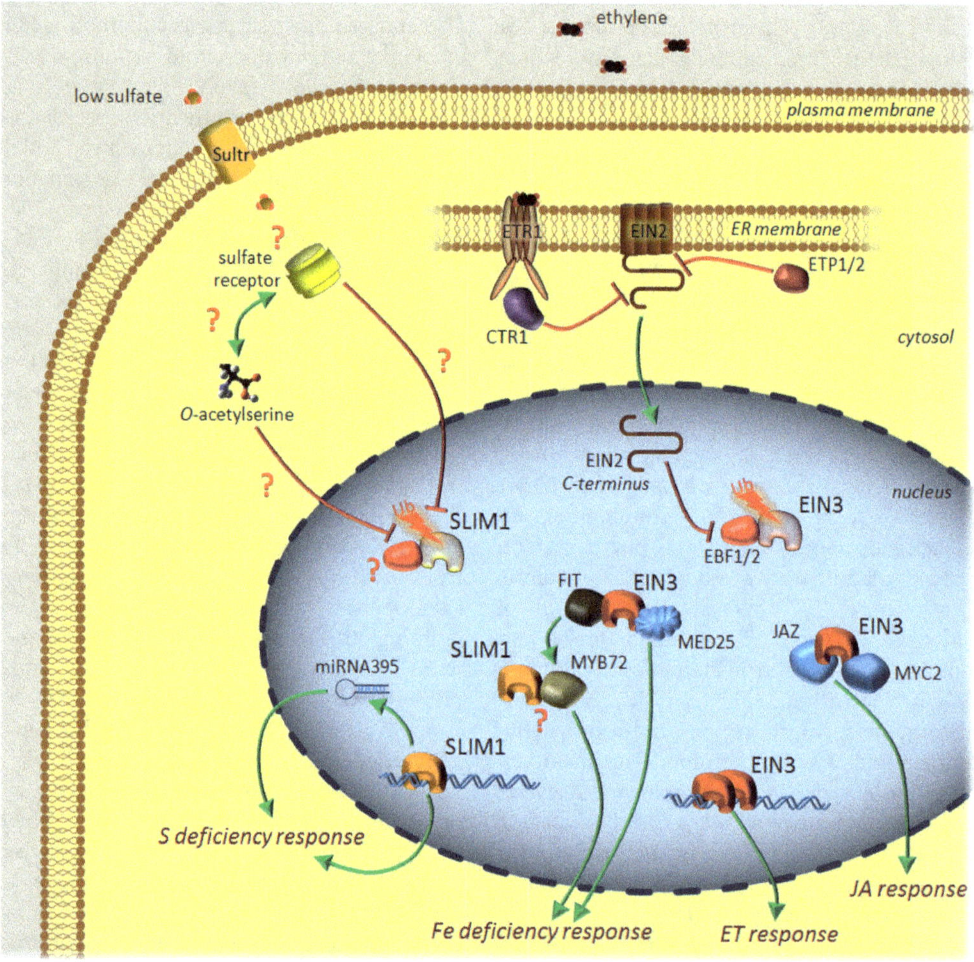

FIGURE 1 | Current model of the ethylene (ET) and sulfur deficiency signaling pathways in *Arabidopsis*. In contrast to ET signaling, sulfur deficiency signaling is poorly characterized. Sulfate is transported to cytosol *via* sulfate transporters of the Sultr family. Low sulfate availability is sensed by an unknown receptor and may depend on *O*-acetylserine level. The low sulfur (LSU) signal is transmitted to the nucleus and putatively stabilizes transcriptional factor SLIM1. SLIM1 induces the transcription of selected genes and miRNA395, thus reprogramming the transcriptional profile to answer the sulfur deficiency conditions. ET is perceived by the receptor proteins (for example, ETR1) present in the ER membrane. When the hormone is absent, the receptors activate a Ser/Thr kinase, CTR1, that dimerizes and suppresses the ET response by inactivating EIN2 through the phosphorylation of its C-terminal end. The EIN2 protein level is negatively regulated by the F-box proteins ETP1 and ETP2 and proteasomal degradation, while two other F-box proteins, EBF1/2 serve for the degradation of the transcription factor EIN3 in the nucleus to shut off the ET response. Upon perception of ET, ETR1 inactivates CTR1 and promotes the cleaving off of the C-terminal end of EIN2 that induces the degradation of EBF1/2 after import to the nucleus. EIN3 dimerizes and activates a transcriptional cascade of ET-responsive genes. Depending on the other environmental factors, EIN3 also interacts with JAZ proteins and transcriptional factor MYC2 to shape the jasmonic acid (JA) response. Another partner of EIN3 is MED25, which is a part of a complex regulating iron homeostasis. Additionally, EIN3 binds to FIT, a central regulator of iron deficiency response affecting the transcription level of many genes, with MYB72 among them. MYB72 can interact with SLIM1; however, the outcome of this interaction is unknown. Positive (green) and negative (red) lines represent activation and downregulation processes, respectively. SLIM1 and EIN3, shown in fading colors with Ub (ubiquitin), correspond to proteins marked for proteasome-mediated degradation. Question marks depict the points that are still waiting to be addressed by researchers.

EIN3/EIL1 can physically interact with FIT, a central regulator of iron acquisition in roots (**Figure 1**). Through this interaction, proteasomal degradation of FIT is reduced and leads to a higher level of expression of the iron acquisition genes. Another factor directly interacting with EIN3 is the Mediator complex subunit MED25 (Yang et al., 2014). Mediator is a conserved multisubunit complex, regulating the transcription by bridging transcription factors with RNA polymerase II. The MED25 subunit tunes up iron homeostasis but is also important for plant disease resistance, flowering, and organ size (Yang et al., 2014). ET is thereby one of the signals that triggers iron deficiency responses at the transcriptional and post-transcriptional levels. Recently, it was evidenced that ET is involved in the sulfur deficiency response with a special highlight on the role of small proteins of unknown function from the LSU/UP9 family (Moniuszko et al., 2013). It's tempting to speculate that similar regulatory mechanisms influencing EIN3 stability and the wealth of interactions refer also to its close homologue, SLIM1.

SLIM1 PROTEIN CONTROL

Not much is known about SLIM1 post-translational modifications or its protein partners. Reportedly, exogenous ET did not affect the expression of any of the *EIL* genes in *Arabidopsis*, tomato, tobacco, and mung bean (Chao et al., 1997; Tieman et al., 2001; Lee and Kim, 2003; Rieu et al., 2003) indicating the regulation at the post-transcriptional level. In contrast, transcriptional induction of *EILs* by ET or mechanical wounding was demonstrated in other plant species, such as petunia, carnation, banana, and rice (Waki et al., 2001; Shibuya et al., 2004; Iordachescu and Verlinden, 2005; Mbeguie et al., 2008; Hiraga et al., 2009). Interestingly, iron deficiency induces the expression of genes involved in ET synthesis and signaling, with *SLIM1* among them, in the *Arabidopsis* roots (Garcia et al., 2010). *SLIM1* is expressed predominantly in vascular tissues and despite the genetically evidenced importance of SLIM1 in sulfur response, its transcription level is not modulated by the changes of sulfur conditions (Maruyama-Nakashita et al., 2006). It is tempting to speculate that SLIM1 may require post-transcriptional mechanisms for the regulation of its performance. Presumably such regulation is accomplished by the highly selective ubiquitin/proteasome system removing SLIM1 protein while its function is not needed. However, unlike EIN3, whose protein level in the nucleus is affected by the ET and carbon status (Guo and Ecker, 2003; Potuschak et al., 2003; Yanagisawa et al., 2003), neither nuclear localization nor abundance of SLIM1 protein were changed by sulfur conditions (Maruyama-Nakashita et al., 2006). Interesting observations come from the recent studies of Aubry et al. (2014) revealing the crucial role of bundle sheath cells in sulfur metabolism. Despite the strong transcriptional upregulation of the whole sulfur assimilation pathway and the glucosinolates metabolism, the representation of *SLIM1* transcript was not increased, again pointing to the SLIM1 protein level control. It cannot be excluded, however, that other factors are controlling sulfur metabolism in this cell type. So far there is only one described protein partner of SLIM1 (**Figure 1**). MYB72, which is involved in induced systemic resistance, has been shown to interact physically with SLIM1 in the yeast two-hybrid assay (Van der Ent et al., 2008). Recognition of the beneficial microbes leads to the induction of *MYB72* and interaction of the protein with SLIM1 to trigger a jasmonate/ET-dependent resistance effective against a broad range of pathogens (Van der Ent et al., 2008). MYB72 together with MYB10 were recently found to be essential for plant survival under iron- deficiency, inducing the nicotianamine synthase gene *NAS4* necessary for proper metal homeostatsis (Palmer et al., 2013). Interestingly, MYB72 has also been described as a direct target of FIT, the root-specific central regulator of iron deficiency (Sivitz et al., 2012). This raises the question of whether the tandem MYB72–SLIM1 plays an additional regulatory role in sulfur deficiency responses. SLIM1, on the other hand, negatively regulates the expression of another MYB family member, ATR1/MYB34, thereby affecting glucosinolate biosynthesis in *Arabidopsis* roots (Maruyama-Nakashita et al., 2006). Another intriguing observation is that SLIM1, in contrast to the homodimers of EIN3, EIL1, and EIL2, exists in the monomeric form (Solano et al., 1998). It was suggested that the dimerization of EIL proteins is important in the stable binding to a pseudo-palindromic DNA sequence present in ET-responsive promoters, although it is still possible for monomeric proteins to bind to a shorter consensus (Yamasaki et al., 2005).

SLIM1 BINDING TO DNA

Solano et al. (1998) have shown that the proteins from the *Arabidopsis* EIL family bind directly to primary ET response DNA elements, which are 28-nt imperfect palindromes found in the promoters of various ET-responsive genes. At the same time the 8-nt consensus binding sequence was defined for tobacco NtEIL1/TEIL, the transcription factor also believed to be involved in ET signaling (Kosugi and Ohashi, 2000). The similarity between those DNA regions is very high, however, the sequence essential for EIN3 binding was bound by TEIL with considerably less affinity than the TEIL binding sequence (*tebs*), showing differences in the binding preference between EIL family members (Kosugi and Ohashi, 2000). On the other hand, it was proved during *in vitro* studies that SLIM1 is able to bind to *tebs*, though the interaction is very unstable and only detectable with surface plasmon resonance but not by electro-mobility shift assay (Yamasaki et al., 2005). *Tebs* are present in the promoters of several sulfur deficiency-induced genes of *Arabidopsis*, the regulation of which is also controlled by SLIM1 (Maruyama-Nakashita et al., 2006). Additionally, the direct interaction of SLIM1 with 20-nt consensus, called the UPE-box was demonstrated (Lewandowska et al., 2010; Wawrzynska et al., 2010). The UPE-box contains two *tebs*, partially overlapping in opposite orientation to each other, and is only present in the promoters of several *Arabidopsis* genes strongly induced by sulfur deficiency (Wawrzynska et al., 2010). Among those genes, are genes encoding proteins from the LSU family, homologues of tobacco UP9C protein. Transgenic tobacco plants with lowered expression of *UP9C* showed the disturbed response of the ET signaling and synthesis pathways during conditions of sulfur deficiency, indicating a crosstalk between ET and sulfur metabolism in plants (Moniuszko et al., 2013). Interestingly, the UPE-boxes were also found in promoters of a co-regulated gene cluster induced by the cysteine precursor *O*-acetylserine, suggesting a potential function for SLIM1 in the sensing of sulfur status (Hubberten et al., 2012). A signaling function of *O*-acetylserine in sulfur assimilation by enteric bacteria had already been stated a long time ago (Ostrowski and Kredich, 1990); however, its role as a sensor of sulfur status in plants is still under debate.

THE *slim1 Arabidopsis* MUTANTS CHARACTERISTICS

The *slim1* mutants are not able to induce expression of the high-affinity sulfate transporter SULTR1;2 and consequently sulfate uptake during sulfur deficiency. SLIM1 inactivation results in a 60% limitation of sulfate uptake rate and a 30% reduction in root length (in comparison to the wild-type plants). The metabolite analysis further suggests that *slim1* mutants may be experiencing the lowered supply of sulfate to the reduction pathway as evidenced by the significant decrease of glutathione content and overaccumulation of *O*-acetylserine in their shoots. Such metabolite profiles are characteristic for sulfur-deficient plants and caused by an insufficient sulfur amount for the cysteine synthesis pathway. Degradation of glucosinolates is another important aspect of sulfur limitation response. Glucosinolates are characteristic compounds for *Brassicales* participating in the defense against herbivores and pathogens (Halkier and Gershenzon, 2006). The *slim1* mutations concomitantly affect the expression of metabolic and regulatory genes of glucosinolate biosynthetic pathways. Consistent with the transcriptional changes, glucosinolates levels were shown to be higher in the *slim1* mutants, even under sulfur deficient conditions (Maruyama-Nakashita et al., 2006). These results provide strong evidence for the function of SLIM1 in the co-regulation of this sulfur recycling process in parallel with sulfate transport systems during sulfur deficiency. However, since the transcriptomic profile of the *slim1* mutants showed alterations in many, but not all, genes responsive to sulfate deficiency, one might expect other factors controlling these processes. Additionally, it was suggested recently that SLIM1 may possess a dual function as an activator at sulfur limitation and a repressor during normal sulfur status (Matthewman et al., 2012). This was evidenced by higher sulfate uptake by the *slim1* mutants on a normal sulfate supply which was consistent with the higher transcript level for *SULTR1;1* (Maruyama-Nakashita et al., 2006).

SLIM1 INTERPLAY WITH microRNA395

Another level of control in gene expression is the regulation by microRNAs (miRNAs), which are a class of naturally occurring, small non-coding RNA molecules. They are partially complementary to one or more mRNA molecules, and their main function is to affect the stability of these molecules in a variety of manners, including translational repression, mRNA cleavage, and deadenylation (Voinnet, 2009). Functionally, miRNAs are involved in a variety of developmental processes in plants, including stress responses with nutrient deficiencies. Among those, miR395 in *Arabidopsis* was identified as being involved in the regulation of sulfate transport and assimilation targeting the mRNAs of three isoforms of ATP sulfurylase and one transporter SULTR2;1 facilitating inter-organ transport of sulfate (Bonnet et al., 2004; Jones-Rhoades and Bartel, 2004). The expression of miR395 is drastically upregulated under sulfur limitation and its induction is directly or indirectly controlled by SLIM1 (Kawashima et al., 2009; **Figure 1**). The cell-type specific expression pattern between miR395 and its target transcripts enables the fine-tuning of the sulfur assimilation rate (Kawashima et al., 2011). Especially interesting is the unexpected positive correlation of expression between miR395 and targeted *SULTR2;1* during sulfur deficiency. It enables restriction of the SULTR2;1 transporter to xylem parenchyma cells, thus together with sulfate reduction shut off, providing for efficient translocation of sulfate from roots to shoots (Kawashima et al., 2009). Additionally, grafting experiments provided convincing evidence that miR395 are phloem-mobile, suggesting their role as long-distance signaling molecules, and underlying the importance of systemic regulation of plant response to varying sulfur levels (Buhtz et al., 2010). Recently, it was demonstrated that the trigger of miR395 accumulation is linked rather to internal sulfate levels and not external sulfate availability (Matthewman et al., 2012), again pointing out to the *O*-acetylserine as an activating signal and to the involvement of SLIM1. Moreover, it was also shown that the redox signaling plays an important role in miR395 induction during sulfur deficiency, placing SLIM1 downstream in the regulatory cascade (Jagadeeswaran et al., 2014). However, whether SLIM1 itself is a target of redox signaling has not yet been determined.

Interestingly, EIN3 also participates in the control of miRNA by integrating different developmental and environmental cues and directly binding to the promoters of miR164 activating leaf senescence processes (Li et al., 2013).

CONCLUSION AND FUTURE PROSPECTS

Due to the sessile life cycle, plants have developed different strategies to adapt to adverse environmental stresses. Plant growth and development is largely impaired by nutrient deficiencies; therefore to maintain good productivity in plant breeding, it is essential to understand fully those mechanisms. In this review, we focused on SLIM1 as the only described transcriptional regulator dedicated to plant response to sulfur deficiency. It belongs to the same protein family of transcription factors as EIN3. Contrary to SLIM1, regulation of EIN3 stability, interaction with other proteins as well as the whole signaling pathway leading to transcriptional response is already well described (**Figure 1**). It is of interest as to whether the same level of complexity can be expected in sulfur deficiency signaling. Still, the exact signaling cascade leading from sensing to activating the expression of the SLIM1-dependent gene set, resulting in sulfur metabolism reprogramming, needs to be clarified and future studies are required to reveal the molecular components, with a special emphasis on the role of *O*-acetylserine (**Figure 1**). Such studies should also concentrate on the investigation of post-transcriptional modifications of SLIM1 influencing its functionality under different sulfur regimes, as well as its direct interaction with specific DNA sequences. We must be cautions, however, in drawing general conclusions and remember that SLIM1 is present mostly in the vascular tissues thus its action might be predominantly connected with the translocation of sulfate between plant parts rather than governing the whole plant sulfur metabolism.

AUTHOR CONTRIBUTIONS

Anna Wawrzyńska drafted the manuscript and prepared the figure; Agnieszka Sirko revised it critically for important intellectual content.

ACKNOWLEDGMENTS

Research in our laboratory is supported by a grant from the Polish National Science Centre (2011/03/B/NZ1/01243 to Anna Wawrzyńska).

REFERENCES

Aubry, S., Smith-Unna, R. D., Boursnell, C. M., Kopriva, S., and Hibberd, J. M. (2014). Transcript residency on ribosomes reveals a key role for the *Arabidopsis thaliana* bundle sheath in sulfur and glucosinolate metabolism. *Plant J.* 78, 659–673. doi: 10.1111/tpj.12502

Benlloch-Gonzalez, M., Romera, J., Cristescu, S., Harren, F., Fournier, J. M., and Benlloch, M. (2010). K^+ starvation inhibits water-stress-induced stomatal closure via ethylene synthesis in sunflower plants. *J. Exp. Bot.* 61, 1139–1145. doi: 10.1093/jxb/erp379

Bie, B. B., Pan, J. S., He, H. L., Yang, X. Q., Zhao, J. L., and Cai, R. (2013). Molecular cloning and expression analysis of the ethylene insensitive3 (EIN3) gene in cucumber (*Cucumis sativus*). *Genet. Mol. Res.* 12, 4179–4191. doi: 10.4238/2013.October.7.4

Binder, B. M., Walker, J. M., Gagne, J. M., Emborg, T. J., Hemmann, G., Bleecker, A. B., et al. (2007). The *Arabidopsis* EIN3 binding F-Box proteins EBF1 and EBF2 have distinct but overlapping roles in ethylene signaling. *Plant Cell* 19, 509–523. doi: 10.1105/tpc.106.048140

Bonnet, E., Wuyts, J., Rouze, P., and Van De Peer, Y. (2004). Evidence that microRNA precursors, unlike other non-coding RNAs, have lower folding free energies than random sequences. *Bioinformatics* 20, 2911–2917. doi: 10.1093/bioinformatics/bth374

Buhtz, A., Pieritz, J., Springer, F., and Kehr, J. (2010). Phloem small RNAs, nutrient stress responses, and systemic mobility. *BMC Plant Biol.* 10:64. doi 10.1186/1471-2229-10-64

Chao, Q., Rothenberg, M., Solano, R., Roman, G., Terzaghi, W., and Ecker, J. R. (1997). Activation of the ethylene gas response pathway in *Arabidopsis* by the nuclear protein ETHYLENE-INSENSITIVE3 and related proteins. *Cell* 89, 1133–1144. doi: 10.1016/S0092-8674(00)80300-1

Dreher, K., and Callis, J. (2007). Ubiquitin, hormones and biotic stress in plants. *Ann. Bot.* 99, 787–822. doi: 10.1093/aob/mcl255

Frugis, G., and Chua, N. H. (2002). Ubiquitin-mediated proteolysis in plant hormone signal transduction. *Trends Cell Biol.* 12, 308–311. doi: 10.1016/S0962-8924(02)02308-5

Gagne, J. M., Smalle, J., Gingerich, D. J., Walker, J. M., Yoo, S. D., Yanagisawa, S., et al. (2004). *Arabidopsis* EIN3-binding F-box 1 and 2 form ubiquitin-protein ligases that repress ethylene action and promote growth by directing EIN3 degradation. *Proc. Natl. Acad. Sci. U.S.A.* 101, 6803–6808. doi: 10.1073/pnas.0401698101

Gallie, D. R., and Young, T. E. (2004). The ethylene biosynthetic and perception machinery is differentially expressed during endosperm and embryo development in maize. *Mol. Genet. Genomics* 271, 267–281. doi: 10.1007/s00438-004-0977-9

Garcia, M. J., Lucena, C., Romera, F. J., Alcantara, E., and Perez-Vicente, R. (2010). Ethylene and nitric oxide involvement in the up-regulation of key genes related to iron acquisition and homeostasis in *Arabidopsis*. *J. Exp. Bot.* 61, 3885–3899. doi: 10.1093/jxb/erq203

Guo, H., and Ecker, J. R. (2003). Plant responses to ethylene gas are mediated by SCF(EBF1/EBF2)-dependent proteolysis of EIN3 transcription factor. *Cell* 115, 667–677. doi: 10.1016/S0092-8674(03)00969-3

Guo, H., and Ecker, J. R. (2004). The ethylene signaling pathway: new insights. *Curr. Opin. Plant Biol.* 7, 40–49. doi: S1369526603001420

Halkier, B. A., and Gershenzon, J. (2006). Biology and biochemistry of glucosinolates. *Annu. Rev. Plant Biol.* 57, 303–333. doi: 10.1146/annurev.arplant.57.032905.105228

Hiraga, S., Sasaki, K., Hibi, T., Yoshida, H., Uchida, E., Kosugi, S., et al. (2009). Involvement of two rice ETHYLENE INSENSITIVE3-LIKE genes in wound signaling. *Mol. Genet. Genomics* 282, 517–529. doi: 10.1007/s00438-009-0483-1.

Hubberten, H. M., Klie, S., Caldana, C., Degenkolbe, T., Willmitzer, L., and Hoefgen, R. (2012). Additional role of O-acetylserine as a sulfur status-independent regulator during plant growth. *Plant J.* 70, 666–677. doi: 10.1111/j.1365-313X.2012.04905.x

Iordachescu, M., and Verlinden, S. (2005). Transcriptional regulation of three EIN3-like genes of carnation (*Dianthus caryophyllus* L. cv. Improved White Sim) during flower development and upon wounding, pollination, and ethylene exposure. *J. Exp. Bot.* 56, 2011–2018. doi: 10.1093/jxb/eri199

Jagadeeswaran, G., Li, Y. F., and Sunkar, R. (2014). Redox signaling mediates the expression of a sulfate-deprivation-inducible microRNA395 in *Arabidopsis*. *Plant J.* 77, 85–96. doi: 10.1111/tpj.12364

Johnson, P. R., and Ecker, J. R. (1998). The ethylene gas signal transduction pathway: a molecular perspective. *Annu. Rev. Genet.* 32, 227–254. doi: 10.1146/annurev.genet.32.1.227

Jones-Rhoades, M. W., and Bartel, D. P. (2004). Computational identification of plant microRNAs and their targets, including a stress-induced miRNA. *Mol. Cell* 14, 787–799. doi: 10.1016/j.molcel.2004.05.027

Kawashima, C. G., Matthewman, C. A., Huang, S., Lee, B. R., Yoshimoto, N., Koprivova, A., et al. (2011). Interplay of SLIM1 and miR395 in the regulation of sulfate assimilation in *Arabidopsis*. *Plant J.* 66, 863–876. doi: 10.1111/j.1365-313X.2011.04547.x

Kawashima, C. G., Yoshimoto, N., Maruyama-Nakashita, A., Tsuchiya, Y. N., Saito, K., Takahashi, H., et al. (2009). Sulphur starvation induces the expression of microRNA-395 and one of its target genes but in different cell types. *Plant J.* 57, 313–321. doi: 10.1111/j.1365-313X.2008.03690.x

Kazan, K., and Manners, J. M. (2012). JAZ repressors and the orchestration of phytohormone crosstalk. *Trends Plant Sci.* 17, 22–31. doi: 10.1016/j.tplants.2011.10.006

Kosugi, S., and Ohashi, Y. (2000). Cloning and DNA-binding properties of a tobacco Ethylene-Insensitive3 (EIN3) homolog. *Nucleic Acids Res.* 28, 960–967. doi: 10.1093/nar/28.4.960

Lee, J. H., Deng, X. W., and Kim, W. T. (2006). Possible role of light in the maintenance of EIN3/EIL1 stability in *Arabidopsis* seedlings. *Biochem. Biophys. Res. Commun.* 350, 484–491. doi: 10.1016/j.bbrc.2006.09.074

Lee, J. H., and Kim, W. T. (2003). Molecular and biochemical characterization of VR-EILs encoding mung bean ETHYLENE INSENSITIVE3-LIKE proteins. *Plant Physiol.* 132, 1475–1488. doi: 10.1104/pp.103.022574

Lewandowska, M., Wawrzynska, A., Moniuszko, G., Lukomska, J., Zientara, K., Piecho, M., et al. (2010). A contribution to identification of novel regulators of plant response to sulfur deficiency: characteristics of a tobacco gene UP9C, its protein product and the effects of UP9C silencing. *Mol. Plant* 3, 347–360. doi: 10.1093/mp/ssq007

Li, J., Li, Z., Tang, L., Yang, Y., Zouine, M., and Bouzayen, M. (2012). A conserved phosphorylation site regulates the transcriptional function of ETHYLENE-INSENSITIVE3-like1 in tomato. *J. Exp. Bot.* 63, 427–439. doi: 10.1093/jxb/err289

Li, Z. H., Peng, J. Y., Wen, X., and Guo, H. W. (2013). Ethylene-insensitive3 is a senescence-associated gene that accelerates age-dependent leaf senescence by directly repressing miR164 transcription in *Arabidopsis*. *Plant Cell* 25, 3311–3328. doi: 10.1105/tpc.113.113340

Lingam, S., Mohrbacher, J., Brumbarova, T., Potuschak, T., Fink-Straube, C., Blondet, E., et al. (2011). Interaction between the bHLH transcription factor FIT and ETHYLENE INSENSITIVE3/ETHYLENE INSENSITIVE3-LIKE1 reveals molecular linkage between the regulation of iron acquisition and ethylene signaling in *Arabidopsis*. *Plant Cell* 23, 1815–1829. doi: 10.1105/tpc.111.084715

Lynch, J., and Brown, K. M. (1997). Ethylene and plant responses to nutritional stress. *Physiol. Plant.* 100, 613–619. doi: 10.1034/j.1399-3054.1997.1000324.x

Mao, C., Wang, S., Jia, Q., and Wu, P. (2006). OsEIL1, a rice homolog of the *Arabidopsis* EIN3 regulates the ethylene response as a positive component. *Plant Mol. Biol.* 61, 141–152. doi: 10.1007/s11103-005-6184-1.

Maruyama-Nakashita, A., Nakamura, Y., Tohge, T., Saito, K., and Takahashi, H. (2006). *Arabidopsis* SLIM1 is a central transcriptional regulator of plant sulfur response and metabolism. *Plant Cell* 18, 3235–3251. doi: 10.1105/tpc.106.046458

Matthewman, C. A., Kawashima, C. G., Huska, D., Csorba, T., Dalmay, T., and Kopriva, S. (2012). miR395 is a general component of the sulfate assimilation regulatory network in *Arabidopsis*. *FEBS Lett.* 586, 3242–3248. doi: 10.1016/j.febslet.2012.06.044

Mbeguie, A. M. D., Hubert, O., Fils-Lycaon, B., Chillet, M., and Baurens, F. C. (2008). EIN3-like gene expression during fruit ripening of Cavendish banana (*Musa acuminata* cv. Grande naine). *Physiol. Plant.* 133, 435–448. doi: 10.1111/j.1399-3054.2008.01083.x

Merchante, C., Alonso, J. M., and Stepanova, A. N. (2013). Ethylene signaling: simple ligand, complex regulation. *Curr. Opin. Plant Biol.* 16, 554–560. doi: 10.1016/j.pbi.2013.08.001

Moniuszko, G., Skoneczny, M., Zientara-Rytter, K., Wawrzynska, A., Glow, D., Cristescu, S. M., et al. (2013). Tobacco LSU-like protein couples sulphur-deficiency response with ethylene signalling pathway. *J. Exp. Bot.* 64, 5173–5182. doi: 10.1093/jxb/ert309

Ostrowski, J., and Kredich, N. M. (1990). In vitro interactions of CysB protein with the cysJIH promoter of *Salmonella* typhimurium: inhibitory effects of sulfide. *J. Bacteriol.* 172, 779–785.

Palmer, C. M., Hindt, M. N., Schmidt, H., Clemens, S., and Guerinot, M. L. (2013). MYB10 and MYB72 are required for growth under iron-limiting conditions. *PLoS Genet.* 9:e1003953. doi: 10.1371/journal.pgen.1003953

Pauwels, L., and Goossens, A. (2011). The JAZ proteins: a crucial interface in the jasmonate signaling cascade. *Plant Cell* 23, 3089–3100. doi: 10.1105/tpc.111.089300

Potuschak, T., Lechner, E., Parmentier, Y., Yanagisawa, S., Grava, S., Koncz, C., et al. (2003). EIN3-dependent regulation of plant ethylene hormone signaling by two *Arabidopsis* F box proteins: EBF1 and EBF2. *Cell* 115, 679–689. doi: 10.1016/S0092-8674(03)00968-1

Rieu, I., Mariani, C., and Weterings, K. (2003). Expression analysis of five tobacco EIN3 family members in relation to tissue-specific ethylene responses. *J. Exp. Bot.* 54, 2239–2244. doi: 10.1093/jxb/erg240

Santner, A., and Estelle, M. (2010). The ubiquitin-proteasome system regulates plant hormone signaling. *Plant J.* 61, 1029–1040. doi: 10.1111/j.1365-313X.2010.04112.x

Shibuya, K., Barry, K. G., Ciardi, J. A., Loucas, H. M., Underwood, B. A., Nourizadeh, S., et al. (2004). The central role of PhEIN2 in ethylene responses throughout plant development in petunia. *Plant Physiol.* 136, 2900–2912. doi: 10.1104/pp.104.046979

Sivitz, A. B., Hermand, V., Curie, C., and Vert, G. (2012). *Arabidopsis* bHLH100 and bHLH101 control iron homeostasis via a FIT-independent pathway. *PLoS ONE* 7:e44843. doi: 10.1371/journal.pone.0044843

Smalle, J., and Vierstra, R. D. (2004). The ubiquitin 26S proteasome proteolytic pathway. *Annu. Rev. Plant Biol.* 55, 555–590. doi: 10.1146/annurev.arplant.55.031903.141801

Solano, R., Stepanova, A., Chao, Q., and Ecker, J. R. (1998). Nuclear events in ethylene signaling: a transcriptional cascade mediated by ETHYLENE-INSENSITIVE3 and ETHYLENE-RESPONSE-FACTOR1. *Genes Dev.* 12, 3703–3714. doi: 10.1101/gad.12.23.3703

Song, S., Huang, H., Gao, H., Wang, J., Wu, D., Liu, X., et al. (2014). Interaction between MYC2 and ETHYLENE INSENSITIVE3 modulates antagonism between jasmonate and ethylene signaling in *Arabidopsis. Plant Cell* 26, 263–279. doi: 10.1105/tpc.113.120394

Tieman, D. M., Ciardi, J. A., Taylor, M. G., and Klee, H. J. (2001). Members of the tomato LeEIL (EIN3-like) gene family are functionally redundant and regulate ethylene responses throughout plant development. *Plant J.* 26, 47–58. doi: 10.1046/j.1365-313x.2001.01006.x

Van der Ent, S., Verhagen, B. W., Van Doorn, R., Bakker, D., Verlaan, M. G., Pel, M. J., et al. (2008). MYB72 is required in early signaling steps of rhizobacteria-induced systemic resistance in *Arabidopsis. Plant Physiol.* 146, 1293–1304. doi: 10.1104/pp.107.113829

Vierstra, R. D. (2003). The ubiquitin/26S proteasome pathway, the complex last chapter in the life of many plant proteins. *Trends Plant Sci.* 8, 135–142. doi: 10.1016/S1360-1385(03)00014-1

Voinnet, O. (2009). Origin, biogenesis, and activity of plant microRNAs. *Cell* 136, 669–687. doi: 10.1016/j.cell.2009.01.046

Waki, K., Shibuya, K., Yoshioka, T., Hashiba, T., and Satoh, S. (2001). Cloning of a cDNA encoding EIN3-like protein (DC-EIL1) and decrease in its mRNA level during senescence in carnation flower tissues. *J. Exp. Bot.* 52, 377–379. doi: 10.1093/jexbot/52.355.377

Wang, K. L., Li, H., and Ecker, J. R. (2002). Ethylene biosynthesis and signaling networks. *Plant Cell* 14, S131–S151. doi: 10.1105/tpc.001768

Wawrzynska, A., Lewandowska, M., and Sirko, A. (2010). Nicotiana tabacum EIL2 directly regulates expression of at least one tobacco gene induced by sulphur starvation. *J. Exp. Bot.* 61, 889–900. doi: 10.1093/jxb/erp356

Yamasaki, K., Kigawa, T., Inoue, M., Yamasaki, T., Yabuki, T., Aoki, M., et al. (2005). Solution structure of the major DNA-binding domain of *Arabidopsis thaliana* ethylene-insensitive3-like3. *J. Mol. Biol.* 348, 253–264. doi: 10.1016/j.jmb.2005.02.065

Yanagisawa, S., Yoo, S. D., and Sheen, J. (2003). Differential regulation of EIN3 stability by glucose and ethylene signalling in plants. *Nature* 425, 521–525. doi: 10.1038/nature01984

Yang, Y., Ou, B., Zhang, J., Si, W., Gu, H., Qin, G., et al. (2014). The *Arabidopsis* Mediator subunit MED16 regulates iron homeostasis by associating with EIN3/EIL1 through subunit MED25. *Plant J.* 77, 838–851. doi: 10.1111/tpj.12440

Yin, X. R., Allan, A. C., Chen, K. S., and Ferguson, I. B. (2010). Kiwifruit EIL and ERF genes involved in regulating fruit ripening. *Plant Physiol.* 153, 1280–1292. doi: 10.1104/pp.110.157081

Yoo, S. D., Cho, Y. H., Tena, G., Xiong, Y., and Sheen, J. (2008). Dual control of nuclear EIN3 by bifurcate MAPK cascades in C2H4 signalling. *Nature* 451, 789–795. doi: 10.1038/nature06543

Zhang, X., Zhu, Z., An, F., Hao, D., Li, P., Song, J., et al. (2014). Jasmonate-Activated MYC2 represses ETHYLENE INSENSITIVE3 activity to antagonize ethylene-promoted apical Hook formation in *Arabidopsis. Plant Cell* 26, 1105–1117. doi: 10.1105/tpc.113.122002

Zhu, Z., An, F., Feng, Y., Li, P., Xue, L., A, M., et al. (2011). Derepression of ethylene-stabilized transcription factors (EIN3/EIL1) mediates jasmonate and ethylene signaling synergy in *Arabidopsis. Proc. Natl. Acad. Sci. U.S.A.* 108, 12539–12544. doi: 10.1073/pnas.1103959108

Conflict of Interest Statement: The authors declare that the research was conducted in the absence of any commercial or financial relationships that could be construed as a potential conflict of interest.

4

Molecular mechanisms of regulation of sulfate assimilation: first steps on a long road

Anna Koprivova and Stanislav Kopriva*

Botanical Institute and Cluster of Excellence on Plant Sciences, Cologne Biocenter, University of Cologne, Cologne, Germany

Edited by:
Dibyendu Talukdar, Raja Peary Mohan College, India

Reviewed by:
Hideki Takahashi, Michigan State University, USA
Ruediger Hell, University of Heidelberg, Germany

****Correspondence:***
Stanislav Kopriva, Botanical Institute and Cluster of Excellence on Plant Sciences, Cologne Biocenter, University of Cologne, Zülpicher Straße 47b, 50674 Cologne, Germany
e-mail: skopriva@uni-koeln.de

The pathway of sulfate assimilation, which provides plants with the essential nutrient sulfur, is tightly regulated and coordinated with the demand for reduced sulfur. The responses of metabolite concentrations, enzyme activities and mRNA levels to various signals and environmental conditions have been well described for the pathway. However, only little is known about the molecular mechanisms of this regulation. To date, nine transcription factors have been described to control transcription of genes of sulfate uptake and assimilation. In addition, other levels of regulation contribute to the control of sulfur metabolism. Post-transcriptional regulation has been shown for sulfate transporters, adenosine 5′phosphosulfate reductase, and cysteine synthase. Several genes of the pathway are targets of microRNA miR395. In addition, protein–protein interaction is increasingly found in the center of various regulatory circuits. On top of the mechanisms of regulation of single genes, we are starting to learn more about mechanisms of adaptation, due to analyses of natural variation. In this article, the summary of different mechanisms of regulation will be accompanied by identification of the major gaps in knowledge and proposition of possible ways of filling them.

Keywords: sulfate assimilation, transcriptional regulation, transcription factors, microRNA, sulfate uptake, adenosine 5′phosphosulfate, glutathione

INTRODUCTION

Sulfur is an essential nutrient for all organisms, found in the amino acids cysteine and methionine, in a large number of cofactors and prosthetic groups, such as FeS centers, thiamine, or S-adenosylmethionine, and in a plethora of primary and secondary metabolites. Plants are able to take up inorganic sulfate from soil, reduce it to sulfide and incorporate into bioorganic compounds. In the pathway of sulfate assimilation sulfate is first activated by ATP sulfurylase (ATPS) to adenosine 5′-phosphosulfate (APS). APS is a branching point in sulfate assimilation, which can proceed by reduction to sulfite catalyzed by APS reductase or by phosphorylation to 3′-phosphoadenosine 5′-phosphosulfate (PAPS) by APS kinase. Sulfite is further reduced to sulfide by sulfite reductase (SiR), followed by incorporation into the amino acid skeleton of O-acetylserine (OAS) to make cysteine, which is the donor of reduced sulfur for all further metabolites (**Figure 1**). PAPS is the donor of activated sulfate for sulfation of peptides and small metabolites primarily in secondary metabolism [reviewed in (Takahashi et al., 2011)].

The biochemistry and physiology of sulfate assimilation and its regulation is well understood (Takahashi et al., 1997; Kopriva et al., 1999, 2001; Koprivova et al., 2000; Vauclare et al., 2002; Yoshimoto et al., 2002; Wirtz et al., 2004, 2012; Kawashima et al., 2005; Maruyama-Nakashita et al., 2006; Heeg et al., 2008; Mugford et al., 2009; Khan et al., 2010; Cao et al., 2013; Yarmolinsky et al., 2013). The pathway is regulated by the demand for reduced sulfur, by sulfur availability, by various environmental factors, or phytohormones, and coordinated with assimilation of carbon and nitrogen (Takahashi et al., 1997; Koprivova et al., 2000, 2008; Kopriva et al., 2002; Hesse et al., 2003; Jost et al., 2005). However, the knowledge of molecular mechanisms of the regulation, transcription factors controlling transcription of sulfate assimilation genes, and further levels of post-transcriptional regulation is still far from sufficient. Therefore, here we will shortly summarize the current knowledge of mechanisms of control of sulfate assimilation and identify the most significant gaps.

CONTROL OF FLUX THROUGH THE SULFATE ASSIMILATION PATHWAY

The quest of finding the mechanisms of control of sulfate assimilation has to start by identification of the steps controlling the flux of sulfur through the pathway. Determination of the flux is (relatively) easily possible by incubating the plants with radioactively labeled sulfate and measuring the label in various sulfur pools (Neuenschwander et al., 1991; Kopriva et al., 1999; Vauclare et al., 2002; Scheerer et al., 2010; Mugford et al., 2011). The flux data can be then used for a control flux analysis to calculate the contribution of individual enzymes to the control of the pathway. In a simple approach, exploiting the feedback inhibition of sulfate assimilation by thiols, two major control points were identified, APS reductase and sulfate transport (Vauclare et al., 2002). When sulfate reduction and incorporation to thiols and protein was analyzed in isolation, i.e., considering internal sulfate as the starting point, APS reductase was responsible for ca. 90% of the control. When the transport of external sulfate was taken into account, it contributed about 50% of the total control (Vauclare et al., 2002). However, it was shown later by a similar control analysis that APS reductase is mainly, but not always the main control

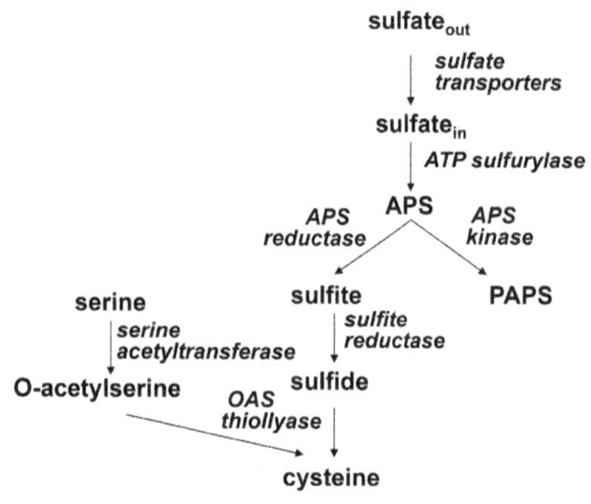

FIGURE 1 | Scheme of sulfate assimilation.

point and contribution of other enzymes was postulated (Scheerer et al., 2010). The flux analysis results thus corroborated the generally accepted view of APS reductase as the key enzyme of the pathway, as demonstrated, e.g., by its strong regulation by environmental factors (Brunold, 1978; Nussbaum et al., 1988; Farago and Brunold, 1990; Neuenschwander et al., 1991; Koprivova et al., 2000). On the other hand, the results were confirmed by analysis of plants with modulated expression of APS reductase, i.e., the accumulation of reduced sulfur compounds in plants overexpressing the genes and reduced flux and reduced tolerance to selenate in mutants of APR2 isoform of APS reductase (Tsakraklides et al., 2002; Grant et al., 2011). In addition, natural variation in *APR2* gene has been shown to cause a variation in sulfate and total sulfur content in several *Arabidopsis* ecotypes (Loudet et al., 2007; Chao et al., 2014).

Analysis of further mutants in the pathway, however, pointed out other genes contributing significantly to the control of flux through sulfate assimilation. Among these genes, three seem to have the highest importance for the reductive part of the pathway. Silencing of mitochondrial isoform of serine acetyltransferase (SAT), the enzyme synthesizing the cysteine precursor OAS, showed a clear correlation between the level of the transcript for this gene and size of the plants (Haas et al., 2008). The same was true for T-DNA insertion mutants in SiR, two knockdown lines showed strong growth inhibition (Khan et al., 2010). In both cases, reduction of cysteine and glutathione (GSH) synthesis rate was observed, but due to use of [^{3}H]serine for the SAT experiments the flux of sulfur was not assessed in these plants (Haas et al., 2008). On the other hand, reduced expression of ATPS1 isoform of ATP sulfurylase (ATPS) leads to reduced flux without growth penalty (Kawashima et al., 2011; Koprivova et al., 2013). The *atps1* mutants of *Arabidopsis*, instead, show an increased accumulation of sulfate in the leaves. While no major alterations of sulfur metabolism in plants overexpressing ATPS1 have been reported, such plants are more tolerant to Se and As and show increased capacity for reduction of selenate (Pilon-Smits et al., 1999; Wangeline et al., 2004). Interestingly, as with *APR2*, natural variation in *ATPS1* contributes to control of variation in sulfate levels in *Arabidopsis* accessions (Koprivova et al., 2013; Herrmann et al., 2014). The flux through reductive sulfate assimilation is, however, altered also due to manipulation of enzymes not directly participating in the pathway. Reduced APS kinase activity in *apk1 apk2* mutants leads to an increased flux through the pathway to cysteine and GSH and to accumulation of reduced sulfur compounds, primarily GSH (Mugford et al., 2009, 2011). In addition, these plants possess low levels of sulfated secondary compounds glucosinolates and are also affected in growth (Mugford et al., 2009, 2010).

The analysis of sulfur fluxes, showing the key role of sulfate transport, APS reductase, and to some extent ATPS and APS kinase in the flux control, thus point to these genes as primary targets for investigations of the molecular mechanisms of regulation of the pathway. Accordingly, promoters of sulfate transporter *SULTR1;2* and *APR3* isoform of APS reductase were used as tools to dissect the regulation of the pathway in several genetic approaches (Maruyama-Nakashita et al., 2006; Koprivova et al., 2010; Lee et al., 2011). However, it is obvious from these results that other mechanisms of the regulation targeting other components of the pathway exist and are important at least for fine tuning of the control. The ways to understand the control of sulfur fluxes is discussed in another contribution to this research topic.

TRANSCRIPTIONAL REGULATION

In the search for molecular mechanisms of regulation of sulfate assimilation, the attention was first focused on the transcriptional regulation (Awazuhara et al., 2002; Maruyama-Nakashita et al., 2004b, 2005, 2006; Falkenberg et al., 2008; Yatusevich et al., 2010; Lee et al., 2011). Indeed, large number of studies showed a clear regulation between the transcript levels of high affinity sulfate transporters and sulfate uptake or between mRNA levels of APS reductase and its protein accumulation, enzyme activity and flux through the pathway suggesting that transcriptional regulation is the main mechanism of control of the pathway (Takahashi et al., 1997, 2000; Kopriva et al., 1999, 2002; Lappartient et al., 1999; Koprivova et al., 2000; Vauclare et al., 2002; Yoshimoto et al., 2002; Hesse et al., 2003; Hartmann et al., 2004; Maruyama-Nakashita et al., 2004b, 2006; Rouached et al., 2008). Transcript levels of the high affinity sulfate transporters *SULTR1;1* and *SULTR1;2* are strongly and specifically upregulated by sulfate starvation and plants expressing GFP under control of promoters of these genes were therefore used in search for factors affecting such regulation. Alternatively, the reporters were expressed under control of synthetic promoter, containing repeats of a 235-bp fragment of a β-conglycinin promoter that confer sulfur starvation response (Awazuhara et al., 2002).

TRANSCRIPTIONAL REGULATION OF SULFATE STARVATION RESPONSE

Increase of sulfate uptake capacity is a characteristic response to sulfate limitation. This increase is primarily triggered by transcriptional regulation of two high affinity sulfate transporters expressed in roots, SULTR1;1 and SULTR1;2 (Takahashi et al., 1997; Yoshimoto et al., 2002). Upon resupply of sulfur, the transcript levels of these transporters are rapidly repressed. Because of robustness of this response, the *SULTR1;1* and *SULTR1;2* genes were used as tools to study the mechanisms of this regulation.

The first gene shown to affect the regulation of *SULTR1;2* by sulfate starvation was the cytokinin receptor *CRE1* (Maruyama-Nakashita et al., 2004b). In a search for factors affecting the sulfate deficiency response of *SULTR1;2* the cytokinin zeatin was found to rapidly repress the induction of the reporter gene. This repression was alleviated in *cre1-1* mutants demonstrating the function of cytokinins as signals in regulating sulfate transport (Maruyama-Nakashita et al., 2004b). However, although a role of cytokinins in the regulatory circuit of sulfate limitation response has been confirmed using the synthetic promoter (Ohkama et al., 2002), CRE1 is not directly involved in control of the transcription of sulfur metabolism genes, and the search thus went further. Using the alternative promoter of *SULTR1;1*, a need for a so far unknown phosphatase in the regulatory circuit has been demonstrated (Maruyama-Nakashita et al., 2004a). The next report got a little closer to the real transcription factors, as a sulfur-responsive SURE *cis* element has been identified in the *SULTR1;1* promoter (Maruyama-Nakashita et al., 2005). Interestingly, the 16 bp element contains an auxin response factor (ARF) binding sequence, which overlaps with the core element GAGAC, as determined by base substitution analysis (Maruyama-Nakashita et al., 2005).

The major breakthrough in the dissection of molecular mechanisms of regulation of sulfur metabolism was made when SULFUR LIMITATION1 (SLIM1) transcription factor has been identified (Maruyama-Nakashita et al., 2006). SLIM1 belongs to the ETHYLENE INSENSITIVE3-LIKE (EIL) family and is also annotated as EIL3. Loss of function of SLIM1 prevents or strongly attenuates sulfate starvation response of most, but not all, transcripts regulated by these conditions, such as sulfate transporters, miR395, or genes involved in glucosinolate synthesis. The exception is the induction of APS reductase, which is SLIM1-independent (**Figure 2**; Maruyama-Nakashita et al., 2006). Somewhat surprisingly, despite the importance of SLIM1 in the regulation of sulfate starvation response, there are many open questions about this factor: primarily the exact binding sequence and the mechanism of action, as *SLIM1* mRNA is not affected by sulfate starvation. A thorough summary of the current knowledge of SLIM1 is provided within this research topic (Wawrzyńska and Sirko, 2014).

Sulfate starvation response has been investigated on many levels, by further genetic screens as well as by systems biology approaches. Interestingly, both lines of research led to pointing out auxin related genes as being involved in regulation of the response. The next gene coming from a genetic screen of altered sulfur limitation response, using the synthetic β-glycinin promoter, was the *BIG* gene, which encodes a protein necessary for the polar transport of auxin (Kasajima et al., 2007). Mutants in the *BIG* gene showed a constitutive upregulation of the reporter gene as well as of some genes upregulated by sulfate deficiency, most interestingly the SLIM1-independent *APR1*. However, since the defect in *BIG* resulted in increase of auxin levels, and auxin itself induces *APR1* and the β-conglycinin expression, this gene is most probably only indirectly related to the sulfate starvation response (Kasajima et al., 2007). At the same time, transcriptomics approach identified several auxin related transcription factors among the genes rapidly responding to sulfate deficiency (Falkenberg et al.,

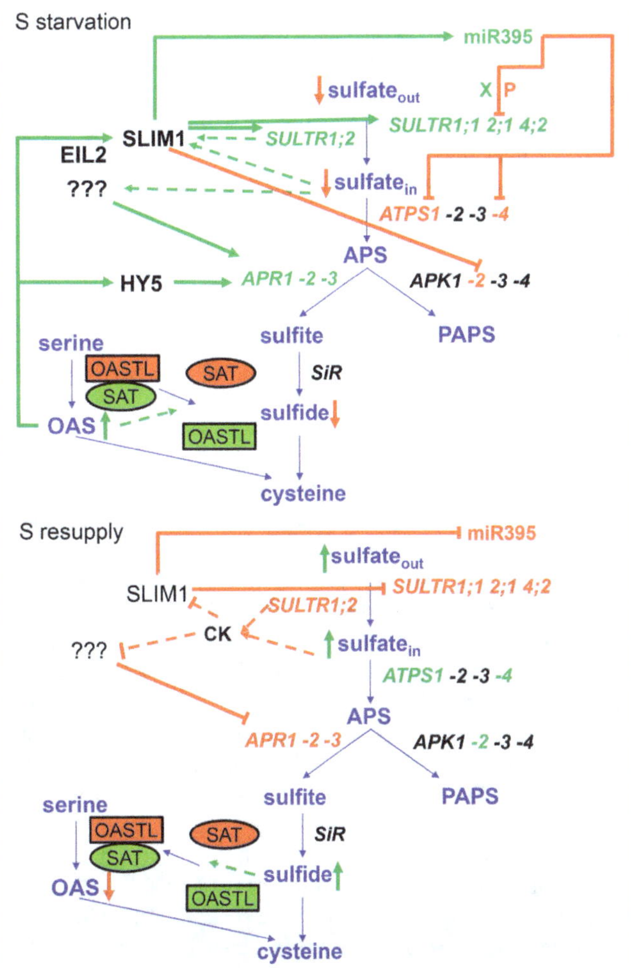

FIGURE 2 | Summary of regulatory mechanisms of response to sulfate starvation and resupply. Green text and arrows depict activation, while red text, arrows and lines represent repression. The interrupted lines symbolise the putative mechanisms of sensing of internal sulfate and transceptor function of SULTR1;2. CK abbreviates cytokinins, X represents xylem, and P phloem.

2008). This has been of special interest also due to the presence of ARF-binding sequence within the SURE cis element (Maruyama-Nakashita et al., 2005). Also in the case of these factors, IAA13, IAA28, and ARF-2, however, their effects on sulfur metabolism seems to be indirect, due to general alteration of auxin signaling (Falkenberg et al., 2008).

Further pieces of the jigsaw have been obtained from studies of tobacco UP9 gene, homologous to *Arabidopsis RESPONSE TO LOW SULFUR* (*LSU*) genes (Lewandowska et al., 2010; Wawrzynska et al., 2010). The gene of unknown function is highly upregulated by sulfate deficiency and so a prime subject of detailed studies. In the promoter of this gene a new sulfur deficiency-responsive motif, named UPE-box, has been identified (Wawrzynska et al., 2010). The UPE-box has no overlap with the SURE motif. It has been found only in eight *Arabidopsis* genes upregulated by sulfate starvation, such as three out of four *LSU* genes and *APR1* and *APR3* isoforms of APS reductase, in many of these together with the SURE element (Wawrzynska et al., 2010). A new transcription factor binding the UPE-box has been characterized as EIL2, belonging to the same family as SLIM1. Interestingly, also SLIM1 was able to drive transcription of a reporter gene from a minimal promoter containing the UPE-box (Wawrzynska et al., 2010). However, since UPE-box is present in only a small subset of genes regulated by sulfate starvation and among them there are promoters of two APR genes that are SLIM1-independent, it is not possible to assign the UPE-box as the prime binding sequence of SLIM1.

TRANSCRIPTIONAL REGULATION OF SULFATE ASSIMILATION GENES

While regulation of sulfate starvation response is an important aspect of sulfur homeostasis, it is by far not the only condition important for control of sulfur metabolism. The transcripts of many sulfate assimilation genes, above all the APS reductase, are upregulated by light, carbohydrates, jasmonic acid, or heavy metals and repressed by nitrogen limitation and reduced sulfur compounds (Kopriva et al., 1999; Koprivova et al., 2000; Vauclare et al., 2002; Hesse et al., 2003; Jost et al., 2005). However, none of these conditions has been reported so far as a basis of genetic screen to find the regulatory factors. The only transcription factor participating in such regulation has been found by a rather indirect approach (Lee et al., 2011). In a genetic screen for defects in GSH homeostasis, using the reduction of root growth by incubation with inhibitor of GSH synthesis buthionine sulfoximine (BSO), the LONG HYPOCOTYL5 (HY5) transcription factor was identified (Lee et al., 2011). The bZIP transcription factor HY5 has been known as a central regulator of photomorphogenesis and is directly binding to promoters of more than 1000 light-inducible genes (Chattopadhyay et al., 1998; Lee et al., 2007). Therefore, the attenuated induction of APS reductase by light in *hy5* mutant was not entirely surprising (Lee et al., 2011). The loss of function of HY5 has, however, further consequences for regulation of the pathway, as the induction of APS reductase by OAS and repression by nitrogen limitation are also attenuated in the mutant (**Figure 3**). The disrupted transcriptional regulation is reflected in altered flux through the pathway and also sulfate uptake. Importantly, HY5 is the only transcription factor for which evidence of direct binding on the corresponding promoters has been obtained and a direct transcriptional regulation can be unequivocally confirmed (Lee et al., 2011). Chromatin immunoprecipitation experiments showed clearly that HY5 binds promoters of *APR1* and *APR2* and also of *SULTR1;2*. This together with HY5's involvement in regulation by light, OAS, and nitrogen limitation positions this factor to a central place in the sulfate assimilation regulatory circuit.

The next transcription factors regulating APS reductase expression have also been found indirectly. The two groups of MYB factors, MYB28, MYB29, and MYB76 controlling synthesis of aliphatic glucosinolates, and MYB51, MYB34, and MYB122 controlling indolic glucosinolates (Gigolashvili et al., 2007a,b, 2008; Hirai et al., 2007; Sonderby et al., 2007; Malitsky et al., 2008), were initially linked to regulation of sulfate assimilation because of the importance of PAPS for glucosinolate synthesis (Mugford et al., 2009; Yatusevich et al., 2010). In the *apk1 apk2* plants, in which PAPS synthesis is low, the reduced levels of glucosinolates trigger a coordinated upregulation of genes involved in synthesis of these metabolites (Mugford et al., 2009). The upregulation is probably controlled by the 6 MYB factors, as their mRNA levels are also elevated. Since (1) the genes of glucosinolate synthesis form a single regulatory network and (2) sufficient PAPS availability is important for glucosinolate synthesis, it was hypothesized that also PAPS synthesis might be part of the network (Yatusevich et al., 2010). Indeed, transactivation assays, in which a reporter gene under control of investigated promoter is co-expressed with transcription factor, demonstrated that *APK1*, *APK2*, and partly *APK3* isoforms of APS kinase and *ATPS1* and *ATPS3* isoforms of ATPS are under the control of all six glucosinolate connected MYB factors (Yatusevich et al., 2010). Interestingly, genes of the dedicated reductive part of sulfate assimilation, APS reductase and SiR were also positive in the transactivation assays and thus regulated by the MYB factors. The results of transactivation assays were confirmed in transgenic plants overexpressing the MYB factors, as in all of them the steady state levels of the ATPS; APK, APR, and SiR genes were elevated. The link of APS reductase with glucosinolate synthesis, although belonging to different pathways, can be explained by the need of reduced sulfur for the thioglucoside bond in the core structure of the glucosinolates as well as the origin of aliphatic glucosinolates from sulfur containing amino acid methionine. It seems, however, that the MYB factors contribute to only part of the regulatory circuits of sulfate assimilation, mainly to those connected to biotic stress in which glucosinolate synthesis is induced. With notable exception of APR and indolic MYBs, the steady state levels of the genes of primary assimilation are not affected in mutants of the 6 MYB factors. The increase transcript levels of APS reductase in, e.g., *myb51* mutant corresponded with increased enzyme activity and accumulation of GSH. This might be an adaptation to low accumulation of indolic glucosinolates in this mutant to increase GSH content as an alternative defense compound (Yatusevich et al., 2010). Function of the MYB factors in general regulation of sulfur homeostasis thus remains rather elusive.

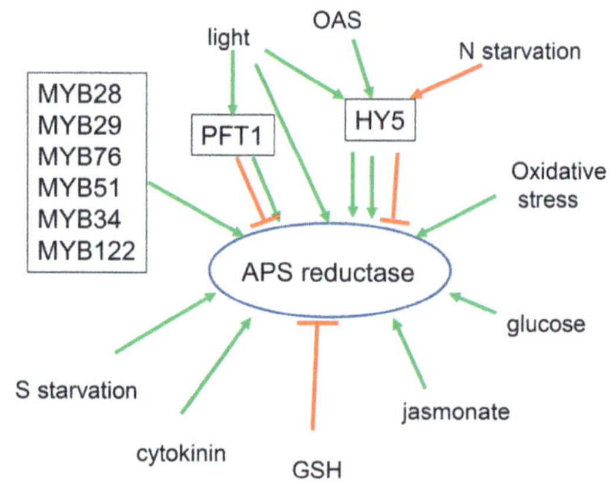

FIGURE 3 | Schematic summary of regulation of APS reductase. Factors known to control *APR* transcription are boxed. Light affects *APR* through HY5, PFT1, and independent from both factors. PFT1 acts as activator for *APR2* but repressor for *APR1* and *APR3* isoforms.

POST-TRANSCRIPTIONAL REGULATION

Although many reports showed a clear correlation between the regulation of transcript levels and activities of the gene products, several exceptions of this pattern have been observed (Bick et al., 2001; Yoshimoto et al., 2007; Koprivova et al., 2008; Kawashima et al., 2011). The more detailed the search for molecular mechanisms has become, the more frequently such post-transcriptional regulation has been observed. It is evident, that without taking post-transcriptional and post-translational regulation of sulfate assimilation into account, the understanding of the pathway control would never be complete. Many different levels of such regulation have been described, but only a few have been sufficiently functionally analyzed to understand the molecular mechanisms.

miR395 IN CONTROL OF SULFATE HOMEOSTASIS

Probably the best understood post-transcriptional regulation of sulfate assimilation pathway is the action of microRNA miR395 (Jones-Rhoades and Bartel, 2004; Kawashima et al., 2009, 2011; Pant et al., 2009; Liang et al., 2010; Matthewman et al., 2012). MicroRNAs are short non-coding molecules that regulate the expression of protein coding genes. Among the first miRNAs to be characterized was the miR395, presumably because its target genes have been easily recognized as genes involved in sulfate assimilation: a low affinity sulfate transporter *SULTR2;1* and three out of four members of the ATPS gene family (ATPS1, 3, and 4; Jones-Rhoades and Bartel, 2004; Allen et al., 2005). MiR395 is strongly induced by sulfate deficiency, and in turn it cleaves mRNAs of its target genes (Jones-Rhoades and Bartel, 2004; Allen et al., 2005; Kawashima et al., 2009). Indeed, three of the targets were confirmed experimentally in both shoots and roots of *Arabidopsis thaliana*, whereas the cleavage of *ATPS3* seems to be restricted to the shoot only (Kawashima et al., 2009). Overexpression of miR395 causes accumulation of sulfate in the leaves, due to increased translocation from the roots (Liang et al., 2010; Kawashima et al., 2011). The increased sulfate translocation seems to be the main mechanism of miR395 function as revealed by analyses of plants with higher and lower miR395 levels because of overexpression or target mimicry, respectively (Kawashima et al., 2011). The higher root-to-shoot transport is achieved through increased translocation rate and reduced flux through sulfate reduction specifically in the roots (Kawashima et al., 2011).

The effects of miR395 on its targets follow different mechanisms. Only the *ATPS4* isoform of ATPS undergoes the canonical regulation, where its transcript levels strongly decrease with increased miR395 accumulation (Jones-Rhoades and Bartel, 2004; Kawashima et al., 2009, 2011; Liang et al., 2010). The *ATPS1* mRNA levels have been reported either to decrease slightly (Jones-Rhoades and Bartel, 2004; Liang et al., 2010) or not to be affected by sulfate deficiency (Hirai et al., 2003; Kawashima et al., 2011). The unexpected response of this miR395 target can be explained by a simultaneous increase in *ATPS1* transcription, as demonstrated by comparison of GFP expression in plants expressing GFP under control of *ATPS1* promoter directly or after fusion with *ATPS1* coding region, and thus targeted for miR395 cleavage (Kawashima et al., 2011). Interestingly, the transcript levels of *SULTR2;1* are actually higher in sulfate deficient roots than in control roots (Kawashima et al., 2009, 2011). The miR395 function is, however, enabled by a non-overlapping cell-specific expression pattern for *SULTR2;1* and the miRNA, which is expressed specifically in phloem companion cells and allows SULTR2;1 in xylem parenchyma cells to remain functional for xylem loading of sulfate (Kawashima et al., 2009). The placement of miR395 in the sulfate deficiency regulatory network was strongly corroborated by showing that the induction of miR395 accumulation is dependent on SLIM1 (Kawashima et al., 2009, 2011) and that miR395 levels are affected by OAS, cysteine, and cadmium (Matthewman et al., 2012; Zhang et al., 2013). Interestingly, miR395 has been found in phloem of S-starved plants pointing to its role as a long-distance signal (Pant et al., 2009). However, the analysis of plants expressing GFP under control of promoters of the six miR395 genes revealed that the expression of miRNA is strongly induced both in shoots and roots, so the significance of the phloem transport is not known (Kawashima et al., 2009).

PROTEIN–PROTEIN INTERACTIONS

Multienzyme complexes often form control points of metabolic pathways as they allow substrate channeling and allosteric modulation of activity. The same is true for sulfate assimilation, where the last enzymatic step, incorporation of sulfide into cysteine, is catalyzed by cysteine synthase (Wirtz and Hell, 2006; Wirtz et al., 2010). The complex is formed by two enzymes, the SAT, which synthesizes OAS, and OAS-(thiol)lyase (OASTL), which uses the OAS and sulfide for synthesis of cysteine. However, the assembly of the two consecutive enzymes does not serve a better channeling of OAS between the two enzymes, but rather strongly modulates their activity, at least in *in vitro* experiments (Droux et al., 1998; Hell et al., 2002; Wirtz et al., 2010). SAT activity is greatly increased in the complex, which also attenuates its feedback inhibition by cysteine. On the other hand, OASTL is inactive in the complex and cysteine is formed by the excess free enzyme only. The stability of the complex is influenced by the substrates OAS and sulfide: whereas sulfide stabilizes the complex, OAS promotes the dissociation of the subunits, as it competes with SAT for the binding site (Berkowitz et al., 2002; Francois et al., 2006; Wirtz et al., 2010). This modulation of complex stability and consequently activity by the two pathway intermediates points to a function in regulating the pathway, particularly during sulfate limitation. In these conditions sulfide availability decreases and OAS accumulates, which leads to dissociation of the complex and reduced synthesis of OAS (Droux et al., 1998; Wirtz and Hell, 2006). The concentration of OAS needed for half-maximal dissociation of the complex is 77 μM and thus within the physiological range in plant cells, which confirms the relevance of this regulation for the control of sulfate assimilation pathway (Berkowitz et al., 2002).

OAS, and OAS-(thiol)lyase take part in another example of protein–protein interactions affecting sulfur metabolism. The cytosolic isoform OASTL-A interacts with a STAS domain of SULTR1;2 transporter and reduces the sulfate uptake rate in yeast heterologous system (Shibagaki and Grossman, 2010). The OASTL is also affected by the interaction and its activity is increased. Interestingly, this modulation of OASTL activity is specific for SULTR1;2, as the same domain from a closely related SULTR1;1

has no effect, despite binding the enzyme (Shibagaki and Grossman, 2010). The repression of sulfate uptake by OASTL-A is more pronounced at high sulfate supply than during sulfate limitation, suggesting a regulatory function of this interaction. The physiological function of this regulation is, however, not very clear and requires further investigation.

An interesting addition to protein–protein interaction in sulfate assimilation has been finding of modulation of chloroplastic SAT activity by interaction with cyclophilin 20-3 (Dominguez-Solis et al., 2008). The cyclophilin has been postulated as a signal in response to oxidative stress, since in wild type plants SAT activity was elevated upon stress treatment, but this activation was strongly attenuated in *cyc20-3* mutants. The increase in SAT activity causes elevated thiol content to combat the oxidative stress (Dominguez-Solis et al., 2008). It has been shown recently, however, that the CYP-SAT interaction is a part of a signaling pathway of the phytohormone (+)-12-oxo-phytodienoic acid (OPDA; Park et al., 2013). OPDA binds to CYP20-3, which increases its affinity for SAT. The CYP-SAT complex facilitates formation of cysteine synthase complex and so increases synthesis of OAS and cysteine, with subsequent alterations of redox potential. The redox changes then modulate expression of at least some OPDA-responsive genes (Park et al., 2013). How far such mechanism contributes to control of sulfate assimilation is currently unclear, but the CYP20-3 does not seem to be necessary for the normal formation of cysteine synthase complex.

REDOX REGULATION

Several enzymes of sulfate assimilation undergo redox regulation. In fact, redox regulation of APS reductase has been the first reported example of post-transcriptional regulatory mechanism of the pathway (Bick et al., 2001). An uncoupling of the regulation of APR transcript levels and enzyme activity was observed in plants under oxidative stress. This could be explained by a redox regulation of the enzyme, which is activated in oxidizing conditions (Bick et al., 2001). This observation *in vivo* agrees with *in vitro* results, which demonstrated inhibition of APS reductase activity by reductants (Kopriva and Koprivova, 2004). Two mechanisms for the regulation have been proposed, a redox-regulated switch between an active protein dimer and inactive monomer or a regulatory cysteine pair (Bick et al., 2001; Kopriva and Koprivova, 2004). Both have been supported by experimental evidence, so it seems that the jury is out until the structure of APS reductase is solved.

However, APS reductase is not the only enzyme of the pathway regulated by changes in redox environment. It has been long known that the first enzyme of GSH synthesis, the γ-glutamylcysteine synthetase (γECS), is feedback regulated by GSH (Hell and Bergmann, 1990). The plant enzyme contains one or two (in Brassicaceae) redox active cysteine pairs and after incubation with reductants, such as GSH, changes its topology from dimer to monomer and loses activity (Jez et al., 2004; Hothorn et al., 2006). The redox regulation thus allows rapid adjustment of the activity and GSH synthesis rate to the redox environment and actual GSH concentration in the cell.

Another enzyme of the pathway regulated by changes in redox potential is APS kinase. The redox control has been unexpectedly discovered after solving the crystal structure of the *Arabidopsis* APK1 isoform (Ravilious et al., 2012). The enzyme contains a redox active disulfide bond within each subunit. Interestingly, in contrast to APS reductase and γECS, this enzyme is activated by the reductants and the reduction also alleviates the otherwise strong substrate inhibition (Ravilious et al., 2012). The opposite redox regulation of APS kinase and APS reductase is particularly relevant as the enzymes use the same substrate. It offers, therefore, an interesting possibility that the partitioning of sulfur between these two enzymes, and so between primary and secondary sulfur metabolism, is at least partly under redox control.

OTHER POST-TRANSCRIPTIONAL REGULATION

Apart of these clearly defined examples of post-transcriptional regulation, other, less well understood observations have been made. The regulation of sulfate transporters by sulfate deficiency includes a post-transcriptional component (Yoshimoto et al., 2007). When *sultr1;1 sultr1;2* mutant was complemented by tagged transporters under control of constitutive 35S promoter, not only the localization in root epidermis was reconstituted, the protein accumulation and sulfate uptake were upregulated by sulfate deficiency. This represent a completely new mechanism of control of sulfate transport (Yoshimoto et al., 2007). However, its relevance *in vivo* remains to be demonstrated, since this mechanism could not complement the loss of SLIM1. It is, however, possible that the components of this post-transcriptional regulation are under SLIM1 control and that SLIM1 is responsible for both transcriptional and post-transcriptional regulation of the transporters.

Another component of the regulatory network affecting sulfate assimilation is PHYTOCHROME AND FLOWERING TIME1 (PFT1). Loss of PFT1 results in altered transcriptional regulation of APR by light, in an isoform specific pattern (Koprivova et al., 2014a). While *APR2* is induced by light to a lesser degree in *pft1* mutants than in wild type, the induction is significantly bigger for *APR1* and *APR3*. This increased response to light is accompanied by increased flux through the pathway (Koprivova et al., 2014a). However, as it is not a transcription factor the effect of PFT1 on APR transcription must be indirect. Indeed, PFT1 is MED25 subunit of the Mediator complex, which facilitates gene transcription by bridging transcription factors with RNA polymerase II complex (Conaway and Conaway, 2011). As part of the Mediator, PFT1 interacts with a number of transcription factors and modulates so their activity (Ou et al., 2011). Mediator, and specifically PFT1 have been shown to affect a large number of processes and may represent the mechanism for integration of various signals into a single response and so for fine tuning of gene expression (Kidd et al., 2009; Elfving et al., 2011; Kim et al., 2011; Inigo et al., 2012). The contribution of Mediator and its individual subunits to control of sulfate assimilation is thus of utmost importance for a deep and full understanding of the processes.

The summary of post-transcriptional regulation of sulfate assimilation would not be complete without mentioning the attempts to dissect the regulation of APS reductase by salt (Koprivova et al., 2008). While in most reports on regulation of the pathway an uncoupling of mRNA and activity was very rare, this study showed a large number of such phenomena. Thus, treatment of *Arabidopsis* with ABA led to a strong decrease of APS reductase enzyme activity without affecting transcript levels of its three isoforms (Koprivova et al., 2008). The largest number of "exceptions" has been observed in the analysis of mutants in signal transduction pathways. For example, in *npr1*, *etr1*, and *jar1*, deficient in salicylate, ethylene, and jasmonate signaling, respectively, salt induced mRNA of all three APR isoforms but not the enzyme activity, whereas in the gibberellin insensitive mutant *gai*, the mRNA was not affected but activity increased (Koprivova et al., 2008). These results imply, that the regulatory network of sulfate assimilation is very complex and well balanced, so that the full extent of the regulation might be seen only after disturbance of the system by multiple factors simultaneously (Koprivova and Kopriva, 2008).

SENSING AND SIGNALING

Regulatory networks are formed not only by transcription factors, miRNAs, or other post-transcriptional mechanisms, another important components are sensors detecting changes in external or internal environment and signaling cascades that transmit the information from sensors to nucleus and trigger the transcriptional response. In higher plants, surprisingly little is known about the sensing mechanisms, how plants recognize sulfur deficiency, what is the sensor of refilled sulfur pools or of excess reduced sulfur. There are two major theories on the sensing of sulfur deficiency/sufficiency, either a receptor monitoring external (or possible apoplastic/vacuolar) sulfate or levels of downstream product(s) of sulfate assimilation. The dissociation of cysteine synthase complex described above might be part of the latter response (Wirtz and Hell, 2006). On the other hand, two observations indicate that sulfate levels are monitored by plants. Firstly, analysis of gene expression in different mutants of sulfur metabolism showed that reduced sulfate content, e.g., in *sultr1;2* and *fry1*, but not low GSH concentration in *cad2* and *rax1*, causes similar changes in gene expression as sulfate deficiency even at normal external sulfate supply (Maruyama-Nakashita et al., 2003; Matthewman et al., 2012). These results thus pointed to internal sulfate being the sensed metabolite. This hypothesis that sulfate is the measure of sulfur status of the plant was corroborated by analysis of new alleles of *sultr1;2* (Zhang et al., 2014). Under normal sulfate supply these mutants showed strongly reduced sulfate levels and activation of genes involved in sulfate limitation response. When the mutants were incubated in high sulfate and the sulfate levels were restored, the expression of sulfate starvation marker genes remained high. These results are best explained by postulating an additional function of SULTR1;2 as sensor of sulfate status (Zhang et al., 2014). When confirmed and mechanistically explained, SULTR1;2 could be considered a transceptor similar to the nitrate transporter NRT1;1 (Ho et al., 2009). Interestingly, in the green alga *Chlamydomonas*, in which the mechanisms of sulfate limitation response are much better understood, a sulfate sensor SAC1 has been identified as a member of sulfate transporter family SLC13 (Davies et al., 1996). It seems, therefore, evident that sulfate is the metabolite used for establishing sulfur status of the plant, but contribution of other systems, such as the cysteine synthase complex, cannot be excluded and may be important for specific parts of the regulatory networks.

Apart of sulfate, other small molecules seem to be integral components of sulfate assimilation regulatory networks. Many metabolites affect individual components of the pathway, such as pathway intermediates OAS, cysteine, glutathione, sugars, or the phytohormones jasmonate, salicylate, ABA, ethylene, nitric oxide, and cytokinins. All these metabolites and many others can potentially be signals in the regulation of the pathway, but on the other hand, their effects may be only indirect and pleiotrophic. Still, several of these metabolites can be considered true signals. The role of cytokinins in repressing the expression of sulfate assimilation genes at sufficient sulfur availability seems to be well established (Maruyama-Nakashita et al., 2004b) and is similar to the role of these hormones in regulation of nitrate assimilation (Sakakibara et al., 2006), making them a good candidate for a true signal.

However, the one compound that immediately comes to mind, when signals are mentioned, is OAS. OAS has been discussed as signal for decades but this role has often been met with controversy. OAS induces transcript levels and activity of sulfate transporters and APS reductase (Neuenschwander et al., 1991; Smith et al., 1997; Koprivova et al., 2000). Incubation with OAS triggers a global response of gene expression, similar to sulfate deficiency (Hirai et al., 2003), including induction of the miR395 (Matthewman et al., 2012). As OAS accumulates during sulfate limitation, it was a logical conclusion to consider OAS as the signal of sulfate starvation, which triggers the changes in gene transcription (Hirai et al., 2003). However, this conclusion has been seriously questioned when a time course experiment shown that the changes in gene expression in sulfur starved plants actually precede the accumulation of OAS (Hopkins et al., 2005). This controversy seems to be resolved by elegant experiments which identified a cluster of genes directly regulated by OAS (Hubberten et al., 2012). The genes were found in a combination of stringent analyses of available -omics datasets, finding correlation between OAS accumulation and gene expression, with analysis of plants with inducible SAT and thus producing a pulse of OAS without changes in other metabolites. The cluster is formed from six genes, which are highly upregulated by sulfate deficiency (Hubberten et al., 2012). Interestingly, it includes both SLIM1-dependent and SLIM1-independent genes (*APR3*). The independent verification of the gene cluster by three methods/datasets establishes OAS as a signal and a direct component of the regulatory network, but the mechanism of its action is still left open (Hubberten et al., 2012). The role of OAS in control of transcription is, however, independent from its effect on the stability of cysteine synthase complex.

Another signal is necessary to transmit the information of sufficient or elevated concentration of reduced sulfur compounds. There are three candidates, H_2S, cysteine, and GSH. Since these metabolites are highly interconnected, feeding of one compound results in increased levels of others, it is not easy to identify the real signal. H_2S is a specific case, since it was recognized as a gaseous signal in human and animal world (Kimura and Kimura, 2004) there are increasing numbers of reports of H_2S being a signal in plants as well, protecting against a large range of stresses and even promoting growth (Lisjak et al., 2010; Dooley et al., 2013; Sun et al., 2013). The story of H_2S and its signaling function is complicated and controversial and is discussed in several recent reviews (Garcia-Mata and Lamattina, 2013; Lisjak et al., 2013; Calderwood and Kopriva, 2014). Both cysteine and GSH have the same effect on gene expression, i.e., repression of sulfate transporters and APS reductase, but since this effect can be attenuated by inhibition of GSH synthesis BSO, GSH is the better candidate for the signal (Lappartient et al., 1999; Vauclare et al., 2002; Hartmann et al., 2004). This seems to be confirmed by the substantial alterations in gene expression in mutants in GSH synthesis (Ball et al., 2004), although in the mutant Cys concentration also differs from wild type. Again, the mechanism of action of GSH as signal is not known, it may be simple redox regulation, glutathionylation of specific transcription factors, or non-covalent binding to a factor and modulation of its function.

CONCLUSIONS AND OPEN QUESTIONS

It is obvious that our knowledge of molecular mechanisms of regulation of sulfate assimilation has been improved. But in many aspects this knowledge is still patchy. We know that SLIM1 is a central regulator of sulfate deficiency response, but do not know the DNA sequence and the complement of promoters it binds. We know that there must be at least one other factor controlling the induction of APR by sulfate limitation, but not the nature of this factor. We know that sulfate assimilation is preferentially localized in bundle sheath cells surrounding the veins in *Arabidopsis* (Aubry et al., 2014), but we do not know the mechanisms and the biological significance. Several signaling molecules have been identified, but we do not know how they transmit the signal. We know that many genes of the pathway are regulated by multiple environmental and metabolic conditions, but we do not know the transcription factors and transduction pathways. We got a first hint of a possible modulation of the transcriptional response by the Mediator complex, but know almost nothing about the contribution of other subunits than PFT1. There are several miRNAs affected by sulfate deficiency (Buhtz et al., 2010), but apart of miR395 nothing is known about their targets and functions. There are many genes highly induced by sulfate limitation, but the functions of most of them are not known. The list of similar questions could be much longer and all of them are important to answer, in order to understand the regulatory networks of the pathway. Or are they?

Many reports applying quantitative genetics and exploiting natural variation to dissect a control of complex traits identified metabolic genes underlying the variation (Loudet et al., 2007; Baxter et al., 2010; Chan et al., 2011; Chao et al., 2012; Koprivova et al., 2013). In QTL analysis of sulfate content *APR2* and *ATPS1* have been found to affect the levels of foliar sulfate. For both genes, substantial variation in amino acid sequence has been found, including those that strongly diminished the enzyme activities (Loudet et al., 2007; Herrmann et al., 2014). Two more independent alleles of *APR2* were found among *Arabidopsis* accessions, associated with high sulfate and total sulfur content (Chao et al., 2014). Genome wide approaches led to identification of gene variants responsible for the large variation in types and amount of glucosinolates (Chan et al., 2011). These natural haplotypes represent sources of alleles that can be directly used for improvement of complex metabolic traits. They also suggest evolutionary adaptations of sulfur metabolism to environment. Whereas for glucosinolates there is a link between the variation of their composition and herbivory (Bidart-Bouzat and Kliebenstein, 2008), such links are not obvious for the APR2 or ATPS1 alleles. There does not seem to be much common between the origins of the three APR2 haplotypes: Middle Asian mountains (Sha), south of Czech Republic (Hod) and northern Sweden (Loudet et al., 2007; Chao et al., 2014). It is, however, possible to speculate that at least for Sha and the Swedish accessions, growth might be restricted due to harsh conditions and the reduction of sulfate assimilation would prevent accumulation of reduced sulfur compound and increasingly reducing cellular environment. The analysis is not limited to *Arabidopsis*, similar approaches have been made directly with crops and similar haplotypes have been identified (Harper et al., 2012; Koprivova et al., 2014b). Thus, modulation of, e.g., sulfate levels, seems to be possible without knowing the regulatory networks, transcription factors, cis elements, or signals controlling sulfate homeostasis. The two approaches and amounts of detail are, however, complementary and together will bring our understanding of sulfur metabolism on the level to know how it is regulated and how we can exploit the knowledge.

REFERENCES

Allen, E., Xie, Z., Gustafson, A. M., and Carrington, J. C. (2005). MicroRNA-directed phasing during trans-acting siRNA biogenesis in plants. *Cell* 121, 207–221. doi: 10.1016/j.cell.2005.04.004

Aubry, S., Smith-Unna, R. D., Boursnell, C. M., Kopriva, S., and Hibberd, J. M. (2014). Transcript residency on ribosomes reveals a key role for the *Arabidopsis thaliana* bundle sheath in sulfur and glucosinolate metabolism. *Plant J.* 78, 659–673. doi: 10.1111/tpj.12502

Awazuhara, M., Kim, H., Goto, D. B., Matsui, A., Hayashi, H., Chino, M., et al. (2002). A 235-bp region from a nutritionally regulated soybean seed-specific gene promoter can confer its sulfur and nitrogen response to a constitutive promoter in aerial tissues of *Arabidopsis thaliana. Plant Sci.* 163, 75–82. doi: 10.1016/S0168-9452(02)00064-X

Ball, L., Accotto, G. P., Bechtold, U., Creissen, G., Funck, D., Jimenez, A., et al. (2004). Evidence for a direct link between glutathione biosynthesis and stress defense gene expression in *Arabidopsis. Plant Cell* 16, 2448–2462. doi: 10.1105/tpc.104.022608

Baxter, I., Brazelton, J. N., Yu, D., Huang, Y. S., Lahner, B., Yakubova, E., et al. (2010). A coastal cline in sodium accumulation in *Arabidopsis thaliana* is driven by natural variation of the sodium transporter AtHKT1;1. *PLoS Genet.* 6:e1001193. doi: 10.1371/journal.pgen.1001193

Berkowitz, O., Wirtz, M., Wolf, A., Kuhlmann, J., and Hell, R. (2002). Use of biomolecular interaction analysis to elucidate the regulatory mechanism of the cysteine synthase complex from *Arabidopsis thaliana. J. Biol. Chem.* 277, 30629–30634. doi: 10.1074/jbc.M111632200

Bick, J. A., Setterdahl, A. T., Knaff, D. B., Chen, Y., Pitcher, L. H., Zilinskas, B. A., et al. (2001). Regulation of the plant-type 5′-adenylyl sulfate reductase by oxidative stress. *Biochemistry* 40, 9040–9048. doi: 10.1021/bi010518v

Bidart-Bouzat, M. G., and Kliebenstein, D. J. (2008). Differential levels of insect herbivory in the field associated with genotypic variation in glucosinolates in *Arabidopsis thaliana. J. Chem. Ecol.* 34, 1026–1037. doi: 10.1007/s10886-008-9498-z

Brunold, C. (1978). Regulation of sulfate assimilation in plants: 7. Cysteine inactivation of adenosine 5′-phosphosulfate sulfotransferase in *Lemna minor* L. *Plant Physiol.* 61, 342–347. doi: 10.1104/pp.61.3.342

Buhtz, A., Pieritz, J., Springer, F., and Kehr, J. (2010). Phloem small RNAs, nutrient stress responses, and systemic mobility. *BMC Plant Biol* 10:64. doi: 10.1186/1471-2229-10-64

Calderwood, A., and Kopriva, S. (2014). Hydrogen sulfide in plants: from dissipation of excess sulfur to signaling molecule. *Nitric Oxide* 41C, 72–78. doi: 10.1016/j.niox.2014.02.005

Cao, M. J., Wang, Z., Wirtz, M., Hell, R., Oliver, D. J., and Xiang, C. B. (2013). SULTR3;1 is a chloroplast-localized sulfate transporter in *Arabidopsis thaliana*. *Plant J.* 73, 607–616. doi: 10.1111/tpj.12059

Chan, E. K., Rowe, H. C., Corwin, J. A., Joseph, B., and Kliebenstein, D. J. (2011). Combining genome-wide association mapping and transcriptional networks to identify novel genes controlling glucosinolates in *Arabidopsis thaliana*. *PLoS Biol.* 9:e1001125. doi: 10.1371/journal.pbio.1001125

Chao, D. Y., Baraniecka, P., Danku, J., Koprivova, A., Lahner, B., Luo, H., et al. (2014). Variation in sulfur and selenium accumulation is controlled by naturally occurring isoforms of the key sulfur assimilation enzyme APR2 across the *Arabidopsis thaliana* species range. *Plant Physiol.* doi: 10.1104/pp.114.247825 [Epub ahead of print].

Chao, D. Y., Silva, A., Baxter, I., Huang, Y. S., Nordborg, M., Danku, J., et al. (2012). Genome-wide association studies identify heavy metal ATPase3 as the primary determinant of natural variation in leaf cadmium in *Arabidopsis thaliana*. *PLoS Genet.* 8:e1002923. doi: 10.1371/journal.pgen.1002923

Chattopadhyay, S., Ang, L. H., Puente, P., Deng, X. W., and Wei, N. (1998). *Arabidopsis* bZIP protein HY5 directly interacts with light-responsive promoters in mediating light control of gene expression. *Plant Cell* 10, 673–683. doi: 10.1105/tpc.10.5.673

Conaway, R. C., and Conaway, J. W. (2011). Function and regulation of the mediator complex. *Curr. Opin. Genet. Dev.* 21, 225–230. doi: 10.1016/j.gde.2011.01.013

Davies, J. P., Yildiz, F. H., and Grossman, A. (1996). Sac1, a putative regulator that is critical for survival of *Chlamydomonas reinhardtii* during sulfur deprivation. *EMBO J.* 15, 2150–2159.

Dominguez-Solis, J. R., He, Z., Lima, A., Ting, J., Buchanan, B. B., and Luan, S. (2008). A cyclophilin links redox and light signals to cysteine biosynthesis and stress responses in chloroplasts. *Proc. Natl. Acad. Sci. U.S.A.* 105, 16386–16391. doi: 10.1073/pnas.0808204105

Dooley, F. D., Nair, S. P., and Ward, P. D. (2013). Increased growth and germination success in plants following hydrogen sulfide administration. *PLoS ONE* 8:e62048. doi: 10.1371/journal.pone.0062048

Droux, M., Ruffet, M. L., Douce, R., and Job, D. (1998). Interactions between serine acetyltransferase and O-acetylserine (thiol) lyase in higher plants – structural and kinetic properties of the free and bound enzymes. *Eur. J. Biochem.* 255, 235–245. doi: 10.1046/j.1432-1327.1998.2550235.x

Elfving, N., Davoine, C., Benlloch, R., Blomberg, J., Brannstrom, K., Muller, D., et al. (2011). The *Arabidopsis thaliana* Med25 mediator subunit integrates environmental cues to control plant development. *Proc. Natl. Acad. Sci. U.S.A.* 108, 8245–8250. doi: 10.1073/pnas.1002981108

Falkenberg, B., Witt, I., Zanor, M. I., Steinhauser, D., Mueller-Roeber, B., Hesse, H., et al. (2008). Transcription factors relevant to auxin signalling coordinate broad-spectrum metabolic shifts including sulphur metabolism. *J. Exp. Bot.* 59, 2831–2846. doi: 10.1093/jxb/ern144

Farago, S., and Brunold, C. (1990). Regulation of assimilatory sulfate reduction by herbicide safeners in *Zea mays* L. *Plant Physiol.* 94, 1808–1812. doi: 10.1104/pp.94.4.1808

Francois, J. A., Kumaran, S., and Jez, J. M. (2006). Structural basis for interaction of O-acetylserine sulfhydrylase and serine acetyltransferase in the *Arabidopsis* cysteine synthase complex. *Plant Cell* 18, 3647–3655. doi: 10.1105/tpc.106.047316

Garcia-Mata, C., and Lamattina, L. (2013). Gasotransmitters are emerging as new guard cell signaling molecules and regulators of leaf gas exchange. *Plant Sci.* 201–202, 66–73. doi: 10.1016/j.plantsci.2012.11.007

Gigolashvili, T., Berger, B., Mock, H. P., Muller, C., Weisshaar, B., and Flugge, U. I. (2007a). The transcription factor HIG1/MYB51 regulates indolic glucosinolate biosynthesis in *Arabidopsis thaliana*. *Plant J.* 50, 886–901. doi: 10.1111/j.1365-313X.2007.03099.x

Gigolashvili, T., Yatusevich, R., Berger, B., Muller, C., and Flugge, U. I. (2007b). The R2R3-MYB transcription factor HAG1/MYB28 is a regulator of methionine-derived glucosinolate biosynthesis in *Arabidopsis thaliana*. *Plant J.* 51, 247–261. doi: 10.1111/j.1365-313X.2007.03133.x

Gigolashvili, T., Engqvist, M., Yatusevich, R., Muller, C., and Flugge, U. I. (2008). HAG2/MYB76 and HAG3/MYB29 exert a specific and coordinated control on the regulation of aliphatic glucosinolate biosynthesis in *Arabidopsis thaliana*. *New Phytol.* 177, 627–642. doi: 10.1111/j.1469-8137.2007.02295.x

Grant, K., Carey, N. M., Mendoza, M., Schulze, J., Pilon, M., Pilon-Smits, E. A., et al. (2011). Adenosine 5′-phosphosulfate reductase (APR2) mutation in *Arabidopsis* implicates glutathione deficiency in selenate toxicity. *Biochem. J.* 438, 325–335. doi: 10.1042/BJ20110025

Haas, F. H., Heeg, C., Queiroz, R., Bauer, A., Wirtz, M., and Hell, R. (2008). Mitochondrial serine acetyltransferase functions as a pacemaker of cysteine synthesis in plant cells. *Plant Physiol.* 148, 1055–1067. doi: 10.1104/pp.108.125237

Harper, A. L., Trick, M., Higgins, J., Fraser, F., Clissold, L., Wells, R., et al. (2012). Associative transcriptomics of traits in the polyploid crop species *Brassica napus*. *Nat. Biotechnol.* 30, 798–802. doi: 10.1038/nbt.2302

Hartmann, T., Honicke, P., Wirtz, M., Hell, R., Rennenberg, H., and Kopriva, S. (2004). Regulation of sulphate assimilation by glutathione in poplars (*Populus tremula* x *P.alba*) of wild type and overexpressing gamma-glutamylcysteine synthetase in the cytosol. *J. Exp. Bot.* 55, 837–845. doi: 10.1093/Jxb/Erh094

Heeg, C., Kruse, C., Jost, R., Gutensohn, M., Ruppert, T., Wirtz, M., et al. (2008). Analysis of the *Arabidopsis* O-acetylserine(thiol)lyase gene family demonstrates compartment-specific differences in the regulation of cysteine synthesis. *Plant Cell* 20, 168–185. doi: 10.1105/tpc.107.056747

Hell, R., and Bergmann, L. (1990). Gamma-glutamylcysteine synthetase in higher-plants – catalytic properties and subcellular-localization. *Planta* 180, 603–612. doi: 10.1007/BF02411460

Hell, R., Jost, R., Berkowitz, O., and Wirtz, M. (2002). Molecular and biochemical analysis of the enzymes of cysteine biosynthesis in the plant *Arabidopsis thaliana*. *Amino Acids* 22, 245–257. doi: 10.1007/s007260200012

Herrmann, J., Ravilious, G. E., Mckinney, S. E., Westfall, C. S., Lee, S. G., Baraniecka, P., et al. (2014). Structure and mechanism of soybean ATP sulfurylase and the committed step in plant sulfur assimilation. *J. Biol. Chem.* 289, 10919–10929. doi: 10.1074/jbc.M113.540401

Hesse, H., Trachsel, N., Suter, M., Kopriva, S., Von Ballmoos, P., Rennenberg, H., et al. (2003). Effect of glucose on assimilatory sulphate reduction in *Arabidopsis thaliana* roots. *J. Exp. Bot.* 54, 1701–1709. doi: 10.1093/Jxb/Erg177

Hirai, M. Y., Fujiwara, T., Awazuhara, M., Kimura, T., Noji, M., and Saito, K. (2003). Global expression profiling of sulfur-starved *Arabidopsis* by DNA macroarray reveals the role of O-acetyl-l-serine as a general regulator of gene expression in response to sulfur nutrition. *Plant J.* 33, 651–663. doi: 10.1046/j.1365-313X.2003.01658.x

Hirai, M. Y., Sugiyama, K., Sawada, Y., Tohge, T., Obayashi, T., Suzuki, A., et al. (2007). Omics-based identification of *Arabidopsis* Myb transcription factors regulating aliphatic glucosinolate biosynthesis. *Proc. Natl. Acad. Sci. U.S.A.* 104, 6478–6483. doi: 10.1073/pnas.0611629104

Ho, C. H., Lin, S. H., Hu, H. C., and Tsay, Y. F. (2009). CHL1 functions as a nitrate sensor in plants. *Cell* 138, 1184–1194. doi: 10.1016/j.cell.2009.07.004

Hopkins, L., Parmar, S., Blaszczyk, A., Hesse, H., Hoefgen, R., and Hawkesford, M. J. (2005). O-acetylserine and the regulation of expression of genes encoding components for sulfate uptake and assimilation in potato. *Plant Physiol.* 138, 433–440. doi: 10.1104/pp.104.057521

Hothorn, M., Wachter, A., Gromes, R., Stuwe, T., Rausch, T., and Scheffzek, K. (2006). Structural basis for the redox control of plant glutamate cysteine ligase. *J. Biol. Chem.* 281, 27557–27565. doi: 10.1074/jbc.M602770200

Hubberten, H. M., Klie, S., Caldana, C., Degenkolbe, T., Willmitzer, L., and Hoefgen, R. (2012). Additional role of O-acetylserine as a sulfur status-independent regulator during plant growth. *Plant J.* 70, 666–677. doi: 10.1111/j.1365-313X.2012.04905.x

Inigo, S., Alvarez, M. J., Strasser, B., Califano, A., and Cerdan, P. D. (2012). PFT1, the MED25 subunit of the plant mediator complex, promotes flowering through CONSTANS dependent and independent mechanisms in *Arabidopsis*. *Plant J.* 69, 601–612. doi: 10.1111/j.1365-313X.2011.04815.x

Jez, J. M., Cahoon, R. E., and Chen, S. (2004). *Arabidopsis thaliana* glutamate-cysteine ligase: functional properties, kinetic mechanism, and regulation of activity. *J. Biol. Chem.* 279, 33463–33470. doi: 10.1074/jbc.M405127200

Jones-Rhoades, M. W., and Bartel, D. P. (2004). Computational identification of plant microRNAs and their targets, including a stress-induced miRNA. *Mol. Cell.* 14, 787–799. doi: 10.1016/j.molcel.2004.05.027

Jost, R., Altschmied, L., Bloem, E., Bogs, J., Gershenzon, J., Hahnel, U., et al. (2005). Expression profiling of metabolic genes in response to methyl jasmonate reveals regulation of genes of primary and secondary sulfur-related pathways in *Arabidopsis thaliana*. *Photosyn. Res.* 86, 491–508. doi: 10.1007/s11120-005-7386-7388

Kasajima, I., Ohkama-Ohtsu, N., Yoko Ide, Y., Hayashi, H., Yoneyama, T., Suzuki, Y., et al. (2007). The BIG gene is involved in regulation of sulfur deficiency-responsive genes in *Arabidopsis thaliana*. *Physiol. Plant.* 129, 351–363. doi: 10.1111/j.1399-3054.2006.00814.x

Kawashima, C. G., Berkowitz, O., Hell, R., Noji, M., and Saito, K. (2005). Characterization and expression analysis of a serine acetyltransferase gene family involved in a key step of the sulfur assimilation pathway in *Arabidopsis*. *Plant Physiol.* 137, 220–230. doi: 10.1104/pp.104.045377

Kawashima, C. G., Matthewman, C. A., Huang, S. Q., Lee, B. R., Yoshimoto, N., Koprivova, A., et al. (2011). Interplay of SLIM1 and miR395 in the regulation of sulfate assimilation in *Arabidopsis*. *Plant J.* 66, 863–876. doi: 10.1111/j.1365-313X.2011.04547.x

Kawashima, C. G., Yoshimoto, N., Maruyama-Nakashita, A., Tsuchiya, Y. N., Saito, K., Takahashi, H., et al. (2009). Sulphur starvation induces the expression of microRNA-395 and one of its target genes but in different cell types. *Plant J.* 57, 313–321. doi: 10.1111/j.1365-313X.2008.03690.x

Khan, M. S., Haas, F. H., Samami, A. A., Gholami, A. M., Bauer, A., Fellenberg, K., et al. (2010). Sulfite reductase defines a newly discovered bottleneck for assimilatory sulfate reduction and is essential for growth and development in *Arabidopsis thaliana*. *Plant Cell* 22, 1216–1231. doi: 10.1105/tpc.110.074088

Kidd, B. N., Edgar, C. I., Kumar, K. K., Aitken, E. A., Schenk, P. M., Manners, J. M., et al. (2009). The mediator complex subunit PFT1 is a key regulator of jasmonate-dependent defense in *Arabidopsis*. *Plant Cell* 21, 2237–2252. doi: 10.1105/tpc.109.066910

Kim, Y. J., Zheng, B., Yu, Y., Won, S. Y., Mo, B., and Chen, X. (2011). The role of mediator in small and long noncoding RNA production in *Arabidopsis thaliana*. *EMBO J.* 30, 814–822. doi: 10.1038/emboj.2011.3

Kimura, Y., and Kimura, H. (2004). Hydrogen sulfide protects neurons from oxidative stress. *FASEB J.* 18, 1165–1167. doi: 10.1096/fj.04-1815fje

Kopriva, S., Buchert, T., Fritz, G., Suter, M., Weber, M., Benda, R., et al. (2001). Plant adenosine 5′-phospho sulfate reductase is a novel iron-sulfur protein. *J. Biol. Chem.* 276, 42881–42886. doi: 10.1074/jbc.M107424200

Kopriva, S., and Koprivova, A. (2004). Plant adenosine 5′-phosphosulphate reductase: the past, the present, and the future. *J. Exp. Bot.* 55, 1775–1783. doi: 10.1093/Jxb/Erh185

Kopriva, S., Muheim, R., Koprivova, A., Trachsel, N., Catalano, C., Suter, M., et al. (1999). Light regulation of assimilatory sulphate reduction in *Arabidopsis thaliana*. *Plant J.* 20, 37–44. doi: 10.1046/j.1365-313X.1999.00573.x

Kopriva, S., Suter, M., Von Ballmoos, P., Hesse, H., Krahenbuhl, U., Rennenberg, H., et al. (2002). Interaction of sulfate assimilation with carbon and nitrogen metabolism in *Lemna minor*. *Plant Physiol.* 130, 1406–1413. doi: 10.1104/Pp.007773

Koprivova, A., Calderwood, A., Lee, B. R., and Kopriva, S. (2014a). Do PFT1 and HY5 interact in regulation of sulfate assimilation by light in *Arabidopsis*? *FEBS Lett.* 588, 1116–1121. doi: 10.1016/j.febslet.2014.02.031

Koprivova, A., Harper, A. L., Trick, M., Bancroft, I., and Kopriva, S. (2014b). Dissection of control of anion homeostasis by associative transcriptomics in *Brassica napus*. *Plant Physiol.* 166, 442–450. doi: 10.1104/pp.114.239947

Koprivova, A., Des Francs-Small, C. C., Calder, G., Mugford, S. T., Tanz, S., Lee, B. R., et al. (2010). Identification of a pentatricopeptide repeat protein implicated in splicing of intron 1 of mitochondrial nad7 transcripts. *J. Biol. Chem.* 285, 32192–32199. doi: 10.1074/jbc.M110.147603

Koprivova, A., Giovannetti, M., Baraniecka, P., Lee, B. R., Grondin, C., Loudet, O., et al. (2013). Natural variation in the ATPS1 Isoform of ATP sulfurylase contributes to the control of sulfate levels in *Arabidopsis*. *Plant Physiol.* 163, 1133–1141. doi: 10.1104/pp.113.225748

Koprivova, A., and Kopriva, S. (2008). Lessons from investigation of regulation of APS reductase by salt stress. *Plant Signal. Behav.* 3, 567–569. doi: 10.4161/psb.3.8.5716

Koprivova, A., North, K. A., and Kopriva, S. (2008). Complex signaling network in regulation of adenosine 5'-phosphosulfate reductase by salt stress in *Arabidopsis* roots. *Plant Physiol.* 146, 1408–1420. doi: 10.1104/pp.107.113175

Koprivova, A., Suter, M., Op Den Camp, R., Brunold, C., and Kopriva, S. (2000). Regulation of sulfate assimilation by nitrogen in *Arabidopsis*. *Plant Physiol.* 122, 737–746. doi: 10.1104/pp.122.3.737

Lappartient, A. G., Vidmar, J. J., Leustek, T., Glass, A. D., and Touraine, B. (1999). Inter-organ signaling in plants: regulation of ATP sulfurylase and sulfate transporter genes expression in roots mediated by phloem-translocated compound. *Plant J.* 18, 89–95. doi: 10.1046/j.1365-313X.1999.00416.x

Lee, B. R., Koprivova, A., and Kopriva, S. (2011). The key enzyme of sulfate assimilation, adenosine 5′-phosphosulfate reductase, is regulated by HY5 in *Arabidopsis*. *Plant J.* 67, 1042–1054. doi: 10.1111/j.1365-313X.2011.04656.x

Lee, J., He, K., Stolc, V., Lee, H., Figueroa, P., Gao, Y., et al. (2007). Analysis of transcription factor HY5 genomic binding sites revealed its hierarchical role in light regulation of development. *Plant Cell* 19, 731–749. doi: 10.1105/tpc.106.047688

Lewandowska, M., Wawrzynska, A., Moniuszko, G., Lukomska, J., Zientara, K., Piecho, M., et al. (2010). A contribution to identification of novel regulators of plant response to sulfur deficiency: characteristics of a tobacco gene UP9C, its protein product and the effects of UP9C silencing. *Mol. Plant* 3, 347–360. doi: 10.1093/mp/ssq007

Liang, G., Yang, F., and Yu, D. (2010). MicroRNA395 mediates regulation of sulfate accumulation and allocation in *Arabidopsis thaliana*. *Plant J.* 62, 1046–1057. doi: 10.1111/j.1365-313X.2010.04216.x

Lisjak, M., Srivastava, N., Teklic, T., Civale, L., Lewandowski, K., Wilson, I., et al. (2010). A novel hydrogen sulfide donor causes stomatal opening and reduces nitric oxide accumulation. *Plant Physiol. Biochem.* 48, 931–935. doi: 10.1016/j.plaphy.2010.09.016

Lisjak, M., Teklic, T., Wilson, I. D., Whiteman, M., and Hancock, J. T. (2013). Hydrogen sulfide: environmental factor or signalling molecule? *Plant Cell Environ.* 36, 1607–1616. doi: 10.1111/pce.12073

Loudet, O., Saliba-Colombani, V., Camilleri, C., Calenge, F., Gaudon, V., Koprivova, A., et al. (2007). Natural variation for sulfate content in *Arabidopsis thaliana* is highly controlled by APR2. *Nat. Genet.* 39, 896–900. doi: 10.1038/Ng2050

Malitsky, S., Blum, E., Less, H., Venger, I., Elbaz, M., Morin, S., et al. (2008). The transcript and metabolite networks affected by the two clades of *Arabidopsis* glucosinolate biosynthesis regulators. *Plant Physiol.* 148, 2021–2049. doi: 10.1104/pp.108.124784

Maruyama-Nakashita, A., Inoue, E., Watanabe-Takahashi, A., Yamaya, T., and Takahashi, H. (2003). Transcriptome profiling of sulfur-responsive genes in *Arabidopsis* reveals global effects of sulfur nutrition on multiple metabolic pathways. *Plant Physiol.* 132, 597–605. doi: 10.1104/pp.102.019802

Maruyama-Nakashita, A., Nakamura, Y., Tohge, T., Saito, K., and Takahashi, H. (2006). *Arabidopsis* SLIM1 is a central transcriptional regulator of plant sulfur response and metabolism. *Plant Cell* 18, 3235–3251. doi: 10.1105/tpc.106.046458

Maruyama-Nakashita, A., Nakamura, Y., Watanabe-Takahashi, A., Inoue, E., Yamaya, T., and Takahashi, H. (2005). Identification of a novel cis-acting element conferring sulfur deficiency response in *Arabidopsis* roots. *Plant J.* 42, 305–314. doi: 10.1111/j.1365-313X.2005.02363.x

Maruyama-Nakashita, A., Nakamura, Y., Watanabe-Takahashi, A., Yamaya, T., and Takahashi, H. (2004a). Induction of SULTR1;1 sulfate transporter in *Arabidopsis* roots involves protein phosphorylation/dephosphorylation circuit for transcriptional regulation. *Plant Cell Physiol.* 45, 340–345. doi: 10.1093/pcp/pch029

Maruyama-Nakashita, A., Nakamura, Y., Yamaya, T., and Takahashi, H. (2004b). A novel regulatory pathway of sulfate uptake in *Arabidopsis* roots: implication of CRE1/WOL/AHK4-mediated cytokinin-dependent regulation. *Plant J.* 38, 779–789. doi: 10.1111/j.1365-313X.2004.02079.x

Matthewman, C. A., Kawashima, C. G., Huska, D., Csorba, T., Dalmay, T., and Kopriva, S. (2012). miR395 is a general component of the sulfate assimilation regulatory network in *Arabidopsis*. *FEBS Lett.* 586, 3242–3248. doi: 10.1016/j.febslet.2012.06.044

Mugford, S. G., Lee, B. R., Koprivova, A., Matthewman, C., and Kopriva, S. (2011). Control of sulfur partitioning between primary and secondary metabolism. *Plant J.* 65, 96–105. doi: 10.1111/j.1365-313X.2010.04410.x

Mugford, S. G., Matthewman, C. A., Hill, L., and Kopriva, S. (2010). Adenosine-5′-phosphosulfate kinase is essential for *Arabidopsis* viability. *FEBS Lett.* 584, 119–123. doi: 10.1016/j.febslet.2009.11.014

Mugford, S. G., Yoshimoto, N., Reichelt, M., Wirtz, M., Hill, L., Mugford, S. T., et al. (2009). Disruption of adenosine-5′-phosphosulfate kinase in *Arabidopsis* reduces levels of sulfated secondary metabolites. *Plant Cell* 21, 910–927. doi: 10.1105/tpc.109.065581

Neuenschwander, U., Suter, M., and Brunold, C. (1991). Regulation of sulfate assimilation by light and O-acetyl-l-serine in *Lemna minor* L. *Plant Physiol.* 97, 253–258. doi: 10.1104/pp.97.1.253

Nussbaum, S., Schmutz, D., and Brunold, C. (1988). Regulation of assimilatory sulfate reduction by cadmium in *Zea mays* L. *Plant Physiol.* 88, 1407–1410. doi: 10.1104/pp.88.4.1407

Ohkama, N., Takei, K., Sakakibara, H., Hayashi, H., Yoneyama, T., and Fujiwara, T. (2002). Regulation of sulfur-responsive gene expression by exogenously applied cytokinins in *Arabidopsis thaliana. Plant Cell Physiol.* 43, 1493–1501. doi: 10.1093/pcp/pcf183

Ou, B., Yin, K. Q., Liu, S. N., Yang, Y., Gu, T., Wing Hui, J. M., et al. (2011). A high-throughput screening system for *Arabidopsis* transcription factors and its application to Med25-dependent transcriptional regulation. *Mol. Plant* 4, 546–555. doi: 10.1093/mp/ssr002

Pant, B. D., Musialak-Lange, M., Nuc, P., May, P., Buhtz, A., Kehr, J., et al. (2009). Identification of nutrient-responsive *Arabidopsis* and rapeseed microRNAs by comprehensive real-time polymerase chain reaction profiling and small RNA sequencing. *Plant Physiol.* 150, 1541–1555. doi: 10.1104/pp.109.139139

Park, S.-W., Li, W., Viehhausrer, A., He, B., Kim, S., Nilsson, A. K., et al. (2013). Cyclophilin 20-3 relays a 12-oxo-phytodienoic acid signal during stress responsive regulation of cellular redox homeostasis. *Proc. Natl. Acad. Sci. U.S.A.* 110, 9559–9564. doi: 10.1073/pnas.1218872110

Pilon-Smits, E. A., Hwang, S., Mel Lytle, C., Zhu, Y., Tai, J. C., Bravo, R. C., et al. (1999). Overexpression of ATP sulfurylase in indian mustard leads to increased selenate uptake, reduction, and tolerance. *Plant Physiol.* 119, 123–132. doi: 10.1104/pp.119.1.123

Ravilious, G. E., Nguyen, A., Francois, J. A., and Jez, J. M. (2012). Structural basis and evolution of redox regulation in plant adenosine-5′-phosphosulfate kinase. *Proc. Natl. Acad. Sci. U.S.A.* 109, 309–314. doi: 10.1073/pnas.1115772108

Rouached, H., Wirtz, M., Alary, R., Hell, R., Arpat, A. B., Davidian, J. C., et al. (2008). Differential regulation of the expression of two high-affinity sulfate transporters, SULTR1.1 *and SULTR*1.2, in *Arabidopsis. Plant Physiol.* 147, 897–911. doi: 10.1104/pp.108.118612

Sakakibara, H., Takei, K., and Hirose, N. (2006). Interactions between nitrogen and cytokinin in the regulation of metabolism and development. *Trends Plant Sci.* 11, 440–448. doi: 10.1016/j.tplants.2006.07.004

Scheerer, U., Haensch, R., Mendel, R. R., Kopriva, S., Rennenberg, H., and Herschbach, C. (2010). Sulphur flux through the sulphate assimilation pathway is differently controlled by adenosine 5′-phosphosulphate reductase under stress and in transgenic poplar plants overexpressing gamma-ECS, SO, or APR. *J. Exp. Bot.* 61, 609–622. doi: 10.1093/Jxb/Erp327

Shibagaki, N., and Grossman, A. R. (2010). Binding of cysteine synthase to the STAS domain of sulfate transporter and its regulatory consequences. *J. Biol. Chem.* 285, 25094–25102. doi: 10.1074/jbc.M110.126888

Smith, F. W., Hawkesford, M. J., Ealing, P. M., Clarkson, D. T., Vanden Berg, P. J., Belcher, A. R., et al. (1997). Regulation of expression of a cDNA from barley roots encoding a high affinity sulphate transporter. *Plant J.* 12, 875–884. doi: 10.1046/j.1365-313X.1997.12040875.x

Sonderby, I. E., Hansen, B. G., Bjarnholt, N., Ticconi, C., Halkier, B. A., and Kliebenstein, D. J. (2007). A systems biology approach identifies a R2R3 MYB gene subfamily with distinct and overlapping functions in regulation of aliphatic glucosinolates. *PLoS ONE* 2:e1322. doi: 10.1371/journal.pone.0001322

Sun, J., Wang, R., Zhang, X., Yu, Y., Zhao, R., Li, Z., et al. (2013). Hydrogen sulfide alleviates cadmium toxicity through regulations of cadmium transport across the plasma and vacuolar membranes in *Populus euphratica* cells. *Plant Physiol. Biochem.* 65, 67–74. doi: 10.1016/j.plaphy.2013.01.003

Takahashi, H., Kopriva, S., Giordano, M., Saito, K., and Hell, R. (2011). Sulfur Assimilation in photosynthetic organisms: molecular functions and regulations of transporters and assimilatory enzymes. *Annu. Rev. Plant Biol.* 62, 157–184. doi: 10.1146/annurev-arplant-042110-103921

Takahashi, H., Watanabe-Takahashi, A., Smith, F. W., Blake-Kalff, M., Hawkesford, M. J., and Saito, K. (2000). The roles of three functional sulphate transporters involved in uptake and translocation of sulphate in *Arabidopsis thaliana. Plant J.* 23, 171–182. doi: 10.1046/j.1365-313x.2000.00768.x

Takahashi, H., Yamazaki, M., Sasakura, N., Watanabe, A., Leustek, T., Engler, J. A., et al. (1997). Regulation of sulfur assimilation in higher plants: a sulfate transporter induced in sulfate-starved roots plays a central role in *Arabidopsis thaliana. Proc. Natl. Acad. Sci. U.S.A.* 94, 11102–11107. doi: 10.1073/pnas.94.20.11102

Tsakraklides, G., Martin, M., Chalam, R., Tarczynski, M. C., Schmidt, A., and Leustek, T. (2002). Sulfate reduction is increased in transgenic *Arabidopsis thaliana* expressing 5′adenylylsulfate reductase from *Pseudomonas aeruginosa. Plant J.* 32, 879–889. doi: 10.1046/j.1365-313X.2002.01477.x

Vauclare, P., Kopriva, S., Fell, D., Suter, M., Sticher, L., Von Ballmoos, P., et al. (2002). Flux control of sulphate assimilation in *Arabidopsis thaliana*: adenosine 5′-phosphosulphate reductase is more susceptible than ATP sulphurylase to negative control by thiols. *Plant J.* 31, 729–740. doi: 10.1046/j.1365-313X.2002.01391.x

Wangeline, A. L., Burkhead, J. L., Hale, K. L., Lindblom, S. D., Terry, N., Pilon, M., et al. (2004). Overexpression of ATP sulfurylase in Indian mustard: effects on tolerance and accumulation of twelve metals. *J. Environ. Qual.* 33, 54–60. doi: 10.2134/jeq2004.5400

Wawrzynska, A., Lewandowska, M., and Sirko, A. (2010). *Nicotiana tabacum* EIL2 directly regulates expression of at least one tobacco gene induced by sulphur starvation. *J. Exp. Bot.* 61, 889–900. doi: 10.1093/jxb/erp356

Wawrzyńska, A., and Sirko, A. (2014). To control and to be controlled – understanding the *Arabidopsis* SLIM1 function in sulfur deficiency through comprehensive investigation of the EIL protein family. *Front. Plant Sci.* 5:575. doi: 10.3389/fpls.2014.00575

Wirtz, M., Beard, K. F. M., Lee, C. P., Boltz, A., Schwarzlander, M., Fuchs, C., et al. (2012). Mitochondrial cysteine synthase complex regulates O-acetylserine biosynthesis in plants. *J. Biol. Chem.* 287, 27941–27947. doi: 10.1074/jbc.M112.372656

Wirtz, M., Birke, H., Heeg, C., Muller, C., Hosp, F., Throm, C., et al. (2010). Structure and function of the hetero-oligomeric cysteine synthase complex in plants. *J. Biol. Chem.* 285, 32810–32817. doi: 10.1074/jbc.M110.157446

Wirtz, M., Droux, M., and Hell, R. (2004). O-acetylserine (thiol) lyase: an enigmatic enzyme of plant cysteine biosynthesis revisited in *Arabidopsis thaliana. J. Exp. Bot.* 55, 1785–1798. doi: 10.1093/Jxb/Erh201

Wirtz, M., and Hell, R. (2006). Functional analysis of the cysteine synthase protein complex from plants: structural, biochemical and regulatory properties. *J. Plant Physiol.* 163, 273–286. doi: 10.1016/j.jplph.2005.11.013

Yarmolinsky, D., Brychkova, G., Fluhr, R., and Sagi, M. (2013). Sulfite reductase protects plants against sulfite toxicity. *Plant Physiol.* 161, 725–743. doi: 10.1104/pp.112.207712

Yatusevich, R., Mugford, S. G., Matthewman, C., Gigolashvili, T., Frerigmann, H., Delaney, S., et al. (2010). Genes of primary sulfate assimilation are part of the glucosinolate biosynthetic network in *Arabidopsis thaliana. Plant J.* 62, 1–11. doi: 10.1111/j.1365-313X.2009.04118.x

Yoshimoto, N., Inoue, E., Watanabe-Takahashi, A., Saito, K., and Takahashi, H. (2007). Posttranscriptional regulation of high-affinity sulfate transporters in *Arabidopsis* by sulfur nutrition. *Plant Physiol.* 145, 378–388. doi: 10.1104/pp.107.105742

Yoshimoto, N., Takahashi, H., Smith, F. W., Yamaya, T., and Saito, K. (2002). Two distinct high-affinity sulfate transporters with different inducibilities mediate uptake of sulfate in *Arabidopsis* roots. *Plant J.* 29, 465–473. doi: 10.1046/j.0960-7412.2001.01231.x

Zhang, B., Pasini, R., Dan, H., Joshi, N., Zhao, Y., Leustek, T., et al. (2014). Aberrant gene expression in the *Arabidopsis* SULTR1;2 mutants suggests a possible regulatory role for this sulfate transporter in response to sulfur nutrient status. *Plant J.* 77, 185–197. doi: 10.1111/tpj.12376

Zhang, L. W., Song, J. B., Shu, X. X., Zhang, Y., and Yang, Z. M. (2013). miR395 is involved in detoxification of cadmium in *Brassica napus. J. Hazard. Mater.* 250–251, 204–211. doi: 10.1016/j.jhazmat.2013.01.053

Conflict of Interest Statement: The authors declare that the research was conducted in the absence of any commercial or financial relationships that could be construed as a potential conflict of interest.

Compartment-specific importance of glutathione during abiotic and biotic stress

Bernd Zechmann *

Center for Microscopy and Imaging, Baylor University, Waco, TX, USA

Edited by:
Agnieszka Sirko, Institute of Biochemistry and Biophysics; Polish Academy of Sciences, Poland

Reviewed by:
Suleyman I. Allakhverdiev, Russian Academy of Sciences, Russia
Markus Wirtz, Centre for Organismal Studies, Germany
Antonio Masi, University of Padova, Italy

****Correspondence:***
Bernd Zechmann, Center for Microscopy and Imaging, Baylor University, One Bear Place #97046, Waco, TX 76798, USA
e-mail: bernd_zechmann@baylor.edu

The tripeptide thiol glutathione (γ-L-glutamyl-L-cysteinyl-glycine) is the most important sulfur containing antioxidant in plants and essential for plant defense against abiotic and biotic stress conditions. It is involved in the detoxification of reactive oxygen species (ROS), redox signaling, the modulation of defense gene expression, and the regulation of enzymatic activities. Even though changes in glutathione contents are well documented in plants and its roles in plant defense are well established, still too little is known about its compartment-specific importance during abiotic and biotic stress conditions. Due to technical advances in the visualization of glutathione and the redox state through microscopical methods some progress was made in the last few years in studying the importance of subcellular glutathione contents during stress conditions in plants. This review summarizes the data available on compartment-specific importance of glutathione in the protection against abiotic and biotic stress conditions such as high light stress, exposure to cadmium, drought, and pathogen attack (*Pseudomonas*, *Botrytis*, tobacco mosaic virus). The data will be discussed in connection with the subcellular accumulation of ROS during these conditions and glutathione synthesis which are both highly compartment specific (e.g., glutathione synthesis takes place in chloroplasts and the cytosol). Thus this review will reveal the compartment-specific importance of glutathione during abiotic and biotic stress conditions.

Keywords: abiotic stress, biotic stress, chloroplasts, drought, glutathione, mitochondria, nuclei, peroxisomes

INTRODUCTION

The subcellular distribution of glutathione in plants is of great importance as this multifunctional metabolite is essential for plant development and growth (Kocsy et al., 2013). It is the most important antioxidant in plants and is a key agent in plant defense against abiotic and biotic stress. It is involved in the detoxification of reactive oxygen species (ROS), either directly through scavenging them or through the ascorbate–glutathione cycle (**Figure 1**). It is also involved in redox signaling, the modulation of gene expression and the regulation of enzymatic activities (extensively reviewed by Noctor, 2006; Foyer and Noctor, 2009, 2011; Noctor et al., 2012; Kocsy et al., 2013; Schmitt et al., 2014). Additionally, glutathione is involved in the detoxification of xenobiotics, herbicides (Edwards et al., 2005; DeRidder and Goldsbrough, 2006; Cummins et al., 2011; Mohsenzadeh et al., 2011), heavy metals such as cadmium (Zawoznik et al., 2007; Ammar et al., 2008; DalCorso et al., 2008; Ducic et al., 2008; Nocito et al., 2008; Tan et al., 2010; Jozefczak et al., 2012; Koffler et al., 2014b), and protects proteins from oxidation by a process called glutathionylation (Dixon et al., 2005; Hurd et al., 2005a,b; Zaffagnini et al., 2012a,b). Therefore subcellular changes in glutathione contents especially during environmental stress situations provide insights into compartment-specific defense reactions and reflect the occurrence of compartment-specific oxidative stress. Such information can be used as a subcellular stress marker and can be very helpful to clarify the importance of the protective roles of glutathione during stress situations in plants on the cellular level.

Synthesis of glutathione takes place in two adenosine triphosphate (ATP)-depending steps and is highly compartment specific as the enzymes that trigger glutathione synthesis are encoded by single genes which are targeted to either chloroplast and/or the cytosol. In *Arabidopsis*, the first step of glutathione synthesis, the formation of γ-glutamylcysteine (γ-EC) out of glutamate and cysteine catalyzed by γ-glutamylcysteine synthetase (GSH1; also referred to as γ-ECS in some literature, EC 6.3.2.2), takes place in chloroplasts as GSH1 is exclusively targeted to chloroplasts (Wachter et al., 2005). The second step catalyzed by glutathione synthetase (GSH2; also referred to as GSHS in some literature, EC 6.3.2.3.), in which glycine is added to γ-glutamylcysteine to form glutathione, takes place in plastids and the cytosol as GSH2 is targeted to both chloroplasts and the cytosol (Noctor et al., 2002; Sugiyama et al., 2004; Wachter et al., 2005). Nevertheless, it has been shown that restricting the final step of glutathione synthesis to the cytosol is sufficient for normal plant development (Pasternak et al., 2008). These results indicate that chloroplasts export γ-EC to and are able to import glutathione from the cytosol through specific transporters (Maughan et al., 2010) as discussed below. In other plant species, the situation is less clear as GSH1 was also detected in the remaining leaf extract in wheat after the isolation of chloroplasts (Noctor et al., 2002) and as

GSH1 is encoded by more than one gene in some plant species (e.g., *Populus trichocarpa*, *Oryza sativa*). Under non-stressed conditions cysteine and subsequently γ-EC are considered to be the rate limiting precursors for glutathione synthesis as it has been shown that both the artificial elevation of cysteine (Gullner et al., 1999; Harms et al., 2000; Bloem et al., 2004, 2007; Zechmann et al., 2007, 2008b) and the overexpression of genes and enzymes involved in cysteine synthesis (Harms et al., 2000; Noji and Saito, 2002; Wirtz and Hell, 2007) increased glutathione contents in plants. The short- and long-term blockage of the first step of glutathione synthesis results in the accumulation of cysteine of up to 300% in some cell compartments also indicating that large amounts of cysteine are used for glutathione synthesis (Koffler et al., 2011). During stress conditions and in the absence of photorespiration glycine can also become the limiting factor for the production of glutathione (Noctor et al., 1997a,b; Zechmann et al., 2008b). The complete absence of glutathione synthesis results in a lethal phenotype (Cairns et al., 2006; Pasternak et al., 2008). Mutants deficient in glutathione synthesis such as the *rml1* mutant which has a single point mutation in the gene that encodes GSH1 develop strong growth defects such as a dwarf phenotype, the lack of a root meristem, short shoots, inflorescence, smaller rosettes, and flowers (Cheng et al., 1995; Vernoux et al., 2000; Cairns et al., 2006). In opposite to *rml1* which shows a reduction of glutathione between 90 and 97% (Vernoux et al., 2000; Cairns et al., 2006) in all cell compartments the *pad2-1* mutant which shows a reduction of glutathione contents of about 80% does not develop a distorted phenotype (Parisy et al., 2007). *pad2-1* mutants are also characterized by a single point mutation of the gene that encodes GSH1 but glutathione contents remain at control levels in mitochondria despite a strong reduction of glutathione in all other cell compartments (Zechmann et al., 2008a; Koffler et al., 2011) which will be further discussed later in this review. Summing up, the ability of plants to synthesize glutathione and the availability of glutathione precursors in glutathione producing organelles are essential for proper plant growth and development and subsequently for defense against abiotic and biotic stress.

Glutathione synthesis is highly compartment specific (e.g., localized in chloroplasts and the cytosol in *Arabidopsis*) but it can be found in different concentrations within the different organelles and accumulates in certain organelles during environmental stress conditions as discussed below. Thus, glutathione transporters must be present in membranes of all organelles in order to facilitate the import and export of glutathione. Whereas the existence of the transport of glutathione through membranes such as the plasma membrane, tonoplast, and the chloroplast envelope is well established (Schneider et al., 1992; Jamai et al., 1996; Foyer et al., 2001; Noctor et al., 2002; Pasternak et al., 2008) the identity and their exact role in glutathione transport still remains unclear especially in plants (Bachhawat et al., 2013). Three low affinity transporters for glutathione have been identified in plants. In *Arabidopsis*, the homologs of the *Plasmodium falciparum* chloroquine resistance transporter (PfCRT) have been described to facilitate the transport of glutathione through the envelope of the chloroplast (Maughan et al., 2010). Three proteins named CLT1, CLT2, and CLT3 were identified to be essential for the transport of glutathione between the chloroplasts and the cytosol (Maughan et al., 2010). Further transporters of glutathione in plants include homologs from the oligopeptide family from yeast. These homologs are mainly associated with the vascular tissue of plants which indicates that they are involved in long distance transport of glutathione rather than transport of glutathione between cell compartments (Koh et al., 2002; Cagnac et al., 2004; Pike et al., 2009). The transport of glutathione conjugates and oxidized glutathione into vacuoles in plants is facilitated by transporters of the ATP-binding cassette (ABC) family (Lu et al., 1998). These transporter might play essential roles in the sequestration of oxidized glutathione in vacuoles in situation of extreme oxidative stress (Queval et al., 2011) as described below.

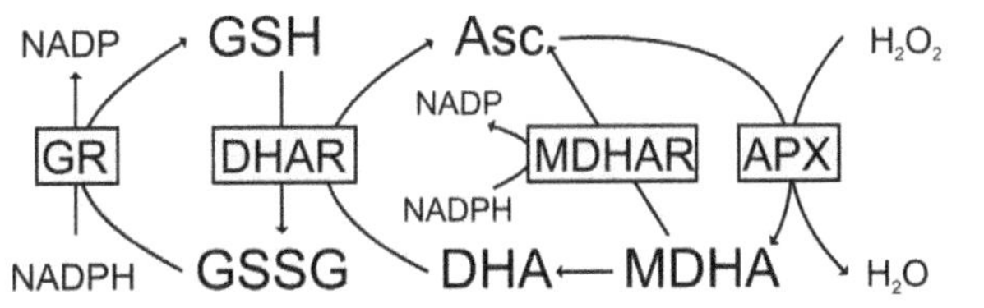

FIGURE 1 | The ascorbate–glutathione cycle in plants. Hydrogen peroxide (H_2O_2) within the plant cell can be detoxified by ascorbate peroxidase (APX). In this reaction, the reduced form of ascorbate (Asc) is oxidized to monodehydroascorbate (MDHA). MDHA is then either reduced by monodehydroascorbate reductase (MDHAR) to Asc or, since very unstable, reacts to dehydroascorbate (DHA). DHA is reduced by dehydroascorbate reductase (DHAR) to Asc. In this reaction, the reduced form of glutathione (GSH) is oxidized to glutathione disulfide (GSSG). GSSG is then reduced by glutathione reductase (GR) to GSH. The electron acceptor NADP is regenerated during the reduction of MDHA and GSSG by the respective enzymes. Asc and GSH are additional able to detoxify reactive oxygen species by direct chemical interaction. Thus, besides the total ascorbate and glutathione level their redox state (reduced vs. oxidized state) which depends on the activity of the described enzymes (gray boxes) is also very important for successful plant protection.

Glutathione degradation is carried out by γ-glutamyl transferase/transpeptidase (GGT, EC 2.3.2.2) which promotes the cleavage of glutamate from glutathione in vacuoles and the apoplast (Ohkama-Ohtsu et al., 2007a,b; Tolin et al., 2013). In *Arabidopsis*, GGT1 and GGT2 are associated with the cell wall and the plasma membrane, respectively, whereas GGT3 does not seem to have a function, and GGT4 occurs in vacuoles (Ohkama-Ohtsu et al., 2007a,b; Ferretti et al., 2009; Destro et al., 2011; Tolin et al., 2013). A further pathway of glutathione degradation is the removal of glycine from glutathione which is carried out by carboxypeptidase located inside vacuoles (Wolf et al., 1996). Metabolization of the remaining dipeptides is then carried out by dipeptidases. Further pathways could involve the degradation of glutathione by γ-glutamyl-cyclotransferase (EC 2.3.2.4) via 5-oxoproline, which would results in the production of free glutamate (Martin and Slovin, 2000; Ohkama-Ohtsu et al., 2008) or by phytochelatin synthase (also known as glutathione γ-glutamylcysteinyltransferase, EC 2.3.2.15) which could facilitate the degradation of glutathione in situations when conjugated glutathione accumulates in the cytosol such as exposure of plants to heavy metals (Grzam et al., 2006; Blum et al., 2007, 2010). As these pathways take place in the cytosol they represent an alternative

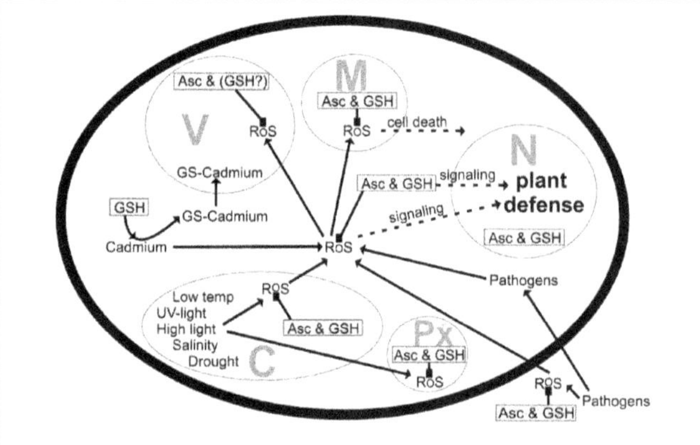

FIGURE 2 | Compartment-specific production of reactive oxygen species (ROS) induced by different stress conditions and possible detoxification and signaling pathways involving ascorbate (Asc) and glutathione (GSH) and ROS. C, chloroplast; M, mitochondrium; N, nucleus; Px, peroxisome; V, vacuole.

pathway of glutathione degradation besides the ones in the vacuole and the apoplast (Noctor et al., 2011). As plants depend on the protection of glutathione and as it is also involved in (redox) signaling and the activation of defense genes, its homeostasis in different organs, tissues, and organelles is essential for plant defense and depends on the control of a complex network of metabolic and environmental factors (Noctor, 2006; Foyer and Noctor, 2009, 2011; Miller et al., 2010; Noctor et al., 2012; Szarka et al., 2012). Considering that stress situations affect cell compartments differently (**Figure 2**) and that glutathione metabolism is highly compartment specific it is essential to investigate subcellular glutathione levels in order to obtain a deeper insight into how glutathione is involved in plant defense.

METHODS TO DETECT SUBCELLULAR GLUTATHIONE CONTENTS

Despite the importance of glutathione within plant cells its detection on the subcellular level is technically challenging as it can be easily washed out and/or redistributed during sample preparation which is also due to the fact that sample preparation itself can be seen as a stress to the plant sample. Currently, there are different methods available that have been used to study the subcellular distribution of glutathione in plants which can be separated into two major approaches: (1) biochemical measurements after the isolation or fractionation of organelles (Jiménez et al., 1997, 1998; Vanacker et al., 1998a,b,c; Kuźniak and Sklodowska, 2001, 2004, 2005a,b; Ohkama-Ohtsu et al., 2007a,b; Krueger et al., 2009) and (2) microscopical investigations after labeling with specific dyes (Meyer and Fricker, 2000; Meyer et al., 2001; Müller et al., 2005), antibodies (Zechmann et al., 2008a; Zechmann and Müller, 2010) or by using ratiometric redox sensitive green fluorescent protein (GFP; Meyer et al., 2007; Gutscher et al., 2008). All methods have advantages and disadvantages but have given valuable insights into the subcellular distribution of glutathione.

With biochemical methods glutathione was localized and measured after organelle isolation or fractionation in mitochondria, chloroplasts, peroxisomes, the apoplast, and vacuoles of different plant species (Jiménez et al., 1997, 1998; Vanacker et al., 1998a,b,c; Kuźniak and Sklodowska, 2001, 2004, 2005a,b; Ohkama-Ohtsu et al., 2007a,b; Krueger et al., 2009). With these methods it was possible to measure glutathione contents in millimolar concentrations and to differentiate between the reduced and oxidized form which gave valuable insights into the redox state of certain cell compartments during different stress conditions. Nevertheless, these methods face the problem that the individual organelles have to be isolated or fractionated from a large amount of plant samples before glutathione can be measured with high performance liquid chromatography or spectrometrically (Jiménez et al., 1997, 1998; Vanacker et al., 1998a,b,c; Kuźniak and Sklodowska, 2001, 2004, 2005a,b; Ohkama-Ohtsu et al., 2007a,b; Krueger et al., 2009). This can lead to contamination of non-organelle-specific glutathione and because of the lengthy procedure it is unclear how well the obtained results reflect the *in vivo* situation as glutathione can be washed out or redistributed between the organelles (Noctor et al., 2002; Chew et al., 2003; Krueger et al., 2009).

With light microscopical methods after monochloro- or monobromobimane staining glutathione could be detected in nuclei and the cytosol (Fricker et al., 2000; Meyer and Fricker, 2000; Meyer et al., 2001; Müller et al., 2005). Nevertheless, light microscopical investigations which allow investigations of the *in vivo* situation are limited by the resolution of the light microscope (about 200 nm), by the ability of the antibodies/dyes to infiltrate the different organelles (Müller et al., 2005) and by their specificity to bind with the respective component. Monochlorobimanes, for example, bind to all thiols (not only to the reduced form of glutathione) in cells and do not infiltrate chloroplasts (Hartmann et al., 2003; Müller et al., 2005; **Figures 3A,B**). Additionally, monochloro- and monobromobimane are toxic to the plant and are transported into the vacuole after complexation with reduced glutathione (Fricker et al., 2000; Meyer and Fricker, 2002). This process can be inhibited by using chemicals that inhibit the transport of glutathione conjugates through the tonoplast such as sodium azide (Fricker et al., 2000). The use of redox sensitive GFP in plants has been proven to overcome some of these problems and has allowed investigations of the glutathione redox potential *in vivo* even in very small cell compartments with fluorescence microscopy (Meyer et al., 2007; Gutscher et al., 2008). As the signal can be targeted specifically to different cell compartments the situation can be even studied in small cell compartments such as the endoplasmic reticulum (ER) which cannot be resolved with the light microscope otherwise (Meyer et al., 2007). Additionally, this method is able to differentiate between the reduced and oxidized form of glutathione and has been used to detect changes in glutathione contents in the cytosol during environmental stress situations (Jubany-Marí et al., 2010; Lim et al., 2014). Nevertheless, *in vivo* investigations with confocal-laser scanning or fluorescence microscopes face the problems that only very thin cell layers, tissues, or organs can be investigated and that the sample preparation (e.g., mechanical separation of the epidermis, or other tissues and cells) and investigations with the microscope (exposure to a strong light source, high temperature, lack of oxygen, water stress) can also be seen as a stress source

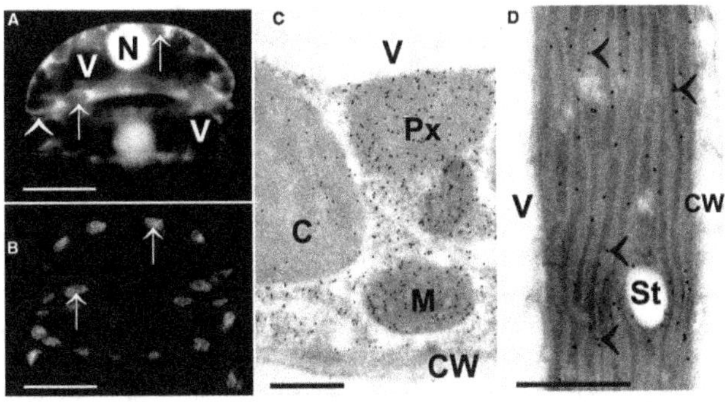

FIGURE 3 | Images show the typical distribution of glutathione. (A) Monochlorobimane staining in guard cells of the upper epidermis of tobacco cells in the light microscope. Fluorescence was observed in cytosol and nuclei (N) but not in vacuoles (V) and cell walls (arrowhead). Additionally, no fluorescence could be observed in chloroplasts (arrows in **A** and **B**) which can be best identified when comparing the autofluorescence of chloroplast **(B)** with monochlorobimane staining **(A)**. Transmission electron micrographs show the subcellular distribution of glutathione **(C,D)** in mesophyll cells of leaves from *Arabidopsis* Col-0 plants. Glutathione-specific labeling could be observed in different concentrations in mitochondria (M), chloroplasts (C), peroxisomes (Px) but not in vacuoles (V) and cell walls (CW). Glutathione-specific labeling was observed in the stroma as well as inside the thylakoid lumen (arrowheads) when plants were exposed to high light intensities of 700 μmol m^{-2} s^{-1}. Bars in **(A,B)** = 10 μm, **(C,D)** = 0.5 μm.

to the sample. Single culture cells are a good alternative but similar to the above-described situation it remains unclear how the situation in single cells, cell types or tissues (e.g., epidermis) correlates with the situation in whole leaves or other tissues. Therefore the labeling results have to be carefully evaluated and the situation in single cells or tissues (e.g., epidermis) cannot always be transferred to the situation in deeper cell layers (e.g., mesophyll, vascular tissue) or the whole plant. Additionally, it was not possible so far with the above-mentioned methods to measure glutathione levels in all cell compartments simultaneously in one experiment.

Another alternative to the above-mentioned techniques is the detection and quantitative evaluation of glutathione after immunogold labeling with computer supported transmission electron microscopy (Zechmann et al., 2008a; Fernandez-Garcia et al., 2009; Zechmann and Müller, 2010; Gao et al., 2012). This method allows the simultaneous detection and quantification of glutathione in all cell compartments of a cell in one experiment (**Figures 3C,D**). Changes in glutathione contents can be resolved even during stress situations on the subcellular level on a high level of resolution (technical resolution limit of the transmission electron microscope is around 0.2 nm). Thus this method allows the detection and quantification of glutathione in even very small cell compartments (e.g., ER, dictyosomes, membranes) that cannot be resolved by light microscopy (Zechmann et al., 2008a; Zechmann and Müller, 2010). Additionally, it is possible with this method to investigate subcellular glutathione levels also in deeper layers and tissues of the leaves (e.g., mesophyll, vascular tissue). As very little plant material has to be used (sample size is usually around 1 mm^2) this method can also achieve a very high spatial resolution which is important if one wants to study local events that can induce oxidative stress, e.g., the penetration site of aphids (Zechmann et al., 2009) or fungi (Simon et al., 2013) or wants to look at very small plant material such as pollen grains (Zechmann et al., 2011). Recently this approach was used to calculate glutathione concentrations in millimolar for the different organelles (Koffler et al., 2013). For this purpose, glutathione-specific labeling density was correlated with biochemical analysis of glutathione concentrations in whole organs (see Queval et al., 2011; Koffler et al., 2013 for details). These methods were used to determine subcellular glutathione concentrations (in millimolar) in different leaf areas in older and younger leaves of *Arabidopsis* wild-type plants and during situations of oxidative stress (Queval et al., 2011; Koffler et al., 2013). Highest glutathione concentrations in the center of older leaves were found in mitochondria (14.8 mM), followed by nuclei (6.4 mM), the cytosol (4.5 mM), peroxisomes (4.4 mM), chloroplasts (1.2 mM), and vacuoles (0.08 mM; **Table 1**).

The main limitations of this method are that the samples have to go through fixation, that the labeling results rely on how well and how close as possible the distribution of glutathione has been preserved to the natural state and on the specificity of the antibodies. Thus, it is impossible to study the distribution of glutathione *in vivo* and the specificity and accuracy of the labeling results have to be carefully and intensively evaluated at the beginning and throughout the use of these methods (Zechmann et al., 2008a, 2011; Zechmann and Müller, 2010). This can be achieved by using different negative controls (e.g., saturation of the antibodies with reduced and oxidized forms of glutathione, using pre-immune serum instead of the antibodies, omission of secondary antibodies), using different fixation methods, and by using mutants that are deficient in or accumulate glutathione (Zechmann et al., 2008a, 2011; Zechmann and Müller, 2010). Unfortunately, the antibody that is currently used for detecting glutathione cannot differentiate between the reduced and oxidized form, thus allowing the compartment-specific detection of the total glutathione status only. Summing up, despite the limitations of these methods they have all contributed toward a deeper understanding of the subcellular distribution of glutathione in plants during abiotic

Table 1 | Compartment-specific glutathione contents in leaves of *Arabidopsis* Col-0.

	Gold particles per μm^2	Calculated concentrations (mM)
Mitochondria	601	14.9
Chloroplasts	47	1.2
Nuclei	260	6.4
Peroxisomes	179	4.4
Cytosol	181	4.5
Vacuoles	3	0.08
Apoplast	Not detected	Not detected

The amount of glutathione in each compartment was obtained by multiplying the amount of total glutathione measured by HPLC (326 nmol/g fresh weight) with the measured fractional contribution of each compartment to overall gold label. From these values, concentrations were calculated using subcellular volumes estimated in leaf sections of Col-0 (for details, see Koffler et al., 2013).

and biotic stress which will be described and discussed in the next sections in connection with their compartment-specific roles during stress conditions.

SUBCELLULAR RESPONSE OF PLANTS TO ABIOTIC STRESS

MITOCHONDRIA

The role of glutathione in mitochondria is quite an important one. It seems that high and stable levels of glutathione especially in mitochondria are essential for proper cell development and cell survival especially in situations when the glutathione pool is depleted (Zechmann et al., 2006, 2008a; Zechmann and Müller, 2010). Glutathione contents within mitochondria of leaves and roots of the glutathione-deficient *Arabidopsis thaliana pad 2-1* mutant remained the same as in the wild-type Col-0 whereas glutathione was strongly decreased in all other cell compartments that contained glutathione of up to −91% (Zechmann et al., 2008a). In the glutathione-deficient mutant *rml1*, glutathione labeling was found to be about 96–98% lower in all cell compartments when compared to the wild-type (Zechmann and Müller, 2010). Interestingly, *rml1* showed the strongest reduction of glutathione in mitochondria (−98%). In opposite to *rml1* which forms very short roots (1–2 mm), small shoots and leaves (Cheng et al., 1995; Vernoux et al., 2000), *pad2-1* mutant has a phenotype similar to the wild-type in non-stressed conditions (Parisy et al., 2007). Thus, the preservation of high levels of glutathione in mitochondria in situations of glutathione deficiency seems to be essential for proper plant development. The *rml1* phenotype can also be achieved by treatment of plants with buthionine sulfoximine (BSO), which inhibits glutathione synthesis (Vernoux et al., 2000). Nevertheless, mild BSO-stress did not affect concentrations of glutathione in mitochondria, even though glutathione was reduced in most other cell compartments in roots and leaves (Zechmann et al., 2006). Additionally, short-term excess light stress led to a massive increase of glutathione in mitochondria of the *pad2-1* mutant of up to 900% whereas glutathione contents in all other cell compartments remained unchanged or decreased (Heyneke et al., 2013). These results demonstrate that high and stable levels of glutathione in mitochondria in plants, especially in situations of glutathione deficiency caused by impaired glutathione synthesis induced by BSO or mutations as described above and during abiotic stress, play an important role for the development and growth of plants, thus allowing a phenotype similar to the wild-type.

CHLOROPLASTS AND PEROXISOMES

Abiotic stress conditions such as excess light, high salinity, and drought (or a combination of all three) represent a unique stress source to the plant that mainly affects metabolic changes in chloroplasts first before other cell compartments are affected (Asada, 2006; Kim et al., 2008; Pfannschmidt et al., 2009). One of the first responses of plants to drought, salt stress and high light conditions is the closure of stomata which limits the gas exchange between the atmosphere and the leaves and decreases the ratio of CO_2 to O_2 (Bhargava and Sawant, 2013; Hernández et al., 2013). This situation leads to oxidative stress in illuminated chloroplasts since low levels of CO_2 in chloroplasts induce disturbances of the Calvin cycle which causes the exhaustion of the primary electron acceptor NADP and the block of the electron transport to NADP. Subsequently electrons will be transferred to O_2 inducing the production of ROS in chloroplasts (Halliwell and Gutteridge, 1999; Tausz et al., 2004; Asada, 2006; Golan et al., 2006; Krieger-Liszkay et al., 2008; Pospisil, 2012; Schmitt et al., 2014). Additionally, this situation favors photorespiration, which leads to the production of phosphoglycolate. The degradation of this toxic component leads to the production of H_2O_2 in peroxisomes (Foyer and Noctor, 2009; Miller et al., 2010; Bhargava and Sawant, 2013; Hernández et al., 2013). Thus it is not surprising that an accumulation of glutathione was observed in chloroplasts and peroxisomes during high light and salt stress conditions (**Table 2**) indicating an increased need of glutathione in chloroplasts to keep ROS under control. Under excess light conditions glutathione also accumulated inside the thylakoid lumen (**Figure 3D**) highlighting the importance of high levels of glutathione in chloroplasts for the protection against ROS produced in these cell compartments during excess light conditions (Heyneke et al., 2013). When the scavenging systems in chloroplasts fail to detoxify ROS, which are produced during excess light conditions, plants suffer photobleaching and eventually cell death. Besides toxic effects of ROS which can directly trigger these events by destroying target components within chloroplasts they additionally seem to be actively governed by the plant as the accumulation of ROS in chloroplasts seems to be involved in the regulation of cell death events (op den Camp et al., 2003; Wagner et al., 2004; Doyle et al., 2010; Schmitt et al., 2014). The generation of singlet oxygen in chloroplasts of *flu* mutants within the first minute of illumination correlated with the inhibition of plant growth and the appearance of necrotic lesions (op den Camp et al., 2003). Based on further investigations of gene expression patterns and studies involving the EXECUTER1 protein (which is a highly conserved protein in plastids) the authors came to the conclusion that changes in growth and development of the *flu* mutants after illumination were not caused by direct toxic effects of singlet oxygen but rather reflected its role as a signal initiator that activated several stress-response pathways (op den Camp et al., 2003; Wagner et al., 2004). Additionally, it has been demonstrated in *Arabidopsis* cell cultures under heat treatment that the interplay of antioxidants (especially glutathione and ascorbate) and ROS production in chloroplasts controlled the number of cells that were subject to apoptosis like cell death and the severity of these events (Doyle et al., 2010). Treatment of glutathione and ascorbate under these conditions decreased ROS levels and therefore inhibited necrosis caused by direct damage of ROS in the tissue but increased apoptosis like programed cell death demonstrating the crucial role of ROS in chloroplasts for signaling cell death events (Doyle et al., 2010). Similar results have been obtained during biotic stress conditions where the accumulation of glutathione contents in chloroplasts was insufficient to prevent the accumulation of ROS and resulted in cell death (Großkinsky et al., 2012; Király et al., 2012; Simon et al., 2013). These results support the conclusion that cell death events during stress conditions are not only caused by direct damaging effects of an excess of ROS in chloroplasts but are also indirectly triggered by signaling events induced by ROS in chloroplast (or other cell compartments). Whereas the accumulation of antioxidants in

Table 2 | Changes in the subcellular distribution of glutathione in *Arabidopsis* Col-0 plants exposed to salt stress (treatment of plants with 100 mM NaCl for 3 days; unpublished data) and excess light conditions (700 μmol m^{-2} s^{-1} light for 4 h and 14 days; data according to Heyneke et al., 2013) when compared to plants grown at 150 μmol m^{-2} s^{-1}.

	Chloroplasts (%)	Peroxisomes (%)
High salinity	26***	84**
Excess light (4 h)	27***	32***
Excess light (14 days)	190***	65***

*Significant differences were calculated using the Mann–Whitney U-test; ** and ***, respectively, indicate significance at the 0.01 and 0.001 level of confidence. n > 20 for peroxisomes and vacuoles and n > 60 for all other cell structures.*

chloroplast under these conditions can reduce direct damaging effects of ROS it cannot prevent the induction of programed cell death.

NUCLEI

The interplay of ROS and glutathione in the nucleus plays an essential role for cell proliferation and subsequently for plant growth and development during abiotic stress (Diaz-Vivancos et al., 2010a,b; Kocsy et al., 2013). Additionally, ROS in nuclei are involved in signaling between nuclei, chloroplasts, and the cytosol especially during stress situations (Galvez-Valdivieso and Mullineaux, 2010; Schmitt et al., 2014). Glutathione contents in nuclei of non-stressed leaves are about double as high than in the cytosol (Koffler et al., 2013) indicating its importance in nuclei. Large amounts of glutathione (up to 80% of the entire cellular pool) have been found to co-localize with nuclear DNA in the early steps of cell proliferation and at the G1 and S phases during the cell cycle (Diaz-Vivancos et al., 2010a). Even the accumulation of H_2O_2 in the cytosol could not prevent the sequestration of glutathione into the nucleus indicating that glutathione is transported into the nucleus even in situations of severe oxidative stress in expense of the cytosolic glutathione pool (Diaz-Vivancos et al., 2010a). A similar situation has been found in animal cells where oxidation events early in G1 phase were essential for the activation of signaling events initiating cell proliferation (Menon et al., 2003). It has been demonstrated for plant cells that the depletion of reduced glutathione by BSO inhibited the transition of cells from G1 to S phase (Vernoux et al., 2000). Additionally, the addition of dehydroascorbate and the inhibition of glutathione synthesis by BSO during G1 phase delayed cell division (Potters et al., 2004). On top low levels of ascorbate in the quiescent center of roots seem to be responsible for keeping these cells in the extended G1 state (Kerk and Feldman, 1995). Thus, it seems likely that high levels of glutathione (and ascorbate) in nuclei during G1 phase are an important strategy in order to allow the cell cycle to be continued (Diaz-Vivancos et al., 2010a). If the redox balance of nuclei is altered, DNA could be damaged which could induce mutations and eventually cell death (Diaz-Vivancos et al., 2010b).

VACUOLES

Recently it has been shown that vacuoles are involved in the protection of plants against ROS during abiotic stress conditions. During high light and drought stress H_2O_2 was found to leak from chloroplast and peroxisomes through the cytosol into vacuoles (Heyneke et al., 2013; Koffler et al., 2014a) which might act as a sink for ROS in situations of extreme stress. Nevertheless, glutathione could not be detected in vacuoles during high light and drought stress indicating that glutathione plays a minor role in the detoxification of H_2O_2 in this cell compartment. In contrast high levels of ascorbate accumulated in vacuoles under these conditions which could help to reduce phenoxyl radicals (created by oxidation of phenols by H_2O_2) and is oxidized to mono- and dehydroascorbic acid which is then transported into the cytosol for reduction (by glutathione through the ascorbate glutathione cycle) to ascorbic acid (Takahama, 2004). The sequestration of oxidized glutathione in vacuoles (10-fold increase in its concentration) seems to be a protective mechanism of the *cat2* mutant in order to avoid proposed negative effects of the accumulation of oxidized glutathione such as lesion formation, dormancy or cell death (Queval et al., 2011). The sequestration of oxidized glutathione and detoxification of H_2O_2 in vacuoles during oxidative stress may be involved in the control of cytosolic redox potential and redox state of target molecules and subsequently in the regulation of cell division, differentiation, and defense. Besides being involved in the detoxification of ROS vacuoles also act as a sink for glutathione conjugates withdrawing large amounts of glutathione from the cytoplasm (Fricker et al., 2000; Koffler et al., 2014b). Cadmium, for example, has a high affinity to thiol groups and forms complexes with reduced glutathione. These complexes are then transported into and sequestered in vacuoles (Rauser, 2001; Maksymiec and Krupa, 2006; Semane et al., 2007; Van Belleghem et al., 2007; DalCorso et al., 2008; Jozefczak et al., 2012) which can lead to the withdrawal of glutathione from the cytosol and other organelles if the demand for fresh glutathione cannot be compensated by glutathione synthesis (Koffler et al., 2014b).

SUBCELLULAR RESPONSE OF PLANTS TO BIOTIC STRESS

MITOCHONDRIA

The accumulation of ROS in mitochondria during biotic stress conditions is involved in the induction of programed cell death (reviewed by Amirsadeghi et al., 2007; Vianello et al., 2007). The depletion of glutathione in mitochondria favors the accumulation of ROS and thus it is not surprising that in *Nicotiana tabacum* plants infected with an incompatible strain of tobacco mosaic virus (TMV) the decrease of glutathione contents in mitochondria was accompanied with the development of necrotic lesions (Király et al., 2012). Similar effects were observed in *Arabidopsis* plants infected with *Botrytis cinerea*. At the infection site the development of necrosis 48 h post inoculation could be correlated with a strong depletion of glutathione contents in mitochondria, whereas glutathione levels in all other cell compartments remained at control levels (Simon et al., 2013). At this stage and also at later stages of infection the breakdown of the antioxidative defense system in mitochondria could also be correlated with a strong accumulation of H_2O_2 in this cell compartment (Simon et al., 2013). In tomato plants infected with *Botrytis cinerea* glutathione contents in mitochondria (besides peroxisomes) were affected the strongest. A strong drop in total glutathione contents accompanied with the accumulation of oxidized glutathione in mitochondria could be observed as early as 48 h post inoculation in this organelle and was accompanied with pathogen induced senescence (Kuźniak and Sklodowska, 2005a). Considering these results the depletion of glutathione contents in mitochondria during pathogen infections seems to favor the accumulation of ROS in this cell compartment and could be the reasons for the induction of programed cell death events. After all it still needs to be clarified if glutathione degradation and the accumulation of oxidized glutathione in mitochondria observed after pathogen infection is actively controlled by the plant or indirectly caused by disturbances in the electron transport chain induced by the invading pathogen.

CHLOROPLASTS AND PEROXISOMES

Compartment-specific changes in glutathione contents in chloroplasts and peroxisomes seem to be involved in the fine tuning of plant defense against pathogens. During *Botrytis cinerea* and *Pseudomonas syringae* infection in *Arabidopsis*, chloroplasts and peroxisomes could be identified as the hotspots of glutathione accumulation at the infection site at the beginning of the infection whereas the breakdown of the antioxidative system in these two cell compartments at the later stages of infection was correlated with the accumulation of ROS and progress of diseases symptoms (Großkinsky et al., 2012; Simon et al., 2013). Similar results were collected in tomato plants infected with the fungal pathogen *Botrytis cinerea* where the collapse of the antioxidative system in peroxisomes was associated with pathogen induced leaf senescence (Kuźniak and Sklodowska, 2005a). In conclusion, it seems that high levels of glutathione in chloroplast and peroxisomes are essential for a successful defense of plants against fungal and bacterial pathogens. The depletion of glutathione in chloroplast and peroxisomes leads to the accumulation of ROS in the tissue and the progression of symptom development.

NUCLEI

High levels of glutathione in nuclei play important roles in the protection of sensitive nuclear components (DNA, proteins, etc.) and also in regulating the expression of genes that are involved in the activation of plant defense (Han et al., 2013). Thus, it is not surprising that changes in glutathione contents are commonly observed in nuclei during pathogen attack. In younger zucchini yellow mosaic virus (ZYMV)-infected leaves of *Cucurbita pepo* plants and TMV-infected *Nicotiana tabacum* plants a strong accumulation of glutathione in nuclei was detected after inoculation (Zechmann et al., 2005, 2007; Király et al., 2012). A similar accumulation of glutathione was observed in *Arabidopsis* plants infected with *Pseudomonas syringae* (Großkinsky et al., 2012) and *Botrytis cinerea* at early stages of infection (Simon et al., 2013). These results confirm the importance of high glutathione levels in nuclei during biotic stress. High levels of reduced glutathione in nuclei could serve to protect DNA and redox-sensitive nuclear proteins from oxidation, as well as driving GRX-related processes which would influence the binding of transcription factors which results in adaptations of gene expression patterns (Mou et al., 2003; Gomez et al., 2004; Green et al., 2006; Diaz-Vivancos et al., 2010a,b). On the other hand, it has been demonstrated that the accumulation of glutathione in nuclei accompanied by its depletion in the cytosol subsequently leads to increased glutathione synthesis and the rapid accumulation of glutathione levels in the whole cell (Diaz-Vivancos et al., 2010a,b). Thus, it could be possible that the accumulation of glutathione in nuclei observed during virus, fungal, and bacterial infections in plants (Zechmann et al., 2005, 2007; Großkinsky et al., 2012; Király et al., 2012; Simon et al., 2013) is used as a signal to activate glutathione synthesis in order to increase cellular glutathione contents. This seems very likely as in TMV and ZYMV infected plants (Zechmann et al., 2005; Király et al., 2012), as well as in *Arabidopsis* plants infected with *Pseudomonas syringae* and *Botrytis cinerea* (Großkinsky et al., 2012; Simon et al., 2013) the increase of glutathione in nuclei was followed by a strong accumulation of glutathione in chloroplasts and the cytosol—which are considered to be the primary cell compartments for glutathione synthesis (Wachter et al., 2005).

APOPLAST

Glutathione contents in the apoplast during non-stressed conditions have been found to be very low or below the level of detection (Vanacker et al., 1998a,b,c, 2000; Zechmann et al., 2008a; Tolin et al., 2013). The reasons therefore can most probably be found in a high degradative activity of GGT1 and GGT2 which degrade glutathione and are located in the cell wall and the plasma membrane, respectively. In *ggt1* knock out mutants levels of glutathione in the apoplast were found to be similar to glutathione contents in plastids (Tolin et al., 2013). These changes were associated with modifications of the proteome that were similar to those found during abiotic and biotic stress conditions. These results indicate that glutathione contents and the redox state in the apoplast are involved in sensing and signaling environmental stress, thus have a key role in the adaption of plants to changing environmental conditions (Tolin et al., 2013). Support for this hypothesis comes from studies during fungal infections in oat and barley plants where a strong accumulation of glutathione was detected in the apoplast which was associated with race- and non-race-specific resistance to *Blumeria graminis* (Vanacker et al., 1998a,b,c, 2000). Decreased amounts of glutathione in the apoplast were found in these plants after infection with a susceptible fungal species where glutathione could not control the accumulation of H_2O_2 in the leaves during the hypersensitive response (Vanacker et al., 1998a,b,c, 2000). Such roles of glutathione could not be verified during viral and bacterial infections (Höller et al., 2010; Großkinsky et al., 2012; Király et al., 2012). In conclusion, it seems that glutathione and/or its redox state (which becomes more oxidized during the hypersensitive response) serves important roles in the apoplast in the response to abiotic and biotic stress. Nevertheless, whether glutathione primarily acts as a signaling agent or as an antioxidant in the apoplast still needs to be clarified.

SUMMARY AND OUTLOOK

Glutathione shows highly compartment-specific changes in plants during abiotic and biotic stress situations, indicating important subcellular roles for plant defense (**Figure 2**). High levels of glutathione in chloroplasts and peroxisomes seem to be of special importance during abiotic stress situations that negatively interfere with photosynthesis such as high light and salt stress. A drop of glutathione levels in these two cell compartments during pathogen attack could be correlated with the accumulation of ROS and with the development of chlorosis and necrosis. A similar situation was found for mitochondria where high and stable levels of glutathione were found to be essential for proper plant and cell development. The breakdown of the antioxidative system and the accumulation of ROS in mitochondria during pathogen attack seems to be involved in the activation of programed cell death. Vacuoles act as a sink for oxidized glutathione in situations of severe oxidative stress and as a sink for glutathione conjugates, for example, during the exposure of heavy metals. On the other side glutathione does not seem to be directly involved in the detoxification of ROS that diffuse into vacuoles during abiotic stress as observed during high light and drought. The roles of glutathione in the apoplast still have to be clearly dissected but seem to involve sensing and signaling stress conditions rather than the plain detoxification of ROS. In nuclei, glutathione fulfills a dual role during abiotic and biotic stress conditions. On the one site it protects nuclear components against oxidation by ROS and on the other site it is involved in the regulation of genes that are involved in plant defense. In conclusion, glutathione is essential for proper plant development, growth, and defense in the individual cell compartments during abiotic and biotic stress conditions. In order to achieve further progress in this field future research should focus:

(1) on the dissection of the functions of glutathione and its redox state in the apoplast which should help to clarify if its functions are related to sensing and signaling stress rather than simply the detoxification of ROS in this cell compartment,
(2) on the identification and investigation of the physiological relevance of glutathione transporters responsible for the import and export of glutathione from cells, tissues, and especially organelles,
(3) on the evaluation of the relevance of glutathione and antioxidants in vacuoles regarding the detoxification of ROS during environmental stress conditions,
(4) on the correlation, combination and progression of current (and possible new) methods available for the detection of subcellular glutathione in order to achieve a combined measurement of the actual glutathione concentration and the redox state in each cell compartment during abiotic and biotic stress conditions, and finally
(5) on combining this data with changes (i) in the subcellular distribution of ROS, (ii) in the transcription of related genes and (iii) changes in the proteome in order to receive a more detailed picture on the physiological relevance and the interplay of ROS and antioxidants especially glutathione in plants during abiotic and biotic stress.

ACKNOWLEDGMENT

This work was supported by the Austrian Science Fund (FWF P22988-B16).

REFERENCES

Ammar, W. B., Mediouni, C., Tray, B., Ghorbel, M. H., and Jemal, F. (2008). Glutathione and phytochelatin contents in tomato plants exposed to cadmium. *Biol. Plant.* 52, 314–320. doi: 10.1007/s10535-008-0065-9

Amirsadeghi, S., Robson, C. A., and Vanlerberghe, G. C. (2007). The role of the mitochondrion in plant responses to biotic stress. *Physiol. Plant.* 129, 253–266. doi: 10.1111/j.1399-3054.2006.00775.x

Asada, K. (2006). Production and scavenging of reactive oxygen species in chloroplasts and their functions. *Plant Physiol.* 141, 391–396. doi: 10.1104/pp.106.082040

Bachhawat, A. K., Thakur, A., Kaur, J., and Zulkifli, M. (2013). Glutathione transporters. *Biochim. Biophys. Acta* 1830, 3154–3164. doi: 10.1016/j.bbagen.2012.11.018

Bhargava, S., and Sawant, K. (2013). Drought stress adaptation: metabolic adjustment and regulation of gene expression. *Plant Breed.* 132, 21–32. doi: 10.1111/pbr.12004

Bloem, E., Haneklaus, S., and Schnug, E. (2007). Sulphur-induced resistance (SIR)—sulphur fertilization as a sustainable strategy for keeping plants healthy. [Schwefel-induzierte Resistenz (SIR)—Schwefeldüngung als nachhaltige Strategie zur Gesunderhaltung von Pflanzen]. *J. Consum. Prot. Food Saf.* 2, 7–12. doi: 10.1007/s00003-006-0107-1

Bloem, E., Riemenschneider, A., Volker, J., Papenbrock, J., Schmidt, A., Salac, I., et al. (2004). Sulphur supply and infection with *Pyrenopeziza brassicae* influence l-cysteine desulphydrase activity in *Brassica napus* L. *J. Exp. Bot.* 55, 2305–2312. doi: 10.1093/jxb/erh236

Blum, R., Beck, A., Korfte, A., Stengel, A., Letzel, T., Lendzian, K., et al. (2007). Function of phytochelatin synthase in catabolism of glutathione-conjugates. *Plant J.* 49, 740–749. doi: 10.1111/j.1365-313X.2006.02993.x

Blum, R., Meyer, K. C., Wünschmann, J., Lendzian, K. J., and Grill, E. (2010). Cytosolic action of phytochelatin synthase. *Plant Physiol.* 153, 159–169. doi: 10.1104/pp.109.149922

Cagnac, O., Bourbouloux, A., Chakrabarty, D., Zhang, M. Y., and Delrot, S. (2004). AtOPT6 transports glutathione derivatives and is induced by primisulfuron. *Plant Physiol.* 135, 1378–1387. doi: 10.1104/pp.104.039859

Cairns, N. G., Pasternak, M., Wachter, A., Cobbett, C. S., and Meyer, A. J. (2006). Maturation of *Arabidopsis* seeds is dependent on glutathione biosynthesis within the embryo. *Plant Physiol.* 141, 446–455. doi: 10.1104/pp.106.077982

Cheng, J. C., Seeley, K. A., and Sung, Z. R. (1995). RML1 and RML2, *Arabidopsis* genes required for cell proliferation at the root tip. *Plant Physiol.* 107, 365–376.

Chew, O., Whelan, J., and Millar, A. H. (2003). Molecular definition of the ascorbate–glutathione cycle in *Arabidopsis* mitochondria reveals dual targeting of antioxidant defenses in plants. *J. Biol. Chem.* 278, 46869–46877. doi: 10.1074/jbc.M307525200

Cummins, I., Dixon, D. P., Freitag-Pohl, S., Skipsey, M., and Edwards, R. (2011). Multiple roles for plant glutathione transferases in xenobiotic detoxification. *Drug Metab. Rev.* 43, 226–280. doi: 10.3109/03602532.2011.552910

DalCorso, G., Farinati, S., Maistri, S., and Furini, A. (2008). How plants cope with cadmium: staking all on metabolism and gene expression. *J. Integr. Plant Biol.* 50, 1268–1280. doi: 10.1111/j.1744-7909.2008.00737.x

DeRidder, B. P., and Goldsbrough, P. B. (2006). Organ-specific expression of glutathione *S*-transferases and the efficacy of herbicide safeners in *Arabidopsis*. *Plant Physiol.* 140, 167–175. doi: 10.1104/pp.105.067199

Destro, T., Prasad, D., Martignago, D., Bernet, I. L., Trentin, A. R., Renu, I. K., et al. (2011). Compensatory expression and substrate inducibility of γ-glutamyl transferase GGT2 isoform in *Arabidopsis thaliana*. *J. Exp. Bot.* 62, 805–815. doi: 10.1093/jxb/erq316

Diaz-Vivancos, P., Dong, Y., Ziegler, K., Markovic, J., Pallardo, F. V., Pellny, T. K., et al. (2010a). Recruitment of glutathione into the nucleus during cell proliferation adjusts whole-cell redox homeostasis in *Arabidopsis thaliana* and lowers the oxidative defence shield. *Plant J.* 64, 825–838. doi: 10.1111/j.1365-313X.2010.04371.x

Diaz-Vivancos, P., Wolff, T., Markovic, J., Pallardo, F. V., and Foyer, C. H. (2010b). A nuclear glutathione cycle within the cell cycle. *Biochem. J.* 431, 169–178. doi: 10.1042/BJ20100409

Dixon, D. P., Skipsey, M., Grundy, N. M., and Edwards, R. (2005). Stress-induced protein *S*-glutathionylation in *Arabidopsis*. *Plant Physiol.* 138, 2233–2244. doi: 10.1104/pp.104.058917

Doyle, S. M., Diamond, M., and McCabe, P. F. (2010). Chloroplast and reactive oxygen species involvement in apoptotic-like programmed cell death in *Arabidopsis* suspension cultures. *J. Exp. Bot.* 61, 473–482. doi: 10.1093/jxb/erp320

Ducic, T., Maksimovic, V., and Radotic, K. (2008). Oxalate oxidase and non-enzymatic compounds of the antioxidative system in young Serbian spruce plants exposed to cadmium stress. *Arch. Biol. Sci.* 60, 67–76. doi: 10.2298/ABS0801067D

Edwards, R., Brazier-Hicks, M., Dixon, D. P., and Cummins, I. (2005). Chemical manipulation of antioxidant defences in plants. *Adv. Bot. Res.* 42, 1–32. doi: 10.1016/S0065-2296(05)42001-7

Fernandez-Garcia, N., Marti, M. C., Jimenez, A., Sevilla, F., and Olmos, E. (2009). Sub-cellular distribution of glutathione in an *Arabidopsis* mutant (vtc1) deficient in ascorbate. *J. Plant Physiol.* 166, 2004–2012. doi: 10.1016/j.jplph.2009.06.006

Ferretti, M., Destro, T., Tosatto, S. C. E., La Rocca, N., Rascio, N., and Masi, A. (2009). Gamma-glutamyl transferase in the cell wall participates in extracellular glutathione salvage from the root apoplast. *New Phytol.* 181, 115–126. doi: 10.1111/j.1469-8137.2008.02653.x

Foyer, C. H., and Noctor, G. (2009). Redox regulation in photosynthetic organisms: signaling, acclimation, and practical implications. *Antioxid. Redox Signal.* 11, 861–905. doi: 10.1089/ars.2008.2177

Foyer, C. H., and Noctor, G. (2011). Ascorbate and glutathione: the heart of the redox hub. *Plant Physiol.* 155, 2–18. doi: 10.1104/pp.110.167569

Foyer, C. H., Theodoulou, F. L., and Delrot, S. (2001). The functions of inter- and intracellular glutathione transport systems in plants. *Trends Plant Sci.* 6, 486–492. doi: 10.1016/S1360-1385(01)02086-6

Fricker, M. D., May, M., Meyer, A. J., Sheard, N., and White, N. S. (2000). Measurement of glutathione levels in intact roots of *Arabidopsis*. *J. Microsc.* 198, 162–173. doi: 10.1046/j.1365-2818.2000.00696.x

Galvez-Valdivieso, G., and Mullineaux, P. M. (2010). The role of reactive oxygen species in signalling from chloroplasts to the nucleus. *Physiol. Plant.* 138, 430–439. doi: 10.1111/j.1399-3054.2009.01331.x

Gao, R., Ng, F. K. L., Liu, P., and Wong, S. M. (2012). Hibiscus chlorotic ringspot virus coat protein upregulates sulfur metabolism genes for enhanced pathogen defense. *Mol. Plant Microbe Interact.* 25, 1574–1583. doi: 10.1094/MPMI-08-12-0203-R

Golan, T., Müller-Moulé, P., and Niyogi, K. K. (2006). Photoprotection mutants of *Arabidopsis thaliana* acclimate to high light by increasing photosynthesis and specific antioxidants. *Plant Cell Environ.* 29, 879–887. doi: 10.1111/j.1365-3040.2005.01467.x

Gomez, L. D., Noctor, G., Knight, M. R., and Foyer, C. H. (2004). Regulation of calcium signalling and expression by glutathione. *J. Exp. Bot.* 55, 1851–1859. doi: 10.1093/jxb/erh202

Green, R. M., Graham, M., O'Donovan, M. R., Chipman, J. K., and Hodges, N. J. (2006). Subcellular compartmentalization of glutathione: correlations with parameters of oxidative stress related to genotoxicity. *Mutagenesis* 21, 383–390. doi: 10.1093/mutage/gel043

Großkinsky, D. K., Koffler, B. E., Roitsch, T., Maier, R., and Zechmann, B. (2012). Compartment-specific antioxidative defense in *Arabidopsis* against virulent and avirulent *Pseudomonas syringae*. *Phytopathology* 102, 662–673. doi: 10.1094/PHYTO-02-12-0022-R

Grzam, A., Tennstedt, P., Clemens, S., Hell, R., and Meyer, A. J. (2006). Vacuolar sequestration of glutathione *S*-conjugates outcompetes a possible degradation of the glutathione moiety by phytochelatin synthase. *FEBS Lett.* 580, 6384–6390. doi: 10.1016/j.febslet.2006.10.050

Gullner, G., Tóbiás, I., Fodor, J., and Kömives, T. (1999). Elevation of glutathione level and activation of glutathione-related enzymes affect virus infection in tobacco. *Free Radic. Res.* 31(Suppl.), S155–S161. doi: 10.1080/10715769900301451

Gutscher, M., Pauleau, A. L., Marty, L., Brach, T., Wabnitz, G. H., Samstag, Y., et al. (2008). Real-time imaging of the intracellular glutathione redox potential. *Nat. Methods* 5, 553–559. doi: 10.1038/nmeth.1212

Halliwell, B., and Gutteridge, J. M. C. (1999). *Free Radicals in Biology and Medicine*. Oxford: Oxford University Press.

Han, Y., Chaouch, S., Mhamdi, A., Queval, G., Zechmann, B., and Noctor, G. (2013). Functional analysis of *Arabidopsis* mutants points to novel roles for glutathione in coupling H_2O_2 to activation of salicylic acid accumulation and signaling. *Antioxid. Redox Signal.* 18, 2106–2121. doi: 10.1089/ars.2012.5052

Harms, K., von Ballmoos, P., Brunold, C., Hofgen, R., and Hesse, H. (2000). Expression of a bacterial serine acetyltransferase in transgenic potato plants leads to increased levels of cysteine and glutathione. *Plant J.* 22, 335–343. doi: 10.1046/j.1365-313x.2000.00743.x

Hartmann, T. N., Fricker, M. D., Rennenberg, H., and Meyer, A. J. (2003). Cell-specific measurement of cytosolic glutathione in poplar leaves. *Plant Cell Environ.* 26, 965–975. doi: 10.1046/j.1365-3040.2003.01031.x

Hernández, I., Cela, J., and Alegre, L. (2013). "Antioxidant defenses against drought stress," in *Plant Responses to Drought Stress*, ed. R. Aroca (Berlin: Springer-Verlag), 231–258. doi: 10.1007/978-3-642-32653-0_9

Heyneke, E., Luschin-Ebengreuth, N., Krajcer, I., Wolkinger, V., Müller, M., and Zechmann, B. (2013). Dynamic compartment specific changes in glutathione and ascorbate levels in *Arabidopsis* plants exposed to different light intensities. *BMC Plant Biol.* 13:104. doi: 10.1186/1471-2229-13-104

Höller, K., Király, L., Künstler, A., Müller, M., Gullner, G., Fattinger, M., et al. (2010). Enhanced glutathione metabolism is correlated with sulfur-induced resistance in tobacco mosaic virus-infected genetically susceptible *Nicotiana tabacum* plants. *Mol. Plant Microbe Interact.* 23, 1448–1459. doi: 10.1094/MPMI-05-10-0117

Hurd, T. R., Filipovska, A., Costa, N. J., Dahm, C. C., and Murphy, M. P. (2005a). Disulphide formation on mitochondrial protein thiols. *Biochem. Soc. Trans.* 33(Pt 6), 1390–1393. doi: 10.1042/BST20051390

Hurd, T. R., Costa, N. J., Dahm, C, C., Beer, S. M., Brown, S. T., Filipovska, A., et al. (2005b). Glutathionylation of mitochondrial proteins. *Antioxid. Redox Signal.* 7, 999–1010. doi: 10.1089/ars.2005.7.999

Jamai, A., Tommasini, R., Martinoia, E., and Delrot, S. (1996). Characterization of glutathione uptake in broad bean leaf protoplasts. *Plant Physiol.* 111, 1145–1152. doi: 10.1104/pp.111.4.1145

Jiménez, A., Hernández, J. A., del Río, L. A., and Sevilla, F. (1997). Evidence for the presence of the ascorbate–glutathione cycle in mitochondria and peroxisomes of pea leaves. *Plant Physiol.* 114, 275–284. doi: 10.1104/pp.114.1.275

Jiménez, A., Hernández, J. A., Pastori, G., del Río, L. A., and Sevilla, F. (1998). Role of the ascorbate–glutathione cycle of mitochondria and peroxisomes in the senescence of pea leaves. *Plant Physiol.* 118, 1327–1335. doi: 10.1104/pp.118.4.1327

Jozefczak, M., Remans, T., Vangronsfeld, J., and Cuypers, A. (2012). Glutathione is a key player in metal-induced oxidative stress defenses. *Int. J. Mol. Sci.* 13, 3145–3175. doi: 10.3390/ijms13033145

Jubany-Marí, T., Munné-Bosch, S., and Alegre, L. (2010). Redox regulation of water stress responses in field-grown plants. Role of hydrogen peroxide and ascorbate. *Plant Physiol. Biochem.* 48, 351–358. doi: 10.1016/j.plaphy.2010.01.021

Kerk, N. M., and Feldman, L. J. (1995). A biochemical model for the initiation and maintenance of the quiescent centre: implications for organization of root meristem. *Development* 121, 2825–2833.

Kim, C., Meskauskiene, R., Apel, K., and Laloi, C. (2008). No single way to understand singlet oxygen signalling in plants. *EMBO Rep.* 9, 435–439. doi: 10.1038/embor.2008.57

Király, L., Künstler, A., Höller, K., Fattinger, M., Juhász, C., Müller, M., et al. (2012). Sulfate supply influences compartment specific glutathione metabolism and confers enhanced resistance to tobacco mosaic virus during a hypersensitive response. *Plant Physiol. Biochem.* 59, 44–54. doi: 10.1016/j.plaphy.2011.10.020

Kocsy, G., Tari, I., Vankova, R., Zechmann, B., Gulyas, Z., Poor, P., et al. (2013). Redox control of plant growth and development. *Plant Sci.* 211, 77–91. doi: 10.1016/j.plantsci.2013.07.004

Koffler, B. E., Bloem, E., Zellnig, G., and Zechmann, B. (2013). High resolution imaging of subcellular glutathione concentrations by quantitative immunoelectron microscopy in different leaf areas of *Arabidopsis*. *Micron* 45, 119–128. doi: 10.1016/j.micron.2012.11.006

Koffler, B. E., Luschin-Ebengreuth, N., Stabentheiner, E., Müller, M., and Zechmann, B. (2014a). Compartment specific response of antioxidants to drought stress in *Arabidopsis*. *Plant Sci.* 227, 133–144. doi: 10.1016/j.plantsci.2014.08.002

Koffler, B. E., Polanschütz, L. M., and Zechmann, B. (2014b). Higher sensitivity of pad2-1 and vtc2-1 mutants to cadmium is related to lower subcellular glutathione rather than ascorbate contents. *Protoplasma* 251, 755–769. doi: 10.1007/s00709-013-0576-x

Koffler, B. E., Maier, R., and Zechmann, B. (2011). Subcellular distribution of glutathione precursors in *Arabidopsis thaliana*. *J. Integr. Plant Biol.* 53, 930–941. doi: 10.1111/j.1744-7909.2011.01085.x

Koh, S., Wiles, A. M., Sharp, J. S., Naider, F. R., Becker, J. M., and Stacey, G. (2002). An oligopeptide transporter gene family in *Arabidopsis*. *Plant Physiol.* 128, 21–29. doi: 10.1104/pp.010332

Krieger-Liszkay, A., Fufezan, C., and Trebst, A. (2008). Singlet oxygen production in photosystem II and related protection mechanism. *Photosynth. Res.* 98, 551–564. doi: 10.1007/s11120-008-9349-3

Krueger, S., Niehl, A., Lopez Martin, M. C., Steinhauser, D., Donath, A., Hildebrandt, T., et al. (2009). Analysis of cytosolic and plastidic serine acetyltransferase mutants and subcellular metabolite distributions suggests interplay of the cellular compartments for cysteine biosynthesis in *Arabidopsis*. *Plant Cell Environ.* 32, 349–367. doi: 10.1111/j.1365-3040.2008.01928.x

Kuźniak, E., and Sklodowska, A. (2001). Ascorbate, glutathione and related enzymes in chloroplasts of tomato leaves infected by *Botrytis cinerea*. *Plant Sci.* 160, 723–731. doi: 10.1016/S0168-9452(00)00457-X

Kuźniak, E., and Sklodowska, A. (2004). Comparison of two methods for preparing mitochondria from tomato leaves to study the ascorbate–glutathione cycle activity. *Biol. Plant.* 48, 537–542. doi: 10.1023/B:BIOP.0000047149.57946.a5

Kuźniak, E., and Sklodowska, A. (2005a). Fungal pathogen-induced changes in the antioxidant systems of leaf peroxisomes from infected tomato plants. *Planta* 222, 192–200. doi: 10.1007/s00425-005-1514-8

Kuźniak, E., and Sklodowska, A. (2005b). Compartment-specific role of the ascorbate–glutathione cycle in the response of tomato leaf cells to *Botrytis cinerea* infection. *J. Exp. Bot.* 56, 921–933. doi: 10.1093/jxb/eri086

Lim, B., Pasternak, M., Meyer, A. J., and Cobbett, C. S. (2014). Restricting glutamylcysteine synthetase activity to the cytosol or glutathione biosynthesis to the plastid is sufficient for normal plant development and stress tolerance. *Plant Biol.* 16, 58–67. doi: 10.1111/plb.12033

Lu, Y. P., Li, Z. S., Drozdowicz, Y. M., Hörtensteiner, S., Martinoia, E., and Rea, P. A. (1998). AtMRP2, an *Arabidopsis* ATP binding cassette transporter able to transport glutathione *S*-conjugates and chlorophyll catabolites: functional comparisons with AtMRP1. *Plant Cell* 10, 267–282.

Maksymiec, W., and Krupa, Z. (2006). The effects of short-term exposition to Cd, excess Cu ions and jasmonate on oxidative stress appearing in *Arabidopsis thaliana*. *Environ. Exp. Bot.* 57, 187–194. doi: 10.1016/j.envexpbot.2005.05.006

Martin, M. N., and Slovin, J. P. (2000). Purified γ-glutamyl transpeptidases from tomato exhibit high affinity for glutathione and glutathione *S*-conjugates. *Plant Physiol.* 122, 1417–1426. doi: 10.1104/pp.122.4.1417

Maughan, S. C., Pasternak, M., Cairns, N., Kiddle, G., Brach, T., Jarvis, R., et al. (2010). Plant homologs of the *Plasmodium falciparum* chloroquine-resistance transporter, PfCRT, are required for glutathione homeostasis and stress responses. *Proc. Natl. Acad. Sci. U.S.A.* 107, 2331–2336. doi: 10.1073/pnas.0913689107

Menon, S. G., Sarsour, E. H., Spitz, D. R., Higashikubo, R., Sturm, M., Zhang, H., et al. (2003). Redox regulation of the G1 to S phase transition in the mouse embryo fibroblast cell cycle. *Cancer Res.* 63, 2109–2117.

Meyer, A. J., Brach, T., Marty, L., Kreye, S., Rouhier, N., Jacquot, J. P., et al. (2007). Redox-sensitive GFP in *Arabidopsis thaliana* is a quantitative biosensor for the redox potential of the cellular glutathione redox buffer. *Plant J.* 52, 973–986. doi: 10.1111/j.1365-313X.2007.03280.x

Meyer, A. J., and Fricker, M. D. (2000). Direct measurement of glutathione in epidermal cells of intact *Arabidopsis* roots by two-photon laser scanning microscopy. *J. Microsc.* 198, 174–181. doi: 10.1046/j.1365-2818.2000.00697.x

Meyer, A. J., and Fricker, M. D. (2002). Control of demand-driven biosynthesis of glutathione in green *Arabidopsis* suspension culture cells. *Plant Physiol.* 130, 1927–1937. doi: 10.1104/pp.008243

Meyer, A. J., May, M. J., and Fricker, M. (2001). Quantitative in vivo measurement of glutathione in *Arabidopsis* cells. *Plant J.* 27, 67–78. doi: 10.1046/j.1365-313x.2001.01071.x

Miller, G., Suzuki, N., Ciftci-Yilmaz, S., and Mittler, R. (2010). Reactive oxygen species homeostasis and signalling during drought and salinity stresses. *Plant Cell Environ.* 33, 453–467. doi: 10.1111/j.1365-3040.2009.02041.x

Mohsenzadeh, S., Esmaeili, M., Moosavi, F., Shahrtash, M., Saffari, B., and Mohabatkar, H. (2011). Plant glutathione *S*-transferase classification, structure and evolution. *Afr. J. Biotechnol.* 43, 8160–8165. doi: 10.5897/AJB11.1024

Mou, Z., Fan, W., and Dong, X. (2003). Inducers of plant systemic acquired resistance regulate NPR1 function through redox changes. *Cell* 27, 935–944. doi: 10.1016/S0092-8674(03)00429-X

Müller, M., Zellnig, G., Urbanek, A., and Zechmann, B. (2005). Recent developments in methods intracellulary localizing glutathione within plant tissues and cells (a minireview). *Phyton* 45, 45–55.

Nocito, F. F., Espen, L., Crema, B., Cocucci, M., and Sacchi, G. A. (2008). Cadmium induces acidosis in maize root cells. *New Phytol.* 179, 700–711. doi: 10.1111/j.1469-8137.2008.02509.x

Noctor, G. (2006). Metabolic signalling in defence and stress: the central roles of soluble redox couples. *Plant Cell Environ.* 29, 409–425. doi: 10.1111/j.1365-3040.2005.01476.x

Noctor, G., Arisi, A. C. M., Jouanin, L., Valadier, M. H., Roux, Y., and Foyer, C. H. (1997a). Light-dependent modulation of foliar glutathione synthesis and associated amino acid metabolism in poplar overexpressing c-glutamylcysteine synthetase. *Planta* 202, 357–369.

Noctor, G., Arisi, A. C. M., Jouanin, L., Valadier, M. H., Roux, Y., and Foyer, C. H. (1997b). The role of glycine in determining the rate of glutathione synthesis in poplar. Possible implications for glutathione production during stress. *Physiol. Plant.* 100, 225–263. doi: 10.1111/j.1399-3054.1997.tb04781.x

Noctor, G., Gomez, L., Vanacker, H., and Foyer, C. H. (2002). Interactions between biosynthesis, compartmentation and transport in the control of glutathione homeostasis and signalling. *J. Exp. Bot.* 53, 1283–1304. doi: 10.1093/jexbot/53.372.1283

Noctor, G., Mhamdi, A., Chaouch, S., Han, Y., Neukermans, J., Marquez-Garcia, B., et al. (2012). Glutathione in plants: an integrated overview. *Plant Cell Environ.* 35, 454–484. doi: 10.1111/j.1365-3040.2011.02400.x

Noctor, G., Queval, G., Mhamdi, A., Chaouch, S., and Foyer, C. H. (2011). Glutathione. *Arabidopsis Book* 9:e0142. doi: 10.1199/tab.0142

Noji, M., and Saito, K. (2002). Molecular and biochemical analysis of serine acetyltransferase and cysteine synthase towards sulfur metabolic engineering in plants. *Amino Acids* 22, 231–243. doi: 10.1007/s007260200011

Ohkama-Ohtsu, N., Oikawa, A., Zhao, P., Xiang, C., Saito, K., and Oliver, D. J. (2008). A γ-glutamyl transpeptidase-independent pathway of glutathione catabolism to glutamate via 5-oxoproline in *Arabidopsis*. *Plant Physiol.* 148, 1603–1613. doi: 10.1104/pp.108.125716

Ohkama-Ohtsu, N., Radwan, S., Peterson, A., Zhao, P., Badr, A. F., Xiang, C., et al. (2007a). Characterization of the extracellular c-glutamyl transpeptidases, GGT1 and GGT2, in *Arabidopsis*. *Plant J.* 49, 865–877. doi: 10.1111/j.1365-313X.2006.03004.x

Ohkama-Ohtsu, N., Zhao, P., Xiang, C., and Oliver, D. J. (2007b). Glutathione conjugates in the vacuole are degraded by γ-glutamyl transpeptidase GGT3 in *Arabidopsis*. *Plant J.* 49, 878–888. doi: 10.1111/j.1365-313X.2006.03005.x

op den Camp, R. G. L., Przybyla, D., Ochsenbein, C., Laloi C., Kim, C., Danon, A., et al. (2003). Rapid induction of distinct stress responses after the release of singlet oxygen in *Arabidopsis*. *Plant Cell* 15, 2320–2332. doi: 10.1105/tpc.014662

Parisy, V., Poinssot, B., Owsianowski, L., Buchala, A., Glazebrook, J., and Mauch, F. (2007). Identification of PAD2 as a γ-glutamylcysteine synthetase highlights the importance of glutathione in disease resistance of *Arabidopsis*. *Plant J.* 49, 159–172. doi: 10.1111/j.1365-313X.2006.02938.x

Pasternak, M., Lim, B., Wirtz, M., Hell, R., Cobbett, C. S., and Meyer, A. (2008). Restricting glutathione biosynthesis to the cytosol is sufficient for normal plant development. *Plant J.* 53, 999–1012. doi: 10.1111/j.1365-313X.2007.03389.x

Pfannschmidt, T., Bräutigam, K., Wagner, R., Dietzel, L., Schröter, Y., Steiner, S., et al. (2009). Potential regulation of gene expression in photosynthetic cells by redox and energy state: approaches towards better understanding. *Ann. Bot.* 103, 599–607. doi: 10.1093/aob/mcn081

Pike, S., Patel, A., Stacey, G., and Gassmann, W. (2009). *Arabidopsis* OPT6 is an oligopeptide transporter with exceptionally broad substrate specificity. *Plant Cell Physiol.* 50, 1923–1932. doi: 10.1093/pcp/pcp136

Pospisil, P. (2012). Molecular mechanisms of production and scavenging of reactive oxygen species by photosystem II. *Biochim. Biophys. Acta* 1817, 218–231. doi: 10.1016/j.bbabio.2011.05.017

Potters, G., Horemans, N., Bellone, S., Caubergs, R. J., Trost, P., Guisez, Y., et al. (2004). Dehydroascorbate influences the plant cell cycle through a glutathione-independent reduction mechanism. *Plant Physiol.* 134, 1479–1487. doi: 10.1104/pp.103.033548

Queval, G., Jaillard, D., Zechmann, B., and Noctor, G. (2011). Increased intracellular H_2O_2 availability preferentially drives glutathione accumulation in vacuoles and chloroplasts. *Plant Cell Environ.* 34, 21–32. doi: 10.1111/j.1365-3040.2010.02222.x

Rauser, W. E. (2001). "The role of glutathione in plant reaction and adaptation to excess metals," in *Significance of Glutathione to Plant Adaptation to the Environment*, eds D. Grill, M. Tausz, and L. J. Cok (Dordrecht: Kluwer), 123–154. doi: 10.1007/0-306-47644-4_6

Schmitt, F. J., Renger, G., Friedrich, T., Kreslavksi, V. D., Zharmukhadmedov, S. K., Los, D. A., et al. (2014). Reactive oxygen species: re-evaluation of generation, monitoring and role in stress-signaling in phototrophic organisms. *Biochim. Biophys. Acta* 1837, 385–848. doi: 10.1016/j.bbabio.2014.02.005

Schneider, A., Martini, N., and Rennenberg, H. (1992). Reduced glutathione (GSH) transport into cultured tobacco cells. *Plant Physiol. Biochem.* 30, 29–38.

Semane, B., Cuypers, A., Smeets, K., Van Belleghem, F., Horemans, N., Schat, H., et al. (2007). Cadmium responses in *Arabidopsis thaliana*: glutathione metabolism and antioxidative defence system. *Physiol. Plant.* 129, 519–528. doi: 10.1111/j.1399-3054.2006.00822.x

Simon, K. U., Polanschütz, L. M., Koffler, B. E., and Zechmann, B. (2013). High resolution imaging of temporal and spatial changes of subcellular ascorbate, glutathione and H_2O_2 distribution during *Botrytis cinerea* infection in *Arabidopsis*. *PLoS ONE* 8:e65811. doi: 10.1371/journal.pone.0065811

Sugiyama, A., Nishimura, J., Mochizuku, Y., Inagaki, K., and Sekiya, J. (2004). Homoglutathione synthesis in transgenic tobacco plants expressing soybean homoglutathione synthetase. *Plant Biotechnol.* 21, 79–83.

Szarka, A., Tomasskovics, B., and Banhegyi, G. (2012). The ascorbate-glutathione-α-tocopherol triad in abiotic stress response. *Int. J. Mol. Sci.* 13, 4458–4483. doi: 10.3390/ijms13044458

Takahama, U. (2004). Oxidation of vacuolar and apoplastic phenolic substrates by peroxidase: physiological significance of the oxidation reactions. *Phytochem. Rev.* 3, 207–219. doi: 10.1023/B:PHYT.0000047805.08470.e3

Tan, J. F., O'Toole, N., Taylor, N. L., and Millar, A. H. (2010). Divalent metal ions in plant mitochondria and their role in interactions with proteins and oxidative stress-induced damage to respiratory function. *Plant Physiol.* 152, 747–761. doi: 10.1104/pp.109.147942

Tausz, M., Šircelj, H., and Grill, D. (2004). The glutathione system as a stress marker in plant ecophysiology: is a stress-response concept valid? *J. Exp. Bot.* 55, 1955–1962. doi: 10.1093/jxb/erh194

Tolin, S., Arrigoni, G., Trentin, A. R., Veljovic-Jovanovic, S., Pivato, M., Zechmann, B., et al. (2013). Biochemical and quantitative proteomics investigations in *Arabidopsis* ggt1 mutant leaves reveal a role for the gamma-glutamyl cycle in plant's adaptation to environment. *Proteomics* 13, 2031–2045. doi: 10.1002/pmic.201200479

Vanacker, H., Carver, T. L. W., and Foyer, C. H. (1998a). Pathogen-induced changes in the antioxidant status of the apoplast in barley leaves. *Plant Physiol.* 117, 1103–1114. doi: 10.1104/pp.117.3.1103

Vanacker, H., Foyer, C. H., and Carver, T. L. W. (1998b). Changes in apoplastic antioxidants induced by powdery mildew attack in oat genotypes with race non-specific resistance. *Planta* 208, 444–452. doi: 10.1007/s004250050581

Vanacker, H., Harbinson, J., Ruisch, J., Carver, T. L. W., and Foyer, C. H. (1998c). Antioxidant defences of the apoplast. *Protoplasma* 205, 129–140. doi: 10.1007/BF01279303

Vanacker, H., Carver, T. L. W., and Foyer, C. H. (2000). Early H_2O_2 accumulation in mesophyll cells leads to induction of glutathione during the hypersensitive response in the barley-powdery mildew interaction. *Plant Physiol.* 123, 1289–1300. doi: 10.1104/pp.123.4.1289

Van Belleghem, F., Cuypers, A., Semane, B., Smeets, K., Vangronsveld, J., D'Haen, J., et al. (2007). Subcellular localization of cadmium in roots and leaves of *Arabidopsis thaliana*. *New Phytol.* 173, 495–508. doi: 10.1111/j.1469-8137.2006.01940.x

Vernoux, T., Wilson, R. C., Seeley, K. A., Reichheld, J. P., Muroy, S., Brown, S., et al. (2000). The root meristemless/cadmium sensitive2 gene defines a glutathione-dependent pathway involved in initiation and maintenance of cell division during postembryonic root development. *Plant Cell* 12, 97–110. doi: 10.1105/tpc.12.1.97

Vianello, A., Zancani, M., Peresson, C., Petrussa, E., Casolo, V., Krajnakova, J., et al. (2007). Plant mitochondrial pathway leading to programmed cell death. *Physiol. Plant.* 129, 242–252. doi: 10.1111/j.1399-3054.2006.00767.x

Wachter, A., Wolf, S., Steininger, H., Bogs, J., and Rausch, T. (2005). Differential targeting of GSH1 and GSH2 is achieved by multiple transcription initiation: implications for the compartmentation of glutathione biosynthesis in the Brassicaceae. *Plant J.* 41, 15–30. doi: 10.1111/j.1365-313X.2004.02269.x

Wagner, D., Przybyla, D., op den Camp, R., Kim, C., Landgraf, F., Lee, K. P., et al. (2004). The genetic basis of singlet oxygen-induced stress responses of *Arabidopsis thaliana*. *Science* 306, 1183–1185. doi: 10.1126/science.1103178

Wirtz, M., and Hell, R. (2007). Dominant-negative modification reveals the regulatory function of the multimeric cysteine synthase protein complex in transgenic tobacco. *Plant Cell* 19, 625–639. doi: 10.1105/tpc.106.043125

Wolf, A. E., Dietz, K. J., and Schröder, P. (1996). Degradation of glutathione *S*-conjugates by a carboxypeptidase in the plant vacuole. *FEBS Lett.* 384, 31–34. doi: 10.1016/0014-5793(96)00272-4

Zaffagnini, M., Bedhomme, M., Lemaire, S. D., and Trost, P. (2012a). The emerging roles of protein glutathionylation in chloroplasts. *Plant Sci.* 185, 86–96. doi: 10.1016/j.plantsci.2012.01.005

Zaffagnini, M., Bedhomme, M., Marchand, C. H., Morisse, S., Trost, P., and Lemaire, S. D. (2012b). Redox regulation in photosynthetic organisms: focus on glutathionylation. *Antioxid. Redox Signal.* 16, 567–586. doi: 10.1089/ars.2011.4255

Zawoznik, M. S., Groppa, M. D., Tomaro, M. L., and Benavides, M. P. (2007). Endogenous salicylic acid potentiates cadmium-induced oxidative stress in *Arabidopsis thaliana*. *Plant Sci.* 173, 190–197. doi: 10.1016/j.plantsci.2007.05.004

Zechmann, B., Koffler, B. E., and Russell, S. D. (2011). Glutathione synthesis is essential for pollen germination in vitro. *BMC Plant Biol.* 11:54. doi: 10.1186/1471-2229-11-54

Zechmann, B., Mauch, F., Sticher, L., and Müller, M. (2008a). Subcellular immunocytochemical analysis detects the highest concentrations of glutathione in mitochondria and not in plastids. *J. Exp. Bot.* 59, 4017–4027. doi: 10.1093/jxb/ern243

Zechmann, B., Müller, M., and Zellnig, G. (2008b). "Modified levels of cysteine affect glutathione metabolism in plant cells," in *Sulfur Assimilation and Abiotic Stress in Plants*, eds N. A. Khan, S. Singh, and S. Umar (Berlin: Springer), 193–206. doi: 10.1007/978-3-540-76326-0_9

Zechmann, B., and Müller, M. (2010). Subcellular compartmentation of glutathione in dicotyledonous plants. *Protoplasma* 246, 15–24. doi: 10.1007/s00709-010-0111-2

Zechmann, B., Müller, M., and Zellnig, G. (2006). Intracellular adaptations of glutathione content in *Cucurbita pepo* L. induced by treatment with reduced glutathione and buthionine sulfoximine. *Protoplasma* 227, 197–209. doi: 10.1007/s00709-005-0129-z

Zechmann, B., Zelený, F., Wonisch, A., and Zelená, E. (2009). Aphid infestation affects subcellular glutathione and cysteine contents in *Beta vulgaris* L. *Belg. J. Bot.* 142, 50–59. doi: 10.2307/20794671

Zechmann, B., Zellnig, G., and Müller, M. (2005). Changes in the subcellular distribution of glutathione during virus infection in *Cucurbita pepo* (L.). *Plant Biol.* 7, 49–57. doi: 10.1055/s-2004-830477

Zechmann, B., Zellnig, G., Urbanek-Krajnc, A., and Müller, M. (2007). Artificial elevation of glutathione affects symptom development in ZYMV-infected *Cucurbita pepo* L. plants. *Arch. Virol.* 152, 747–762. doi: 10.1007/s00705-006-0880-2

Conflict of Interest Statement: The author declares that the research was conducted in the absence of any commercial or financial relationships that could be construed as a potential conflict of interest.

6

Predictive sulfur metabolism – a field in flux

Alexander Calderwood[1]*, Richard J. Morris[1] and Stanislav Kopriva[2]

[1] Department of Computational and Systems Biology, John Innes Centre, Norwich, UK
[2] Botanical Institute and Cluster of Excellence on Plant Sciences, University of Cologne, Cologne Biocenter, Cologne, Germany

Edited by:
Stanislaus Francis D'Souza, Bhabha Atomic Research Centre, India

Reviewed by:
Guillaume Pilot, Virginia Tech, USA
Hideki Takahashi, Michigan State University, USA

***Correspondence:**
Alexander Calderwood, Department of Computational and Systems Biology, John Innes Centre, Norwich Research Park, Colney Lane, Norwich NR14 6AG, UK
e-mail: alexander.calderwood@jic.ac.uk

The key role of sulfur metabolites in response to biotic and abiotic stress in plants, as well as their importance in diet and health has led to a significant interest and effort in trying to understand and manipulate the production of relevant compounds. Metabolic engineering utilizes a set of theoretical tools to help rationally design modifications that enhance the production of a desired metabolite. Such approaches have proven their value in bacterial systems, however, the paucity of success stories to date in plants, suggests that challenges remain. Here, we review the most commonly used methods for understanding metabolic flux, focusing on the sulfur assimilatory pathway. We highlight known issues with both experimental and theoretical approaches, as well as presenting recent methods for integrating different modeling strategies, and progress toward an understanding of flux at the whole plant level.

Keywords: sulfur, flux, genome scale, flux balance analysis, kinetic modeling, metabolic control analysis

INTRODUCTION

Sulfur is an essential nutrient; available in the soil as sulfate, plants are able to reduce inorganic sulfur, for use in a large number of primary and secondary metabolites.

Unsurprisingly, the study of reductive sulfur assimilation by plants is often pragmatically motivated; *Brassicaceae* especially have large sulfur requirements, and the quality and yield of oilseed rape is known to be affected by low sulfur availability (De Pascale et al., 2008). Furthermore, the importance of sulfur metabolites in diet and health (Sekiz et al., 1975; Tawfiq et al., 1995; Tripathi and Mishra, 2007; Traka and Mithen, 2011), their intrinsic economic value (Li et al., 2004), and conferred tolerance to abiotic or biotic stresses in the plant (Gatehouse, 2002; Bednarek et al., 2009; Yadav, 2010) has led to an interest in manipulating their production.

A tenet of metabolic engineering is that with sufficient understanding of the components, reactions, and fluxes through a pathway we can rationally design modifications that improve, for instance, the production of a desired metabolite. This interest in, and comparative lack of examples of successful sulfur pathway engineering in plants suggests that approaches thus far have failed to significantly advance our understanding of sulfur assimilation at some level.

For plants, all is flux; all biological responses are ultimately to direct the movement of molecules and energy through the metabolic network in the most appropriate way, often acting to buffer changes in metabolite levels (Mugford et al., 2011). Consequently, understanding the control of flux is a pre-requisite for successful metabolic engineering. Unfortunately, this dynamic property is comparatively difficult to measure and interpret, and thus requires the integrated involvement of theoretical biology. Mathematical modeling has developed a number of approaches to understand control of flux through metabolism, ranging from theoretical frameworks to integrate experimental results, to highly detailed kinetic models of small fragments of a pathway, to constraint-based methods which can encompass the entire reactome.

Here, we review the most commonly used methods for studying flux, focusing on the sulfur assimilatory pathway, not just because of the commercial and scientific importance of sulfur, but because it illustrates well the more general challenges and weaknesses of each approach. Starting with the difficulties of experimental attempts to partition control of flux among the enzymes of the pathway, we then consider small scale kinetic models of several pathway branches, and flux balance analysis (FBA), as well as recent approaches to integrate different modeling strategies, and progress toward an understanding of flux at the whole plant level.

SULFUR ASSIMILATORY PATHWAY

The sulfur assimilatory pathway has been recently reviewed (Takahashi et al., 2011). In summary; sulfate is taken up from the environment, facilitated by specialized transporters. A large fraction of the sulfate is stored in the vacuole, while sulfate in chloroplasts or the cytosol is activated by ATP sulfurylase, forming adenosine 5′-phosphosulfate (APS). APS may then either be further phosphorylated by APS Kinase (APK) or reduced by APS Reductase (APR). Phoshorylation of APS forms 3′-phosphoadenosine 5′-phosphosulphate (PAPS), which acts as a promiscuous donor of activated sulfate, and is involved in the modification of a variety of proteins, saccharides, and secondary metabolites, including desulfo-glucosinolates. In primary assimilation, APS is instead reduced in the plastid to sulfite by APR, and then to sulfide by sulfite reductase (SiR). Sulfide in chloroplasts, mitochondria and the cytosol, is incorporated into *O*-acetylserine (OAS) to form cysteine, the precursor of all organic compounds containing reduced sulfur. Cysteine in the plastid may be converted into glutathione (GSH) via γ-glutamyl-cysteine, or reacts with phosphohomoserine, to form cystathionine, which can then be converted to methionine via homocysteine. Excess sulfite can be oxidized in the peroxisome back to sulfate by sulfite oxidase (**Figure 1**).

SULFUR FLUX CONTROL – MEASURE BY MEASURE

Efforts to experimentally characterize flux through the assimilation pathway are based on accumulation of radiolabelled ^{35}S, from $^{35}SO_4$, into various metabolite pools (Koprivova et al., 2000; Vauclare et al., 2002; Mugford et al., 2011). Assuming that over the timescale considered there is no significant turnover of the most downstream metabolites measured, this allows calculation of sulfur flux from SO_4 through the pathway. By measuring alterations to flux distribution under genetically (Khan et al., 2010; Mugford et al., 2011), and environmentally (Koprivova et al., 2000; Vauclare et al., 2002; Scheerer et al., 2010) perturbed conditions it was hoped that insights could be gained into the control of flux through the sulfur assimilation pathway, and into various organic molecules.

CONTROL IS DISTRIBUTED

To quantify control of flux, Vauclare et al. (2002) applied the metabolic control analysis (MCA) framework (for a comprehensive introduction to MCA, see Fell, 1992). Based on flux correlation with decreased APR activity, they calculated that APR has a large proportion of the total control of flux through the assimilatory pathway. From this, and several qualitative studies (Tsakraklides et al., 2002; Loudet et al., 2007), the hypothesis arose that APR is the key enzyme, controlling flux through the reductive assimilation pathway (Vauclare et al., 2002; Yoshimoto et al., 2007; Davidian and Kopriva, 2010; Scheerer et al., 2010). Consistent with this idea, APR has been shown to be highly regulated by demand for reduced sulfur products (Lappartient et al., 1999; Kopriva, 2006; Davidian and Kopriva, 2010; Takahashi et al., 2011), internal sulfate levels (Lee et al., 2012), and other environmental signals (Jost et al., 2005; Koprivova et al., 2008; Lee et al., 2011; Huseby et al., 2013).

More recently, however, a number of different enzymes have also been implicated in altered flux through the sulfur reduction pathway; Khan et al. (2010) found that SiR knockdown plants have a strongly reduced flux to thiols, variation in ATPS has been shown to cause altered flux of sulfur into primary metabolism (Koprivova et al., 2013), and reduction in APK increased flux through primary

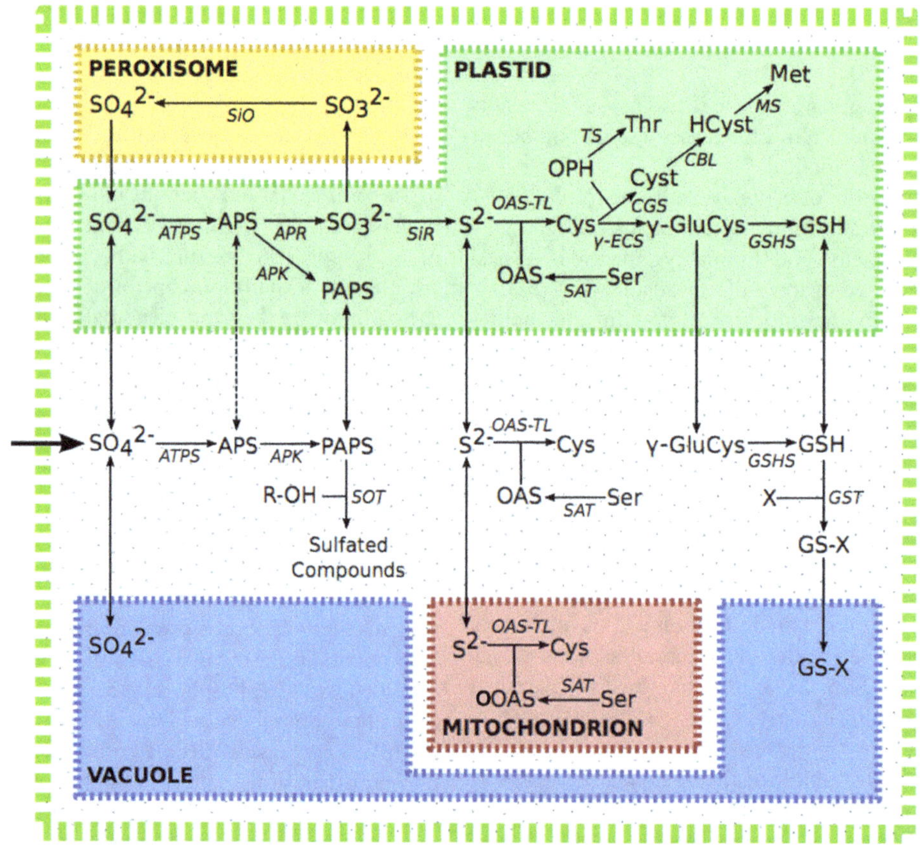

FIGURE 1 | The sulfur assimilation pathway. Dashed line indicates putative transport of APS. Metabolite abbreviations; APS, adenosine 5′-phosphosulfate; Cys, cysteine; Cyst, cystathionine; γ-GluCys, γ-glutamyl-cysteine; GSH, glutathione; GS-X, glutathione conjugate; Hcy, homocysteine; Met, methionine; OAS, *O*-acetylserine; OPH, *O*-phosphohomoserine; PAPS, 3′-phosphoadenosine 5′-phosphosulfate; R-OH, hydroxylated precursor; Ser, serine; Thr, threonine. Enzyme abbreviations; APK, APS kinase; APR, APS reductase, ATPS, ATP sulfurylase; CBL, cystathionine β-lyase; CGS, cystathionine γ-synthase; γ-ECS, γ-glutamyl-cysteine synthetase; GSHS, glutathione synthetase; GST, glutathione-*S*-transferase; MS, methionine synthase; OAS-TL, OAS(thiol)lyase; SAT, serine acetyltransferase; SiO, sulphite oxidase; SiR, sulphite reductase; TS, threonine synthase.

assimilation (Mugford et al., 2011). These results suggest that flux control is more complicated than had been previously thought, and extends beyond the APR enzyme. This distribution of control among multiple enzymes is a common feature of metabolic pathways (Thomas and Fell, 1998).

DIFFICULTY OF APPLYING MCA FRAMEWORK TO EXPERIMENTS

Distributed control of flux means that a quantitative understanding, as attempted by Vauclare et al. (2002), becomes increasingly important for successfully engineering overproduction of metabolites. However, the results of this kind of perturbation experiment, in which the activity of an enzyme is artificially increased or decreased, are difficult to interpret within the MCA framework due to regulatory interactions, and non-linear changes in control coefficients with enzyme activity.

Metabolic control analysis defines flux control coefficients (FCCs) as the sensitivity of flux through the pathway to an infinitesimal change in a given enzyme activity from the reference state (**Figure 2**). These coefficients can be interpreted as a measure of the 'rate limitingness' of the enzyme to flux though the pathway, and potentially used to identify targets for overexpression to increase flux to metabolites of interest.

To determine FCCs experimentally, by measuring the effect of a genetic perturbation on flux through the pathway requires the assumption that no other enzyme activities change in compensation (Fell, 1992). However, demonstrating that this is the case is challenging (Vauclare et al., 2002; Scheerer et al., 2010), and given the complex regulation of the pathway by a number of metabolites (Takahashi et al., 2011) is generally unlikely to be true. Although for some simple purposes this may not matter, it does hinder an understanding the root causes of changes of flux through the network, limiting the applicability of any findings.

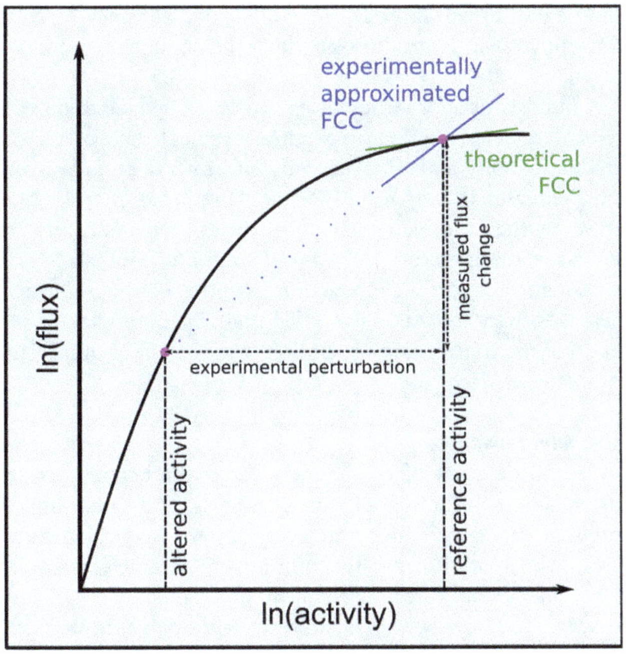

FIGURE 2 | Flux control coefficient varies non-linearly with enzyme activity, confounding experimental approximations. Flux control coefficients (FCCs) can be computed from the gradient of the ln(flux) versus ln(activity) curve for a given enzyme. Most experimental attempts to approximate the FCC at the reference state measure the change in flux through the pathway with change in an enzyme activity. Larger experimental perturbations result in less accurate estimates, and knock down experiments can be expected to overestimate FCC at the reference state. Ideally multiple perturbations should be made, and the curve fitted to the data, allowing a point estimate of FCC at the reference state.

Flux control coefficients vary non-linearly with enzyme activity. Therefore although a large experimental change in enzyme activity may result in significantly altered flux through the pathway, this does not mean that the enzyme actually has a high control coefficient in the unperturbed state (**Figure 2**). Ideally several magnitudes of perturbation would be made, and used to estimate the control coefficient in unperturbed conditions, however, this has not been done to date within the sulfur community (Vauclare et al., 2002; Khan et al., 2010). Furthermore, this means that experimentally approximated control coefficients cannot be directly compared to each other, as genetic perturbations vary in magnitude (Vauclare et al., 2002; Khan et al., 2010). It is therefore still not quantitatively clear which reactions have how much control of sulfur assimilation, even under controlled experimental conditions.

This non-linearity also means that MCA is not a robust predictive framework for engineering; control coefficients at the reference state are not necessarily likely to reflect control coefficients under genetically altered conditions. Interestingly, Scheerer et al. (2010) found that distribution of FCCs through the sulfur assimilation pathway varied with environmental conditions and organism, likely due to altered enzyme expression levels. This highlights the importance of a predictive understanding of the controlling steps through the pathway, not only to more robustly predict the effect of genetic alterations, but also due to the impracticality of experimentally determining control distributions under all environments of interest.

SMALL IS BEAUTIFUL – KINETIC MODELING

Although experimental investigations into control of flux through the pathway have been useful in qualitatively identifying important enzymes, this approach is limited, as multiple controlling enzymes and non-linear dynamics make predicting behavior away from measured conditions difficult. This was clearly seen in the analysis of poplar roots, where despite an increase in APR activity in many conditions, only few resulted in higher flux (Scheerer et al., 2010). Furthermore, the data generated experimentally is not easily integrated into a formal framework for analysis. In contrast, the MCA framework is easily applied to kinetic models of the pathway, which can be used not only to calculate control coefficients at the reference state more accurately than is possible experimentally, but also to simulate altered conditions. The difficulty lies, however, in producing an appropriate model.

Kinetic modeling of metabolic pathways is well established (see Curien et al., 2014 for a practical introduction). Models comprise a coupled system of ordinary non-linear, differential equations, functions of metabolite concentrations and kinetic parameters, which specify the rate of a reaction. A given pathway system can be mapped onto these equations, and solved numerically using a range of freely available software (Copeland et al., 2012).

In sulfur metabolism, this has allowed not only dissection of flux control distribution at several points (Curien et al., 2003; Mendoza-Cózatl and Moreno-Sánchez, 2006), but also predictions about how environmental perturbation changes control of flux to GSH (Mendoza-Cózatl and Moreno-Sánchez, 2006), and suggested engineering interventions to modify levels of sulfur metabolites. For example Curien et al. (2003) were able to predict that overexpressing cystathionine-γ-synthase would allow overproduction of methionine, without compromising threonine production, and that this was therefore a better strategy than knocking down threonine synthase (TS).

However, although kinetic models have yielded useful insights into flux through the sulfur assimilation pathway, the rarity of models published in this area hints at the difficulties of the approach.

PROBLEMS WITH KINETIC MODELS

Kinetic models require detailed understanding of the biological pathway under study, at the structural, thermodynamic, and kinetic levels. In many instances the pathway structure is well known, and thermodynamic data are either available (Goldberg et al., 2004), or can be calculated approximately (Jankowski et al., 2008), however, incomplete knowledge of enzyme kinetic parameters remains as the biggest hurdle to model building, particularly given that isoenzymes in different tissues or compartments often display different kinetics. Strategies for determining parameter values can be broadly split into measurement, and estimation approaches.

Parameter measurement

For small models, it may be possible to measure all kinetic parameters required. Curien et al. (2003) were able to measure the kinetic parameters of TS and cystathionine-γ-synthase for their model of the branch point of methionine and threonine biosynthesis *in vitro*, however, the large experimental effort required (Stitt and Gibon, 2014; Tummler et al., 2014; van Eunen and Bakker, 2014) makes this a comparatively rare example; it is more common to search the literature to recover the majority of parameters required (Rohwer, 2014). Several databases (Schomburg et al., 2004; Wittig et al., 2012) facilitate the search for previously determined kinetic parameters, however, generally poor coverage, particularly for allosteric regulation, means that it is accepted practice to use whichever parameters are available, either from experiments under differing conditions, or from orthologous proteins (Rohwer, 2014). The validity of transferring parameters in this way is generally unclear (Stitt and Gibon, 2014), the exception being enzyme activity parameters, which are acknowledged to vary so greatly with environment, that they should be measured under the condition of interest (Curien et al., 2014). There has been some speculation that advances in robotics, and microfluidics could lead to 'omics style investigations' into enzyme kinetics (Gibon et al., 2004; Sjostrom et al., 2013), however, a reliable high throughput pipeline has not to our knowledge been developed, and poor coverage is likely to remain a problem in the immediate future.

Kinetics databases highlight a further shortcoming of kinetic parameters measured *in vitro*; the assay conditions used are typically far from the *in vivo* environment seen by the enzyme. This problem of non-physiological *in vitro* assay media can be seen in Curien et al. (2003), where the use of high phosphohomoserine media, likely contributed to a poor initial model fit to data. Initiatives to design more *in vivo* like *in vitro* media, are underway for several microorganisms (Garcia-Contreras et al., 2012; Goel et al., 2012; Leroux et al., 2013), but to the best of our knowledge, no such effort has been reported in plants, where the problem is exacerbated by the presence of multiple subcellular compartments, each with a unique environment. In the sulfur assimilation pathway, three out of the five reactions converting sulfate to cysteine occur in multiple compartments (Takahashi et al., 2011) and thus likely require multiple sets of kinetic parameters.

Parameter estimation

The difficulties of obtaining experimentally measured kinetic parameters mean that in the vast majority of published models, at least some parameters are fitted by minimizing the difference between model predictions (e.g. of flux through the path), and experimental measurements (Tummler et al., 2014). Aside from the experimental difficulties of acquiring data, especially within subcellular compartments, one problem with this approach is overfitting; assigning parameter values to fit the data more precisely than is justified. As a result, many models parameterized using a top down approach lose predictive accuracy as conditions move away from those at which the parameters were fitted (Hawkins, 2004).

A number of approaches have been developed based on sensitivity of model predictions to parameter values to analyze parameter identifiability, and calculate confidence intervals for parameters and predictions (Cotten and Reed, 2013; Kravaris et al., 2013). However, these problems have not always been rigorously considered in the literature. Mendoza-Cózatl and Moreno-Sánchez (2006) ignored possible interaction terms between parameters in their sensitivity analysis, and Curien et al. (2003) provide no indication of the robustness of their predictions to error in measured parameter values.

Parameter reduction

Given the problem of estimating a large number of unknown parameters with limited data, most models tend to use lumped, empirical rate laws, which aim to capture the salient kinetic features, whilst minimizing the number of parameters required, rather than complicated, mechanistic laws (Heijnen, 2005; Curien et al., 2014; Rohwer, 2014). However, although this simplification can be useful (Costa et al., 2011), it inevitably leads to a loss in model fidelity. Curien et al. (2003) found that even replacing a ping-pong rate law with relatively complex Michaelis–Menten kinetics led to their model losing the experimentally identified insensitivity to cysteine concentration of cystathionine γ-synthase.

SMALL MAY NOT BE SUFFICIENT

A minimal model of a subsystem should include everything that affects the internal variables of the model (Curien et al., 2014), however, in practice, lack of biological knowledge can make it difficult to know what has to be included. Within the sulfur assimilation pathway, it is still often unclear which metabolites regulate enzyme activity allosterically, although it seems that many potentially can (Vauclare et al., 2002; Hopkins et al., 2005; Rouached et al., 2009; Hubberten et al., 2012; Lee et al., 2012). This means that, for example, even a small model of the APR, APK branch point must be large enough to consider the reduction pathway at least as far as GSH production, as this feeds back to regulate APR (Vauclare et al., 2002; Hacham et al., 2014) and possibly APK through changes in redox environment (Ravilious et al., 2012). This results in the requirement for a large number of kinetic parameters.

The extent to which models can be simplified and still remain useful is unclear, as an overly reduced model system can result in inaccurate predictions. For example ignoring phosphohomoserine production meant that Curien et al. (2003) were only able to identify a subset of the intervention steps that have since been experimentally shown to increase methionine production Lee et al. (2005) and Mendoza-Cózatl and Moreno-Sánchez (2006) demonstrated the importance of considering demand for GSH, as well as its production by finding that including demand results in large changes in the control coefficients of synthesizing reactions.

It is possible that a much larger metabolic network has to be considered when modeling sulfur assimilation than just the pathway itself, for example, GSH production at night is limited by availability of glycine, as its major source is photorespiration (Noctor et al., 1999). Integration of sulfur assimilation within the wider metabolic network is demonstrated by the tight coordination of sulfur uptake with nitrogen and carbon availability (Koprivova et al., 2000; Kopriva and Rennenberg, 2004; Nero et al., 2009), and the broad range of conditions which have been shown to alter enzyme activities in the pathway (Kopriva et al., 1999; Koprivova et al., 2008; Huseby et al., 2013). For instance cysteine links both nitrogen and carbon metabolism to sulfur assimilation via OAS. OAS availability is likely a dominant factor in regulating the production of cysteine by controlling formation of the cysteine synthase complex (Birke et al., 2012), and so its availability has to be considered in models of cysteine synthesis. Furthermore, as at least under some conditions, cysteine availability limits production of downstream metabolites such as methionine and GSH (Noctor et al., 1996), and these downstream metabolites can regulate upstream components (Vauclare et al., 2002; Hacham et al., 2014), therefore this link to wider metabolism should be acknowledged whichever part of the sulfur pathway is being studied.

THE DIFFICULTY WITH LARGER MODELS

Unfortunately, as model size increases, the problems of unknown parameters, and rate laws become extremely difficult to overcome. To generate a large kinetic model, simplifying assumptions about parameter values (Smallbone and Mendes, 2013), and rate laws (Alves et al., 2008), are frequently made, but this often results in poor model quality away from the fitted conditions (Chakrabarti et al., 2013) and so is of limited predictive value.

Other kinetic modeling frameworks acknowledge the inherently greater unknowns of a large system, and use the available data to define a cadre of related models, or sample feasible parameter space, reflecting either structural, or parameter and rate law uncertainty in their predictions (Famili et al., 2005; Steuer et al., 2006; Tran et al., 2008; Miskovic and Hatzimanikatis, 2011). Some of these approaches have resulted in the production of large kinetic models, in the order of 200 metabolites and reactions (Khodayari et al., 2014), but do not scale well to bigger models. As model size increases, parameter space expands enormously (Zamora-Sillero et al., 2011), resulting in prohibitive computational requirements (Link et al., 2014). As such, kinetic modeling currently does not scale to the size that is likely to be required to gain a holistic understanding of flux through sulfur related pathways.

BIGGER IS BETTER – GENOME SCALE MODELS

In contrast, constraint-based modeling provides a number of hugely scalable, largely parameter free methods for understanding flux through large metabolic networks (Lewis et al., 2012; Bordbar et al., 2014a). Here we focus on FBA as the most commonly used constraint-based method, the only method currently applicable to the genome scale, and a foundation for of many closely related variants.

Flux balance analysis (Varma and Palsson, 1994) is a powerful technique to estimate internal flux distributions within a large-scale network using only the structure of the reaction network, an objective function, and a small number of measured nutrient uptake fluxes as constraints, (for a practical introduction, to the method, see Grafahrend-Belau et al., 2014). By assuming metabolic steady state, and that fluxes are distributed so as to maximize some cellular objective, feasible flux space is reduced, and a subset of biologically likely internal flux distributions are predicted.

Flux balance analysis can be directly used in a number of areas, including understanding metabolic efficiency (Chen and Shachar-Hill, 2012), interpreting 'omics data' (Toepfer et al., 2013; Simons et al., 2014b), and predicting novel metabolic pathways (Hay and Schwender, 2011; Bordbar et al., 2014b). Furthermore extensions to the method interpret structural properties related to control of flux (Notebaart et al., 2008; Sajitz-Hermstein and Nikoloski, 2013), predict how flux distribution changes in response to genetic and environmental changes (Segre et al., 2002; Cheung et al., 2014), and suggest optimal intervention strategies to engineer metabolite production (Zomorrodi et al., 2012; Tomar and De, 2013; Ohno et al., 2014), as well as having a number of other applications (Papp et al., 2011; Bordbar et al., 2014a). This array of methods has been recently reviewed (Lewis et al., 2012), but continues to rapidly expand.

PROBLEMS WITH GENOME SCALE MODELS

In spite of a profligacy of analytical methods and well-documented metabolic engineering case studies in microbes, application of FBA based methods to plants has been limited to date. This is likely due to challenges in genome scale model construction, and the assumptions of the FBA method, as will be discussed below.

Model construction

Despite a large number of available tools (Kim et al., 2012), construction of a genome scale model is not a facile task, and particularly in plants remains a laborious undertaking (Saha et al., 2014; Simons et al., 2014a). Here we highlight some of the difficulties biological unknowns cause in creating even single cell type models of plants.

Although primary metabolism is well understood, the generally poorer understanding of the huge plant secondary metabolism (Shachar-Hill, 2013), is reflected in the focus of models published to date (Poolman et al., 2009; Williams et al., 2010; Cheung et al., 2013). The problem of unknown metabolites was recently highlighted for sulfur metabolites in particular (Glaeser et al., 2014), and the potentially large numbers of missing reactions suggested by the large proportion of genome content with unknown function (Seaver et al., 2012) could also adversely affect prediction quality.

All published genome scale models of plant metabolism include compartmentalization to some extent, but the problem of biological unknowns again raises concern over the quality of some of the assignations. Wide variation between models in which compartments reactions occur (Poolman et al., 2009; Masakapalli et al., 2010; Mintz-Oron et al., 2012) suggests that despite databases of subcellular enzyme location (Heazlewood et al., 2007; Sun et al., 2009), and parsimony based methods for extending database coverage (Mintz-Oron et al., 2012), the number of reactions which can be confidently assigned to particular compartments, and in particular to the vacuole, is probably much lower than occur in reality (Krueger et al., 2011). This reflects the current difficulty of experimentally determining subcellular reaction location.

Additionally, transport between compartments is often poorly understood; even in well-studied parts of metabolism, it is not always clear which metabolites can move between compartments, and the energetic costs of transport reactions are rarely known. This is shown in the sulfur assimilatory pathway by the only recent identification of PAPS transport between the plastid and cytosol (Gigolashvili et al., 2012).

FBA analysis

Objective functions. In addition to defining network structure, some biological knowledge of the system is required to choose an appropriate objective function. One commonly used objective is maximization of biomass production (Feist and Palsson, 2010), which is equivalent to finding the most efficient way of generating biomass from nutrients taken up by the cell (Zarecki et al., 2014). Although maximization of biomass is generally accepted as a good objective function for bacteria in log phase, the accuracy of fluxes predicted using it vary with environment, growth phase, and species, suggesting that this is not always appropriate (Schuster et al., 2008; Feist and Palsson, 2010).

A number of other objective functions have been considered in the literature, most often tied either explicitly or implicitly to efficiency in some regard (Chen and Shachar-Hill, 2012), although recently other objectives have been proposed which either aim to maximize growth rate (Zarecki et al., 2014) or minimize conflict with 'omic data' (Becker and Palsson, 2008; Collins et al., 2012). Much literature assessing the performance of these various different objective functions in correctly predicting observed growth, gene essentiality or flux states in bacteria (Burgard and Maranas, 2003; Schuetz et al., 2007; Feist and Palsson, 2010), and plants (Cheung et al., 2013) has been produced, however, it is not clear that organisms act to optimize a single objective function, even under constant conditions (Nagrath et al., 2007; Schuetz et al., 2012; Harcombe et al., 2013).

Combined objectives give the most accurate predicted flux distributions, and both in bacteria (Schuetz et al., 2012), and Eukaryotes (Nagrath et al., 2007), cell fluxes apparently occupy a Pareto surface, at which several objectives trade-off against each other, as further increase in one objective leads to a decrease in another. It is not obvious how this problem of competing objectives can be addressed by FBA; defining an appropriate objective function becomes much more difficult, because although frameworks for optimizing multiple objectives are well established (Oh et al., 2009; Zomorrodi and Maranas, 2012; Zomorrodi et al., 2014), the lack of a predictive understanding of which trade-offs are likely to apply, limits their application, as it is likely to vary with species, environment, and developmental state.

Degeneracy. Another problem is degeneracy, FBA is often unable to distinguish between a number of flux distributions, which all maximize the objective function. Although this degeneracy of predicted distributions is often considered undesirable (Pozo et al., 2014), it is in fact likely to reflect biological reality. Degenerate optimal solutions are consistent with robustness, which seems to be a common feature of biological networks (Kitano, 2004), and a population of cells is unlikely to be adequately described by a single flux distribution (Labhsetwar et al., 2013).

The real difficulty associated with degenerate flux distributions is that experimentally measured fluxes in bacteria often actually exist in suboptimal regions, which allow large flux variation (Schuetz et al., 2012; Harcombe et al., 2013; Roman et al., 2014) without further compromising the best combination of assumed objectives. Although the extent to which apparent sub-optimal distributions arise through the averaging of measured fluxes in a heterogeneous population, rather than sub-optimality in a single cell is unclear, the FBA assumption that flux is distributed in order to maximize an objective function may only be a useful approximation in specific cases. FBA based methods are beginning to appear that address the need to consider only partially optimized distributions (Wintermute et al., 2013), but sub-optimal distributions are a major challenge for the FBA framework, which given the sophistication and size of plant metabolic networks, and numerous differentiated cell types, is likely to be particularly relevant to their study.

In spite of these difficulties, an increasing number of studies have accurately predicted flux distributions in plants cells using FBA (Dal'Molin et al., 2010a; Williams et al., 2010; Hay and Schwender, 2011; Saha et al., 2011; Cheung et al., 2013). Precise external flux measurements have been shown to be more important for accurately predicting internal fluxes than the objective function used (Cheung et al., 2013), and although current approaches to use transcriptomics data to improve flux prediction accuracy have been recently questioned (Machado and Herrgard, 2014), fluxomics data generated by metabolic flux analysis [recently reviewed from a sulfur perspective by Rennenberg and Herschbach (2014)] can be used to add additional constraints, and further improve prediction accuracy (Hay and Schwender, 2011).

It seems only a matter of time before FBA is used to facilitate engineering outcomes in plant cell cultures.

SULFUR AND BEYOND – TOWARD WHOLE PLANT FLUX MODELS

There is great interest in bridging the gap between the long tradition of eco-physiological agronomic models (Keurentjes et al., 2013), and molecular models. Combined, these two approaches could provide an integrated understanding of control of economically important traits (Baldazzi et al., 2012; Poorter et al., 2013). This fusion requires the ability to model differentiated tissue, at least at the organ level, and consider dynamic changes to flux.

Genome scale plant models to date have generally focused on cell cultures grown in suspension (Williams et al., 2010; Cheung et al., 2013) and so bypassed the problem of differentiation, but a particular challenge in whole plant models is the large number of cell types present. Although attempts to address this remain fairly crude, and restricted to models of only a few cell types or organs, the framework, in which proteomics data is integrated into constraint-based models to generate tissue, or organ specific sub-models (Mintz-Oron et al., 2012), which can then be coupled together, and used to predict fluxes through heterogeneous material (Dal'Molin et al., 2010b; Grafahrend-Belau et al., 2013) is established.

The greatest limitation of the FBA method is that it can only consider a steady state snapshot of flux distribution, especially as the extent to which metabolism at the whole plant level is ever in steady state remains unclear. Although recent work in plants has studied responses to light–dark cycles using a purely FBA approach (Cheung et al., 2014), and FBA metrics have been identified which correlate with metabolite concentration dynamics (Reznik et al., 2013), the application of a purely FBA based methodology to study dynamic systems is limited. In bacteria, there has been a movement toward integrating FBA models within a kinetic model. This is used to dynamically modify the exchange reaction constraints while FBA is repeatedly performed, allowing internal flux dynamics to be approximated (Varma and Palsson, 1994; Mahadevan et al., 2002). Output from the FBA simulation may then be fed back, to modify the kinetic model, or not (Feng et al., 2012). This hybrid modeling approach allows prediction of genome scale fluxes over time (Vargas et al., 2011; Jouhten et al., 2012), with only a few parameters required to capture input and output fluxes, biomass prerequisites, and maintenance costs.

This dynamic flux balance analysis (dFBA) approach makes a pseudo steady state assumption, that intracellular metabolism equilibrates several orders of magnitude faster than extracellular changes. Although it is likely that the pseudo steady state assumption might not be justified in all aspects of plant metabolism, a recent extension to the method potentially relaxes this requirement (Birch et al., 2014). Grafahrend-Belau et al. (2013) recently used dFBA to link a multi organ model of barley metabolism with an agronomic model, and provide insight into the dynamic interaction of source and sink organs in relation to senescence. Although fairly unsophisticated in the models used and interaction framework considered, the first steps toward a whole plant model have been taken, and dFBA is likely to find wider application in the coming years.

CONCLUDING REMARKS

Flux is perhaps the most important metric to determine for a practical, applied understanding of plant biology. Through the study of sulfur metabolism we have seen that to understand flux requires a fusion of experimental and modeling approaches, but that to date, no integration of the two satisfactorily solves the problem to provide an accurate predictive framework. Each approach considered continues to advance independently both theoretically and experimentally, but currently perhaps most promising is the joining of kinetic and constraint-based approaches, which although an immature field, has the potential to finally deliver a useful facet of the famous 'virtual plant.'

ACKNOWLEDGMENTS

This work was supported by the Biotechnology and Biological Sciences Research Council (BBSRC) grant BB/J004561/1 and the John Innes Foundation.

REFERENCES

Alves, R., Vilaprinyo, E., Hernandez-Bermejo, B., and Sorribas, A. (2008). Mathematical formalisms based on approximated kinetic representations for modeling genetic and metabolic pathways. *Biotechnol. Genet. Eng. Rev.* 25, 1–40. doi: 10.5661/bger-25-1

Baldazzi, V., Bertin, N., De Jong, H., and Genard, M. (2012). Towards multiscale plant models: integrating cellular networks. *Trends Plant Sci.* 17, 728–736. doi: 10.1016/j.tplants.2012.06.012

Becker, S. A., and Palsson, B. O. (2008). Context-specific metabolic networks are consistent with experiments. *PLoS Comput. Biol.* 4:e1000082. doi: 10.1371/journal.pcbi.1000082

Bednarek, P., Pislewska-Bednarek, M., Svatos, A., Schneider, B., Doubsky, J., Mansurova, M., et al. (2009). A glucosinolate metabolism pathway in living plant cells mediates broad-spectrum antifungal defense. *Science* 323, 101–106. doi: 10.1126/science.1163732

Birch, E. W., Udell, M., and Covert, M. W. (2014). Incorporation of flexible objectives and time-linked simulation with flux balance analysis. *J. Theor. Biol.* 345, 12–21. doi: 10.1016/j.jtbi.2013.12.009

Birke, H., Mueller, S. J., Rother, M., Zimmer, A. D., Hoernstein, S. N. W., Wesenberg, D., et al. (2012). The relevance of compartmentation for cysteine synthesis in phototrophic organisms. *Protoplasma* 249, 147–155. doi: 10.1007/s00709-012-0411-9

Bordbar, A., Monk, J. M., King, Z. A., and Palsson, B. O. (2014a). Constraint-based models predict metabolic and associated cellular functions. *Nat. Rev. Genet.* 15, 107–120. doi: 10.1038/nrg3643

Bordbar, A., Nagarajan, H., Lewis, N. E., Latif, H., Ebrahim, A., Federowicz, S., et al. (2014b). Minimal metabolic pathway structure is consistent with associated biomolecular interactions. *Mol. Syst. Biol.* 10:737. doi: 10.15252/msb.20145243

Burgard, A. P., and Maranas, C. D. (2003). Optimization-based framework for inferring and testing hypothesized metabolic objective functions. *Biotechnol. Bioeng.* 82, 670–677. doi: 10.1002/bit.10617

Chakrabarti, A., Miskovic, L., Soh, K. C., and Hatzimanikatis, V. (2013). Towards kinetic modeling of genome-scale metabolic networks without sacrificing stoichiometric, thermodynamic and physiological constraints. *Biotechnol. J.* 8, 1043–1057. doi: 10.1002/biot.201300091

Chen, X., and Shachar-Hill, Y. (2012). Insights into metabolic efficiency from flux analysis. *J. Exp. Bot.* 63, 2343–2351. doi: 10.1093/jxb/ers057

Cheung, C. Y. M., Poolman, M. G., Fell, D. A., Ratcliffe, R. G., and Sweetlove, L. J. (2014). A diel flux balance model captures interactions between light and dark metabolism during day-night cycles in C-3 and crassulacean acid metabolism leaves. *Plant Physiol.* 165, 917–929. doi: 10.1104/pp.113.234468

Cheung, C. Y. M., Williams, T. C. R., Poolman, M. G., Fell, D. A., Ratcliffe, R. G., and Sweetlove, L. J. (2013). A method for accounting for maintenance costs in flux balance analysis improves the prediction of plant cell metabolic phenotypes under stress conditions. *Plant J.* 75, 1050–1061. doi: 10.1111/tpj.12252

Collins, S. B., Reznik, E., and Segre, D. (2012). Temporal expression-based analysis of metabolism. *PLoS Comput. Biol.* 8:e1002781. doi: 10.1371/journal.pcbi.1002781

Copeland, W. B., Bartley, B. A., Chandran, D., Galdzicki, M., Kim, K. H., Sleight, S. C., et al. (2012). Computational tools for metabolic engineering. *Metab. Eng.* 14, 270–280. doi: 10.1016/j.ymben.2012.03.001

Costa, R. S., Machado, D., Rocha, I., and Ferreira, E. C. (2011). Critical perspective on the consequences of the limited availability of kinetic data in metabolic dynamic modelling. *IET Syst. Biol.* 5, 157–163. doi: 10.1049/iet-syb.2009.0058

Cotten, C., and Reed, J. L. (2013). Mechanistic analysis of multi-omics datasets to generate kinetic parameters for constraint-based metabolic models. *BMC Bioinformatics* 14:32. doi: 10.1186/1471-2105-14-32

Curien, G., Cardenas, M. L., and Cornish-Bowden, A. (2014). Analytical kinetic modeling: a practical procedure. *Methods Mol. Biol.* 1090, 261–280. doi: 10.1007/978-1-62703-688-7_16

Curien, G., Ravanel, S., and Dumas, R. (2003). A kinetic model of the branch-point between the methionine and threonine biosynthesis pathways in *Arabidopsis thaliana*. *Eur. J. Biochem.* 270, 4615–4627. doi: 10.1046/j.1432-1033.2003.03851.x

Dal'Molin, C. G. D. O., Quek, L.-E., Palfreyman, R. W., Brumbley, S. M., and Nielsen, L. K. (2010a). AraGEM, a genome-scale reconstruction of the primary metabolic network in *Arabidopsis*. *Plant Physiol.* 152, 579–589. doi: 10.1104/pp.109.148817

Dal'Molin, C. G. D. O., Quek, L.-E., Palfreyman, R. W., Brumbley, S. M., and Nielsen, L. K. (2010b). C4GEM, a genome-scale metabolic model to study C-4 plant metabolism. *Plant Physiol.* 154, 1871–1885. doi: 10.1104/pp.110.166488

Davidian, J. C., and Kopriva, S. (2010). Regulation of sulfate uptake and assimilation-the same or not the same? *Mol. Plant* 3, 314–325. doi: 10.1093/mp/ssq001

De Pascale, S., Maggio, A., Orsini, F., Bottino, A., and Barbieri, G. (2008). Sulphur fertilisation affects yield and quality in friarielli (*Brassica rapa* L. subsp. sylvestris L. Janch. var. esculenta Hort.) grown in a floating system. *J. Hortic. Sci. Biotechnol.* 83, 743–748.

Famili, I., Mahadevan, R., and Palsson, B. O. (2005). k-cone analysis: determining all candidate values for kinetic parameters on a network scale. *Biophys. J.* 88, 1616–1625. doi: 10.1529/biophysj.104.050385

Feist, A. M., and Palsson, B. O. (2010). The biomass objective function. *Curr. Opin. Microbiol.* 13, 344–349. doi: 10.1016/j.mib.2010.03.003

Fell, D. A. (1992). Metabolic control analysis – a survey of its theoretical and experimental development. *Biochem. J.* 286, 313–330.

Feng, X., Xu, Y., Chen, Y., and Tang, Y. J. (2012). Integrating flux balance analysis into kinetic models to decipher the dynamic metabolism of *Shewanella oneidensis* MR-1. *PLoS Comput. Biol.* 8:e1002376. doi: 10.1371/journal.pcbi.1002376

Garcia-Contreras, R., Vos, P., Westerhoff, H. V., and Boogerd, F. C. (2012). Why in vivo may not equal in vitro – new effectors revealed by measurement of enzymatic activities under the same in vivo-like assay conditions. *FEBS J.* 279, 4145–4159. doi: 10.1111/febs.12007

Gatehouse, J. A. (2002). Plant resistance towards insect herbivores: a dynamic interaction. *New phytol.* 156, 145–169. doi: 10.1046/j.1469-8137.2002.00519.x

Gibon, Y., Blaesing, O. E., Hannemann, J., Carillo, P., Hohne, M., Hendriks, J. H. M., et al. (2004). A robot-based platform to measure multiple enzyme activities in *Arabidopsis* using a set of cycling assays: comparison of changes of enzyme activities and transcript levels during diurnal cycles and in prolonged darkness. *Plant Cell* 16, 3304–3325. doi: 10.1105/tpc.104.025973

Gigolashvili, T., Geier, M., Ashykhmina, N., Frerigmann, H., Wulfert, S., Krueger, S., et al. (2012). The *Arabidopsis* thylakoid ADP/ATP carrier TAAC has an additional role in supplying plastidic phosphoadenosine 5′-phosphosulfate to the cytosol. *Plant Cell* 24, 4187–4204. doi: 10.1105/tpc.112.101964

Glaeser, K., Kanawati, B., Kubo, T., Schmitt-Kopplin, P., and Grill, E. (2014). Exploring the *Arabidopsis* sulfur metabolome. *Plant J.* 77, 31–45. doi: 10.1111/tpj.12359

Goel, A., Santos, F., Vos, W. M., Teusink, B., and Molenaar, D. (2012). Standardized assay medium to measure *Lactococcus lactis* enzyme activities while mimicking intracellular conditions. *Appl. Environ. Microbiol.* 78, 134–143. doi: 10.1128/aem.05276-11

Goldberg, R. N., Tewari, Y. B., and Bhat, T. N. (2004). Thermodynamics of enzyme-catalyzed reactions – a database for quantitative biochemistry. *Bioinformatics* 20, 2874–2877. doi: 10.1093/bioinformatics/bth314

Grafahrend-Belau, E., Junker, A., Eschenroeder, A., Mueller, J., Schreiber, F., and Junker, B. H. (2013). Multiscale metabolic modeling: dynamic flux balance analysis on a whole-plant scale. *Plant Physiol.* 163, 637–647. doi: 10.1104/pp.113.224006

Grafahrend-Belau, E., Junker, A., Schreiber, F., and Junker, B. H. (2014). Flux balance analysis as an alternative method to estimate fluxes without labeling. *Methods Mol. Biol.* 1090, 281–299. doi: 10.1007/978-1-62703-688-7_17

Hacham, Y., Koussevitzky, S., Kirma, M., and Amir, R. (2014). Glutathione application affects the transcript profile of genes in *Arabidopsis* seedling. *J. Plant Physiol.* 171, 1444–1451. doi: 10.1016/j.jplph.2014.06.016

Harcombe, W. R., Delaney, N. F., Leiby, N., Klitgord, N., and Marx, C. J. (2013). The ability of flux balance analysis to predict evolution of central metabolism scales with the initial distance to the optimum. *PLoS Comput. Biol.* 9:e1003091. doi: 10.1371/journal.pcbi.1003091

Hawkins, D. M. (2004). The problem of overfitting. *J. Chem. Inf. Comput. Sci.* 44, 1–12. doi: 10.1021/ci0342472

Hay, J., and Schwender, J. (2011). Computational analysis of storage synthesis in developing *Brassica napus* L. (oilseed rape) embryos: flux variability analysis in relation to C-13 metabolic flux analysis. *Plant J.* 67, 513–525. doi: 10.1111/j.1365-313X.2011.04611.x

Heazlewood, J. L., Verboom, R. E., Tonti-Filippini, J., Small, I., and Millar, A. H. (2007). SUBA: the *Arabidopsis* subcellular database. *Nucleic Acids Res.* 35, D213–D218. doi: 10.1093/nar/gkl863

Heijnen, J. J. (2005). Approximative kinetic formats used in metabolic network modeling. *Biotechnol. Bioeng.* 91, 534–545. doi: 10.1002/bit.20558

Hopkins, L., Parmar, S., Baszczyk, A., Hesse, H., Hoefgen, R., and Hawkesford, M. J. (2005). *O*-acetylserine and the regulation of expression of genes encoding components for sulfate uptake and assimilation in potato. *Plant Physiol.* 138, 433–440. doi: 10.1104/pp.104.057521

Hubberten, H.-M., Klie, S., Caldana, C., Degenkolbe, T., Willmitzer, L., and Hoefgen, R. (2012). Additional role of O-acetylserine as a sulfur status-independent regulator during plant growth. *Plant J.* 70, 666–677. doi: 10.1111/j.1365-313X.2012.04905.x

Huseby, S., Koprivova, A., Lee, B.-R., Saha, S., Mithen, R., Wold, A.-B., et al. (2013). Diurnal and light regulation of sulphur assimilation and glucosinolate biosynthesis in *Arabidopsis*. *J. Exp. Bot.* 64, 1039–1048. doi: 10.1093/jxb/ers378

Jankowski, M. D., Henry, C. S., Broadbelt, L. J., and Hatzimanikatis, V. (2008). Group contribution method for thermodynamic analysis of complex metabolic networks. *Biophys. J.* 95, 1487–1499. doi: 10.1529/biophysj.107.124784

Jost, R., Altschmied, L., Bloem, E., Bogs, J., Gershenzon, J., Hahnel, U., et al. (2005). Expression profiling of metabolic genes in response to methyl jasmonate reveals regulation of genes of primary and secondary sulfur-related pathways in *Arabidopsis thaliana*. *Photosynth. Res.* 86, 491–508. doi: 10.1007/s11120-005-7386-8

Jouhten, P., Wiebe, M., and Penttila, M. (2012). Dynamic flux balance analysis of the metabolism of *Saccharomyces cerevisiae* during the shift from fully respirative or respirofermentative metabolic states to anaerobiosis. *FEBS J.* 279, 3338–3354. doi: 10.1111/j.1742-4658.2012.08649.x

Keurentjes, J. J. B., Molenaar, J., and Zwaan, B. J. (2013). Predictive modelling of complex agronomic and biological systems. *Plant Cell Environ.* 36, 1700–1710. doi: 10.1111/pce.12156

Khan, M. S., Haas, F. H., Samami, A. A., Gholami, A. M., Bauer, A., Fellenberg, K., et al. (2010). Sulfite reductase defines a newly discovered bottleneck for assimilatory sulfate reduction and is essential for growth and development in *Arabidopsis thaliana*. *Plant Cell* 22, 1216–1231. doi: 10.1105/tpc.110.074088

Khodayari, A., Zomorrodi, A. R., Liao, J. C., and Maranas, C. D. (2014). A kinetic model of *Escherichia coli* core metabolism satisfying multiple sets of mutant flux data. *Metab. Eng.* 25, 50–62. doi: 10.1016/j.ymben.2014.05.014

Kim, T. Y., Sohn, S. B., Kim, Y. B., Kim, W. J., and Lee, S. Y. (2012). Recent advances in reconstruction and applications of genome-scale metabolic models. *Curr. Opin. Biotechnol.* 23, 617–623. doi: 10.1016/j.copbio.2011.10.007

Kitano, H. (2004). Biological robustness. *Nat. Rev. Genet.* 5, 826–837. doi: 10.1038/nrg1471

Kopriva, S. (2006). Regulation of sulfate assimilation in *Arabidopsis* and beyond. *Ann. Bot.* 97, 479–495. doi: 10.1093/aob/mcl006

Kopriva, S., Muheim, R., Koprivova, A., Trachsel, N., Catalano, C., Suter, M., et al. (1999). Light regulation of assimilatory sulphate reduction in *Arabidopsis thaliana*. *Plant J.* 20, 37–44. doi: 10.1046/j.1365-313X.1999.00573.x

Kopriva, S., and Rennenberg, H. (2004). Control of sulphate assimilation and glutathione synthesis: interaction with N and C metabolism. *J. Exp. Bot.* 55, 1831–1842. doi: 10.1093/jxb/erh203

Koprivova, A., Giovannetti, M., Baraniecka, P., Lee, B.-R., Grondin, C., Loudet, O., et al. (2013). Natural variation in the ATPS1 Isoform of ATP sulfurylase contributes to the control of sulfate levels in *Arabidopsis*. *Plant Physiol.* 163, 1133–1141. doi: 10.1104/pp.113.225748

Koprivova, A., North, K. A., and Kopriva, S. (2008). Complex signaling network in regulation of adenosine 5′-phosphosulfate reductase by salt stress in *Arabidopsis* roots. *Plant Physiol.* 146, 1408–1420. doi: 10.1104/pp.107.113175

Koprivova, A., Suter, M., Op Den Camp, R., Brunold, C., and Kopriva, S. (2000). Regulation of sulfate assimilation by nitrogen in *Arabidopsis. Plant Physiol.* 122, 737–746. doi: 10.1104/pp.122.3.737.

Kravaris, C., Hahn, J., and Chu, Y. (2013). Advances and selected recent developments in state and parameter estimation. *Comput. Chem. Eng.* 51, 111–123. doi: 10.1016/j.compchemeng.2012.06.001

Krueger, S., Giavalisco, P., Krall, L., Steinhauser, M.-C., Buessis, D., Usadel, B., et al. (2011). A topological map of the compartmentalized *Arabidopsis thaliana* leaf metabolome. *PLoS ONE* 6:e17806, doi: 10.1371/journal.pone.0017806

Labhsetwar, P., Cole, J. A., Roberts, E., Price, N. D., and Luthey-Schulten, Z. A. (2013). Heterogeneity in protein expression induces metabolic variability in a modeled *Escherichia coli* population. *Proc. Natl. Acad. Sci. U.S.A.* 110, 14006–14011. doi: 10.1073/pnas.1222569110

Lappartient, A. G., Vidmar, J. J., Leustek, T., Glass, A. D. M., and Touraine, B. (1999). Inter-organ signaling in plants: regulation of ATP sulfurylase and sulfate transporter genes expression in roots mediated by phloem-translocated compound. *Plant J.* 18, 89–95. doi: 10.1046/j.1365-313X.1999.00416.x

Lee, B.-R., Huseby, S., Koprivova, A., Chetelat, A., Wirtz, M., Mugford, S. T., et al. (2012). Effects of fou8/fry1 mutation on sulfur metabolism: is decreased internal sulfate the trigger of sulfate starvation response? *PLoS ONE* 7:e39425. doi: 10.1371/journal.pone.0039425

Lee, B.-R., Koprivova, A., and Kopriva, S. (2011). The key enzyme of sulfate assimilation, adenosine 5′-phosphosulfate reductase, is regulated by HY5 in *Arabidopsis. Plant J.* 67, 1042–1054. doi: 10.1111/j.1365-313X.2011.04656.x

Lee, M., Martin, M. N., Hudson, A. O., Lee, J., Muhitch, M. J., and Leustek, T. (2005). Methionine and threonine synthesis are limited by homoserine availability and not the activity of homoserine kinase in *Arabidopsis thaliana. Plant J.* 41, 685–696. doi: 10.1111/j.1365-313X.2004.02329.x

Leroux, A. E., Haanstra, J. R., Bakker, B. M., and Krauth-Siegel, R. L. (2013). Dissecting the catalytic mechanism of *Trypanosoma brucei* trypanothione synthetase by kinetic analysis and computational modeling. *J. Biol. Chem.* 288, 23751–23764. doi: 10.1074/jbc.M113.483289

Lewis, N. E., Nagarajan, H., and Palsson, B. O. (2012). Constraining the metabolic genotype-phenotype relationship using a phylogeny of in silico methods. *Nat. Rev. Microbiol.* 10, 291–305. doi: 10.1038/nrmicro2737

Li, Y., Wei, G. Y., and Chen, J. (2004). Glutathione: a review on biotechnological production. *Appl. Microbiol. Biotechnol.* 66, 233–242. doi: 10.1007/s00253-004-1751-y

Link, H., Christodoulou, D., and Sauer, U. (2014). Advancing metabolic models with kinetic information. *Curr. Opin. Biotechnol.* 29, 8–14. doi: 10.1016/j.copbio.2014.01.015

Loudet, O., Saliba-Colombani, V., Camilleri, C., Calenge, F., Gaudon, V., Koprivova, A., et al. (2007). Natural variation for sulfate content in *Arabidopsis thaliana* is highly controlled by APR2. *Nat. Genet.* 39, 896–900. doi: 10.1038/ng2050

Machado, D., and Herrgard, M. (2014). Systematic evaluation of methods for integration of transcriptomic data into constraint-based models of metabolism. *PLoS Comput. Biol.* 10:e1003580. doi: 10.1371/journal.pcbi.1003580

Mahadevan, R., Edwards, J. S., and Doyle, F. J. III. (2002). Dynamic flux balance analysis of diauxic growth in *Escherichia coli. Biophys. J.* 83, 1331–1340. doi: 10.1016/s0006-3495(02)73903-9

Masakapalli, S. K., Le Lay, P., Huddleston, J. E., Pollock, N. L., Kruger, N. J., and Ratcliffe, R. G. (2010). Subcellular flux analysis of central metabolism in a heterotrophic *Arabidopsis* cell suspension using steady-state stable isotope labeling. *Plant Physiol.* 152, 602–619. doi: 10.1104/pp.109.151316

Mendoza-Cózatl, D. G., and Moreno-Sánchez, R. (2006). Control of glutathione and phytochelatin synthesis under cadmium stress. Pathway modeling for plants. *J. Theor. Biol.* 238, 919–936. doi: 10.1016/j.jtbi.2005.07.003

Mintz-Oron, S., Meir, S., Malitsky, S., Ruppin, E., Aharoni, A., and Shlomi, T. (2012). Reconstruction of *Arabidopsis* metabolic network models accounting for subcellular compartmentalization and tissue-specificity. *Proc. Natl. Acad. Sci. U.S.A.* 109, 339–344. doi: 10.1073/pnas.1100358109

Miskovic, L., and Hatzimanikatis, V. (2011). Modeling of uncertainties in biochemical reactions. *Biotechnol. Bioeng.* 108, 413–423. doi: 10.1002/bit.22932

Mugford, S. G., Lee, B.-R., Koprivova, A., Matthewman, C., and Kopriva, S. (2011). Control of sulfur partitioning between primary and secondary metabolism. *Plant J.* 65, 96–105. doi: 10.1111/j.1365-313X.2010.04410.x

Nagrath, D., Avila-Elchiver, M., Berthiaume, F., Tilles, A. W., Messac, A., and Yarmush, M. L. (2007). Integrated energy and flux balance based multiobjective framework for large-scale metabolic networks. *Ann. Biomed. Eng.* 35, 863–885. doi: 10.1007/s10439-007-9283-0

Nero, D., Krouk, G., Tranchina, D., and Coruzzi, G. M. (2009). A system biology approach highlights a hormonal enhancer effect on regulation of genes in a nitrate responsive "biomodule." *BMC Syst. Biol.* 3:59. doi: 10.1186/1752-0509-3-59

Noctor, G., Arisi, A. C. M., Jouanin, L., and Foyer, C. H. (1999). Photorespiratory glycine enhances glutathione accumulation in both the chloroplastic and cytosolic compartments. *J. Exp. Bot.* 50, 1157–1167. doi: 10.1093/jexbot/50.336.1157

Noctor, G., Strohm, M., Jouanin, L., Kunert, K. J., Foyer, C. H., and Rennenberg, H. (1996). Synthesis of glutathione in leaves of transgenic poplar overexpressing gamma-glutamylcysteine synthetase. *Plant Physiol.* 112, 1071–1078.

Notebaart, R. A., Teusink, B., Siezen, R. J., and Papp, B. (2008). Co-regulation of metabolic genes is better explained by flux coupling than by network distance. *PLoS Comput. Biol.* 4:e26. doi: 10.1371/journal.pcbi.0040026

Oh, Y.-G., Lee, D.-Y., Lee, S. Y., and Park, S. (2009). Multiobjective flux balancing using the NISE method for metabolic network analysis. *Biotechnol. Prog.* 25, 999–1008. doi: 10.1002/btpr.193

Ohno, S., Shimizu, H., and Furusawa, C. (2014). FastPros: screening of reaction knockout strategies for metabolic engineering. *Bioinformatics* 30, 981–987. doi: 10.1093/bioinformatics/btt672

Papp, B., Notebaart, R. A., and Pal, C. (2011). Systems-biology approaches for predicting genomic evolution. *Nat. Rev. Genet.* 12, 591–602. doi: 10.1038/nrg3033

Poolman, M. G., Miguet, L., Sweetlove, L. J., and Fell, D. A. (2009). A genome-scale metabolic model of *Arabidopsis* and some of its properties. *Plant Physiol.* 151, 1570–1581. doi: 10.1104/pp.109.141267

Poorter, H., Anten, N. P. R., and Marcelis, L. F. M. (2013). Physiological mechanisms in plant growth models: do we need a supra-cellular systems biology approach? *Plant Cell Environ.* 36, 1673–1690. doi: 10.1111/pce.12123

Pozo, C., Miró, A., Guillén-Gosálbez, G., Sorribas, A., Alves, R., and Jiménez, L. (2014). Gobal optimization of hybrid kinetic/FBA models via outer-approximation. *Comput. Chem. Eng.* 72, 325–333. doi: 10.1016/j.compchemeng.2014.06.011

Ravilious, G. E., Amelia, N., Francois, J. A., and Jez, J. M. (2012). Structural basis and evolution of redox regulation in plant adenosine-5′-phosphosulfate kinase. *Proc. Natl. Acad. Sci. U.S.A.* 109, 309–314. doi: 10.1073/pnas.1115772108

Rennenberg, H., and Herschbach, C. (2014). A detailed view on sulphur metabolism at the cellular and whole-plant level illustrates challenges in metabolite flux analyses. *J. Exp. Bot.* 65, 5711–5724. doi: 10.1093/jxb/eru315

Reznik, E., Mehta, P., and Segre, D. (2013). Flux imbalance analysis and the sensitivity of cellular growth to changes in metabolite pools. *PLoS Comput. Biol.* 9:e1003195. doi: 10.1371/journal.pcbi.1003195

Rohwer, J. M. (2014). Applications of kinetic modeling to plant metabolism. *Methods Mol. Biol.* 1083, 275–286. doi: 10.1007/978-1-62703-661-0_16

Roman, M. S., Cancela, H., and Acerenza, L. (2014). Source and regulation of flux variability in *Escherichia coli. BMC Syst. Biol.* 8:67. doi: 10.1186/1752-0509-8-67

Rouached, H., Secco, D., and Arpat, A. B. (2009). Getting the most sulfate from soil: regulation of sulfate uptake transporters in *Arabidopsis. J. Plant Physiol.* 166, 893–902. doi: 10.1016/j.jplph.2009.02.016

Saha, R., Chowdhury, A., and Maranas, C. D. (2014). Recent advances in the reconstruction of metabolic models and integration of omics data. *Curr. Opin. Biotechnol.* 29, 39–45. doi: 10.1016/j.copbio.2014.02.011

Saha, R., Suthers, P. F., and Maranas, C. D. (2011). *Zea mays* iRS1563: a comprehensive genome-scale metabolic reconstruction of maize metabolism. *PLoS ONE* 6:e21784. doi: 10.1371/journal.pone.0021784

Sajitz-Hermstein, M., and Nikoloski, Z. (2013). Structural control of metabolic flux. *PLoS Comput. Biol.* 9:e1003368. doi: 10.1371/journal.pcbi.1003368

Scheerer, U., Haensch, R., Mendel, R. R., Kopriva, S., Rennenberg, H., and Herschbach, C. (2010). Sulphur flux through the sulphate assimilation pathway is differently controlled by adenosine 5′-phosphosulphate reductase under stress and in transgenic poplar plants overexpressing gamma-ECS, SO, or APR. *J. Exp. Bot.* 61, 609–622. doi: 10.1093/jxb/erp327

Schomburg, I., Chang, A., Ebeling, C., Gremse, M., Heldt, C., Huhn, G., et al. (2004). BRENDA, the enzyme database: updates and major new developments. *Nucleic Acids Res.* 32, D431–D433. doi: 10.1093/nar/gkh081

Schuetz, R., Kuepfer, L., and Sauer, U. (2007). Systematic evaluation of objective functions for predicting intracellular fluxes in *Escherichia coli*. *Mol. Syst. Biol.* 3:119. doi: 10.1038/msb4100162

Schuetz, R., Zamboni, N., Zampieri, M., Heinemann, M., and Sauer, U. (2012). Multidimensional optimality of microbial metabolism. *Science* 336, 601–604. doi: 10.1126/science.1216882

Schuster, S., Pfeiffer, T., and Fell, D. A. (2008). Is maximization of molar yield in metabolic networks favoured by evolution? *J. Theor. Biol.* 252, 497–504. doi: 10.1016/j.jtbi.2007.12.008

Seaver, S. M. D., Henry, C. S., and Hanson, A. D. (2012). Frontiers in metabolic reconstruction and modeling of plant genomes. *J. Exp. Bot.* 63, 2247–2258. doi: 10.1093/jxb/err371

Segre, D., Vitkup, D., and Church, G. M. (2002). Analysis of optimality in natural and perturbed metabolic networks. *Proc. Natl. Acad. Sci. U.S.A.* 99, 15112–15117. doi: 10.1073/pnas.232349399

Sekiz, S. S., Scott, M. L., and Nesheim, M. C. (1975). The effect of methionine deficiency on body weight, food and energy utilization in the chick. *Poult. Sci.* 54, 1184–1188. doi: 10.3382/ps.0541184

Shachar-Hill, Y. (2013). Metabolic network flux analysis for engineering plant systems. *Curr. Opin. Biotechnol.* 24, 247–255. doi: 10.1016/j.copbio.2013.01.004

Simons, M., Misra, A., and Sriram, G. (2014a). Genome-scale models of plant metabolism. *Methods Mol. Biol.* 1083, 213–230. doi: 10.1007/978-1-62703-661-0_13

Simons, M., Saha, R., Guillard, L., Clement, G., Armengaud, P., Canas, R., et al. (2014b). Nitrogen-use efficiency in maize (*Zea mays* L.): from 'omics' studies to metabolic modelling. *J. Exp. Bot.* 65, 5657–5671. doi: 10.1093/jxb/eru227

Sjostrom, S. L., Joensson, H. N., and Svahn, H. A. (2013). Multiplex analysis of enzyme kinetics and inhibition by droplet microfluidics using picoinjectors. *Lab Chip* 13, 1754–1761. doi: 10.1039/c3lc41398e

Smallbone, K., and Mendes, P. (2013). Large-scale metabolic models: from reconstruction to differential equations. *Ind. Biotechnol.* 9, 179–184. doi: 10.1089/ind.2013.0003

Steuer, R., Gross, T., Selbig, J., and Blasius, B. (2006). Structural kinetic modeling of metabolic networks. *Proc. Natl. Acad. Sci. U.S.A.* 103, 11868–11873. doi: 10.1073/pnas.0600013103

Stitt, M., and Gibon, Y. (2014). Why measure enzyme activities in the era of systems biology? *Trends Plant Sci.* 19, 256–265. doi: 10.1016/j.tplants.2013.11.003

Sun, Q., Zybailov, B., Majeran, W., Friso, G., Olinares, P. D. B., and Van Wijk, K. J. (2009). PPDB, the plant proteomics database at cornell. *Nucleic Acids Res.* 37, D969–D974. doi: 10.1093/nar/gkn654

Takahashi, H., Kopriva, S., Giordano, M., Saito, K., and Hell, R. (2011). "Sulfur assimilation in photosynthetic organisms: molecular functions and regulations of transporters and assimilatory enzymes," in *Annual Review of Plant Biology*, Vol. 62, eds S. S. Merchant, W. R. Briggs, and D. Ort (Palo Alto, CA: Annual Reviews), 157–184.

Tawfiq, N., Heaney, R. K., Plumb, J. A., Fenwick, G. R., Musk, S. R., and Williamson, G. (1995). Dietary glucosinolates as blocking agents against carcinogenesis: glucosinolate breakdown products assessed by induction of quinone reductase activity in murine hepa1c1c7 cells. *Carcinogenesis* 16, 1191–1194. doi: 10.1093/carcin/16.5.1191

Thomas, S., and Fell, D. A. (1998). The role of multiple enzyme activation in metabolic flux control. *Adv. Enzyme Regul.* 38, 65–85. doi: 10.1016/S0065-2571(97)00012-5

Toepfer, N., Caldana, C., Grimbs, S., Willmitzer, L., Fernie, A. R., and Nikoloski, Z. (2013). Integration of genome-scale modeling and transcript profiling reveals metabolic pathways underlying light and temperature acclimation in *Arabidopsis*. *Plant Cell* 25, 1197–1211. doi: 10.1105/tpc.112.108852

Tomar, N., and De, R. K. (2013). Comparing methods for metabolic network analysis and an application to metabolic engineering. *Gene* 521, 1–14. doi: 10.1016/j.gene.2013.03.017

Traka, M. H., and Mithen, R. F. (2011). Plant science and human nutrition: challenges in assessing health-promoting properties of phytochemicals. *Plant Cell* 23, 2483–2497. doi: 10.1105/tpc.111.087916

Tran, L. M., Rizk, M. L., and Liao, J. C. (2008). Ensemble modeling of metabolic networks. *Biophys. J.* 95, 5606–5617. doi: 10.1529/biophysj.108.135442

Tripathi, M., and Mishra, A. (2007). Glucosinolates in animal nutrition: a review. *Anim. Feed Sci. Technol.* 132, 1–27. doi: 10.1016/j.anifeedsci.2006.03.003

Tsakraklides, G., Martin, M., Chalam, R., Tarczynski, M. C., Schmidt, A., and Leustek, T. (2002). Sulfate reduction is increased in transgenic *Arabidopsis thaliana* expressing 5′-adenylylsulfate reductase from *Pseudomonas aeruginosa*. *Plant J.* 32, 879–889. doi: 10.1046/j.1365-313X.2002.01477.x

Tummler, K., Lubitz, T., Schelker, M., and Klipp, E. (2014). New types of experimental data shape the use of enzyme kinetics for dynamic network modeling. *FEBS J.* 281, 549–571. doi: 10.1111/febs.12525

van Eunen, K., and Bakker, B.M. (2014). The importance and challenges of in vivo-like enzyme kinetics. *Perspect. Sci.* 1, 126–130. doi: 10.1016/j.pisc.2014.02.011

Vargas, F. A., Pizarro, F., Ricardo Perez-Correa, J., and Agosin, E. (2011). Expanding a dynamic flux balance model of yeast fermentation to genome-scale. *BMC Syst. Biol.* 5:75. doi: 10.1186/1752-0509-5-75

Varma, A., and Palsson, B. O. (1994). Stoichiometric flux balance models quantitatively predict growth and metabolic by-product secretion in wild-type *Escherichia coli* w3110. *Appl. Environ. Microbiol.* 60, 3724–3731.

Vauclare, P., Kopriva, S., Fell, D., Suter, M., Sticher, L., Von Ballmoos, P., et al. (2002). Flux control of sulphate assimilation in *Arabidopsis thaliana*: adenosine 5′-phosphosulphate reductase is more susceptible than ATP sulphurylase to negative control by thiols. *Plant J.* 31, 729–740. doi: 10.1046/j.1365-313X.2002.01391.x

Williams, T. C. R., Poolman, M. G., Howden, A. J. M., Schwarzlander, M., Fell, D. A., Ratcliffe, R. G., et al. (2010). A genome-scale metabolic model accurately predicts fluxes in central carbon metabolism under stress conditions. *Plant Physiol.* 154, 311–323. doi: 10.1104/pp.110.158535

Wintermute, E. H., Lieberman, T. D., and Silver, P. A. (2013). An objective function exploiting suboptimal solutions in metabolic networks. *BMC Syst. Biol.* 7:98. doi: 10.1186/1752-0509-7-98

Wittig, U., Kania, R., Golebiewski, M., Rey, M., Shi, L., Jong, L., et al. (2012). SABIO-RK-database for biochemical reaction kinetics. *Nucleic Acids Res.* 40, D790–D796. doi: 10.1093/nar/gkr1046

Yadav, S. (2010). Heavy metals toxicity in plants: an overview on the role of glutathione and phytochelatins in heavy metal stress tolerance of plants. *S. Afr. J. Bot.* 76, 167–179. doi: 10.1016/j.sajb.2009.10.007

Yoshimoto, N., Inoue, E., Watanabe-Takahashi, A., Saito, K., and Takahashi, H. (2007). Posttranscriptional regulation of high-affinity sulfate transporters in *Arabidopsis* by sulfur nutrition. *Plant Physiol.* 145, 378–388. doi: 10.1104/pp.107.105742

Zamora-Sillero, E., Hafner, M., Ibig, A., Stelling, J., and Wagner, A. (2011). Efficient characterization of high-dimensional parameter spaces for systems biology. *BMC Syst. Biol.* 5:142. doi: 10.1186/1752-0509-5-142

Zarecki, R., Oberhardt, M. A., Yizhak, K., Wagner, A., Segal, E. S., Freilich, S., et al. (2014). Maximal sum of metabolic exchange fluxes outperforms biomass yield as a predictor of growth rate of microorganisms. *PLoS ONE* 9:e98372. doi: 10.1371/journal.pone.0098372

Zomorrodi, A. R., Islam, M. M., and Maranas, C. D. (2014). d-OptCom: dynamic multi-level and multi-objective metabolic modeling of microbial communities. *ACS Synth. Biol.* 3, 247–257. doi: 10.1021/sb4001307

Zomorrodi, A. R., and Maranas, C. D. (2012). OptCom: a multi-level optimization framework for the metabolic modeling and analysis of microbial communities. *PLoS Comput. Biol.* 8:e1002363. doi: 10.1371/journal.pcbi.1002363

Zomorrodi, A. R., Suthers, P. F., Ranganathan, S., and Maranas, C. D. (2012). Mathematical optimization applications in metabolic networks. *Metab. Eng.* 14, 672–686. doi: 10.1016/j.ymben.2012.09.005

Conflict of Interest Statement: The authors declare that the research was conducted in the absence of any commercial or financial relationships that could be construed as a potential conflict of interest.

7

The role of bacteria and mycorrhiza in plant sulfur supply

Jacinta Gahan and Achim Schmalenberger*

Department of Life Sciences, University of Limerick, Limerick, Ireland

Edited by:
Stanislav Kopriva, University of Cologne, Germany

Reviewed by:
Tamara Gigolashvili, University of Cologne, Germany
Michael Kertesz, University of Sydney, Australia

Correspondence:
Achim Schmalenberger, Department of Life Sciences, University of Limerick, Castletroy, Limerick, Ireland
e-mail: achim.schmalenberger@ul.ie

Plant growth is highly dependent on bacteria, saprophytic, and mycorrhizal fungi which facilitate the cycling and mobilization of nutrients. Over 95% of the sulfur (S) in soil is present in an organic form. Sulfate-esters and sulfonates, the major forms of organo-S in soils, arise through deposition of biological material and are transformed through subsequent humification. Fungi and bacteria release S from sulfate-esters using sulfatases, however, release of S from sulfonates is catalyzed by a bacterial multi-component mono-oxygenase system. The *asfA* gene is used as a key marker in this desulfonation process to study sulfonatase activity in soil bacteria identified as *Variovorax*, *Polaromonas*, *Acidovorax*, and *Rhodococcus*. The rhizosphere is regarded as a hot spot for microbial activity and recent studies indicate that this is also the case for the mycorrhizosphere where bacteria may attach to the fungal hyphae capable of mobilizing organo-S. While current evidence is not showing sulfatase and sulfonatase activity in arbuscular mycorrhiza, their effect on the expression of plant host sulfate transporters is documented. A revision of the role of bacteria, fungi and the interactions between soil bacteria and mycorrhiza in plant S supply was conducted.

Keywords: sulfonate desulfurization, sulfate esters, mycorrhizal fungi, plant–microbe interactions, asf gene cluster, sulfatases, mycorrhizosphere

INTRODUCTION

Sulfur (S), an essential macro-element required for growth, is increasingly becoming limiting to crop yield and quality as a result of a reduction in atmospheric S levels and crop varieties removing S from soil more rapidly (Fowler et al., 2005). S present in soil is approximately 95% organically bound largely in one of two major forms; sulfate-esters and sulfonates (**Figure 1**; Autry and Fitzgerald, 1990; Kertesz and Mirleau, 2004). These forms of organo-S are not directly available to plants which rely upon microbes in soil and rhizosphere for organo-S mobilization (Kertesz et al., 2007). Plant root activity impacts the physicochemical properties of the soil through the release of organic compounds (rhizodeposition) which accounts for 15–30% of photosynthetically produced carbon (C; Russell, 1977). This process provides soil organisms with an energy source that enables them to fulfill their respective functional roles (Lynch and Whipps, 1990; Farrar et al., 2003).

Many bacteria and fungi in soil are capable of mineralizing S from sulfate-esters (Klose et al., 1999). In contrast, an exclusively bacterial multicomponent mono-oxygenase enzyme complex is necessary to mobilize sulfonates, the dominant organo-S source in soil (Vermeij et al., 1999; Kertesz and Mirleau, 2004). In fact, soil S cycling may involve complex interactions between several free living and symbiotic root associated microbial populations. Arbuscular mycorrhizal (AM) fungi form symbiosis with 80% of land plant species which depend upon them for growth (Wang and Qiu, 2006). AM fungal symbiosis is characterized by fungal penetration of root cortical cells forming microscopic branched structures called arbuscules that increase efficiency of plant-fungus metabolite exchange (Smith and Read, 1997). Extraradicular AM hyphae provide surfaces for functional bacterial populations to colonize. A number of studies have reported interactions between AM fungi and phosphorus (P) and nitrogen (N) mobilizing bacteria (Richardson et al., 2009; Hodge and Storer, 2014), and the impact of AM on bacterial community structures (Bianciotto and Bonfante, 2002; Toljander et al., 2007). Like S, both N, and P exist predominantly inaccessible to plants which rely on interactions with mycorrhizal fungi and associated microbes to facilitate their mobilization (Richardson et al., 2009).

SULFUR FOR PLANT GROWTH

S owes its importance as a component of the (i) proteinaceous amino acids cysteine and methionine, (ii) non-protein amino acids including cystine, lanthionine, and ethionine (iii) tripeptide glutathione, and (iv) components including vitamins thiamine and biotin, phytochelatins, chlorophyll, coenzyme A, S-adenosyl-methionin and sulfolipids (Scherer, 2001). S plays critical structural roles in cells as disulphide bonds in proteins, is involved in enzyme regulation (redox control), provides protection from oxidative stress via glutathione, and its derivatives are involved in heavy metal stress mediation (Leustek and Saito, 1999). Plant S also plays an important role in disease protection and defense response as a component of glucosinolates and allin compounds (Jones et al., 2004; Brader et al., 2006). Various plant species prevent fungal infection via deposition of elemental S in the xylem parenchyma (Cooper and Williams, 2004).

Plant S demand is dependent on species and stage of development, with increased demand observed during periods of vegetative growth and seed development (Leustek and Saito, 1999). Inorganic sulfate (SO_4^{2-}) is the dominant plant available source of S, while to a lesser extent atmospheric reduced S may be utilized (Leustek et al., 2000). Regulation of SO_4^{2-} uptake involves

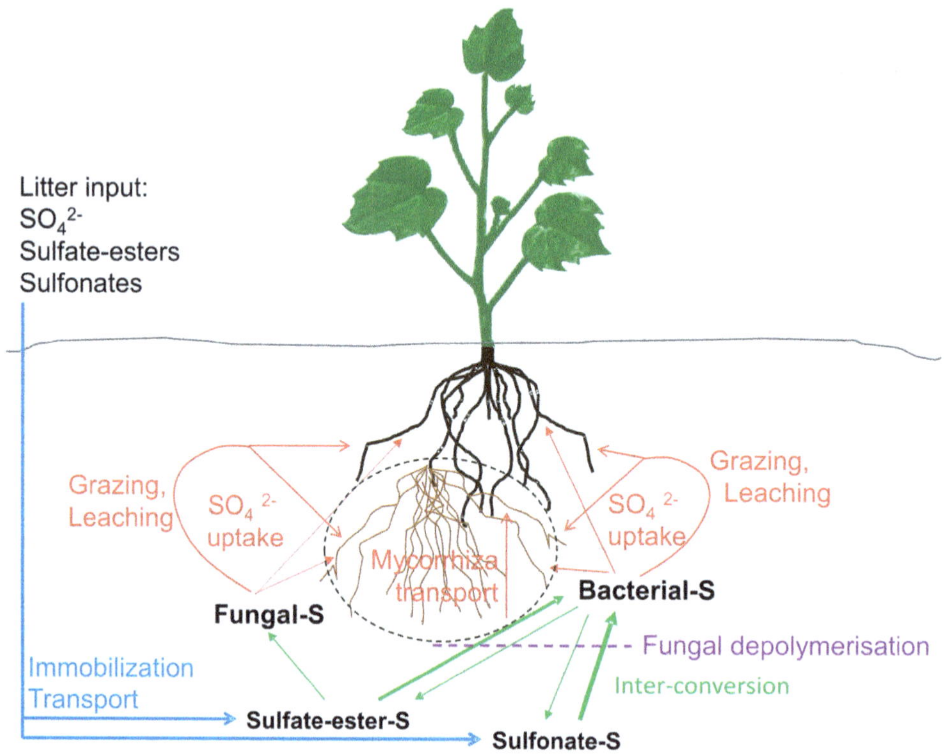

FIGURE 1 | Illustration of the sulfur cycle in soil with plant cover. Major sulfur (S) inputs to soils originate from organic litter deposition and animal droppings (blue lines). Most of this deposited S is organically bound (organo-S). Atmospheric deposition of inorganic S has greatly declined in Europe, America and elsewhere, thus is often only a minor source for plants. Organo-S (sulfate-esters and sulfonates) can be transformed by soil microbes between the two major organo-S pools or mineralized to inorganic S (green lines, thickness suggests main direction of pathway). At the same time, inorganic S can be immobilized into organo-S (green lines). While the sulfate-ester pool is largely available to both fungi and bacteria, sulfonates are primarily accessible to bacteria only and aromatic sulfonates are only available to a particular functional clade of bacteria. Bacterial sulfonate desulfurization via the mono-oxygenase multi-enzyme pathway may occur intracellular, thus polymeric sulfonate may need depolymerisation, e.g., by saprophytic fungi prior to uptake (dotted purple line). Organo-S mineralised by fungi and bacteria need to be made available for plant uptake in the form of sulfate. This may happen via sulfate uptake by mycorrhizal fungal hyphae as an intermediate step (dashed gray line). In the absence of any direct evidence of a sulfate transport system from fungus or bacterium to the plant root or symbiotic mycorrhizal hyphae, release of mineralised S through autolysis and grazing by protists and microscopic nematodes may play an import role in inorganic sulfate release and plant sulfate uptake (red lines).

multiple transport steps and a large family of SO_4^{2-} transporters have been characterized (Hawkesford, 2003). Assimilation of SO_4^{2-} to cysteine occurs primarily in the chloroplasts of young leaves, while cysteine and methionine can also be synthesized in roots and seeds (Leustek and Saito, 1999). S starvation has been shown to negatively impact plant vitality when the P and N status is adequate (Sieh et al., 2013). During S limitation plant SO_4^{2-} transporters are up-regulated for rapid SO_4^{2-} up-take from the rhizosphere leading to a zone of SO_4^{2-} depletion (Buchner et al., 2004). In this zone, bacterial desulfurization of organo-S is induced to mineralize organo-S, thus indirectly regulating plant S uptake (Kertesz and Mirleau, 2004). However, S-deficiency in plants can result in reduced root exudation (Alhendawi et al., 2005) or alteration of root exudates (Astolfi et al., 2010) which can influence bacterial communities seeking exudates as source of carbon.

X-ray absorption near edge structure (XANES) spectroscopy has revealed that sulfonates and sulfate-esters compose 30–70% and 20–60% of the organo-S in soil, respectively (Zhao et al., 2006). Directly plant available SO_4^{2-} constitutes less than 5% of the total soil S (Autry and Fitzgerald, 1990). Organo-S compounds arise through deposition of biological material containing S, including plant and animal residues, and are subsequently incorporated into organic molecules through complex humification processes (Guggenberger, 2005). Animal residues are particularly high in organo-S with sheep dung comprising ~80% of S as sulfonates, and while SO_4^{2-} is rapidly leached from soil, organo-S can persist for longer time periods (Haynes and Williams, 1993). Additionally, soil-S pools are not static but rapidly interconverted between forms by soil microbial activity (Freney et al., 1975; Kertesz et al., 2007). Sulfonates were found to be mineralized more rapidly than other S-fractions and accounted for the majority of S released in short term incubation studies (Zhao et al., 2003, 2006). These findings indicate that C-bound S in soils may be of greatest importance (Ghani et al., 1992).

MICROBIAL MINERALIZATION OF ORGANO-S

Microbial mineralization of organo-S is undertaken to access carbon, energy or S, with the latter also vital for plant growth (Ghani et al., 1992; Cook et al., 1998; Cook and Denger, 2002).

Sulfate-ester mineralization is catalyzed by sulfatases of the esterase class (Deng and Tabatabai, 1997). Arylsulfatase enzymes act on aromatic sulfate-esters by splitting the O-S bond while alkylsulfatase enzymes act on aliphatic sulfate-esters by splitting the C-O bond (Kertesz, 1999). Both reactions release sulfate and are common in rhizospheric soil (Kertesz and Mirleau, 2004). Bacterial arylsulfatase activity is induced during S starvation and repressed in the presence of SO_4^{2-} in *Pseudomonas aeruginosa,* while in a *Streptomyces* strain, a membrane bound sulfatase was also induced independently via substrate presence (Hummerjohann et al., 2000; Cregut et al., 2013). The ability to mobilize sulfate-esters has been observed in a range of bacteria including *Pseudomonas, Klebsiella, Salmonella, Enterobacter, Serratia,* and *Comamonas* (Hummerjohann et al., 2000). Additionally, arylsulfatase activity is influenced by various external factors including soil temperature, moisture content, vegetative cover, and crop rotation (Tabatabai and Bremner, 1970).

Fungi play an important role in the rhizosphere as plant symbionts or as free living saprotrophs. Soil filamentous fungi were reported to be important in mobilization of sulfate-esters (Omar and Abd-Alla, 2000; Baum and Hrynkiewicz, 2006), where enhanced arylsulfatase activity was found under S-limiting conditions (Fitzgerald, 1976; Marzluf, 1997). Likewise, wood-rotting fungi utilized sulfate-esters and thiols from wood (Schmalenberger et al., 2011).

The most abundant organo-S source in soil is present as aliphatic or aromatic sulfonates (Autry and Fitzgerald, 1990; Zhao et al., 2006). The ability to mobilize S from aliphatic sulfonates is widespread among soil bacteria with over 90% of morphologically distinct isolates capable of C2-sulfonate utilization (King and Quinn, 1997). However, aromatic sulfonates have been shown to be of greater importance for S nutrition and the ability to mobilize these sulfonates has been associated with plant growth promotion (PGP) of tomato (Kertesz and Mirleau, 2004) and *Arabidopsis* (Kertesz et al., 2007).

The desulfonating ability of the sewage sludge bacterial isolate *Pseudomonas putida* S-313 has been widely studied across a broad substrate range (Kertesz et al., 1994; Cook et al., 1998; Vermeij et al., 1999; Kahnert et al., 2000). Mobilization of SO_4^{2-} from aromatic and aliphatic sulfonates is catalyzed by a $FMNH_2$-dependent monooxygenase enzyme complex encoded in the *ssu* gene cluster (Eichhorn et al., 1999). The monooxygenase SsuD cleaves sulfonates to their corresponding aldehydes and the reduced flavin for this process is provided by the FMN-NADPH reductase SsuE. Although its function is unknown, *ssuF* from the *ssu* gene cluster was found to be essential for sulfonate desulfurization as well. For aromatic desulfonation the *asfRABC* gene cluster is required as an additional 'tool-kit' to complement *ssu.* The *asf* gene cluster includes a substrate binding protein, an ABC type transporter, a reductase/ferredoxin electron transport system involved in electron transfer and energy provision during oxygenation of the C-S bond, and a LysR-type regulatory protein, which activates the system during SO_4^{2-} limitation (Vermeij et al., 1999). Transposon mutagenesis in the *asfA* gene of sewage isolate *P. putida* S-313 resulted in mutants without the capability to utilize aromatic sulfonates, while the utilization of aliphatic sulfonates was unchanged (Vermeij et al., 1999). This mutant was used in a plant growth experiment alongside its wild type, where the PGP effect was directly attributed to an functioning *asfA* gene (Kertesz and Mirleau, 2004). This particular type of bacterium has recently been isolated from the hyphae of symbiotic mycorrhizal fungi (Gahan and Schmalenberger, 2014). Various recent studies on the bacterial phylogeny of aromatic sulfonate mobilizing bacteria have expanded the diversity to the Beta-Proteobacteria; *Variovorax, Polaromonas, Hydrogenophaga, Cupriavidus, Burkholderia,* and *Acidovorax,* the Actinobacteria; *Rhodococcus* and the Gamma-Proteobacteria; *Pseudomonas* (**Figure 2**; Schmalenberger and Kertesz, 2007; Schmalenberger et al., 2008, 2009; Fox et al., 2014). Additionally, *Stenotrophomonas* and *Williamsia* species, isolated from hand-picked AM hyphae, have recently been added to these groups (Gahan and Schmalenberger, 2014).

Until now, there has been little evidence to suggest fungal catalysis of sulfonate desulfurization (Kertesz et al., 2007; Schmalenberger et al., 2011). Indeed, while some saprotrophic fungi appear to breakdown some sulfonated molecules they do not release inorganic S in the process, for example, the white rot fungus *Phanerochaete chrysporium* transforms the aromatic alkylbenzene sulfonate but does so exclusively on its side chain without S-release (Yadav et al., 2001). Cultivation of fungi *in vitro* suggested that sulfonates could be utilized as an S source by wood degrading fungus *Geophyllum trabeum,* however, XANES spectra taken from wood accessible solely to the fungus displayed no evidence of sulfonate mobilization (Schmalenberger et al., 2011). Other cultivation experiments indicated a use of aliphatic sulfonates by various strains of yeasts via a putative 2-oxoglutarate dependent dioxygenase pathway (Uria-Nickelsen et al., 1993; Linder, 2012). However, this desulfurization capability may be limited to certain C4–C6 alkanesulfonates as this is the case for the taurine dioxygenase (Kertesz, 1999). Thus, the importance of bacteria and fungi with a dioxygenase pathway for sulfonate desulfurization is still somewhat unclear. As aforementioned, bacterial desulfonation based on the monooxygenase pathway occurs intracellularly and, as such, availability of sulfonates of different molecular size may be of importance. Therefore, saprotrophic fungi, including several genera of the Basidomycota, may play a role in sulfonate mobilization by secreting enzymes such as laccases and peroxidases in order to depolymerize large organic compounds in the soil (**Figure 1**; Muralikrishna and Renganathan, 1993; Tuor et al., 1995; Heinzkill et al., 1998). Lignolytic degradation of large organic complexes releases mono and oligomeric sulfonates which can be further mobilized by functional bacterial guilds as described above (Kertesz et al., 2007).

THE ROLE OF ARBUSCULAR MYCORRHIZA IN SULFUR SUPPLY

Arbuscular mycorrhizal fungi are the most common form of mycorrhizal association and their evolution can be dated back 460 million years (Smith and Read, 1997). They form symbiosis with 77% of angiosperms, 45% of 84 species of gymnosperms and 52% of 400 species of fern and lycopod (Wang and Qiu, 2006). The defining characteristic structure, the arbuscule, acts as an efficient site for plant-fungus metabolite exchange (Smith and Read, 1997). AM intra-radicular hyphae (IRH) provide the means for fungal extension within the host plant's cortical region (Morton

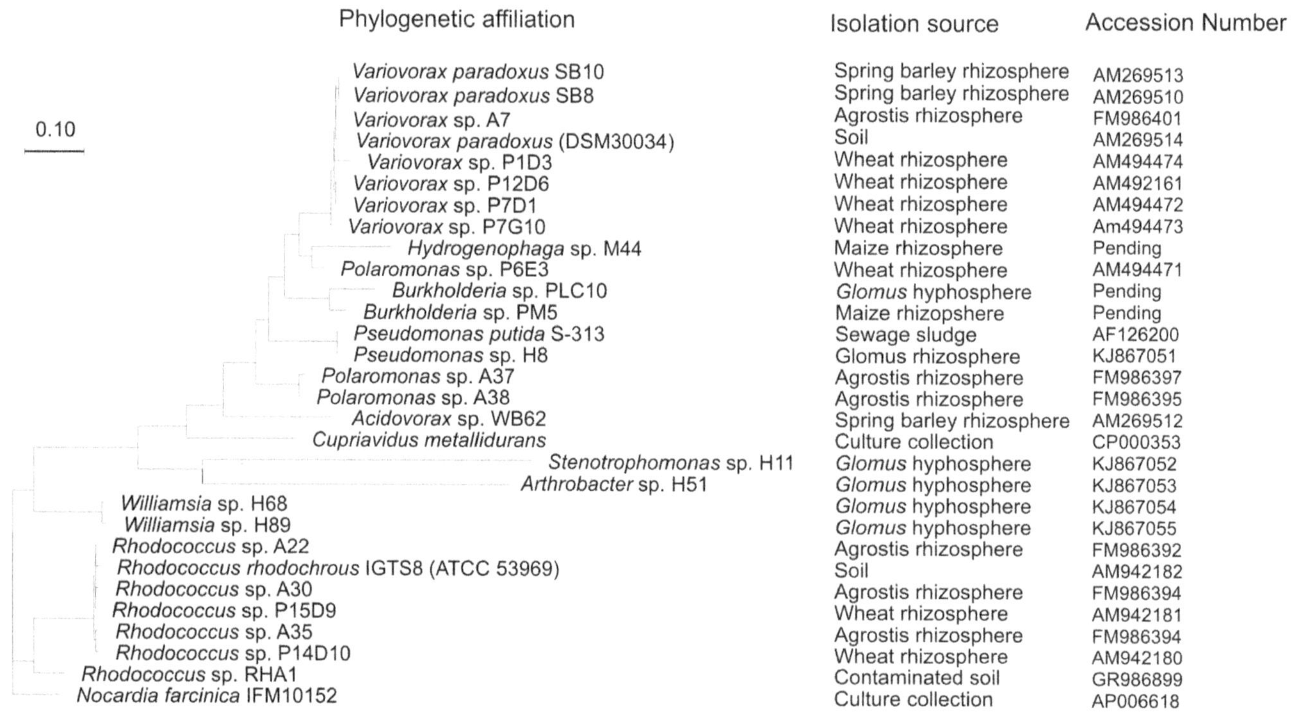

FIGURE 2 | Randomized axelerated maximum likelihood tree from truncated AsfA sequences obtained from aromatic sulfonate desulfurizing bacteria isolated from soil, rhizosphere, or hyphosphere alongside strains from culture collections.

and Benny, 1990), while extra-radicular hyphae (ERH) have three primary functions – nutrient acquisition, infection of host plants, and production of fertile spores (Nagahashi and Douds, 2000).

Available studies on the effects of AM colonization on uptake of S have presented equivocal results (Gray and Gerdemann, 1973; Cooper and Tinker, 1978; Rhodes and Gerdemann, 1978). However, studies have shown that the presence of AM fungi enhances S uptake for maize, clover (Gray and Gerdemann, 1973) and tomato (Cavagnaro et al., 2006). More recently, AM fungus *G. intraradices* on transformed carrot roots demonstrated uptake of reduced forms of S *in vitro* (Allen and Shachar-Hill, 2009). Rates of this uptake and transfer of reduced S were comparable to that of SO_4^{2-} when the latter was largely absent. Soil to root SO_4^{2-} translocation is demand driven, with strongly induced SO_4^{2-} absorption under conditions of S limitation. This rapid uptake of SO_4^{2-} in the rhizosphere leads to a zone of SO_4^{2-} depletion similar to that observed with P (Buchner et al., 2004). The AM fungal ERH could extend out past this zone of SO_4^{2-} depletion and may play an important role in provision of S under conditions of S limitation (Kertesz et al., 2007). Recent investigations revealed that AM fungi can influence the expression of plant sulfate transporters and as a consequence improve the S nutritional status of the host plant (Giovannetti et al., 2014). This is important for all hyphospheric and rhizospheric soil microbes as lack of readily available sulfate in soil can lead to a reduction in plant exudates (Alhendawi et al., 2005) and as a consequence can affect soil microbial activity due to reduced availability of photosynthate as a source of carbon.

Extra-radicular hyphae are surrounded by complex bacterial and fungal communities that interact with the plant-mycorrhiza partnership and sustain its metabolic functioning (Frey-Klett and Garbaye, 2005). AM formation effects microbial communities in the rhizosphere via alteration of root exudates and translocation of energy rich C compounds to the extended soil environment for instance in the form of hyphal exudates (Barea et al., 2002; Boer et al., 2005). AM hyphae have a surface area several orders of magnitude greater than the plant roots which provides a niche for functional microbial interactions essential for nutrient cycling (Gryndler et al., 2000). Diverse soil microbial communities are essential for soil fertility and plant vitality (Gianinazzi and Schüepp, 1994; Siciliano et al., 2014) and AM hyphae have been shown to host a larger community of sulfonate desulfurizing bacteria than bulk soil (Gahan and Schmalenberger, 2014). Sulfonate desulfurization has been found to be characteristically rhizo- and hyphospheric in nature (**Figure 2**) and dominant sulfonate desulfurizing hyphospheric bacteria were found to be able to putatively attach and migrate with hyphae (Gahan and Schmalenberger, 2014). Inoculation of *Lolium perenne* soil microcosms with AM fungi significantly increased percentage root colonization and the quantity of cultivable sulfonate mobilizing bacteria (Gahan and Schmalenberger, 2013). Increased abundance of desulfonating bacteria as a result of elevated AM root colonization may be beneficial for plant-S supply. Likewise, addition of 2-(N-morpholine)-ethanesulfonic acid (MES) to soil putatively stimulated sulfonate mobilizing bacteria whose metabolites may have been responsible for the enhanced ERH growth of *Glomus* *intraradices* (Vilarino et al., 1997). This is important for maximizing S uptake as enhanced hyphal growth stemming from sulfonate mobilizing bacterial metabolites may further stimulate the proliferation of this community in a potential positive feedback loop. AM fungi may, therefore, play an increasingly important role in plant S metabolism not only through uptake and up-regulation of plant sulfate transporters but also through interaction with organo-S mobilizing microbes.

The hyphosphere of AM fungi can be regarded as a zone of increased bacterial abundance and activity, similar to the rhizosphere (Linderman, 1988; Andrade et al., 1998). Recent studies on the hyphosphere of ectomycorrhizae found that bacteria were co-migrating with the hyphae *in vitro*, putatively using a type III secretion system (T3SS) encoded infection needle for attachment (Warmink and van Elsas, 2008). This T3SS was also recently found to be present in aromatic sulfonate desulfurizing bacteria from the AM hyphosphere (Gahan and Schmalenberger, 2014), thus co-migration with ERH of AM fungi may be established via deployment of such an infection needle. While various pathogens are known to utilize T3SS for toxin injection into the host cells, nothing is known about any potential transfer of plant nutrients via such an infection needle to the mycorrhizal hyphae.

Currently, there is a profound knowledge gap when it comes to transfer of S from associated microbes to the plant host and its fungal symbiont. Extracellular sulfatases release S into soil solution which is then available to plant roots, mycorrhizal hyphae and various microbes, the release of S from sulfonates is potentially more complicated. While the possibility exists of a targeted transfer of S to the plant host via the ERH of AM fungi, there is currently no direct evidence provided in the literature. However, indirect release of S from sulfonate desulfurizing bacteria is a possibility. These bacteria may be turned over through grazing by microscopic predators such as nematodes and protozoa in the microbial loop (Bonkowski, 2004; Irshad et al., 2011). Indeed, soil amendments with biochar resulted not only in a significant increase in aromatic sulfonate desulfurizing bacteria but also in a significant increase in bacteria feeding nematodes (Fox et al., 2014), thus nematode activity may enhance the release of sulfonate desulfurized S in the rhizosphere and mycorrhizosphere/hyphosphere (**Figure 1**).

In conclusion, as a result of the limited nature of plant available S in soil it is increasingly necessary to understand the pathways and interactions required to mobilize the sulfate-esters and sulfonates that dominate the soil S pool. Saprotrophic fungi can depolymerize large humic material releasing sulfate-esters to bacteria and fungi, and sulfonates to specialist bacteria in possession of a monooxygenase enzyme complex. Desulfurizing microbial populations have been shown to be enriched in the rhizosphere and hyphosphere, however, released SO_4^{2-} is quickly assimilated leaving an S depleted zone in the rhizosphere. AM fungi can extend past this zone, and indeed, are stimulated by organo-S mobilizing bacterial metabolites to expand their hyphal networks, increasing the area of soil and volume of S available to the plant. Additionally, inoculation with AM fungi has been shown to increase both percentage root colonization and the magnitude of the sulfonate mobilizing bacterial community. Inoculation practices, therefore, have huge potential to sustainably increase crop yield in areas where S is becoming a limiting factor to growth.

REFERENCES

Alhendawi, R. A., Kirkby, E. A., and Pilbeam, D. J. (2005). Evidence that sulfur deficiency enhances molybdenum transport in xylem sap of tomato plants. *J. Plant Nutr.* 28, 1347–1353. doi: 10.1081/PLN-200067449

Allen, J. W., and Shachar-Hill, Y. (2009). Sulfur transfer through an arbuscular mycorrhiza. *Plant Physiol.* 149, 549–560. doi: 10.1104/pp.108.129866

Andrade, G., Mihara, K., Linderman, R., and Bethlenfalvay, G. (1998). Soil aggregation status and rhizobacteria in the mycorrhizosphere. *Plant Soil* 202, 89–96. doi: 10.1023/A:1004301423150

Astolfi, S., Zuchi, S., Hubberten, H. M., Pinton, R., and Hoefgen, R. (2010). Supply of sulphur to S-deficient young barley seedlings restores their capability to cope with iron shortage. *J. Exp. Bot.* 61, 799–806. doi: 10.1093/jxb/erp346

Autry, A. R., and Fitzgerald, J. W. (1990). Sulfonate S: a major form of forest soil organic sulfur. *Biol. Fertil. Soils* 10, 50–56. doi: 10.1007/BF00336124

Barea, J.-M., Azcón, R., and Azcón-Aguilar, C. (2002). Mycorrhizosphere interactions to improve plant fitness and soil quality. *Antonie Van Leeuwenhoek* 81, 343–351. doi: 10.1023/A:1020588701325

Baum, C., and Hrynkiewicz, K. (2006). Clonal and seasonal shifts in communities of saprotrophic microfungi and soil enzyme activities in the mycorrhizosphere of *Salix* spp. *J. Plant Nutr. Soil Sci.* 169, 481–487. doi: 10.1002/jpln.200521922

Bianciotto, V., and Bonfante, P. (2002). Arbuscular mycorrhizal fungi: a specialised niche for rhizospheric and endocellular bacteria. *Antonie Van Leeuwenhoek* 81, 365–371. doi: 10.1023/A:1020544919072

Boer, W. D., Folman, L. B., Summerbell, R. C., and Boddy, L. (2005). Living in a fungal world: impact of fungi on soil bacterial niche development. *FEMS Microbiol. Rev.* 29, 795–811. doi: 10.1016/j.femsre.2004.11.005

Bonkowski, M. (2004). Protozoa and plant growth: the microbial loop in soil revisited. *New Phytol.* 162, 617–631. doi: 10.1111/j.1469-8137.2004.01066.x

Brader, G., Mikkelsen, M. D., Halkier, B. A., and Tapio Palva, E. (2006). Altering glucosinolate profiles modulates disease resistance in plants. *Plant J.* 46, 758–767. doi: 10.1111/j.1365-313X.2006.02743.x

Buchner, P., Takahashi, H., and Hawkesford, M. J. (2004). Plant sulphate transporters: co-ordination of uptake, intracellular and long-distance transport. *J. Exp. Bot.* 55, 1765–1773. doi: 10.1093/jxb/erh206

Cavagnaro, T. R., Jackson, L. E., Six, J., Ferris, H., Goyal, S., Asami, D., et al. (2006). Arbuscular mycorrhizas, microbial communities, nutrient availability, and soil aggregates in organic tomato production. *Plant Soil* 282, 209–225. doi: 10.1007/s11104-005-5847-7

Cook, A. M., and Denger, K. (2002). Dissimilation of the C2 sulfonates. *Arch. Microbiol.* 179, 1–6. doi: 10.1007/s00203-002-0497-0

Cook, A. M., Laue, H., and Junker, F. (1998). Microbial desulfonation. *FEMS Microbiol. Rev.* 22, 399–419. doi: 10.1111/j.1574-6976.1998.tb00378.x

Cooper, K. M., and Tinker, P. (1978). Translocation and transfer of nutrients in vesicular arbuscular mycorrhizas. *New Phytol.* 88, 327–339. doi: 10.1111/j.1469-8137.1981.tb01728.x

Cooper, R. M., and Williams, J. S. (2004). Elemental sulphur as an induced antifungal substance in plant defence. *J. Exp. Bot.* 55, 1947–1953. doi: 10.1093/jxb/erh179

Cregut, M., Piutti, S., Slezack-Deschaumes, S., and Benizri, E. (2013). Compartmentalization and regulation of arylsulfatase activities in *Streptomyces sp., Microbacterium sp.* and *Rhodococcus sp.* soil isolates in response to inorganic sulfate limitation. *Microbiol. Res.* 168, 12–21. doi: 10.1016/j.micres.2012.08.001

Deng, S., and Tabatabai, M. (1997). Effect of tillage and residue management on enzyme activities in soils: III. Phosphatases and arylsulfatase. *Biol. Fertil. Soils* 24, 141–146. doi: 10.1007/s003740050222

Eichhorn, E., Van Der Ploeg, J. R., and Leisinger, T. (1999). Characterization of a two-component alkanesulfonate monooxygenase from *Escherichia coli*. *J. Biol. Chem.* 274, 26639–26646. doi: 10.1074/jbc.274.38.26639

Farrar, J., Hawes, M., Jones, D., and Lindow, S. (2003). How roots control the flux of carbon to the rhizosphere. *Ecology* 84, 827–837. doi: 10.1890/0012-9658(2003)084[0827:HRCTFO]2.0.CO;2

Fitzgerald, J. W. (1976). Sulfate ester formation and hydrolysis: a potentially important yet often ignored aspect of the sulfur cycle of aerobic soils. *Bacteriol. Rev.* 40:698.

Fowler, D., Smith, R., Muller, J., Hayman, G., and Vincent, K. (2005). Changes in the atmospheric deposition of acidifying compounds in the UK between 1986 and 2001. *Environ. Pollut.* 137, 15–25. doi: 10.1016/j.envpol.2004.12.028

Fox, A., Kwapinski, W., Griffiths, B. S., and Schmalenberger, A. (2014). The role of sulfur and phosphorus mobilizing bacteria in biochar induced growth promotion of *Lolium perenne*. *FEMS Microbiol. Ecol.* 90, 78–91. doi: 10.1111/1574-6941.12374

Freney, J., Melville, G., and Williams, C. (1975). Soil organic matter fractions as sources of plant-available sulphur. *Soil Biol. Biochem.* 7, 217–221. doi: 10.1016/0038-0717(75)90041-3

Frey-Klett, P., and Garbaye, J. (2005). Mycorrhiza helper bacteria: a promising model for the genomic analysis of fungal–bacterial interactions. *New Phytol.* 168, 4–8. doi: 10.1111/j.1469-8137.2005.01553.x

Gahan, J., and Schmalenberger, A. (2013). "Bacterial and fungal communities in the mycorrhizospheres of *Agrostis*, *Lolium* and *Plantago* respond to inoculation with arbuscular mycorrhizal fungi," in *Agricultural Research Forum 2014*, ed. M. G. Diskin (Tullamore: Teagasc), 4.

Gahan, J., and Schmalenberger, A. (2014). Arbuscular mycorrhizal hyphae in grassland select for a diverse and abundant hyphospheric bacterial community involved in sulfonate desulfurization. *Appl. Soil Ecol.*

Ghani, A., Mclaren, R., and Swift, R. (1992). Sulphur mineralisation and transformations in soils as influenced by additions of carbon, nitrogen and sulphur. *Soil Biol. Biochem.* 24, 331–341. doi: 10.1016/0038-0717(92)90193-2

Gianinazzi, S., and Schüepp, H. (1994). *Impact of Arbuscular Mycorrhizas on Sustainable Agriculture and Natural Ecosystems.* Berlin: Birkhäuser (Springer). doi: 10.1007/978-3-0348-8504-1

Giovannetti, M., Tolosano, M., Volpe, V., Kopriva, S., and Bonfante, P. (2014). Identification and functional characterization of a sulfate transporter induced by both sulfur starvation and mycorrhiza formation in *Lotus japonicus*. *New Phytol.* 204, 609–619. doi: 10.1111/nph.12949

Gray, L., and Gerdemann, J. (1973). Uptake of sulphur-35 by vesicular-arbuscular mycorrhizae. *Plant Soil* 39, 687–689. doi: 10.1007/BF00264184

Gryndler, M., Hršelová, H., and Stříteská, D. (2000). Effect of soil bacteria on hyphal growth of the arbuscular mycorrhizal fungus *Glomus claroideum*. *Folia Microbiol.* 45, 545–551. doi: 10.1007/BF02818724

Guggenberger, G. (2005). "Humification and mineralization in soils," in *Microorganisms in Soils: Roles in Genesis and Functions*, eds F. Buscot and A. Varma (Berlin: Springer), 85–106. doi: 10.1007/3-540-26609-7_4

Hawkesford, M. J. (2003). Transporter gene families in plants: the sulphate transporter gene family—redundancy or specialization? *Physiol. Plant.* 117, 155–163. doi: 10.1034/j.1399-3054.2003.00034.x

Haynes, R., and Williams, P. (1993). Nutrient cycling and soil fertility in the grazed pasture ecosystem. *Adv. Agron.* 49, 119–199. doi: 10.1016/S0065-2113(08)60794-4

Heinzkill, M., Bech, L., Halkier, T., Schneider, P., and Anke, T. (1998). Characterization of laccases and peroxidases from wood-rotting fungi (family Coprinaceae). *Appl. Environ. Microbiol.* 64, 1601–1606.

Hodge, A., and Storer, K. (2014). Arbuscular mycorrhiza and nitrogen: implications for individual plants through to ecosystems. *Plant Soil* (in press). doi: 10.1007/s11104-014-2162-1

Hummerjohann, J., Laudenbach, S., Rétey, J., Leisinger, T., and Kertesz, M. A. (2000). The sulfur-regulated arylsulfatase gene cluster of *Pseudomonas aeruginosa*, a new member of the cys regulon. *J. Bacteriol.* 182, 2055–2058. doi: 10.1128/JB.182.7.2055-2058.2000

Irshad, U., Villenave, C., Brauman, A., and Plassard, C. (2011). Grazing by nematodes on rhizosphere bacteria enhances nitrate and phosphorus availability to *Pinus pinaster* seedlings. *Soil Biol. Biochem.* 43, 2121–2126. doi: 10.1016/j.soilbio.2011.06.015

Jones, M. G., Hughes, J., Tregova, A., Milne, J., Tomsett, A. B., and Collin, H. A. (2004). Biosynthesis of the flavour precursors of onion and garlic. *J. Exp. Bot.* 55, 1903–1918. doi: 10.1093/jxb/erh138

Kahnert, A., Vermeij, P., Wietek, C., James, P., Leisinger, T., and Kertesz, M. A. (2000). The ssu locus plays a key role in organosulfur metabolism in *Pseudomonas putida* S-313. *J. Bacteriol.* 182, 2869–2878. doi: 10.1128/JB.182.10.2869-2878.2000

Kertesz, M. A. (1999). Riding the sulfur cycle – metabolism of sulfonates and sulfate esters in Gram-negative bacteria. *FEMS Microbiol. Rev.* 24, 135–175.

Kertesz, M. A., Cook, A. M., and Leisinger, T. (1994). Microbial metabolism of sulfur and phosphorus-containing xenobiotics. *FEMS Microbiol. Rev.* 15, 195–215. doi: 10.1111/j.1574-6976.1994.tb00135.x

Kertesz, M. A., Fellows, E., and Schmalenberger, A. (2007). Rhizobacteria and plant sulfur supply. *Adv. Appl. Microbiol.* 62, 235–268. doi: 10.1016/S0065-2164(07)62008-5

Kertesz, M. A., and Mirleau, P. (2004). The role of microbes in plant sulphur supply. *J. Exp. Bot.* 55, 1939–1945. doi: 10.1093/jxb/erh176

King, J., and Quinn, J. (1997). The utilization of organosulphonates by soil and freshwater bacteria. *Lett. Appl. Microbiol.* 24, 474–478. doi: 10.1046/j.1472-765X.1997.00062.x

Klose, S., Moore, J. M., and Tabatabai, M. A. (1999). Arylsulfatase activity of microbial biomass in soils as affected by cropping systems. *Biol. Fertil. Soils* 29, 46–54. doi: 10.1007/s003740050523

Leustek, T., Martin, M. N., Bick, J. A., and Davies, J. P. (2000). Pathways and regulation of sulfur metabolism revealed through molecular and genetic studies. *Ann. Rev. Plant Biol.* 51, 141–165. doi: 10.1146/annurev.arplant.51.1.141

Leustek, T., and Saito, K. (1999). Sulfate transport and assimilation in plants. *Plant Physiol.* 120, 637–644. doi: 10.1104/pp.120.3.637

Linder, T. (2012). Genomics of alternative sulfur utilization in ascomycetous yeasts. *Microbiology* 158, 2585–2597. doi: 10.1099/mic.0.060285-0

Linderman, R. (1988). Mycorrhizal interactions with the rhizosphere microflora: the mycorrhizosphere effect. *Phytopathology* 78, 366–371.

Lynch, J., and Whipps, J. (1990). Substrate flow in the rhizosphere. *Plant Soil* 129, 1–10. doi: 10.1007/BF00011685

Marzluf, G. A. (1997). Molecular genetics of sulfur assimilation in filamentous fungi and yeast. *Annu. Rev. Microbiol.* 51, 73–96. doi: 10.1146/annurev.micro.51.1.73

Morton, J. B., and Benny, G. L. (1990). Revised classification of arbuscular mycorrhizal fungi (Zygomycetes): a new order, *Glomales*, two new suborders, Glomineae and Gigasporineae, and two new families, Acaulosporaceae and Gigasporaceae, with an emendation of Glomaceae. *Mycotaxon* 37, 471–491.

Muralikrishna, C., and Renganathan, V. (1993). Peroxidase-catalyzed desulfonation of 3, 5-dimethyl-4-hydroxy and 3, 5-dimethyl-4-aminobenzenesulfonic acids. *Biochem. Biophys. Res. Commun.* 197, 798–804. doi: 10.1006/bbrc.1993.2549

Nagahashi, G., and Douds, D. D. (2000). Partial separation of root exudate components and their effects upon the growth of germinated spores of AM fungi. *Mycol. Res.* 104, 1453–1464. doi: 10.1017/S0953756200002860

Omar, S., and Abd-Alla, M. (2000). Physiological aspects of fungi isolated from root nodules of faba bean (*Vicia faba* L.). *Microbiol. Res.* 154, 339–347. doi: 10.1016/S0944-5013(00)80008-7

Rhodes, L., and Gerdemann, J. (1978). Influence of phosphorus nutrition on sulfur uptake by vesicular-arbuscular mycorrhizae of onion. *Soil Biol. Biochem.* 10, 361–364. doi: 10.1016/0038-0717(78)90058-5

Richardson, A. E., Barea, J.-M., Mcneill, A. M., and Prigent-Combaret, C. (2009). Acquisition of phosphorus and nitrogen in the rhizosphere and plant growth promotion by microorganisms. *Plant Soil* 321, 305–339. doi: 10.1007/s11104-009-9895-2

Russell, R. S. (1977). *Plant Root Systems: Their Function and Interaction with Soil.* London: McGraw-Hill Book Company.

Scherer, H. (2001). Sulphur in crop production—invited paper. *Eur. J. Agron.* 14, 81–111. doi: 10.1016/S1161-0301(00)00082-4

Schmalenberger, A., Hodge, S., Bryant, A., Hawkesford, M. J., Singh, B. K., and Kertesz, M. A. (2008). The role of *Variovorax* and other Comamonadaceae in sulfur transformations by microbial wheat rhizosphere communities exposed to different sulfur fertilization regimes. *Environ. Microbiol.* 10, 1486–1500. doi: 10.1111/j.1462-2920.2007.01564.x

Schmalenberger, A., Hodge, S., Hawkesford, M. J., and Kertesz, M. A. (2009). Sulfonate desulfurization in *Rhodococcus* from wheat rhizosphere communities. *FEMS Microbiol. Ecol.* 67, 140–150. doi: 10.1111/j.1574-6941.2008.00602.x

Schmalenberger, A., and Kertesz, M. A. (2007). Desulfurization of aromatic sulfonates by rhizosphere bacteria: high diversity of the asfA gene. *Environ. Microbiol.* 9, 535–545. doi: 10.1111/j.1462-2920.2006.01172.x

Schmalenberger, A., Pritzkow, W., Ojeda, J. J., and Noll, M. (2011). Characterization of main sulfur source of wood-degrading Basidiomycetes by S K-edge X-ray absorption near edge spectroscopy (XANES). *Int. Biodeterior. Biodegradation* 65, 1215–1223. doi: 10.1016/j.ibiod.2011.08.013

Siciliano, S. D., Palmer, A. S., Winsley, T., Lamb, E., Bissett, A., Brown, M. V., et al. (2014). Soil fertility is associated with fungal and bacterial richness, whereas pH is associated with community composition in polar soil microbial communities. *Soil Biol. Biochem.* 78, 10–20. doi: 10.1016/j.soilbio.2014.07.005

Sieh, D., Watanabe, M., Devers, E. A., Brueckner, F., Hoefgen, R., and Krajinski, F. (2013). The arbuscular mycorrhizal symbiosis influences sulfur starvation responses of *Medicago truncatula*. *New Phytol.* 197, 606–616. doi: 10.1111/nph.12034

Smith, S. E., and Read, D. J. (1997). *Mycorrhizal Symbiosis*. Oxford: Academic Press.

Tabatabai, M., and Bremner, J. (1970). Arylsulfatase activity of soils. *Crop Sci. Soc. Am. J.* 34, 225–229. doi: 10.2136/sssaj1970.03615995003400020016x

Toljander, J. F., Lindahl, B. D., Paul, L. R., Elfstrand, M., and Finlay, R. D. (2007). Influence of arbuscular mycorrhizal mycelial exudates on soil bacterial growth and community structure. *FEMS Microbiol. Ecol.* 61, 295–304. doi: 10.1111/j.1574-6941.2007.00337.x

Tuor, U., Winterhalter, K., and Fiechter, A. (1995). Enzymes of white-rot fungi involved in lignin degradation and ecological determinants for wood decay. *J. Biotechnol.* 41, 1–17. doi: 10.1016/0168-1656(95)00042-O

Uria-Nickelsen, M. R., Leadbetter, E. R., and Godchaux, W. III. (1993). Sulfonate-sulfur assimilation by yeasts resembles that of bacteria. *FEMS Microbiol. Lett.* 114, 73–77. doi: 10.1111/j.1574-6968.1993.tb06553.x

Vermeij, P., Wietek, C., Kahnert, A., Wüest, T., and Kertesz, M. A. (1999). Genetic organization of sulphur-controlled aryldesulphonation in *Pseudomonas putida* S-313. *Mol. Microbiol.* 32, 913–926. doi: 10.1046/j.1365-2958.1999.01398.x

Vilarino, A., Frey, B., and Shüepp, H. (1997). MES [2-(N-morpholine)-ethane sulphonic acid] buffer promotes the growth of external hyphae of the arbuscular mycorrhizal fungus *Glomus intraradices* in an alkaline sand. *Biol. Fertil. Soils* 25, 79–81. doi: 10.1007/s003740050284

Wang, B., and Qiu, Y. L. (2006). Phylogenetic distribution and evolution of mycorrhizas in land plants. *Mycorrhiza* 16, 299–363. doi: 10.1007/s00572-005-0033-6

Warmink, J. A., and van Elsas, J. D. (2008). Selection of bacterial populations in the mycosphere of *Laccaria proxima*: is type III secretion involved. *ISME J.* 2, 887–900. doi: 10.1038/ismej.2008.41

Yadav, J. S., Lawrence, D. L., Nuck, B. A., Federle, T. W., and Reddy, C. A. (2001). Biotransformation of linear alkylbenzene sulfonate (LAS) by *Phanerochaete chrysosporium*: oxidation of alkyl side-chain. *Biodegradation* 12, 443–453. doi: 10.1023/A:1015056530264

Zhao, F., Knights, J., Hu, Z., and Mcgrath, S. (2003). Stable sulfur isotope ratio indicates long-term changes in sulfur deposition in the Broadbalk experiment since 1845. *J. Environ. Qual.* 32, 33–39. doi: 10.2134/jeq2003.3300

Zhao, F., Lehmann, J., Solomon, D., Fox, M., and Mcgrath, S. (2006). Sulphur speciation and turnover in soils: evidence from sulphur K-edge XANES spectroscopy and isotope dilution studies. *Soil Biol. Biochem.* 38, 1000–1007. doi: 10.1016/j.soilbio.2005.08.013

Conflict of Interest Statement: The authors declare that the research was conducted in the absence of any commercial or financial relationships that could be construed as a potential conflict of interest.

8

The impact of sulfate restriction on seed yield and quality of winter oilseed rape depends on the ability to remobilize sulfate from vegetative tissues to reproductive organs

Alexandra Girondé[1,2,3†], Lucie Dubousset[1,2,3†], Jacques Trouverie[1,2,3], Philippe Etienne[1,2,3] and Jean-Christophe Avice[1,2,3*]

[1] Normandie University, Caen, France
[2] Université de Caen Basse Normandie, UMR INRA–UCBN 950 Ecophysiologie Végétale, Agronomie and Nutritions N.C.S., Caen, France
[3] Institut National de la Recherche Agronomique, UMR INRA–UCBN 950 Ecophysiologie Végétale, Agronomie and Nutritions N.C.S., Caen, France

Edited by:
Stanislav Kopriva, University of Cologne, Germany

Reviewed by:
Ruediger Hell, University of Heidelberg, Germany
Karl H. Muehling, Kiel University, Germany

****Correspondence:***
Jean-Christophe Avice, UMR INRA–UCBN 950 EVA, Ecophysiologie Végétale, Agronomie and Nutritions N.C.S., Université de Caen Basse-Normandie, Esplanade de la Paix, F-14032 Caen, France
e-mail: jean-christophe.avice@unicaen.fr

† These authors have contributed equally to this work.

Our current knowledge about sulfur (S) management by winter oilseed rape to satisfy the S demand of developing seeds is still scarce, particularly in relation to S restriction. Our goals were to determine the physiological processes related to S use efficiency that led to maintain the seed yield and quality when S limitation occurred at the bolting or early flowering stages. To address these questions, a pulse-chase $^{34}SO_4^{2-}$ labeling method was carried out in order to study the S fluxes from uptake and remobilization at the whole plant level. In response of S limitation at the bolting or early flowering stages, the leaves are the most important source organ for S remobilization during reproductive stages. By combining ^{34}S-tracer with biochemical fractionation in order to separate sulfate from other S-compounds, it appeared that sulfate was the main form of S remobilized in leaves at reproductive stages and that tonoplastic SULTR4-type transporters were specifically involved in the sulfate remobilisation in case of low S availability. In response to S limitation at the bolting stage, the seed yield and quality were dramatically reduced compared to control plants. These data suggest that the increase of both S remobilization from source leaves and the root proliferation in order to maximize sulfate uptake capacities, were not sufficient to maintain the seed yield and quality. When S limitation occurred at the early flowering stage, oilseed rape can optimize the mobilization of sulfate reserves from vegetative organs (leaves and stem) to satisfy the demand of seeds and maintain the seed yield and quality. Our study also revealed that the stem may act as a transient storage organ for remobilized S coming from source leaves before its utilization by seeds. The physiological traits (S remobilization, root proliferation, transient S storage in stem) observed under S limitation could be used in breeding programs to select oilseed rape genotypes with high S use efficiency.

Keywords: *Brassica napus* L., sulfate restriction, ^{34}S labeling, S flux, sulfate transporter, seed yield, seed quality

INTRODUCTION

Oilseed rape (*Brassica napus* L.) is a crop that produces seeds with high contents of oil and proteins for human and animal nutrition as well as non-edible uses. To provide an adequate amount of S for oilseed rape culture, the recommendations are about 20–30 kg S.ha^{-1}. Oil and protein concentrations in seeds have been shown to increase with S fertilization (Ahmad et al., 2007; Malhi et al., 2007). In addition, the application of S fertilizers also improves N-use efficiency and maintains a sufficient oil level and fatty acid quality in oilseed rape (Schnug et al., 1993; Fismes et al., 2000; Dubousset et al., 2010). In modern-grown oilseed rape varieties (i.e., double low varieties with zero erucic acid and a reduced content of seed glucosinolates), the S harvest index (S content in seeds divided by the total S in the whole crop) is typically only 20% (McGrath and Zhao, 1996). This indicates that a large proportion of S taken up by the crop is retained in the vegetative tissues and pods.

In order to increase crop competitiveness, the oil yield needs to be improved while limiting fertilizer inputs. Compared with other crops such as cereals, oilseed rape is particularly sensitive to S deficiency because it has a high demand for S (Zhao et al., 1997). The reduction in atmospheric deposition of S has increased the incidence of S deficiency in oilseed rape (McGrath and Zhao, 1996). S deprivation in oilseed rape leads to reduced seed yields and oil quality (Janzen and Bettany, 1984; Scherer, 2001). Recent work based on proteomics approaches in mature seeds obtained from winter oilseed rape plants grown under low sulfate applied at the bolting, early flowering or start of pod filling stages (D'Hooghe

Abbreviations: DM, Dry Matter; FAME, fatty acid methyl ester; GC, gas chromatography; GS, growth stage; HS, High S; LS, Low S; NIRS, Near Infrared Spectroscopy; S, Sulfur; V-CDT, Vienna-Canyon Diablo Troilite.

et al., 2014) have revealed that the protein quality of seeds was reduced depending on the severity of S limitation and was associated with a reduction in S-rich seed storage protein accumulation (such as Cruciferin Cru4) which favored S-poor seed storage protein (such as Cruciferin BnC1). Nevertheless, our knowledge about the stages of development the more sensitive to S limitation or the physiological processes that are involved in S management by oilseed rape subjected to sulfate restriction remains largely unclear (Dubousset et al., 2010).

In oilseed rape, S is mainly taken up by the roots as sulfate (SO_4^{2-}) and transported *via* the xylem to the leaves by specific transporters (Hawkesford and De Kok, 2006; Takahashi et al., 2011). Sulfate is mainly reduced to Cysteine (Cys) in leaves, and either converted to Methionine or incorporated into proteins, Cys-containing peptides such as glutathione, or numerous secondary metabolites involved in plant defense (Sulfur-containing Defense Compounds, SDC) such as glucosinolates. The root uptake and subsequent distribution of SO_4^{2-} to the leaves is closely related to growth demand and mineral S availability in the soil. Sulfate stored in vacuoles is the main form of S released by the mesophyll cells under low S nutrition conditions (Blake-Kalff et al., 1998; Scherer, 2001; Matula and Pechová, 2002; Parmar et al., 2007). Indeed, a reduction in the sulfate supply leads to an up-regulation of genes encoding for specific transporters involved in (i) sulfate uptake by roots, and (ii) sulfate remobilization at the level of roots and leaves, especially *BnSultr4;1* and *BnSultr4;2*, which are involved in vacuolar efflux of sulfate (Parmar et al., 2007; Dubousset et al., 2009; Abdallah et al., 2010).

The leaves play a crucial role in recycling foliar compounds to sustain seed filling during the reproductive stages and therefore contribute to the maintenance of seed yield of oilseed rape. Noquet et al. (2004) have reported that ablation of 50% of the leaves present at the bolting stage results in a 30% decrease in seed yield in oilseed rape. In the case of a transient mineral S limitation occurring at the rosette stage, winter oilseed rape is able to maintain its growth by optimization of the recycling of endogenous foliar S compounds (particularly sulfate) from old and mature leaves without any acceleration of leaf senescence processes (Dubousset et al., 2009; D'Hooghe et al., 2013). Although mobilization of S from vegetative tissues is likely to be crucial for seed filling, very little is known about the dynamics, the efficiency and the contribution of S mobilization from vegetative tissues to seeds in oilseed rape.

In this study we investigated the S management of winter oilseed rape in response to S restriction applied at two crucial growth stages (bolting or early flowering). The impact of sulfate limitation was studied on seed yield and quality. To evaluate the level of S remobilization from vegetative parts using stable isotope as tracer, a greenhouse experiment was carried out for long-term steady state ^{34}S-SO_4^{2-} pulse labeling. This method is particularly appropriate for studying S management at the whole plant level and for showing the source/sink relationships for S from uptake or remobilisation. In addition, fluxes of S remobilization were studied in relation to the dynamics of mobilization of S compounds (sulfate vs. S-reduced compounds) as well as the gene expression of sulfate transporters (SULTR) of group 4 in response to sulfate availability in the main source organs i.e., mature leaves.

MATERIALS AND METHODS

EXPERIMENTAL TREATMENTS, MINERAL S RESTRICTION AND ^{34}S LABELING

The experimental design was previously described by Dubousset et al. (2010). Briefly, seedling of winter oilseed rape (*Brassica napus* cv Capitol) were first grown for 36 days on vermiculite under a thermoperiod of 20°C (day-16 h) and 15°C (night-8 h). To initiate flowering and the reproductive stages of development, a period of vernalization was applied to plants for 46 days, consisting of a thermoperiod of 10°C (day 10 h) and 4°C (night 14 h). During pre-culture and vernalization, light was supplied by High Pressure Sodium Lamps (Philips, MASTER GreenPower T400W) with a PAR (Photosynthetically Active Radiations) of 400 μmol photon.s^{-1}.m^{-2} at the top of the canopy, and plants were supplied with 25% Hoagland nutrient solution (1.25 mM $Ca(NO_3)_2$,$4H_2O$, 1.25 mM KNO_3, 0.5 mM $MgSO_4$, 0.25 mM KH_2PO_4, 0.2 mM EDTA 2NaFe,$3H_2O$, 14 μM H_3BO_3, 5 μM $MnSO_4$, 3 μM $ZnSO_4$, 0.7 μM $(NH_4)_6Mo_7O_{24}$, 0.7 μM $CuSO_4$, 0.1 $CoCl_2$) renewed twice a week.

After vernalization, plants were individually transferred into 2.5 L pots containing mixed vermiculite (1V) and perlite (2V). Greenhouse conditions were maintained at a thermoperiod of 20°C (day-16 h) and 15°C (night-8 h). Plants grew under natural light conditions and received 25% Hoagland nutrient solution as described above. In order to determine the endogenous and exogenous S fluxes at the whole plant level, a long term pulse-chase ^{34}S labeling ($^{34}SO_4^{2-}$ at 1 atom% excess) was applied from the end of the vernalization period up to the beginning of the S limitation treatments. S restriction (Low S: LS) equivalent to 8.7 μM sulfate was applied at two different stages of development: at GS32 (bolting stage, LS32 plants) or GS53 (early flowering, LS53 plants). When S restriction started, ^{34}S-labeling was stopped and unlabeled-sulfate was supplied in the nutrient solution until the final stage of development (GS99: mature seeds). For each S limitation treatment, control plants (High S treatment: HS) supplied with 508,7 μM sulfate were also submitted to the same period of pulse-chase ^{34}S labeling in order to compare the S fluxes between S-limited and non-limited (control) plants. LS and HS plants were harvested at GS32, GS53, GS70 (first pods), GS81 (seed coloring) and GS99 (mature seeds). Plants were separated into roots, leaves, stems, floral stems, pod walls and seeds. The age of leaves were determined by the relative chlorophyll concentration using a non-destructive SPAD (Soil Plant Analysis Development) chlorophyll meter (Minolta, SPAD-502 model) and the measurement of leaf area using a LI-COR 300 area meter (LI-COR, Lincoln, NE, USA). Based on leaf area and chlorophyll level, two mature leaves (leaf ranks 7 and 8) that became source for S during the experiment, were chosen for further biochemical and molecular analyses (sulfate, residual S in dead leaves, expression of sulfate transporters of group 4) in relation to S remobilization. Each organ was freeze-dried, weighed and ground to a fine powder to be analyzed. Aliquots of fresh matter were also stored at −80°C until further gene expression analyses.

DETERMINATION OF TOTAL S AMOUNT, ^{34}S AMOUNT AND S FLUXES

For each organ and each date of harvest, S and ^{34}S analyses were performed with an elemental analyser (EA3000, EuroVector,

Milan, Italy) linked to a continuous flow isotope mass spectrometer (IRMS, IsoPrime GV instruments, Manchester, UK). The long-term period of isotope labeling described previously allowed a homogenous incorporation of tracers into the whole plant. Therefore, the amount of ^{34}S in excess was proportional to the amount of S and allowed identification of the "sink" or "source" status of organs. As such, an organ was considered as "a sink" if a gain in the ^{34}S amount was observed for a period Δt. Inversely, a "source" organ was characterized by a loss in the amount of ^{34}S for a period Δt. For the determination of S fluxes (Salon et al., 2014), it is necessary to quantify the ^{34}S amount in excess ($Q^{34}S_{in\ excess}$) in each organ. For this, the value of isotope abundance (A%) given by the IRMS was used as follows:

$$Q^{34}S_{in\ excess}(mg) = [(A\%_{sample} - A\%_{natural\ standard})/100] \times QS$$

where $A\%_{natural\ standard}$ corresponds to natural ^{34}S abundance i.e., 4.2549%, QS is the amount of total S in a given organ in mg, and $A\%_{sample}$ corresponds to ^{34}S isotope abundance in a given organ. A% in a sample or natural standard is calculated as:

$$A\% = 100 \times \left[^{34}S/\left(^{34}S + ^{32}S\right)\right]$$

For every organ on each date, ^{34}S amounts were normalized ($Q^{34}S_{normalized}$) with the following calculation at every date (t):

$$Q^{34}S_{normalized} = \left(Q^{34}St_{organ} \times \text{Average } Q^{34}S \text{ whole plant}_{(all\ dates\ included)}\right) / Q^{34}St_{whole\ plant}$$

where $Q^{34}St_{organ}$ indicates the ^{34}S amount in excess in a given organ at the date t, Average $Q^{34}S$ whole plant $_{(all\ dates\ included)}$ represents the average ^{34}S amount in the whole plant at every study date and $Q^{34}St_{whole\ plant}$ indicates the ^{34}S amount in the whole plant at the date t.

For source organs (decline in the ^{34}S amount between t and $t+\Delta t$), the S amount remobilized (QSR_{source}) was defined as:

$$QSR_{source} = \left[QS_{total}t \times \left(Q^{34}St - Q^{34}St + \Delta t\right)\right] / Q^{34}St$$

where $QS_{total}t$ indicates total S amount in the source organ at the date t, $Q^{34}St$ represents the ^{34}S amount in excess in the source organs at the date t and $Q^{34}St + \Delta t$ indicates the ^{34}S amount in excess in the source organs at $t+\Delta t$.

For sink organs (gain of ^{34}S amount between t and $t+\Delta t$), the S amount remobilized (QSR_{sink}) corresponded to:

$$QSR_{sink} = \left(Q^{34}St + \Delta t - Q^{34}St\right) / \left[\left(\Sigma Q^{34}S_{source\ organs}t\right) - \left(\Sigma Q^{34}S_{source\ organs}t + \Delta t\right) / QS_{total} \text{ remobilized}\right]$$

where $\Sigma Q^{34}S_{source\ organs}t$ indicates the total ^{34}S amount in excess from source organs at t, $\Sigma Q^{34}S_{source\ organs}t + \Delta t$ represents the total ^{34}S amount in excess from source organs at $t+\Delta t$ and QS_{total} remobilized indicates the total S amount in excess remobilized (from all source organs).

The inflow of S taken up was also calculated between t and $t+\Delta t$.

For source organs, the S amount taken up (QSI_{source}) was defined as:

$$QSI_{source} = (QSt + \Delta t - QSt) + QSR_{source}$$

For sink organs, the S amount taken up (QSI_{sink}) corresponded to:

$$QSI_{sink} = (QSt + \Delta t - QSt) - QSR_{sink}$$

where QSR_{source} indicates the S amount from the studied source organ and QSR_{sink} indicates the S amount from the studied sink organ.

SULFATE AMOUNT AND DETERMINATION OF ^{34}S IN SULFATE

Forty-five milligrams of lyophilized powder from source leaves were used to extract sulfate after two successive incubations with 2 mL of 50% ethanol at 45°C for 1 h, centrifugation at 10,000 g for 20 min and two incubations with water at 95°C for 1 h, ending with centrifugation at 10,000 g for 20 min. The supernatants were pooled and evaporated under vacuum (Concentrator Evaporator RC 10.22, Jouan, Saint-Herblain, France). The dry residue was re-suspended in 2 mL of ultra-pure water and the sulfate concentration was determined by high performance liquid chromatography (HPLC; DX100, DIONEX Corp, Sunnyvale, CA, USA). This extract contains sulfate and other soluble S compounds. Therefore, in order to determine the contribution of sulfate to the S remobilization from source leaves it was necessary to purify sulfate and quantify the changes in S-sulfate and ^{34}S-sulfate. Thus, this extract passed through a DOWEX 50W (H+) column (4 cm long, 1 cm of diameter) and was eluted with 7.5 mL of water. The resulting eluted fraction contained purified sulfate. A part of this fraction was evaporated under vacuum and finally re-suspended with ultra-pure water in a volume to obtain an S concentration of around 0.85 $\mu g\ S.\mu L^{-1}$. Thirty microliters were placed in a tin capsule and the S and ^{34}S amounts were determined with an elemental analyser coupled with IRMS as described above. The rest of the fraction (about 1.5 mL) was used to determine the sulfate concentration by HPLC (as described above) in order to verify the purity of the sulfate fraction after comparison with the IRMS data.

DETERMINATION OF OIL AND GLUCOSINOLATE CONTENTS BY NIRS

The seed samples were scanned on a monochromator NIR system model 6500 (FOSS NIRSystems Inc, Silver Spring, MD, USA) equipped with the transport module, in the reflectance mode. Intact seeds (about 5 g) were placed in a standard ring cup and scanned. The results were predicted from an external calibration established for oil and total glucosinolate content (CRAW, Gembloux, Belgium). Three determinations were performed for each sample. The results were given in % of oil per DM seed and in μmol of total glucosinolates per g of DM seed.

DETERMINATION OF OIL CONTENT AND LIPID COMPOSITION OF SEEDS

The method for determination of seed oil content was based on direct methylation of fatty acids. Briefly, 10 mature dried seeds were ground in a microtube tube containing 3 inox balls using a Tissue Lyser system (Qiagen, Chatsworth, CA, USA). For each sample, three aliquots of 10 mg were weighed and transferred into glass tubes with Teflon-lined screw caps containing 1.32 mL of a freshly prepared solution of methanol/sulfuric acid/toluene (100/2.5/30, v/v/v) with 400 μg.mL^{-1} of heptadecanoic acid as an internal standard. The mixture was shaken vigorously for 30 s and then heated at 95°C for 1 h. After cooling on ice for 10–15 min, fatty acid methyl esters (FAMEs) were extracted into 500 μL of hexane following the addition of 1 mL of water. After vigorous hand shaking (15 s) and centrifugation (5 min, 2000 rpm), 10 μL of the upper organic phase was analyzed by gas chromatography (GC). If necessary, extracts were evaporated under nitrogen and dissolved into 50 μL of hexane before GC analysis. FAMEs were separated on a DB-WAX column (30 m l. × 0.25 mm i.d., 25 μm film, J&W Scientific Columns from Agilent Technologies Co., Palo Alto, CA, USA) and quantified with a flame ionization detector. The GC conditions were as follows: split mode injection (1:100), injector temperature at 250°C, oven temperature fixed at 220°C for 15 min and detector temperature at 260°C, with hydrogen as the carrier gas at a pressure of 100 kPa and a flow rate of 1 mL.min^{-1}, in a 6890N GC GLC system (Agilent Technologies). To determine the mass of each fatty acid, the peak area was compared to the internal standard peak area. Results were given in μg of fatty acid per mg of DM seed and were expressed as % of control (HS plants).

TRANSCRIPTOMIC ANALYSIS

Total RNA was extracted from 200 mg of fresh matter previously ground to a powder in a mortar containing liquid nitrogen and immersed in a hot mixture (80°C) containing 750 μL of phenol pH 4.3 and 750 μL of extraction buffer (0.1 M LiCl, 0.1 M Tris Base, 10 mM EDTA, 1% SDS (w/v), pH 8). Samples were vortexed for 40 s and 750 μL of chloroform/isoamylalcohol (24/1 v/v) were added. Samples were vortexed for 40 s again and centrifuged at 1500 g (4°C, 5 min). The supernatant was transferred to 750 μL of LiCl (4M) and incubated overnight at 4°C. After incubation, samples were centrifuged at 1500 g (4°C, 20 min). The supernatant was discarded and the pellet was suspended in 100 μL of sterile water. Then, RNA purification was made by Qiagen® RNeasy Kit according to the manufacturer's recommendation. RNA quantity was measured by a spectrophotometer (BioPhotometer, Eppendorf®) at 260 nm and their quality was check on an agarose gel (1.2% (w/v)).

A reverse transcription (RT) was performed on the purified RNA. For this, 1 μg of RNA was converted to cDNA using an "iScript cDNA synthesis kit" (Bio-Rad, Marne-La-Coquette, France) according to the manufacturer's recommendation. The obtained cDNA was used to perform Q-PCR using specific primers for *Brassica napus BnSultr4;1* (Accession no: AJ416461; forward primer: 5′-GACCAGACCCGTTAAGGTCA-3′, reverse primer: 5′TTGGAATCCATGTGGAAGCAA-3′) and *BnSultr4;2* (Accession no: FJ688133; forward primer: 5′-AGCAAGATCAGG GATTGTGG-3′, reverse primer: 5′-TGCAACATTTGTGGGT GTCT-3′) (Dubousset et al., 2009). Two internal control genes were used in this experiment: *EF1-α* (Accession no: DQ312264; forward primer: 5′-GCCTGGTATGGTTGTGACCT-3′, reverse primer: 5′-GAAGTTAGCAGCACCCTTGG-3′) and *18S RNA* (Accession no: X16077; Forward primer: 5′-CGGATAACCGTA GTAATTCTAG-3′, reverse primer: 5′-GTACTCATTCCAATTA CCAGAC-3′). The Q-PCR was performed with Qbiogene Taq polymerase (MP Biomedicals, Illkirch, France) according to the manufacturer's protocol, and the amplification program was: 1 cycle at 95°C for 3 min, 40 cycles including one step at 95°C for 15 s, one at 60°C for 60 s and fluorescence reading. Lastly, a thermal decoupling was executed to obtain a melting curve to verify the specificity of the Q-PCR. To normalize our quantitative cycle (Cq) of the genes of interest (*BnSultr4;1* and *4;2*), a calculation was made as:

ΔCq = Quantitative cycle of the gene of interest (Cq) – (geometric average of quantitative cycle of internal control genes)

Then, the $\Delta\Delta$Cq method is applied as:

$\Delta\Delta Cq = \Delta Cq$ treated sample – ΔCq control sample,

and the relative expression (RE) was calculated as:

$RE = (1 + E)^{-[\Delta\Delta Cq]} \approx 2^{-[\Delta\Delta Cq]}$

where E = Q-PCR efficiency.

STATISTICS

All data were performed with MINITAB13 on Windows® (Minitab Inc., State College, PA, USA). The normality of the data was studied with the Ryan–Joiner test at 95%. Analysis of variance (ANOVA) and the Tukey test were used to compare the means. When the normality law of data was not respected, the non-parametric test of Kruskal–Wallis was carried out. Statistical significance was postulated at $P < 0.05$. Four biological repetitions were used ($n = 4$) and all the data presented here are expressed as the mean value ± standard error (SE).

RESULTS

IMPACT OF S LIMITATION OCCURRING AT THE BOLTING STAGE (LS32 PLANTS)

Growth, S amount, seed yield and seed quality

When S limitation occurred at the bolting stage (LS32 plants), the dry matter of roots and inflorescences was significantly higher than in control (HS plants) with an increase of +42.3% (roots) and +59.7% (inflorescences) (**Figure 1A**). At the beginning of pod development (GS70), root dry matter was still significantly higher than in HS plants (+34%). At GS81 (seed coloring), an increase in dry matter was observed for leaves (+52.4%) and pod walls (+20.7%) in LS32 plants compared to control. As expected, a decrease in seed dry matter was observed at GS81 in response to S restriction (−34.3% compared to control HS plants; **Figure 1A**).

At the early flowering stage (GS53), the S amount increased only in roots (+60.8%) in response to sulfate limitation occurring at the bolting stage. As expected, at GS81, S limitation led to a decline in S amount in all aerial organs compared with the HS plants (**Figure 1B**). Thereafter, compared to HS plants, S amount decreased for all organs in later growing stages in LS32 plants.

At the mature seed stage (**Figure 1C**), the seed dry matter per plant was significantly reduced in LS32 plants (−54.3% compared to HS plants). Seed quality was also highly impacted by S limitation at the bolting stage. Protein ($p < 0.05$), glucosinolate ($p < 0.01$) and oil ($p < 0.01$) contents were lower in LS32

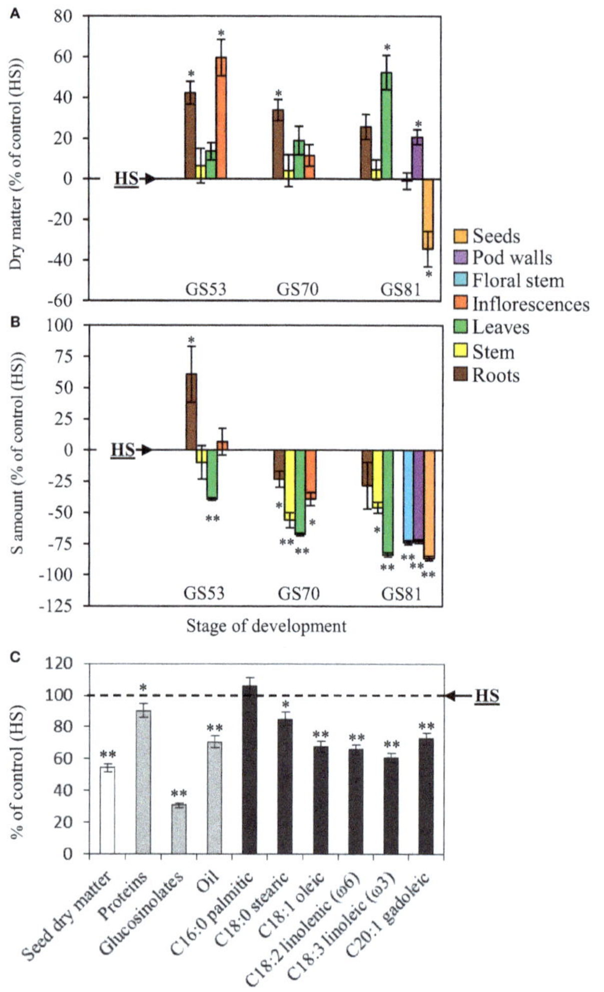

FIGURE 1 | Evolution of the dry matter (A), changes in the S amount (B) and seed yield and composition (C) of oilseed rape subjected to limitation of sulfate at the bolting stage. S limitation (LS32 plants supplied with 8.7 μM sulfate) was compared to control plants (HS plants supplied with 508.7 μM sulfate). **(A)** Variations in dry matter of LS32 plants (as % of HS plants) between GS53 (early flowering), GS70 (start of pod filling) and GS81 (seed coloring). **(B)** Variations in the S amount in different organs of LS32 plants (as % of HS plants) between GS53 and GS81. A positive or a negative value indicates that the dry matter or the S amount is increased or reduced compared with HS plants, respectively. **(C)** Seed yield and composition of seeds (proteins, glucosinolates, oil, fatty acids) in LS32 plants (as % of HS plants) at mature seed stage (GS99). Significant differences between treatments are indicated with asterisks ($n = 4$; *$P < 0.05$; **$P < 0.01$).

plants compared to HS plants. Oil composition was also significantly affected. Indeed, except for palmitic acid (C16:0) which did not differ from HS seeds, there was a decrease in all fatty acids, especially linoleic acid (C18:3 belonging to omega-3) and linolenic acid (C18:2 belonging to omega-6).

S management at the whole plant level: uptake, remobilization and loss by fallen leaves

The ^{34}S labeling pulse-chase method used in this experiment was performed to determine the exogenous S flux (from S uptake) and endogenous S flux (from remobilization of S reserves) at different stages of development (**Figure 2**).

Between bolting (GS32) and the early flowering (GS53) stages (**Figure 2A**), the S taken up from the soil in control (HS) plants was mainly allocated to leaves (64.1% of total S uptake) with nearly equal amounts going to roots and stems (11.3 and 12.3%, respectively). A significant part of this S is lost through dead leaves (11.3%). Despite the low level of S fertilization, a low but significant S uptake was observed in LS32 plants and this S taken up was mainly allocated to leaves, but in a smaller proportion (40.2%) than in HS plants (**Figure 2A**). The determination of S remobilization fluxes (**Figures 2B,C**) indicated that leaves were the only source organs in both sulfate treatments. The main sink for S remobilized from source leaves was the stems in HS plants (31.3%) and roots in LS32 plants (20%) (**Figure 2C**). This large redistribution to roots in LS32 plants can be explained by the increase in root dry matter (**Figure 1A**) and can explain the increase in total S in this organ (**Figure 1B**). Surprisingly, more than a half of the remobilized S was lost through dead leaves in HS plants (58.7% of total remobilized S). By contrast, only 23.3% of remobilized S was lost due to leaf fall in LS32 plants (**Figure 2C**).

Between the early flowering stage (GS53) and the start of pod filling (GS70) (**Figure 2E**), leaves were the main sink organs for S taken up in HS plants (42.2%) and the remainder was allocated to inflorescences (20%), stem (17.3%) and roots (10.8%). The proportion of S lost by dead leaves was the same as in previous growing stages (about 10%). For LS32 plants, exogenous S was still taken up between GS53 and GS70 and was equally distributed between leaves (25%), roots (22%) and stem (21.2%). As before, the increase in root sink strength in response to LS32 treatment was probably due to the increase in dry matter (**Figure 1A**). The S remobilization fluxes (**Figures 2F,G**) revealed that (i) leaves were the only source organs (**Figure 2F**) and (ii) the inflorescences were the main sink organs in both S treatments (60.8% for HS and 41.3% for LS32 plants; **Figure 2G**). S was highly remobilized to stems in response to LS32 treatment (22.7%) compared to HS plants (4.5%; **Figure 2G**). The same proportion of remobilized S was lost by dead leaves in both treatments (22.73 and 20% for HS and LS32 plants, respectively) (**Figure 2G**).

Unlike HS plants, there was no S uptake by LS32 plants between the onset of pod filling (GS70) and the seed coloring (GS81) stages (**Figure 2I**). In HS plants, the main sinks for S taken up were pod walls (33.3% of total S taken up) and seeds (28%). The S taken up that was lost by dead leaves at this time was lower than in previous stages of development but was still significant (7.5%; **Figure 2I**). The leaves were the main source organs for S remobilization (57.8 and 48.5% for HS and LS32 plants, respectively; **Figure 2I**). The contribution of roots and inflorescences as source organs for remobilized S in LS32 plants was stronger in response to S limitation (23.3% for roots; 25.6% for inflorescences) than in HS plants (10% for roots; 16.6% for inflorescences) (**Figure 2J**). In response to LS32 treatment a larger part of the remobilized S was distributed to pod walls (30.3% compared to 17.4% for HS plants) to the detriment of seeds (21.5% compared to 43.2% for HS plants; **Figure 2K**). As previously observed between GS53 and GS70, the proportion of S lost

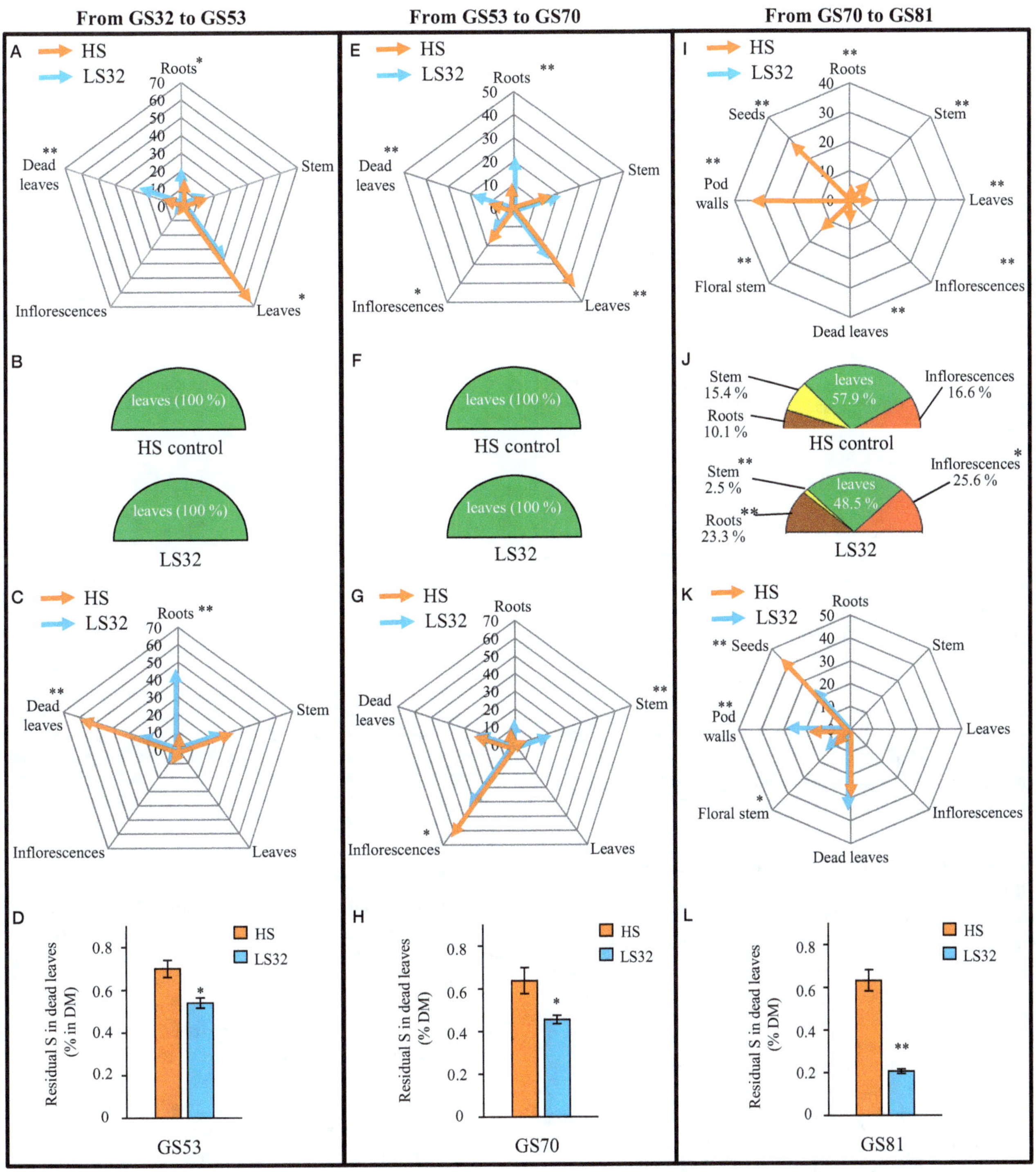

FIGURE 2 | Sulfur fluxes (A–C, E–G, I–K) and residual S in dead leaves (D,H,L) of oilseed rape subjected to limitation of sulfate at the bolting stage. S limitation (LS32 plants supplied with 8.7 μM sulfate) was compared to control plants (HS plants supplied with 508.7 μM sulfate) between GS32 (bolting stage) and GS81 (seeds coloring). Flux of S taken up from the soil allocated to the sink organs are indicated as % of total S taken up between **(A)** GS32 and GS53 (early flowering), **(E)** GS53 and GS70 (start of pod filling) and **(I)** GS70 and GS81 (seeds coloring). The diagrams **(B,F,J)** represent the contribution of source organs to the S remobilization in HS and LS32 plants between GS32 and GS53, GS53 and GS70, and GS70 and GS81, respectively (data are expressed as % of total S remobilized). Flux of S remobilized from the source organs toward the sink organs are mentioned as % of total remobilized S between **(C)** GS32 and GS53, **(G)** GS53 and GS70, and **(K)** GS70 and GS81. Residual S in dead leaves (as % of dry matter) of HS and LS32 plants are indicated at **(D)** GS53, **(H)** GS70 and **(L)** GS81. Significant differences between treatments are indicated with asterisks ($n = 4$; $*P < 0.05$; $**P < 0.01$).

by dead leaves was the same in both S treatments between GS70 and GS81 (**Figure 2K**).

The residual S percentage in fallen leaves did not change during the whole experiment in HS plants (0.67 ± 0.03% in average; **Figures 2D,H,L**). On the contrary, in LS32 plants, a decrease in residual S in dead leaves was observed from GS53 and it is lower than in HS plants at every growing stage (0.54, 0.45 and 0.20% at GS53, GS70 and GS81, respectively; **Figures 2D,H,L**).

Involvement of sulfate and vacuolar sulfate transporters in S remobilization of source leaves

Analysis of S fluxes has revealed that leaf S remobilization was improved in response to an S deficiency. Therefore, the dynamics of sulfate and S-reduced compounds (including S-amino acids, glutathione, proteins and other S-organic compounds) in source leaves were investigated in relation to (i) flux of leaf S and ^{34}S remobilization and (ii) transcript levels of the *BnSULTR4;1* and *4;2* transporters involved in vacuolar efflux of sulfate. In response to S restriction occurring at the bolting stage, the decrease in S in source leaves was mainly related to a strong decline of sulfate while S-reduced compounds slightly decreased at the early flowering stage (GS53) or remained constant at the onset of pod filling (GS70; **Figure 3A**). Moreover, a decrease of ^{34}S in the sulfate fraction was observed at GS53 and GS70 for LS32 plants, while the ^{34}S in S-reduced compounds remained constant at these growth stages (**Figure 3B**). In parallel to this decrease in sulfate, an induction of *BnSULTR4;1* transcripts was observed in source leaves of LS32 plants (3-fold and 2.5-fold higher than in HS plants at GS53 and GS70, respectively; **Figures 4A,B**). Compared to HS plants, the transcript level of *BnSULTR4;2* in source leaves of LS32 plants was highly up-regulated (9 and 60-fold higher at GS53 and GS70, respectively; **Figures 4C,D**).

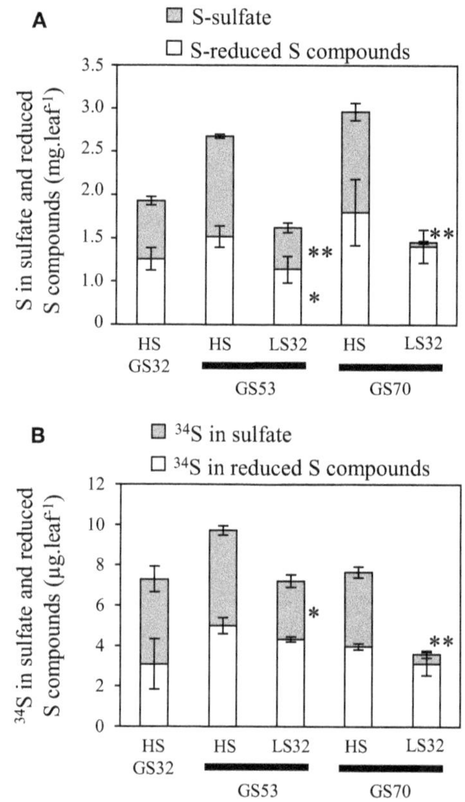

FIGURE 3 | Changes in S (A) and ^{34}S (B) amount determined in the sulfate and reduced S-compounds fractions in source leaves of oilseed rape subjected to limitation of sulfate at the bolting stage. S limitation (LS32 plants supplied with 8.7 μM sulfate) was compared to control plants (HS plants supplied with 508.7 μM sulfate) between GS32 (bolting stage), GS53 (early flowering) and GS70 (start of pod filling). Significant differences between treatments are indicated with asterisks ($n = 4$; $^{*}P < 0.05$; $^{**}P < 0.01$).

IMPACT OF S LIMITATION OCCURRING AT THE EARLY FLOWERING STAGE (LS53 PLANTS)

Growth, S amount, seed yield and seed quality

Compared to control (HS plants), a limitation of sulfate occurring at the early flowering stage (LS53 plants) had no effect on dry matter of all organs except for dry matter of leaves at GS81 (-61.3% compared to HS, **Figure 5A**). The S amount of LS53 plants is lower than HS plants in all organs at every growing stage, except for stem at GS70 and roots at GS81, for which there were no significant differences compared to HS plants (**Figure 5B**). The seed dry matter per plant (**Figure 5C**) was not affected by S limitation. In addition, protein content was not altered, in contrast to the glucosinolate content, which decreased by 82% compared to HS plants (**Figure 5C**). Oil content and oil quality were also slightly reduced in LS plants ($P < 0.05$) due to the decrease in linoleic and linolenic acid content (−9.4 and −11.5%, respectively, compared to HS plants) (**Figure 5C**).

S management at the whole plant level: uptake, remobilization and loss by fallen leaves

Between early flowering (GS53) and the onset of pod filling (GS70), the main sink organs for S taken up in LS53 plants were leaves (34% of total S from the soil; **Figure 6A**). The main difference between HS and LS53 plants was the sink strength of stems. Indeed, the allocation of S taken up into the stem increased for LS53 plants (25.3% compared to 17.3% for HS plants). In both S treatments, leaves were the main source organs for remobilized S between GS53 and GS70 (**Figure 6B**). A larger proportion of the remobilized S was redistributed to stem in LS53 plants (47.4%) compared to HS plants (4.5%) (**Figure 6C**). These results explain the lack of difference between the total S amount in stems of HS and LS53 plants at GS70. Moreover, in response to LS53 treatment, stem dry matter slightly increased, but it was not significantly different to HS plants. Therefore, the increase in S taken up and the redistribution of remobilized S to stems cannot be explained by an increase of growth. As expected for LS53 plants, a smaller proportion of remobilized S was lost through dead leaves (13.55% for LS53 plants vs. 22.73% for HS plants, **Figure 6C**).

Between the beginning of pod filling (GS70) and the seed coloring stage (GS81), there was no S uptake for LS53 plants while the S taken up in HS plants was essentially allocated to pods and seeds (**Figure 6E**). Leaves were the major source organs for remobilized S in LS53 and HS plants (**Figure 6F**). In addition,

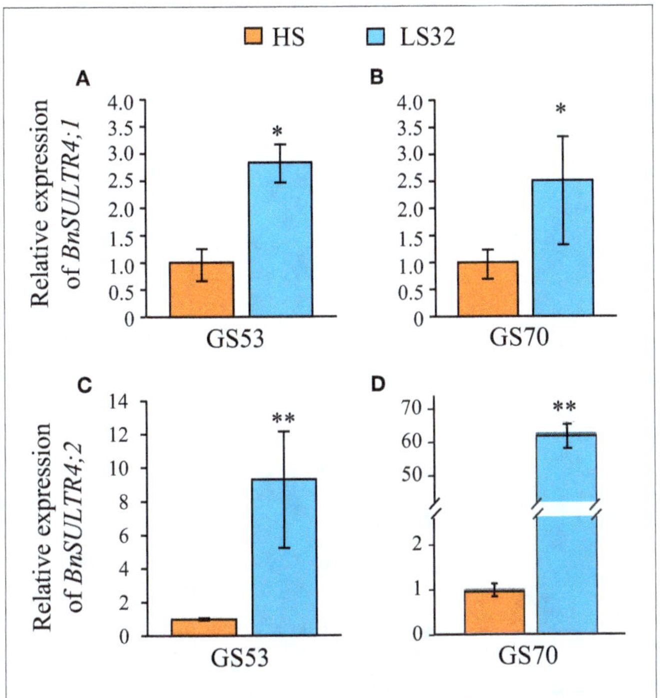

FIGURE 4 | Relative expression of *BnSultr4;1* (A,B) and *BnSultr4;2* (C,D) genes observed at GS53 (early flowering) and GS70 (start of pod filling) in source leaves of oilseed rape subjected to limitation of sulfate at the bolting stage. S limitation (LS32 plants supplied with 8.7 μM sulfate) was compared to control plants (HS plants supplied with 508.7 μM sulfate). For each stage of development, the value 1 is attributed for the level of transcripts observed in source leaves of HS plants. Significant differences between treatments are indicated with asterisks ($n = 4$; $^{*}P < 0.05$; $^{**}P < 0.01$).

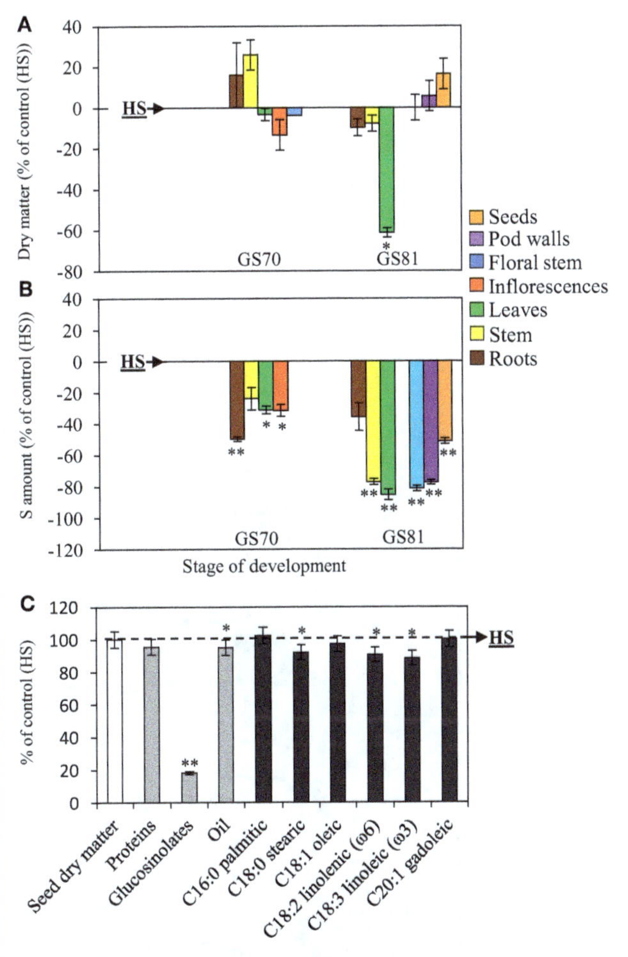

FIGURE 5 | Evolution of the dry matter (A), changes in the S amount (B) and seed yield and composition (C) of oilseed rape subjected to limitation of sulfate at the early flowering stage. S limitation (LS53 plants supplied with 8.7 μM sulfate) was compared to control plants (HS plants supplied with 508.7 μM sulfate). **(A)** Variations in dry matter of LS53 plants (as % of HS plants) between GS70 (start of pod filling) and GS81 (seed coloring). **(B)** Variations in the S amount in different organs of LS53 plants (as % of HS plants) between GS70 and GS81. A positive or a negative value indicates that the dry matter or the S amount is increased or reduced compared with HS plants, respectively. **(C)** Seed yield and composition of seeds (proteins, glucosinolates, oil, fatty acids) in LS53 plants (as % of HS plants) at mature seed stage (GS99). Significant differences between treatments are indicated with asterisks ($n = 4$; $^{*}P < 0.05$; $^{**}P < 0.01$).

stems represented a more important source for S remobilization in LS53 plants (25% of total remobilized S) than HS plants (15.4%; **Figure 6F**). The main sink organs for remobilized S were seeds for both S treatments but especially for LS53 plants with 53.7% of remobilized S redistributed to seeds vs. 43.2% in HS plants (**Figure 6G**). This larger S allocation and remobilization toward seeds could explain why LS53 plants had a similar seed yield to HS plants despite S restriction. At the onset of pod filling (GS70) the residual S percentage in dead leaves of LS53 plants was slightly decreased compared to HS plants, but not significantly (0.64 and 0.56%, respectively; **Figure 6D**). In contrast, at the seed coloring stage (GS81), a significant decrease was observed (0.64% and 0.31% for HS and LS53 plants, respectively; **Figure 6H**).

Involvement of sulfate and vacuolar sulfate transporters in S remobilization of source leaves

At GS70, a decrease in the amount of sulfate and ^{34}S-sulfate was noted in source leaves of LS53 plants while the S and ^{34}S in S-reduced compounds remained constant (**Figures 7A,B**). This decrease in sulfate was concomitant with up-regulation of *BnSULTR4;1* (2.3-fold) and *BnSULTR4;2* (8-fold) expression in source leaves of LS53 plants compared to HS plants (**Figures 8A,B**).

DISCUSSION

S SUPPLY BEFORE THE FLOWERING STAGE IS CRUCIAL FOR SEED YIELD AND SEED QUALITY IN WINTER OILSEED RAPE

The link between S availability, seed yield and seed quality of cereals and oilseed crops has been described previously (Janzen and Bettany, 1984; Zhao et al., 1997; Hawkesford, 2000; Scherer, 2001; Ahmad et al., 2007). Nevertheless, the stages of development that are the more sensitive to sulfate limitation and the source/sink relationships for S, and more particularly the contribution of S from absorption or remobilization in the reallocation of S to seeds, remain largely unclear in winter oilseed rape. The aims of

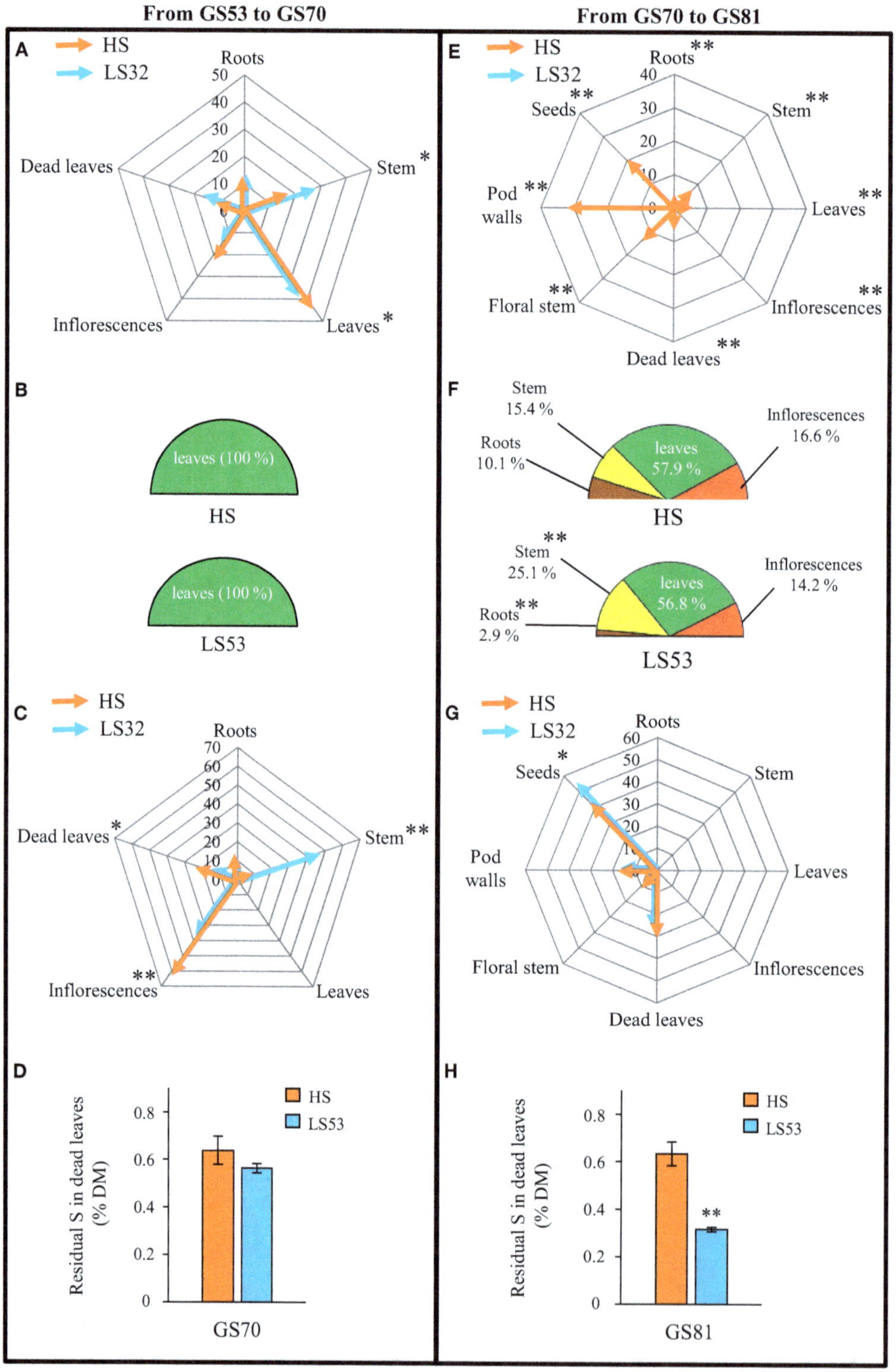

FIGURE 6 | Sulfur fluxes (A–C, E–G) and residual S in dead leaves (D,H) of oilseed rape subjected to limitation of sulfate at the early flowering stage. S limitation (LS53 plants supplied with 8.7 μM sulfate) was compared to control plants (HS plants supplied with 508.7 μM sulfate) between GS53 (early flowering stage) and GS81 (seeds coloring). Flux of S taken up from the soil allocated to the sink organs are indicated as % of total S taken up between **(A)** GS53 and GS70 (start of pod filling) and **(E)** GS70 and GS81 (seeds coloring). The diagrams **(B,F)** represent the contribution of source organs to the S remobilization in HS and LS53 plants between GS53 and GS70, and GS70 and GS81, respectively (data are expressed as % of total S remobilized). Flux of S remobilized from the source organs toward the sink organs are mentioned as % of total remobilized S between **(C)** GS53 and GS70, and **(G)** GS70 and GS81. Residual S in dead leaves (as % of dry matter) of HS and LS53 plants are indicated at **(D)** GS70 and **(H)** GS81. Significant differences between treatments are indicated with asterisks ($n = 4$; *: $P < 0.05$; **: $P < 0.01$).

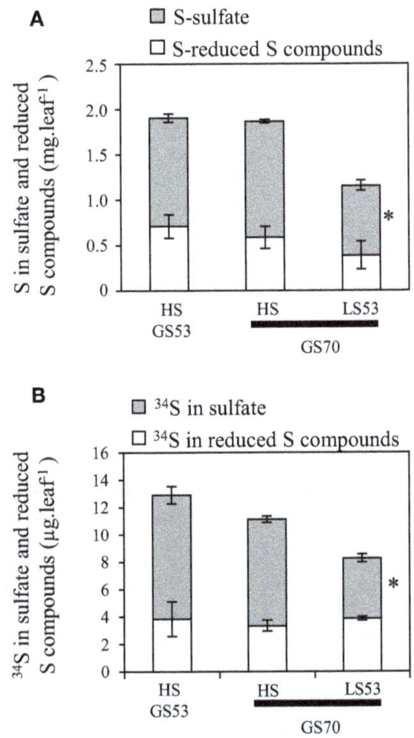

FIGURE 7 | Changes in S (A) and ^{34}S (B) amounts determined in the sulfate and reduced S-compounds fractions in source leaves of oilseed rape subjected to limitation of sulfate at the early flowering stage. S limitation (LS53 plants supplied with 8.7 μM sulfate) was compared to control plants (HS plants supplied with 508.7 μM sulfate) between GS53 (early flowering) and GS70 (start of pod filling). Significant differences between treatments are indicated with asterisks ($n = 4$; $^{*}P < 0.05$; $^{**}P < 0.01$).

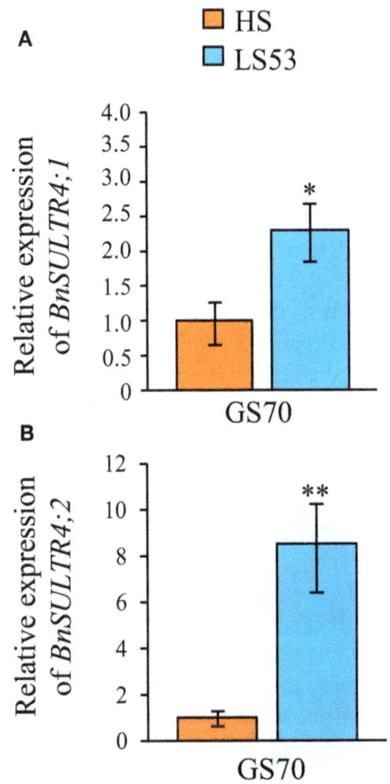

FIGURE 8 | Relative expression of *BnSultr4;1* (A) and *BnSultr4;2* (B) genes observed at GS70 (start of pod filling) in source leaves of oilseed rape subjected to limitation of sulfate at the early flowering stage. S limitation (LS53 plants supplied with 8.7 μM sulfate) was compared to control plants (HS plants supplied with 508.7 μM sulfate). For each stage of development, the value 1 is attributed for the level of transcripts observed in source leaves of HS plants. Significant differences between treatments are indicated with asterisks ($n = 4$; $^{*}P < 0.05$; $^{**}P < 0.01$).

this study were to determine how oilseed rape was able to manage its S reserves in order to maintain the seed yield and seed quality when sulfate limitation (LS) occurred at bolting stage (LS32 treatment) or early flowering stage (LS53 treatment). A sulfate limitation applied at the bolting stage (LS32 plants) affects seed yield, and seed quality with lower proteins and oil content (**Figure 2C**). Moreover, oil quality was affected leading to a decline in many fatty acids especially linoleic and linolenic acids which are particularly required in human nutrition. When S limitation occurred at the early flowering stage (LS53 plants), seed yield was the same as in control plants, but linoleic and linolenic acid contents were affected too. On the contrary, as shown by Dubousset et al. (2010), S restriction applied at the onset of pod filling did not affect seed yield and oil quality. The same results on grain quality were obtained in wheat by Anderson and Fitzgerald (2001) when S restriction occurred after anthesis. These results show that the S supply before the anthesis is crucial to sustain S demand for seed growth and to guarantee an optimal seed yield and seed quality. Moreover, the glucosinolate content in mature seeds declined when S limitation was applied at bolting or early flowering stages (**Figures 1C, 5C**). These secondary metabolites may represent a problem for meal production because a high level of some glucosinolates can affect palatability (Robbelen and Thies, 1980). On the other hand, glucosinolates may serve, *via* the myrosinase activity, as precursor of isothiocyanates, thiocyanates, nitriles, or epithionitriles that are important in defense mechanisms against pathogens (Halkier and Gershenzon, 2006) as demonstrated in leaves of oilseed rape (Dubuis et al., 2005). Interestingly, D'Hooghe et al. (2014) have reported that the mature seeds of LS53 and LS32 plants were characterized by a low myrosinase abundance. Even if the low level of glucosinolates and myrosinase observed in mature seeds in response to S limitation applied before anthesis may improve meal palatability, these data suggest that these S-limited seeds may be more susceptible to pathogens attacks.

S UPTAKE AND S REMOBILIZATION EFFICIENCY ARE DIFFERENTLY IMPACTED BY THE SULFATE LIMITATION OCCURRING AT THE BOLTING OR THE EARLY FLOWERING STAGES

In response to S limitation applied at the bolting stage (LS32 plants), seed yield and quality were drastically affected meaning that the S management by the LS32 plants was not efficient to compensate the low sulfate availability. However, our study revealed that LS32 plants significantly increased their root dry matter (+42.3% compared to HS plants, **Figure 1A**). It was also well established that the two high-affinity sulfate transporters (*Sultr1;1* and *Sultr1;2*) were involved in the major part of sulfate uptake (Davidian and Kopriva, 2010), and that gene expression of *BnSultr1;1* and *BnSultr1;2* was up-regulated in roots of oilseed rape subjected to S restriction (Buchner et al., 2004a,b; Parmar et al., 2007; Dubousset et al., 2009). In our experiment, the enhancement of root dry matter was concomitant with the increase in root sink strength for S (**Figures 2A,C**) indicating an increase in the sulfate uptake capacities as previously reported in oilseed rape at the vegetative stage (Abdallah et al., 2010) or in *Arabidopsis* (Hoefgen and Nikiforova, 2008). In our experimental conditions (culture in pot containing perlite and vermiculite with a very low sulfate availability), the increase of sulfate absorption capacities observed in LS32 plants did not lead to sustain the S demand for growth of reproductive organs. Although the remobilization of S from leaves was strongly induced in LS32 plants to sustain the root growth between the bolting and early flowering stages (**Figure 2C**), this was not sufficient to satisfy the S requirements for seed development and filling. However, under field conditions, the improvement of S absorption efficiency in response to limitation of S fertilizers *via* an increase of root proliferation and the induction of sulfate transporters in roots may help to explore a more important volume of soil and consequently maximize uptake of mineral S (Hawkesford, 2000). These physiological traits (increase of S remobilization from leaves to sustain the root proliferation and the S uptake efficiency) may limit the negative impacts on the seed yield and quality in case of low availability of sulfate occurring before the bolting stage.

When S limitation occurred at the early stages of flowering (LS53 plants), the S use efficiency of oilseed rape was improved to maintain seed yield and quality (**Figure 5C**). To understand the compensation mechanisms that occur in LS53 plants, the S fluxes were investigated and demonstrated that no matter what growing stages are considered, leaves are the most important source organs for S. Except at the onset of pod development (GS70), where the S amount was identical between LS53 and HS plants (**Figure 6D**), the residual S amount in leaves was lower in LS53 plants than in HS plants, which implies a better foliar S remobilization in S-limited plants (**Figure 6H**). Between the early flowering and the beginning of pod development, the remobilized S from leaves was mainly redistributed toward the stem in LS53 plants (**Figure 6C**). Between the onset of pod development and the seed coloring stages, the remobilization of S toward seeds was greater in LS53 plants than HS plants (**Figure 6G**). In parallel, the contribution of source leaves to S remobilisation was similar between LS53 and HS plants while the stem represented a major source organ for S remobilization in LS53 plants (25% of the total remobilized S vs. 15.4% for HS plants; **Figure 6F**). These results support the hypothesis that, in response of S restriction occurring at the flowering stages, the stem is a transient storage organ for S and may have a pivotal function in the case of asynchronism between S remobilization from source leaves and S use by reproductive organs.

SULFATE IS THE MAIN S COMPOUND REMOBILIZED IN SOURCE LEAVES

Once taken up from the soil, S in the form of sulfate is assimilated or transiently stored in vacuoles of roots or leaves (Davidian and Kopriva, 2010). Sulfate is the main storage form of S in Brassicacea. Indeed, Blake-Kalff et al. (1998) showed that in leaves of oilseed rape well supplied with S, sulfate can represent 70–90% of the total S amount. Several studies have suggested that vacuolar sulfate is a major form for S remobilization in response to S restriction at the vegetative stages in *Arabidopsis* or oilseed rape (Hawkesford, 2000; Kataoka et al., 2004; Parmar et al., 2007; Dubousset et al., 2009).

For LS32 and LS53 plants, a diminution in the total amount of S was observed in parallel to a decrease in the amount of sulfate while the amount of S-reduced compounds remained constant or slightly decreased (**Figures 3A, 7A**). Moreover, the ^{34}S-sulfate decreases but not the ^{34}S in S-reduced compounds. These results indicate that sulfate is the main S compound remobilized in leaves of plants subjected to an S restriction that is applied at the bolting and early flowering stages. In addition, an up-regulation of *BnSULTR4;1* and *BnSULTR4;2* gene expression was observed alongside the decrease in sulfate. After the bolting stage, sulfate destined for remobilization is probably derived from sulfate previously stored in vacuoles of source leaves. The increase in gene expression of these two SULTR4-type transporters has been shown at the vegetative stage in old and mature leaves (Dubousset et al., 2009) and in roots (Parmar et al., 2007) of oilseed rape. Moreover, Parmar et al. (2007) have reported that the transcript abundance of *BnSULTR4;2* is inversely correlated with sulfate concentration trends in tissues of *Brassica napus* at the vegetative stage. Although other transporters are supposed to play a significant role in sulfate remobilization, including transport from cytoplasm to phloem such as the transporters of groups 1 or 2, our results highlight the contribution of tonoplastic SULTR4-type transporters in the efflux of sulfate from the vacuole of source leaves in response to S restriction occurring at the reproductive stages (Buchner et al., 2004b).

In conclusion, our findings clearly indicate that leaves are the most important source organs for S during reproductive stages of oilseed rape. By combining ^{34}S-labeling with biochemical fractionation in order to separate sulfate from other S-compounds, the present study shows that sulfate is the main form of S remobilized in leaves at reproductive stages and that tonoplastic SULTR4-type transporters are particularly involved in the sulfate remobilisation in case of low S availability. Nevertheless, further experiments need to be performed to determine if sulfate itself is transported by the phloem to young pods and seeds or if it is reduced to other transportable S compounds. Our investigations on S fluxes at whole plant level also reveal that (i) the induction of root proliferation may help to maximize the sulfate uptake capacities especially when S limitation appeared at the bolting stage and (ii) the stem may serve as a transient storage organ for S in response to S limitation occurred at the early flowering stage. Consequently, these physiological traits (sulfate remobilization, root proliferation and/or transient S storage in stem) could be used in breeding programs to select oilseed rape genotypes with high S use efficiency or able to limit the impact of mineral S limitation on seed yield and quality.

AUTHOR CONTRIBUTIONS

All authors contributed to the experimental design, to plant growth and tissue sampling and have been involved in revising the article critically for important intellectual content. Alexandra Girondé, Lucie Dubousset, Jacques Trouverie and Jean-Christophe Avice carried out the S flux calculations. Alexandra Girondé and Jean-Christophe Avice performed the biochemical fractionation. Alexandra Girondé and Philippe Etienne performed the molecular analyses for determination of gene expression of SULTR4-type transporters. Alexandra Girondé and Lucie Dubousset made other measurements and analyses, including statistical analyses, interpretation of data and drafting the article.

ACKNOWLEDGMENTS

This work was supported by the French National Research Agency (program ANR-COSMOS n°ANR-05-JC05-51097) and by a Ph.D. grant for Mrs Lucie Dubousset from the French Ministry of Research. The authors thank Ms Marie-Paule Bataillé and Mr Raphaël Ségura for their technical help in S and ^{34}S analyses (Plateform PLATIN, SFR ICORE, University of Caen Basse-Normandie). The authors wish to acknowledge Dr. Nathalie Nési for the analysis of seed quality, and Anne-Françoise Ameline for her valuable help in the greenhouse experiment.

REFERENCES

Abdallah, M., Dubousset, L., Meuriot, F., Etienne, P., Avice, J. C., and Ourry, A. (2010). Effect of mineral sulphur availability on nitrogen and sulphur uptake and remobilization during the vegetative growth of *Brassica napus* L. *J. Exp. Bot.* 61, 2635–2646. doi: 10.1093/jxb/erq096

Ahmad, G., Jan, A., Arif, M., Jan, M. T., and Khattak, R. A. (2007). Influence of nitrogen and sulfur fertilization on quality of canola (*Brassica napus* L.) under rainfed conditions. *J. Shejiang Univ. Sci. B* 8, 731–737. doi: 10.1631/jzus.2007.B0731

Anderson, J. W., and Fitzgerald, M. A. (2001). Physiological and metabolic origin of sulphur for the synthesis of seed storage proteins. *J. Plant Physiol.* 158, 447–456. doi: 10.1078/0176-1617-00356

Blake-Kalff, M. M. A., Harrison, K. R., Hawkesford, M. J., Zhao, F. J., and McGrath, S. P. (1998). Distribution of sulfur within oilseed rape leaves in response to sulfur deficiency during vegetative growth. *Plant Physiol.* 118, 1337–1344. doi: 10.1104/pp.118.4.1337

Buchner, P., Elisabeth, C., Stuiver, E., Westerman, S., Wirtz, M., Hell, R., et al. (2004a). Regulation of sulfate uptake and expression of sulfate transporter genes in *Brassica oleracea* as affected by atmospheric H_2S and pedospheric sulfate nutrition. *Plant Physiol.* 136, 3396–3408. doi: 10.1104/pp.104.046441

Buchner, P., Takajashi, H., and Hawkesford, M. J. (2004b). Plant sulphate transporters; co-ordination of uptake, intracellular and long-distance transport. *J. Exp. Bot.* 55, 1765–1773. doi: 10.1093/jxb/erh206

D'Hooghe, P., Dubousset, L., Gallardo, K., Kopriva, S., Avice, J. C., and Trouverie, J. (2014). Evidence for proteomic and metabolic adaptations associated to alterations of seed yield and quality in sulphur-limited *Brassica napus* L. *Mol. Cell. Proteomics* 13, 1165–1183. doi: 10.1074/mcp.M113.034215

Davidian, J. C., and Kopriva, S. (2010). Regulation of sulfate uptake and assimilation – The same or not the same? *Mol. Plant* 3, 314–325. doi: 10.1093/mp/ssq001

D'Hooghe, P., Escamez, S., Trouverie, J., and Avice, J. C. (2013). Sulphur limitation provokes physiological and leaf proteome changes in oilseed rape that lead to perturbation of sulphur, carbon and oxidative metabolisms. *BMC Plant Biology* 13:23. doi: 10.1186/1471-2229-13-23

Dubousset, L., Abdallah, M., Desfeux, A. S., Etienne, P., Meuriot, F., Hawkesford, M. J., et al. (2009). Remobilization of leaf S compounds and senescence in response to restricted sulphate supply during the vegetative stage of oilseed rape are affected by mineral N availability. *J. Exp. Bot.* 60, 3239–3253. doi: 10.1093/jxb/erp172

Dubousset, L., Etienne, P., and Avice, J. C. (2010). Is the remobilization of S and N reserves for seed filling of winter oilseed rape modulated by sulphate restrictions occurring at different growth stages? *J. Exp. Bot.* 61, 4313–4324. doi: 10.1093/jxb/erq233

Dubuis, P. H., Marazzi, C., Städler, E., and Mauch, F. (2005). Sulphur deficiency causes a reduction in antimicrobial potential and leads to increased disease susceptibility of oilseed rape. *J. Phytopathol.* 153, 27–36. doi: 10.1111/j.1439-0434.2004.00923.x

Fismes, J., Vong, P. C., Guckert, A., and Frossard, E. (2000). Influence of sulfur on apparent N-use efficiency, yield and quality of oilseed rape (*Brassica napus* L.) grown on a calcareous soil. *Eur. J. Agron.* 12, 127–141. doi: 10.1016/S1161-0301(99)00052-0

Halkier, B. A., and Gershenzon, J. (2006). Biology and biochemistry of glucosinolates. *Annu. Rev. Plant Biol.* 57, 303–333. doi: 10.1146/annurev.arplant.57.032905.105228

Hawkesford, M. J. (2000). Managing sulphur metabolism in plants. *Plant Cell Environ.* 29, 382–395. doi: 10.1111/j.1365-3040.2005.01470.x

Hawkesford, M. J., and De Kok, L. J. (2006). Managing sulphur metabolism in plants. *Plant Cell Environ.* 29, 382–395. doi: 10.1111/j.1365-3040.2005.01470.x

Hoefgen, R., and Nikiforova, V. J. (2008). Metabolomics integrated with transcriptomics: assessing systems response to sulfur-deficiency stress. *Physiol. Plant.* 132, 190–198. doi: 10.1111/j.1399-3054.2007.01012.x

Janzen, H. H., and Bettany, J. R. (1984). Sulfur nutrition of rapeseed: I. Influence of fertilizer nitrogen and sulfur rates. *Soil Sci. Soc. Am. J.* 48, 100–107. doi: 10.2136/sssaj1984.03615995004800010019x

Kataoka, T., Watanabe-Takahashi, A., Hayashi, N., Ohnishi, M., Mimura, T., Buchner, P., et al. (2004). Vacuolar sulfate transporters are essential determinants controlling internal distribution of sulfate in *Arabidopsis*. *Plant Cell* 16, 2693–2704. doi: 10.1105/tpc.104.023960

Malhi, S. S., Gan, Y., and Raney, J. P. (2007). Yield, seed quality, and sulfur uptake of *Brassica* oilseed crops in response to sulfur fertilization. *Agron. J.* 99, 570–577. doi: 10.2134/agronj2006.0269

Matula, J., and Pechová, M. (2002). Sulfur and sulfate concentrations in leaves of oilseed rape under field conditions. *Rostlinná Vıroba* 48, 433–440.

McGrath, S. P., and Zhao, F. J. (1996). Sulphur uptake, yield response and the interactions between N and S in winter oilseed rape (*Brassica napus*). *J. Agric. Sci.* 126, 53–62. doi: 10.1017/S0021859600088808

Noquet, C., Avice, J. C., Rossato, L., Beauclair, P., Henry, M. P., and Ourry, A. (2004). Effects of altered source-sink relationships on N allocation and vegetative storage protein accumulation in *Brassica napus* L. *Plant Sci.* 166, 1007–1018. doi: 10.1016/j.plantsci.2003.12.014

Parmar, S., Buchner, P., and Hawkesford, M. J. (2007). Leaf developmental stage affects sulfate depletion and specific sulfate transporter expression during sulfur deprivation in *Brassica napus* L. *Plant Biol.* 9, 647–653. doi: 10.1055/s-2007-965428

Robbelen, G., and Thies, W. (1980). "Biology and Breeding," in *Brassica Crops and Wild Allies*, eds S. Tsunoda, H. Hinata, G. Gomez-Campo (Tokyo: Japan Sci. Soc. Press), 285–300.

Salon, C., Bataillé, M. P., Gallardo, K., Jeudy, C., Santoni, A. L., Trouverie, J., et al. (2014). ^{34}S and ^{15}N labelling to model S and N flux in plants and determine the different components of N and S use efficiency. *Methods Mol. Biol.* 1090, 335–346. doi: 10.1007/978-1-62703-688-7_20

Scherer, H. W. (2001). Sulfur in crop production. *Eur. J. Agron.* 14, 81–111. doi: 10.1078/0176-1617-00356

Schnug, E., Haneklaus, S., and Murphy, D. (1993). Impact of sulphur fertilization on fertilizer nitrogen efficiency. *Sulphur Agric.* 17, 8–12.

Takahashi, H., Buchner, P., Yoshimoto, N., Hawkesford, M. J., and Shiu, S.-H. (2011). Evolutionary relationships and functional diversity of plant sulfate transporters. *Front. Plant Sci.* 2:119. doi: 10.3389/fpls.2011.00119

Zhao, F. J., Bilsborrow, P. E., Evans, E. J., and McGrath, S. P. (1997). Nitrogen to sulphur ratio in rapeseed and in rapeseed protein and its use in diagnosing sulphur deficiency. *J. Plant Nutr.* 20, 549–558. doi: 10.1080/01904169709365273

Conflict of Interest Statement: The Associate Editor Stanislav Kopriva declares that, despite having collaborated with authors Lucie Dubousset, Jacques Trouverie and Jean-Christophe Avice, the review process was handled objectively and no conflict of interest exists. The authors declare that the research was conducted in the absence of any commercial or financial relationships that could be construed as a potential conflict of interest.

9

Are changes in sulfate assimilation pathway needed for evolution of C_4 photosynthesis?

Silke C. Weckopp and Stanislav Kopriva*

Botanical Institute and Cluster of Excellence on Plant Sciences, Cologne Biocenter, University of Cologne, Cologne, Germany

Edited by:
Ruediger Hell, University of Heidelberg, Germany

Reviewed by:
Cornelia Herschbach, Albert Ludwigs University of Freiburg, Germany
Sascha Offermann, Leibniz University Hannover, Germany

****Correspondence:***
Stanislav Kopriva, Botanical Institute and Cluster of Excellence on Plant Sciences, Cologne Biocenter, University of Cologne, Zülpicher Str. 47b, 50674 Cologne, Germany
e-mail: skopriva@uni-koeln.de

C_4 photosynthesis characteristically features a cell-specific localization of enzymes involved in CO_2 assimilation in bundle sheath cells (BSC) or mesophyll cells. Interestingly, enzymes of sulfur assimilation are also specifically present in BSC of maize and many other C_4 species. This localization, however, could not be confirmed in C_4 species of the genus *Flaveria*. It was, therefore, concluded that the bundle sheath localization of sulfate assimilation occurs only in C_4 monocots. However, recently the sulfate assimilation pathway was found coordinately enriched in BSC of *Arabidopsis*, opening new questions about the significance of such cell-specific localization of the pathway. In addition, next generation sequencing revealed expression gradients of many genes from C_3 to C_4 species and mathematical modeling proposed a sequence of adaptations during the evolutionary path from C_3 to C_4. Indeed, such gradient, with higher expression of genes for sulfate reduction in C_4 species, has been observed within the genus *Flaveria*. These new tools provide the basis for reexamining the intriguing question of compartmentalization of sulfur assimilation. Therefore, this review summarizes the findings on spatial separation of sulfur assimilation in C_4 plants and *Arabidopsis*, assesses the information on sulfur assimilation provided by the recent transcriptomics data and discusses their possible impact on understanding this interesting feature of plant sulfur metabolism to find out whether changes in sulfate assimilation are part of a general evolutionary trajectory toward C_4 photosynthesis.

Keywords: sulfate assimilation, C_4 photosynthesis, bundle sheath cells, *Flaveria*, glutathione

INTRODUCTION

Sulfur is an important element in living organisms where it possesses a plethora of functions. As a component of the amino acids cysteine and methionine, sulfur is incorporated in peptides and proteins. The formation of disulfide bridges by cysteine residues of proteins is important for conformational and regulatory processes: however, sulfur in proteins possesses also catalytic and electrochemical functions and participates in electron transport in iron–sulfur clusters, is part of the catalytic sites of various enzymes and coenzymes. Furthermore, the sulfhydryl group of the tripeptide glutathione (GSH) is involved in redox reactions which protect the cell from oxidative stress (Leustek and Saito, 1999; Foyer and Noctor, 2009).

Plants have the ability to incorporate reduced sulfur components from the atmosphere in form of sulfur dioxide or hydrogen sulfide, but the majority of sulfur in higher plants derives from sulfate (Leustek et al., 2000; Saito, 2004). Sulfate (SO_4^{2-}) is present in the soil as the most oxidized and accessible form of sulfur for plants, algae and many microorganisms. Sulfate uptake into the plant and its distribution within the cells, occurs via various sulfate transporters which differ in their properties and functions (Leustek and Saito, 1999; Buchner et al., 2004; Takahashi et al., 2011). Within plant cells, sulfate can either be stored in the vacuole or directly be incorporated in organic compounds. For synthesis of reduced sulfur compounds, sulfate needs to be reduced and assimilated to cysteine. The initial step of sulfur assimilation is the activation of sulfate, an inert and stable compound, by ATP sulfurylase (ATPS) to adenosine-5′-phosphosulfate (APS). APS can be reduced to sulfite by APS reductase (APR) or phosphorylated to the common sulfo-group donor 3′-phosphoadenosine-5′-phosphosulfate (PAPS) by APS kinase. APR uses two molecules of reduced GSH for the reduction of APS. Sulfite reductase (SIR) then reduces sulfite to sulfide by transferring electrons from the iron–sulfur protein ferredoxin. Finally, sulfide can be incorporated in the amino acid backbone of *O*-acetylserine (OAS) by OAS (thiol)lyase (OASTL). OAS is derived from serine by acetylation mediated by serine acetyltransferase (SAT; Leustek et al., 2000; Saito, 2004; Kopriva, 2006). Serine biosynthesis is closely linked to carbon and nitrogen assimilation, the synthesis of cysteine thus merges the three assimilatory pathways. Accordingly, light, carbon, and nitrogen compounds regulate sulfate assimilation (Kopriva et al., 1999, 2002; Koprivova et al., 2000). Using thiol- and stress-dependent regulation of APR expression as a tool, the reduction of activated sulfate was found to be a key step in sulfur assimilation (Vauclare et al., 2002; Scheerer et al., 2010).

Reduced sulfur is preferentially stored and transported in the form of GSH, the most abundant low molecular weight thiol in plants with a number of different functions. It is substantially involved in the detoxification of reactive oxygen species (ROS) and heavy metals, serves as sulfur donor in catalytic processes and is involved in redox signaling (Foyer and Noctor, 2009; Takahashi et al., 2011). Two enzymes catalyze the ATP-dependent reactions to form GSH from its constituent amino acids. γ-glutamylcysteine synthetase (γ-ECS) catalyzes the reaction of cysteine and glutamate to form γ-glutamylcysteine (γ-EC). Subsequently, glycine is added to γ-EC by GSH synthetase to form GSH.

The traditional view on the whole plant sulfur metabolism suggests that sulfate reduction predominantly takes place in leaves and reduced sulfur compounds are subsequently distributed to sink tissues (Herschbach and Rennenberg, 2001). However, although GSH is widely used to store and transport reduced sulfur, most plant tissues are capable of sulfate reduction and thus are able to cover their needs of reduced sulfate. Indeed, according to available microarray data in the Genevestigator database, APR and SIR transcripts are present in all *Arabidopsis* tissues, including reproductive organs (Zimmermann et al., 2004). The presence of sulfate reduction could also be detected in roots and developing seeds (Brunold and Suter, 1989; Kopriva et al., 2001; Tabe and Droux, 2002).

There are, however, exceptions and some plant tissues seem to have lost the capacity to reduce sulfate. Most notable among these tissues are the mesophyll cells (MC) in C_4 plants, as in these plants the pathway is confined to a specific tissue, the bundle sheath cells (BSC; Gerwick et al., 1980; Passera and Ghisi, 1982; Burnell, 1984; Schmutz and Brunold, 1984; Burgener et al., 1998). The intercellular distribution of sulfate (and nitrate) assimilation in C_4 plants is one of the enigmatic open questions of sulfur metabolism, not just in terms of mechanisms but mainly of its biological significance. This question has been very recently revived by Aubry et al. (2014) who showed that genes involved in the pathway are mainly expressed in the BSC of *Arabidopsis*. Therefore, here we review what is known about the localization of sulfate assimilation in C_4 plants, and discuss how the current progress in studies of C_4 photosynthesis may contribute to understanding this interesting feature of plant sulfur metabolism.

C_4 PHOTOSYNTHESIS

During photosynthesis, carbon fixation takes place in the Calvin–Benson cycle and is mediated by the dual-specific enzyme Ribulose 1,5-bisphosphate carboxylase/oxygenase (Rubisco). The carboxylation reaction yields 3-phosphoglycerate (3-PGA) which is reduced to carbohydrates. In contrast, the oxygenation reaction yields 2-phosphoglycolate (2-PG), a toxic compound that enters photorespiration to be metabolized. The photorespiratory pathway releases one molecule of previously fixed CO_2 and regenerates the other into 3-PGA. Although Rubisco shows higher affinity toward CO_2 than O_2, one third of the assimilated carbon is lost during photorespiration, increasing so the energy costs of photosynthesis. An efficient mechanism to minimize the oxygenation reaction of Rubisco and photorespiration is C_4 photosynthesis. C_4 plants avoid such loss of assimilated CO_2 by splitting the CO_2 fixing reactions among two differentiated cell types, MC and BSC. MC and BSC differ from each other by their morphological properties, distribution within the plant tissue and cell-specific localization of many enzymes. BSC build up a radial pattern around the vascular tissue. They exhibit thickened cell-walls and many starch containing chloroplasts. BSC are surrounded by an adjacent layer of MC with small and randomly distributed chloroplasts. The radial arrangement of BSC and MC around the vascular tissue is known as Kranz anatomy (Laetsch, 1974).

The primary CO_2 fixation takes place in MC. This reaction, catalyzed by phosphoenolpyruvate carboxylase (PEPCase), results in oxaloacetate, a C_4 compound. Oxaloacetate is immediately converted to malate or aspartate, C_4 acids that enter BSC by diffusion. Here, the C_4 acids are decarboxylated, resulting in enriched CO_2 concentrations in the BSC. C_4 photosynthesis can be divided into three subtypes defined by the enzymes which catalyze the decarboxylation reaction: NADP-dependent malic enzyme, NAD-dependent malic enzyme or phosphoenolpyruvate carboxykinase. The released CO_2 enters the Calvin–Benson cycle within the BSC (Edwards and Huber, 1981; Edwards and Walker, 1983; Hatch, 1987). Besides Rubisco, additional genes involved in the carbon cycle are solely expressed in the BSC (Sheen, 1999; Edwards et al., 2001). The separation of PEPCase and Rubisco, as well as the Kranz anatomy and a lack of photosystem II (PSII) in BSC of most C_4 plants (Hatch and Osmond, 1976) support more effective carbon assimilation. The CO_2 enrichment in the BSC minimizes the rate of the oxygenation reaction. Any CO_2, lost during residual photorespiration, is separated from the aerial space by the thick cell walls of the BSC and by the adjacent layer of MC, where it is directly re-assimilated. In addition, the lack of PSII in BSC restricts the emergence of oxygen by water splitting to MC chloroplasts. Paired with a BSC-specific expression of glycine decarboxylase (GDC), these mechanisms generate a low photosynthetic CO_2 compensation point in C_4 plants. The CO_2 compensation point describes the concentration at which the photosynthetic CO_2 uptake equals the rate of respiration.

The last decade has seen a revival of interest in C_4 photosynthesis, as a potential route to increase crop productivity (Hibberd et al., 2008). Particular interest has been given the evolutionary route from C_3 to C_4 plants as this could identify steps needed to engineer C_4 photosynthesis in C_3 plants (Heckmann et al., 2013; Mallmann et al., 2014). C_4 photosynthesis evolved not less than 62 times independently from C_3 ancestors (Sage et al., 2011). Genera containing C_3 and C_4 as well as C_3–C_4 intermediate species, such as *Flaveria*, are especially suited to give information about the evolution from C_3 to C_4 photosynthesis (Kopriva et al., 1996; Heckmann et al., 2013; Williams et al., 2013; Mallmann et al., 2014). C_3–C_4 species are characterized by a CO_2 compensation point intermediate of C_3 and C_4 species (Apel and Maass, 1981; Bauwe et al., 1987). While there are many different ways to achieve such CO_2 compensation points (Bassüner et al., 1984; Bauwe, 1984; Monson et al., 1986), the underlying photosynthetic mechanism is universally based on a photorespiratory CO_2 pump. Shortly, GDC activity is lost in the MC, resulting in the transport of photorespiratory glycine to BSC for decarboxylation, which causes a moderate increase in BSC CO_2 concentration and a boost of carboxylation over oxygenation reaction. Since the two-carbon compound glycine serves as C transporter, this process is known as C_2 photosynthesis (Bauwe and Kolukisaoglu, 2003; Sage, 2004;

Sage et al., 2012; Schulze et al., 2013). C_2 photosynthesis was postulated to be a prerequisite to C_4 metabolism (Sage et al., 2012). This conclusion was confirmed by analysis of transcriptomes of nine *Flaveria* species and by flux balance analysis (Mallmann et al., 2014). However, the translocation of glycine decarboxylation from MC to BSC results in the shift of carbon and nitrogen balance. This disbalance can be counterbalanced by a basal activity of the C_4 cycle, acting as an efficient ammonium recirculation pathway. Thus, the C_2 photosynthesis triggers development of a basic C_4 cycle which further leads to evolution of a full C_4 photosynthesis. In another mathematical approach, a model of fitness landscape describes the evolutionary trajectory from C_3 to C_4 photosynthesis as a process of 30 individual steps, each of them gradually yielding a gain of biochemical fitness (Heckmann et al., 2013). This model positions the known C_3–C_4 intermediates as real transitory states in the evolution from C_3 to C_4 photosynthesis and explains the ease with which C_4 photosynthesis evolved independently, making it feasible to be recreated by genetic engineering.

However, one major aspect of C_4 photosynthesis remains largely unexplained. Besides the cell-specific distribution of enzymes involved directly in photosynthesis and photorespiration, intercellular differences could be detected in the localization of enzymatic reactions involved in nitrogen and sulfur assimilation in C_4 plants. The reduction of nitrate and nitrite is restricted to MC, whereas the incorporation of reduced nitrogen into the amino acids glutamate and glutamine takes place in the BSC or in MC and BSC (Rathnam and Edwards, 1976; Moore and Black, 1979; Becker et al., 2000) which accords with the translocation of GDC into BSC. C_4 species also exhibit higher nitrogen use efficiency, presumably because of the concentration of Rubisco into BSC and so decreasing the amount of this protein per leaf area (Brown, 1978; Ghannoum et al., 2011). Whether the cell-specific localization of nitrate reduction contributes to the improved nitrogen nutrition is not clear yet. Furthermore, since spatial separation was reported for the assimilation of carbon and nitrogen, two nutrients that are taken up into the plant in their oxidized form (CO_2 and NO_3^-) and since the translocation of GDC to the BSC restricts production of photorespiratory serine, the precursor of OAS and cysteine, to these cells, the question of intercellular distribution of sulfate assimilation in C_4 plants has long been of major interest.

INTERCELLULAR COMPARTMENTATION OF SULFATE ASSIMILATION IN C_4 PLANTS

The question of intercellular compartmentation of sulfate assimilation in C_4 species was first addressed by Gerwick and Black (1979) in the crabgrass *Digitaria sanguinalis*. They focused on the initial step of sulfur assimilation, the activation of sulfate by ATPS, and showed that more than 90% of total leaf ATPS activity was restricted to BSC in crabgrass. Similar results could be obtained for 17 additional C_4 monocot species, representing each of the three C_4 subtypes (Gerwick et al., 1980; Passera and Ghisi, 1982; Burnell, 1984; Schmutz and Brunold, 1984; Burgener et al., 1998). Analyses of the spatial separation of sulfate assimilation in maize leaves could additionally show restriction of APR activity to BSC

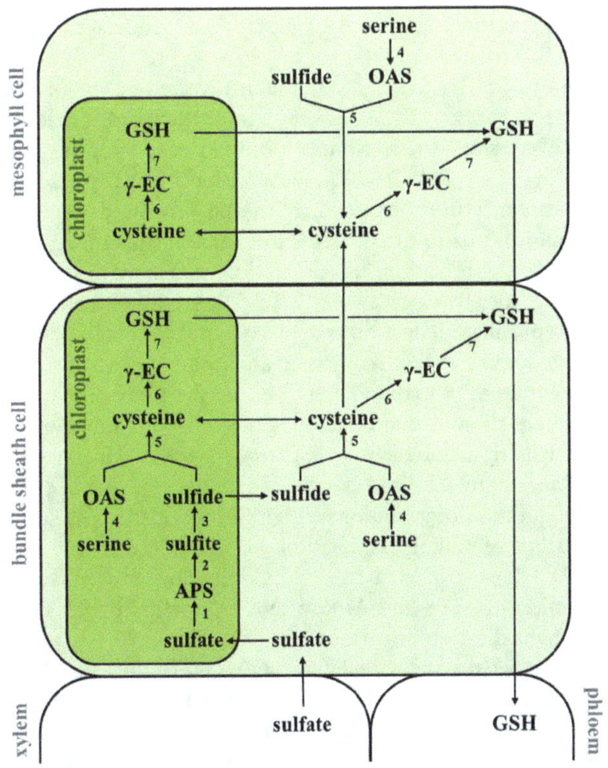

FIGURE 1 | Intercellular compartmentation of sulfate assimilation and glutathione biosynthesis in maize. Sulfate is taken up from the soil and transported to the bundle sheath cells (BSC) through the xylem. The reduction of sulfate takes place exclusively in the plastids of BSC and is mediated by ATP sulfurylase (1), APS reductase (2), and sulfite reductase (3). Sulfide is further incorporated into the amino acid backbone of OAS by OAS (thiol)lyase (5) to form cysteine in chloroplasts, cytosol and mitochondria (not included) of BSC. OAS is derived from serine by serine acetyltransferase-mediated acetylation (4). Reduced sulfur is transported in form of cysteine to mesophyll cells where glutathione (GSH) synthesis is predominantly localized. GSH synthesis is driven by γ-EC synthetase (6) and GSH synthethase (7). APS, adenosine-5′-phosphosulfate; GSH, glutathione; OAS, *O*-acetylserine; γ-EC, γ-glutamylcysteine.

(**Figure 1**; Schmutz and Brunold, 1984; Burgener et al., 1998), whereas activity of SIR and OASTL was equally detected in MC and BSC (Passera and Ghisi, 1982; Schmutz and Brunold, 1984, 1985).

Corresponding to these findings, transcripts of ATPS and APR were found exclusively in the BSC and OASTL mRNA was present in MC and BSC in maize leaves (Kopriva et al., 2001). These findings were supported by recent analyses of the maize MC and BSC transcriptomes which showed a predominant expression of all four ATPS as well as two APR genes in the BSC and localization of OASTL transcripts in MC and BSC (Chang et al., 2012). The identical localization of APR, ATPS, and OASTL mRNA and their enzymes activity, indicates a transcriptional regulation of genes involved in sulfate assimilation. However, APR mRNA but not enzyme activity could also be detected in MC of maize leaves after exposure to chilling stress. This finding reveals participation of post-transcriptional processes in the compartmentalization of sulfate assimilation in maize (Kopriva et al., 2001). Interestingly, post-transcriptional mechanisms seem to be responsible also for the MC-specific localization of glutathione reductase (GR) in maize (Pastori et al., 2000).

Although SIR activity was originally detected in MC of maize leaves (Schmutz and Brunold, 1984), in further studies SIR transcripts were shown to accumulate only in BSC (Kopriva et al., 2001; Chang et al., 2012). The localization of SIR, particularly the discrepancy in protein and transcript, thus needs revisiting, although it is possible that in MC SIR fulfills a role independent from sulfate assimilation, e.g., protection against SO_2 (Yarmolinsky et al., 2013). Burgener et al. (1998) were able to show that all steps of sulfate reduction take place in BSC. The restriction of sulfate reduction to BSC requires an efficient transport of reduced sulfate compounds to the adjacent MC. Feeding experiments with $[^{35}S]$sulfate showed that cysteine is the transport metabolite in maize. It is transported from BSC to MC where GSH synthesis is predominantly localized (Burgener et al., 1998).

The question of evolutionary significance of the BSC-specific localization of sulfate assimilation was addressed in the genus *Flaveria* (Koprivova et al., 2001). Comparing the cell-specific localization of ATPS and APR mRNA by northern analysis and *in situ* hybridization in various *Flaveria* species, Koprivova et al. (2001) expected an accumulation of the transcripts in the BSC in C_4 species and ubiquitous expression of ATPS and APR in all photosynthetic cells of C_3 species. Remarkably, they found comparable transcript levels of both genes in each species, independent of the photosynthetic mechanism. Immunogold electron microscopy confirmed a similar distribution of APR protein in chloroplasts of BSC and MC in all species analyzed (Koprivova et al., 2001).

These findings contradicted the previously postulated compartmentation of sulfate assimilation in C_4 plants (Gerwick et al., 1980; Passera and Ghisi, 1982; Burnell, 1984; Schmutz and Brunold, 1984; Burgener et al., 1998). Admittedly, earlier studies were conducted in maize and 17 other C_4 species, all belonging to the group of monocotelydons. *F. trinervia* and *F. australasica* were the first C_4 dicots analyzed for the intercellular separation of sulfate assimilation (Koprivova et al., 2001). However, BSC-specific localization of sulfate assimilation is not a monocot-specific trait. In wheat, a C_3 monocot, ATPS and APR activities were found at equivalent levels in MC and BSC (Schmutz and Brunold, 1984). In addition, recently, Aubry et al. (2014) described the preferential expression of genes associated with sulfate assimilation in the BSC of the C_3 dicot *Arabidopsis thaliana*. Thus, BSC-specific localization of sulfate assimilation might no longer be seen as a C_4-associated trait but possibly as species-specific characteristic. Therefore it is even more imperative to understand the biological reasons and/or consequences for this localization.

Indeed, the physiological significance of spatial separation in sulfate assimilation in maize has been widely discussed, unfortunately without a clear conclusion. The BSC of maize lack PSII and with it the water-splitting complex (Sheen and Bogorad, 1988; Pfundel et al., 1996). Burgener et al. (1998) considered that the consequential reduction in O_2 concentration might be the reason for BSC-specific expression of sulfate assimilation in maize. The low levels of O_2 might prevent the oxidation of sulfite and sulfide, the intermediates of sulfate assimilation, and so increase the efficiency of the pathway. However, such oxidative events, except the enzymatic sulfite oxidation by sulfite oxidase, have not been reported for C_3 plants which possess PSII in the chloroplasts of all photosynthetic cells. As a precursor of cysteine, serine is necessary for sulfate reduction. In C_3 plants, photorespiration is the main source of serine production (Douce et al., 2001; Ros et al., 2014). GDC and serine hydroxymethyltransferase, key enzymes in photorespiration and serine biosynthesis, are localized in BSC of C_4 plants (Ohnishi and Kanai, 1983; Gardeström et al., 1985; Becker et al., 1993). Therefore, Burgener et al. (1998) hypothesized that the localization of sulfate reduction in maize coincides with the site of photorespiration because of serine availability. Several physiological aspects, however, challenge this hypothesis. C_4 species of the genus *Flaveria* do not show BSC-specific localization of sulfur reduction, although GDC activity is restricted to this cell type. Moreover, serine would need to be transported from BSC to MC to participate, e.g., in protein biosynthesis. Indeed, an alternative pathway of serine biosynthesis exists and is vital for plants (Benstein et al., 2013), so that photorespiratory serine cannot be the link. The BSC localization of sulfate assimilation in *Arabidopsis* (Aubry et al., 2014) also contradicts the link to serine synthesis. The *Arabidopsis* data are derived from comparison of BSC expression vs. whole leaf expression, therefore the spatial separation may not be complete as it is in C_4 monocots. The BSC expression of sulfate assimilation is linked with BSC expression of genes for synthesis of glucosinolates, sulfur-containing secondary metabolites. As these compounds are important in the vasculature for defense against insects (Shroff et al., 2008) the need for glucosinolate synthesis might drive the BSC specific expression of sulfate assimilation genes. This hypothesis, even though it has to be tested yet, seems to strengthen the view that the spatial distribution of sulfate assimilation may be a species-specific adaptation to diverse environmental conditions. Obviously, more research is needed to understand the biological reason(s) and consequences for plant fitness of such intercellular separation of sulfate assimilation, if there are any.

SULFUR ASSIMILATION PATHWAY FROM C_3 TO C_4 PLANTS

The analysis of sulfate assimilation in different *Flaveria* species revealed another interesting result. The activity of APR as well as cysteine and GSH levels were significantly higher in leaves of C_4 and C_4-like species compared to C_3 and C_3–C_4 species (Koprivova et al., 2001). Enzyme activity, cysteine, and GSH content correlated with the degree of expression of C_4 characteristics, considering the photosynthetic CO_2 compensation point as a quantitative measure of C_4 photosynthetic traits (Kopriva and Koprivova, 2005). The intriguing results of analyses of sulfate assimilation in *Flaveria* were unfortunately not followed by further studies. The question of intercellular localization of sulfur assimilation, and nitrate assimilation at the same time, seemed to be forgotten for more than a decade. However, with the increased interest in C_4 photosynthesis as a means to improve plant productivity and with the more frequent use of next generation sequencing in plant science, these questions can be addressed from a different angle.

Indeed, the transcriptomes of nine *Flaveria* species were analyzed recently to assess the influence of photorespiration on the evolution of C_4 photosynthesis (Mallmann et al., 2014). To obtain the transcriptomic data, the authors performed four independent cultivations during different seasons of two C_3, four C_3–C_4, one C_4-like, and two C_4 species of the genus *Flaveria* (**Figure 2**). The analysis showed clear gradients in expression of many genes across the *Flaveria* species, the photorespiratory enzymes being expressed to lower degree with increasing degree of C_4 photosynthesis, whereas the genes of C_4 pathway being expressed more. Data mining of this transcriptome dataset focusing on genes associated with sulfate assimilation provides an interesting insight into the expression of ATPS, APR and SIR, depending on the photosynthetic mechanism and seasonal changes (**Figure 2A**). APS1 and APS2 are two isoforms of ATPS in *Arabidopsis thaliana*, with transcript sequences matching *Flaveria* ATPS mRNA. Whereas APS1 is exclusively localized in chloroplasts, APS2 can also be found in the cytosol. The expression of the ATPS genes does not show a recognizable correlation with the photosynthetic mechanism, but seems to follow seasonal changes. Direct comparison of APS1 and APS2 signals in each species indicate preferential expression of one of the isoforms for at least the C_3 species, *F. pringlei* and *F. robusta*, and the C_4-like species *F. brownii* (**Figure 2B**). Overall, the highest expression of ATPS genes can be detected in *Flaveria* C_3 species.

The transcript level of SIR was least variable among the different species and time points of harvest, with a tendency to higher levels in the C_3 species (**Figures 2A,B**). This agrees with the low level of transcriptional regulation of this gene in the model species *Arabidopsis* (Takahashi et al., 2011). On the other hand, the differences in mRNA levels of genes similar to the *Arabidopsis* APR1 and APR2 sequences were striking (**Figures 2A,B**). Both isoforms show similar expression patterns, which indicates either a coordinate expression or a presence of a single APR isoform in the *Flaveria* species. Strongest expression of APR can be detected in the *Flaveria* C_4 species *F. bidentis* and *F. trinervia*. The C_3–C_4 species *F. pubescens* constitutes an exception with rather high APR mRNA levels. The increased transcript levels of APR in C_4 species correspond with the elevated APR enzyme activity in C_4 and C_4-like *Flaveria*, reported by Koprivova et al. (2001), and indicate transcriptional regulation of APR activity depending on the photosynthetic mechanism. To test whether this trait can be extended to other C_4 species, we analyzed the comparative adult leaf transcriptomes of closely related C_3 and C_4 *Cleomaceae* species, *Cleome gynandra* (C_4) and *Cleome spinosa* (C_3; Brautigam et al., 2011). The ratio of ATPS and APR mRNA levels between the two species were comparable to transcript level relations in *Flaveria* C_3 and C_4 species (**Figure 2C**). The relative abundance of ATPS transcripts is higher in *C. spinosa*, whereas *C. gynandra* exhibits increased abundance of APR mRNA.

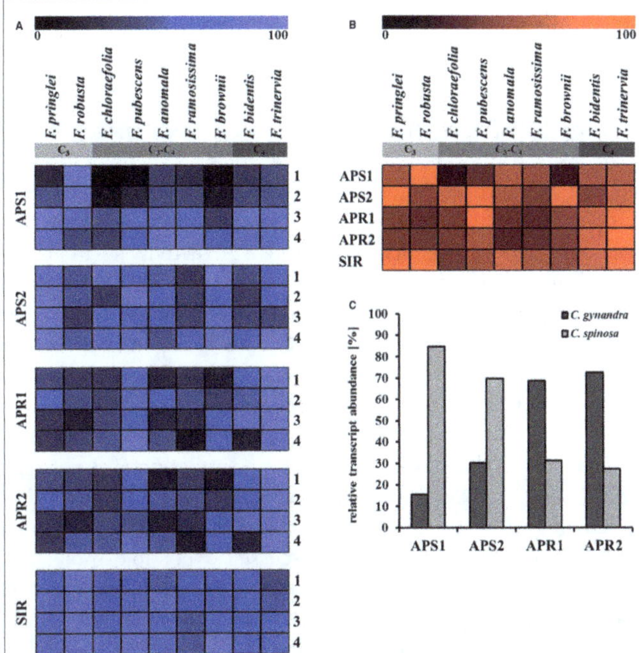

FIGURE 2 | Transcript abundance of genes involved in sulfate assimilation in leaves of C_3, C_4, and C_3–C_4 intermediate plants. Based on the transcriptomic data of nine *Flaveria* species by Mallmann et al. (2014) obtained in four independent cultivations harvested at different seasons (1, September 2009; 2, June 2010; 3, October 2010; 4, April 2011), the transcript levels of the individual cultivations **(A)** and mean values **(B)** were normalized and plotted in heat maps. C_3: *F. pringlei*, *F. robusta*; C_3–C_4: *F. chloraefolia*, *F. pubescens*, *F. anomala*, *F. ramosissima*; C_4-like: *F. brownii*; C_4: *F. bidentis*, *F. trinervia*. **(C)** The relative transcript abundances of genes participating in sulfate assimilation in the *Cleome* species *C. gynandra* (C_4) and *C. spinosa* (C_3) are based on the Brautigam et al. (2011). Transcripts were annotated by allocation to their *Arabidopsis* homologues. APR1 and APR2, APS reductase; APS1 and APS2, ATP sulfurylase; SIR, sulfite reductase.

Adenosine-5′-phosphosulfate reductase has been reported to be the key enzyme of sulfate assimilation with strong regulation by many environmental factors (Kopriva et al., 2001; Vauclare et al., 2002; Scheerer et al., 2010). Among others, sulfur starvation and increased need for cysteine, required for GSH synthesis, induce APR transcription (Vauclare et al., 2002). Hence, increased transcript levels of APR in two independent C_4 species compared to their close C_3 relatives might indicate the need of enhanced levels of reduced sulfur in C_4 plants. Indeed, C_4 plants are particularly adapted to dry and warm habitats, environmental conditions which cause oxidative stress by formation of ROS. To protect the cells from damage, ROS is detoxified in the GSH–ascorbate cycle in which GSH serves as reductant of dehydroascorbate (Noctor and Foyer, 1998). Thus, exposure to oxidative stress increases the demand for GSH. Cysteine, required for GSH synthesis, is provided by sulfate reduction. Adaptation to exceptional habitats could explain the high levels of GSH and cysteine in C_4 *Flaveria* described by Koprivova et al. (2001) as well as the increased APR activity and transcript levels. The C_4 monocot maize is sensitive to chilling, another oxidative stress causing condition. Here as well exposure to chilling stress increases the mRNA levels and activity of APR and ATPS (Brunner et al., 1995; Kopriva et al., 2001). Remarkably, the species *Zea mays* also contains chilling-tolerant genotypes which show higher levels of GSH and cysteine as well as increased activity of APR and GR, the enzyme that reduces oxidized GSH (Kocsy et al., 1996, 1997). Kocsy et al. (2000) demonstrated a direct link between GSH synthesis and chilling tolerance by inhibiting γ-ECS using buthionine sulfoximine (BSO). BSO-treated maize plants lost their chilling tolerance which could be restored by supplementation with GSH or γ-EC.

Overall, these findings in maize fit the general hypothesis of demand-driven regulation of sulfate assimilation formulated from studies with C_3 Brassicaceae species (Lappartient and Touraine, 1996). Sulfur starvation leads to increased transcript levels of ATPS, APR, and sulfate transporters in maize (Bolchi et al., 1999; Hopkins et al., 2004), whereas supplementation of reduced sulfur compounds results in repression of ATPS expression (Bolchi et al., 1999). Besides oxidative stress, cadmium has been reported to enhance ATPS and APR activity (Nussbaum et al., 1988). The detoxification of heavy metals requires the synthesis of phytochelatins which bind to the metals and enable their relocation to the vacuole. As GSH is indispensable for the formation of phytochelatins (Cobbett and Goldsbrough, 2002), exposure to heavy metals increases the demand for reduced sulfur. In Brassicacea species GSH was identified as the molecular regulator of demand-driven sulfate assimilation (Lappartient and Touraine, 1996; Vauclare et al., 2002). In contrast, cysteine directly regulates sulfate reduction in maize (Bolchi et al., 1999). Kopriva and Koprivova (2005) discussed a possible molecular background to the findings of Bolchi et al. (1999). They assumed that cysteine as regulatory molecule is a consequence of the BSC-specific localization of sulfate assimilation in maize. Whereas GSH can be synthesized in maize BSC and MC, cysteine synthesis is restricted to BSC and cysteine was shown to be transported from BSC protoplasts (Burgener et al., 1998; Gomez et al., 2004). This led to the conclusion, that the cysteine pools of MC and BSC are permanently connected and enable a rapid reaction to changes in cysteine concentration (Kopriva and Koprivova, 2005).

Regulation of sulfate assimilation in different *Flaveria* species has not been addressed yet. The elucidation of such regulation might give an insight view on the molecular mechanism causing a constitutionally higher activity of the sulfate assimilatory pathway in C_4 species compared to C_3 plants.

CONCLUSION AND OUTLOOK

The BSC localization of sulfate assimilation and the expression gradient from C_3 to C_4 species are interesting aspects of both sulfur metabolism and C_4 photosynthesis. It is important to find out whether this localization and difference in expression give the plants an evolutionary advantage, which has to be considered in attempts to engineer C_4 photosynthesis in C_3 crops as well as to improve sulfur use efficiency of crop plants. The wealth of genomics and transcriptomics resources in C_4 plants offers new ways to approach these questions, based not on a single model. The findings of preferential localization of sulfate assimilation in BSC in *Arabidopsis* could be the right trigger to assess the consequences of this localization for the general sulfur metabolism. All in all, while we cannot answer the question of importance of the changes in expression and localization of sulfate assimilation for evolution of C_4 photosynthesis yet it seems that this enigmatic question of plant sulfur research might be answered in near future.

ACKNOWLEDGMENT

Research in Stanislav Kopriva's lab is supported by Deutsche Forschungsgemeinschaft (EXC 1028).

REFERENCES

Apel, P., and Maass, L. (1981). Photosynthesis in species of *Flaveria*. CO_2 compensation concentration, O_2 influence on photosynthetic gas exchange and d13C values in species of *Flaveria* (Asteraceae). *Biochem. Physiol. Pflanz.* 176, 396–399. doi: 10.1016/S0015-3796(81)80052-2

Aubry, S., Smith-Unna, R. D., Boursnell, C. M., Kopriva, S., and Hibberd, J. M. (2014). Transcript residency on ribosomes reveals a key role for the *Arabidopsis thaliana* bundle sheath in sulphur and glucosinolate metabolism. *Plant J.* 78, 659–673. doi: 10.1111/tpj.12502

Bassüner, B., Keerberg, O., Bauwe, H., Pyarnik, T., and Keerberg, H. (1984). Photosynthetic CO_2 metabolism in C_3–C_4 intermediate and C_4 species of *Flaveria* (Asteraceae). *Biochem. Physiol. Pflanz.* 179, 631–634. doi: 10.1016/S0015-3796(84)80039-6

Bauwe, H. (1984). Photosynthetic enzyme activities and immunofluorescence studies on the localization of ribulose-1, 5-bisphosphate carboxylase/oxygenase in leaves of C_3, C_4, and C_3–C_4 intermediate species of *Flaveria* (Asteraceae). *Biochem. Physiol. Pflanz.* 179, 253–268. doi: 10.1016/S0015-3796(84)80041-4

Bauwe, H., Keerberg, O., Bassuner, R., Parnik, T., and Bassuner, B. (1987). Reassimilation of carbon dioxide by *Flaveria* (Asteraceae) species representing different types of photosynthesis. *Planta* 172, 214–218. doi: 10.1007/BF00394590

Bauwe, H., and Kolukisaoglu, U. (2003). Genetic manipulation of glycine decarboxylation. *J. Exp. Bot.* 54, 1523–1535. doi: 10.1093/jxb/erg171

Becker, T. W., Carrayol, E., and Hirel, B. (2000). Glutamine synthetase and glutamate dehydrogenase isoforms in maize leaves: localization, relative proportion and their role in ammonium assimilation or nitrogen transport. *Planta* 211, 800–806. doi: 10.1007/s004250000355

Becker, T. W., Perrot-Rechenmann, C., Suzuki, A., and Hirel, B. (1993). Subcellular and immunocytochemical localization of the enzymes involved in ammonia assimilation in mesophyll and bundle-sheath cells of maize leaves. *Planta* 191, 129–136. doi: 10.1007/BF00240904

Benstein, R. M., Ludewig, K., Wulfert, S., Wittek, S., Gigolashvili, T., Frerigmann, H., et al. (2013). *Arabidopsis* phosphoglycerate dehydrogenase1 of the phosphoserine pathway is essential for development and required for ammonium assimilation and tryptophan biosynthesis. *Plant Cell* 25, 5011–5029. doi: 10.1105/tpc.113.118992

Bolchi, A., Petrucco, S., Tenca, P. L., Foroni, C., and Ottonello, S. (1999). Coordinate modulation of maize sulfate permease and ATP sulfurylase mRNAs in response to variations in sulfur nutritional status: stereospecific down-regulation by L-cysteine. *Plant Mol. Biol.* 39, 527–537. doi: 10.1023/A:1006148815106

Brautigam, A., Kajala, K., Wullenweber, J., Sommer, M., Gagneul, D., Weber, K. L., et al. (2011). An mRNA blueprint for C_4 photosynthesis derived from comparative transcriptomics of closely related C_3 and C_4 species. *Plant Physiol.* 155, 142–156. doi: 10.1104/pp.110.159442

Brown, R. H. (1978). A difference in N use efficiency in C_3 and C_4 plants and its implications in adaptation and evolution. *Crop Sci.* 18, 93–98.

Brunner, M., Kocsy, G., Rftegsegger, A., Schmutz, D., and Brunold, C. (1995). Effect of chilling on assimilatory sulfate reduction and glutathione synthesis in maize. *J. Plant Physiol.* 146, 743–747. doi: 10.1016/S0176-1617(11)81942-5

Brunold, C., and Suter, M. (1989). Localization of enzymes of assimilatory sulfate reduction in pea roots. *Planta* 179, 228–234. doi: 10.1007/BF00393693

Buchner, P., Takahashi, H., and Hawkesford, M. J. (2004). Plant sulphate transporters: co-ordination of uptake, intracellular and long-distance transport. *J. Exp. Bot.* 55, 1765–1773. doi: 10.1093/jxb/erh206

Burgener, M., Suter, M., Jones, S., and Brunold, C. (1998). Cyst(e)ine is the transport metabolite of assimilated sulfur from bundle-sheath to mesophyll cells in maize leaves. *Plant Physiol.* 116, 1315–1322. doi: 10.1104/pp.116.4.1315

Burnell, J. N. (1984). Sulfate assimilation in C_4 plants: intercellular and intracellular location of ATP sulfurylase, cysteine synthase, and cystathionine β-lyase in maize leaves. *Plant Physiol.* 75, 873–875. doi: 10.1104/pp.75.3.873

Chang, Y. M., Liu, W. Y., Shih, A. C., Shen, M. N., Lu, C. H., Lu, M. Y., et al. (2012). Characterizing regulatory and functional differentiation between maize mesophyll and bundle sheath cells by transcriptomic analysis. *Plant Physiol.* 160, 165–177. doi: 10.1104/pp.112.203810

Cobbett, C., and Goldsbrough, P. (2002). Phytochelatins and metallothioneins: roles in heavy metal detoxification and homeostasis. *Annu. Rev. Plant Biol.* 53, 159–182. doi: 10.1146/annurev.arplant.53.100301.135154

Douce, R., Bourguignon, J., Neuburger, M., and Rébeillé, F. (2001). The glycine decarboxylase system: a fascinating complex. *Trends Plant Sci.* 6, 167–176. doi: 10.1016/S1360-1385(01)01892-1

Edwards, G. E., Franceschi, V. R., Ku, M. S., Voznesenskaya, E. V., Pyankov, V. I., and Andreo, C. S. (2001). Compartmentation of photosynthesis in cells and tissues of C_4 plants. *J. Exp. Bot.* 52, 577–590. doi: 10.1093/jexbot/52.356.577

Edwards, G. E., and Huber, S. C. (1981). "The C_4 pathway," in *Biochemistry of Plants*, eds M. D. Hatch and N. K. Boardman (New York: Academic Press), 238–281.

Edwards, G. E., and Walker, D. A. (1983). *C_3, C_4: Mechanisms and Cellular and Environmental Regulation of Photosynthesis.* Oxford: Blackwell Scientific.

Foyer, C. H., and Noctor, G. (2009). Redox regulation in photosynthetic organisms: signaling, acclimation, and practical implications. *Antioxid. Redox Signal.* 11, 861–905. doi: 10.1089/ars.2008.2177

Gardeström, P., Edwards, G. E., Henricson, D., and Ericson, I. (1985). The localization of serine hydroxymethyltransferase in leaves of C_3 and C_4 species. *Physiol. Plant.* 64, 29–33. doi: 10.1111/j.1399-3054.1985.tb01208.x

Gerwick, B. C., and Black, C. C. (1979). Sulfur assimilation in C_4 plants. *Plant Physiol.* 64, 590–593.

Gerwick, B. C., Ku, S. B., and Black, C. C. (1980). Initiation of sulfate activation: a variation in C_4 photosynthesis plants. *Science* 209, 513–515. doi: 10.1126/science.209.4455.513

Ghannoum, O., Evans, J., and von Caemmerer, S. (2011). "Nitrogen and water use efficiency of C_4 plants," in *Advances in Photosynthesis and Respiration*, Vol. 32, *C_4 photosynthesis and related CO_2 concentrating mechanisms*, eds A. S. Raghavendra and R. F. Sage (Dordrecht: Springer), 109–128.

Gomez, L. D., Vanacker, H., Buchner, P., Noctor, G., and Foyer, C. H. (2004). Intercellular distribution of glutathione synthesis in maize leaves and its response to short-term chilling. *Plant Physiol.* 134, 1662–1671. doi: 10.1104/pp.103.033027

Hatch, M. D. (1987). C_4 Photosynthesis: a unique blend of modified biochemistry, anatomy, and ultrastructure. *Biochim. Biophys. Acta* 895, 81–106. doi: 10.1016/S0304-4173(87)80009-5

Hatch, M. D., and Osmond, C. B. (1976). "Compartmentation and transport in C_4 photosynthesis," in *Transport in Plants III*, eds C. R. Stocking and U. Heber (Berlin: Springer), 144–184.

Heckmann, D., Schulze, S., Denton, A., Gowik, U., Westhoff, P., Weber, A. P., et al. (2013). Predicting C_4 photosynthesis evolution: modular, individually adaptive steps on a Mount Fuji fitness landscape. *Cell* 153, 1579–1588. doi: 10.1016/j.cell.2013.04.058

Herschbach, C., and Rennenberg, H. (2001). "Significance of phloem-translocated organic sulfur compounds for the regulation of sulfur nutrition," in *Progress in Botany*, eds K. Esser, U. Lüttge, J. W. Kadereit, and W. Beyschlag (Berlin: Springer), 177–193.

Hibberd, J. M., Sheehy, J. E., and Langdale, J. A. (2008). Using C_4 photosynthesis to increase the yield of rice-rationale and feasibility. *Curr. Opin. Plant Biol.* 11, 228–231. doi: 10.1016/j.pbi.2007.11.002

Hopkins, L., Parmar, S., Bouranis, D. L., Howarth, J. R., and Hawkesford, M. J. (2004). Coordinated expression of sulfate uptake and components of the sulfate assimilatory pathway in maize. *Plant Biol. (Stuttg.)* 6, 408–414. doi: 10.1055/s-2004-820872

Kocsy, G., Brunner, M., Rüegsegger, A., Stamp, P., and Brunold, C. (1996). Glutathione synthesis in maize genotypes with different sensitivities to chilling. *Planta* 198, 365–370. doi: 10.1007/BF00620052

Kocsy, G., Owttrim, G., Brander, K., and Brunold, C. (1997). Effect of chilling on the diurnal rhythm of enzymes involved in protection against oxidative stress in a chilling-tolerant and a chilling-sensitive maize genotype. *Physiol. Plant.* 99, 249–254. doi: 10.1111/j.1399-3054.1997.tb05409.x

Kocsy, G., Von Ballmoos, P., Suter, M., Ruegsegger, A., Galli, U., Szalai, G., et al. (2000). Inhibition of glutathione synthesis reduces chilling tolerance in maize. *Planta* 211, 528–536. doi: 10.1007/s004250000308

Kopriva, S. (2006). Regulation of sulfate assimilation in *Arabidopsis* and beyond. *Ann. Bot.* 97, 479–495. doi: 10.1093/aob/mcl006

Kopriva, S., Chu, C. C., and Bauwe, H. (1996). Molecular phylogeny of *Flaveria* as deduced from the analysis of nucleotide sequences encoding the H-protein of the glycine cleavage system. *Plant Cell Environ.* 19, 1028–1036. doi: 10.1111/j.1365-3040.1996.tb00209.x

Kopriva, S., Jones, S., Koprivova, A., Suter, M., Ballmoos, P., Brander, K., et al. (2001). Influence of chilling stress on the intercellular distribution of assimilatory sulfate reduction and thiols in *Zea mays*. *Plant Biol.* 3, 24–31. doi: 10.1055/s-2001-11745

Kopriva, S., and Koprivova, A. (2005). Sulfate assimilation and glutathione synthesis in C_4 plants. *Photosynth. Res.* 86, 363–372. doi: 10.1007/s11120-005-3482-z

Kopriva, S., Muheim, R., Koprivova, A., Trachsel, N., Catalano, C., Suter, M., et al. (1999). Light regulation of assimilatory sulphate reduction in *Arabidopsis thaliana*. *Plant J.* 20, 37–44. doi: 10.1046/j.1365-313X.1999.00573.x

Kopriva, S., Suter, M., Von Ballmoos, P., Hesse, H., Krahenbuhl, U., Rennenberg, H., et al. (2002). Interaction of sulfate assimilation with carbon and nitrogen metabolism in *Lemna minor*. *Plant Physiol.* 130, 1406–1413. doi: 10.1104/pp.007773

Koprivova, A., Melzer, M., Von Ballmoos, P., Mandel, T., Brunold, C., and Kopriva, S. (2001). Assimilatory sulfate reduction in C_3, C_3–C_4, and C_4 species of *Flaveria*. *Plant Physiol.* 127, 543–550. doi: 10.1104/pp.010144

Koprivova, A., Suter, M., Den Camp, R. O., Brunold, C., and Kopriva, S. (2000). Regulation of sulfate assimilation by nitrogen in *Arabidopsis*. *Plant Physiol.* 122, 737–746. doi: 10.1104/pp.122.3.737

Laetsch, W. M. (1974). The C_4 syndrome: a structural analysis. *Annu. Rev. Plant Physiol. Plant Mol. Biol.* 25, 27–52. doi: 10.1146/annurev.pp.25.060174.000331

Lappartient, A. G., and Touraine, B. (1996). Demand-driven control of root ATP sulfurylase activity and SO_4^{2-} uptake in intact canola (the role of phloem-translocated glutathione). *Plant Physiol.* 111, 147–157.

Leustek, T., Martin, M. N., Bick, J. A., and Davies, J. P. (2000). Pathways and regulation of sulfur metabolism revealed through molecular and genetic studies. *Annu. Rev. Plant Physiol. Plant Mol. Biol.* 51, 141–165. doi: 10.1146/annurev.arplant.51.1.141

Leustek, T., and Saito, K. (1999). Sulfate transport and assimilation in plants. *Plant Physiol.* 120, 637–644. doi: 10.1104/pp.120.3.637

Mallmann, J., Heckmann, D., Brautigam, A., Lercher, M. J., Weber, A. P., Westhoff, P., et al. (2014). The role of photorespiration during the evolution of C_4 photosynthesis in the genus *Flaveria*. *Elife* 3, e02478. doi: 10.7554/eLife.02478

Monson, R. K., Moore, B. D., Ku, M. S. B., and Edwards, G. E. (1986). Co-function of C_3-and C_4-photosynthetic pathways in C_3, C_4 and C_3–C_4 intermediate *Flaveria* species. *Planta* 168, 493–502. doi: 10.1007/BF00392268

Moore, R., and Black, C. C. (1979). Nitrogen assimilation pathways in leaf mesophyll and bundle sheath cells of C_4 photosynthesis plants formulated from comparative studies with *Digitaria sanguinalis* (L.) Scop. *Plant Physiol.* 64, 309–313. doi: 10.1104/pp.64.2.309

Noctor, G., and Foyer, C. H. (1998). ASCORBATE AND GLUTATHIONE: keeping active oxygen under control. *Annu. Rev. Plant Physiol. Plant Mol. Biol.* 49, 249–279. doi: 10.1146/annurev.arplant.49.1.249

Nussbaum, S., Schmutz, D., and Brunold, C. (1988). Regulation of assimilatory sulfate reduction by cadmium in *Zea mays* L. *Plant Physiol.* 88, 1407–1410. doi: 10.1104/pp.88.4.1407

Ohnishi, J.-I., and Kanai, R. (1983). Differentiation of photorespiratory activity between mesophyll and bundle sheath cells of C_4 plants I. Glycine oxidation by mitochondria. *Plant Cell Physiol.* 24, 1411–1420.

Passera, C., and Ghisi, R. (1982). ATP sulphurylase and *O*-acetylserine sulphydrylase in isolated mesophyll protoplasts and bundle sheath strands of S-deprived maize leaves. *J. Exp. Bot.* 33, 432–438. doi: 10.1093/jxb/33.3.432

Pastori, G. M., Mullineaux, P. M., and Foyer, C. H. (2000). Post-transcriptional regulation prevents accumulation of glutathione reductase protein and activity in the bundle sheath cells of maize. *Plant Physiol.* 122, 667–675. doi: 10.1104/pp.122.3.667

Pfundel, E., Nagel, E., and Meister, A. (1996). Analyzing the light energy distribution in the photosynthetic apparatus of C_4 plants using highly purified mesophyll and bundle-sheath thylakoids. *Plant Physiol.* 112, 1055–1070.

Rathnam, C. K., and Edwards, G. E. (1976). Distribution of nitrate-assimilating enzymes between mesophyll protoplasts and bundle sheath cells in leaves of three groups of C_4 plants. *Plant Physiol.* 57, 881–885. doi: 10.1104/pp.57.6.881

Ros, R., Munoz-Bertomeu, J., and Krueger, S. (2014). Serine in plants: biosynthesis, metabolism, and functions. *Trends Plant Sci.* 19, 564–569. doi: 10.1016/j.tplants.2014.06.003

Sage, R. F. (2004). The evolution of C_4 photosynthesis. *New phytol.* 161, 341–370. doi: 10.1111/j.1469-8137.2004.00974.x

Sage, R. F., Christin, P. A., and Edwards, E. J. (2011). The C_4 plant lineages of planet Earth. *J. Exp. Bot.* 62, 3155–3169. doi: 10.1093/jxb/err048

Sage, R. F., Sage, T. L., and Kocacinar, F. (2012). Photorespiration and the evolution of C_4 photosynthesis. *Annu. Rev. Plant Biol.* 63, 19–47. doi: 10.1146/annurev-arplant-042811-105511

Saito, K. (2004). Sulfur assimilatory metabolism. The long and smelling road. *Plant Physiol.* 136, 2443–2450. doi: 10.1104/pp.104.046755

Scheerer, U., Haensch, R., Mendel, R. R., Kopriva, S., Rennenberg, H., and Herschbach, C. (2010). Sulphur flux through the sulphate assimilation pathway is differently controlled by adenosine 5′-phosphosulphate reductase under stress and in transgenic poplar plants overexpressing *γ-ECS, SO*, or *APR. J. Exp. Bot.* 61, 609–622. doi: 10.1093/jxb/erp327

Schmutz, D., and Brunold, C. (1984). Intercellular localization of assimilatory sulfate reduction in leaves of *Zea mays* and *Triticum aestivum. Plant Physiol.* 74, 866–870. doi: 10.1104/pp.74.4.866

Schmutz, D., and Brunold, C. (1985). Localization of nitrite and sulfite reductase in bundle sheath and mesophyll cells of maize leaves. *Physiol. Plant.* 64, 523–528. doi: 10.1111/j.1399-3054.1985.tb08533.x

Schulze, S., Mallmann, J., Burscheidt, J., Koczor, M., Streubel, M., Bauwe, H., et al. (2013). Evolution of C_4 photosynthesis in the genus *Flaveria*: establishment of a photorespiratory CO_2 pump. *Plant Cell* 25, 2522–2535. doi: 10.1105/tpc.113.114520

Sheen, J. (1999). C_4 gene expression. *Annu. Rev. Plant Physiol. Plant Mol. Biol.* 50, 187–217. doi: 10.1146/annurev.arplant.50.1.187

Sheen, J. Y., and Bogorad, L. (1988). Differential expression in bundle sheath and mesophyll cells of maize of genes for photosystem II components encoded by the plastid genome. *Plant Physiol.* 86, 1020–1026. doi: 10.1104/pp.86.4.1020

Shroff, R., Vergara, F., Muck, A., Svatos, A., and Gershenzon, J. (2008). Nonuniform distribution of glucosinolates in *Arabidopsis thaliana* leaves has important consequences for plant defense. *Proc. Natl. Acad. Sci. U.S.A.* 105, 6196–6201. doi: 10.1073/pnas.0711730105

Tabe, L. M., and Droux, M. (2002). Limits to sulfur accumulation in transgenic lupin seeds expressing a foreign sulfur-rich protein. *Plant Physiol.* 128, 1137–1148. doi: 10.1104/pp.010935

Takahashi, H., Kopriva, S., Giordano, M., Saito, K., and Hell, R. (2011). Sulfur assimilation in photosynthetic organisms: molecular functions and regulations of transporters and assimilatory enzymes. *Annu. Rev. Plant Biol.* 62, 157–184. doi: 10.1146/annurev-arplant-042110-103921

Vauclare, P., Kopriva, S., Fell, D., Suter, M., Sticher, L., Von Ballmoos, P., et al. (2002). Flux control of sulphate assimilation in *Arabidopsis thaliana*: adenosine 5′-phosphosulphate reductase is more susceptible than ATP sulphurylase to negative control by thiols. *Plant J.* 31, 729–740. doi: 10.1046/j.1365-313X.2002.01391.x

Williams, B. P., Johnston, I. G., Covshoff, S., and Hibberd, J. M. (2013). Phenotypic landscape inference reveals multiple evolutionary paths to C_4 photosynthesis. *Elife* 2, e00961. doi: 10.7554/eLife.00961

Yarmolinsky, D., Brychkova, G., Fluhr, R., and Sagi, M. (2013). Sulfite reductase protects plants against sulfite toxicity. *Plant Physiol.* 161, 725–743. doi: 10.1104/pp.112.207712

Zimmermann, P., Hirsch-Hoffmann, M., Hennig, L., and Gruissem, W. (2004). GENEVESTIGATOR. *Arabidopsis* microarray database and analysis toolbox. *Plant Physiol.* 136, 2621–2632. doi: 10.1104/pp.104.046367

Conflict of Interest Statement: The authors declare that the research was conducted in the absence of any commercial or financial relationships that could be construed as a potential conflict of interest.

10

Differential response to sulfur nutrition of two common bean genotypes differing in storage protein composition

Sudhakar Pandurangan[1,2], Mark Sandercock[3], Ronald Beyaert[2], Kenneth L. Conn[2], Anfu Hou[3] and Frédéric Marsolais[1,2] *

[1] Department of Biology, University of Western Ontario, London, ON, Canada
[2] Genomics and Biotechnology, Southern Crop Protection and Food Research Centre, Agriculture and Agri-Food Canada, London, ON, Canada
[3] Cereal Research Centre Morden, Agriculture and Agri-Food Canada, Morden, MB, Canada

Edited by:
Stanislav Kopriva, University of Cologne, Germany

Reviewed by:
Rachel Amir, Migal – Tel Hai, Israel
Christian Zörb, University of Hohenheim, Germany

***Correspondence:**
Frédéric Marsolais, Genomics and Biotechnology, Southern Crop Protection and Food Research Centre, Agriculture and Agri-Food Canada, 1391 Sandford Street, London, Ontario, N5V 4T3, Canada
e-mail: frederic.marsolais@agr.gc.ca

It has been hypothesized that the relatively low concentration of sulfur amino acids in legume seeds might be an ecological adaptation to nutrient poor, marginal soils. SARC1 and SMARC1N-PN1 are genetically related lines of common bean (dry bean, *Phaseolus vulgaris*) differing in seed storage protein composition. In SMARC1N-PN1, the lack of phaseolin and major lectins is compensated by increased levels of sulfur-rich proteins, resulting in an enhanced concentration of cysteine and methionine, mostly at the expense of the abundant non-protein amino acid, *S*-methylcysteine. To identify potential effects associated with an increased concentration of sulfur amino acids in the protein pool, the response of the two genotypes to low and high sulfur nutrition was evaluated under controlled conditions. Seed yield was increased by the high sulfate treatment in SMARC1N-PN1. The seed concentrations of sulfur, sulfate, and *S*-methylcysteine were altered by the sulfur treatment in both genotypes. The concentration of total cysteine and extractible globulins was increased specifically in SMARC1N-PN1. Proteomic analysis identified arcelin-like protein 4, lipoxygenase-3, albumin-2, and alpha amylase inhibitor beta chain as having increased levels under high sulfur conditions. Lipoxygenase-3 accumulation was sensitive to sulfur nutrition only in SMARC1N-PN1. Under field conditions, both SARC1 and SMARC1N-PN1 exhibited a slight increase in yield in response to sulfur treatment, typical for common bean.

Keywords: sulfur nutrition, seed yield, sulfur amino acids, globulin, lipoxygenase, albumin, lectin, common bean

INTRODUCTION

Common bean (dry bean, *Phaseolus vulgaris*) is an important source of protein and fiber in human diets. Like other grain legumes, its protein quality is sub-optimal, being limited by the levels of the essential sulfur amino acids, methionine, and cysteine. During the past decades, a lot of effort has been dedicated to improving protein quality in grain legumes, primarily using transgenic approaches (Krishnan, 2005; Ufaz and Galili, 2008; Amir et al., 2012; Galili and Amir, 2013). Transgenic expression of foreign proteins can be limited by the supply of sulfur, and often results in a shift of sulfur away from endogenous, sulfur-rich proteins (Streit et al., 2001; Tabe and Droux, 2002; Chiaiese et al., 2004). **Table 1** lists the different experiments that were performed involving transgenic expression of sulfur-rich proteins in legumes and their outcomes. In *Vicia narbonensis*, co-expression of Brazil nut 2S albumin and a feedback-insensitive, bacterial aspartate kinase was associated with increased sulfur concentration in seed (Demidov et al., 2003). A common concern with these approaches is the potential allergenicity of the foreign proteins (Nordlee et al., 1996; Kelly and Hefle, 2000; Krishnan et al., 2010). A possible solution to this problem is the expression of a *de novo* synthetic protein, MB-16. An alternative approach involves the transgenic manipulation of sulfur amino acid pathways. Overexpression of cytosolic serine acetyltransferase in developing soybean seed led to a 70% increase in total cysteine concentration (Kim et al., 2012). Expression of a feedback-insensitive *Arabidopsis* cystathionine γ-synthase (AtD-CGS), encoding a protein lacking 30 amino acids in the N-terminal domain, raised total methionine concentration by 1.8 to 2.3-fold, with an overall increase in seed protein concentration (Song et al., 2013). By contrast, expression of the feedback-insensitive *mto1-1* allele, harboring a point mutation, led to elevated levels of free methionine, but not total methionine in soybean, whereas in azuki bean, the levels of cystathionine were raised while total methionine concentration was actually decreased (Hanafy et al., 2013a,b). A completely different approach proposed to improve protein quality in common bean involves the introduction of highly digestible phaseolin types from wild accessions by conventional breeding (Montoya et al., 2010). Based on *in vitro* protein digestibility corrected amino acid score, genotypes having highly digestible phaseolin types could increase bioavailable sulfur amino acids by approximately 30% as compared with S type phaseolin present in Mesoamerican cultivars.

Grain yield in legumes has a low heritability due to environmental variables. Consequently, agronomic practices combined with proper fertilizer management heavily influence yield. Sulfur, which has long been known to play a major role in plant metabolism (Takahashi et al., 2011), increases yield in common bean (Malavolta et al., 1987) and influences seed quality via the

Table 1 | Attempts to improve sulfur amino acid concentration in legumes by transgenic expression of sulfur-rich proteins.

Crop plant	Foreign protein	Increase in sulfur amino acids	Reference
Soybean	Brazil nut 2S albumin	Methionine 26%	Townsend and Thomas (1994)
Soybean	15 kDa δ-zein	Methionine by 20% and cysteine by 35%	Dinkins et al. (2001)
Soybean	27 kDa γ-zein	Methionine by 19% and cysteine by 30%	Li et al. (2005)
Soybean	11 kDa δ-zein	Methionine (alcohol soluble fraction)	Kim and Krishnan (2004)
Common bean	Brazil nut 2S albumin	Methionine by 20%	Aragão et al. (1999)
Lupin	Sunflower seed albumin	Methionine by 90%	Molvig et al. (1997), Tabe and Droux (2002)
Chickpea	Sunflower seed albumin	Methionine by 90%	Chiaiese et al. (2004)
Vicia narbonensis	Brazil nut 2S albumin and feedback-insensitive aspartate kinase	Methionine by 100% and cysteine by 20%	Demidov et al. (2003)
Soybean	MB-16	Methionine by 16% and cysteine by 66%	Zhang et al. (2014)

proportion of sulfur containing amino acids, cysteine, and methionine. Sulfate is the most significant and readily mobilized form of sulfur. Sulfate taken up by the roots is reduced to sulfide and further incorporated into cysteine. Cysteine is converted to methionine or incorporated into glutathione and proteins. Sulfate, and/or organic forms of sulfur, such as glutathione (Anderson and Fitzgerald, 2001) or *S*-methylmethionine (Bourgis et al., 1999; Lee et al., 2008; Tan et al., 2010), is transported through the phloem, followed by uptake by transporters into the developing embryo and translocation between seed tissues (Zuber et al., 2010). Delivery of adequate sulfur to seed tissues is needed for maximizing production and to improve protein quality (Hawkesford and De Kok, 2006). Nutrient status of the plant regulates the uptake and assimilation of sulfate (Smith et al., 1995; Buchner et al., 2004). Studies have shown that a decrease in sulfate availability results in a several-fold enhanced expression of sulfate transporter genes, which enhances the capacity for sulfate uptake (Hawkesford, 2000, 2003). Sulfur fertilization favorably affects protein quality by increasing the expression of proteins rich in sulfur amino acids. Control of seed protein accumulation by the sulfur status has been well documented in several legumes, including globulins in soybean and lupine (Blagrove et al., 1976; Gayler and Sykes, 1985), and globulins, and albumins in pea (Chandler et al., 1983, 1984; Higgins et al., 1987). Reduced expression of pea legumin and albumin 1 genes in response to sulfur deficiency was further confirmed in transgenic tobacco (Rerie et al., 1991; Morton et al., 1998). In general, high sulfur stimulates the expression of sulfur-rich globulins and albumins while sulfur deficiency increases the expression of sulfur-poor globulins. In soybean, the accumulation of the sulfur-poor β-subunit of β-conglycinin is repressed by exogenous methionine (Holowach et al., 1984, 1986). This was confirmed in transgenic *Arabidopsis* (Naito et al., 1995). The immediate metabolic precursor of cysteine, *O*-acetylserine, seems involved in the up-regulation of the β-subunit of β-conglycinin under conditions of sulfur deficiency (Kim et al., 1999). A high nitrogen to sulfur ratio not only increases the accumulation of the β-subunit of β-conglycinin, but also reduces the levels of sulfur-rich Bowman-Birk inhibitor (Krishnan et al., 2005). Recent research has focused on adaptation of legumes to sulfur deficiency, highlighting the possible role of a vacuolar sulfate transporter in *Medicago truncatula* (Zuber et al., 2013). This research is relevant to improvement of sulfur use efficiency (De Kok et al., 2011).

Crop plants mitigate the effect of silencing or deficiency in storage proteins through rebalancing of the seed proteome (Marsolais et al., 2010; Herman, 2014; Wu and Messing, 2014). SARC1 and SMARC1N-PN1 are related genotypes of common bean differing in seed protein composition (Osborn et al., 2003). They share 87.5 and 83.6% of the recurrent, Sanilac parental background, respectively. SARC1 integrates the lectin arcelin-1 from a wild accession. SMARC1N-PN1 lacks phaseolin and major lectins, through introgressions from a *P. coccineus* accession and Great Northern 1140, respectively. These changes are associated with an increased concentration of methionine and cysteine, by 10 and 70%, respectively, concomitant with 70% decrease in S-methylcysteine concentration (Taylor et al., 2008). Proteomic and transcript profiling indicated that several sulfur-rich proteins have increased levels in SMARC1N-PN1, including the 11S globulin legumin, albumin-2, defensin D1, Bowman-Birk type proteinase inhibitor 2, albumin-1, basic 7S globulin, and Kunitz trypsin protease inhibitor (Marsolais et al., 2010; Yin et al., 2011; Liao et al., 2012). SARC1 and SMARC1N-PN1 offer a unique system to investigate how related legume genotypes, harboring natural genetic variation in storage protein composition, respond to sulfur deficiency. The presence of an endogenous sink for sulfur in SMARC1N-PN1 is associated with an increased plasticity of the seed composition in response to sulfur nutrition.

MATERIALS AND METHODS

PLANT MATERIALS AND GROWTH CONDITIONS

SARC1 and SMARC1N-PN1 were evaluated for their response to sulfur nutrition by fertilizing with a nutrient solution containing low sulfur (LS) or high sulfur (HS) as described in previous work with common bean (Sánchez et al., 2002) and chickpea (Chiaiese et al., 2004), with modifications (Pandurangan et al., submitted to *Sulfur Metabolism in Plants. Molecular Physiology and Ecophysiology of Sulfur. Proceedings of the International Plant Sulfur Workshop*). Seeds were sown in small trays containing vermiculite for better germination. Ten day old seedlings were transplanted to pots (17 cm× 20 cm) containing sand, perlite, and vermiculite in a 2:1:1 ratio. The experimental unit consisted in a pot with two plants. There were two groups (LS and HS) of five pots for each genotype. For the initial establishment, the transplanted seedlings were fertilized once with 20:20:20 (N:P:K; Plant Prod, Brampton, ON, Canada) before the actual sulfur treatment. The nutrient solutions for the treatment were made fresh from stock solutions and applied once weekly. The LS nutrient solution contained 0.2 mM K_2SO_4 and 1.8 mM $MgCl_2$; HS contained 0.2 mM K_2SO_4 and 1.8 mM $MgSO_4$. Other nutrients included 4.5 mM $Ca(NO_3)_2$, 1.7 mM K_2HPO_4, 4 μM $MnSO_4.H_2O$, 5 μM H_3BO_3, 10 μM Fe-EDTA, 0.25 μM $CuSO_4.5H_2O$, 1 μM $ZnSO_4.7H_2O$, and 0.2 μM $Na_2MoO_4.2H_2O$. Plants were grown in cabinets (Conviron E8H, Winnipeg, MB, Canada) with 16 h light (300–400 μmol photons m^{-2} s^{-1}) and 8 h dark, with a temperature cycling between 18 and 24°C (Pandurangan et al., 2012).

FIELD TRIAL

The response of SARC1 and SMARC1N-PN1 genotypes to sulfur fertilization was assessed in a field trial conducted at the Cereal Research Centre Morden, MB, Canada, in 2012. Soil was sampled in the fall 2011 and analyzed at Exova, Calgary, AB, Canada, to determine the amount of nutrients needed for the treatment. Nutrient analysis found 24:129:1345 kg ha^{-1} as nitrogen, phosphorus, potassium, and 47 kg ha^{-1} as sulfur. Crops were either grown with or without applied sulfur (30 kg ha^{-1}) as gypsum ($CaSO_4 \cdot 2H_2O$). Recommended seed rate (250,000 seeds ha^{-1} = 25 seed m^{-2}) and cultural practices were used at all plots. Plot size was 1 m × 5.5 m trimmed to 5.0 m^2 with spacing of two rows at 0.5 m between plots. All plots were planted in a randomized complete block design with four replications for each treatment, each replication consisting of two rows of 5.5 m long accounting for 550 seeds per treatment. Two adjacent rows represented one replicate. A post emergent herbicide, Basagran (BASF Canada, Mississauga, ON, Canada) was applied at the rate of 2.2 l ha^{-1}. Fertilizer added in all the treatment plots was 120 kg N ha^{-1}. Dry mature seeds from the net area of each plot were harvested separately, weighed, and recorded as seed yield (kg ha^{-1}).

AMINO ACID ANALYSIS

Extraction and quantification of sulfur amino acids from mature seed tissue was performed as previously described, using HPLC after derivatization with phenylisothiocyanate (Hernández-Sebastià et al., 2005; Taylor et al., 2008). Cysteine was quantified separately as cysteic acid after oxidation with performic acid.

ALBUMIN AND GLOBULIN EXTRACTION AND QUANTIFICATION

Albumin and globulin fractions were extracted from mature seed as described by Rolletschek et al. (2005). Protein in the extracts was quantified using the Bio-Rad Protein Assay reagent (Mississauga, ON, Canada) with bovine serum albumin as standard. Protein concentration was normalized according to the volume of extract recovered. A volume of sample equivalent to the same weight of tissue extracted was subjected to SDS-PAGE on a 12% polyacrylamide gel. Following staining with Coomassie R-250, band intensities in globulin extracts were measured with Quantity One 4.2.1 (Bio-Rad). Quantity One is very tolerant of an assortment of electrophoretic artifacts, and can measure total and average quantities, determine relative and actual amounts of protein. Prior to quantification the image acquired from scanning the gels was optimized by the software by performing lane background subtraction to reduce any noise or background density while maintaining image quality followed by filtering to remove small noise features while leaving larger features relatively unaffected. The software was used for identifying lanes and defining, quantitating, and matching bands. Lane-based quantitation used to calculate intensity of similar bands across lanes involves calculating the average intensity of pixels across the band width and integrating over the band height. The quantity of a band as measured by the area under its intensity profile curve is expressed as intensity × mm. Apparent molecular mass was calculated based on standards using the same software.

SAMPLE PREPARATION AND MASS SPECTROMETRY

Proteomic experiments were performed at the London Regional Proteomics Centre of the University of Western Ontario. Sample preparation was carried out at the Functional Proteomics Facility. Protein bands of interest identified by band intensities in the globulin extracts were excised by the robotic Ettan Spot Picker (GE Healthcare Life Sciences, Baie d'Urfé, QC, Canada) and suspended in 50% methanol and 5% acetic acid for digestion. Trypsin digestion was performed using the MassPREP automated digester (Waters, Mississauga, ON, Canada). Gel pieces were destained using 50 mM ammonium bicarbonate and 50% acetonitrile followed by protein reduction with 10 mM dithiothreitol, alkylation with 55 mM iodoacetamide and tryptic digestion. Peptides were extracted using a solution of 1% formic acid and 2% acetonitrile and lyophilized. Peptides were dissolved in a solution of 30% acetonitrile and 0.1% trifluoroacetic acid mixed with α-cyano-4-hydroxycinnamic acid in 50% acetonitrile, 12.5 mM ammonium citrate, 0.1% trifluoroacetic acid, and analyzed on a 4700 Proteomics Discovery System (Life Technologies, Burlington, ON, Canada) at the MALDI-MS facility. MS analysis was carried out in an *m/z* range of 500–3500 and mass tolerance of 50 ppm. Data acquisition and processing were done using 4000 Series Explorer and Data Explorer (Life Technologies). The instrument was equipped with a 355 nm Nd:YAG laser and the laser rate was 200 Hz. Reflectron and linear positive ion modes were used. Each mass spectrum was collected as a sum of 1000 shots. Samples from protein bands no. 1, 4, and 5 were further analyzed by LC–MS–MS at the Biological Mass Spectrometry Laboratory. They were reconstituted in 18 μl of 0.1% formic acid in water and 10 μl was injected into the UPLC-MS/MS system. The system was comprised of a Waters nanoAcquity UPLC with a Waters C18 trapping and Waters 25 cm analytical column coupled to a Waters QToF Ultima Global Mass Spectrometer. The sample was run at a flow rate of 0.3 μl/min. Solvent A was water:formic acid 0.1% and solvent B was acetonitrile:formic acid 0.1%. Solvent B was set to go from 5% to 60% in 40 min and then reach 95% by 42.5 min. B was kept at 95% for 5 min and brought back to 5% at 50 min. The column was re-equilibrated for 25 min prior to the following injection. Sample loading took 3 min with a flow rate of 10 μl/min at 99% A and 1% B. MS survey scan was performed at a cone voltage of 35 V and set to 1.4 s with 0.1 s interscan and recorded from 300 to 1800 *m/z*. In a given survey scan, all doubly and triply charged ions with intensities greater than 40 counts were considered candidate to undergo MS/MS fragmentation. MS/MS acquisition would stop as soon as the total ion current would reach 25000 counts per second or after a maximum time of 6 s. MS/MS scan was acquired from 50 to 1800 *m/z* for 1.4 s with an interscan time of 0.1 s. Selected ions were fragmented with a collision energy of 30 eV.

Peptide mass fingerprint data were analyzed by searching peptide mass values against a translated version of the preliminary release of the common bean genome (June 26, 2012; Schmutz et al., 2014) using MASCOT (Matrix Science, Boston, MA, USA). The following parameters were used: 1 missed cleavage; fixed carbamidomethyl alkylation of cysteine; variable oxidation of methionine; peptide mass tolerance: $\pm$ 1.2 Da; peptide charge state: +1, significant threshold: $p < 0.05$. For MS-MS, raw data were converted to mgf files using PEAKS 5.3 (Bioinformatics Solutions Inc., Waterloo, ON, Canada). MS/MS ion search was performed with MASCOT against the same database, as well the Mascot database (MSDB, August 31, 2006) using similar cleavage and post-translational modification parameters.

SULFATE ANALYSIS

Replicate samples (~50 seeds) were ground to a fine powder in a Kleco Ball Mill (Visalia, CA, USA) and lyophilized. Approximately 100 mg of ground tissue was used for sulfate analysis by chemical suppression ion chromatography and conductivity detection using a Dionex DX-600 Ion Chromatograph (Thermo Fisher Scientific, Sunnyvale, CA, USA), as described in Herschbach et al. (2000) with modifications. Approximately 100 mg of tissue was extracted in 0.5 ml of deionized water. The suspension was centrifuged at 16,000 × *g* for 10 min at 4°C. A 300 μl aliquot of the cleared supernatant was transferred to an ion chromatography vial for testing using an IonPac anion-exchange column (AS14A, 4 mm; Thermo Fisher Scientific) and eluted with a mixture of 3.5 mM sodium hydrogen carbonate, and 1.0 mM sodium carbonate. A 10 μl aliquot of the solution contained in vials was injected into the eluent stream and background conductivity of eluents reduced by a suppressor (Anion Self-Regenerating Suppressor Ultra, 4 mm). An AS50 auto sampler equipped with a refrigerated chamber was used to house the vials and Dionex Peaknet 6.0 software was employed to track and analyze data.

ELEMENTAL ANALYSIS

Approximately 500 mg of ground seed tissue was submitted to elemental analysis which was performed by dry combustion with a CNS-2000 Elemental Analyzer (LECO Instruments ULC, Mississauga, ON, Canada) as described by Taylor et al. (2008).

STATISTICAL ANALYSIS

Analysis of variance was performed using SAS version 9.2 (Toronto, ON, Canada). Homogeneity of the variances was inspected by residual graphic analysis.

ACCESSION NUMBERS

Accession numbers for proteins in this study are as follows: arcelin-like 4 [Uniprot:Q8RVX7]; lipoxygenase-3 [Phytozome:Phvul.005g 157000.1]; albumin-2 [Phytozome:Phvul007g275800]; α-amylase inhibitor β-subunit [Uniprot:Q9S9E1]; α-amylase inhibitor 1 [Uniprot:Q6J2U4].

RESULTS

INCREASED YIELD OF SMARC1N-PN1 IN RESPONSE TO HIGH SULFUR UNDER CONTROLLED CONDITIONS

To determine whether differences in response to sulfur nutrition are associated with the presence of an additional, endogenous sink for sulfur in SMARC1N-PN1, an experiment was performed under controlled conditions with two levels of sulfate fertilization. Treatment conditions were designed so that the LS conditions correspond to a sulfur deficiency at the reproductive stage. The LS condition was found to be non-limiting for vegetative growth (**Figure 1**). The nitrogen levels selected are non-limiting (Sánchez et al., 2002; Chiaiese et al., 2004). The two genotypes were compared for their agronomic parameters. The following variables were evaluated: number of seeds, seed weight, and seed yield (**Table 2**). The fact that SARC1 and SMARC1N-PN1 are not completely isogenic explains the occurrence of genotypic differences for some of these characteristics. There were significant interactions between factors for seed weight and yield. Whereas the average seed weight decreased under HS for SARC1, it actually increased for SMARC1N-PN1 (G × T; $p \leq 0.01$). This was associated with increased yield, specifically in SMARC1N-PN1, by 8% (G × T; $p \leq 0.05$). A trial was performed to determine if the differences observed under controlled conditions would be replicated in the field. The large difference in yield between genotypes indicates that SMARC1N-PN1 is not well adapted to agronomic conditions in Manitoba. Both genotypes exhibited a limited yield response to sulfate fertilization, by 3–15% (**Table 3**). This response is typical for common bean and other legume crops.

INCREASED SEED CONCENTRATION OF SULFUR AND SULFATE IN RESPONSE TO HIGH SULFATE TREATMENT

To determine if the sulfur treatment effectively altered seed composition and particularly the concentration of sulfur and its metabolites, mature seeds were analyzed for total carbon, nitrogen,

FIGURE 1 | Plants grown under controlled conditions with low sulfur (LS) and high sulfur (HS), 60 days after germination and 26 days after flowering. Vegetative growth appeared similar between genotypes.

Table 2 | Effect of sulfur on number of seeds, seed weight, and yield under controlled conditions.

Genotype	Treatment	Number of seeds	Seed weight (mg)	Yield (g)
SARC1	LS	134 ± 16	214.6 ± 10.1	28.6 ± 2.1
	HS	143 ± 9	196.9 ± 8.9	28.2 ± 1.5
SMARC1N-PN1	LS	152 ± 12	195.5 ± 9.0	29.6 ± 1.2
	HS	157 ± 9	203.8 ± 11.8	32.0 ± 0.7
Source of variation	**d. f.**		**ANOVA *p*-value**	
Genotype (G)	1	0.008	n. s.	0.002
Treatment (T)	1	n. s.	n. s.	n. s.
G × T	1	n. s.	0.01	0.05
Error	15			

Values are the average ± SD; n = 5; LS, low sulfur (0.2 mM SO_4); HS, high sulfur (2 mM SO_4); d. f., degrees of freedom; n. s., not significant.

Table 3 | Seed yield in sulfur deficient and -sufficient field conditions.

Genotype	Yield without sulfur (kg ha^{-1})	Yield with sulfur (kg ha^{-1})
SARC1	1736 ± 407	1995 ± 564
SMARC1N-PN1	667 ± 330	689 ± 211

Values are the average ± SD; n = 4.

and sulfur by elemental analysis and for sulfate concentration by ion analysis. Previously, SARC1 and SMARC1N-PN1 were shown to have similar nitrogen concentration in seed (Hartweck and Osborn, 1997), and Taylor et al. (2008) reported a similar seed concentration of carbon, nitrogen, and sulfur. The sulfur treatment did not change carbon and nitrogen concentration, but had a significant effect on sulfur and sulfate concentration (**Table 4**). Sulfur concentration was raised by the HS treatment by approximately 15 to 20% in both genotypes. Sulfate concentration was increased by 17% in SARC1 and 38% in SMARC1N-PN1. The differences in sulfur and sulfate concentrations indicate that treatment conditions are suitable to investigate whether the two genotypes respond differently to sulfur nutrition.

INCREASED CONCENTRATION OF CYSTEINE AND GLOBULINS IN SMARC1N-PN1 UNDER HIGH SULFATE CONDITIONS

To evaluate whether sulfur nutrition has an effect on the total concentration of sulfur amino acids, methionine, cysteine, and the non-protein amino acid, *S*-methylcysteine were quantified after acid hydrolysis of ground seed tissue. As expected, the concentration of these three amino acids was different between genotypes, methionine, and cysteine being higher, and *S*-methylcysteine lower in SMARC1N-PN1 than in SARC1 (**Table 5**), as previously reported (Taylor et al., 2008). HS increased the levels of *S*-methylcysteine by approximately 40% in both genotypes. Cysteine concentration was raised in response to the HS treatment specifically in SMARC1N-PN1, by 16% (G × T; $p \leq 0.03$). On average, the combined levels of methionine and cysteine were elevated by 13% in SMARC1N-PN1, while they were decreased by 2% in SARC1 in response to the HS treatment.

To investigate whether the differences in sulfur nutrient allocation influenced seed storage protein composition, an important determinant of seed quality, albumins, and globulins were sequentially extracted and their concentration quantified (**Table 6**). The concentration of extractible albumins was unchanged by the treatment. However, the concentration of extractible globulins increased specifically in SMARC1N-PN1, by 24% (G × T; $p \leq 0.008$).

The globulin extracts were analyzed by SDS-PAGE. A volume of sample equal to a similar weight of seed tissue extracted was separated by electrophoresis (**Figure 2**). The volume of five protein bands appeared to be increased by the sulfur treatment. This was confirmed by image analysis with Quantity One (**Table 7**). The fold change and statistical significance of the changes was confirmed by analyzing replicate extracts of each genotype on separate gels (Supplementary Figure S1; Supplementary Tables S1 and S2). The protein bands were excised, digested with trypsin and analyzed by MALDI-MS or LC–MS–MS and identified by MASCOT search against the MASCOT database, or a translated database of the common bean genome (Schmutz et al., 2014). Protein bands no. 2 and 3 could be identified by MALDI-MS, whereas protein bands no. 1, 4, and 5 required more sensitive LC–MS–MS analysis. **Tables 8** and **9** provide information about protein identifications and list the number of methionine, cysteine, and total residues in each protein. In all cases, the apparent molecular mass measured by electrophoresis matched the predicted molecular mass relatively closely.

Protein band no. 2 was identified as a lipoxygenase, which was named lipoxygenase-3, based on its similarity to the corresponding soybean and pea proteins. Its baseline levels were higher in SMARC1N-PN1 than in SARC1 by approximately twofold. These results are consistent with the prior identification of this lipoxygenase as elevated in SMARC1N-PN1 as compared with SARC1, both by spectral counting (1.6-fold) and immunoblotting (2.5-fold; Marsolais et al., 2010). This protein had been identified on the basis of the soybean lipoxygenase-3 sequence (Yenofsky et al., 1988). The common bean accession shares 88% identity with soybean lipoxygenase-3, and 84% identity with pea lipoxygenase-3 (Ealing and Casey, 1988). Lipoxygenase-3 levels were increased by the sulfur treatment by approximately twofold. This was observed exclusively in SMARC1N-PN1. The apparent molecular mass determined for lipoxygenase-3 is consistent with the fact that the pea lipoxygenase A1 polypeptide, whose N-terminal sequence matched the deduced amino acid sequence of the lipoxygenase-3 cDNA, had an apparent molecular mass greater than 97.4 kDa (Domoney et al., 1990). Lipoxygenase-3 is particularly rich in sulfur amino acids with 1.7% of its residues as methionine and 0.5% as cysteine.

Protein band no. 3 was identified as albumin-2. Its baseline levels were sixfold higher in SMARC1N-PN1 than in SARC1. These results are consistent with the prior identification of albumin-2

Table 4 | Elemental and sulfate concentrations in mature seed.

Genotype	Treatment	C (%)	N (%)	S (%)	SO_4^{2-} (nmol/mg)
SARC1	LS	46.4 ± 0.2	3.90 ± 0.11	0.20 ± 0.12	0.18 ± 0.03
	HS	46.4 ± 0.1	3.91 ± 0.17	0.23 ± 0.01	0.21 ± 0.03
SMARC1N-PN1	LS	46.0 ± 0.1	3.63 ± 0.20	0.19 ± 0.01	0.21 ± 0.05
	HS	46.0 ± 0.1	3.71 ± 0.14	0.23 ± 0.01	0.29 ± 0.02
Source of variation	**d. f.**			**ANOVA *p*-value**	
Genotype (G)	1	0.0001	0.006	n. s.	0.002
Treatment (T)	1	n. s.	n. s.	0.0001	0.001
G × T	1	n. s.	n. s.	n. s.	n. s.
Error	15				

Values are the average ± SD; n = 5.

Table 5 | Sulfur amino acid concentration in mature seed.

Genotype	Treatment	Methionine (nmol per mg)	Cysteine (nmol per mg)	*S*-Methylcysteine (nmol per mg)
SARC1	LS	16.7 ± 0.2	22.3 ± 3.0	18.2 ± 2.0
	HS	16.7 ± 1.4	21.6 ± 2.2	23.6 ± 3.8
SMARC1N-PN1	LS	17.2 ± 1.0	24.4 ± 1.1	9.1 ± 0.4
	HS	19.0 ± 1.1	28.2 ± 1.9	12.8 ± 0.9
Source of variation	**d. f.**		**ANOVA *p*-value**	
Genotype (G)	1	0.04	0.0004	0.0001
Treatment (T)	1	n. s.	n. s.	0.007
G × T	1	n. s.	0.03	n. s.
Error	15			

Values are the average ± SD; n = 3; n = 5 for cysteine.

as being elevated by about 10-fold in SMARC1N-PN1 as compared with SARC1 according to two-dimensional electrophoresis based proteomics (Marsolais et al., 2010). This protein had been identified by *de novo* sequencing, on the basis of its similarity with pea albumin-2 (Higgins et al., 1987) and mung bean seed albumin [Uniprot:Q43680], and its full-length cDNA had been subsequently cloned (Yin et al., 2011). Its transcript levels were elevated in developing seeds of SMARC1N-PN1 as compared with SARC1 (Liao et al., 2012). Albumin-2 levels were increased by the HS treatment in both genotypes, by approximately 20–30%. The sequence of albumin-2 is relatively rich in sulfur amino acids, with 0.9% of its residues as methionine and 1.3% as cysteine.

Protein band no. 1 was identified as arcelin-like protein 4 (Lioi et al., 2003). It was only present in the SARC1 extracts. This is consistent with previous spectral counting and two-dimensional electrophoresis based proteomic data (Marsolais et al., 2010). Arcelin-like protein 4 contains 0.4% its residues as methionine and 0.8% as cysteine. Protein bands no. 4 and 5 were only observed in SMARC1N-PN1. They were identified as α-amylase inhibitor β subunit (Kasahara et al., 1996). The levels of this protein had been shown to be elevated by 20-fold in SMARC1N-PN1 as compared with SARC1 by spectral counting, and this had been validated by two-dimensional electrophoresis based proteomics (Marsolais et al., 2010). The apparent molecular mass measured for protein band no. 4 is in agreement with the results obtained with α-amylase inhibitor purified from Great Northern beans (Furuichi et al., 1993). A protein band of 15.5 kDa had been identified as the β subunit by N-terminal sequencing and appeared not to be glycosylated. Protein band no. 5 is likely be a minor form of the α-amylase inhibitor β subunit lacking one or more residues at the C-terminal. Indeed, a peptide containing the sequence of the N-terminus was detected for this band. In the work by Furuichi et al. (1993) a protein band of 13.5 kDa band was tentatively assigned as the α subunit and appeared to be glycosylated. Here, the proteomic data were unambiguous and did not identify a match to the α subunit. Interestingly, the α subunit was conspicuously absent from the SMARC1N-PN1 proteome determined by spectral counting (Marsolais et al., 2010). Purification of the β subunit yielded a complex with the α-amylase inhibitor like protein, with a predicted molecular mass of 25 kDa, in which the α subunit was

Table 6 | Concentration of extractible albumins and globulins in mature seed.

Genotype	Treatment	Albumins (%)	Globulins (%)
SARC1	LS	1.79 ± 0.25	3.01 ± 0.51
	HS	1.88 ± 0.10	2.71 ± 0.21
SMARC1N-PN1	LS	1.95 ± 0.11	3.59 ± 0.57
	HS	1.92 ± 0.24	4.47 ± 0.31
Source of variation	**d. f.**	**ANOVA *p*-value**	
Genotype (G)	1	n. s.	0.0001
Treatment (T)	1	n. s.	n. s.
G × T	1	n. s.	0.008
Error	15		

Values are the average ± SD; n = 5.

absent (Yin and Marsolais, unpublished results). The apparent molecular mass of this complex measured by size exclusion chromatography was equal to 41.2 kDa. In SDS-PAGE, the purified fraction contained a minor band of approximately 13.8 kDa, similar to the results in **Figure 2**. The α-amylase inhibitor β chain is devoid of sulfur amino acids. However, the corresponding α-amylase-1 precursor contains 1.6% of its residues as methionine without any cysteine (Prescott et al., 2005). The four methionines are located at the N-terminus of the polypeptide precursor. Sulfur amino acid residues are neither present in the α subunit (Kasahara et al., 1996).

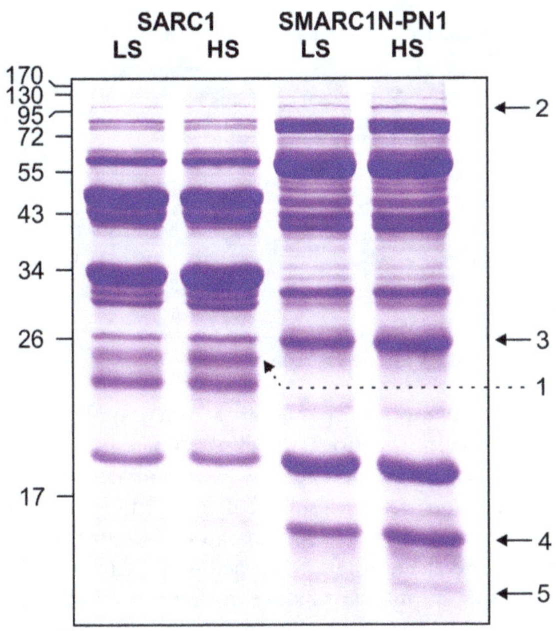

FIGURE 2 | SDS-PAGE of globulin fractions extracted from mature seeds of SARC1 and SMARC1N-PN1 grown under LS or HS. Size of molecular mass markers is indicated on the left. Bands quantified by image analysis and excised for proteomic identification are numbered and indicated by arrows. A volume of sample equivalent to the same weight of tissue extracted was subjected to SDS-PAGE on a 12% polyacrylamide gel. The total volume of bands in each lane measured with Quantity One was as follows, SARC1-LS: 632; SARC1-HS: 636; SMARC1N-PN1-LS: 945; and SMARC1N-PN1-HS: 1007.

DISCUSSION

The most important finding reported from this study is that the protein pool of SMARC1N-PN1 was able to accommodate an increase in sulfur amino acids, particularly cysteine, in response to enhanced sulfate nutrition, whereas this was not the case in SARC1. This property is associated with the presence of an endogenous sink for sulfur in SMARC1N-PN1. The increase in cysteine concentration was associated with a specific increase in the concentration of extractible globulins and seed yield in SMARC1N-PN1. It has been reported that sulfur nutrition influences the response of chickpea plants to the transgenic expression of sunflower seed albumin (Chiaiese et al., 2004). An increase in the seed concentration of reduced sulfur was more pronounced in the transgenic line than in wild-type under conditions of low nitrogen and high sulfur nutrition. Further results were suggestive of a higher methionine concentration in the transgenic line in response to high sulfur and high nitrogen, although this could not be analyzed statistically due to a lack of replication. Recently, Kim et al. (2014) reported that adequate sulfate nutrition is required to maximize the accumulation of maize δ-zein in transgenic soybean lines where expression of the endogenous β-conglycinin was suppressed by RNA interference.

The seed concentration of sulfate reported here for common bean is much lower than the concentration of oxidized sulfur in lupin or pea, by approximately 100-fold (Tabe and Droux, 2002; Chiaiese et al., 2004). Whereas in lupin or pea, sulfate represents a reserve of sulfur, whose levels can be significantly reduced upon transgenic expression of a sulfur-rich protein, *S*-methylcysteine plays a similar role in common bean. This is supported by the fact that the *S*-methylcysteine concentration was reduced by approximately twofold in SMARC1N-PN1 as compared with SARC1. However, the effect of sulfur nutrition on the concentration of *S*-methylcysteine was similar between genotypes. Likewise in chickpea, nitrogen and sulfur treatments had a similar effect on oxidized sulfur concentration in wild-type and a transgenic line expressing sunflower seed albumin (Chiaiese et al., 2004).

The increase in the extractible globulin fraction associated with enhanced levels of cysteine stimulated an analysis of sulfur-responsive proteins in the extracts. Two lectins, uniquely present in either genotype, arcelin-like protein 4, and the β subunit of the α-amylase inhibitor, were identified as sulfur-responsive. In the case of the α-amylase inhibitor, sulfur is needed for the accumulation of the polypeptide precursor but not the mature β subunit. Albumin-2 was found to be increased by a similar fold change in response to HS in both genotypes, although its baseline levels were higher in SMARC1N-PN1. In pea, albumin-2 levels were initially found to be reduced in response to severe sulfur deficiency (Randall et al., 1979). In later work, protein levels were found to be relatively unchanged in response to moderate sulfur deficiency (Higgins et al., 1987). Albumin-2 is characterized by the presence of four repeats of the hemopexin domain. The results

Table 7 | Quantification and apparent molecular mass of sulfur-responsive protein bands identified in Figure 2.

Protein band no.	Apparent molecular mass (kDa)	SARC1			SMARC1N-PN1		
		Volume		Increase (%)	Volume		Increase (%)
		LS	HS		LS	HS	
1	25.0	29.7	44	48	–	–	–
2	105.4	6.30	6.54	4	12.4	24.9	101
3	26.5	24.5	29.1	19	142	189	33
4	15.7	–	–	–	83.2	110	32
5	13.8	–	–	–	9.64	13.1	36

Table 8 | Identification of protein bands in Figure 2 by MALDI-MS and MASCOT search following trypsin digestion.

Protein band no.	Sample	Accession	Predicted mass (kDa)	Protein score	*p*-value	Matches	Protein coverage (%)	Residues vs. Met vs. Cys
2	SMARC1N-PN1-LS	Lox-3	97527	68	0.03	46	33	860/15/4
2	SMARC1N-PN1-HS	Lox-3	97527	85	0.0006	27	25	860/15/4
3	SMARC1N-PN1-LS	Alb-2	25574	83	0.0009	21	73	227/2/3
3	SMARC1N-PN1-HS	Alb-2	25574	82	0.001	19	70	227/2/3

Lox-3, lipoxygenase-3; Alb-2, albumin-2.

Table 9 | Identification of protein bands in Figure 2 by MS-MS and MASCOT search following trypsin digestion.

Band no.	Sample	Accession	Predicted mass (kDa)	Total ions score	Matches	Sequences	Protein coverage (%)	Residues vs. Met vs. Cys
1	SARC1-LS	ARL4	26585	56	1	1	4	240/1/2
1	SARC1-HS	ARL4	26585	241	4	4	20	240/1/2
4	SMARC1N-PN1-LS	α-AI β subunit	15395	628	46	8	66	137/0/0
4	SMARC1N-PN1-LS	α-AI β subunit	15395	645	42	8	66	137/0/0
5	SMARC1N-PN1-LS	α-AI β subunit	15395	373	32	5	47	137/0/0
5	SMARC1N-PN1-HS	α-AI β subunit	15395	484	21	6	56	137/0/0

ARL4, Arcelin-like protein 4; α-AI, α-amylase inhibitor. The α-amylase inhibitor polypeptide precursor has a total of 246 residues with four methionines and no cysteine.

of crystallization studies have revealed that these proteins bind spermine (Gaur et al., 2010, 2011). Binding of heme and spermine was found to be mutually exclusive in grasspea hemopexin. A pea mutant lacking albumin-2 had altered levels of polyamines, and this was associated with increased seed protein concentration (Vigeolas et al., 2008). Whether SARC1 and SMARC1N-PN1 differ in their level of polyamines could be the subject of future investigation.

The identification of lipoxygenase-3 as a sulfur-responsive protein is particularly interesting, because this response was observed in only one of the genotypes. The present result strongly suggests that the sulfur-responsive albumin protein of 95 kDa identified in pea is actually lipoxygenase-3 (Higgins et al., 1987). We speculate that differences in sulfur-responsiveness between the two common bean genotypes are probably determined by polymorphisms in the promoter or 3′-untranslated region of the lipoxygenase-3 gene. These polymorphisms must arise from recombination at the lipoxygenase-3 locus between the different parents. If this is true, sequence comparisons between the two genotypes might lead to the identification of a much sought after cis-acting regulatory element determining a positive response to sulfur nutrition in higher plants. To date, the only known cis-acting regulatory motif determining a transcriptional response to sulfur, the SURE motif, is involved in the up-regulation of sulfate transporter and assimilatory genes in response to sulfur deficiency (Maruyama-Nakashita et al., 2005).

The present results have special implications for the agronomic management of common bean, if storage protein deficiency is used as a trait for the improvement of protein quality through conventional breeding. Although SARC1 and SMARC1N-PN1 responded equally to sulfate fertilization under field conditions, the results obtained under controlled conditions suggest that adequate sulfur nutrition is required to maximize the concentration of sulfur amino acids and therefore protein quality in genotypes lacking phaseolin and major lectins like SMARC1N-PN1. As deposition of sulfur due to atmospheric pollution decreases, sulfate fertilization might become necessary for common bean production in Southwestern Ontario. It is currently an integral part of agronomic production in the plains of Manitoba.

ACKNOWLEDGMENTS

We thank staff at the Southern Crop Protection and Food Research Centre, Igor Lalin for sulfate analysis, Tim McDowell for elemental analysis, Patrick Chapman for bioinformatics, and Alex Molnar for graphics. We are grateful to staff at the London Regional Proteomics Centre, Paula Pittock and Kristina Jurcic. We acknowledge Rey Interior, at the SPARC BioCentre, Hospital for Sick Children, for help with amino acid analyses. We thank Larry Stitt of LW Stitt Statistical Services for help with statistical interpretations. We are indebted to Cynthia Grant at the Brandon Research Centre of Agriculture and Agri-Food Canada, for advice on field sulfur trials, and to Derek Lydiate at the Saskatoon Research Centre of Agriculture and Agri-Food Canada, for discussions that stimulated this work. Steven Karcz of the Saskatoon Research Centre, who passed away in 2012, helped with MASCOT. This work was supported in part by a grant from the Ontario Research Fund for Research Excellence, Round 4 to FM (project 043), and grants from the Manitoba Pulse Growers Association, and Pulse Science Cluster of the AgriInnovation Program of Agriculture and Agri-Food Canada to AH and FM.

REFERENCES

Amir, R., Han, T., and Ma, F. (2012). Bioengineering approaches to improve the nutritional values of seeds by increasing their methionine content. *Mol. Breed.* 29, 915–924. doi: 10.1007/s11032-011-9690-7

Anderson, J. W., and Fitzgerald, M. A. (2001). Physiological and metabolic origin of sulphur for the synthesis of seed storage proteins. *J. Plant Physiol.* 158, 447–456. doi: 10.1078/0176-1617-00356

Aragão, F. J. L., Barros, L. M. G., De Sousa, M. V., Grossi De Sá, M. F., Almeida, E. R. P., Gander, E. S., et al. (1999). Expression of a methionine-rich storage albumin from the Brazil nut (*Bertholletia excelsa* HBK Lecythidaceae) in transgenic bean plants (*Phaseolus vulgaris* L. *Fabaceae). Genet. Mol. Biol.* 22, 445–449. doi: 10.1590/S1415-47571999000300026

Blagrove, R. J., Gillesie, J. M., and Randall, P. J. (1976). Effect of sulphur supply on the seed globulin composition of *Lupinus angustifolius. Aust. J. Plant Physiol.* 3, 173–184. doi: 10.1071/PP9760173

Bourgis, F., Roje, S., Nuccio, M. L., Fisher, D. B., Tarczynski, M. C., Li, C. J., et al. (1999). S-Methylmethionine plays a major role in phloem sulfur transport and is synthesized by a novel type of methyltransferase. *Plant Cell* 11, 1485–1497. doi: 10.1105/tpc.11.8.1485

Buchner, P., Stuiver, C. E., Westerman, S., Wirtz, M., Hell, R., Hawkesford, M. J., et al. (2004). Regulation of sulfate uptake and expression of sulfate transporter genes in *Brassica oleracea* as affected by atmospheric H2S and pedospheric sulfate nutrition. *Plant Physiol.* 136, 3396–3408. doi: 10.1104/pp.104.046441

Chandler, P. M., Higgins, T. J., Randall, P. J., and Spencer, D. (1983). Regulation of legumin levels in developing pea seeds under conditions of sulfur deficiency: rates of legumin synthesis and levels of legumin mRNA. *Plant Physiol.* 71, 47–54. doi: 10.1104/pp.71.1.47

Chandler, P. M., Spencer, D., Randall, P. J., and Higgins, T. J. (1984). Influence of sulfur nutrition on developmental patterns of some major pea seed proteins and their mRNAs. *Plant Physiol.* 75, 651–657. doi: 10.1104/pp.75.3.651

Chiaiese, P., Ohkama-Ohtsu, N., Molvig, L., Godfree, R., Dove, H., Hocart, C., et al. (2004). Sulphur and nitrogen nutrition influence the response of chickpea seeds to an added, transgenic sink for organic sulphur. *J. Exp. Bot.* 55, 1889–1901. doi: 10.1093/jxb/erh198

De Kok, L. J., Stulen, I., and Hawkesford, M. J. (2011). "Sulfur nutrition in crop plants," in *The Molecular and Physiological Basis of Nutrient Use Efficiency in Crops*, eds M. J. Hawkesford and P. Barraclough (Chichester: Wiley-Blackwell), 295–309. doi: 10.1002/9780470960707.ch14

Demidov, D., Horstmann, C., Meixner, M., Pickardt, T., Saalbach, I., Galili, G., and Müntz, K. (2003). Additive effects of the feed-back insensitive bacterial aspartate kinase and the Brazil nut 2S albumin on the methionine content of transgenic narbon bean (*Vicia narbonensis* L.). *Mol. Breed.* 11, 187–201. doi: 10.1023/A:1022814506153

Dinkins, R. D., Reddy, M. S. S., Meurer, C. A., Yan, B., Trick, H., Thibaud-Nissen, F., et al. (2001). Increased sulfur amino acids in soybean plants over expressing the maize 15 kDa zein protein. *In vitro Cell. Dev. Biol. Plant* 37, 742–747. doi: 10.1007/s11627-001-0123-x

Domoney, C., Firmin, J. L., Sidebottom, C., Ealing, P. M., Slabas, A., and Casey, R. (1990). Lipoxygenase heterogeneity in *Pisum sativum. Planta* 181, 35–43. doi: 10.1007/BF00202322

Ealing, P. M., and Casey, R. (1988). The complete amino acid sequence of a pea (*Pisum sativum*) seed lipoxygenase predicted from a near full-length cDNA. *Biochem. J.* 253, 915–918.

Furuichi, Y., Takemura, M., Uesaka, N., Kamemura, K., Shimada, S., Komada, H., et al. (1993). Some characteristics of an α-amylase inhibitor from *Phaseolus vulgaris* (cultivar Great Northern) seeds. *Biosci. Biotechnol. Biochem.* 57, 147–148. doi: 10.1271/bbb.57.147

Galili, G., and Amir, R. (2013). Fortifying plants with the essential amino acids lysine and methionine to improve nutritional quality. *Plant Biotechnol. J.* 11, 211–222. doi: 10.1111/pbi.12025

Gaur, V., Chanana, V., Jain, A., and Salunke, D. M. (2011). The structure of a haemopexin-fold protein from cow pea (*Vigna unguiculata*) suggests functional diversity of haemopexins in plants. *Acta Crystallogr. Sect. F. Struct. Biol. Cryst. Commun.* 67, 193–200. doi: 10.1107/s1744309110051250

Gaur, V., Qureshi, I. A., Singh, A., Chanana, V., and Salunke, D. M. (2010). Crystal structure and functional insights of hemopexin fold protein from grass pea. *Plant Physiol.* 152, 1842–1850. doi: 10.1104/pp.109.150680

Gayler, K. R., and Sykes, G. E. (1985). Effects of nutritional stress on the storage proteins of soybeans. *Plant Physiol.* 78, 582–585. doi: 10.1104/pp.78.3.582

Hanafy, M. S., Rahman, S. M., Nakamoto, Y., Fujiwara, T., Naito, S., Wakasa, K., et al. (2013a). Differential response of methionine metabolism in two grain legumes, soybean and azuki bean, expressing a mutated form of *Arabidopsis* cystathionine γ-synthase. *J. Plant Physiol.* 170, 338–345. doi: 10.1016/j.jplph.2012.10.018

Hanafy, M. S., Rahman, S. M., Nakamoto, Y., Fujiwara, T., Naito, S., Wakasa, K., et al. (2013b). Erratum to differential response of methionine metabolism in two grain legumes, soybean, and azuki bean, expressing a mutated form of *Arabidopsis* cystathionine γ-synthase. *J. Plant Physiol.* 170, 1469. doi: 10.1016/j.jplph.2013.06.003

Hartweck, L. M., and Osborn, T. C. (1997). Altering protein composition by genetically removing phaseolin from common bean seeds containing arcelin or phytohemagglutinin. *Theor. Appl. Genet.* 95, 1012–1017. doi: 10.1007/s001220050655

Hawkesford, M. J. (2000). Plant responses to sulphur deficiency and the genetic manipulation of sulphate transporters to improve S-utilization efficiency. *J. Exp. Bot.* 51, 131–138. doi: 10.1093/jexbot/51.342.131

Hawkesford, M. J. (2003). Transporter gene families in plants: the sulphate transporter gene family - redundancy or specialization? *Physiol. Plant.* 117, 155–163. doi: 10.1034/j.1399-3054.2003.00034.x

Hawkesford, M. J., and De Kok, L. J. (2006). Managing sulphur metabolism in plants. *Plant Cell Environ.* 29, 382–395. doi: 10.1111/j.1365-3040.2005.01470.x

Herman, E. M. (2014). Soybean seed proteome rebalancing. *Front. Plant Sci.* 5:437. doi: 10.3389/fpls.2014.00437

Hernández-Sebastià, C., Marsolais, F., Saravitz, C., Israel, D., Dewey, R. E., and Huber, S. C. (2005). Free amino acid profiles suggest a possible role for asparagine in the control of storage-product accumulation in developing seeds of low- and high-protein soybean lines. *J. Exp. Bot.* 56, 1951–1963. doi: 10.1093/jxb/eri191

Herschbach, C., Van Der Zalm, E., Schneider, A., Jouanin, L., De Kok, L. J., and Rennenberg, H. (2000). Regulation of sulfur nutrition in wild-type and transgenic poplar over-expressing γ-glutamylcysteine synthetase in the cytosol as affected by atmospheric H2S. *Plant Physiol.* 124, 461–473. doi: 10.1104/pp.124.1.461

Higgins, T. J. V., Beach, L. R., Spencer, D., Chandler, P. M., Randall, P. J., Blagrove, R. J., et al. (1987). cDNA and protein sequence of a major pea seed albumin (PA 2: Mr≈26 000). *Plant Mol. Biol.* 8, 37–45. doi: 10.1007/BF00016432

Holowach, L. P., Madison, J. T., and Thompson, J. F. (1986). Studies on the mechanism of regulation of the mRNA level for a soybean storage protein subunit by exogenous L-methionine. *Plant Physiol.* 80, 561–567. doi: 10.1104/pp.80.2.561

Holowach, L. P., Thompson, J. F., and Madison, J. T. (1984). Effects of exogenous methionine on storage protein composition of soybean cotyledons cultured in vitro. *Plant Physiol.* 74, 576–583. doi: 10.1104/pp.74.3.576

Kasahara, K., Hayashi, K., Arakawa, T., Philo, J. S., Hara, S., and Yamaguchi, H. (1996). Complete sequence, subunit structure, and complexes with pancreatic α-amylase of an α-amylase inhibitor from *Phaseolus vulgaris* white kidney beans. *J. Biochem. (Tokyo)* 120, 177–183. doi: 10.1093/oxfordjournals.jbchem.a021381

Kelly, J. D., and Hefle, S. L. (2000). 2S Methionine-rich protein (SSA) from sunflower seed is an IgE-binding protein. *Allergy* 55, 556–560. doi: 10.1034/j.1398-9995.2000.00498.x

Kim, H., Hirai, M. Y., Hayashi, H., Chino, M., Naito, S., and Fujiwara, T. (1999). Role of O-acetyl-L-serine in the coordinated regulation of the expression of a soybean seed storage-protein gene by sulfur and nitrogen nutrition. *Planta* 209, 282–289. doi: 10.1007/s004250050634

Kim, W. S., Chronis, D., Juergens, M., Schroeder, A. C., Hyun, S. W., Jez, J. M., et al. (2012). Transgenic soybean plants overexpressing O-acetylserine sulfhydrylase accumulate enhanced levels of cysteine and Bowman-Birk protease inhibitor in seeds. *Planta* 235, 13–23. doi: 10.1007/s00425-011-1487-8

Kim, W., Jez, J. M., and Krishnan, H. B. (2014). Effects of proteome rebalancing and sulfur nutrition on the accumulation of methionine rich δ-zein in transgenic soybeans. *Front. Plant Sci.* 5:633 doi: 10.3389/fpls.2014.00633

Kim, W. S., and Krishnan, H. B. (2004). Expression of an 11 kDa methionine-rich delta-zein in transgenic soybean results in the formation of two types of novel protein bodies in transitional cells situated between the vascular tissue and storage parenchyma cells. *Plant Biotechnol. J.* 2, 199–210. doi: 10.1111/j.1467-7652.2004.00063.x

Krishnan, H. B. (2005). Engineering soybean for enhanced sulfur amino acid content. *Crop Sci.* 45, 454–461. doi: 10.2135/cropsci2005.0454

Krishnan, H. B., Bennett, J. O., Kim, W., Krishnan, H. A., and Mawhinney, T. P. (2005). Nitrogen lowers the sulfur amino acid content of soybean (*Glycine max* [L.] Merr.) by regulating the accumulation of Bowman-Birk protease inhibitor. *J. Agric. Food Chem.* 53, 6347–6354. doi: 10.1021/jf050510i

Krishnan, H. B., Kerley, M. S., Allee, G. L., Jang, S., Kim, W. S., and Fu, C. J. (2010). Maize 27 kDa gamma-zein is a potential allergen for early weaned pigs. *J. Agric. Food Chem.* 58, 7323–7328. doi: 10.1021/jf100927u

Lee, M. S., Huang, T. F., Toro-Ramos, T., Fraga, M., Last, R. L., and Jander, G. (2008). Reduced activity of *Arabidopsis thaliana* HMT2, a methionine biosynthetic enzyme, increases seed methionine content. *Plant J.* 54, 310–320. doi: 10.1111/j.1365-313X.2008.03419.x

Li, Z., Meyer, S., Essig, J. S., Liu, Y., Schapaugh, M. A., Muthukrishnan, S., et al. (2005). High-level expression of maize gamma-zein protein in transgenic soybean (*Glycine max*). *Mol. Breed.* 16, 11–20. doi: 10.1007/s11032-004-7658-6

Liao, D., Pajak, A., Karcz, S. R., Chapman, B. P., Sharpe, A. G., Austin, R. S., et al. (2012). Transcripts of sulphur metabolic genes are co-ordinately regulated in developing seeds of common bean lacking phaseolin and major lectins. *J. Exp. Bot.* 63, 6283–6295. doi: 10.1093/jxb/ers280

Lioi, L., Sparvoli, F., Galasso, I., Lanave, C., and Bollini, R. (2003). Lectin-related resistance factors against bruchids evolved through a number of duplication events. *Theor. Appl. Genet.* 107, 814–822. doi: 10.1007/s00122-003-1343-8

Malavolta, E., Vitti, G. C., Rosolem, C. A., Fageria, N. K., and Guimarães, P. T. G. (1987). Sulfur responses of Brazilian crops. *J. Plant Nutr.* 10, 2153–2158. doi: 10.1080/01904168709363766

Marsolais, F., Pajak, A., Yin, F., Taylor, M., Gabriel, M., Merino, D. M., et al. (2010). Proteomic analysis of common bean seed with storage protein deficiency reveals up-regulation of sulfur-rich proteins and starch and raffinose metabolic enzymes, and down-regulation of the secretory pathway. *J. Proteomics* 73, 1587–1600. doi: 10.1016/j.jprot.2010.03.013

Maruyama-Nakashita, A., Nakamura, Y., Watanabe-Takahashi, A., Inoue, E., Yamaya, T., and Takahashi, H. (2005). Identification of a novel cis-acting element conferring sulfur deficiency response in *Arabidopsis* roots. *Plant J.* 42, 305–314. doi: 10.1111/j.1365-313X.2005.02363.x

Molvig, L., Tabe, L. M., Eggum, B. O., Moore, A. E., Craig, S., Spencer, D., et al. (1997). Enhanced methionine levels and increased nutritive value of seeds of transgenic lupins (*Lupinus angustifolius* L.) expressing a sunflower seed albumin gene. *Proc. Natl. Acad. Sci. U.S.A.* 94, 8393–8398. doi: 10.1073/pnas.94.16.8393

Montoya, C. A., Lallès, J. P., Beebe, S., and Leterme, P. (2010). Phaseolin diversity as a possible strategy to improve the nutritional value of common beans (*Phaseolus vulgaris*). *Food Res. Int.* 43, 443–449. doi: 10.1016/j.foodres.2009.09.040

Morton, R. L., Ellery, A. J., and Higgins, T. J. (1998). Downstream elements from the pea albumin 1 gene confer sulfur responsiveness on a reporter gene. *Mol. Gen. Genet.* 259, 309–316. doi: 10.1007/s004380050817

Naito, S., Hirai, M. Y., Inaba-Higano, K., Nambara, E., Fujiwara, T., Hayashi, H., et al. (1995). Expression of soybean seed storage protein genes in transgenic plants and their response to sulfur nutritional conditions. *J. Plant Physiol.* 145, 614–619. doi: 10.1016/S0176-1617(11)81272-1

Nordlee, J. A., Taylor, S. L., Townsend, J. A., Thomas, L. A., and Bush, R. K. (1996). Identification of a Brazil-nut allergen in transgenic soybeans. *N. Engl. J. Med.* 334, 688–692. doi: 10.1056/NEJM199603143341103

Osborn, T. C., Hartweck, L. M., Harmsen, R. H., Vogelzang, R. D., Kmiecik, K. A., and Bliss, F. A. (2003). Registration of *Phaseolus vulgaris* genetic stocks with altered seed protein compositions. *Crop Sci.* 43, 1570–1571. doi: 10.2135/cropsci2003.1570

Pandurangan, S., Pajak, A., Molnar, S. J., Cober, E. R., Dhaubhadel, S., Hernandez-Sebastia, C., et al. (2012). Relationship between asparagine metabolism and protein concentration in soybean seed. *J. Exp. Bot.* 63, 3173–3184. doi: 10.1093/jxb/ers039

Prescott, V. E., Campbell, P. M., Moore, A., Mattes, J., Rothenberg, M. E., Foster, P. S., et al. (2005). Transgenic expression of bean α-amylase inhibitor in peas results in altered structure and immunogenicity. *J. Agric. Food Chem.* 53, 9023–9030. doi: 10.1021/jf050594v

Randall, P. J., Thomson, J. A., and Schroeder, H. E. (1979). Cotyledonary storage proteins in *Pisum sativum. IV.* Effects of sulfur, phosphorus, potassium and magnesium deficiencies. *Aust. J. Plant Physiol.* 6, 11–24. doi: 10.1071/PP9790011

Rerie, W. G., Whitecross, M., and Higgins, T. J. (1991). Developmental and environmental regulation of pea legumin genes in transgenic tobacco. *Mol. Gen. Genet.* 225, 148–157. doi: 10.1007/BF00282653

Rolletschek, H., Hosein, F., Miranda, M., Heim, U., Götz, K. P., Schlereth, A., et al. (2005). Ectopic expression of an amino acid transporter (VfAAP1) in seeds of *Vicia narbonensis* and pea increases storage proteins. *Plant Physiol.* 137, 1236–1249. doi: 10.1104/pp.104.056523

Sánchez, E., Ruiz, J. M., and Romero, L. (2002). Proline metabolism in response to nitrogen toxicity in fruit of French Bean plants (*Phaseolus vulgaris* L. cv Strike). *Sci. Hortic.* 93, 225–233. doi: 10.1016/S0304-4238(01)00342-9

Schmutz, J., Mcclean, P. E., Mamidi, S., Wu, G. A., Cannon, S. B., Grimwood, J., et al. (2014). A reference genome for common bean and genome-wide analysis of dual domestications. *Nat. Genet.* 46, 707–713. doi: 10.1038/ng.3008

Smith, F. W., Ealing, P. M., Hawkesford, M. J., and Clarkson, D. T. (1995). Plant members of a family of sulfate transporters reveal functional subtypes. *Proc. Natl. Acad. Sci. U.S.A.* 92, 9373–9377. doi: 10.1073/pnas.92.20.9373

Song, S., Hou, W., Godo, I., Wu, C., Yu, Y., Matityahu, I., et al. (2013). Soybean seeds expressing feedback-insensitive cystathionine γ-synthase exhibit a higher content of methionine. *J. Exp. Bot.* 64, 1917–1926. doi: 10.1093/jxb/ert053

Streit, L. G., Beach, L. R., Register, J. C., Jung, R., and Fehr, W. R. (2001). Association of the Brazil nut protein gene and Kunitz trypsin inhibitor alleles with soybean protease inhibitor activity and agronomic traits. *Crop Sci.* 41, 1757–1760. doi: 10.2135/cropsci2001.1757

Tabe, L. M., and Droux, M. (2002). Limits to sulfur accumulation in transgenic lupin seeds expressing a foreign sulfur-rich protein. *Plant Physiol.* 128, 1137–1148. doi: 10.1104/pp.010935

Takahashi, H., Kopriva, S., Giordano, M., Saito, K., and Hell, R. (2011). Sulfur assimilation in photosynthetic organisms: molecular functions and regulations of transporters and assimilatory enzymes. *Annu. Rev. Plant Biol.* 62, 157–184. doi: 10.1146/annurev-arplant-042110-103921

Tan, Q., Zhang, L., Grant, J., Cooper, P., and Tegeder, M. (2010). Increased phloem transport of S-methylmethionine positively affects sulfur and nitrogen metabolism and seed development in pea plants. *Plant Physiol.* 154, 1886–1896. doi: 10.1104/pp.110.166389

Taylor, M., Chapman, R., Beyaert, R., Hernández-Sebastià, C., and Marsolais, F. (2008). Seed storage protein deficiency improves sulfur amino acid content in common bean (*Phaseolus vulgaris* L.): redirection of sulfur from γ-glutamyl-S-methyl-cysteine. *J. Agric. Food Chem.* 56, 5647–5654. doi: 10.1021/jf800787y

Townsend, J. A., and Thomas, L. A. (1994). Factors which influence the *Agrobacterium*-mediated transformation of soybean. *J. Cell Biochem.* (Suppl. 18A), 78.

Ufaz, S., and Galili, G. (2008). Improving the content of essential amino acids in crop plants: goals and opportunities. *Plant Physiol.* 147, 954–961. doi: 10.1104/pp.108.118091

Vigeolas, H., Chinoy, C., Zuther, E., Blessington, B., Geigenberger, P., and Domoney, C. (2008). Combined metabolomic and genetic approaches reveal a link between the polyamine pathway and albumin 2 in developing pea seeds. *Plant Physiol.* 146, 74–82. doi: 10.1104/pp.107.111369

Wu, Y., and Messing, J. (2014). Proteome balancing of the maize seed for higher nutritional value. *Front. Plant Sci.* 5:240. doi: 10.3389/fpls.2014.00240

Yenofsky, R. L., Fine, M., and Liu, C. (1988). Isolation and characterization of a soybean (*Glycine max*) lipoxygenase-3 gene. *Mol. Gen. Genet.* 211, 215–222. doi: 10.1007/BF00330597

Yin, F., Pajak, A., Chapman, R., Sharpe, A., Huang, S., and Marsolais, F. (2011). Analysis of common bean expressed sequence tags identifies sulfur metabolic pathways active in seed and sulfur-rich proteins highly expressed in the absence of phaseolin and major lectins. *BMC Genomics* 12:268. doi: 10.1186/1471-2164-12-268

Zhang, Y., Schernthaner, J., Labbé, N., Hefford, M. A., Zhao, J., and Simmonds, D. H. (2014). Improved protein quality in transgenic soybean expressing a *de novo* synthetic protein, MB-16. *Transgenic Res.* 3, 1–13. doi: 10.1007/s11248-013-9777-5

Zuber, H., Davidian, J. C., Aubert, G., Aimé, D., Belghazi, M., Lugan, R., et al. (2010). The seed composition of *Arabidopsis* mutants for the group 3 sulfate transporters indicates a role in sulfate translocation within developing seeds. *Plant Physiol.* 154, 913–926. doi: 10.1104/pp.110.162123

Zuber, H., Poignavent, G., Le Signor, C., Aime, D., Vieren, E., Tadla, C., et al. (2013). Legume adaptation to sulfur deficiency revealed by comparing nutrient allocation and seed traits in *Medicago truncatula*. *Plant J.* 76, 982–996. doi: 10.1111/tpj.12350

Conflict of Interest Statement: The authors declare that the research was conducted in the absence of any commercial or financial relationships that could be construed as a potential conflict of interest.

11

Sulfate transporters in the plant's response to drought and salinity: regulation and possible functions

Karine Gallardo[1]*, Pierre-Emmanuel Courty[2], Christine Le Signor[1], Daniel Wipf[3] and Vanessa Vernoud[1]

[1] *Institut National de la Recherche Agronomique, UMR1347 Agroécologie, Dijon, France*
[2] *Zurich-Basel Plant Science Center, Department of Environmental Sciences, Botany, University of Basel, Basel, Switzerland*
[3] *Université de Bourgogne, UMR1347 Agroécologie, Dijon, France*

Edited by:
Ruediger Hell, University of Heidelberg, Germany

Reviewed by:
Chengbin Xiang, University of Science and Technology of China, China
Peter Buchner, Rothamsted Research, UK

***Correspondence:**
Karine Gallardo, Institut National de la Recherche Agronomique, UMR1347 Agroécologie, 17 rue de Sully, BP 86510, Dijon, France
e-mail: karine.gallardo@dijon.inra.fr

Drought and salinity are two frequently combined abiotic stresses that affect plant growth, development, and crop productivity. Sulfate, and molecules derived from this anion such as glutathione, play important roles in the intrinsic responses of plants to such abiotic stresses. Therefore, understanding how plants facing environmental constraints re-equilibrate the flux of sulfate between and within different tissues might uncover perspectives for improving tolerance against abiotic stresses. In this review, we took advantage of genomics and post-genomics resources available in *Arabidopsis thaliana* and in the model legume species *Medicago truncatula* to highlight and compare the regulation of sulfate transporter genes under drought and salt stress. We also discuss their possible function in the plant's response and adaptation to abiotic stresses and present prospects about the potential benefits of mycorrhizal associations, which by facilitating sulfate uptake may assist plants to cope with abiotic stresses. Several transporters are highlighted in this review that appear promising targets for improving sulfate transport capacities of crops under fluctuating environmental conditions.

Keywords: sulfate, transporters, abiotic stresses, *M. truncatula*, *Arabidopsis*

INTRODUCTION

Drought, the incidence of which is expected to increase with climatic changes, is one of the major abiotic constraints on agricultural productivity. Because drought is often associated with salinity, one challenge for sustainable agriculture is to breed crops for enhanced tolerance to both stresses. This requires an understanding of the adaptive mechanisms allowing plants to survive in low-water and high-salt environments. Sulfur is a key component in helping plants to cope with such abiotic stresses (for review, see Chan et al., 2013). For example, sulfur is used for the synthesis of glutathione, which acts in the maintenance of the cellular redox balance and mitigates damage caused by reactive oxygen species. Most of the sulfur taken up by plants is in the form of sulfate, and several studies point to a role of this anion in the plant response to drought and salinity in relation to the phytohormone abscisic acid (ABA), a major regulator of leaf stomatal conductance (Wilkinson and Davies, 2002). It was proposed that sulfate acts as a primary signal to enhance the anti-transpirant effect of ABA reaching the stomata in leaves (Ernst et al., 2010). More recently, Cao et al. (2014) provided evidence for a significant co-regulation of sulfur and ABA metabolisms in *Arabidopsis* that may help to combat environmental stresses. Such metabolic adjustments undoubtedly rely on the plant's ability to absorb and distribute sulfate to the different organs in amounts sufficient to fulfill requirements.

Major advances have been made toward identifying and characterizing the transporters involved in the uptake, distribution, or efflux of sulfate from the vacuoles, especially in *Arabidopsis* (Buchner et al., 2004 and references therein). The investigation of the contribution of sulfate transporters (SULTR) to abiotic stress tolerance has begun more recently. Cao et al. (2014) proposed a role for SULTR3;1 in helping plants to cope with environmental stresses by providing sulfate for the synthesis of cysteine that serves as a sulfur donor during ABA biosynthesis. With the advances made over the last decade in the integration of "omics" data, gene expression atlases are now available for several species, giving access to the regulation of any gene of interest in different conditions. In this review, we took advantage of these resources to highlight the regulation of *SULTR* genes in response to drought and salinity. We focus on *Arabidopsis* and *M. truncatula*, the latter being a wild legume species originating from the Mediterranean basin that makes use of symbiotic associations to obtain nutrients and that has evolved to develop a tolerance to extreme environmental conditions including drought and salinity (Friesen et al., 2010). After a search of the *SULTR* sequences in *M. truncatula* and of their closest homologs in *Arabidopsis*, we discuss and compare their regulation and possible contribution to protection against unfavorable environmental conditions. We also highlight the potential benefit of using arbuscular mycorrhizal (AM) fungi to improve sulfate uptake.

COMPARATIVE ANALYSIS OF *SULTR* GENE FAMILIES BETWEEN *Arabidopsis* AND *M. truncatula*

Medicago truncatula is an annual forage species adopted in 2001 as a model for legumes because of its small genome, compared to crop legumes such as pea, and its ability to perform symbiotic interactions with nitrogen-fixing rhizobia and AM fungi, like most legume species (Frugoli and Harris, 2001). The close relationship of the *M. truncatula* genome with that of pea (*Pisum sativum* L.) facilitates the transfer of information to the crop, and molecular

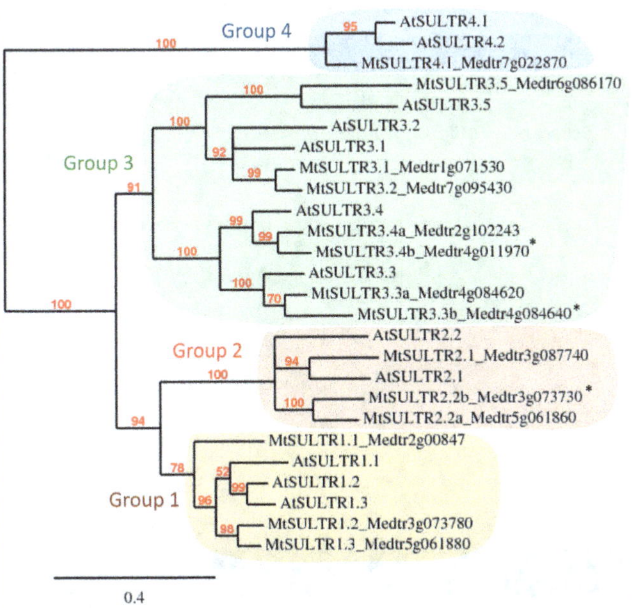

FIGURE 1 | Phylogenetic tree of *Medicago truncatula* and *Arabidopsis* sulfate transporters. The Maximum likelihood phylogenetic tree was generated using all SULTR amino acid sequences available in the *Arabidopsis* and *M. truncatula* (Mt4.0v1) genomic resources. *, sequences for which there was no corresponding probe in the *Medicago* Gene Expression Atlas.

markers have been developed for translational genomics between the two species (Bordat et al., 2011). *M. truncatula* is native to the arid and semi-arid environments of the Mediterranean. It is thus adapted to this climate, making it a good model to identify adaptation processes to low-water or high-salt stresses. Genomic resources were developed for this species that we used here to retrieve *SULTR* genes (*MtSULTR*). Fourteen genes homologous to the *Arabidopsis SULTR* genes (*AtSULTR*) were identified in the last *Medicago* genome version 4.0v1[1]. Phylogenetic analysis using SULTR full length amino-acid sequences allowed us to re-annotate the MtSULTRs and to refine their phylogenetic relationship with AtSULTRs (**Figure 1**). The corresponding neighbor-joining tree divided into four clusters matching the four groups described in *Arabidopsis* (Buchner et al., 2004), as previously observed by Casieri et al. (2013). Three MtSULTRs clustered with the three *Arabidopsis* transporters of high-affinity belonging to group 1, involved in sulfate uptake (SULTR1;1 and 1;2, Yoshimoto et al., 2007; Barberon et al., 2008) or in its distribution to sink organs (SULTR1;3, Yoshimoto et al., 2003). Three others MtSULTRs clustered with the two *Arabidopsis* members of group 2 that deliver sulfate to aerial parts and developing tissues (Takahashi et al., 2000; Awazuhara et al., 2005). Group 3 is the largest group, with seven members in *M. truncatula* compared to five in *Arabidopsis*. They play multiple roles, such as facilitating sulfate transport to aerial parts or controlling cysteine level in seeds and seedlings in tight interaction with ABA metabolism (Kataoka et al., 2004a; Zuber et al., 2010; Cao et al., 2014). One member of this group, SULTR3;1, is responsible for sulfate transport into chloroplasts (Cao et al., 2013). Within group 4, unlike *Arabidopsis* which contains two *SULTR4* genes, there was only one *M. truncatula* gene. It encodes a protein with high homology to AtSULTR4;1 which plays a major role in the efflux of sulfate from the vacuole lumen to the cytosol (Kataoka et al., 2004b). This suggests a unique function for MtSULTR4;1 in remobilizing the stored sulfate. This may apply to other species as there is only one transporter of group 4 with high homology to AtSULTR4;1 in pea (RNAseq data, Burstin J, personal communication) and rice (Kumar et al., 2011).

The recent transcriptome analysis of *M. truncatula* subjected to progressive drought (Zhang et al., 2014a) allowed us to investigate the transcriptional regulation of the *MtSULTR* gene family in response to this abiotic stress and in comparison with a salt stress response (Li et al., 2009). Data were downloaded from the Gene Expression Atlas (MtGEA)[2], and expression fold-change between treated and non-treated samples was calculated (cutoff of 2.0, **Table 1**). Expression of three of the 14 *MtSULTR* genes (*MtSULTR2;2b*, *MtSULTR3;3b*, and *MtSULTR3;4b*, **Figure 1**) could not be investigated as there was no corresponding probe set in the Affymetrix chip used to build the MtGEA. To compare *SULTR* gene regulation between *M. truncatula* and *Arabidopsis*, we used transcriptomic data available in *Arabidopsis* for drought and salt stress experiments (Kilian et al., 2007; Huang et al., 2008; Perera et al., 2008; Nishiyama et al., 2012; Geng et al., 2013; Pandey et al., 2013; Wang et al., 2013; Ha et al., 2014). The studies showing the most substantial regulation of *SULTR* genes are included in **Table 1**. Results are discussed in the light of functional data available, mainly in *Arabidopsis*.

SULTR OF GROUP 3 ARE STRONGLY REGULATED BY ABIOTIC STRESSES IN ROOTS

Of particular interest is the up-regulation of the *SULTR3;1* gene in roots of both species subjected to drought and salt stress. Interestingly, the expression of *AtSULTR3;1* is enhanced by ABA and required for cysteine synthesis (Cao et al., 2014). Cysteine, whose precursor is sulfate, plays a key role in ABA synthesis as it serves as sulfur donor for the sulfuration of molybdenum, a co-factor needed in its sulfurylated form for the last reaction in the pathway (Xiong et al., 2001). The cysteine formed may also serve for the synthesis of the stress-defense compound glutathione. Cao et al. (2014) proposed that sulfur metabolism and ABA biosynthesis interplay to ensure sufficient cysteine for ABA production under abiotic stresses. From these data and the reported plastid-localization of AtSULTR3;1 (Cao et al., 2013), it is tempting to speculate on a role for this transporter in directing the flux of sulfate toward cysteine biosynthesis in the root plastids that may further be used for ABA production in response to both abiotic stresses. In *M. truncatula*, SULTR3;1 has not been functionally characterized. However, the gene is up-regulated in response to both abiotic stresses (**Table 1**) and co-localizes with quantitative trait loci (QTL) regions for salt tolerance (Friesen et al., 2010; Arraouadi et al., 2012), as also observed for *AtSULTR3;1* (El-Soda et al., 2014; Zhang et al., 2014c). This

[1]http://www.jcvi.org/medicago/

[2]http://mtgea.noble.org/v3/

Table 1 | Regulation of *SULTR* gene expression in *Medicago truncatula* and *Arabidopsis* subjected to drought and salt stress.

Medicago truncatula

Group	Gene	Probeset ID[g]	DROUGHT[a] ROOT Mild	DROUGHT[a] ROOT Moderate	DROUGHT[a] ROOT Severe	DROUGHT[a] SHOOT Mild	DROUGHT[a] SHOOT Moderate	DROUGHT[a] SHOOT Severe	SALINITY[b] ROOT Early	SALINITY[b] ROOT Late
1	*SULTR1;1*	Mtr.12106.1.S1_at	-3,7	-13,4	-15,7	ns	ns	ns	ns	-2,9
	SULTR1;2	Mtr.28489.1.S1_at	ns	ns	ns	ns	ns	ns	3,7	1,7
	SULTR1;3	Mtr.5111.1.S1_at	-3,9	ns	-1,7	ns	-2,6	-3,2	1,9	1,7
2	*SULTR2;1*	Mtr.11734.1.S1_at	-2,5	ns	ns	-5,4	-4,6	-6,0	ns	ns
	SULTR2;2a	Mtr.45143.1.S1_at	ns	ns	1,8	ns	ns	ns	ns	-1,5
3	*SULTR3;1*	Mtr.18757.1.S1_at	4,7	14,4	27,9	-1,9	-3,1	-2,8	2,5	3,0
	SULTR3;2	Mtr.41982.1.S1_at	ns	ns	ns	-2,1	-4,0	-4,3	7,4	-2,4
	SULTR3;3a	Mtr.41524.1.S1_at	2,1	1,6	2,0	ns	-2,1	-2,3	ns	ns
	SULTR3;4a	Mtr.31749.1.S1_at	3,8	5,5	5,8	2,5	2,5	1,5	-1,9	ns
	SULTR3;5	Mtr.37708.1.S1_at	-2,0	-2,1	-3,3	-1,7	-1,6	-1,8	25,1	78,6
4	*SULTR4;1*	Mtr.45139.1.S1_at	ns	ns	ns	1,6	1,5	ns	1,7	2,9

Arabidopsis thaliana

Group	Gene	Accession Nb[h]	DROUGHT[c,d] ROOT[c]	DROUGHT[c,d] LEAVES c	DROUGHT[c,d] LEAVES d	SALINITY[e,f] ROOT[e]	SALINITY[e,f] SHOOT/LEAVES e	SALINITY[e,f] SHOOT/LEAVES f
1	*SULTR1;1*	AT4G08620	ns	ns	ns	ns	ns	1,2
	SULTR1;2	AT1G78000	2,1	2,2	ns	1,7	ns	ns
	SULTR1;3	AT1G22150	ns	ns	ns	ns	ns	ns
2	*SULTR2;1*	AT5G10180	ns	ns	-2,1	1,9	ns	ns
	SULTR2;2	AT1G77990	ns	ns	-4,4	ns	ns	ns
3	*SULTR3;1*	AT3G51895	8,7	ns	2,4	7,1	ns	2,4
	SULTR3;2	AT4G02700	ns	-1,2	ns	ns	ns	ns
	SULTR3;3	AT1G23090	ns	ns	ns	1,4	ns	ns
	SULTR3;4	AT3G15990	4,9	ns	ns	3,8	1,7	2,9
	SULTR3;5	AT5G19600	ns	ns	-3,1	-2,6	-2,1	-4,6
4	*SULTR4;1*	AT5G13550	1,4	2,0	5,8	1,6	2,2	2,1
	SULTR4;2	AT3G12520	ns	4,0	ns	2,2	2,3	1,8

Gene expression in response to drought or salinity: Up-regulated / Down-regulated

SULTR gene regulation from: (a) Zhang et al. (2014a): mild, moderate or severe water stress (corresponding to 7, 10, or 14 days of water withdrawal, respectively) applied on 24 day-old M. truncatula plants. (b) Li et al. (2009): young seedlings (2 days) treated with 180 mM of NaCl for 6 h (early response) or 2 days (late reponse); (c) Ha et al. (2014): aerial portions of 24 day-old plants detached and exposed to dehydration on paper towels for 4 h; (d) Pandey et al. (2013): 3 week-old plantlets grown for 9 days on soil with a moisture level below 30%; (e) Kilian et al. (2007): 150 mM NaCl applied to Arabidopsis seedlings in vitro; (f) Wang et al. (2013): 10 day-old seedlings grown for 4 days on a medium supplemented with 100 mM NaCl. For each MtSULTR gene, data for the correponding probeset ID (g) were downloaded from the Medicago Gene Expression Atlas at http://mtgea.noble.org/v3/experiments. (h) Genbank accession number of the Arabidopsis SULTR genes. The values refer to gene expression fold change between treated and non-treated samples. Changes in gene expression of at least twofold are highlighted using a color scale; ns, non-significant change in gene expression in response to drought or salt stress.

makes *MtSULTR3;1* a potential target for modulating the abiotic stress response in legumes. In addition, *MtSULTR3;1* expression is higher at late stages of water stress, i.e., severe water stress in **Table 1**, known to be associated with ABA biosynthesis in roots (Goodger and Schachtman, 2010), suggesting that MtSULTR3;1 could be closely linked in its action with ABA production, as is the case in *Arabidopsis* (Cao et al., 2014). Another gene of group 3 (*AtSULTR3;4, MtSULTR3;4a*) is co-expressed in roots with *SULTR3;1* in response to drought in the two species and in response to salt stress in *Arabidopsis* (**Table 1**). The reduced ABA content in seedlings for the two mutants *Atsultr3;1* and *Atsultr3;4* suggests a role for both genes in relation to ABA production. The subcellular localization of SULTR3;4 is unknown. Investigating spatial and subcellular localizations in roots for both transporters might help to decipher whether they can have a coordinated function or a functional redundancy in this tissue. It should be noted that in contrast to *Arabidopsis*, *MtSULTR3;1* and *MtSULTR3;4a* are differentially regulated in response to salt stress (only *MtSULTR3;1* is up-regulated) and that a second *MtSULTR3;4* gene (*MtSULTR3;4b*, **Figure 1**) exists whose response to salt stress is currently unknown.

In *M. truncatula*, the expression of another group 3 *SULTR* (*MtSULTR3;5*) is strongly up-regulated in roots subjected to salt stress (up to 78-fold; **Table 1**). Its closest *Arabidopsis* homolog,

AtSULTR3;5, shows opposite trends of expression in roots with a consistent down-regulation in response to salinity. This suggests distinct roles or transcriptional regulation of *SULTR3;5* between the two species. In the legume species *Lotus japonicus*, the *SULTR3;5* homolog *SST1 (Symbiotic Sulfate Transporter 1)* is necessary for nodule formation and essential for the symbiotic supply of sulfur to the bacteria (Krusell et al., 2005). In this connection, Varin et al. (2010) identified sulfur supply as necessary for proper accumulation of nitrogenase and leghaemoglobin, two proteins rich in sulfur amino acids and needed for nitrogen fixation. This highlights the importance of maintaining efficient sulfate transport systems in nodules to exploit the nitrogen-fixing capacity of legume plants in agroecological systems. *MtSULTR3;5* is strongly expressed in nodules (Roux et al., 2014) and studies are ongoing to understand the function of MtSULTR3;5 in nodules and to deciphering its contribution to the salt stress response.

RE-EQUILIBRATION OF SULFATE FLUX IN AERIAL PARTS IN RESPONSE TO ABIOTIC STRESSES

In contrast to the functional SST1 (Krusell et al., 2005), AtSULTR3;5 is a non-functional transporter by itself (Kataoka et al., 2004a). This transporter forms a complex with AtSULTR2;1, thus enhancing its sulfate import activity into cells of root vascular tissues for loading into the xylem and transfer to aerial parts, especially when sulfur availability is limited (Takahashi et al., 2000; Kataoka et al., 2004a). The flux of sulfur from roots to shoots is in part controlled by microRNA(Mir)395, which limits expression of *SULTR2;1* to xylem parenchyma, thus enhancing sulfate translocation to aerial parts (Kawashima et al., 2011). Interestingly, Mir395 is up-regulated in response to drought stress in rice (Zhou et al., 2010) and under high salinity conditions in maize (*Zea mays* L.; Ding et al., 2009), suggesting it participates in abiotic stress responses, presumably by maintaining the flux of sulfur toward aerial parts. In roots, the expression of AtSULTR2;1 is not affected by salinity and drought, whereas that of AtSULTR3;5 decreased significantly in response to salt stress (**Table 1**). Owing to the co-activator function of AtSULTR3;5, this may slow the allocation of sulfate to aerial parts. It is therefore possible that *Arabidopsis* adjusts the level of sulfate in roots under salt stress by modulating *AtSULTR3;5* expression. This could be part of the adaptive mechanisms used by *Arabidopsis* to load sulfate into xylem vessels while ensuring that sufficient sulfate remains in roots when uptake is limited due to high salt concentrations in soils. In *M. truncatula*, the *SULTR2;1* gene is not significantly regulated in roots in response to salt stress, but down-regulated in this tissue at early stages of water stress. The function of this transporter has not been reported yet, but if we assume a similar role to its *Arabidopsis* homolog, the down-regulation observed is likely to reflect a need to maintain sulfate in roots at these stages.

A continued loading of sulfate into xylem vessels is of paramount importance for maintaining the synthesis of sulfur molecules in aerial parts. Moreover, sulfate from the xylem acts as a chemical signal for ABA-dependent stomatal closure in leaves during early stages of water stress when ABA biosynthesis is restricted to leaves (Ernst et al., 2010). Several *SULTR* genes in **Table 1** that are regulated in shoots or leaves are good candidates for re-equilibrating the flux of sulfate in aerial parts in response to abiotic stresses. First, *SULTR2;1* is significantly down-regulated in leaves of *Arabidopsis* and *M. truncatula* subjected to drought. *AtSULTR2;1* has been shown to be not only expressed in the xylem parenchyma cells but also in the phloem cells of mature leaves, where it participates in the translocation of sulfate to young leaves (Takahashi et al., 2000). Hence, the down-regulation of *SULTR2;1* suggests a decreased flux of sulfate to young leaves, presumably to save sulfate for protection mechanisms, such as those involving ABA. Second, in *M. truncatula* subjected to drought, one *SULTR3* gene, *MtSULTR3;4*, is significantly up-regulated in aerial parts and more strongly at early stages of water stress (mild and moderate in **Table 1**). It would be of particular interest to investigate whether this transporter could play a role in leaves in controlling their early response to water stress in strong connection with ABA biosynthesis. In *Arabidopsis*, *AtSULTR3;1* and *3;4* are both significantly up-regulated in leaves subjected to salt stress, reinforcing the hypothesis raised in the previous section that both transporters could act in concert to mitigate the effect of salt stress.

Interestingly, the expression of both vacuolar *AtSULTR4* genes is significantly enhanced in leaves by drought and salinity. Moreover, *AtSULTR4;1* and *AtSULTR4;2* fall in QTL regions for tolerance to both stresses (Juenger et al., 2005; McKay et al., 2008). They are thus good candidates for multiple stress tolerance. The only *SULTR4* gene in *M. truncatula* is also up-regulated in shoots in response to drought with a statistically significant but lower fold-change compared to *Arabidopsis*. Because in *Arabidopsis*, the SULTR4 transporters were shown to enable the mobilization of the sulfate stored in the vacuoles, they may play a critical role in ensuring sulfur metabolism in plant cells when sulfate uptake is limited due to environmental constraints. Furthermore, efflux of sulfate from the vacuole may contribute to osmotic adjustments that play a fundamental role in water and salt stress responses. The role of SULTR4 (Kataoka et al., 2004b) has been investigated in roots but their involvement in shoots merits further investigations in relation to abiotic stress tolerance.

REGULATION OF GENES INVOLVED IN SULFATE UPTAKE UNDER ABIOTIC STRESS CONDITIONS

The capacity of roots to take up nutrients generally declines in salt- and water-stressed plants, which may explain the changes in expression of SULTR genes belonging to groups 2, 3, and 4 under these conditions to rebalance sulfate flux between affected tissues. By examining the regulation of the two *SULTR1* genes known to control sulfate uptake in *Arabidopsis*, we observed a contrasted pattern for both genes (**Table 1**). *MtSULTR1;1* appeared down-regulated in roots subjected to both abiotic stresses, whereas *MtSULTR1;2* and *AtSULTR1;2* were up-regulated in response to salinity and drought, respectively. Barberon et al. (2008) demonstrated that *SULTR1;1* and *SULTR1;2* display unequal functional redundancy in *Arabidopsis* and left open the possibility for the *SULTR1;1* gene to display an additional function besides its role in sulfate membrane transport. Recent findings also proposed a supplementary role for AtSULTR1;2 in the regulatory or sensing/signaling pathways related to sulfur metabolism (Zhang et al., 2014b). Further studies are needed to better understand their additional function(s) and contribution to abiotic stress responses.

AM FUNGI, A PROMISING PERSPECTIVE FOR IMPROVING SULFATE UPTAKE IN FLUCTUATING ENVIRONMENTS?

The emerging role of sulfate in plant adaptation to abiotic stresses reinforces the need to sustain proper sulfate uptake and use in cultures that face environmental stresses. One specific feature of legumes, compared to *Arabidopsis*, is their ability to perform symbiotic interactions with AM fungi. This mutualistic association is known to increase plant tolerance to drought (Augé, 2001), an abiotic stress limiting the absorption of ions, including sulfate, by roots. Recent studies in *M. truncatula* revealed that AM fungi improve sulfur nutrition in low-sulfate environments (Casieri et al., 2012; Sieh et al., 2013), probably through their capacity to take up and translocate sulfate to the root (Gray and Gerdemann, 1973; Rhodes and Gerdemann, 1978a,b; Allen and Shachar-Hill, 2009). To date, there is no information available on the regulation of plant sulfate uptake or plant sulfate transporter genes in the presence of AM fungi under drought conditions. However, because drought is associated with reduced sulfate availability, the *SULTR* genes up-regulated at low sulfate concentrations in roots colonized with AM fungi (Casieri et al., 2012; Sieh et al., 2013) might help the plant partner to survive in such environments. This is the case for *MtSULTR1;1* and *MtSULTR1;2*, both up-regulated in roots of AM symbiotic plants, especially at low sulfate concentrations (Casieri et al., 2012). Recently, Giovannetti et al. (2014) demonstrated the induction of the *LjSULTR1;2* gene during the *Lotus japonicus/Rhizophagus irregularis* mutualistic interaction, and the specific expression of this transporter in arbuscule-containing cells, strongly suggesting AM-specific sulfate transport. Investigating the regulation of such genes during AM symbiosis in response to abiotic stresses might help to decipher the roles played by these transporters in fluctuating environments.

CONCLUSION

Several *SULTR* genes regulated by drought and/or salinity were highlighted in this review that may contribute to adjust sulfur distribution in plants subjected to abiotic stresses. We discussed their possible roles using information available in *Arabidopsis*, for which considerable advances have been made in the last two decades toward understanding SULTR functions, more recently in response to salinity (Cao et al., 2014). *SULTR* genes similarly regulated in *Arabidopsis* and *M. truncatula* are promising targets for improving sulfate transport capacities under fluctuating environmental conditions. Among these are group 3 *SULTR*, also in the list of abiotic stress-responsive genes shared between *Arabidopsis* and *M. truncatula* of Hyung et al. (2014). Group 1 SULTR are other potential targets for enhancing sulfate uptake in fluctuating environmental conditions. Members of this group were found to be up-regulated by drought stress and by AM fungi associations that increased significantly the root uptake of sulfate in low-sulfate environments, as it is the case in drought conditions. Broad collections of ecotypes and TILLING mutants are available in *M. truncatula* and in the pea crop (Dalmais et al., 2008; Le Signor et al., 2009; Deulvot et al., 2010) that can be used to study and confirm *SULTR* genes as relevant candidates for discovering favorable alleles for abiotic stress tolerance.

ACKNOWLEDGMENTS

Present work of the authors on sulfate transport is done in the frame of the European Community's Seventh Framework Programme (FP7/2007-2013) under the grant agreement number FP7-613551, LEGATO project. Pierre-Emmanuel Courty gratefully acknowledges his current support by the Swiss National Science Foundation through an AMBIZIONE fellowship (grant No. PZ00P3_136651).

REFERENCES

Allen, J. W., and Shachar-Hill, Y. (2009). Sulfur transfer through an arbuscular mycorrhiza. *Plant Physiol.* 149, 549–560. doi: 10.1104/pp.108.129866

Arraouadi, S., Badri, M., Abdelly, C., Huguet, T., and Aouani, M. E. (2012). QTL mapping of physiological traits associated with salt tolerance in *Medicago truncatula* Recombinant Inbred Lines. *Genomics* 99, 118–125. doi: 10.1016/j.ygeno.2011.11.005

Augé, R. M. (2001). Water relations, drought and vesicular-arbuscular mycorrhizal symbiosis. *Mycorrhiza* 11, 3–42. doi: 10.1007/s005720100097

Awazuhara, M., Fujiwara, T., Hayashi, H., Watanabe-Takahashi, A., Takahashi, H., and Saito, K. (2005). The function of SULTR2;1 sulfate transporter during seed development in *Arabidopsis thaliana*. *Plant Physiol.* 125, 95–105. doi: 10.1111/j.1399-3054.2005.00543.x

Barberon, M., Berthomieu, P., Clairotte, M., Shibagaki, N., Davidian, J.-C., and Gosti, F. (2008). Unequal functional redundancy between the two *Arabidopsis thaliana* high-affinity sulphate transporters SULTR1;1 and SULTR1;2. *New Phytol.* 180, 608–619. doi: 10.1111/j.1469-8137.2008.02604.x

Bordat, A., Savois, V., Nicolas, M., Salse, J., Chauveau, A., Bourgeois, M., et al. (2011). Translational genomics in legumes allowed placing in silico 5460 unigenes on the pea functional map and identified candidate genes in *Pisum sativum* L. *G3* (Bethesda) 1, 93–103. doi: 10.1534/g3.111.000349

Buchner, P., Takahashi, H., and Hawkesford, M. J. (2004). Plant sulphate transporters: co-ordination of uptake, intracellular and long-distance transport. *J. Exp. Bot.* 55, 1765–1773. doi: 10.1093/jxb/erh206

Cao, M. J., Wang, Z., Wirtz, M., Hell, R., Oliver, D. J., and Xiang, C. B. (2013). SULTR3;1 is a chloroplast-localized sulfate transporter in *Arabidopsis thaliana*. *Plant J.* 73, 607–616. doi: 10.1111/tpj.12059

Cao, M. J., Wang, Z., Zhao, Q., Mao, J. L., Speiser, A., Wirtz, M., et al. (2014). Sulfate availability affects ABA levels and germination response to ABA and salt stress in *Arabidopsis thaliana*. *Plant J.* 77, 604–615. doi: 10.1111/tpj.12407

Casieri, L., Ait Lahmidi, N., Doidy, J., Veneault-Fourrey, C., Migeon, A., Bonneau, L., et al. (2013). Biotrophic transportome in mutualistic plant-fungal interactions. *Mycorrhiza* 23, 597–625. doi: 10.1007/s00572-013-0496-9

Casieri, L., Gallardo, K., and Wipf, D. (2012). Transcriptional response of *Medicago truncatula* sulphate transporters to arbuscular mycorrhizal symbiosis with and without sulphur stress. *Planta* 235, 1431–1447. doi: 10.1007/s00425-012-1645-7

Chan, K. X., Wirtz, M., Phua, S. Y., Estavillo, G. M., and Pogson, B. J. (2013). Balancing metabolites in drought: the sulfur assimilation conundrum. *Trends Plant Sci.* 18, 18–29. doi: 10.1016/j.tplants.2012.07.005

Dalmais, M., Schmidt, J., Le Signor, C., Moussy, F., Burstin, J., Savois, V., et al. (2008). UTILLdb, a *Pisum sativum* in silico forward and reverse genetics tool. *Genome Biol.* 9, R43. doi: 10.1186/gb-2008-9-2-r43

Deulvot, C., Charrel, H., Marty, A., Jacquin, F., Donnadieu, C., Lejeune-Hénaut, I., et al. (2010). Highly-multiplexed SNP genotyping for genetic mapping and germplasm diversity studies in pea. *BMC Genomics* 11:468. doi: 10.1186/1471-2164-11-468

Ding, D., Zhang, L., Wang, H., Liu, Z., Zhang, Z., and Zheng, Y. (2009). Differential expression of miRNAs in response to salt stress in maize roots. *Ann. Bot.* 103, 29–38. doi: 10.1093/aob/mcn205

El-Soda, M., Kruijer, W., Malosetti, M., Koornneef, M., and Aarts, M. G. (2014). Quantitative trait loci and candidate genes underlying genotype by environment interaction in the response of *Arabidopsis thaliana* to drought. *Plant Cell Environ.* doi: 10.1111/pce.12418 [Epub ahead of print].

Ernst, L., Goodger, J. Q., Alvarez, S., Marsh, E. L., Berla, B., Lockhart, E., et al. (2010). Sulphate as a xylem-borne chemical signal precedes the expression of ABA biosynthetic genes in maize roots. *J. Exp. Bot.* 61, 3395–3405. doi: 10.1093/jxb/erq160

Friesen, M. L., Cordeiro, M. A., Penmetsa, R. V., Badri, M., Huguet, T., Aouani, M. E., et al. (2010). Population genomic analysis of Tunisian *Medicago truncatula* reveals candidates for local adaptation. *Plant J.* 63, 623–635. doi: 10.1111/j.1365-313X.2010.04267.x

Frugoli, J., and Harris, J. (2001). *Medicago truncatula* on the move! *Plant Cell.* 13, 458–463. doi: 10.1105/tpc.13.3.458

Geng, Y., Wu, R., Wee, C. W., Xie, F., Wei, X., Chan, P. M., et al. (2013). A spatio-temporal understanding of growth regulation during the salt stress response in *Arabidopsis. Plant Cell* 25, 2132–2154. doi: 10.1105/tpc.113.112896

Giovannetti, M., Tolosano, M., Volpe, V., Kopriva, S., and Bonfante, P. (2014). Identification and functional characterization of a sulfate transporter induced by both sulfur starvation and mycorrhiza formation in *Lotus japonicus. New Phytol.* 204, 609–619. doi: 10.1111/nph.12949

Goodger, J. Q., and Schachtman, D. P. (2010). Re-examining the role of ABA as the primary long-distance signal produced by water-stressed roots. *Plant Signal. Behav.* 5, 1298–1301. doi: 10.4161/psb.5.10.13101

Gray, L. E., and Gerdemann, J. W. (1973). Uptake of sulphur-35 by vesicular-arbuscular mycorrhizae. *Plant Soil* 39, 687–689. doi: 10.1007/BF00264184

Ha, C. V., Leyva-González, M. A., Osakabe, Y., Tran, U. T., Nishiyama, R., Watanabe, Y., et al. (2014). Positive regulatory role of strigolactone in plant responses to drought and salt stress. *Proc. Natl. Acad. Sci. U.S.A.* 111, 851–856. doi: 10.1073/pnas.1322135111

Huang, D., Wu, W., Abrams, S. R., and Cutler, A. J. (2008). The relationship of drought-related gene expression in *Arabidopsis thaliana* to hormonal and environmental factors. *J. Exp. Bot.* 59, 2991–3007. doi: 10.1093/jxb/ern155

Hyung, D., Lee, C., Kim, J. H., Yoo, D., Seo, Y. S., Jeong, S. C., et al. (2014). Cross-family translational genomics of abiotic stress-responsive genes between *Arabidopsis* and *Medicago truncatula. PLoS ONE* 9:e91721. doi: 10.1371/journal.pone.0091721

Juenger, T. E., McKay, J. K., Hausmann, N., Keurentjes, J. J. B., Sen, S., Stowe, K. A., et al. (2005). Identification and characterization of QTL underlying whole-plant physiology in *Arabidopsis thaliana*: Delta C-13, stomatal conductance and transpiration efficiency. *Plant Cell Environ.* 28, 697–708. doi: 10.1111/j.1365-3040.2004.01313.x

Kataoka, T., Hayashi, N., Yamaya, T., and Takahashi, H. (2004a). Root-to-shoot transport of sulfate in *Arabidopsis.* Evidence for the role of SULTR3;5 as a component of low-affinity sulfate transport system in the root vasculature. *Plant Physiol.* 136, 4198–4204. doi: 10.1104/pp.104.045625

Kataoka, T., Watanabe-Takahashi, A., Hayashi, N., Ohnishi, M., Mimura, T., Buchner, P., et al. (2004b). Vacuolar sulfate transporters are essential determinants controlling internal distribution of sulfate in *Arabidopsis. Plant Cell* 16, 2693–2704. doi: 10.1105/tpc.104.023960

Kawashima, C. G., Matthewman, C. A., Huang, S., Lee, B. R., Yoshimoto, N., Koprivova, A., et al. (2011). Interplay of SLIM1 and miR395 in the regulation of sulfate assimilation in *Arabidopsis. Plant J.* 66, 863–876. doi: 10.1111/j.1365-313X.2011.04547.x

Kilian, J., Whitehead, D., Horak, J., Wanke, D., Weinl, S., Batistic, O., et al. (2007). The AtGenExpress global stress expression data set: protocols, evaluation and model data analysis of UV-B light, drought and cold stress responses. *Plant J.* 50, 347–363. doi: 10.1111/j.1365-313X.2007.03052.x

Krusell, L., Krause, K., Ott, T., Desbrosses, G., Krämer, U., Sato, S., et al. (2005). The sulfate transporter SST1 is crucial for symbiotic nitrogen fixation in Lotus japonicus root nodules. *Plant Cell* 17, 1625–1636. doi: 10.1105/tpc.104.030106

Kumar, S., Asif, M. H., Chakrabarty, D., Tripathi, R. D., and Trivedi, P. K. (2011). Differential expression and alternative splicing of rice sulphate transporter family members regulate sulphur status during plant growth, development and stress conditions. *Funct Integr Genomics* 11, 259–273. doi: 10.1007/s10142-010-0207-y

Le Signor, C., Savois, V., Aubert, G., Verdier, J., Nicolas, M., Pagny, G., et al. (2009). Optimizing TILLING populations for reverse genetics in *Medicago truncatula. Plant Biotechnol. J.* 7, 430–441. doi: 10.1111/j.1467-7652.2009.00410.x

Li, D., Su, Z., Dong, J., and Wang, T. (2009). An expression database for roots of the model legume *Medicago truncatula* under salt stress. *BMC Genomics* 10:517. doi: 10.1186/1471-2164-10-517

McKay, J. K., Richards, J. H., Nemali, K. S., Sen, S., Mitchell-Olds, T., Boles, S., et al. (2008). Genetics of drought adaptation in *Arabidopsis thaliana* II. QTL analysis of a new mapping population, Kas-1 × Tsu-1. *Evolution* 62, 3014–3026. doi: 10.1111/j.1558-5646.2008.00474.x

Nishiyama, R., Le, D. T., Watanabe, Y., Matsui, A., Tanaka, M., Seki, M., et al. (2012). Transcriptome analyses of a salt-tolerant cytokinin-deficient mutant reveal differential regulation of salt stress response by cytokinin deficiency. *PLoS ONE* 7:e32124. doi: 10.1371/journal.pone.0032124

Pandey, N., Ranjan, A., Pant, P., Tripathi, R. K., Ateek, F., Pandey, H. P., et al. (2013). CAMTA 1 regulates drought responses in *Arabidopsis thaliana. BMC Genomics* 14:216. doi: 10.1186/1471-2164-14-216

Perera, I. Y., Hung, C. Y., Moore, C. D., Stevenson-Paulik, J., and Boss, W. F. (2008). Transgenic *Arabidopsis* plants expressing the type 1 inositol 5-phosphatase exhibit increased drought tolerance and altered abscisic acid signaling. *Plant Cell* 20, 2876–2893. doi: 10.1105/tpc.108.061374

Rhodes, L. H., and Gerdemann, J. W. (1978a). Hyphal translocation and uptake of sulfur by vesicular-arbuscular mycorrhizae of onion. *Soil Biol. Biochem.* 10, 355–360. doi: 10.1016/0038-0717(78)90057-3

Rhodes, L. H., and Gerdemann, J. W. (1978b). Influence of phosphorus nutrition on sulphur uptake by vesicular arbuscular. *Soil Biol. Biochem.* 10, 361–364. doi: 10.1016/0038-0717(78)90058-5

Roux, B., Rodde, N., Jardinaud, M. F., Timmers, T., Sauviac, L., Cottret, L., et al. (2014). An integrated analysis of plant and bacterial gene expression in symbiotic root nodules using laser-capture microdissection coupled to RNA sequencing. *Plant J.* 77, 817–837. doi: 10.1111/tpj.12442

Sieh, D., Watanabe, M., Devers, E. A., Brueckner, F., Hoefgen, R., and Krajinski, F. (2013). The arbuscular mycorrhizal symbiosis influences sulfur starvation responses of *Medicago truncatula. New Phytol.* 197, 606–616. doi: 10.1111/nph.12034

Takahashi, H., Watanabe-Takahashi, A., Smith, F. W., Blake-Kalff, M., Hawkesford, M. J., and Saito, K. (2000). The roles of three functional sulphate transporters involved in uptake and translocation of sulphate in *Arabidopsis thaliana. Plant J.* 23, 171–182. doi: 10.1046/j.1365-313x.2000.00768.x

Varin, S., Cliquet, J. B., Personeni, E., Avice, J.-C., and Lemauviel-Lavenant, S. (2010). How does sulphur availability modify N acquisition of white clover (Trifolium repens L.)? *J. Exp. Bot.* 61, 225–234. doi: 10.1093/jxb/erp303

Wang, Y., Yang, L., Zheng, Z., Grumet, R., Loescher, W., Zhu, J. K., et al. (2013). Transcriptomic and physiological variations of three *Arabidopsis* ecotypes in response to salt stress. *PLoS ONE* 8:e69036. doi: 10.1371/journal.pone.0069036

Wilkinson, S., and Davies, W. J. (2002). ABA-based chemical signaling: the co-ordination of responses to stress in plants. *Plant Cell Environ.* 25, 195–210. doi: 10.1046/j.0016-8025.2001.00824.x

Xiong, L., Ishitani, M., Lee, H., and Zhu, J. K. (2001). The *Arabidopsis* LOS5/ABA3 locus encodes a molybdenum cofactor sulfurase and modulates cold stress- and osmotic stress-responsive gene expression. *Plant Cell* 13, 2063–2083. doi: 10.1105/tpc.13.9.2063

Yoshimoto, N., Inoue, E., Saito, K., Yamaya, T., and Takahashi, H. (2003). Phloem-localizing sulfate transporter, Sultr1;3, mediates re-distribution of sulfur from source to sink organs in *Arabidopsis. Plant Physiol.* 131, 1511–1517. doi: 10.1104/pp.014712

Yoshimoto, N., Inoue, E., Watanabe-Takahashi, A., Saito, K., and Takahashi, H. (2007). Posttranscriptional regulation of high-affinity sulfate transporters in *Arabidopsis* by sulfur nutrition. *Plant Physiol.* 145, 378–388. doi: 10.1104/pp.107.105742

Zhang, J. Y., Cruz, D. E., Carvalho, M. H., Torres-Jerez, I., Kang, Y., Allen, S. N., et al. (2014a). Global reprogramming of transcription and metabolism in *Medicago truncatula* during progressive drought and after rewatering. *Plant Cell Environ.* In press (free access online). doi: 10.1111/pce.12328 [Epub ahead of print].

Zhang, B., Pasini, R., Dan, H., Joshi, N., Zhao, Y., Leustek, T., et al. (2014b). Aberrant gene expression in the *Arabidopsis* SULTR1;2 mutants suggests a possible regulatory role for this sulfate transporter in response to sulfur nutrient status. *Plant J.* 77, 185–197. doi: 10.1111/tpj.12376

Zhang, X., Lu, G., Long, W., Zou, X., Li, F., and Nishio, T. (2014c). Recent progress in drought and salt tolerance studies in *Brassica* crops. *Breed Sci.* 64, 60–73. doi: 10.1270/jsbbs.64.60

Zhou, L., Liu, Y., Liu, Z., Kong, D., Duan, M., and Luo, L. (2010). Genome-wide identification and analysis of drought-responsive microRNAs in Oryza sativa. *J. Exp. Bot.* 61, 4157–4168. doi: 10.1093/jxb/erq237

Zuber, H., Davidian, J.-C., Aubert, G., Aimé, D., Belghazi, M., Lugan, R., et al. (2010). The seed composition of *Arabidopsis* mutants for the group 3 sulfate transporters indicates a role in sulfate translocation within developing seeds. *Plant Physiol.* 154, 913–926. doi: 10.1104/pp.110.162123

Conflict of Interest Statement: The authors declare that the research was conducted in the absence of any commercial or financial relationships that could be construed as a potential conflict of interest.

12

Garlic γ-glutamyl transpeptidases that catalyze deglutamylation of biosynthetic intermediate of alliin

Naoko Yoshimoto[1], Ayami Yabe[1], Yuka Sugino[1], Soichiro Murakami[1], Niti Sai-ngam[1], Shin-ichiro Sumi[2], Tadamitsu Tsuneyoshi[3] and Kazuki Saito[1,4]*

[1] Graduate School of Pharmaceutical Sciences, Chiba University, Chiba, Japan
[2] Research Planning Department, Wakunaga Pharmaceutical Company, Akitakata, Japan
[3] Central Research Institute, Wakunaga Pharmaceutical Company, Akitakata, Japan
[4] RIKEN Center for Sustainable Resource Science, Yokohama, Japan

Edited by:
Stanislav Kopriva, University of Cologne, Germany

Reviewed by:
Naoko Ohtsu, Tokyo University of Agriculture and Technology, Japan
Pete Rose, University of Lincoln, UK

***Correspondence:**
Kazuki Saito, Graduate School of Pharmaceutical Sciences, Chiba University, Inohana 1-8-1, Chuo-ku, Chiba 260-8675, Japan
e-mail: ksaito@faculty.chiba-u.jp

S-Alk(en)yl-L-cysteine sulfoxides are pharmaceutically important secondary metabolites produced by plants that belong to the genus *Allium*. Biosynthesis of *S*-alk(en)yl-L-cysteine sulfoxides is initiated by *S*-alk(en)ylation of glutathione, which is followed by the removal of glycyl and γ-glutamyl groups and *S*-oxygenation. However, most of the enzymes involved in the biosynthesis of *S*-alk(en)yl-L-cysteine sulfoxides in *Allium* plants have not been identified. In this study, we identified three genes, *AsGGT1*, *AsGGT2*, and *AsGGT3*, from garlic (*Allium sativum*) that encode γ-glutamyl transpeptidases (GGTs) catalyzing the removal of the γ-glutamyl moiety from a putative biosynthetic intermediate of *S*-allyl-L-cysteine sulfoxide (alliin). The recombinant proteins of AsGGT1, AsGGT2, and AsGGT3 exhibited considerable deglutamylation activity toward a putative alliin biosynthetic intermediate, γ-glutamyl-*S*-allyl-L-cysteine, whereas these proteins showed very low deglutamylation activity toward another possible alliin biosynthetic intermediate, γ-glutamyl-*S*-allyl-L-cysteine sulfoxide. The deglutamylation activities of AsGGT1, AsGGT2, and AsGGT3 toward γ-glutamyl-*S*-allyl-L-cysteine were elevated in the presence of the dipeptide glycylglycine as a γ-glutamyl acceptor substrate, although these proteins can act as hydrolases in the absence of a proper acceptor substrate, except water. The apparent K_m values of AsGGT1, AsGGT2, and AsGGT3 for γ-glutamyl-*S*-allyl-L-cysteine were 86 μM, 1.1 mM, and 9.4 mM, respectively. Subcellular distribution of GFP-fusion proteins transiently expressed in onion cells suggested that AsGGT2 localizes in the vacuole, whereas AsGGT1 and AsGGT3 possess no apparent transit peptide for localization to intracellular organelles. The different kinetic properties and subcellular localizations of AsGGT1, AsGGT2, and AsGGT3 suggest that these three GGTs may contribute differently to the biosynthesis of alliin in garlic.

Keywords: γ-glutamyl transpeptidase, deglutamylation, alliin, secondary metabolism, garlic

INTRODUCTION

Production of cysteine-derived secondary metabolites, *S*-alk(en)yl-L-cysteine sulfoxides, is a pharmaceutically important characteristic of plants that belong to the genus *Allium*. These compounds are hydrolyzed by the endogenous vacuolar enzyme alliinase (EC. 4.4.1.4) upon tissue disruption to yield highly reactive alk(en)ylsulfenic acids that are spontaneously converted to various sulfur-containing compounds with diverse pharmacological activities, including antibacterial, antifungal, antivirus, immunostimulating, antioxidant, anticarcinogenic, antithrombotic, cholesterol- and triglyceride-lowering, and hypotensive effects (Jones et al., 2004; Rose et al., 2005; Iciek et al., 2009). To date, four major *S*-alk(en)yl-L-cysteine sulfoxides, *S*-allyl-L-cysteine sulfoxide (alliin), *S*-methyl-L-cysteine sulfoxide (methiin), *S*-*trans*-1-propenyl-L-cysteine sulfoxide (isoalliin), and *S*-propyl-L-cysteine sulfoxide (propiin), have been identified and isolated from *Allium* plants (Jones et al., 2004; Rose et al., 2005).

Biosynthesis of *S*-alk(en)yl-L-cysteine sulfoxides in *Allium* plants has previously been proposed to proceed via glutathione *S*-conjugates, according to the results of precursor feeding and pulse radiolabeling experiments (Suzuki et al., 1962; Turnbull et al., 1980; Lancaster and Shaw, 1989). In the proposed pathway, glutathione is *S*-alk(en)ylated at the cysteine residue, followed by the removal of a glycyl group to form a biosynthetic intermediate, γ-glutamyl-*S*-alk(en)yl-L-cysteine. This γ-glutamylated sulfide compound is further deglutamylated and *S*-oxygenated to yield *S*-alk(en)yl-L-cysteine sulfoxide (**Figure 1**). Although the results of pulse radiolabeling suggest that *S*-oxygenation may likely occur before deglutamylation in onion (*Allium cepa*; Lancaster and Shaw, 1989), the order of *S*-oxygenation and deglutamylation in other *Allium* plants remains unclear.

γ-Glutamyl transpeptidase (GGT; EC 2.3.2.2), also known as γ-glutamyl transferase, is the enzyme that catalyzes the transfer of the γ-glutamyl moiety of γ-glutamyl compounds to amino acids, short peptides (transpeptidation), or water (hydrolysis; Tate

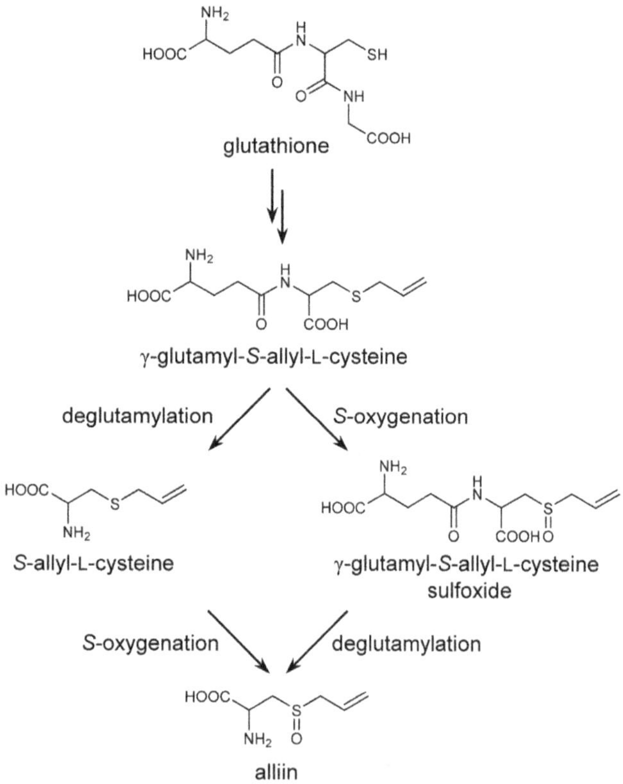

FIGURE 1 | Proposed biosynthetic pathway for alliin in garlic.

and Meister, 1981). The physiological role of GGT, commonly observed in bacteria, yeast, animals, and plants, is the catabolism of glutathione. *Saccharomyces cerevisiae* has a vacuolar GGT protein responsible for the degradation of glutathione in the vacuole during nitrogen starvation in order to supply the constituent amino acids of glutathione to the starved cell (Mehdi and Penninckx, 1997), whereas GGTs in *Escherichia coli*, mammals, and plants function in the breakdown of extracellular glutathione (Suzuki et al., 1999; Storozhenko et al., 2002; Dominici et al., 2005; Martin et al., 2007; Ohkama-Ohtsu et al., 2007a). GGT is also responsible for degrading glutathione-related compounds. For example, GGT is involved in the biosynthesis of the phytoalexin camalexin by removing a γ-glutamyl group from glutathione-indole-3-acetonitrile in *Arabidopsis* (Su et al., 2011), in the conversion of the endogenous glutathione *S*-conjugate leukotriene C4 to leukotriene D4 in rats (Anderson et al., 1982), and in the glutathione-mediated detoxification of xenobiotics in both animals and plants (Zhang et al., 2005; Grzam et al., 2007; Ohkama-Ohtsu et al., 2007b). Given that the removal of a γ-glutamyl group from the biosynthetic intermediate γ-glutamyl-*S*-alk(en)yl-L-cysteine is required for the biosynthesis of *S*-alk(en)yl-L-cysteine sulfoxides in the genus *Allium*, the involvement of GGTs in the biosynthesis of *S*-alk(en)yl-L-cysteine sulfoxides as deglutamylation enzymes has been proposed. The fact that the levels of biosynthetic intermediate γ-glutamyl peptides were decreased while GGT activity was increased during sprouting in onion bulbs also supports this idea (Lancaster and Shaw, 1991). To date, several efforts have been made to identify and characterize GGTs in *Allium* plants. A GGT partially purified from onion showed high substrate specificity toward γ-glutamyl compounds that are putative intermediates of *S*-alk(en)yl-L-cysteine sulfoxide biosynthesis, strongly suggesting the involvement of this GGT in the biosynthesis of *S*-alk(en)yl-L-cysteine sulfoxides (Lancaster and Shaw, 1994). Recently, a GGT protein was purified to homogeneity from sprouting onion bulbs, and a partial cDNA for this GGT, *AcGGT*, was cloned; however, in contrast to the previously partially purified onion GGT (Lancaster and Shaw, 1994), the purified AcGGT protein showed high affinity for glutathione and glutathione *S*-conjugates but could not utilize γ-glutamyl-*trans*-*S*-1-propenyl-L-cysteine sulfoxide as a good γ-glutamyl donor substrate, suggesting that AcGGT is not the major enzyme catalyzing deglutamylation in the biosynthesis of *S*-alk(en)yl-L-cysteine sulfoxides in onion (Shaw et al., 2005). A partial cDNA of *AsGGT*, which has high sequence homology to *AcGGT*, was isolated from garlic (*Allium sativum*), and its mRNA expression patterns suggested that AsGGT may play a role in synthesizing *S*-alk(en)yl-L-cysteine sulfoxides in garlic cloves during cold storage (Cho et al., 2012).

In this study, we cloned three genes encoding GGTs, *AsGGT1*, *AsGGT2*, and *AsGGT3*, that are suggested to be involved in the biosynthesis of alliin in garlic. The substrate preferences of AsGGT1, AsGGT2, and AsGGT3 suggest that a key biosynthetic intermediate, γ-glutamyl-*S*-allyl-L-cysteine, is deglutamylated by these GGTs prior to being *S*-oxygenated during alliin biosynthesis in garlic.

MATERIALS AND METHODS

PLANT MATERIALS AND REAGENTS

Total RNA was extracted from the bulbs of *A. sativum* L. 'Fukuchi-howaito'. *S*-Allyl-L-cysteine was purchased from Tokyo Chemical Industry (Tokyo, Japan). Alliin [$(R_C S_S)$-*S*-allyl-L-cysteine sulfoxide] was synthesized and purified according to previously reported methods (Yu et al., 1994; Kubec et al., 1999; Kubec and Dadáková, 2008). γ-Glutamyl-*S*-allyl-L-cysteine was synthesized as follows. A mixture of 2.59 g of *N*-phthaloyl-L-glutamic anhydride and 1.93 g of *S*-allyl-L-cysteine in 10 mL of acetic acid was stirred at 60°C for 2 h. The solvent in the reaction mixture was removed in vacuo, and the residue was suspended in ethyl acetate and washed with brine. The organic layer was dried using sodium sulfate, and the solvent was removed in vacuo. After 1.5 g of residue was dissolved in 10 mL of methanol, 0.15 mL of hydrazine monohydrate was added, and this mixture was refluxed at 80°C for 1 h. Solvent was removed in vacuo, and the residue was washed with ethanol. The residue was recrystallized in a mixture of ethanol and water. The crystalline powder was applied to Dowex^{TM} 50Wx8 (The Dow Chemical Company, USA), and the column eluate and rinsing were combined and lyophilized. γ-Glutamyl-*S*-allyl-L-cysteine sulfoxide was synthesized as follows. γ-Glutamyl-*S*-allyl-L-cysteine was dissolved in water, and 1.1 equimolar of hydrogen peroxide was added. The mixture was stirred at room temperature, and the solvent was removed in vacuo. The residue was dried under reduced pressure with phosphorus (V) oxide at room temperature. Structures of synthesized γ-glutamyl-*S*-allyl-L-cysteine and γ-glutamyl-*S*-allyl-L-cysteine sulfoxide were confirmed using ^{13}C-NMR and ^{1}H-NMR. All other chemicals were of analytical grade and were purchased from Sigma (St. Louis, MO, USA), Nacalai Tesque (Kyoto, Japan), or Wako Pure Chemical Industries (Osaka, Japan).

CLONING OF *AsGGT1*, *AsGGT2*, AND *AsGGT3* FROM GARLIC

Molecular biological experiments were performed according to the standard protocols (Sambrook et al., 1989), unless otherwise specified. Total RNA was extracted from garlic cloves by using the RNeasy plant mini kit (Qiagen, Valencia, CA, USA) and treated with DNase I (Life Technologies, Carlsbad, CA, USA). Reverse transcription (RT) was performed using SuperScript II reverse transcriptase (Life Technologies) and oligo-d(T)$_{12-18}$. Partial cDNAs of *AsGGT1* and *AsGGT2* were amplified by PCR using *Ex*Taq DNA polymerase (Takara, Tokyo, Japan) and oligonucleotide primers designed from the nucleotide sequences of two garlic EST clones, i.e., EPP005LLAA12S004013 and EPP005LLAA12S003688 in GarlicESTdb (Kim et al., 2009[1]): AsGGT1-Core-F (5′-ATCGCCACTTCATATGAACC-3′) and AsGGT1-Core-R (5′-GATAATGCTAGATATGGCTC-3′) for *AsGGT1*; AsGGT2-Core-F (5′-CTCCTCCACATTAATGGAAC-3′) and AsGGT2-Core-R (5′-AAGTGGTCCCAAACATTTGTC-3′) for *AsGGT2*. For the amplification of a partial region of *AsGGT3* cDNA, degenerate primers designed based on the sequences of conserved regions of known GGTs, GGT-degenerate-F (5′-ATHGTNYTNAAYAAYGARATG-3′) and GGT-degenerate-R (5′-CCNCCYTTNCKNGGRTC-3′), were used. Rapid amplification of cDNA ends (RACE) was performed using 5′-Full RACE Core Set (Takara) and 3′-Full RACE Core Set (TaKaRa), according to the manufacturer's protocols. 5′-RACE was performed using the following primers: AsGGT1-5′-RACE-RT (5′-[Phos]TCTTCTGAACCG-3′), AsGGT1-5′-RACE-F1(5′-TGCTCTCACCACTCTGTTC-3′), AsGGT1-5′-RACE-F2 (5′-GACTCCATCTCTCATCAGTTC-3′), AsGGT1-5′-RACE-R1 (5′-TCACGAACGATGAGCGATG-3′), and AsGGT1-5′-RACE-R2 (5′-CCAGTTTCTGATCAGAAGAAGC-3′) for *AsGGT1*; AsGGT2-5′-RACE-RT (5′-[Phos]TGAGCTCGTAAACTC-3′), AsGGT2-5′-RACE-F1 (5′-TGTGCGACGGTATCCGATCA-3′), AsGGT2-5′-RACE-F2 (5′-CTCAATCCAATTCAACCTAGAC-3′), AsGGT2-5′-RACE-R1 (5′-CATTGTGCAGCGGACGATAG-3′), and AsGGT2-5′-RACE-R2 (5′-GGTTCCATTAATGTGGAGGAG-3′) for *AsGGT2*; AsGGT3-5′-RACE-RT (5′-[Phos]GTATCCATCGGGAAT-3′), AsGGT3-5′-RACE-F1 (5′-TGAAAAAGAAAGGGCAGCTC-3′), AsGGT3-5′-RACE-F2 (5′-GGTTTAGGGATTGCAAATGG-3′), AsGGT3-5′-RACE-R1 (5′-CCTCCACTTGCGCCTAGAG-3′), and AsGGT3-5′-RACE-R2 (5′-GGTGGCGGCATATTGTTATT-3′) for *AsGGT3*. 3′-RACE was performed using 3 sites adaptor primer (5′-CTGATCTAGAGGTACCGGATCC-3′) and the following gene-specific primers: AsGGT1-3′-RACE-F1 (5′-AGCTGGTCTACATGCTGCATGG-3′) and AsGGT1-3′-RACE-F2 (5′-TCCCATGGAAGTCACTTTTCG-3′) for *AsGGT1*; AsGGT2-3′-RACE-F1 (5′-GCTTTTGATGCTAGAGAGACTGC-3′) and AsGGT2-3′-RACE-F2 (5′-ATCACTCCGACAAATGTTTG-3′) for *AsGGT2*; AsGGT3-3′-RACE-F (5′-TGAAAAAGAAAGGGCAGCTC-3′) for *AsGGT3*. cDNA clones of *AsGGT1*, *AsGGT2*, and *AsGGT3* were re-isolated by RT-PCR using KOD plus DNA polymerase (Toyobo, Osaka, Japan) and the following primers: AsGGT1-F (5′-TCATATTCTGACGCAGATTCCACAG-3′) and AsGGT1-R (5′-TGTTCAATCATATTTTGTACAAATAGAC-3′) for *AsGGT1*; AsGGT2-F (5′-CGAGCAAATTAATTCATTTTGGCTCAC-3′) and AsGGT2-R (5′-GCATACCAATCGCCACAAACTC-3′) for *AsGGT2*; AsGGT3-F (5′-GTTAACAACAGGATTGGTCAATGCTC-3′) and AsGGT3-R (5′-CAGCAAACAACGCACTATTCAGTTTCTG-3′) for *AsGGT3*.

[1]http://garlicdb.kribb.re.kr

HETEROLOGOUS EXPRESSION OF AsGGT1, AsGGT2, AND AsGGT3 IN YEAST

The coding regions of *AsGGT1*, *AsGGT2*, and *AsGGT3* were amplified by PCR using the cloned cDNA fragments described above, KOD plus DNA polymerase (Toyobo), and the following gene-specific primers: AsGGT1-FKpn3A (5′-<u>GGTACC</u>AAAATGAACCAAATGGCGCCGGCTTC-3′) and AsGGT1-stop-RXh (5′-<u>CTCGAG</u>CTATACACAAGCAGGACTTCCATC-3′) for *AsGGT1*; AsGGT2-FKpn3A (5′-<u>GGTACC</u>AAAATGGAACCGGCGCATGATGACTTAG-3′) and AsGGT2-stop-RXh (5′-<u>CTCGAG</u>TCACACACATGCAGGACTTCCATC-3′) for *AsGGT2*; AsGGT3-FKpn3A (5′-<u>GGTACC</u>AAAATGCTAATTAATTCATACCCTGC-3′) and AsGGT3-stop-RXh (5′-<u>CTCGAG</u>TCAGTATCCATCGGGAATACC-3′) for *AsGGT3*. The underlined sequences in the primers correspond to *Kpn*I and *Xho*I restriction sites for subcloning. The amplified fragments were cloned into the pGEM-T easy vector (Promega, Madison, WI, USA). After their nucleotide sequences were confirmed, the coding regions of *AsGGT1*, *AsGGT2*, and *AsGGT3* were cut out as *Kpn*I-*Xho*I fragments and were inserted between the *Kpn*I and *Xho*I sites in the yeast expression vector pYES2 (Life Technologies). The resulting plasmids, pYES2-AsGGT1, pYES2-AsGGT2, and pYES2-AsGGT3, and pYES2 empty vector were transformed into the *Saccharomyces cerevisiae* mutant strain BJ2168 (*MATa, prb1-1122, prc1-407, pep4-3, ura3-52, leu2, trp1*; Nippon Gene, Tokyo, Japan) by using the lithium acetate method (Gietz et al., 1992). The transformants were selected on SD minimal medium (Sherman, 1991) containing no uracil. For the induction of recombinant proteins, the yeast cells grown in SD minimal medium without uracil at 28°C for 1 days were transferred to 10 volumes of uracil-less SD medium containing 2% (w/v) galactose instead of glucose to activate the *GAL1* promoter on pYES2, and cultured at 28°C for 1 days. The cells were harvested and disrupted at 4°C with 425–600-μm (diameter) glass beads in buffer G [10 mM Tris-HCl (pH 7.5), 300 mM sorbitol, 100 mM NaCl, 5 mM $MgCl_2$, 1 mM EDTA, and 1 μM pepstatin A]. The lysate was centrifuged at 10,000 × *g* for 5 min, and the supernatant was collected. Buffer G of the supernatant was subsequently replaced with 50 mM Tris-HCl (pH 8.0) by using the Sephadex column PD Mini Trap G-25 (GE Healthcare, Uppsala, Sweden), according to the manufacturer's protocol. The eluted yeast crude proteins were used for the enzymatic activity assay described below. Protein concentrations were determined using the Bio-Rad protein assay (Bio-Rad, CA, USA) based on the Bradford method (Bradford, 1976), using bovine serum albumin as the standard.

ASSAYS OF GGT ENZYME ACTIVITIES

Assays of GGT enzyme activities were performed by analyzing the amount of deglutamylated compounds produced from γ-glutamylated compounds by yeast crude proteins in 6 h at 37°C. The amount of deglutamylated compounds increased linearly over the 6-h incubation period.

Deglutamylation activities using γ-glutamyl-*p*-nitroanilide as the substrate were determined spectrophotometrically according to a previously described method (Orlowski and Meister, 1963), with slight modifications, as follows: the reaction mixture, which consisted of 0.0125 $\mu g\ \mu l^{-1}$ yeast crude protein, 50 mM Tris-HCl (pH 8.0), 10 mM glycylglycine, and 1 mM γ-glutamyl-*p*-nitroanilide, was incubated for 6 h at 37°C, and *p*-nitroaniline released from γ-glutamyl-*p*-nitroanilide was monitored at 412 nm.

For the analysis of deglutamylation activities toward γ-glutamyl-*S*-allyl-L-cysteine and γ-glutamyl-*S*-allyl-L-cysteine sulfoxide, the enzyme assay reaction mixture consisted of 0.67 $\mu g\ \mu l^{-1}$ yeast crude protein, 50 mM Tris-HCl (pH 8.0), 10 mM glycylglycine, and 1 mM γ-glutamyl-*S*-allyl-L-cysteine or γ-glutamyl-*S*-allyl-L-cysteine sulfoxide was incubated for 6 h at 37°C. For the determination of the enzyme activity in a pH range of 6.0–7.0, the reaction mixture containing 50 mM 2-(*N*-morpholino)ethanesulfonic acid buffer, instead of 50 mM Tris-HCl buffer, was used. For the determination of the enzyme activity in a pH range of 7.0–9.0, 50 mM Tris-HCl buffer at pH values from 7.0 to 9.0, instead of 50 mM Tris-HCl (pH 8.0), was used. For the analysis of the effects of glycylglycine as a γ-glutamyl acceptor, deglutamylation activity was determined in the reaction mixture with or without 10 mM glycylglycine. The reaction was initiated by the addition of yeast crude proteins. After incubation at 37°C, proteins in the reaction mixture were removed using a centrifugal ultrafiltration device (molecular weight cutoff, 10 kD; Kurabo, Osaka, Japan). *S*-Allyl-L-cysteine and alliin in the ultrafiltrated solution were quantified using high-performance liquid chromatography (HPLC). For the kinetic analysis, assays were carried out with γ-glutamyl-*S*-allyl-L-cysteine concentrations ranging from 12.5 to 1000 μM for AsGGT1, 0.5–8 mM for AsGGT2, and 1–25 mM for AsGGT3. K_m values were calculated from triplicate date sets according to the Michaelis-Menten equation.

ANALYSIS OF SULFUR-CONTAINING METABOLITES BY USING HPLC

The enzymatic products were analyzed quantitatively by using HPLC (Hitachi, Tokyo, Japan) with the cation-exchange column (TSKgel Aminopak, Tosoh, Tokyo, Japan). For the determination of *S*-allyl-L-cysteine, a mobile phase composed of 67 mM sodium citrate, 8% (v/v) ethanol, and 0.01% (v/v) octanoic acid (pH 3.26) was used for an isocratic elution. The column temperature was 35°C. For the determination of alliin, the mobile phase consisted of 22 mM trisodium citrate and 80 mM citric acid and the column temperature was 40°C. After separation, *S*-allyl-L-cysteine and alliin were fluorescently derivatized using *o*-phthalaldehyde and were detected using a fluorescence detector (excitation 340 nm, emission 455 nm). Identification of *S*-allyl-L-cysteine and alliin in the enzymatic reaction mixture was based on comparisons of retention times of the synthesized standards.

SUBCELLULAR LOCALIZATION ANALYSIS

For the construction of the fusion gene constructs of *35S*$_{pro}$*:AsGGT1*$_{N100}$*:GFP*, *35S*$_{pro}$*:AsGGT2*$_{N100}$*:GFP*, and *35S*$_{pro}$*:AsGGT3*$_{N100}$*:GFP*, partial coding regions of *AsGGT1*, *AsGGT2*, and *AsGGT3* that encode the N-terminal 100 amino acid residues were amplified by PCR using KOD plus DNA polymerase (Toyobo) and the following gene-specific primers: AsGGT1-FSal (5′-GTCGACATGAACCAAATGGCGCCGGCTTCTTC-3′) and AsGGT1-N100-RNco (5′-CCATGGAACCACCACCACCACCACCACCTTTTCTCAGAACTGAAGCTCC-3′) for *AsGGT1*; AsGGT2-FSal (5′-GTCGACATGGAACCGGCGCATGATGACTTAG-3′) and AsGGT2-N100-RNco (5′-CCATGGAACCACCACCACCACCACCTAAAGCGTCCACAGCATGACC-3′) for *AsGGT2*; AsGGT3-FSal (5′-GTCGACATGCTAATTAATTCATACCCTGCATATC-3′) and AsGGT3-N100-RNco (5′-CCATGGAACCACCACCACCACCACCACTAACAACTCCTAAACAAAATG-3′) for *AsGGT3*. The sequence encoding hexa-Gly residues was generated downstream of the sequence of *AsGGT1*, *AsGGT2*, and *AsGGT3* by the PCR. The underlined sequences in the primers correspond to *Sal*I and *Nco*I restriction sites for subcloning. The amplified DNA fragments were cloned into the pGEM-T easy vector (Promega) to confirm the nucleotide sequence. Partial *AsGGT1*, *AsGGT2*, and *AsGGT3*, fused with the sequence encoding hexa-Gly residues, were cut out as *Sal*I-*Nco*I fragments and were inserted between the *Sal*I and *Nco*I sites in pTH2 (Chiu et al., 1996). Each of the resulting plasmids was co-introduced with pDsRed plasmid (Kitajima et al., 2009) into onion epidermal cells by particle bombardment at 150 psi, using a Helios gene gun (Bio-Rad). After bombardment, onion peels were incubated for 26–47 h on B5 medium (Gamborg et al., 1968) in the dark at 25°C. GFP and DsRed fluorescence in the onion cells were observed using a LSM710 confocal laser-scanning microscope (Carl Zeiss, Oberkochen, Germany).

PHYLOGENETIC ANALYSIS

Phylogenetic analysis was performed using MEGA version 6 software (Tamura et al., 2013) based on the ClustalW multiple alignment. A phylogenetic tree was generated using the neighbor-joining method.

RESULTS

IDENTIFICATION OF THREE GENES ENCODING γ-GLUTAMYL TRANSPEPTIDASES IN GARLIC

We found two garlic EST clones, EPP005LLAA12S004013 and EPP005LLAA12S003688, that show high sequence homology with known GGTs in GarlicESTdb (Kim et al., 2009[2]). Utilizing the sequence information on these EST clones, we obtained two different full-length cDNA clones by 5′- and 3′-RACE and RT-PCR from the RNA of garlic cloves and designated them as *AsGGT1* (GenBank Accession No. LC008010) and *AsGGT2* (GenBank Accession No. LC008011). In addition, we amplified one garlic cDNA fragment using degenerate primers designed based on the conserved regions of known plant GGTs. A full-length cDNA clone was obtained by RACE and RT-PCR, and was designated as *AsGGT3* (GenBank Accession No. LC008012). The cDNAs of *AsGGT1*, *AsGGT2*, and *AsGGT3* coded for polypeptides of 627, 622, and

[2]http://garlicdb.kribb.re.kr

605 amino acids, respectively. The deduced amino acid sequences of *AsGGT1* and *AsGGT2* shared 69% identity, whereas the amino acid sequence identity of *AsGGT3* with *AsGGT1* and *AsGGT2* was 46 and 43%, respectively. The amino acid sequence of *AsGGT3* showed 99% sequence identity with that of a partial sequence of garlic *AsGGT* (Cho et al., 2012) in their 158 aa overlapped region and showed 92% sequence identity with that of a partial sequence of onion *AcGGT* (Shaw et al., 2005) in their 543 aa overlapped region. *Arabidopsis thaliana* has three functional GGTs, AtGGT1, AtGGT2, and AtGGT4. Among them, AtGGT4 is known to have a long N-terminal sequence that determines vacuolar localization, compared to AtGGT1 and AtGGT2 that localize in the extracellular space (Grzam et al., 2007; Ohkama-Ohtsu et al., 2007b). As in *Arabidopsis* AtGGT4, the deduced amino acid sequences of *AsGGT1*, *AsGGT2*, and *AsGGT3* had longer N-terminal sequences than those of *Arabidopsis* AtGGT1 and AtGGT2, suggesting the presence of the N-terminal signal sequences for targeting to cellular organelles in AsGGT1, AsGGT2, and AsGGT3. N-terminal regions of the deduced amino acid sequences of *AsGGT1*, *AsGGT2*, and *AsGGT3* were not highly similar to each other or to that of AtGGT4, despite the high sequence similarities among the rest of their regions.

A phylogenic tree was generated by the neighbor-joining method, using the amino acid sequences of known GGTs from plants, yeast, bacteria, and humans (**Figure 2**). All plant GGTs were classified into the same branch, which was further divided into two distinct subgroups. AsGGT1 and AsGGT2 belonged to the subgroup containing *Arabidopsis* AtGGT4 that functions in the degradation of glutathione *S*-conjugates in the vacuole (Grzam et al., 2007; Ohkama-Ohtsu et al., 2007b), whereas AsGGT3 belonged to the subgroup containing *Arabidopsis* AtGGT1 and AtGGT2 that function in the breakdown of extracellular glutathione (Martin et al., 2007; Ohkama-Ohtsu et al., 2007a) together with onion AcGGT (Shaw et al., 2005).

Generally, in bacteria, yeast, plants, and mammals, GGT is a heteromeric protein consisting of large and small subunits, both of which are generated from a common inactive precursor polypeptide by autoprocessing (Penninckx and Jaspers, 1985; Storozhenko et al., 2002; Suzuki and Kumagai, 2002; Ikeda and Taniguchi, 2005; Shaw et al., 2005; Nakano et al., 2006; Boanca et al., 2007). Some plants, such as tomato, onion, and radish, are suggested to have GGT proteins consisting of a single polypeptide, although their sequence information remains unknown (Lancaster and Shaw, 1994; Martin and Slovin, 2000; Nakano et al., 2006). The deduced amino acid sequences of *AsGGT1*, *AsGGT2*, and *AsGGT3* possessed the conserved threonine residue required for autocatalytic processing and the amino acid residues necessary for GGT activity, which were previously identified by biochemical and structural analyses of GGTs from humans and *E. coli* (Ikeda et al., 1993, 1995a,b; Okada et al., 2006).

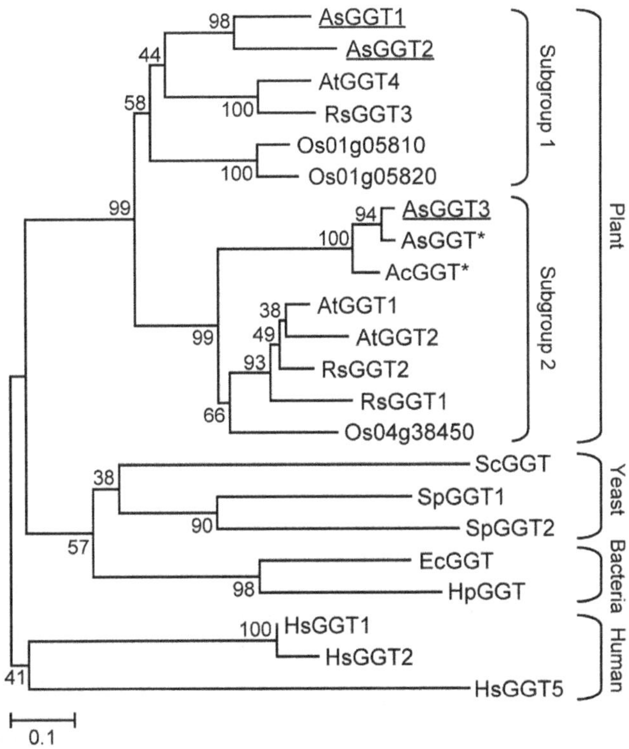

FIGURE 2 | Phylogenetic tree for the γ-glutamyl transpeptidases. An unrooted tree was constructed using MEGA version 6 software based on the ClustalW multiple alignment. Bootstrap values (1000 replicates) are shown next to the branches. Plant GGTs are classified into two subgroups. Garlic AsGGT1, AsGGT2, and AsGGT3 analyzed in this study are underlined. Asterisks indicate partial amino acid sequences. Abbreviations for species are: Ac, *Allium cepa*; As, *Allium sativum*; At, *Arabidopsis thaliana*; Ec, *Escherichia coli*; Hp, *Helicobacter pylori*; Hs, *Homo sapiens*; Os, *Oryza sativa*; Rs, *Raphanus sativus*; Sc, *Saccharomyces cerevisiae*; Sp, *Schizosaccharomyces pombe*. The GenBank accession numbers for the sequences are shown in parentheses: AcGGT (AAL61611); AsGGT1 (LC008010); AsGGT2 (LC008011); AsGGT3 (LC008012); AtGGT1 (AEE87097); AtGGT2 (AEE87099); AtGGT4 (AEE85602); EcGGT (AAA23869); HpGGT (AAD08162); HsGGT1 (AAH25927); HsGGT2 (XP_006724458); HsGGT5 (AAH73999); Os01g05810 (BAD61112); Os01g05820 (BAD61113); Os04g38450 (CAD40892); RsGGT1 (BAC45233); RsGGT2 (BAC56855); RsGGT3 (BAD22536); ScGGT (DAA09609); SpGGT1 (AAN01227); SpGGT2 (AAQ57121).

IN VITRO CHARACTERIZATION OF RECOMBINANT AsGGT1, AsGGT2, AND AsGGT3

Recombinant proteins of AsGGT1, AsGGT2, and AsGGT3 were independently expressed in budding yeast, and the crude protein extracts were used for the *in vitro* enzymatic activity assays. To confirm whether recombinant proteins of AsGGT1, AsGGT2, and AsGGT3 were expressed as mature GGT enzymes in yeast cells, we first examined deglutamylation activities of these recombinant proteins by using the standard procedure that utilizes γ-glutamyl-*p*-nitroanilide, a common synthetic γ-glutamyl donor substrate for known GGTs (Orlowski and Meister, 1963). For a γ-glutamyl acceptor substrate, we used dipeptide glycylglycine. Crude protein extracts from control yeast carrying the empty vector converted γ-glutamyl-*p*-nitroanilide to *p*-nitroaniline (**Table 1**), showing that yeast endogenous GGT could utilize γ-glutamyl-*p*-nitroanilide as a γ-glutamyl donor substrate, as reported previously (Payne and Payne, 1984). The amounts of *p*-nitroaniline released from γ-glutamyl-*p*-nitroanilide in assays using crude protein extracts from yeast expressing AsGGT1, AsGGT2, and AsGGT3, respectively, were significantly higher than that in assays using crude protein extracts from control yeast (**Table 1**), indicating that the

Substrate	GGT activity (pmol μg^{-1} protein hr^{-1})			
	Empty vector	AsGGT1	AsGGT2	AsGGT3
γ-glutamyl-*p*-nitroanilide	52.0 ± 3.6	135.9 ± 10.7	103.6 ± 2.2	130.4 ± 6.0
γ-glutamyl-*S*-allyl-L-cysteine	ND	30.5 ± 0.2	17.4 ± 1.2	35.6 ± 2.5
γ-glutamyl-*S*-allyl-L-cysteine sulfoxide	ND	1.6 ± 0.0	ND	1.0 ± 0.0

Each γ-glutamyl donor substrate was used at a concentration of 1 mM. Activities were measured in the presence of glycylglycine as the γ-glutamyl acceptor substrate. Data represent the mean ± SD (n = 4 for γ-glutamyl-p-nitroanilide, n = 3 for γ-glutamyl-S-allyl-L-cysteine, and n = 3 for γ-glutamyl-S-allyl-L-cysteine sulfoxide). ND means no detectable activity.

recombinant proteins of AsGGT1, AsGGT2, and AsGGT3 were successfully expressed and folded to form mature functional GGT proteins that can utilize γ-glutamyl-*p*-nitroanilide as a γ-glutamyl donor substrate in yeast cells.

Next, we examined the enzymatic activities of AsGGT1, AsGGT2, and AsGGT3 toward γ-glutamyl-*S*-allyl-L-cysteine and γ-glutamyl-*S*-allyl-L-cysteine sulfoxide, which are two possible biosynthetic intermediates in alliin biosynthesis, as potential γ-glutamyl donor substrates (**Figures 3** and **4**). Activities were measured in the presence of glycylglycine as a γ-glutamyl acceptor substrate. When γ-glutamyl-*S*-allyl-L-cysteine was used as a γ-glutamyl donor substrate, *S*-allyl-L-cysteine was not formed at a detectable level in assays using crude protein extracts from control yeast (**Figure 3**; **Table 1**), indicating that yeast endogenous GGT could not use γ-glutamyl-*S*-allyl-L-cysteine as a γ-glutamyl donor substrate. By contrast, considerable amounts of *S*-allyl-L-cysteine were detected in assays using crude protein extracts prepared from yeast cells expressing AsGGT1, AsGGT2, and AsGGT3 (**Figure 3**; **Table 1**), demonstrating that the recombinant proteins of AsGGT1, AsGGT2, and AsGGT3 can convert γ-glutamyl-*S*-allyl-L-cysteine to *S*-allyl-L-cysteine. The deglutamylation activities of AsGGT1, AsGGT2, and AsGGT3 toward γ-glutamyl-*S*-allyl-L-cysteine were decreased in the absence of glycylglycine (**Figure 5**), indicating that these garlic GGTs can catalyze transpeptidation more effectively than hydrolysis. In the presence of glycylglycine, the activities of AsGGT2 were higher with lower pH (**Figure 5**). Kinetic characterization of the recombinant AsGGT1, AsGGT2, and AsGGT3 exhibited typical Michaelis-Menten behavior, and the apparent K_m values of AsGGT1, AsGGT2, and AsGGT3 for γ-glutamyl-*S*-allyl-L-cysteine were 86 μM, 1.1 mM, and 9.4 mM, respectively, in the presence of glycylglycine (**Figure 6**).

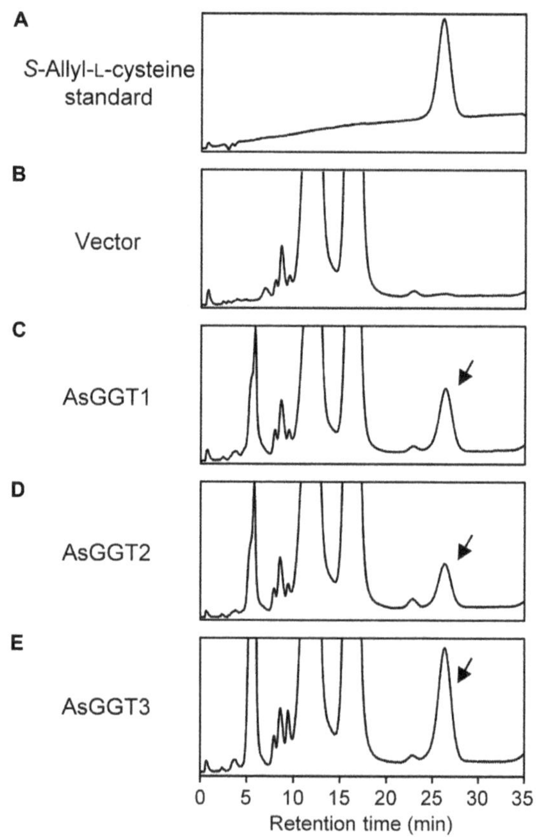

FIGURE 3 | Deglutamylation activities of recombinant AsGGT1, AsGGT2, and AsGGT3 toward γ-glutamyl-*S*-allyl-L-cysteine. High-performance liquid chromatography (HPLC) elution profiles of the *S*-allyl-L-cysteine standard **(A)** and the reaction products from γ-glutamyl-*S*-allyl-L-cysteine by the crude protein extracts of yeast carrying empty vector **(B)** or yeast expressing AsGGT1 **(C)**, AsGGT2 **(D)**, and AsGGT3 **(E)** are shown. Arrows indicate peaks of *S*-allyl-L-cysteine in the reaction products.

When γ-glutamyl-*S*-allyl-L-cysteine sulfoxide was used as a γ-glutamyl donor substrate, only small amounts of alliin were detected in assays using crude protein extracts from yeast expressing AsGGT1 or AsGGT3 (**Figure 4**; **Table 1**), suggesting the recombinant AsGGT1 and AsGGT3 exhibit a weak activity to deglutamylate γ-glutamyl-*S*-allyl-L-cysteine sulfoxide. However, under the conditions we examined, the activities of AsGGT1 and AsGGT3 to deglutamylate γ-glutamyl-*S*-allyl-L-cysteine sulfoxide were too weak to perform further enzymatic characterization. AsGGT2 exhibited no detectable deglutamylation activity toward γ-glutamyl-*S*-allyl-L-cysteine sulfoxide (**Figure 4**; **Table 1**). These results indicate that AsGGT1, AsGGT2, and AsGGT3 are the functional GGT proteins with a preference for γ-glutamyl-*S*-allyl-L-cysteine as a γ-glutamyl donor substrate over γ-glutamyl-*S*-allyl-L-cysteine sulfoxide.

SUBCELLULAR LOCALIZATION OF AsGGT1, AsGGT2, AND AsGGT3

The probable subcellular localization of AsGGT1, AsGGT2, and AsGGT3 was computationally analyzed using the program TargetP

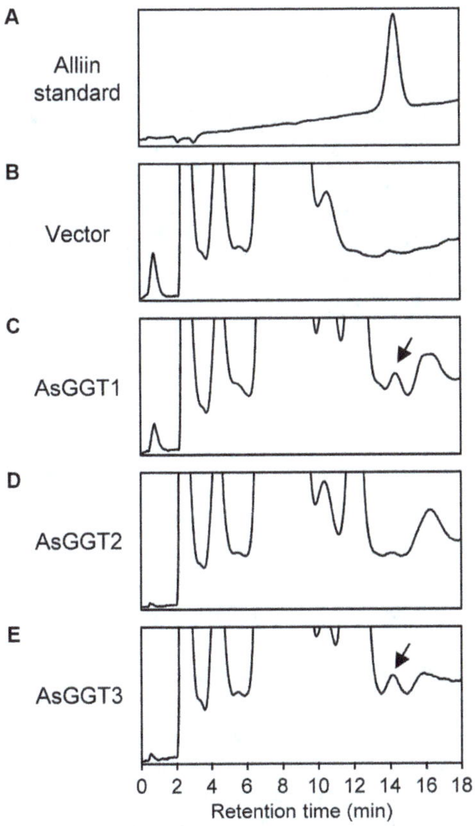

FIGURE 4 | Deglutamylation activities of recombinant AsGGT1, AsGGT2, and AsGGT3 toward γ-glutamyl-*S*-allyl-L-cysteine sulfoxide. HPLC elution profiles of the alliin standard **(A)** and the reaction products from γ-glutamyl-*S*-allyl-L-cysteine sulfoxide by the crude protein extracts of yeast carrying empty vector **(B)** or yeast expressing AsGGT1 **(C)**, AsGGT2 **(D)**, and AsGGT3 **(E)** are shown. Arrows indicate peaks of alliin in the reaction products.

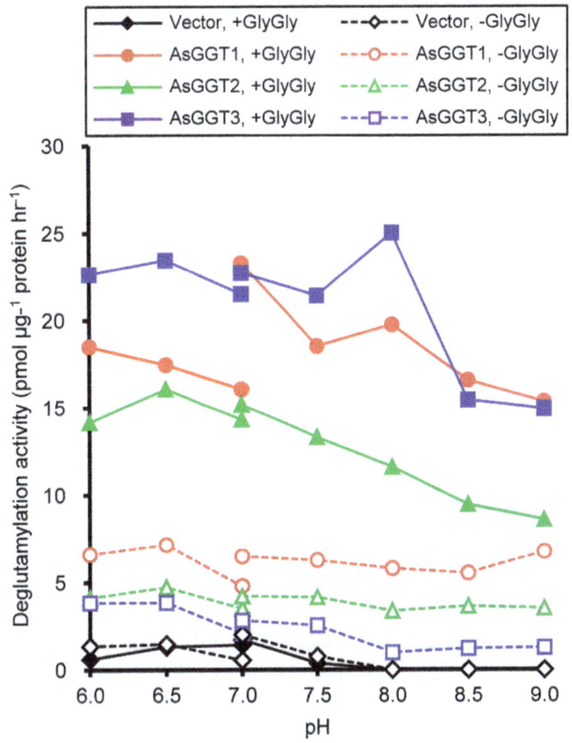

FIGURE 5 | pH dependence of transpeptidation and hydrolysis by AsGGT1, AsGGT2, and AsGGT3. Production of *S*-allyl-L-cysteine from γ-glutamyl-*S*-allyl-L-cysteine was analyzed in the presence or absence of glycylglycine as the γ-glutamyl acceptor substrate. 2-(*N*-morpholino)ethanesulfonic acid buffer was used to cover the pH range from 6.0 to 7.0, and Tris-HCl buffer was used to cover the pH range from 7.0 to 9.0.

v1.1[3] and WoLF PSORT[4]. Both programs predicted that these three garlic GGT proteins lack signal peptides for secretion or localization to cellular organelles.

To determine subcellular localization of AsGGT1, AsGGT2, and AsGGT3, the green fluorescent protein (GFP)-fusion constructs of AsGGT1, AsGGT2, and AsGGT3 were transiently expressed in onion epidermal cells under the control of the cauliflower mosaic virus 35S RNA promoter by using the particle bombardment method. Since N-terminal sequences of AsGGT1, AsGGT2, and AsGGT3 are long and may contain the signal sequence for secretion or targeting to cellular organelles, three types of fusion proteins, GFP C-terminally fused to the N-terminal 100-amino acid residues of GGT ($AsGGT1_{N100}$-GFP, $AsGGT2_{N100}$-GFP, and $AsGGT3_{N100}$-GFP), GFP C-terminally fused to the N-terminal 300-amino acid residues of GGT ($AsGGT1_{N300}$-GFP, $AsGGT2_{N300}$-GFP, and $AsGGT3_{N300}$-GFP), and GFP C-terminally fused to the full-length GGT protein ($AsGGT1_{Full}$-GFP, $AsGGT2_{Full}$-GFP, and $AsGGT3_{Full}$-GFP), were analyzed. As a control of cytosolic localization, DsRed protein was simultaneously expressed with each GFP-fusion protein. The green fluorescent signals derived from $AsGGT1_{N100}$-GFP and $AsGGT3_{N100}$-GFP overlapped with the red fluorescence of DsRed (**Figures 7A,C**), suggesting that N-terminal regions of AsGGT1 and AsGGT3 have no signal sequence for secretion or targeting to cellular organelles. Similarly, the green fluorescent signals from $AsGGT1_{N300}$-GFP, $AsGGT1_{Full}$-GFP, $AsGGT3_{N300}$-GFP, and $AsGGT3_{Full}$-GFP were observed in the cytosol, although the fluorescent signal intensities were much weaker (data not shown). By contrast, the green fluorescence from $AsGGT2_{N100}$-GFP was observed predominantly in the vacuole (**Figure 7B**), indicating that AsGGT2 has a signal sequence for targeting to the vacuole within its N-terminal 100 amino acid residues. When $AsGGT2_{N300}$-GFP and $AsGGT2_{Full}$-GFP were expressed, weak fluorescent signals were detected both in the vacuole and cytosol (data not shown).

DISCUSSION

In this study, we identified three novel genes encoding GGTs, *AsGGT1*, *AsGGT2*, and *AsGGT3*, from garlic by utilizing their partial sequence information found in a publicly available EST database or by utilizing sequence information of conserved regions of known plant GGTs. The deduced amino acid sequences of

[3]http://www.cbs.dtu.dk/services/TargetP/
[4]http://wolfpsort.org/

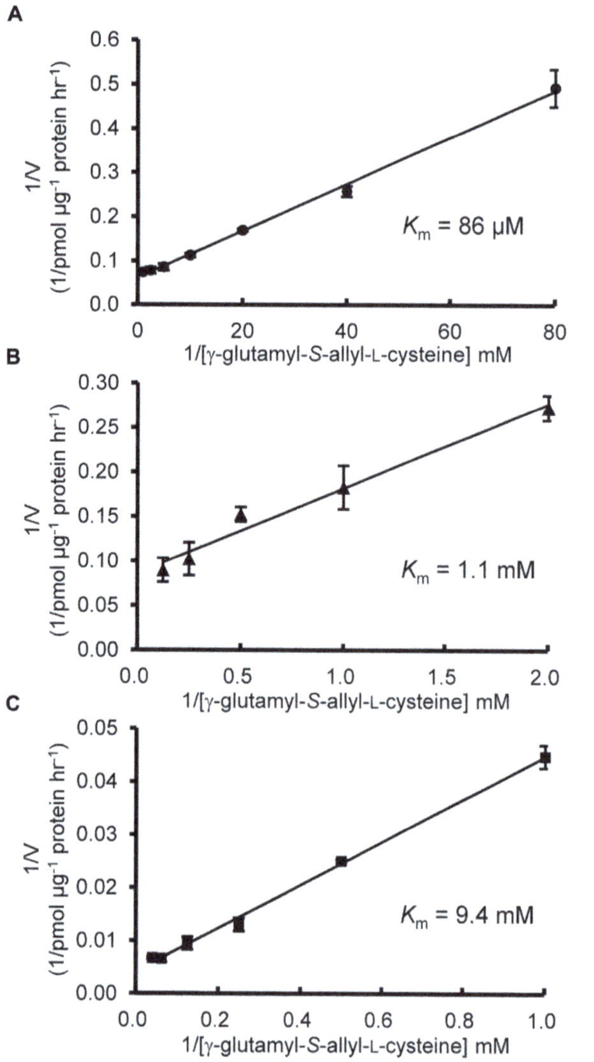

FIGURE 6 | Kinetic characterization of AsGGT1, AsGGT2, and AsGGT3. Lineweaver-Burk plots between deglutamylation activity and γ-glutamyl-*S*-allyl-L-cysteine concentrations. Assays were carried out with γ-glutamyl-*S*-allyl-L-cysteine concentrations ranging from 12.5 to 1000 µM for AsGGT1 **(A)**, 0.5–8 mM for AsGGT2 **(B)**, and 1–25 mM for AsGGT3 **(C)**. Each data point represents the mean ± SD (n = 3).

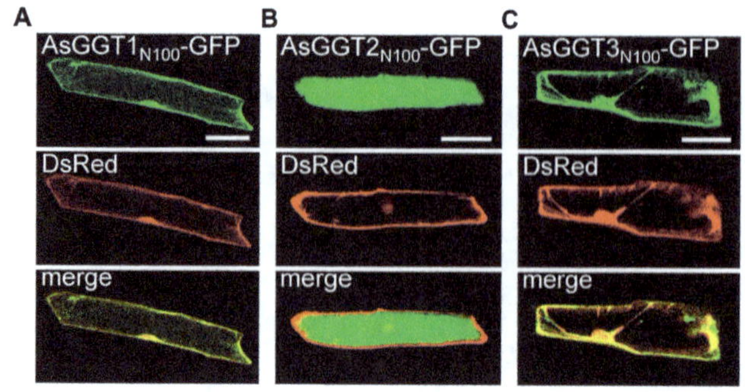

FIGURE 7 | Transient expression of GFP fusion proteins of AsGGT1, AsGGT2, and AsGGT3 in onion epidermal cells. GFP fusion proteins, AsGGT1$_{N100}$-GFP **(A)**, AsGGT2$_{N100}$-GFP **(B)**, or AsGGT3$_{N100}$-GFP **(C)**, were transiently co-expressed with DsRed in onion epidermal cells. DsRed was used as a control for cytosolic localization. Bars = 100 µm.

AsGGT1, *AsGGT2*, and *AsGGT3* contained threonine residues required for autoprocessing and the residues required for GGT activity (Ikeda et al., 1993, 1995a,b; Okada et al., 2006), suggesting that *AsGGT1*, *AsGGT2*, and *AsGGT3* all encode functional GGT proteins. The cDNA sequence of *AsGGT3* and the previously identified partial cDNA sequence of garlic *AsGGT* (Cho et al., 2012) were almost identical in their overlapping region, suggesting that these two cDNAs were derived from a single gene encoding GGT. The high sequence homology between garlic *AsGGT3* and onion *AcGGT* (Shaw et al., 2005) may indicate that they are orthologs. By contrast, *AsGGT1* and *AsGGT2* showed relatively low sequence similarity with garlic *AsGGT* (Cho et al., 2012) and onion *AcGGT* (Shaw et al., 2005). Phylogenetic analysis revealed that AsGGT3 belongs to a subgroup different from that containing AsGGT1 and AsGGT2 (**Figure 2**), suggesting that the biochemical characteristics of AsGGT3 might be somewhat different from those of AsGGT1 and AsGGT2.

The deglutamylation activities of AsGGT1, AsGGT2, and AsGGT3 toward alliin biosynthetic intermediates were demonstrated by *in vitro* biochemical assays, using recombinant proteins expressed in yeast. In the hypothetical alliin biosynthetic pathway, two different routes from the intermediate γ-glutamyl-*S*-allyl-L-cysteine to alliin are possible, according to differences in the order of deglutamylation and *S*-oxygenation reactions (**Figure 1**): a potential route via deglutamylation of γ-glutamyl-*S*-allyl-L-cysteine to yield *S*-allyl-L-cysteine that is further *S*-oxygenated to alliin, and an alternative route via *S*-oxygenation of γ-glutamyl-*S*-allyl-L-cysteine to form γ-glutamyl-*S*-allyl-L-cysteine sulfoxide that is further deglutamylated to yield alliin. Our results demonstrated that AsGGT1, AsGGT2, and AsGGT3 actively deglutamylate γ-glutamyl-*S*-allyl-L-cysteine, whereas these GGTs have almost no deglutamylation activity toward γ-glutamyl-*S*-allyl-L-cysteine sulfoxide (**Figures 3** and **4**; **Table 1**). It can be speculated that the intermediate γ-glutamyl-*S*-allyl-L-cysteine is mainly deglutamylated prior to being *S*-oxygenated in alliin biosynthesis in garlic. This hypothesis is also supported by our recent study on flavin-dependent *S*-oxygenase, which preferably utilizes *S*-allyl-L-cysteine, rather than γ-glutamyl-*S*-allyl-L-cysteine, as the substrate (unpublished results). The presence of dipeptide glycylglycine as a γ-glutamyl acceptor increased the deglutamylation activities of AsGGT1, AsGGT2, and AsGGT3 (**Figure 5**), showing that these GGTs catalyze transpeptiation more efficiently than hydrolysis, as in GGTs from *Arabidopsis* and onion (Storozhenko et al., 2002; Shaw et al., 2005).

There are two characteristic differences among the AsGGT1, AsGGT2, and AsGGT3 proteins. One is in their affinity for γ-glutamyl-*S*-allyl-L-cysteine (**Figure 6**). The apparent K_m values of AsGGT1 and AsGGT2 for γ-glutamyl-*S*-allyl-L-cysteine determined in this study (86 µM and 1.1 mM, respectively) are lower than or comparable to those of partially purified onion GGT for γ-glutamyl-*S*-propenyl-L-cysteine (K_m = 1.68 mM) and for γ-glutamyl-*S*-methyl-L-cysteine ($K_m = 0.55$ mM; Lancaster and Shaw, 1994). By contrast, AsGGT3 exhibited a relatively low affinity for γ-glutamyl-*S*-allyl-L-cysteine ($K_m = 9.4$ mM). This is in agreement with the results of a previous study that onion AcGGT, which shares high sequence homology with AsGGT3, could not utilize γ-glutamyl-*trans*-*S*-1-propenyl-L-cysteine sulfoxide (a major γ-glutamylated biosynthetic intermediate in onion) as a good γ-glutamyl donor substrate (Shaw et al., 2005). In garlic, the content of alliin is increased dramatically before and during the maturation of bulbs (Ueda et al., 1991). Alliin is largely found in leaves before the formation of bulbs and in the initial stage of bulb maturation, whereas it is found predominantly in bulbs in the later stage of bulb formation (Ueda et al., 1991; Koch and Lawson, 1996). It is suggested that leaves of garlic actively biosynthesize alliin before the formation of bulbs and in the initial stage of bulb maturation. However, in leaves at the same stages, γ-glutamyl-*S*-allyl-L-cysteine is present in trace levels (0.01 mg g^{-1} fresh weight; Matsuura et al., 1996). The concentration of γ-glutamyl-*S*-allyl-L-cysteine in cells of these tissues is calculated to be approximately 38 μM, when the content of water in tissues is estimated to be 90%. By contrast, the content of γ-glutamyl-*S*-allyl-L-cysteine in mature bulbs is ~5 mg g^{-1} fresh weight (Matsuura et al., 1996; Ichikawa et al., 2006a,b), and the concentration of γ-glutamyl-*S*-allyl-L-cysteine in cells of bulbs is calculated to be 26 mM, when the content of water in tissues is estimated to be 65%. The highly accumulated γ-glutamyl-*S*-allyl-L-cysteine in bulbs is stored during dormancy of bulbs at −3°C, while it is rapidly converted to alliin when bulb dormancy is broken at 4°C (Ichikawa et al., 2006a). Based on these observations, we hypothesize that AsGGT1 and AsGGT2, which exhibit high-affinity for γ-glutamyl-*S*-allyl-L-cysteine, would contribute to the biosynthesis of alliin in leaves during the formation and maturation of bulbs, while AsGGT3 may contribute to alliin biosynthesis in bulbs during dormancy-breaking or, alternatively, the main *in vivo* function of AsGGT3 may not be the deglutamylation of γ-glutamyl-*S*-allyl-L-cysteine. The other major difference observed among AsGGT1, AsGGT2, and AsGGT3 is in their subcellular localization. Transient expression analyses of GFP-fused AsGGT2 proteins in onion cells suggested that AsGGT2 is predominantly localized in the vacuole *in vivo*. In contrast to the almost exclusive localization of $AsGGT2_{N100}$-GFP in the vacuole (**Figure 7B**), the green fluorescence signals from $AsGGT2_{N300}$-GFP and $AsGGT2_{Full}$-GFP were detected both in the vacuole and cytosol (data not shown), suggesting that a part of $AsGGT2_{N300}$-GFP and $AsGGT2_{Full}$-GFP polypeptides was not properly processed and/or assembled and thus was not sorted to the vacuole in the heterologous expression system we used. Alternatively, AsGGT2 may localize both in the vacuole and cytosol in garlic cells. The signal sequence for targeting to the vacuole of AsGGT2 is located within its N-terminal 100 amino acids, as in *Arabidopsis* AtGGT4 (Grzam et al., 2007; Ohkama-Ohtsu et al., 2007b). To date, several sequence motifs for vacuolar targeting have been identified from plants (Xiang et al., 2013). However, we could not identify the potential motif for vacuolar targeting that is conserved between AsGGT2 and AtGGT4. Future studies are needed to determine the sequence motif and the mechanism for their targeting to the vacuole. Consistent with vacuolar localization of AsGGT2, the deglutamylation activities of AsGGT2 toward γ-glutamyl-*S*-allyl-L-cysteine were increased under weakly acidic conditions (**Figure 5**). AsGGT2 is suggested to contribute alliin biosynthesis mainly in the vacuole. In addition, AsGGT2 may function in the breakdown of glutathione *S*-conjugates in the vacuole in a similar manner as *Arabidopsis* AtGGT4 (Grzam et al., 2007; Ohkama-Ohtsu et al., 2007b). In contrast, GFP-fusion proteins of AsGGT1 and AsGGT3 were retained in the cytosol (**Figures 7A,C**), suggesting that AsGGT1 and AsGGT3 have no apparent signal sequence for targeting to the cellular organelles in their N-terminal peptides. To the best of our knowledge, there have been no reports of GGT proteins localizing in the cytosol. Future investigations will reveal whether AsGGT1 and AsGGT3 are new types of GGT proteins that localize and function in the cytosol *in vivo* or not. To date, the subcellular distribution of alliin biosynthetic intermediates and enzymes in garlic remains largely unclear, although the previous cell fractionation experiment suggested that γ-glutamyl peptides and *S*-alk(en)yl-L-cysteine sulfoxides are mainly located in the cytosol in onion (Lancaster et al., 1989). Our results suggest that AsGGT1, AsGGT2, and AsGGT3 contribute differently to alliin biosynthesis, according to differences in their kinetic properties and localization patterns. Recently, five γ-glutamyl peptidases (GGPs), which have similar catalytic functions but no sequence homology with GGTs, were identified from *Arabidopsis*. Among these, GGP1 and GGP3 were shown to be cytosolic proteins that play major roles in the removal of γ-glutamyl groups from glutathione *S*-conjugates in the biosynthesis of glucosinolates and camalexins (Geu-Flores et al., 2009, 2011). Although identification of GGPs from garlic has not been reported to date, it is likely that GGPs exist and function in the deglutamylation reaction in alliin biosynthesis, perhaps together with GGTs.

In the present study, we succeeded in identifying three garlic GGTs, AsGGT1, AsGGT2, and AsGGT3, that can deglutamylate an alliin biosynthetic intermediate, γ-glutamyl-*S*-allyl-L-cysteine. Future investigations of the *in vivo* functions of AsGGT1, AsGGT2, and AsGGT3 will provide a better understanding of the molecular mechanisms underlying the biosynthesis of alliin in garlic, which can be applied to future metabolic engineering of plants.

ACKNOWLEDGMENTS

We are grateful to Dr. Toshiaki Mitsui (Niigata University) for providing the pDsRed vector. This work was supported in part by JSPS KAKENHI Grant Number 23770036 and 26460118 (to Naoko Yoshimoto), and Grant Number 22108008 (to Kazuki Saito) of Japan.

REFERENCES

Anderson, M. E., Allison, R. D., and Meister, A. (1982). Interconversion of leukotrienes catalyzed by purified γ-glutamyl transpeptidase: concomitant formation of leukotriene D4 and γ-glutamyl amino acids. *Proc. Natl. Acad. Sci. U.S.A.* 79, 1088–1091. doi: 10.1073/pnas.79.4.1088

Boanca, G., Sand, A., Okada, T., Suzuki, H., Kumagai, H., Fukuyama, K., et al. (2007). Autoprocessing of *Helicobacter pylori* γ-glutamyltranspeptidase leads to the formation of a threonine-threonine catalytic dyad. *J. Biol. Chem.* 282, 534–541. doi: 10.1074/jbc.M607694200

Bradford, M. M. (1976). A rapid and sensitive method for the quantitation of microgram quantities of protein utilizing the principle of protein-dye binding. *Anal. Biochem.* 72, 248–254. doi: 10.1016/0003-2697(76)90527-3

Chiu, W. L., Niwa, Y., Zeng, W., Hirano, T., Kobayashi, H., and Sheen, J. (1996). Engineered GFP as a vital reporter in plants. *Curr. Biol.* 6, 325–330. doi: 10.1016/S0960-9822(02)00483-9

Cho, J., Park, M., Choi, D., and Lee, S. K. (2012). Cloning and expression of γ-glutamyl transpeptidase and its relationship to greening in crushed garlic (*Allium sativum*) cloves. *J. Sci. Food Agric.* 92, 253–257. doi: 10.1002/jsfa.4610

Dominici, S., Paolicchi, A., Corti, A., Maellaro, E., and Pompella, A. (2005). Prooxidant reactions promoted by soluble and cell-bound γ-glutamyltransferase activity. *Methods Enzymol.* 401, 484–501. doi: 10.1016/S0076-6879(05)01029-3

Gamborg, O. L., Miller, R. A., and Ojima, K. (1968). Nutrient requirements of suspension cultures of soybean root cells. *Exp. Cell Res.* 50, 151–158. doi: 10.1016/0014-4827(68)90403-5

Geu-Flores, F., Møldrup, M. E., Böttcher, C., Olsen, C. E., Scheel, D., and Halkier, B. A. (2011). Cytosolic γ-glutamyl peptidases process glutathione conjugates in the biosynthesis of glucosinolates and camalexin in *Arabidopsis. Plant Cell* 23, 2456–2469. doi: 10.1105/tpc.111.083998

Geu-Flores, F., Nielsen, M. T., Nafisi, M., Møldrup, M. E., Olsen, C. E., Motawia, M. S., et al. (2009). Glucosinolate engineering identifies a γ-glutamyl peptidase. *Nat. Chem. Biol.* 5, 575–577. doi: 10.1038/nchembio.185

Gietz, D., Jean, A. S., Woods, D. A., and Schiestl, R. H. (1992). Improved method for high efficiency transformation of intact yeast cells. *Nucleic Acids Res.* 20, 1425. doi: 10.1093/nar/20.6.1425

Grzam, A., Martin, M. N., Hell, R., and Meyer, A. J. (2007). γ-Glutamyl transpeptidase GGT4 initiates vacuolar degradation of glutathione *S*-conjugates in *Arabidopsis. FEBS Lett.* 581, 3131–3138. doi: 10.1016/j.febslet.2007.05.071

Ichikawa, M., Ide, N., and Ono, K. (2006a). Changes in organosulfur compounds in garlic cloves during storage. *J. Agric. Food Chem.* 54, 4849–4854. doi: 10.1021/jf060083

Ichikawa, M., Ide, N., Yoshida, J., Yamaguchi, H., and Ono, K. (2006b). Determination of seven organosulfur compounds in garlic by high-performance liquid chromatography. *J. Agric. Food Chem.* 54, 1535–1540. doi: 10.1021/jf051742k

Iciek, M., Kwiecień, I., and Włodek, L. (2009). Biological properties of garlic and garlic-derived organosulfur compounds. *Environ. Mol. Mutagen.* 50, 247–265. doi: 10.1002/em.20474

Ikeda, Y., Fujii, J., Anderson, M. E., Taniguchi, N., and Meister, A. (1995a). Involvement of Ser-451 and Ser-452 in the catalysis of human γ-glutamyl transpeptidase. *J. Biol. Chem.* 270, 22223–22228. doi: 10.1074/jbc.270.38.22223

Ikeda, Y., Fujii, J., Taniguchi, N., and Meister, A. (1995b). Human γ-glutamyl transpeptidase mutants involving conserved aspartate residues and the unique cysteine residue of the light subunit. *J. Biol. Chem.* 270, 12471–12475. doi: 10.1074/jbc.270.21.12471

Ikeda, Y., Fujii, J., and Taniguchi, N. (1993). Significance of Arg-107 and Glu-108 in the catalytic mechanism of human γ-glutamyl transpeptidase. *J. Biol. Chem.* 268, 3980–3985.

Ikeda, Y., and Taniguchi, N. (2005). Gene expression of γ-glutamyltranspeptidase. *Methods Enzymol.* 401, 408-425. doi: 10.1016/S0076-6879(05)01025-6

Jones, M. G., Hughes, J., Tregova, A., Milne, J., Tomsett, A. B., and Collin, H. A. (2004). Biosynthesis of the flavor precursors of onion and garlic. *J. Exp. Bot.* 55, 1903–1918. doi: 10.1093/jxb/erh138

Kim, D.-W., Jung, T.-S., Nam, S.-H., Kwon, H.-R., Kim, A., Chae, S.-H., et al. (2009). GarlicESTdb: an online database and mining tool for garlic EST sequences. *BMC Plant Biol.* 9:61. doi: 10.1186/1471-2229-9-61

Kitajima, A., Asatsuma, S., Okada, H., Hamada, Y., Kaneko, K., Nanjo, Y., et al. (2009). The rice α-amylase glycoprotein is targeted from the Golgi apparatus through the secretory pathway to the plastids. *Plant Cell* 21, 2844–2858. doi: 10.1105/tpc.109.068288

Koch, H. P., and Lawson, L. D. (1996). *Garlic: The Science and Therapeutic Application of Allium sativum L. and Related Species*, 2nd Edn. Baltimore, MD: Williams & Wilkins.

Kubec, R., and Dadáková, E. (2008). Quantitative determination of *S*-alk(en)ylcysteine-*S*-oxides by micellar electrokinetic capillary chromatography. *J. Chromatogr. A* 1212, 154–157. doi: 10.1016/j.chroma.2008.10.024

Kubec, R., Svobodová, M., and Velíšek, J. (1999). Gas chromatographic determination of *S*-alk(en)ylcysteine sulfoxides. *J. Chromatogr. A* 862, 85–94. doi: 10.1016/S0021-9673(99)00902-4

Lancaster, J. E., Reynolds, P. H. S., Shaw, M. L., Dommisse, E. M., and Munro, J. (1989). Intra-cellular localization of the biosynthetic pathway to flavor precursors in onion. *Phytochemistry* 28, 461–464. doi: 10.1016/0031-9422(89)80032-9

Lancaster, J. E., and Shaw, M. L. (1989). γ-Glutamyl peptides in the biosynthesis of *S*-alk(en)yl-L-cysteine sulphoxides (flavor precursors) in *Allium. Phytochemistry* 28, 455–460. doi: 10.1016/0031-9422(89)80031-7

Lancaster, J. E., and Shaw, M. L. (1991). Metabolism of γ-glutamyl peptides during development, storage and sprouting of onion bulbs. *Phytochemistry* 30, 2857–2859. doi: 10.1016/S0031-9422(00)98212-8

Lancaster, J. E., and Shaw, M. L. (1994). Characterization of purified γ-glutamyl transpeptidase in onions: evidence for *in vivo* role as a peptidase. *Phytochemistry* 36, 1351–1358. doi: 10.1016/S0031-9422(00)89723-X

Martin, M. N., Saladores, P. H., Lambert, E., Hudson, A. O., and Leustek, T. (2007). Localization of members of the γ-glutamyl transpeptidase family identified sites of glutathione and glutathione *S*-conjugate hydrolysis. *Plant Physiol.* 144, 1715–1732. doi: 10.1104/pp.106.094409

Martin, M. N., and Slovin, J. P. (2000). Purified γ-glutamyl transpeptidases from tomato exhibit high affinity for glutathione and glutathione *S*-conjugates. *Plant Physiol.* 122, 1417–1426. doi: 10.1104/pp.122.4.1417

Matsuura, H., Inagaki, M., Maeshige, K., Ide, N., Kajimura, Y., and Itakura, Y. (1996). Changes in contents of γ-glutamyl peptides and fructan during growth of *Allium sativum. Plant Med.* 62, 70–71. doi: 10.1055/s-2006-957805

Mehdi, K., and Penninckx, M. J. (1997). An important role for glutathione and γ-glutamyltranspeptidase in the supply of growth requirements during nitrogen starvation of the yeast *Saccharomyces cerevisiae. Microbiology* 143, 1885–1889. doi: 10.1099/00221287-143-6-1885

Nakano, Y., Okawa, S., Yamauchi, T., Koizumi, Y., and Sekiya, J. (2006). Purification and properties of soluble and bound γ-glutamyltransferases from radish cotyledon. *Biosci. Biotechnol. Biochem.* 70, 369–376. doi: 10.1271/bbb.70.369

Ohkama-Ohtsu, N., Radwan, S., Peterson, A., Zhao, P., Badr, A. F., Xiang, C., et al. (2007a). Characterization of the extracellular γ-glutamyl transpeptidases, GGT1 and GGT2, in *Arabidopsis. Plant J.* 49, 865–877. doi: 10.1111/j.1365-313X.2006.03004.x

Ohkama-Ohtsu, N., Zhao, P., Xiang, C., and Oliver, D. J. (2007b). Glutathione conjugates in the vacuole are degraded by γ-glutamyl transpeptidase GGT3 in *Arabidopsis. Plant J.* 49, 878–888. doi: 10.1111/j.1365-313X.2006.03005.x

Okada, T., Suzuki, H., Wada, K., Kumagai, H., and Fukuyama, K. (2006). Crystal structures of γ-glutamyltranspeptidase from *Escherichia coli*, a key enzyme in glutathione metabolism, and its reaction intermediate. *Proc. Natl. Acad. Sci. U.S.A.* 103, 6471–6476. doi: 10.1073/pnas.0511020103

Orlowski, M., and Meister, A. (1963). γ-Glutamyl-*p*-nitroanilide: a new convenient substrate for determination and study of L- and D-γ-glutamyltranspeptidase activities. *Biochim. Biophys. Acta* 73, 679–681. doi: 10.1016/0926-6569(63)90197-4

Payne, G. M., and Payne, J. W. (1984). γ-Glutamyltransferase is not involved in the bulk uptake of amino acids, peptides or γ-glutamyl-amino acids in yeast (*Saccharomyces cerevisiae*). *Biochem. J.* 218, 147–155.

Penninckx, M. J., and Jaspers, C. J. (1985). Molecular and kinetic properties of purified γ-glutamyl transpeptidase from yeast (*Saccharomyces cerevisiae*). *Phytochemistry* 24, 1913–1918. doi: 10.1016/S0031-9422(00)83091-5

Rose, P., Whiteman, M., Moore, P. K., and Zhu, Y. Z. (2005). Bioactive *S*-alk(en)yl cysteine sulfoxide metabolites in the genus *Allium*: the chemistry of potential therapeutic agents. *Nat. Prod. Rep.* 22, 351–368. doi: 10.1039/B417639C

Sambrook, J., Fritsch, E. F., and Maniatis, T. (1989). *Molecular Cloning: a Laboratory Manual*, 2nd Edn. New York: Cold Spring Harbor Laboratory Press.

Shaw, M. L., Pither-Joyce, M. D., and McCallum, J. A. (2005). Purification and cloning of a γ-glutamyl transpeptidase from onion (*Allium cepa*). *Phytochemistry* 66, 515–522. doi: 10.1016/j.phytochem.2005.01.017

Sherman, F. (1991). Getting started with yeast. *Methods Enzymol.* 194, 4–21. doi: 10.1016/0076-6879(91)94004-V

Storozhenko, S., Belles-Boix, E., Babiychuk, E., Hérouart, D., Davey, M. W., Slooten, L., et al. (2002). γ-Glutamyl transpeptidase in transgenic tobacco plants. Cellular localization, processing, and biochemical properties. *Plant Physiol.* 128, 1109–1119. doi: 10.1104/pp.010887

Su, T., Xu, J., Li, Y., Lei, L., Zhao, L., Yang, H., et al. (2011). Glutathione-indole-3-acetonitrile in required for camalexin biosynthesis in *Arabidopsis thaliana. Plant Cell* 23, 364–380. doi: 10.1105/tpc.110.079145

Suzuki, H., Hashimoto, W., and Kumagai, H. (1999). Glutathione metabolism in *Escherichia coli. J. Mol. Catal. B Enzymat.* 6, 175–184. doi: 10.1016/S1381-1177(98)00116-7

Suzuki, H., and Kumagai, H. (2002). Autocatalytic processing of γ-glutamyltranspeptidase. *J. Biol. Chem.* 277, 43536–43543. doi: 10.1074/jbc.M207680200

Suzuki, T., Sugii, M., and Kakimoto, T. (1962). Incorporation of L-valine-[^{14}C] into *S*-(2-carboxypropyl)glutathione and *S*-(2-carboxypropyl)cysteine in garlic. *Chem. Pharm. Bull.* 10, 328–331. doi: 10.1248/cpb.10.328

Tamura, K., Stecher, G., Peterson, D., Filipski, A., and Kumar, S. (2013). MEGA6: molecular evolutionary genetics analysis version 6.0. *Mol. Biol. Evol.*30, 2725–2729. doi: 10.1093/molbev/mst197

Tate, S. S., and Meister, A. (1981). γ-Glutamyl transpeptidase: catalytic, structural and functional aspects. *Mol. Cell. Biochem.* 39, 357–368. doi: 10.1007/BF00232585

Turnbull, A., Galpin, I. J., and Collin, H. A. (1980). Comparison of the onion plant (*Allium cepa*) and onion tissue culture. III. Feeding of ^{14}C labeled precursors of the flavor precursor compounds. *New Phytol.* 85, 483–487. doi: 10.1111/j.1469-8137.1980.tb00762.x

Ueda, Y., Kawajiri, H., Miyamura, N., and Miyajima, R. (1991). Content of some sulfur-containing components and free amino acids in various strains of garlic. *Nippon Shokuhin Kogyo Gakkaishi* 38, 429–434. doi: 10.3136/nskkk1962.38.429

Xiang, L., Etxeberria, E., and Van den Ende, W. (2013). Vacuolar protein sorting mechanisms in plants. *FEBS J.* 280, 979–993. doi: 10.1111/febs.12092

Yu, T.-H., Wu, C.-M., Rosen, R. T., Hartman, T. G., and Ho, C.-T. (1994). Volatile compounds generated from thermal degradation of alliin and deoxyalliin in an aqueous solution. *J. Agric. Food Chem.* 42, 146–153. doi: 10.1021/jf00037a026

Zhang, H., Forman, H. J., and Choi, J. (2005). γ-Glutamyl transpeptidase in glutathione biosynthesis. *Methods Enzymol.* 401, 468–483. doi: 10.1016/S0076-6879(05)01028-1

Conflict of Interest Statement: The authors declare that the research was conducted in the absence of any commercial or financial relationships that could be construed as a potential conflict of interest.

13

Responsibility of regulatory gene expression and repressed protein synthesis for triacylglycerol accumulation on sulfur-starvation in *Chlamydomonas reinhardtii*

Atsushi Sato[1,2], Rie Matsumura[1], Naomi Hoshino[1], Mikio Tsuzuki[1,2] and Norihiro Sato[1,2]*

[1] School of Life Sciences, Tokyo University of Pharmacy and Life Sciences, Hachioji, Japan
[2] Japan Science and Technology Agency, Core Research for Evolutionary Science and Technology, Chiyoda-ku, Japan

Edited by:
Stanislaus Francis D'Souza, BMG BhabhaAtomic Research Centre, India

Reviewed by:
David Gonzalez-Ballester, Universidad de Cordoba, Spain
Stanislaus Francis D'Souza, BMG BhabhaAtomic Research Centre, India

***Correspondence:**
Norihiro Sato, School of Life Sciences, Tokyo University of Pharmacy and Life Sciences, Horinouchi 1432-1, Hachioji, Tokyo 192-0392, Japan
e-mail: nsato@ls.toyaku.ac.jp

Triacylglycerol (TG) synthesis is induced for energy and carbon storage in algal cells under nitrogen(N)-starved conditions, and helps prevent reactive oxygen species (ROS) production through fatty acid synthesis that consumes excessive reducing power. Here, the regulatory mechanism for the TG content in sulfur(S)-starved cells of *Chlamydomonas reinhardtii* was examined, in comparison to that in N- or phosphorus(P)-starved cells. S- and N- starved cells exhibited markedly increased TG contents with up-regulation of mRNA levels of diacylglycerol acyltransferase (DGAT) genes. S-Starvation also induced expression of the genes for phosphatidate synthesis. In contrast, P-starved cells exhibited little alteration of the TG content with almost no induction of these genes. The results implied deficient nutrient-specific regulation of the TG content. An *arg9* disruptant defective in arginine synthesis, even without nutritional deficiencies, exhibited an increased TG content upon removal of supplemented arginine, which repressed protein synthesis. Repression of protein synthesis thus seemed crucial for TG accumulation in S- or N- starved cells. Meanwhile, the results of inhibitor experiments involving cells inferred that TG accumulation during S-starvation is supported by photosynthesis and de novo fatty acid synthesis. During S-starvation, *sac1* and *snrk2.2* disruptants, which are defective in the response to the ambient S-status, accumulated TG at lower and higher levels, respectively, than the wild type. The *sac1* and *snrk2.2* disruptants showed no or much greater up-regulation of DGAT genes, respectively. In conclusion, TG synthesis would be activated in S-starved cells, through the diversion of metabolic carbon-flow from protein to TG synthesis, and simultaneously through up-regulation of the expression of a particular set of genes for TG synthesis at proper levels through the actions of *SAC1* and *SNRK2.2*.

Keywords: acyltransferases, *Chlamydomonas reinhardtii*, lipid droplets, *SAC1*, sulfur-starvation, *SNRK2.2*, protein synthesis, triacylglycerol

INTRODUCTION

Triacylglycerol (TG), which is one of the neutral lipids, is ubiquitous in eukaryotes and also present in a limited group of prokaryotes (Athenstaedt and Daum, 2006). Distinct from polar lipids that are predominantly membrane components, TG is localized in a specific organelle designated as lipid droplets. TG has been considered as a storage compound: e.g., in oil seed plants, fatty acids esterified to TG in seeds undergo β-oxidation for energy production and gluconeogenesis, which supports the seedling growth after germination (Quettier and Eastmond, 2009). However, recent studies indicated that TG also participates in the synthesis of membrane lipids as an intermediate metabolite by supplying fatty acids in actively growing cells of yeast, *Saccharomyces cerevisiae*, and thus is critical for maintenance of lipid homeostasis (Rajakumari et al., 2010; Kohlwein and Henry, 2011). On the other hand, from industrial aspects, TG is important as a food oil and has been recently expected to be a source of biodiesel fuel (BDF), which is produced through its chemical conversion into methyl or ethyl esters of fatty acids. Extensive attention has been paid to BDF production with photosynthetic organisms in particular, in terms of the carbon neutrality concept (Durrett et al., 2008; Hu et al., 2008).

Many algal species have several advantages over terrestrial plants as to the production of biomaterials, including high annual biomass productivity on an area basis that is ensured by their high growth rates (Hu et al., 2008). Eukaryotic algae in general contain TG at a low level during optimal growth conditions, but the content remarkably increases with culture aging (Dunstan et al., 1993) or under ambient stress conditions such as nutritional limitation, high salinity (Siaut et al., 2011), or high light (Khotimchenko and Yakovleva, 2004). It has been consistently shown that, as to nutrients, nitrogen deficiency is markedly effective for induction of accumulation of TG in numerous algal species including green and red algae, diatoms, golden algae, haptophytes, eustigmatophytes, dinophytes, and yellow-green algae (see a review by Hu et al., 2008). Meanwhile, the effects of deficiencies of other nutrients on the TG content were investigated, especially in a green alga, *Chlamydomonas reinhardtii*, distinct results concerning sulfur-starved cells being reported in the literature: Boyle et al. (2012) reported marked induction of TG accumulation whereas Fan et al. (2012) observed little induction. Phosphorus (P)-deficiency seemed to have no or only a minor effect, if any, on the increase in the TG content (Boyle et al., 2012; Fan et al., 2012).

Study of the mechanism by which the accumulation of TG is stimulated, e.g., identification of key genes as to the synthesis of TG and proteomic analysis of lipid droplets, is indispensable for enhancement of the productivity of TG, but this has only just begun for algal species (Nguyen et al., 2011; Boyle et al., 2012; Deng et al., 2012; Msanne et al., 2012). The genes responsible for the terminal step of the synthesis of TG include those for diacylglycerol acyltransferases (DGATs), which could be important players determining the accumulated level of TG. The genome of *C. reinhardtii* contains one homolog of *DGAT1* and five homologs of *DGAT2* designated as *DGTT1-5* (Merchant et al., 2012). Expression of *DGAT1* and *DGTT1-4* was induced at the transcript level in response to N- or S-starvation, but was little affected by P-starvation (Miller et al., 2010; Boyle et al., 2012; Blaby et al., 2013; Ramanan et al., 2013).

Sulfoquinovosyl diacylglycerol (SQDG) is one of the membrane lipids specific to chloroplasts, and is responsible for the structural and functional integrity of the photosystem II complex in *C. reinhardtii* (Sato et al., 1995, 2003a; Minoda et al., 2003; Sato, 2004). Intriguingly, we recently reported that cells of *C. reinhardtii* degrade SQDG for utilization of it as a major intracellular sulfur (S)-source for the synthesis of protein, especially at an early phase of S-starvation (Sugimoto et al., 2007, 2008, 2010). In the course of that study, we noticed that S-deficiency stress as well as N-limiting stress induces pronounced accumulation of TG. *C. reinhardtii* is suitable for gene manipulation, and abundant mutants defective in a variety of physiological processes are available (Rochaix, 1995). Moreover, its whole genome has been successfully sequenced (Merchant et al., 2007). *C. reinhardtii* would thus be a strong biological tool for elucidation of the mechanism by which the accumulated level of TG is enhanced.

S is one of the macronutrients, and is incorporated into plants mainly as sulfate. A greater part of study on plant behavior in response to the ambient S-status has been restricted to green plants such as a seed plant, *Arabidopsis thaliana*, and a green alga, *C. reinhardtii*. These two organisms when exposed to S-limitation exhibit similar up-regulation of expression of the genes for acquisition of external S such as that of the sulfate transporter, and also for primary S-assimilation including the synthesis of cysteine (Nikiforova et al., 2003; Zhang et al., 2004; Toepel et al., 2011; Hubberten et al., 2012). In *C. reinhardtii*, in particular, regulatory components including the SAC1 (Sulfur Acclimation 1; Davies et al., 1996) and SNRK2.2 (SNF1-related protein kinase 2.2, previously known as SAC3; Davies et al., 1999; Moseley et al., 2009) proteins have been identified as components of the signaling pathway responsible for the cellular response to the ambient S-status. The SAC1 protein is homologous to anion transporters of other organisms such as the mammal Na^+/SO_4^{2-} transporter whereas SAC1 like transporters designated as SLT1 and 2 seem to function as high-affinity sulfate transporters at the plasma membrane in *C. reinhardtii* (Davies et al., 1996; Pootakham et al., 2010). SAC1 is thus postulated to sense a shortage of an external S-source, and then to transduce the signal for up-regulation of the transcript levels of a special set of genes (Davies et al., 1996; Zhang et al., 2004; Moseley et al., 2009). On the other hand, the SNRK2.2 protein belongs to the serine/threonine kinase group, and seems to either positively or negatively regulate physiological responses related to the ambient S-status (Davies et al., 1999; Zhang et al., 2004; Moseley et al., 2009). The degradation of SQDG induced by S-starvation was performed by the use of both the SAC1 and SAC3 proteins as positive regulators (Sugimoto et al., 2010). It is thus of interest whether or not, or how these factors are involved in the above-mentioned TG accumulation under S-starved conditions.

Here, we investigated alterations in the TG content and the expression levels of the genes for the synthesis of TG in cells of *C. reinhardtii* during S-starvation, and compared them with those during N- or P-starvation, to reveal the deficient-nutrient dependent regulatory mechanism for the level of TG accumulation. The mechanism by which the accumulation of TG is stimulated under S-starved conditions will be discussed in view of the actions of the *SAC1* and *SNRK2.2* genes, and repression of global protein synthesis.

MATERIALS AND METHODS

STRAINS AND GROWTH CONDITIONS

The *C. reinhardtii* strains used were CC-125 as the wild type, and three disruptants as to the *SAC1* (CC-3794), *SNRK2.2* (CC-3799), and *ARG9* (CC-4440) genes, and the respective complemented strains (CC-3795, CC-3798, CC-4441). These disruptants and complemented strains were purchased from the *Chlamydomonas* Resource Center (http://chlamycollection.org/). *Chlorella kessleri* 11 h was also used (Sato et al., 2003b). Cells were cultured with continuous illumination at 30°C in TAP medium (Gorman and Levine, 1965) for mixotrophic growth of *C. reinhartdii* cells in a flask on a rotary shaker, or in 3/10 HSM (Sueoka, 1960) for photoaoutotrophic growth of *C. reinhardtii* and *C. kessleri* cells in an oblong glass vessel with aeration. A disruptant as to the *ARG9* gene for N-acetyl ornithine aminotransferase in chloroplasts (*arg9*), which is deficient in the synthesis of arginine, was cultured with supplementation of arginine (0.57 mM; Remacle et al., 2009). For transfer to nutritionally starved conditions, cells grown to the mid-logarithmic phase in TAP medium or ones grown to the mid-linear phase in 3/10 HSM were harvested by centrifugation, washed twice and then resuspended in the corresponding S-, N-, or P-free medium. The S- and P-free media were prepared by replacing sulfate and phosphate, respectively, with chloride, whereas the N-free medium was prepared by substituting potassium for ammonium. Growth of the cells was monitored by determination of the optical density at 730 nm with a spectrophotometer, DU640 (Beckman, USA). When needed, the flask, in which cells of *C. reinhardtii* were mixotrophically cultured, was completely covered with aluminum foil to obtain dark conditions.

INHIBITORS OF PROTEIN SYNTHESIS, FATTY ACID SYNTHESIS, AND PHOTOSYNTHESIS

An inhibitor, such as chloramphenicol (CAP, 100 μg ml^{-1}), cycloheximide (CHI, 8 μg ml^{-1}), cerulenin (10 μM), or 3-(3,4-dichlorophenyl)-1,1-dimethylurea (DCMU, 50 μM), was added at the respective final concentrations, immediately after the change from nutrient-repleted to –depleted medium. Methanol was used as a carrier of CAP (final concentration, 1%). CAP and CHI are inhibitors of de novo protein synthesis on chloroplast and mitochondrial 70S ribosomes, and cytoplasmic 80S ribosomes, respectively. Cerulenin is an inhibitor of de novo synthesis of fatty acids, whereas DCMU inhibits photosynthesis.

EXTRACTION OF LIPIDS, PROTEINS, AND CHLOROPHYLL (Chl) AND THEIR QUANTITATIVE ANALYSIS

Total lipids were extracted from cells of *C. reinhardtii* or *C. kessleri*, according to the method of Bligh and Dyer (1959), and thereafter separated into individual neutral lipid classes by TLC on precoated silica gel plates (Merck 5721) with a solvent system of hexane/diethylether/acetate (70:30:1, v/v/v). The spots of lipids were visualized by illumination with UV light after spraying with primulin (0.01% in 80% acetone, w/v). Fatty acid methyl esters were prepared from the total lipids, TG, and free fatty acids (FFA) by heating at 95°C with 5% anhydrous methanolic HCl, and thereafter analyzed by capillary GLC, as described previously (Sato et al., 1995). The fatty acid content of each fraction was estimated with arachidonic acid as an internal standard. The TG content relative to total lipids was expressed as mol% on the basis of included fatty acids. On the other hand, whole-cell extracts were prepared through disruption of cells in an extraction buffer (50 mM Tris-HCl, pH 7.5, 0.5% Triton X-100) with a Beads Crusher mT-12 (TAITEC, Japan), and thereafter the protein contents were measured with a BCA assay kit (Pierce, USA). Chl was extracted from cells with methanol and its content was measurement by spectroscopy, according to the method of Grimme and Boardman (1972).

MICROSCOPIC OBSERVATION OF LIPID DROPLETS AND STARCH GRANULES

A Nile red solution (0.25 mg·ml^{-1} in acetone) was added to cell suspensions of *C. reinhardtii* and *C. kessleri* (1:50, v/v), and the stained cells were observed under a fluorescence microscope (BX-FLA; Olympus Optical Co., Tokyo, Japan) with the use of a 520–550 nm excitation filter. Meanwhile, a solution of 0.1% KI containing 0.1% I_2 was added to a cell suspension of *C. reinhardtii* (1:3, v/v) for staining of starch granules, which were thereafter observed microscopically, as described by Izumo et al. (2011).

SEMI-QUANTITATIVE DETERMINATION OF TRANSCRIPT LEVELS BY REVERSE TRANSCRIPTASE (RT)-PCR

Total RNA was extracted and purified by phenol–chloroform extraction, as described by Los et al. (1997), and then used for the synthesis of cDNA by reverse-transcription with random primers (Tabei et al., 2007). The cDNA synthesized was used as a template for semi-quantitative RT-PCR (Tabei et al., 2007). The specific forward (F) and reverse (R) primers specific to the respective genes are shown in the supplementary material, with available ID numbers for the transcripts.

The primer sets for *DGAT1* and *DGTT1-4*, *GPDH1-3*, and 18 rRNA were the same as those reported by Msanne et al. (2012), Herrera-Valencia et al. (2012), and Teramoto et al. (2002), respectively. The primer sets for *GPAT1*, *GPAT2*, and *LPAAT* were designed on the basis of the corresponding cDNA sequences. The amplified DNA fragments were subjected to agarose gel electrophoresis, and a fluorescent image of a gel after staining with ethidium bromide was obtained by photography. The fluorescence intensities of DNA bands were quantified with ImageJ (http://rsbweb.nih.gov/ij/). As regards the respective genes, the values were estimated relative to that of 18S rRNA as an internal control.

RESULTS

ENHANCED ACCUMULATION OF TG UNDER S-STARVED CONDITIONS IN *C. RENHARDTII* AND *C. KESSLERI*

We first examined whether or not S-deficiency stress has an impact on the contents of non-polar lipids. The mixotrophic growth of cells of *C. reinhardtii* was retarded upon exposure to S-starvation, which finally resulted in an only 2.1-fold increase at 120 h (cf. a 4.9-fold increase under normal conditions, **Figure 1A**). TLC analysis of total lipids from *C. reinhardtii* cells showed the separation of non-polar lipid classes, TG and FFA, on a plate, from polar lipid classes that migrated much more slowly as a group with chlorophyll (**Figure 1B**). TG and FFA accounted for merely 5–10 mol% (see 0 h in **Figure 1C**) and 2–4 mol% (data not shown), respectively, relative to total lipids on the basis of fatty acids, under normal growth conditions. However, after the onset of S-starvation, the TG content began to increase as early as at 4 h, and thereafter accounted maximally for 40.3 mol% at 72 h, relative to total lipids, on the basis of fatty acids (**Figures 1B,C**). Accordingly, the TG content in the culture finally reached a level corresponding to 77 μM fatty acids (**Figure 1D**). In contrast, S-starvation had only a minor effect on the content of FFA (**Figure 1B**). Consistent with the enhanced accumulated level of TG, the S-starved cells eventually contained a substantial amount of lipid droplets, which appeared as intracellular globules that emitted yellow fluorescence against the background of red autofluorescence of Chl (**Figure 1E**, see –S for *C. reinhardtii*). Simultaneously, we observed that S-starved cells, but not S-repleted ones, markedly accumulated starch granules on staining of them with I_2 (**Figure 1F**). S-Starvation under photoautotrophic growth conditions also caused drastic accumulation of TG to a level corresponding to 123 μM fatty acids in the culture or 56.9 mol% relative to total lipids on the basis of fatty acids (**Figure 1D**).

The effects of depletion of other macronutrients such as N and phosphorus (P) on the TG content were then examined. N-Starvation, as compared with S-starvation, retarded mixotrophic cell growth more severely (**Figure 1A**), with more pronounced elevation of the TG content to 56.6 mol% relative to total lipids, on the basis of fatty acids, or 162 μM fatty acids in the culture (**Figures 1C,D**). P-Starved mixotrophic cells showed similar delayed growth to S-starved ones, at least for 96 h (**Figure 1A**), however, the TG content increased little (**Figure 1C**). Compatible with these observations, the content of lipids droplets was greatly increased in N-starved cells, but was maintained at a low level

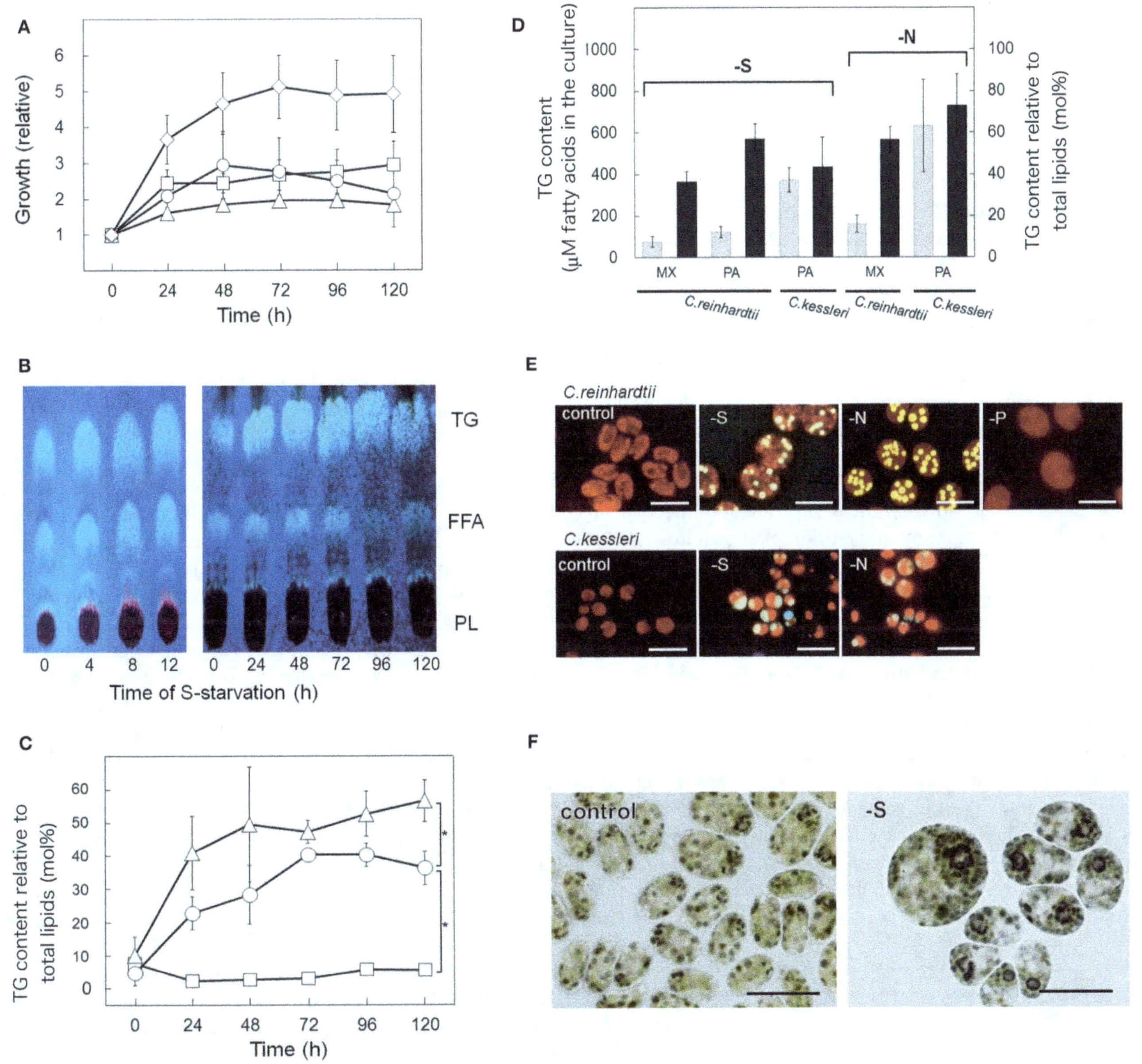

FIGURE 1 | Accumulation of TG in *C. reinhardtii* and *C. kessleri* cells. **(A)** Cell growth of *C. reinhardtii* in TAP (open diamonds), TAP-S (open circles), TAP-N (open triangles), and TAP-P (open squares) medium. OD_{730} values relative to those at 0 h are shown. **(B)** A TLC profile of lipids in *C. reinhardtii* cells during S-starvation for a short term (up to 12 h, left panel) or long term (up to 120 h, right panel). Total lipids isolated from cells in 30-mL culture were loaded. FFA, free fatty acids. PL, polar lipids. **(C)** Changing patterns of TG contents relative to total lipids on the basis of fatty acids in cells of *C. reinhardtii* starved of S, N, or P. The symbols are the same as in **(A)**. **(D)** Accumulation of TG caused by S- or N-starvation for 120 h in *C. reinhardtii* and *C. kessleri*. The TG content of cells was expressed as that in the cultures (gray bars), and that relative to total lipids (black bars), respectively, on the basis of fatty acids. MX and PA indicate mixotrophic and photoautotrophic culturing conditions, respectively. **(E)** Marked accumulation of Nile-red stained lipid droplets that emitted yellow fluorescence in *C. reinhardtii* and *C. kessleri* cells starved of S or N for 120 h. **(F)** Induction of I_2-stained starch accumulation in *C. reinhardtii* cells by S-starvation for 24 h. The values shown in **(A–D)** are the averages ± SE for three distinct groups of data. The significance of differences was evaluated by means of Student's *t*-test. $^*P < 0.05$.

in P-starved ones (**Figure 1E**). Overall, it was concluded that S-starvation is a strong induction factor for the accumulation of TG in *C. reinhardtii*, irrespective of whether the cells are mixotrophically or photoautotrophically grown, although the effect is less outstanding than that of N-starvation. Similarly, S-starvation caused *C. kessleri* cells to accumulate abundant TG with the appearance of lipid droplets (**Figures 1D,E**, see –S for *C. kessleri*), however, the TG accumulation was lower than with N-starvation (**Figure 1D**, compare –S with –N for *C. kessleri*).

In *C. reinhardtii*, the constituent fatty acids of TG that accumulated under S-starved conditions were composed of C16 and C18 acids, which respectively accounted for ca. 44 mol% and

56 mol% (**Figure 2**). C16 acids consisted of 16:0 (29 mol%) and mono- to tetra-unsaturated acids (totally, 15.2 mol%), whereas C18 acids were composed of 18:0 (3.6 mol%) and mono- to tetra-unsaturated acids (totally, 52.0 mol%). The fatty acid profile of TG in N-starved cells was found to be almost identical to that in S-starved cells.

EFFECTS OF INHIBITORS OF SOME PHYSIOLOGICAL PROCESSES, LIGHT CONDITIONS, AND RESPECTIVE MUTATIONS THAT CAUSE A DEFICIENCY IN ARGININE SYNTHESIS AND AN ABERRANT CELLULAR RESPONSE TO THE AMBIENT S-STATUS ON THE ACCUMULATED LEVEL OF TG IN CELLS OF *C. REINHARDTII*

We then characterized the induction system for the accumulation of TG under S-starved conditions in cells of *C. reinhardtii*, with the use of the inhibitors of protein synthesis, fatty acid synthesis, and photosynthesis, and light conditions (**Figure 3A**). Application of CHI resulted in almost complete repression of the accumulation of TG, whereas CAP had no effect. It thus was likely that the enhanced accumulation of TG under S-starved conditions depends on de novo synthesis of protein(s) encoded on the nuclear genome, and not on the chloroplast or mitochondria genome. On the other hand, as cerulenin prevented the elevation in the TG content, the elevation was implied to require de novo synthesis of fatty acids. Dark conditions and DCMU, respectively, repressed the increase in the TG content, inferring that light as an energy source for photosynthesis is indispensable for the enhanced accumulation of TG.

The Chl content, which reflects the level of the PSI and PSII complexes, was decreased at 24 h to 0.7- and 0.5-fold of the initial level in S- and N-starved cells, respectively (**Figure 3B**, inset). In contrast, the Chl content was increased in P-starved cells, although the increased level (1.3-fold of the initial level, **Figure 3B**, inset) was lower than in normal cells (a 3.6-fold increase, data not shown). Simultaneously observed was severe repression in accumulation of total cellular proteins in S- or N-starved cells (**Figure 3B**, see –S or –N, cf. Control). These results indicated that global protein synthesis was so severely repressed in S- and N-starved cells such that degradation of the PS complexes became necessary, but not in P-starved cells, which could still continue to construct the PS complexes. The stimulating effects of S- and N-starvation, but not of P-starvation, on the accumulation of TG might explain that the severe repression of global protein synthesis is a prerequisite for the stimulation of TG synthesis. To examine this possibility, first, we utilized the *arg9* disruptant of *C. reinhardtii* (Remacle et al., 2009). The *arg9* cells were grown in advance with supplementation of arginine, and thereafter were transferred to arginine-free medium for further growth for 24 h for repression of protein synthesis (**Figure 3B**, see –Arg, cf. +*Arg*, for *arg9*). This transfer simultaneously induced substantial accumulation of lipid droplets with an increase in the TG content from 6.6 to 24.7 mol%, relative to total lipids, on the basis of fatty acids (**Figure 3C**). Second, wild-type cells were cultured in the presence of CHI, which exhibited dose-dependent accumulation of lipid droplets (**Figure 3D**), i.e., 0.9 ± 0.7 and 3.7 ± 1.5 lipid droplets on a cell basis in untreated cells and cells treated with 10 $\mu g \cdot ml^{-1}$ CHI, respectively. Overall, the repression of two quite different metabolic processes for protein synthesis (arginine synthesis and the peptidyl transferase step in 70S ribosomes) was correlated with increased accumulated levels of TG and/or lipid droplets. These findings thus strongly supported the novel notion that repression of global protein synthesis is at least one of the key factors for the stimulation of accumulation of TG under S- or N-starved conditions.

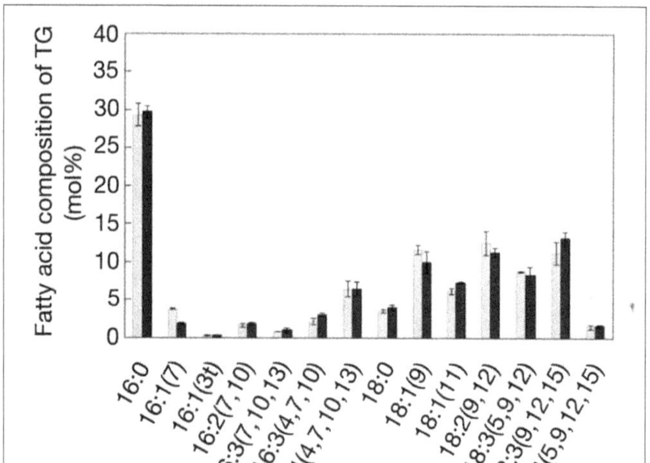

FIGURE 2 | The composition of constituent fatty acids of TG isolated from cells starved of S (gray bars) or N (black bars). The values are the averages ± SE for three distinct groups of data.

We then investigated whether or not the *SAC1* and *SNRK2.2* genes are involved in the stimulation of the TG accumulation by S-starvation. Cells of the two disruptants as to the *SAC1* and *SNRK2.2* genes (*sac1* and *snrk2.2*, respectively), when mixotrophically grown in TAP medium under S-repleted conditions, contained 3 to 5 mol% TG relative to total lipids, on the basis of fatty acids, which was similar to the level in the case of the wild type (**Figure 3E**, see 0 h). Although S-starvation led to accumulation of lipid droplets at substantial levels in the disruptants of *sac1 and snrk2.2* (**Figure 3F**), characteristic mutational phenotypes as regards the TG content became evident on quantitative analysis of lipids (**Figure 3E**). The TG content on the basis of fatty acids was increased in the *sac1* disruptant to only 31.6 mol% relative to total lipids at 72 h, i.e., to a lower level than in WT (40.3 mol%, **Figure 1C**), with no further increase at 120 h. The effect of the *sac1* mutation of repression of TG accumulation could be demonstrated more definitely when the *sac1* disruptant was compared with the *SAC1/sac1* strain, which had accumulated TG to 53.2 mol% at 72 h, and thereafter maintained the increased content until at 120 h. Meanwhile, the TG content increased in the *snrk2.2* disruptant to 35.8 mol% at 72 h, and further to as high as 57.9 mol% at 120 h, i.e., to a higher level than in WT or the *SNRK2.2/snrk2.2* strain. Hence, the *snrk2.2* mutation, in contrast to the *sac1* one, had an impact of stimulation of TG accumulation.

EFFECTS OF NUTRITIONAL DEFICIENCY ON THE EXPRESSION LEVELS OF THE GENES FOR TG SYNTHESIS IN *C. REINHARDTII*

Here, the possibility was investigated that expression of the genes that constitute the Kennedy pathway for the biosynthesis of TG

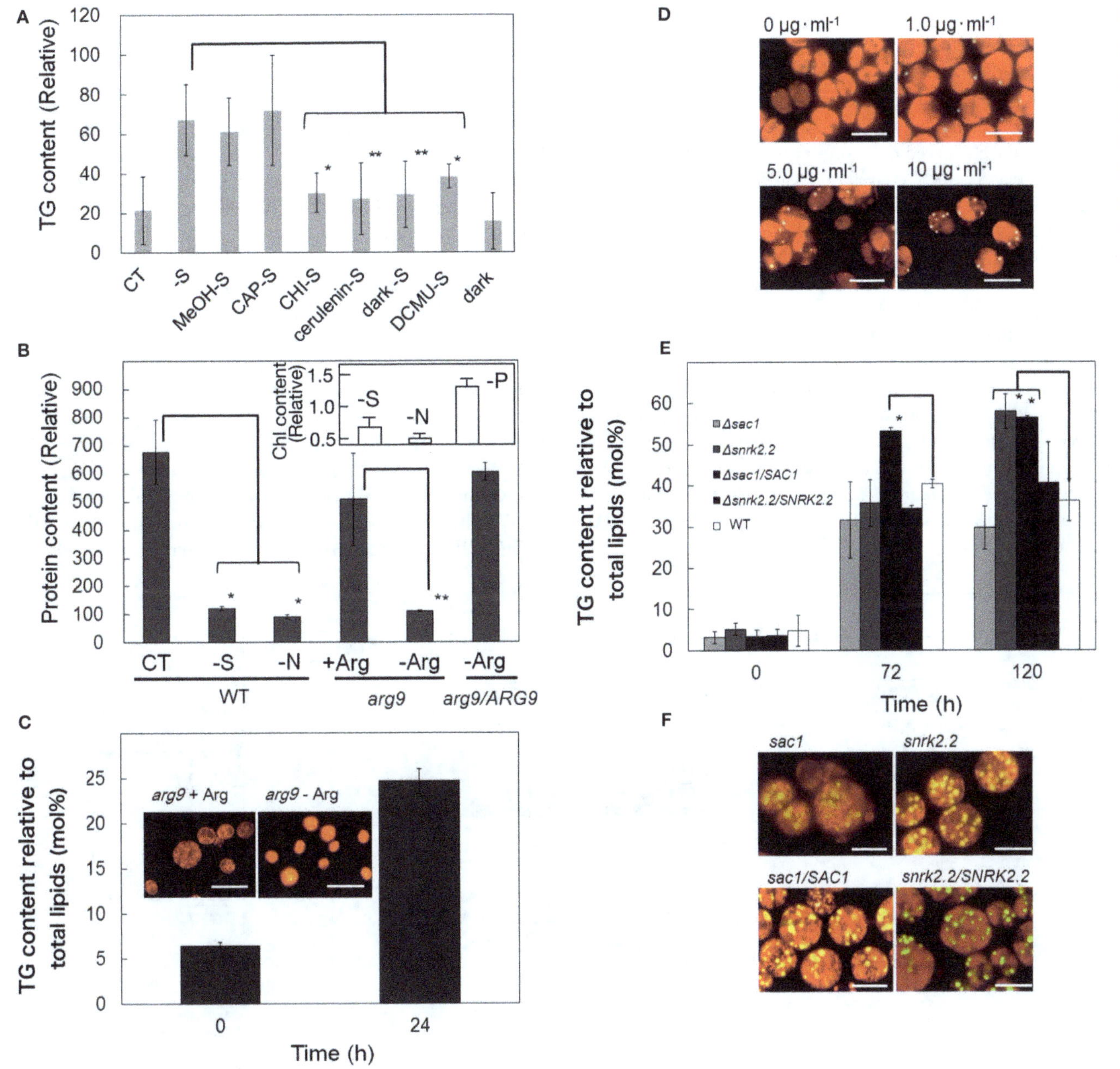

FIGURE 3 | Effects of inhibitors, light conditions, and mutations on the cellular content of TG in *C. reinhardtii*. **(A)** TG contents in cells that were treated with inhibitors or placed under dark conditions, and then starved of S for 24 h. CT, control. MeOH, methanol used as a carrier of CAP. The values were estimated as fatty acid contents of TG per OD_{730} values. **(B)** Contents of total cellular proteins in cells of WT, an *arg9* disruptant, and a complemented strain as to *arg9* (*arg9/ARG9*) after 24-h culturing under normal or stress conditions. The values were estimated relative to those at 0 h. CT, –S, and -N indicate normal, and S- and N-starved conditions for the WT cells, respectively. +Arg and –Arg represent supplementation and non-supplementation of arginine, respectively, for the cultures of *arg9* and *arg9/ARG9* cells. Inset, the Chl content in WT cells starved for S, N, or P for 24 h. The values were estimated relative to those at 0 h. **(C)** An increase in the content of TG or lipid droplets in the *arg9* disruptant after 24-h culture in arginine-free medium. The values were estimated as TG content relative to total lipids on the basis of fatty acids. **(D)** Marked accumulation of Nile-red stained lipid droplets in cells upon treatment with CHI (1–10 $\mu g \cdot mL^{-1}$) for 24 h. S-Starvation induced accumulation of TG **(E)** or Nile-red stained lipid droplets **(F)** in mixotrophically grown cells of the *sac1* and *snrk2.2* disruptants, and complemented strains of *sac1/SAC1* and *snrk2.2/SNRK2.2*. The values **(E)** were estimated as TG content relative to total lipids on the basis of fatty acids. The values for the wild type **(E)** were the same as those in **Figure 1C**. The values shown in **(A–C,E)** are the averages ± SE for three distinct groups of data. The significance of differences was evaluated by means of Student's *t*-test. $^{*}P < 0.05$. $^{**}P < 0.1$.

is regulated under S-starved conditions (**Figure 4**), with the first focus on the *DGAT1* and *DGAT2* genes (Deng et al., 2012). Since the TG content increased steadily in wild-type cells of *C. reinhardtii* from the early to mid-phase of S-starvation (4–72 h, **Figures 1B,C**), the effects of S-starvation for 8 h on the transcript levels of the *DGAT1* and *DGAT2* genes were investigated by semi-quantitative RT-PCR (**Figure 4**). On S-starvation, *DGAT1* had little effect on the mRNA level. However, the *DGAT2* family

(*DGTT1* to *4*), respectively, showed a higher mRNA level in S-starved cells than in control ones, the exception being *DGTT5* mRNA, which was below the detectable level irrespective of the ambient S-status (data not shown). The *DGAT2* family thus seemed to participate more positively than *DGAT1* in the increase of TG, at least during this short term S-starvation. Meanwhile, it was found that N-starved cells, similar to S-starved ones, exhibited an increased mRNA level of *DGTT1-4*, as a short-term response, with the level of *DGAT1* mRNA being hardly altered (**Figure 4**). In contrast, P-starved cells, as compared with control cells, exhibited little difference in the mRNA level of either *DGAT1* or *DGTT1-4* as a short-term response (**Figure 4**).

We then examined the transcript levels of the genes for the synthesis of phosphatidate (PA) from glycerol 3-phosphate (G3P) in the Kennedy pathway (**Figure 4**). This pathway starts with the first acylation of G3P at the *sn*-1 position, yielding lysophosphatidate (LPA), by G3P acyltransferase (GPAT), followed by the second acylation of LPA at the *sn*-2 position by lysophosphatidate acyltransferase (LPAAT), for construction of PA. It is predicted that *C. reinhardtii* possesses two genes that encode chloroplast- and ER-associated GPAT [designated here as *GPAT1* (a homolog of plastid *GPAT* of *A. thaliana*; Nishida et al., 1993) and *GPAT2* (a homolog of mammalian *GPAT3* or *A. thaliana GPAT9*; Nguyen et al., 2011), respectively]. Concerning the *LPAAT* gene, a homolog of plastid *LPAAT* of *A. thaliana* (Yu et al., 2004), but not of the ER-associated one, has been found. S-Starved cells showed higher mRNA levels for *GPAT1* and *LPAAT*, respectively, than control ones, with little effect on the level of *GPAT2* mRNA. N-Starved cells, distinct from S-starved ones, showed there was little impact on the mRNA level of either acyltransferase, while P-starved cells exhibited a higher level of only *GPAT1* mRNA compared to control cells.

Glycerol 3-phosphate (G3P), the first substrate in the Kennedy pathway (**Figure 4**), is generated by G3P dehydrogenase (GPDH) through reduction of dihydroxyacetone phosphate (DHAP). GPDH might thus be another key enzyme for induction of accumulation of TG. S-Starved cells, as compared with control ones, showed higher mRNA levels of CrGPDH2 and CrGPDH3, which are putative chloroplast isozymes (Herrera-Valencia et al., 2012). N-Starved cells, relative to the control ones, exhibited higher and lower levels of CrGPDH2 and CrGPDH3 mRNA, respectively, while P-starved cells showed remarkably lower mRNA levels of both chloroplast isozymes. On the other hand, mRNA of CrGPDH1, a probable cytoplasmic isozyme (Herrera-Valencia

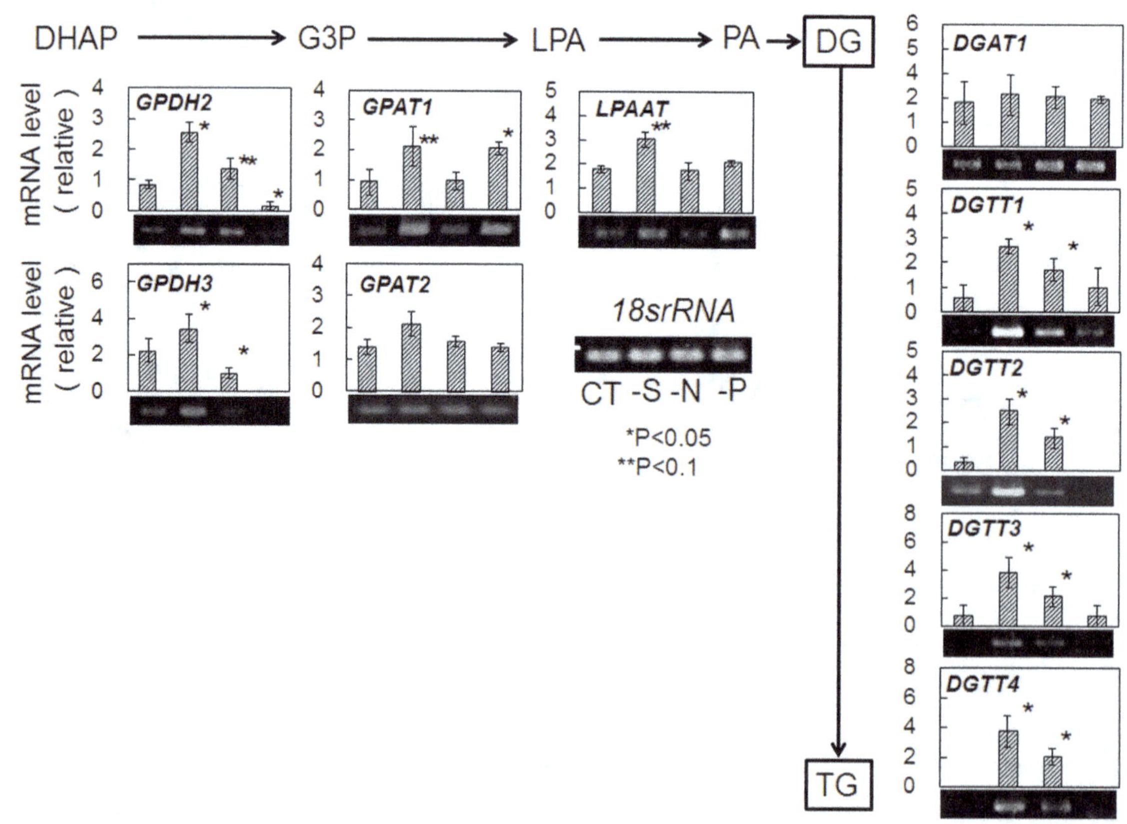

FIGURE 4 | Semi-quantitative RT-PCR analysis of mRNA levels of the genes for TG synthesis in *C. reinhardtii* cells starved of S, N or P for 8 h. CT indicates normal conditions. The values were estimated from the intensities of the DNA bands corresponding to mRNAs of the respective genes for TG synthesis relative to that of 18s rRNA, as described under Materials and Methods. DHAP, dihydroxyacetone phosphate; G3P, glycerol 3-phosphate; LPA, lysophosphatidate; PA, phosphatidate; GPDH, G3P dehydrogenase; GPAT, G3P acyltransferase; LPAAT, lysophosphatidate acyltransferase; DGAT, diacylglycerol acyltransferase type I; DGTT, diacylglycerol acyltransferase type II. The values are the averages ± SE for three distinct groups of data. The significance of differences was evaluated by means of Student's *t*-test. $^{*}P < 0.05$. $^{**}P < 0.1$.

et al., 2012), was undetectable in our study, irrespective of the growth conditions (data not shown). Overall, it was found that the expression levels of the genes for TG synthesis in **Figure 4**, with the exceptions of *DGAT1* and *GPAT2*, are up-regulated under S-starved conditions, but that the genes subjected to up-regulation are limited to the *DGTT1-4* and *GPDH2* ones under N-starved conditions (**Table 1**).

The wild-type cells, when starved of S, showed induction of expression of the genes for DGTT1-4 at particularly high levels. The expression levels of *DGTT1-4* under S-starved conditions were then examined in *sac1* and *snrk2.2* disruptants (**Figure 5**). The *sac1* disruptant, which was repressed as to TG accumulation, was defective in the up-regulation of the mRNA levels of *DGTT1-4*. In contrast, the *snrk2.2* disruptant, which was stimulated as to TG accumulation, showed much stronger up-regulation of the levels of the *DGTT1-4* mRNAs. On the other hand, the *snrk2.2* mutation had little impact on the mRNA level of *DGAT1*. However, the *sac1* mutation brought about an increased level of *DGAT1* mRNA, implying some compensatory mechanism for the failure in the induction system as to expression of the *DGTT1-4* genes.

DISCUSSION

ESTABLISHMENT OF S-STARVATION AS A STRESS CONDITION THAT INDUCES TG ACCUMULATION IN GREEN ALGAE

S-Starvation, distinct from N-starvation, has hardly attracted researchers' attention as a stress condition for algal TG production. *Chlorella ellipsoidea* cells showed an increase in the content of total lipids during S-starvation, however, the responsible lipid class(es) were not specified (Otsuka, 1961). The TG content was increased in *C. reinhardtii* cells under combinational stress conditions of S-starvation and anaerobicity, which obscured the causal environmental factor for the TG accumulation (Matthew et al., 2009). During the course of our study, two reports regarding the effects of S-starvation on the accumulated level of TG in *C. reinhardtii* were published, however, the results were incompatible: Fan et al. (2012) found little stimulation of TG accumulation after 3-day starvation of S, whereas Boyle et al. (2012) reported induction of TG accumulation after 2-day starvation. Cakmak et al. (2012) reported that S-starved cells of *C. reinhardtii* accumulated TG, and that, in contrast to our results, its attained level was higher than in N-starved cells. In the last study, TG was quantified with the use of a spectrum band by Fourier transform infrared spectroscopy (FTIR), which was attributed to ester group vibration of TG. However, they analyzed the dried cells containing the other cellular lipids, ester group vibration of which would interfere with that of TG. In this context, our detailed time-scale study, although consistent with the work by Boyle et al. (2012), is the first that demonstrated conclusively that S-starvation causes TG to steadily accumulate in *C. reinhartdii*, and in another green alga, *C. kessleri*, as well (**Figures 1B–E**). Moreover, our study is novel in definitely demonstrating that stimulation of TG accumulation is a specific response to S- or N-starvation, but not a general response to nutritional deficiency (**Figure 1C**).

Table 1 | Summary of effects of S-, N-, or P-starvation on the levels of transcripts for the genes for TG synthesis.

	GPDH2	*GPDH3*	*GPAT1*	*GPAT2*	*LPAAT*	*DGAT1*	*DGTTs*
-S	↑	↑	↑	LE[a]	↑	LE	↑
-N	↑	↓	LE	LE	LE	LE	↑
-P	↓	↓	↑	LE	LE	LE	LE

Red and blue arrows indicate increases or decreases in the mRNA level.
[a]LE, little effect.

METABOLIC ALTERATION FOR STIMULATION OF TG SYNTHESIS IN *C. REINHARDTII* CELLS DURING S-STARVATION

Prevention of TG accumulation by cerulenin (**Figure 3A**) would suggest that acyl-groups for TG synthesis are supplied predominantly through de novo synthesis of fatty acids in S-starved cells, as was previously reported for N-starved cells (Fan et al., 2011). In line with this idea, the content of glycerate 3-phosphate, a precursor of acetyl-CoA that provides building blocks for fatty acid synthetase, was 3.7-fold elevated in S-starved cells of *C. reinhardtii* (Bölling and Fiehn, 2005). Simultaneously, the cellular content of G3P increased by 3.3-fold with a decrease in the DHAP content to 0.6-fold (Bölling and Fiehn, 2005). S-Starvation thus seemed to stimulate G3P synthesis through enhancement of the transcript levels of CrGPDH2 and 3 (**Figure 4**), and, accordingly, the activity of GPDH. S-Starved cells must perform photosynthesis to accumulate TG (**Figure 3A**), which would reflect requirement of photosynthesis for production of chemical energy and fixed carbon for synthesis of acyl-groups and G3P.

As reflected in the delayed growth of S-starved cells (**Figure 1A**), *C. reinhardtii* cells starved of S are known to exhibit down-regulated photosynthesis (Wykoff et al., 1998), which would help repress the production of reactive oxygen species (ROS; González-Ballester et al., 2010). Simultaneously, activated synthesis of acyl-groups and G3P would also contribute to repressed production of ROS by facilitating the consumption of excessive reducing power and the resultant prevention of over-reduction of cells in a redox state (Hu et al., 2008; Li et al., 2012). It is likely that starch accumulation (**Figure 1F**) also contributes to repression of ROS production. Such regulation of the energy balance would meet the algal requirements for TG and starch, as intracellular carbon- and energy-sources, just when algal cells resume growth upon relief from S-starvation. The metabolic regulation for allocating carbon between TG and starch syntheses is a future subject.

Meanwhile, *in C. reinhardtii* cells under normal conditions, the cellular content of total protein relative to dry cell weight reached as high as 48% (w/w), cf., lipids and carbohydrates that only represent 21 and 17%, respectively (Becker, 2007). TG accumulation stimulated by the *arg9* mutation or application of CHI (**Figures 3C,D**) could reflect diversion of a predominant metabolic carbon-flux from protein synthesis to the synthesis of

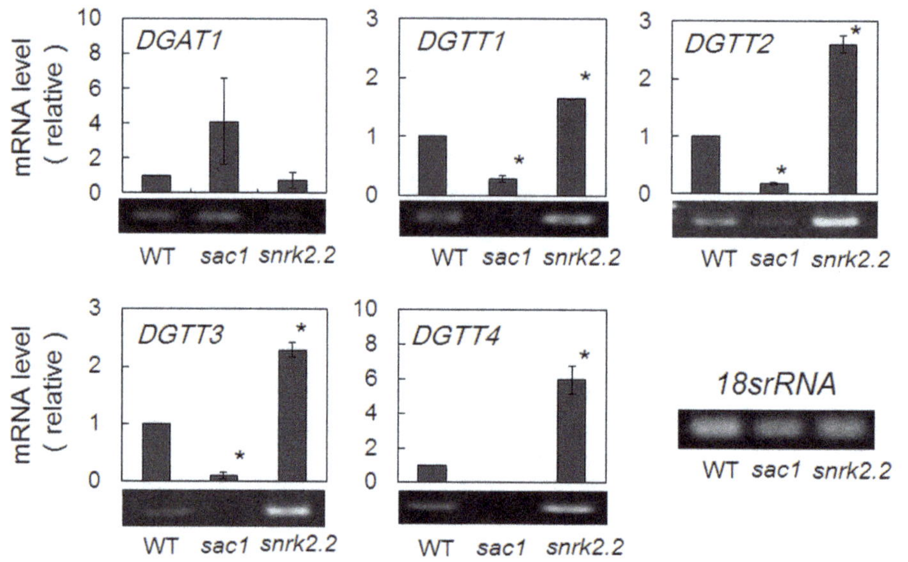

FIGURE 5 | Semi-quantitative RT-PCR analysis of mRNA levels of the DGAT1-4 genes in WT, and *sac1* and *snrk2.2* cells after S-starvation for 8 h. The values were estimated, as described in the legend to **Figure 5**, and are the averages ± SE for three distinct groups of data. The significance of differences was evaluated by means of Student's *t*-test. $^{*}P < 0.05$.

carbon-storage compounds including TG. Repression of global protein synthesis might thus be one of the key factors in the stimulation of TG synthesis under S- or N-starved conditions. These observations in turn indicated that pre-existing enzymes for TG synthesis, by themselves, could contribute to TG synthesis. Compatible with this idea, the *sac1* disruptant that was unable to up-regulate the expression levels of *DGTT1-4* could still elevate the TG content (**Figures 3E, 5**). In this context, it seemed that the complete repression of TG accumulation in S-starved cells of *C. reinhardtii* on the application of CHI (**Figure 3A**) was caused by some secondary effect of CHI, which would become apparent only under severe S-starved conditions, i.e., not under normal nutritional ones. Similar interpretation might be necessary, concerning inhibition of TG accumulation in N-starved *C. reinhardtii* cells (Fan et al., 2011). Again, it should be emphasized that repression of protein synthesis could trigger TG accumulation in S-or N-starved cells of *C. reinhardtii*. Induction of TG accumulation by S-starvation would be conserved in a wide range of algal species in terms of their substantial protein contents, i.e., equal to or more than half of the dry cell weight (Becker, 2007).

REGULATION OF EXPRESSION LEVELS OF THE GENES RESPONSIBLE FOR THE ELEVATION OF TG SYNTHESIS IN *C. REINHARDTII* CELLS UNDER S-STARVED CONDITIONS

Expression patterns of genes for TG synthesis have been extensively examined in N-starved *C. reinhardtii* cells (see below), but infrequently in S-starved ones. Positive participation of the *DGAT2* family, but not of *DGAT1*, was suggested in TG accumulation during short term S-starvation (**Figure 5** and **Table 1**). Long-term S-starvation for 2 days induced up-regulation of the respective mRNA levels of *DGTT1* and *DGAT1* in *C. reinhardtii*, although the effect was much less remarkable on that of *DGAT1* (Boyle et al., 2012). It can thus be proposed that the induction of *DGTT1* starts at an early stage of starvation (at 8 h at latest), and that a time-dependent alteration in the level of *DGAT1* mRNA should be investigated in more detail to understand regulation of the expression of *DGAT1*.

The elevation of the transcript levels of the genes for chloroplast-type GPDH, GPAT, and LPAAT, and ER-type DGTT1-4 in S-starved cells (**Figure 4** and **Table 1**) implied enhancement of the corresponding enzymatic activities for facilitation of PA synthesis in chloroplasts and subsequent TG synthesis at the ER. Intriguingly, C16 acids exclusively occupied the *sn*-2 position of TG in N-starved cells of *C. reinhardtii*, which inferred predominant synthesis of a DG moiety for TG synthesis in chloroplasts (Fan et al., 2011). The similar fatty acid profile of TG for S- starved cells to that for N-starved ones (**Figure 2**) would also imply synthesis of a DG moiety for TG synthesis mainly at chloroplasts in S-starved cells.

The *sac1* and *snrk2.2* mutations led to repression and facilitation, respectively, in up-regulation of the *DGTT1-4* genes under S-starved conditions (**Figure 5**). The arylsulfatase gene (*ARS1*) previously showed regulatory properties similar to those of the *DGAT1-4* genes: the *ARS1* mRNA level is elevated in cells of *C. reinhardtii* during S-starvation, and the elevated level becomes lower and higher with the *sac1* and *snrk2.2* mutations, respectively (Davies et al., 1996; Ravina et al., 2002). It is likely that the SAC1 protein acts as a positive regulator under S-starved conditions, causing stimulated expression of the *DGTT1 to 4* genes as well as the *ARS1* one. The pronouncedly high accumulation of TG in *sac1/SAC1* (**Figure 3E**) might be caused by some mechanism that elevates the expression level of *SAC1*, which has yet to be investigated. Contrarily, it might be that the SNRK2.2 protein acts as a negative regulator for repression of expression of the *DGTT1-4* genes, as well as of that of the *ARS1* one. It should be emphasized that these actions of *SAC1* and *SNRK2.2* could be directly related to the accumulation level of TG in S-starved cells (**Figure 3E**). Collectively, it was newly found that, in S-starved cells, a series of genes for chloroplast PA synthesis, similar to those for DGTTs, are up-regulated in expression levels, and that *SAC1* and *SNRK2.2* are greatly responsible for the up-regulated levels of *DGTT1-4* expression, and consequently of TG.

Numerous groups have attempted to enhance the production level of algal TG to one suitable for industrial application, through manipulation of the genes related to TG synthesis, including those of DGAT and acetyl-CoA carboxylase, which catalyze the rate-limiting step of fatty acid synthesis (e.g., Deng et al., 2012). However, this kind of strategy has limitations in the level of stimulation of TG synthesis through overexpresssion of individual genes. Moreover, the notion of the establishment of a predominant metabolic carbon-flow to TG synthesis has been overlooked. In this context, this study is novel in the discovery that regulatory genes, *SAC1* and *SNRK2.2*, can be used to simultaneously manipulate the expression levels of a series of *DGAT2* genes, and that this aberrant regulation takes place under S-starved conditions, where a prominently large carbon flux into protein synthesis could be deviated to TG synthesis.

DIFFERENT ASPECTS OF TG ACCUMULATION UNDER N-OR P-STARVED CONDITIONS FROM THOSE UNDER S-STARVED CONDITIONS

It seems likely that N-starved cells as well as S-starved ones are repressed in protein synthesis for stimulation of TG synthesis (**Figure 3B**). Moreover, up-regulation as to mRNA levels of DGTT1-4 and GPDH2 was observed also in N-starved cells (**Figure 4** and **Table 1**), as was previously reported (Miller et al., 2010; Boyle et al., 2012; Msanne et al., 2012; Blaby et al., 2013; Ramanan et al., 2013). However, N-starved cells of *C. reinhardtii*, distinct from S-starved ones, showed no increases in the mRNA levels of *GPDH3*, *GPAT1*, and *LPAAT* (**Figure 4** and **Table 1**). Blaby et al. (2013) also reported no significant increases for *GPDH3* and *GPAT1*. Thus, we could interpret that N-starved cells, despite their even higher TG content than that of S-starved ones, exhibited no positive evidence of reinforcement of these three enzyme functions in PA synthesis through confirmation of previous reports (*GPDH3*, *GPAT1*) and with a new information (*LPAAT*). These results might reflect the possible greater availability of PA for TG synthesis in N-starved cells than in S-starved ones. Photosynthetic organisms show decreased synthesis of phospholipids and SQDG under P- and S-starved conditions, respectively, because of a shortage of nutrients for their synthesis (Sato, 2004). N-Containing lipids such as phosphatidylethanolamine and diacylglyceryltrimethylhomoserine totally occupy as much as 30 mol% of total lipids in cells of *C. reinhardtii* (Sugimoto et al., 2008). Repression of the synthesis of N-containing lipids at ER would support preferential metabolic flow of G3P and acyl-groups into PA synthesis at chloroplasts in N-starved cells. This scenario, however, needs experimental verification.

P-Starvation for a short term had little effect on the mRNAs of *DGTT1-4* (**Figure 4** and **Table 1**). It was previously found that, even after long-term P-starvation for 2 days, the *DGTT1* mRNA level hardly changed (Boyle et al., 2012). It thus seemed likely that P-starvation has little impact on *DGAT2* expression throughout a P-starvation period, in line with no stimulated accumulation of TG (**Figure 1C**). Meanwhile, P-starvation caused the mRNA levels of *GPAT1*, and *GPDH 2* and *3* to increase and decrease, respectively (**Figure 4** and **Table 1**). The decreased levels of *GPDH2* and *3* mRNAs would reflect the low requirement of P-starved cells for G3P, owing to repressed phospholipid synthesis. The increased level of *GPAT1* mRNA would lead to enhancement of GPAT activity, thereby drawing G3P from its cellular pool preferentially into the pathway of polar lipid synthesis for cell proliferation. We can thus suggest deficient-nutrient specific regulatory mechanisms, as to the expression levels of the genes for TG synthesis.

Different regulation of gene expression for TG synthesis between S- and N-starved *C. reinhardtii* cells was previously reported for phospholipid:diacylglycerol acyltransferase (PDAT). The level of PDAT mRNA was little affected by S-starvation, but increased during N-starvation. Consistently, disruption of the *PDAT* gene had no impact on TG accumulation in S-starved cells, whereas it caused a 25% decrease in the accumulated level of TG in N-starved cells (Boyle et al., 2012).

Overall, our study demonstrated two novel key points that contribute to stimulation of TG synthesis under S-starved conditions. The first is repression of protein synthesis by S-starvation, probably to make the metabolic pool larger, as to fixed carbon and chemical energy for the synthesis of other organic compounds including TG. The second is induction of expression of a series of genes for TG synthesis at properly high levels through the actions of the *SAC1* and *SNRK2.2* genes, which thereby draws metabolic carbon flow preferentially into TG synthesis. This study has provided a basic architecture for a comprehensive understanding of the regulatory mechanism, from both metabolic and molecular aspects, by which TG synthesis is stimulated by S-starvation, and would further contribute to elucidation of the regulatory mechanisms under other stress conditions such as N-starvation. Future study along these lines will allow us to develop a strategy for the production of algal TG on a large scale that is practical for industrial application.

REFERENCES

Athenstaedt, K., and Daum, G. (2006). The life cycle of neutral lipids: synthesis, storage and degradation. *Cell. Mol. Life Sci.* 63, 1355–1369. doi: 10.1007/s00018-006-6016-8

Becker, E. W. (2007). Micro-algae as a source of protein. *Biotechnol. Adv.* 25, 207–210. doi: 10.1016/j.biotechadv.2006.11.002

Blaby, I. K., Glaesener, A. G., Mettler, T., Fitz-Gibbon, S. T., Gallaher, S. D., Liu, B., et al. (2013). Systems-level analysis of nitrogen starvation-induced modifications of carbon metabolism in a *Chlamydomonas reinhardtii* starchless mutant. *Plant Cell* 25, 4305–4323. doi: 10.1105/tpc.113.117580

Bligh, E. G., and Dyer, W. J. (1959). A rapid method of total lipid extraction and purification. *Can. J. Biochem. Physiol.* 37, 911–917. doi: 10.1139/o59-099

Bölling, C., and Fiehn, O. (2005). Metabolite profiling of *Chlamydomonas reinhardtii* under nutrient deprivation. *Plant Physiol.* 139, 1995–2005. doi: 10.1104/pp.105.071589

Boyle, N. R., Page, M. D., Liu, B., Blaby, I. K., Casero, D., Kropat, J., et al. (2012). Three acyltransferases and nitrogen-responsive regulator are implicated in nitrogen starvation-induced triacylglycerol accumulation in *Chlamydomonas. J. Biol. Chem.* 287, 15811–15825. doi: 10.1074/jbc.M111.334052

Cakmak, T., Angun, P., Demiray, Y. E., Ozkan, A. D., Elibol, Z., and Tekinay, T. (2012). Differential effects of nitrogen and sulfur deprivation on growth and biodiesel feedstock production of *Chlamydomonas reinhardtii. Biotechnol. Bioeng.* 109, 1947–1957. doi: 10.1002/bit.24474

Davies, J. P., Yildiz, F. H., and Grossman, A. (1996). *Sacl*, a putative regulator that is critical for survival of *Chlamydomonas reinhardtii* during sulfur deprivation. *EMBO J.* 15, 2150–2159.

Davies, J. P., Yildiz, F. H., and Grossman, A. R. (1999). *Sac3*, an *Snf1*-like serine/threonine kinase that positively and negatively regulates the responses of *Chlamydomonas* to sulfur limitation. *Plant Cell* 11, 1179–1190.

Deng, X.-D., Gu, B., Li, Y.-J., Hu, X.-W., Guo, J.-C., and Fei, X.-W. (2012). The roles of acyl-CoA: diacylglycerol acyltransferase 2 genes in the biosynthesis of triacylglycerols by the green algae *Chlamydomonas reinhardtii. Mol. Plant* 5, 945–947. doi: 10.1093/mp/sss040

Dunstan, G. A., Volkman, J. K., Barrett, S. M., and Garland, C. D. (1993). Changes in the lipid composition and maximisation of the polyunsaturated fatty acid content of three microalgae grown in mass culture. *J. Appl. Phycol.* 5, 71–83.

Durrett, T. P., Benning, C., and Ohlrogge, J. (2008). Plant triacylglycerols as feedstocks for the production of biofuels. *Plant J.* 54, 593–607. doi: 10.1111/j.1365-313X.2008.03492.x

Fan, J., Andre, C., and Xu, C. (2011). A chloroplast pathway for the de novo biosynthesis of triacylglycerol in *Chlamydomonas reinhardtii. FEBS Lett.* 585, 1985–1991. doi: 10.1016/j.febslet.2011.05.018

Fan, J., Yan, C., Andre, C., Shanklin, J., Schwender, J., and Xu, C. (2012). Oil accumulation is controlled by carbon precursor supply for fatty acid synthesis in *Chlamydomonas reinhardtii. Plant Cell Physiol.* 53, 1380–1390. doi: 10.1093/pcp/pcs082

González-Ballester, D., Casero, D., Cokus, S., Pellegrini, M., Merchant, S. S., and Grossman, A. R. (2010). RNA-Seq analysis of sulfur-deprived *Chlamydomonas* cells reveals aspects of acclimation critical for cell survival. *Plant Cell* 22, 2058–2084. doi: 10.1105/tpc.109.071167

Gorman, D. S., and Levine, R. P. (1965). Cytochrome f and plastocyanin: their sequence in the photosynthetic electron transport chain of *Chlamydomonas reinhardi. Proc. Natl. Acad. Sci. U.S.A.* 54, 1665–1669.

Grimme, L. H., and Boardman, N. K. (1972). Photochemical activities of a particle fraction P_1 obtained from the green alga *Chlorella fusca. Biochem. Biophys. Res. Commun.* 49, 1617–1623.

Herrera-Valencia, V. A., Macario-González, L. A., Casais-Molina, M. L., Beltran-Aguilar, A. G., and Peraza-Echeverría, S. (2012). In silico cloning and characterization of the glycerol-3-phosphate dehydrogenase (GPDH) gene family in the green microalga *Chlamydomonas reinhardtii. Curr. Microbiol.* 64, 477–485. doi: 10.1007/s00284-012-0095-6

Hu, Q., Sommerfeld, M., Jarvis, E., Ghirardi, M., Posewitz, M., Seibert, M., et al. (2008). Microalgal triacylglycerols as feedstocks for biofuel production: perspectives and advances. *Plant J.* 54, 621–639. doi: 10.1111/j.1365-313X.2008.03492.x

Hubberten, H. M., Drozd, A., Tran, B. V., Hesse, H., and Hoefgen, R. (2012). Local and systemic regulation of sulfur homeostasis in roots of *Arabidopsis thaliana. Plant J.* 72, 625–635. doi: 10.1111/j.1365-313X.2012.05105.x

Izumo, A., Fujiwara, S., Sakurai, T., Ball, S. G., Ishii, Y., Ono, H., et al. (2011). Effects of granule-bound starch synthase 1-defective mutant on the morphology and structure of pyrenoidal starch in *Chlamydomonas. Plant Sci.* 180, 238–245. doi: 10.1016/j.plantsci.2010.08.014

Khotimchenko, S. V., and Yakovleva, I. M. (2004). Effect of solar irradiance on lipids of the green alga *Ulva fenestrata Postels et Ruprecht. Bot. Mar.* 47, 395–401. doi: 10.1515/BOT.2004.050

Kohlwein, S. D., and Henry, S. A. (2011). Coordination of storage lipid synthesis and membrane biogenesis: evidence for cross-talk between triacylglycerol metabolism and phosphatidylinositol synthesis. *J. Biol. Chem.* 286, 1696–1708. doi: 10.1074/jbc.M110.172296

Li, X., Moellering, E. R., Liu, B., Johnny, C., Fedewa, M., Sears, B. B., et al. (2012). A galactoglycerolipid lipase is required for triacylglycerol accumulation and survival following nitrogen deprivation in *Chlamydomonas reinhardtii. Plant Cell* 24, 4670–4686. doi: 10.1105/tpc.112.105106

Los, D. A., Ray, M. K., and Murata, N. (1997). Differences in the control of the temperature-dependent expression of four genes for desaturases in *Synechocystis* sp. PCC 6803. *Mol. Microbiol.* 25, 1167–1175. doi: 10.1046/j.1365-2958.1997.5641912.x

Matthew, T., Zhou, W., Rupprecht, J., Lim, L., Thomas-Hall, S. R., Doebbe, A., et al. (2009). The metabolome of *Chlamydomonas reinhardtii* following induction of anaerobic H_2 production by sulfur depletion. *J. Biol. Chem.* 284, 23415–23425. doi: 10.1074/jbc.M109.003541

Merchant, S. S., Kropat, J., Liu, B., Shaw, J., and Warakanont, J. (2012). TAG, You're it! Chlamydomonas as a reference organism for understanding algal triacylglycerol accumulation. *Curr. Opin. Biotechnol.* 23, 352–363. doi: 10.1016/j.copbio.2011.12.001

Merchant, S. S., Prochnik, S. E., Vallon, O., Harris, E. H., Karpowicz, S. J., Witman, G. B., et al. (2007). The *Chlamydomonas* genome reveals the evolution of key animal and plant functions. *Science* 318, 245–250. doi: 10.1126/science.1143609

Miller, R., Wu, G., Deshpande, R. R., Vieler, A., Gärtner, K., Li, X., et al. (2010). Changes in transcript abundance in *Chlamydomonas reinhardtii* following nitrogen deprivation predict diversion of metabolism. *Plant Physiol.* 154, 1737–1752. doi: 10.1104/pp.110.165159

Minoda, A., Sonoike, K., Okada, K., Sato, N., and Tsuzuki, M. (2003). Decrease in the efficiency of the electron donation to tyrosine Z of photosystem II in an SQDG-deficient mutant of *Chlamydomonas. FEBS Lett.* 553, 109–112. doi: 10.1016/S0014-5793(03)00981-5

Moseley, J. L., Gonzalez-Ballester, D., Pootakham, W., Bailey, S., and Grossman, A. R. (2009). Genetic interactions between regulators of *Chlamydomonas* phosphorus and sulfur deprivation responses. *Genetics* 181, 889–905. doi: 10.1534/genetics.108.099382

Msanne, J., Xu, D., Konda, A. R., Casas-Mollano, J. A., Awada, T., Cahoon, E. B., et al. (2012). Metabolic and gene expression changes triggered by nitrogen deprivation in the photoautotrophically grown microalgae *Chlamydomonas reinhardtii* and *Coccomyxa* sp. C-169. *Phytochemistry* 75, 50–59. doi: 10.1016/j.phytochem.2011.12.007

Nguyen, H. M., Baudet, M., Cuine, S., Adriano, J.-M., Barthe, D., Billon, E., et al. (2011). Proteomic profiling of oil bodies isolated from the unicellular green microalga *Chlamydomonas reinhardtii*: with focus on proteins involved in lipid metabolism. *Proteomics* 11, 4266–4273. doi: 10.1002/pmic.201100114

Nikiforova, V., Freitag, J., Kempa, S., Adamik, M., Hesse, H., and Hoefgen, R. (2003). Transcriptome analysis of sulfur depletion in *Arabidopsis thaliana*: interlacing of biosynthetic pathways provides response specificity. *Plant J.* 33, 633–650. doi: 10.1046/j.1365-313X.2003.01657.x

Nishida, I., Tasaka, Y., Shiraishi, H., and Murata, N. (1993). The gene and the RNA for the precursor to the plastid-located glycerol-3-phosphate acyltransferase of *Arabidopsis thaliana. Plant Mol. Biol.* 21, 267–277. doi: 10.1007/BF00019943

Otsuka, H. (1961). Changes of lipid and carbohydrate contents of *Chlorella* cells during the sulfur starvation, as studied by the technique of synchronous culture. *J. Gen. Appl. Microbiol.* 7, 72–77. doi: 10.2323/jgam.7.72

Pootakham, W., Gonzalez-Ballester, D., and Grossman, A. R. (2010). Identification and regulation of plasma membrane sulfate transporters in *Chlamydomonas. Plant Physiol.* 153, 1653–1668. doi: 10.1104/pp.110.157875

Quettier, A. L., and Eastmond, P. J. (2009). Storage oil hydrolysis during early seedling growth. *Plant Physiol. Biochem.* 47, 485–490. doi: 10.1016/j.plaphy.2008.12.005

Rajakumari, S., Rajasekharan, R., and Daum, G. (2010). Triacylglycerol lipolysis is linked to sphingolipid and phospholipid metabolism of the yeast, *Saccharomyces cerevisiae. Biochim. Biophys. Acta* 1801, 1314–1322. doi: 10.1016/j.bbalip.2010.08.004

Ramanan, R., Kim, B. H., Cho, D. H., Ko, S. R., Oh, H. M., and Kim, H. S. (2013). Lipid droplet synthesis is limited by acetate availability in starchless mutant of *Chlamydomonas reinhardtii. FEBS Lett.* 587, 370–377. doi: 10.1016/j.febslet.2012.12.020

Ravina, C. G., Chang, C.-I., Tsakraklides, G. P., McDermott, J. P., Vega, J. M., Leustek, T., et al. (2002). The *sac* mutants of *Chlamydomonas reinhardtii* reveal transcriptional and posttranscriptional control of cysteine biosynthesis. *Plant Physiol.* 130, 2076–2084. doi: 10.1104/pp.012484

Remacle, C., Cline, S., Boutaffala, L., Gabilly, S., Larosa, V., Barbieri, M. R., et al. (2009). The *ARG9* gene encodes the plastid-resident N-acetyl ornithine aminotransferase in the green alga *Chlamydomonas reinhardtii. Eukaryot. Cell* 8, 1460–1463. doi: 10.1128/EC.00108-09

Rochaix, J. D. (1995). *Chlamydomonas reinhardtii* as the photosynthetic yeast. *Annu. Rev. Genet.* 29, 209–230. doi: 10.1146/annurev.ge.29.120195.001233

Sato, N. (2004). Roles of the acidic lipids sulfoquinovosyl diacylglycerol and phosphatidylglycerol in photosynthesis: their specificity and evolution. *J. Plant Res.* 117, 495–505. doi: 10.1007/s10265-004-0183-1

Sato, N., Aoki, M., Maru, Y., Sonoike, K., Minoda, A., and Tsuzuki, M. (2003a). Involvement of sulfoquinovosyl diacylglycerol in the structural integrity and heat-tolerance of photosystem II. *Planta* 217, 245–251. doi: 10.1007/s00425-003-0992-9

Sato, N., Tsuzuki, M., and Kawaguchi, A. (2003b). Glycerolipid synthesis in *Chlorella kessleri* 11h. I. Existence of a eukaryotic pathway. *Biochim. Biophys. Acta* 1633, 27–34. doi: 10.1016/S1388-1981(03)00069-6

Sato, N., Tsuzuki, M., Matsuda, Y., Ehara, T., Osafune, T., and Kawaguchi, A. (1995). Isolation and characterization of mutants affected in lipid metabolism of *Chlamydomonas reinhardtii. Eur. J. Biochem.* 230, 987–993. doi: 10.1111/j.1432-1033.1995.0987g.x

Siaut, M., Cuiné, S., Cagnon, C., Fessler, B., Nguyen, M., Carrier, P., et al. (2011). Oil accumulation in the model green alga *Chlamydomonas reinhardtii*: characterization, variability between common laboratory strains and relationship with starch reserves. *BMC Biotechnol.* 11:7. doi: 10.1186/1472-6750-11-7

Sueoka, N. (1960). Mitotic replication of deoxyribonucleic acid in *Chlamydomonas reinhardti. Proc. Natl. Acad. Sci. U.S.A.* 46, 83–89.

Sugimoto, K., Midorikawa, T., Tsuzuki, M., and Sato, N. (2008). Upregulation of PG synthesis on sulfur-starvation for PSI in *Chlamydomonas. Biochem. Biophys. Res. Commun.* 369, 660–665. doi: 10.1016/j.bbrc.2008.02.058

Sugimoto, K., Sato, N., and Tsuzuki, M. (2007). Utilization of a chloroplast membrane sulfolipid as a major internal sulfur source for protein synthesis in the early phase of sulfur starvation in *Chlamydomonas reinhardtii. FEBS Lett.* 581, 4519–4522. doi: 10.1016/j.febslet.2007.08.035

Sugimoto, K., Tsuzuki, M., and Sato, N. (2010). Regulation of synthesis and degradation of a sulfolipid under sulfur-starved conditions and its physiological significance in *Chlamydomonas reinhardtii. New Phytol.* 185, 676–686. doi: 10.1111/j.1469-8137.2009.03115.x

Tabei, Y., Okada, K., and Tsuzuki, M. (2007). Sll1330 controls the expression of glycolytic genes in *Synechocystis* sp. PCC 6803. *Biochem. Biophys. Res. Commun.* 355, 1045–1050. doi: 10.1016/j.bbrc.2007.02.065

Teramoto, H., Nakamori, A., Minagawa, J., and Ono, T. (2002). Light-intensity-dependent expression of Lhc gene family encoding light-harvesting chlorophyll-a/b proteins of photosystem II in *Chlamydomonas reinhardtii. Plant Physiol.* 130, 325–333. doi: 10.1104/pp.004622

Toepel, J., Albaum, S. P., Arvidsson, S., Goesmann, A., la Russa, M., Rogge, K., et al. (2011). Construction and evaluation of a whole genome microarray of *Chlamydomonas reinhardtii. BMC Genomics* 25:579. doi: 10.1186/1471-2164-12-579

Wykoff, D. D., Davies, J. P., Melis, A., and Grossman, A. R. (1998). The regulation of photosynthetic electron transport during nutrient deprivation in *Chlamydomonas reinhardtii. Plant Physiol.* 117, 129–139. doi: 10.1104/pp.117.1.129

Yu, B., Wakao, S., Fan, J., and Benning, C. (2004). Loss of plastidic lysophosphatidic acid acyltransferase causes embryo-lethality in *Arabidopsis. Plant Cell Physiol.* 45, 503–510. doi: 10.1093/pcp/pch064

Zhang, Z., Shrager, J., Jain, M., Chang, C.-W., Vallon, O., and Grossman, A. R. (2004). Insights into the survival of *Chlamydomonas reinhardtii* during sulfur starvation based on microarray analysis of gene expression. *Eukaryot. Cell* 3, 1331–1348. doi: 10.1128/EC.3.5.1331-1348.2004

Conflict of Interest Statement: The authors declare that the research was conducted in the absence of any commercial or financial relationships that could be construed as a potential conflict of interest.

14

The family of LSU-like proteins

Agnieszka Sirko*, Anna Wawrzyńska, Milagros Collados Rodríguez and Paweł Sęktas

Institute of Biochemistry and Biophysics, Polish Academy of Sciences, Warsaw, Poland

Edited by:
Stanislav Kopriva, University of Cologne, Germany

Reviewed by:
Stanislav Kopriva, University of Cologne, Germany
Rainer Hoefgen, Max Planck Institute of Molecular Plant Physiology, Germany

****Correspondence:***
Agnieszka Sirko, Institute of Biochemistry and Biophysics, Polish Academy of Sciences, Pawinskiego 5A, 02-106 Warsaw, Poland
e-mail: asirko@ibb.waw.pl

The plant response to sulfur deficiency includes extensive metabolic changes which can be monitored at various levels (transcriptome, proteome, metabolome) even before the first visible symptoms of sulfur starvation appear. Four members of the plant-specific *LSU* (response to Low SUlfur) gene family occur in *Arabidopsis thaliana* (*LSU1-4*). Variable numbers of *LSU* genes occur in other plant species but they were studied only in *Arabidopsis* and tobacco. Three out of four of the *Arabidopsis LSU* genes are induced by sulfur deficiency. The *LSU*-like genes in tobacco were characterized as *UP9* (UPregulated by sulfur deficit 9). LSU-like proteins do not have characteristic domains that provide clues to their function. Despite having only moderate primary sequence conservation they share several common features including small size, a coiled–coil secondary structure and short conserved motifs in specific positions. Although the precise function of LSU-like proteins is still unknown there is some evidence that members of the LSU family are involved in plant responses to environmental challenges, such as sulfur deficiency, and possibly in plant immune responses. Various bioinformatic approaches have identified LSU-like proteins as important hubs for integration of signals from environmental stimuli. In this paper we review a variety of published data on *LSU* gene expression, the properties of *lsu* mutants and features of LSU-like proteins in the hope of shedding some light on their possible role in plant metabolism.

Keywords: *Arabidopsis*, tobacco, coiled coil, SALK mutants, gene expression, OAS, ethylene, UP9

INTRODUCTION

The first global analyzes of gene expression profiles under sulfur deficiency stress in *Arabidopsis* appeared in Hirai et al. (2003), Maruyama-Nakashita et al. (2003) and Nikiforova et al. (2003), however, these studies focused on genes encoding proteins with known functions. Two years later *LSU1* (At3g49580) and *LSU2* (At5g24660) were identified as two out of 15 sulfur-responsive genes which were significantly up-regulated in roots as early as 2 h (*LSU1*) or 4 h (*LSU2*) after plants were transferred to sulfur-free medium; a sulfur-responsive element (SURE) was identified in their promoter regions (Maruyama-Nakashita et al., 2005). In the same year the At3g49580 gene appeared on the list of important network elements identified in a pioneering study involving reconstruction of the gene-metabolite network involved in the plant response to sulfur deficiency stress (Nikiforova et al., 2005). At the same time the tobacco *UP9* gene was independently shown to be strongly and specifically up-regulated by sulfur deficiency (-S) using an unbiased suppression subtractive hybridization approach (Wawrzyńska et al., 2005). Since then rather few studies focusing on *LSU*-like genes and proteins have been published; however, several reports presented results of high throughput experiments which included also data related to the regulation of expression and phenotypes of the *Arabidopsis lsu* mutants. The systematic review of available data presented below provides clear evidence of the importance of this family of proteins and, hopefully contributes to uncovering their function.

LSU/UP9 GENES AND THEIR EXPRESSION

LSU GENES IN *Arabidopsis*

Arabidopsis thaliana contains four *LSU* genes (*LSU1–LSU4*) which are localized in pairs of direct repeats on two chromosomes (**Figure 1**). The nucleotide sequences of chromosome III corresponding to *LSU1* and *LSU3* transcripts are separated by about 2250 bp; the distance between *LSU2* and *LSU4* is slightly shorter (about 2060 bp). The open reading frames (ORFs) are relatively small and consist of about 280 bp. Most *LSU* genes have no introns; however, a spliced variant of *LSU1* (At3g49580.2) encoding a protein with internal deletion of 19 amino acids was reported [http://www.arabidopsis.org]. Searches of publicly available microarrays using the Genvestigator platform (Zimmermann et al., 2004, 2008) showed that *LSU1* and *LSU2* are strongly expressed under -S but *LSU4* appears not to be induced by -S. *LSU3* was not included in these microarrays. Expression of *LSU1* and *LSU2* is induced not only by -S but also by other stressful environmental conditions such as salt stress and $AgNO_3$ treatment.

Several global analyzes of *Arabidopsis* gene expression in various growth conditions and developmental stages provide valuable information about expression of *LSU*s. Most of these data relate to *LSU2*, suggesting that this member of the family is preferentially involved in the plant response to certain stresses or certain processes. Expression of *LSU2* is induced during oxidative stress (Davletova et al., 2005) and at the beginning of an extended night, which may indicate that it is induced by carbon starvation and in response to sugar (Usadel et al., 2008). Recently it has been

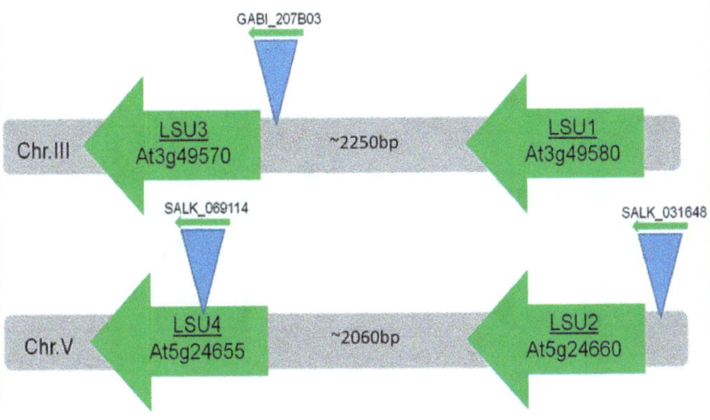

FIGURE 1 | Localization of 'response to Low Sulfur' (LSU) genes in the *Arabidopsis* genome. Number of base pairs (bp) between LSU open reading frames is indicated. Positions of T-DNA inserts are shown schematically.

shown that expression of *LSU2* is induced by a combination of light and plastid signaling (Ruckle et al., 2012); these authors identified *LSU2* as one of seven so-called *END* (enhanced de-etiolation) genes. They went on to characterize some of the *end* mutants, including *lsu2* (SALK_031648), and showed that expression of some photosynthesis-related genes (*Lhcb1.4*, *PsbS*, *RbcS1*, and *CHS*) was attenuated in them. The mechanisms responsible for the regulation of transcription in *end* mutants remain unclear; the expression of *END* genes is regulated by a variety of signals besides light and plastid signals, so it may be different for different mutants. Ruckle et al. (2012) concluded that the products of *END* genes contribute to a complex network responsible for optimization of chloroplast function during chloroplast biogenesis, and probably during periods of chloroplast dysfunction. The link between *LSU2* and chloroplasts was also emphasized in a recent report, where *LSU2* was identified as one of 39 genes that were differentially expressed in six independent microarray experiments using plants with the provoked retrograde signaling in response to disturbances of chloroplast performance by chemical treatment or mutation of some metabolic pathways (Glasser et al., 2014).

In addition *LSU2* was tentatively identified as one of the genes involved in the crosstalk between several signals (nitrate, sulfur, iron, and hormones) from analysis of transcriptome data for *Arabidopsis* plants grown under sulfur and iron restriction, and various nitrate and stress hormone treatments (Omranian et al., 2012).

Somewhat surprisingly expression of *LSU1* was found to be repressed during infection with cabbage leaf curl virus (CaLCuV), whereas *LSU2* expression was apparently unaffected (Ascencio-Ibanez et al., 2008). The *LSU1* gene was also shown to be constitutively (phase-independently) expressed during pollen germination and tube growth (Wang et al., 2008).

Analysis of publicly available data from two sets of high-throughput experiments led to the identification of *LSU1* as a member of a six-gene cluster responding to O-acetylserine (OAS) levels in shoots (Hubberten et al., 2012b). One set of data was from experiments on diurnal oscillations of genes and metabolites (Espinoza et al., 2010); the second set was from studies of plants during the light-dark transition (Caldana et al., 2011). OAS was one of the compounds most affected by changes in conditions in both studies. Hubberten et al. (2012b) confirmed that regardless of temperature (20 or 4°C), the level of OAS (and the expression of the above-mentioned genes) increased during the night and decreased during the day. Increased expression of *LSU1* (and the other five genes) was also observed following induction of the chemically inducible ectopic copy of *SERAT* (encodes serine acetyltransferase, which is involved in OAS synthesis) in sulfur-sufficient transgenic plants (Hubberten et al., 2012b). The same group used a split-root approach to explore further the role of OAS in the regulation of plant S-status in *Arabidopsis*. One half of the root was exposed to -S, whilst the other half of the root of the same plant was grown in sulfur-sufficient conditions. OAS levels were low in both halves of the split root, and expression of previously mentioned OAS-responsive genes, including *LSU1* was also low (Hubberten et al., 2012a).

It has recently been reported that expression of *LSU1* (and *BGLU28* [At2g44460], *SDI1* [At5g48850] and *SULTR4;2* [At3g12520]) is much less affected by S availability in the *sultr1;2* mutants than in the wild type (Zhang et al., 2014). This observation is not strictly related to the function of LSU/UP9 proteins, nevertheless it is worth noting because it makes an important contribution to understanding the plant mechanisms responsible for sensing S availability and thus also S status-dependent regulation of gene expression.

The results of *in silico* analysis of the promoter regions of the *LSU*s are shown in **Figure 2**. Analyzes of the 500 bp upstream transcription start site (TSS) demonstrate the potential for differential expression of each *LSU* gene. In all but the *LSU4* promoter, there is an element specific for induction in -S, UPE-box (Wawrzyńska et al., 2010). Additional sulfur-responsive elements (SURE boxes) which are not included in **Figure 2** have previously been identified in the promoter regions of *LSU1* and *LSU2* (Maruyama-Nakashita et al., 2005). The *LSU1* promoter contains the largest number of potential regulatory elements. Only *LSU1* has consensuses for FUSCA3 and OPAQUE2-like factors, both of which are essential for seed-specific expression (Moreno-Risueno et al., 2008) and the *cis*-elements related to response to dehydration and sucrose. The binding site for the bZIP transcription factors (TFs; G-box) is present in the *LSU1* and *LSU3* promoters, whilst the consensus for binding the WRKY TFs is present in *LSU1* and *LSU4*. The promoter regions of *LSU2*, *LSU3*, and *LSU4* (but not *LSU1*) have sequences for binding the light-responsive factors. The *LSU2* promoter contains sequences responsive to auxins and jasmonic acid as well as sequences for APETALA2 and SQUAMOSA promoter-binding protein (SBP), indicating that *LSU2* may play an important role in ontogenesis. The *LSU3* promoter contains a specific sequence which binds the INDETERMINATE1 (ID domain) responsible for the transition to flowering; the *LSU4* promoter has a *cis*-acting sequence responsive to abscisic acid (ABA). The putative roles of these elements in the regulation of *LSU* gene expression should be verified experimentally.

UP9 GENES IN TOBACCO

The tobacco LSU-like proteins were grouped into six clusters (UP9A to UP9F); however, the exact number of such genes in

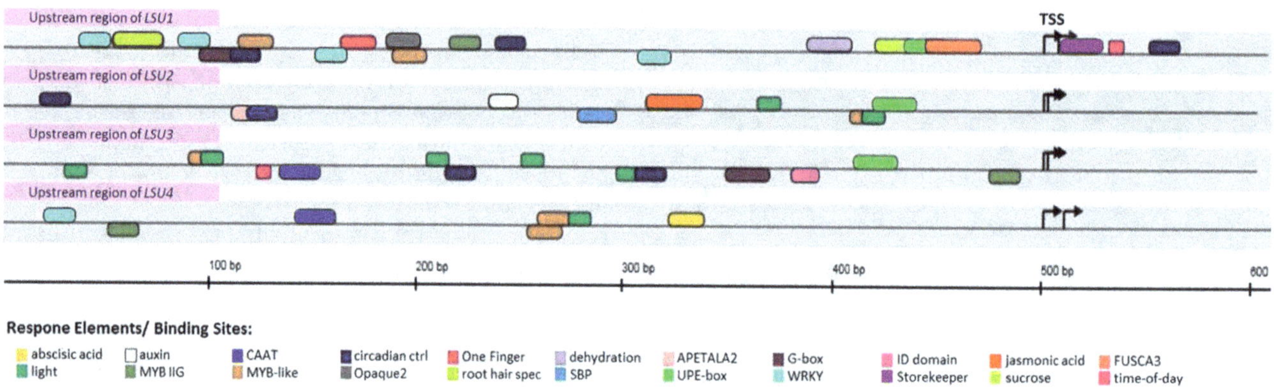

FIGURE 2 | Depiction of transcription factor (TF) binding sites found in the upstream regions of *LSU1-4*. The sequences were scanned for matches to TFs binding sites using MatInspector, part of the Genomatix Software Suite (www.genomatix.de). A match is represented by a round-ended rectangle. Matches with the positive and negative strands are depicted above or below the sequence line, respectively. The arrow symbol on the sequence stands for a transcription start site (TSS); note that there are several putative TSSs for each *LSU*.

tobacco remains unclear (Lewandowska et al., 2010). Only one of these genes, *UP9C*, has been investigated further. An increase in the level of the *UP9C* transcript was observed just 2 days after transferring plants from sulfur-sufficient to sulfur-deficient medium in all parts tested (roots, young leaves, mature leaves, stems) and a further increase in transcript level was observed after additional days under -S. Analysis of the promoter region of *UP9C* indicated that it has only one TSS located 109 bp upstream of the translational start site (Wawrzyńska et al., 2010). The same study also reported the presence of an interesting motif, UPE-box, in the *UP9C* promoter. The authors used the DNA fragment containing UPE-box (from the promoter region of *UP9C*) in a yeast-one-hybrid experiment and identified NtEIL2, a tobacco member of the EIL family, as a TF which bound to the UPE-box (Wawrzyńska et al., 2010). Transient expression assays in *Nicotiana benthamiana* plants indicated that NtEIL2 was responsible for the UPE-box-dependent up-regulation of the reporter gene in -S conditions. Interestingly, an *Arabidopsis* homolog of NtEIL2, SLIM1, which has been identified earlier as a critical transcriptional regulator of plant sulfur response and sulfur metabolism (Maruyama-Nakashita et al., 2006), was also able to bind to *UP9C* promoter containing UPE-box. Mutations in UPE-box affect the binding of both factors, NtEIL2 and AtSLIM1; however, in the presence of SLIM1 the promoter was constitutively active, regardless of the plants' sulfur status (Wawrzyńska et al., 2010). In conclusion, *UP9C* seems to be regulated directly by NtEIL2, in a sulfur-dependent manner. Some as yet unidentified species-specific factors guarantee the specificity of the NtEIL2-dependent up-regulation of the *UP9C* gene (and possibly other genes containing UPE-box) in -S conditions. Further *in silico* analysis of the promoter region showed that the *UP9C* promoter has elements which are potentially responsive to light, salt stress and phytohormones such as ABA, ethylene and cytokines as well as the above-mentioned SURE located 350 bp upstream of the start codon. The biological significance of these cis-factors is unknown.

UPE-box is also present in the promoters of several *Arabidopsis* genes (Wawrzyńska et al., 2010). A search of the genome sequence revealed that it was present in the promoter regions of *LSU1* (At3g49580), *LSU2* (At5g24660), and *LSU3* (At3g49570; but not *LSU4*) and also in several other genes which are up-regulated in -S. Interestingly the set of genes containing UPE-box in promoter appears to be very similar to the OAS cluster genes (Hubberten et al., 2012b).

PHENOTYPES OF THE MUTANTS

ANALYSIS OF *Arabidopsis* SALK MUTANTS

One of the difficulties in determining the function of proteins from the LSU family is that information about the phenotypes of knock-out (KO) and knock-down (KD) mutants is scarce. There are T-DNA insertional mutants for *LSU2* (e.g., SALK_31648, SALK_070105), *LSU3* (e.g., GABI_207B03), and *LSU4* (e.g., SALK_069114) but not for *LSU1* (**Figure 1**). The high probability of functional overlap makes it desirable to test multiple *lsu* KO or KD mutants, but so far no data have been published. Most available data relate to *lsu2* mutants, for example an interesting report on the functional characterization of abiotic stress response proteins with unknown function was published recently (Luhua et al., 2013). These authors tested the response to treatments such as salinity, oxidative, osmotic, heat, cold, and hypoxia stress of 1007 T-DNA insertional mutants in genes with unknown function. The *lsu2* mutant (SALK_31648C) was one of 69 genes with an unknown function that seemed to be more tolerant of osmotic stress than the wild type; responses to other stresses did not appear to be altered. Another study reported that *lsu2* mutants (SALK_031648, SALK_070105) exhibited enhanced susceptibility to two evolutionarily distinct pathogens, *Pseudomonas syringae* and *Hyaloperonospora arabidopsidis* (Mukhtar et al., 2011). According to the authors, LSU2 (and other proteins, for example JAZ3) has some effect on the functioning of the NB-LRR (nucleotide binding leucine-rich repeat) intracellular immune receptors with particular emphasis on the RPS2 (Resistance to *Pseudomonas syringae* 2) protein. Activation of NB-LRR proteins is responsible for robust disease-resistance responses such as host cell death and systemic defense signaling.

Defects in flower and inflorescence development were observed in the insertion mutant *lsu4* (SALK_069114) when grown under short-day conditions (Myakushina et al., 2009). Mutation of the *LSU4* gene caused delayed flowering and disturbances in the formation of flower organs. There were also significant changes in the expression of many regulatory genes, including down-regulation of *LEAFY* (*LFY*), *APETALA1* (*AP1*), *APETALA3* (*AP3*), *PISTILLATA* (*PI*) and *SEPALLATA3* (*SEP3*) and up-regulation of *APETALA2* (*AP2*), *AGAMOUS* (*AG*) and *SEPALLATA* (*SEP2*). It is worth mentioning that the authors noted that *LSU4* expression increased two to threefold under phosphorus, nitrogen, potassium, or iron deficiency.

SILENCING OF *UP9* IN TOBACCO

Analysis of the tobacco antisense *UP9C* transformants (KD) revealed no evidence of phenotypic differences from the wild type, although the KD transformants did have a different metabolite profile from wild type plants (Lewandowska et al., 2010). The metabolite profiles of KDs grown in -S were more similar to the profiles of parental line plants grown in sulfur-sufficient conditions, suggesting that the KD lines failed to adjust their metabolism to the -S conditions. In addition the level of non-protein thiols (consisting mostly of glutathione) in mature leaves and roots, but not in young leaves, was different in KD plants. Wild type plants showed the expected reduction in glutathione levels in mature leaves 2 days after transfer to -S, but there was no change in the KDs, which had a high level of glutathione in the mature leaves regardless of the conditions. The mutants did, however, have low levels of glutathione in the roots, particularly under -S; mutants also had lower levels of sulfur in the roots under -S than the wild type. Another interesting observation was that under -S several genes were misregulated in the mutants; usually the level of transcription was lower in the KDs than the wild type. It must be remembered, however, that only a limited selection of genes was tested and no high-throughput analysis was performed in this study.

Ethylene plays a very important role in plant response to several stresses and regulates many processes (Adie et al., 2007; Lin et al., 2009). In -S conditions ethylene levels increase in wild type tobacco. KD plants have lower levels of ethylene than wild type plants in -S conditions (Moniuszko et al., 2013), but in sulfur-sufficient conditions the *UP9C*-silenced line produced slightly more ethylene than the wild type. Transcriptome analysis revealed significant changes in the gene expression pattern of the KD line relative to the wild type; only 130 of the 360 genes up-regulated in the wild type in -S were also up-regulated in the mutants and only 14 of 91 genes down-regulated in the wild type were also down-regulated in the mutant. Some genes were regulated in the mutant but not in the wild type. Differences in the expression profiles of the mutants and wild type may provide clues to function. Gene ontology (GO) analysis indicated clearly that UP9C does not participate in sulfur deficiency-dependent regulation of genes encoding isoforms of APS reductase (APR) or genes encoding S-adenosylmethionine synthase (SAMS) as these genes were induced in -S in both the mutant and the wild type. Genes from several categories including 'response to hormone stimulus,' 'signal transduction,' 'defense response' and 'regulation of transcription' genes were, however, misregulated in the mutant. Although many genes had different expression profiles in the KD several genes related to ethylene signaling (homologs of *Arabidopsis* EIN3-BINDING F BOX PROTEIN 1 (EBF1), ETHYLENE INSENSITIVE 4 (EIN4) and ETHYLENE RESPONSE SENSOR 1 (ERS1)) and ABA- and cytokine-mediated signaling (homologs of ARABIDOPSIS THALIANA HOMEOBOX 7 (ATHB-7) and HISTIDINE-CONTAINING PHOSPHOTRANSMITTER 1 (AHP1)) attracted particular attention (Moniuszko et al., 2013). The expression of these genes was slightly higher in the KD than in the wild type in sulfur-sufficient medium, but the most interesting effect was the very low expression of these genes in the KD line when plants were transferred to -S conditions. In *Arabidopsis* EBF1 is important for proteosomal degradation of ETHYLENE-INSENSITIVE3 (EIN3), the positive regulator of ethylene-responsive genes, whilst EIN4 and ERS1 are genes for ethylene receptors (Wang et al., 2006). These observations, along with the reduced ethylene level in the mutant grown in -S, prompted the authors to hypothesize that UP9C is involved in modulation of the ethylene signaling pathway, which is important in plant response to -S conditions. The main conclusion to be drawn from this work is that one of the functions of UP9C - and possibly also other LSU-like proteins - in plant response to -S may be related to the involvement of LSU-like proteins in tuning up 'hormone stimulus' signals induced by -S conditions. Although the authors focused on ethylene it is likely that other hormone signaling systems, possibly those involved in -S response, are also affected in the mutant.

LSU-LIKE PROTEINS AND THEIR POTENTIAL INTERACTING PARTNERS

LSU/UP9 family proteins are small (10–13 kDa) and consist of about 100 amino acids (**Figure 3**). A BLAST (blastp) search of non-redundant protein sequences revealed multiple homologs of LSU in various plant species, both monocotyledons and dicotyledons, including *Solanum lycopersicum* (4 homologs), *Solanum tuberosum* (4), *Glycine max* (3), *Populus trichocarpa* (3), *Zea mays* (3), *Hordeum vulgare* (2), *Oryza sativa* (3), *Beta vulgaris* (2) and many others. The LSU-like proteins are also present in gymnosperms, like *Pinus* sp. We believe that so far only *Arabidopsis* LSUs and tobacco UP9s have been analyzed. Computer analysis and the circular dichroism spectra indicated that UP9C has an alpha-helical structure (Lewandowska et al., 2010). The presence of two stranded coiled–coil regions in UP9C (Lewandowska et al., 2010) is strongly suggestive of multimer formation; UP9C-UP9C interactions were observed in yeast two-hybrid (Y2H) experiments. Interestingly, despite relatively weak conservation of the primary sequence, both homologous UP9C-UP9C and heterologous LSU-UP9C (cross-species) interactions were observed. A potential nuclear localization signal was found in UP9C using the MOTIFSCAN program; according to PSORT UP9C has a cytosol-nuclear localization. No nuclear localization motifs have been identified in *Arabidopsis* LSU proteins. Nuclear localization of UP9C was reported (Lewandowska et al., 2010), but more recent experimental data suggest that it is present in both cytoplasm and nucleus (Moniuszko et al., 2013). Because they are small proteins it is likely that LSU-like proteins can cross the nuclear pores without

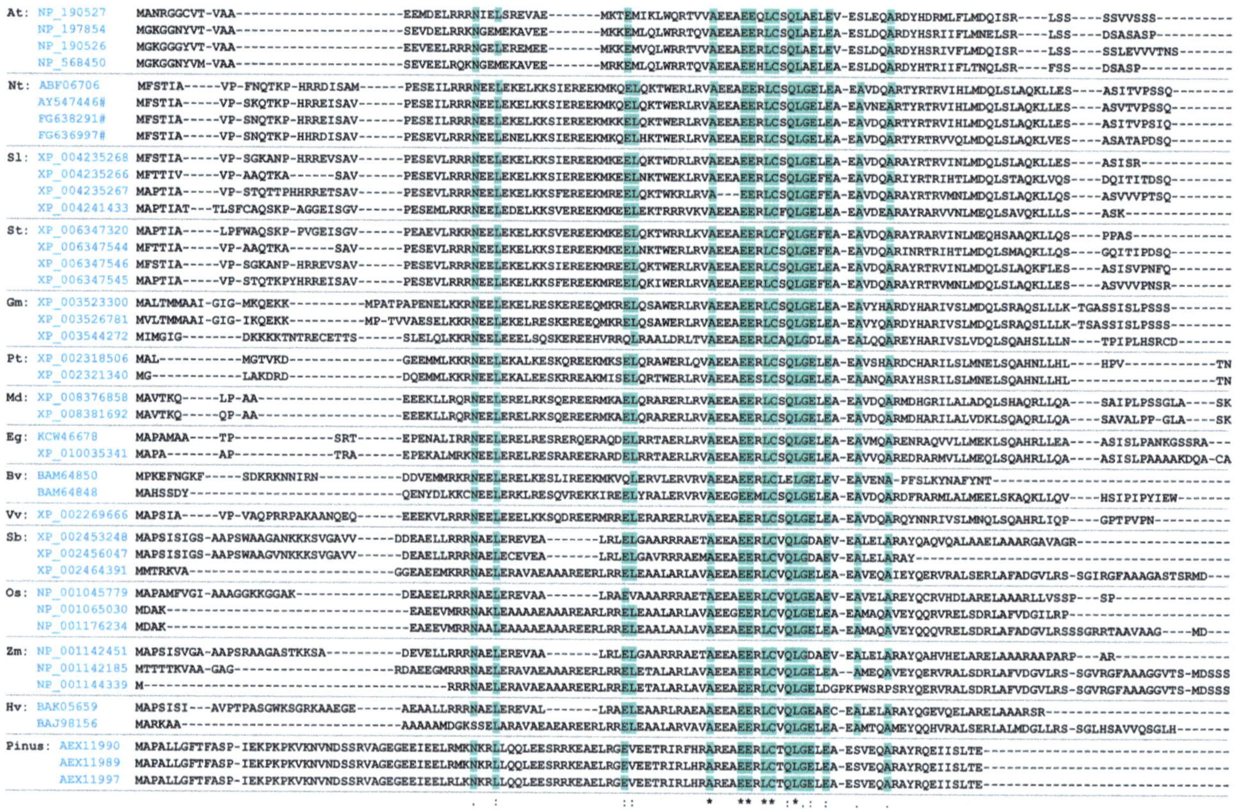

FIGURE 3 | Alignment of the selected LSU-like proteins. The evolutionary conserved amino acids identified by the MAFT alignment software [http://mafft.cbrc.jp] are highlighted. The accession numbers of the protein sequences are provided. # denotes the accession number to the corresponding nucleotide sequence; At, *Arabidopsis thaliana*; Nt, *Nicotiana tabacum*; Sl, *Solanum lycopersicum*; St, *Solanum tuberosum*; Gm, *Glycine max*; Pt, *Populus trichocarpa*; Md, *Malus domesticus*; Eg, *Eucalyptus grandis*; Bv, *Beta vulgaris*; Vv, *Vitis vinifera*; Sb, *Sorghum bicolor*; Os, *Oryza sativa*; Zm, *Zea mays*; Hv, *Hordeum vulgare*; Pinus, *Pinus taeda*.

a specific transport mechanism. There are no specific motifs or domains in LSU/UP9 proteins that suggest their function. The significance of the short, strongly evolutionarily conserved region in the members of this family (**Figure 3**) is unknown.

The LSU/UP9 proteins seem to be involved in multiple protein–protein contacts (**Table 1**; **Figure 4**). Data from tobacco are limited; however, some of the interacting partners identified by the Y2H approach have been confirmed using other methods. For example, UP9C interacts with ACO2A, an enzyme which converts 1-aminocyclopropane-1-carboxylate (ACC) to ethylene; it was therefore proposed that ethylene production might be controlled by UP9C through its interactions with ACO2A (Moniuszko et al., 2013). Joka2/NBR1 functions as a cargo receptor in selective autophagy (Zientara-Rytter et al., 2011) and is another protein which is unquestionably involved in interactions with UP9/LSU; however, the biological significance of these interactions is as yet unexplained.

Mapping of the *Arabidopsis* interactome based on the Y2H system (Arabidopsis Interactome Mapping Consortium, 2011) has revealed numerous partners of LSU1 and LSU2; unfortunately LSU3 and LSU4 were not included in the experiments. The lists of proteins which potentially interact with LSU1 or LSU2 are quite long (80 and 37 proteins, respectively) and include 17 elements common to both proteins (**Figure 4**). Functional categorization of potential interacting partners using GO analysis indicated some changes in the distribution of gene product locations, molecular functions and biological processes relative to those for the genome as a whole (**Figure 5**). Both groups (LSU1 and LSU2 interacting partners) were more likely than average to be located in the nucleus, chloroplasts (plastids) or ribosomes. Nuclear proteins which are LSU1 or LSU2 interacting partners include members of the JAZ family of repressors. It is worth noting that it has been demonstrated that the tobacco homolog of JAZ interacts with UP9C (**Table 1**). Molecular Functions GO categories such as 'DNA or RNA binding,' 'protein binding' and 'TF

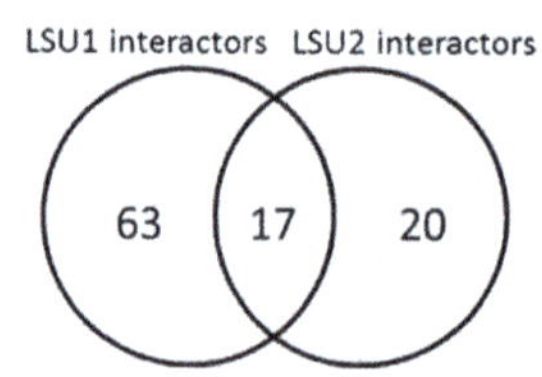

FIGURE 4 | Venn diagram of potential LSU1 and LSU2 interacting partners (*Arabidopsis* Interactome Mapping Consortium, 2011).

Table 1 | Tobacco proteins found as interacting with tobacco UP9C.

Accession number	Clone name	Number of amino acids in the clone	Identification/Function	Corresponding *A. thaliana* gene
Library from *Nicotiana plumbaginifolia* seedlings grown in S-sufficient conditions				
ABF06703	NpJoka2	467	NBR1-like, cargo receptor of selective autophagy	At4g24690
ABF06705	NpJoka8	360	HLH superfamily; bHLH66	At2g24260
ABF06704	NpJoka20	161	Ribosomal L7/L12	At3g27850 At4g36420 At4g37660
Library from 3-month-old *Nicotiana tabacum* plants transfered for 2 days into S-deficient conditions				
GU066878	Joka 31A	117	ACC oxidase	At1g05010
GU066879	Joka 31B	56	ACC oxidase	At1g05010
GU066880	Joka 32	253	PRP11; ZnF-U1 – splicing factor	At2g32600
GU066881	Joka 33	245	TIM50 (mt-inner membrane)	At1g55900
GU066882	Joka 34	376	RING-finger-containing E3 ubiquitin ligase	At3g58030
GU066883	Joka 35	147	RING-finger-containing E3 ubiquitin ligase	At3g16720
GU066884	Joka 36	314	Apetala 2-like (transcription factor)	At2g28550 At4g36920 At5g67180 At5g60120
GU066885	Joka 37	110	Function unknown; Involucin repeat; phosphoenolopyruvate carboxylase; E2-enzyme	At2g28540 At3g55720
GU066886	Joka 38	144	DUF248/methyltransferase	At4g18030 At1g26850
GU066887	Joka 39	119	DUF632/Function unknown, leucine zipper	At2g27090
GU066888	Joka 40	515	Function unknown, nucleoporin-like	At4g37130
GU066889	Joka 41	99	Poly A binding	At1g49760 At4g34110 At2g23350 At1g22760 At1g71770
GU066890	Joka 42	77	FtsH protease	At2g26140
GU066891	Joka 43	128	Unknown	At3g24506 At2g17240
GU066892	Joka 44	75	Microtubule-associated MAP65-1a	At5g55230 At4g26760
GU066893	Joka 46	184	CHORD, PBS2, RAR1, interacts with SGT1; Rar1/TMV resistance	At5g51700
GU066894	Joka 47	200	JAZ1 (transcription factor)	At1g19180

activity' are over-represented among LSU1 and LSU2 interacting partners, whereas categories related to some enzymatic activities are under-represented.

We have also noticed that the group of LSU1 interacting partners includes a relatively high proportion of proteins from the Molecular Function GO category 'structural molecule activity' (all are ribosomal proteins). Overrepresentation of any Biological Process GO category was less apparent; 'cell organization and biogenesis' and 'DNA-dependent transcription' and perhaps the 'protein metabolism' and 'transcription, DNA-dependent' categories were only slightly overrepresented among LSU1 partners. The category of 'cell organization and biogenesis' proteins which interact with LSU1 includes some ribosomal proteins, chaperones and members of RING superfamily (potential E3 ubiquitin ligases). LSU1 partners include, amongst others, members of the ERF/AP2, bHLH and myb-like HTH families of transcriptional factors.

It has also been demonstrated that LSU2 protein interacts with the pathogenic effectors of two different plant pathogens, the bacterium *P. syringae* and the oocyte *Hyaloperonospora arabidopsidis*

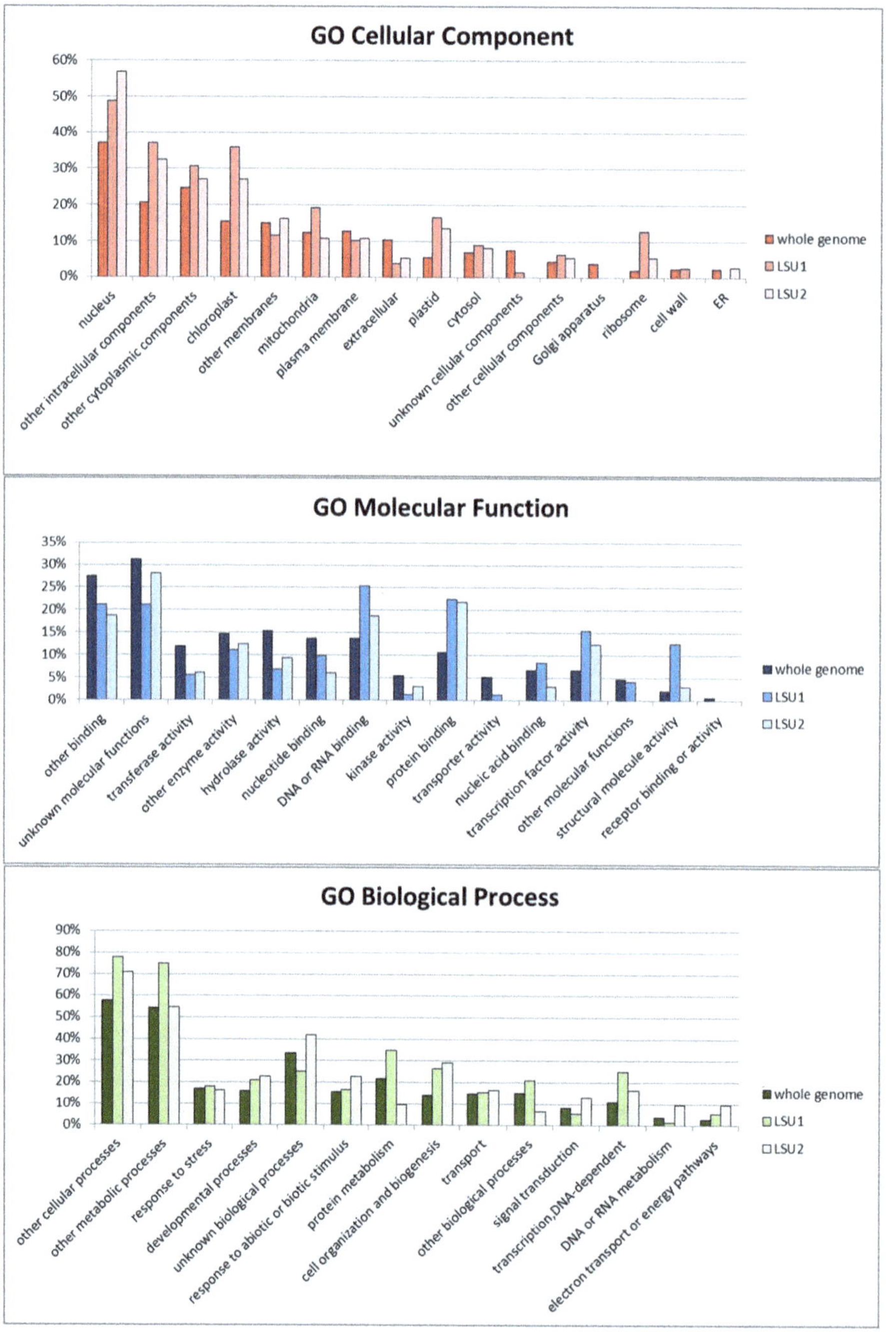

FIGURE 5 | Functional categorization of the potential LSU1 and LSU2 interacting partners for GO Cellular Component, GO Molecular Function and GO Biological Process. The analysis was done using the Gene ontology tools available at TAIR [http://www.arabidopsis.org].

(Mukhtar et al., 2011). The involvement of LSU2 in the immune response to these pathogens was verified by the same authors through the demonstration of enhanced susceptibility in *lsu2* mutants (see also above).

CONCLUDING REMARKS

It is unclear why plants have several isoforms of LSU. The proteins have probably partially overlapping functions; however, the data reported above suggest some functional specificity. *LSU1, LSU2,* and *LSU3* genes from *Arabidopsis* are induced by sulfur deficiency; however, only *LSU2* has been shown to be involved in retrograde signaling associated with chloroplast malfunction. The molecular role of LSU-like family members remains unclear although an increasing amount of evidence links the family with complex intracellular regulatory functions and coordination of organellar and cytosolic metabolism. It is possible that LSU/UP9 proteins modulate degradation of some specific "strategic" targets (such as TFs) in response to environmental stresses or are (directly or indirectly) involved in regulation of cellular degradation machinery. Although there is no clear evidence that LSU-like family members play such roles their interactions with presumed E3 ubiquitin ligases, chaperons (DnaJ-domain, Hsp60) and particularly with NBR1 (a selective autophagy cargo receptor) make the hypothesis plausible.

AUTHOR CONTRIBUTIONS

Agnieszka Sirko and Anna Wawrzyńska drafted the manuscript. Milagros Collados Rodríguez and Paweł Sęktas contributed to the writing process and preparation of figures. All authors were involved in preparing the final version.

ACKNOWLEDGMENTS

Research in our laboratory is supported by grants from the Polish Ministry of Science and Higher Education (W16/7.PR/2011), the Polish National Science Centre (2011/03/B/NZ1/01243) and EC FP7-PEOPLE (BIONUT-ITN; project ref.: 264296).

REFERENCES

Adie, B., Chico, J. M., Rubio-Somoza, I., and Solano, R. (2007). Modulation of plant defenses by ethylene. *J. Plant Growth Regul.* 26, 160–177. doi: 10.1007/s00344-007-0012-6

Arabidopsis Interactome Mapping Consortium. (2011). Evidence for network evolution in an *Arabidopsis* interactome map. *Science* 333, 601–607. doi: 10.1126/science.1203877

Ascencio-Ibanez, J. T., Sozzani, R., Lee, T. J., Chu, T. M., Wolfinger, R. D., Cella, R., et al. (2008). Global analysis of *Arabidopsis* gene expression uncovers a complex array of changes impacting pathogen response and cell cycle during geminivirus infection. *Plant Physiol.* 148, 436–454. doi: 10.1104/pp.108.121038

Caldana, C., Degenkolbe, T., Cuadros-Inostroza, A., Klie, S., Sulpice, R., Leisse, A., et al. (2011). High-density kinetic analysis of the metabolomic and transcriptomic response of *Arabidopsis* to eight environmental conditions. *Plant J.* 67, 869–884. doi: 10.1111/j.1365-313X.2011.04640.x

Davletova, S., Schlauch, K., Coutu, J., and Mittler, R. (2005). The zinc-finger protein Zat12 plays a central role in reactive oxygen and abiotic stress signaling in *Arabidopsis. Plant Physiol.* 139, 847–856. doi: 10.1104/pp.105.068254

Espinoza, C., Degenkolbe, T., Caldana, C., Zuther, E., Leisse, A., Willmitzer, L., et al. (2010). Interaction with diurnal and circadian regulation results in dynamic metabolic and transcriptional changes during cold acclimation in *Arabidopsis. PLoS ONE* 5:e14101. doi: 10.1371/journal.pone.0014101

Glasser, C., Haberer, G., Finkemeier, I., Pfannschmidt, T., Kleine, T., Leister, D., et al. (2014). Meta-analysis of retrograde signaling in *Arabidopsis thaliana* reveals a core module of genes embedded in complex cellular signaling networks. *Mol. Plant* 7, 1167–1190. doi: 10.1093/mp/ssu042

Hirai, M. Y., Fujiwara, T., Awazuhara, M., Kimura, T., Noji, M., and Saito, K. (2003). Global expression profiling of sulfur-starved *Arabidopsis* by DNA macroarray reveals the role of O-acetyl-l-serine as a general regulator of gene expression in response to sulfur nutrition. *Plant J.* 33, 651–663. doi: 10.1046/j.1365-313X.2003.01658.x

Hubberten, H. M., Drozd, A., Tran, B. V., Hesse, H., and Hoefgen, R. (2012a). Local and systemic regulation of sulfur homeostasis in roots of *Arabidopsis thaliana. Plant J.* 72, 625–635. doi: 10.1111/j.1365-313X.2012.05105.x

Hubberten, H. M., Klie, S., Caldana, C., Degenkolbe, T., Willmitzer, L., and Hoefgen, R. (2012b). Additional role of O-acetylserine as a sulfur status-independent regulator during plant growth. *Plant J.* 70, 666–677. doi: 10.1111/j.1365-313X.2012.04905.x

Lewandowska, M., Wawrzynska, A., Moniuszko, G., Lukomska, J., Zientara, K., Piecho, M., et al. (2010). A contribution to identification of novel regulators of plant response to sulfur deficiency: characteristics of a tobacco gene UP9C, its protein product and the effects of UP9C silencing. *Mol. Plant* 3, 347–360. doi: 10.1093/mp/ssq007

Lin, Z., Zhong, S., and Grierson, D. (2009). Recent advances in ethylene research. *J. Exp. Bot.* 60, 3311–3336. doi: 10.1093/jxb/erp204

Luhua, S., Hegie, A., Suzuki, N., Shulaev, E., Luo, X., Cenariu, D., et al. (2013). Linking genes of unknown function with abiotic stress responses by high-throughput phenotype screening. *Physiol. Plant.* 148, 322–333. doi: 10.1111/ppl.12013

Maruyama-Nakashita, A., Inoue, E., Watanabe-Takahashi, A., Yamaya, T., and Takahashi, H. (2003). Transcriptome profiling of sulfur-responsive genes in *Arabidopsis* reveals global effects of sulfur nutrition on multiple metabolic pathways. *Plant Physiol.* 132, 597–605. doi: 10.1104/pp.102.019802

Maruyama-Nakashita, A., Nakamura, Y., Tohge, T., Saito, K., and Takahashi, H. (2006). *Arabidopsis* SLIM1 is a central transcriptional regulator of plant sulfur response and metabolism. *Plant Cell* 18, 3235–3251. doi: 10.1105/tpc.106.046458

Maruyama-Nakashita, A., Nakamura, Y., Watanabe-Takahashi, A., Inoue, E., Yamaya, T., and Takahashi, H. (2005). Identification of a novel cis-acting element conferring sulfur deficiency response in *Arabidopsis* roots. *Plant J.* 42, 305–314. doi: 10.1111/j.1365-313X.2005.02363.x

Moniuszko, G., Skoneczny, M., Zientara-Rytter, K., Wawrzynska, A., Glow, D., Cristescu, S. M., et al. (2013). Tobacco LSU-like protein couples sulphur-deficiency response with ethylene signalling pathway. *J. Exp. Bot.* 64, 5173–5182. doi: 10.1093/jxb/ert309

Moreno-Risueno, M. A., Gonzalez, N., Diaz, I., Parcy, F., Carbonero, P., and Vicente-Carbajosa, J. (2008). FUSCA3 from barley unveils a common transcriptional regulation of seed-specific genes between cereals and *Arabidopsis. Plant J.* 53, 882–894. doi: 10.1111/j.1365-313X.2007.03382.x

Mukhtar, M. S., Carvunis, A. R., Dreze, M., Epple, P., Steinbrenner, J., Moore, J., et al. (2011). Independently evolved virulence effectors converge onto hubs in a plant immune system network. *Science* 333, 596–601. doi: 10.1126/science.1203659

Myakushina, Y. A., Milyaeva, E. L., Romanov, G. A., and Nikiforova, V. Y. (2009). Mutation in LSU4 gene affects flower development in *Arabidopsis thaliana. Doklady Biochem. Biophys.* 428, 257–260. doi: 10.1134/S1607672909050093

Nikiforova, V. J., Daub, C. O., Hesse, H., Willmitzer, L., and Hoefgen, R. (2005). Integrative gene-metabolite network with implemented causality deciphers informational fluxes of sulphur stress response. *J. Exp. Bot.* 56, 1887–1896. doi: 10.1093/jxb/eri179

Nikiforova, V., Freitag, J., Kempa, S., Adamik, M., Hesse, H., and Hoefgen, R. (2003). Transcriptome analysis of sulfur depletion in *Arabidopsis thaliana*: interlacing of biosynthetic pathways provides response specificity. *Plant J.* 33, 633–650. doi: 10.1046/j.1365-313X.2003.01657.x

Omranian, N., Mueller-Roeber, B., and Nikoloski, Z. (2012). PageRank-based identification of signaling crosstalk from transcriptomics data: the case of *Arabidopsis thaliana. Mol. Biosyst.* 8, 1121–1127. doi: 10.1039/c2mb05365a

Ruckle, M. E., Burgoon, L. D., Lawrence, L. A., Sinkler, C. A., and Larkin, R. M. (2012). Plastids are major regulators of light signaling in *Arabidopsis. Plant Physiol.* 159, 366–390. doi: 10.1104/pp.112.193599

Usadel, B., Blasing, O. E., Gibon, Y., Retzlaff, K., Hohne, M., Gunther, M., et al. (2008). Global transcript levels respond to small changes of the carbon status during progressive exhaustion of carbohydrates in *Arabidopsis* rosettes. *Plant Physiol.* 146, 1834–1861. doi: 10.1104/pp.107.115592

Wang, W., Esch, J. J., Shiu, S. H., Agula, H., Binder, B. M., Chang, C., et al. (2006). Identification of important regions for ethylene binding and signaling in the transmembrane domain of the ETR1 ethylene receptor of *Arabidopsis. Plant Cell* 18, 3429–3442. doi: 10.1105/tpc.106.044537

Wang, Y., Zhang, W. Z., Song, L. F., Zou, J. J., Su, Z., and Wu, W. H. (2008). Transcriptome analyses show changes in gene expression to accompany pollen germination and tube growth in *Arabidopsis. Plant Physiol.* 148, 1201–1211. doi: 10.1104/pp.108.126375

Wawrzyńska, A., Lewandowska, M., Hawkesford, M. J., and Sirko, A. (2005). Using a suppression subtractive library-based approach to identify tobacco genes regulated in response to short-term sulphur deficit. *J. Exp. Bot.* 56, 1575–1590. doi: 10.1093/jxb/eri152

Wawrzyńska, A., Lewandowska, M., and Sirko, A. (2010). *Nicotiana tabacum* EIL2 directly regulates expression of at least one tobacco gene induced by sulphur starvation. *J. Exp. Bot.* 61, 889–900. doi: 10.1093/jxb/erp356

Zhang, B., Pasini, R., Dan, H., Joshi, N., Zhao, Y., Leustek, T., et al. (2014). Aberrant gene expression in the *Arabidopsis* SULTR1;2 mutants suggests a possible regulatory role for this sulfate transporter in response to sulfur nutrient status. *Plant J.* 77, 185–197. doi: 10.1111/tpj.12376

Zientara-Rytter, K., Łukomska, J., Moniuszko, G., Gwozdecki, R., Surowiecki, P., Lewandowska, M., et al. (2011). Identification and functional analysis of Joka2, a tobacco member of the family of selective autophagy cargo receptors. *Autophagy* 7, 1145–1158. doi: 10.4161/auto.7.10.16617

Zimmermann, P., Hirsch-Hoffmann, M., Hennig, L., and Gruissem, W. (2004). GENEVESTIGATOR. *Arabidopsis* microarray database and analysis toolbox. *Plant Physiol.* 136, 2621–2632. doi: 10.1104/pp.104.046367

Zimmermann, P., Laule, O., Schmitz, J., Hruz, T., Bleuler, S., and Gruissem, W. (2008). Genevestigator transcriptome meta-analysis and biomarker search using rice and barley gene expression databases. *Mol. Plant* 1, 851–857. doi: 10.1093/mp/ssn048

Conflict of Interest Statement: The Associate Editor Stanislav Kopriva declares that, despite having edited a Research Topic with the author Agnieszka Sirko, the review process was handled objectively and no conflict of interest exists. The authors declare that the research was conducted in the absence of any commercial or financial relationships that could be construed as a potential conflict of interest.

15

Effects of proteome rebalancing and sulfur nutrition on the accumulation of methionine rich δ-zein in transgenic soybeans

Won-Seok Kim[1], Joseph M. Jez[2] and Hari B. Krishnan[1]*

[1] Plant Genetics Research Unit, Agricultural Research Service, U.S. Department of Agriculture, University of Missouri, Columbia, MO, USA
[2] Department of Biology, Washington University, St. Louis, MO, USA

Edited by:
Stanislav Kopriva, University of Cologne, Germany

Reviewed by:
Karine Gallardo, National Institute for Agronomic Research, France
Jean-Christophe Avice, University of Caen Basse-Normandie and National Institute for Agronomic Research, France

***Correspondence:**
Hari B. Krishnan, USDA-ARS, Plant Genetics Research Unit, University of Missouri, 108 Curtis Hall, Columbia, MO 65211, USA
e-mail: hari.krishnan@ars.usda.gov

Expression of heterologous methionine-rich proteins to increase the overall sulfur amino acid content of soybean seeds has been only marginally successful, presumably due to low accumulation of transgenes in soybeans or due to gene silencing. Proteome rebalancing of seed proteins has been shown to promote the accumulation of foreign proteins. In this study, we have utilized RNAi technology to suppress the expression of the β-conglycinin, the abundant 7S seed storage proteins of soybean. Western blot and 2D-gel analysis revealed that β-conglycinin knockdown line (SAM) failed to accumulate the α′, α, and β-subunits of β-conglycinin. The proteome rebalanced SAM retained the overall protein and oil content similar to that of wild-type soybean. We also generated transgenic soybean lines expressing methionine-rich 11 kDa δ-zein under the control of either the glycinin or β-conglycinin promoter. The introgression of the 11 kDa δ-zein into β-conglycinin knockdown line did not enhance the accumulation of the 11 kDa δ-zein. However, when the same plants were grown in sulfur-rich medium, we observed 3- to 16-fold increased accumulation of the 11 kDa δ-zein. Transmission electron microscopy observation revealed that seeds grown in sulfur-rich medium contained numerous endoplasmic reticulum derived protein bodies. Our findings suggest that sulfur availability, not proteome rebalancing, is needed for high-level accumulation of heterologous methionine-rich proteins in soybean seeds.

Keywords: *Glycine max* (L.) Merr., 7S globulin, RNA interference, proteome rebalancing, sulfur assimilation, δ-zein

INTRODUCTION

Due to their high protein content and relatively low cost, soybeans (*Glycine max* (L.) Merr.) are used as an animal feed throughout the world. Protein accounts for about 40% of the dry weight of soybean seeds. The abundant storage proteins of soybean are the salt-soluble globulins, the 7S β-conglycinin and the 11S glycinin, which together could account for about 70% of the total seed proteins (Nielsen, 1996; Nielsen and Nam, 1999; Krishnan, 2000). Both these classes of proteins are encoded by multiple gene families (Harada et al., 1989; Nielsen et al., 1989). Glycinins are hexameric proteins that are classified into two groups based on DNA sequence similarities (Nielsen et al., 1989). Group-1 glycinins contain three genes, *Gy1*, *Gy2*, and *Gy3*, while group-2 glycinins include two genes, *Gy4* and *Gy5*. In addition, two genes, *Gy7* and *Gy6*, resembling group-1 glycinins are also present (Beilinson et al., 2002). The expression of *Gy7* gene is significantly lower than other *Gy* genes while the Gy6 has been identified as a pseudogene (Beilinson et al., 2002). Each of the glycinins is synthesized as a precursor protein that is subsequently cleaved into acidic and basic subunits (Staswick et al., 1984). β-conglycinin consists α′, α, and β subunits (Coates et al., 1985). Unlike the glycinins, the β-conglycinins are glycoproteins (Thanh and Shibasaki, 1978). Because of their abundance, the 7S and 11S globulins are mainly responsible for the nutritional quality of soybeans.

Monogastric animals including humans are unable to synthesize essential amino acids and are dependent on their feed/food to meet the essential amino acid requirements. Even though soybeans are an excellent source of high quality protein, the sulfur amino acid content of soybean seed proteins is not optimal for the formulation of animal feed. This deficiency necessitates supplementation of animal feeds with synthetic methionine that adds additional cost to the livestock producers. Consequently, there have been numerous attempts to improve the sulfur amino acid content of legume seed proteins utilizing different approaches including genetic engineering (Tabe and Higgins, 1998; Krishnan, 2005; Amir et al., 2012; Galili and Amir, 2013). One common approach involves the expression of heterologous seed proteins in transgenic soybeans (Townsend and Thomas, 1994; Dinkins et al., 2001; Kim and Krishnan, 2004; Krishnan, 2005). We have previously reported the expression a methionine-rich δ-zein in soybean (Kim and Krishnan, 2004). The δ-zein stably accumulated with endoplasmic derived protein bodies in transgenic soybean seeds. However, the accumulation of the δ-zein was less than 0.5% of the total seed protein and did not increase the overall methionine content of soybean seed (Kim and Krishnan, 2004).

Similarly, attempts by others to substantially increase the methionine content of soybeans by expressing heterologous proteins rich in methionine have not been successful (reviewed in Krishnan, 2005). Low gene expression of methionine-rich protein in soybeans can be a major contributing factor for the marginal increase in the methionine content in transgenic soybeans.

Several approaches to increase the amount of heterologous protein production in plants have been proposed (Streatfield, 2007). These approaches focus on boosting the heterologous gene replication, transcription, and translation and message stabilization (Streatfield, 2007). Another approach to enhance foreign protein production exploits the high protein synthesis capacity of legumes such as soybean (Schmidt and Herman, 2008; Herman, 2014). By suppressing the production of endogenous seed storage proteins, one could redirect the available protein synthesis capacity to the synthesis of introduced foreign proteins. To test this hypothesis, a green fluorescent protein (GFP)-kdel reporter was introgressed in β-conglycinin suppressed transgenic soybeans. This approach resulted in four-fold increase in GFP-kdel accumulation in transgenic soybean suggesting that proteome rebalancing can enhance foreign protein accumulation (Schmidt and Herman, 2008). Since our previous attempt to increase the sulfur amino acid content of soybean seed by expressing methionine-rich δ-zein was marginally successful due to low accumulation of the δ-zein, we wanted to explore if proteome rebalancing could be exploited to elevate the δ-zein accumulation in soybean.

In this study, we have created β-conglycinin knockdown transgenic soybean lines expressing methionine-rich 11 kDa δ-zein. Interestingly, the expression the 11 kDa δ-zein in β-conglycinin knockdown lines did not elevate the accumulation of the methionine-rich protein. However, when the same plants were grown in sulfur-rich medium, a drastic increase in the 11 kDa δ-zein accumulation was observed. Our results indicate that the availability of sulfur, not proteome rebalancing, is more critical for high-level accumulation of methionine-rich proteins in soybean seeds.

MATERIALS AND METHODS

GENERATION OF Β-CONGLYCININ SUPPRESSED TRANSGENIC SOYBEANS

To simultaneously suppress the expression of all three subunits of β-conglycinin we constructed an RNAi cassette by the following procedure. First, the intron from pKannibal (Wesley et al., 2001) was amplified employing the primer pair: 5′ primer *XbNdXh*INTRi (5′-**TCTAGA** A**CATATG**GTC**CTCGAG**AGTTACTAGTACCCCA-3′) containing *Xba*I, *Nde*I and *Xho*I sites (bold) and 3′ primer INTRi*RVSXb* (5- **TCTAGAGGTCGAC**CG**GATATC**CGCTTTGTTATATTAGC-3) containing XbaI, *Sal*I, and *Eco*RV sites (bold). The amplified intron product was cloned into pGEM-T easy to create pKINTRi. The intron was excised by digestion with *Xba*I and cloned into pHK vector. Primer pair SNd2AplhaRi (5′-GTCGAC**CATATG** TACAGGAACCAAGCATGCCAC-3′; introduced restriction sites *Sal*I and *Nde*I are underlined and bold, respectively) and Alpha′314XhRV (5′-GATATC**CTCGAG**TGAGGTTGGTG TGGGCGTGGG-3′; introduced restriction sites *EcoRV* and *Xho*I are underlined and bold, respectively) was used to amplify a 314 bp coding region of α′-subunit of the β-conglycinin. This 314 bp product was cloned into pGEM-Teasy vector (Promega, WI) and named as pBCON and sequenced at the University of Missouri DNA Core Facility. DNA prepared from pBCON was subjected to restriction enzyme digests with *Nde*I and *Xho*I or *Eco*RV and *Sal*I to facilitate cloning the 314 bp fragments in the opposite orientation. The *Nde*I and *Xho*I excised fragment was ligated into the 5′ end of the pHK-intron while the *Eco*RV and *Sal*I fragement was ligated into the 3′ end pHK-intron. This procedure resulted in a cassette containing a 314 bp α′-subunit of the β-conglycinin hairpin flanking the pHK-intron. Following digestion with *Xba*I the entire β-conglycinin hairpin was cloned into the corresponding restriction site of pZPlapha′P binary vector (Kim and Krishnan, 2004). This final vector places the β-conglycinin hairpin under the regulatory control of the α′-subunit of β-conglycinin promoter and the terminator of the potato proteinase inhibitor gene (*Pin*II). This vector also contains the *bar*-coding region under the regulatory control of cauliflower mosaic virus 35S promoter and 3′-region of the nopaline synthase gene (**Figure 1**). The final RNAi construct was mobilized into *Agrobacterium tumefaciens* (strain EHA105) by triparental mating (Friedman et al., 1982). Soybean cultivar Maverick was transformed by *Agrobacterium*-cotyledonary node method (Hinchee et al., 1988). Regenerated transgenic soybean plants were screened for tolerance to herbicide Liberty by a leaf-painting assay as described earlier (Zhang et al., 1999). The knockout of the target proteins in glufosinate-resistance plants

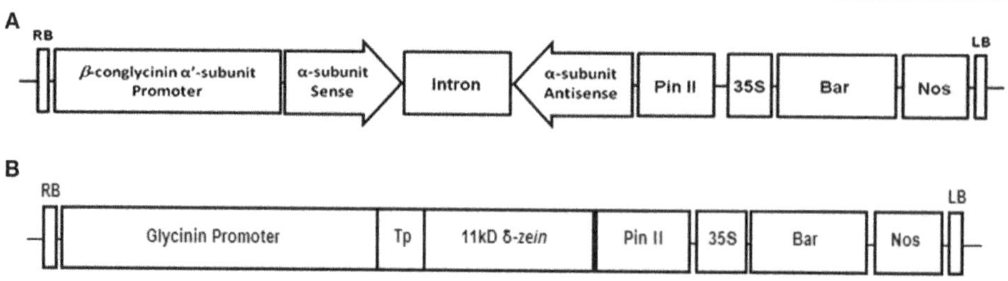

FIGURE 1 | Schematic diagram of the constructs used for the suppression of β-conglycinin soybean seeds (A) and expression of 11 kDa δ-zein (B). The RNAi cassette contains a 314 bp region of the α′ subunit of the β-conglycinin cloned in an inverted repeat orientation and separated by the intron from pKannibal. The RNAi construct is under the control of the soybean β-conglycinin α′-promoter while the expression of 11 kDa δ-zein is under the control of soybean glycinin promoter (Kim and Krishnan, 2004). The constructs also contain a gene expression cassette that includes the cauliflower mosaic virus 35S promoter, the bar-coding region and the 3′ region of the nopaline synthase gene (nos).

was confirmed by PCR and SDS-PAGE. For all comparisons soybean cultivar Maverick was used and referred as wild-type.

RT-PCR ANALYSIS

Total RNA was extracted from transgenic soybean seed at the mid-maturation stage (seed size 8–10 mm) using a Trizol reagent (Invitrogen, Grand Island, NY, USA) following the manufacture's protocol. RNA was quantified by measuring the A260/A280 ratio using a spectrophotometer. One μg of DNase I treated RNA was used as template for RT-PCR. Primers used for alpha′ subunit were Alpha′F, 5-ATGATGAGAGC GCGGTTCCCATTACTG-3 and Alpha′R, 5-TCAGTAAAAAG CCCTCAAAATTGAAGAC-3. Primers used for alpha subunit were Alpha F, 5-ATGATGAGAGCACGGTTCCCATTACTG-3 and Alpha′R. Primers used for beta subunit were BetaF, 5-ATGAT GAGAGTGCGGTTTCCTTTGTTGG-3 and BetaR, 5- TCAGTAG AGAGCACCTAAGATTGAAG-3. Primers used for glycinin 4 were SoyGy4F, 5- ATGGGGAAGCCCTTCACTCTCTCTCTTTC-3 and SoyGy4R, 5- TTATGCGACTTTAACACGGGGTGAGC-3. The cycling condition of RT-PCR were 50°C for 30 min for cDNA production, 95°C for 15 min for inactivation of cDNA production, then 35 cycles of 94°C for 60 s, 60°C for 60 s, 72°C, for 90 s with a final 72°C for 240 s extension steps.

DETERMINATION OF SEED PROTEIN AND OIL CONTENT

Soybean protein content was measured using the Leco model FP-428 nitrogen analyzer (LECO Corporation, Michigan, USA). The oil content was quantified by near-infrared reflectance (NIR) spectroscopy (Tecator AB, Hoganas, Sweden). The fatty acid profiles of soybean were determined by gas chromatograph as described previously (Lee et al., 2009). Briefly, crushed seeds were extracted overnight with 5 mL of chloroform: hexane: methanol (8:5:2, v/v/v). Fatty acids from 100 μL aliquots of the extract were methylated with 75 μL of methanolic sodium methoxide:petroleum ether:ethyl ether (1:5:2, v/v/v). Fatty acids were separated utilizing Agilent Series 6890 capillary gas chromatograph (Palo Alto, CA, USA) that was fitted with an AT-Silar capillary column (Alltech Associates, Deerfield, IL, USA). Standard fatty acid mixtures were used for determining relative amounts of each fatty acid. Four replicates of "Maverick" (MAV) and β-conglycinin knockdown line (SAM) were compared using the t test function within JMP® software Version 9 (SAS Institute Inc., Cary, NC). Significantly different means are indicated by *, **, or *** ($p \leq 0.05, p \leq 0.01, p \leq 0.001$, respectively) and insignificant differences are indicated by "NS."

PROTEIN ISOLATION AND IMMUNOBLOT ANALYSIS

Total soybean seed proteins were extracted from 10 mg of dry seed powder with 1 ml of SDS sample extract buffer (2% SDS, 60 mM Tris-HCl, pH 6.8, 5% β-mecaptoethanol), followed by boiling at 100°C for 5 min. After maximum speed centrifuge the supernatant was used for total seed protein fraction. The total seed protein fraction was electrophoresed with 8 or 15% SDS-PAGE and visualized by staining with Coomassie Brilliant Blue. For western blot analysis, the total seed proteins were transferred to a nitrocellulose membrane after resolved by SDS-PAGE. After blocking with TBS (10 mM Tris-HCl, pH 7.5, 500 mM NaCl) containing 5% non-fat dry milk, the membrane was incubated over-night with 7S globulin storage protein or the 11 kDa δ-zein antibodies (Kim and Krishnan, 2003) that had been diluted 1:5000 in TBST (TBS with 3% non-fat dry milk containing 0.2% Tween 20). After several washes with TBST, the membrane was incubated with goat anti-rabbit IgG-horseradish peroxidase conjugate. Immunoreactive polypeptide signals were detected by according to the SuperSignal West Pico kit's instruction (Pierce, Rockford, Il, USA).

TWO-DIMENSIONAL GEL ELECTROPHORESIS, IMAGE ANALYSIS AND QUANTIFICATION OF SPOT VOLUME

Two-dimensional gel electrophoresis of soybean seed proteins were performed as described earlier (Krishnan et al., 2009). Coomassie stained gels were destained with multiple changes of ultrapure water to remove background and scanned using an Epson Perfection V700 scanner controlled through Adobe Photoshop. Images were analyzed for proteome differences using Delta2D image analysis software. Delta2D parameters were set to maximize spot detection using global image warping and exact spot matching. Background subtraction parameters were identical for all gels and eliminated a large percentage of spots that were less than 0.05% of the total normalized spot volumes. A total of 732 remained on the fusion image of all gels included in the analysis, after background subtraction, and were used for the total volume ratio normalization. A total of 34 spots were chosen as a subset of this total, for major seed storage protein quantification, comprising nearly 50% of the total volume ratio normalized spot volume. This subset of 2-D resolved proteins were previously identified (Krishnan et al., 2009; Krishnan and Nelson, 2011) using peptide mass fingerprinting (MALDI-TOF MS) and then categorized as shown in **Figure 5**. The exact same spot regions were chosen from software generated fused image of all gels used in the analysis. Spot % volume quantities were calculated within each comparison individually; Maverick ($n = 3$) and SAM ($n = 3$).

SULFUR TREATMENT AND PLANT GROWTH CONDITIONS

Soybeans were grown in an environmentally controlled growth chamber at 26 ± 2°C and 50% humidity. The plants received 14 h of daylight for the duration of the experiment. Each individual plant was grown in its own modified hydroponics container, consisting of a pot containing a 1:1 mixture of perlite:vermiculite, with a lower 2 L reservoir for the nutrient solution. Cotton wicks were placed in the lower portion of each pot to deliver nutrient solution to the plant prior to the root growth eventually reaching the lower reservoir. The nutrient solution composition was: 1.25 mM $Ca(NO_3)_2$ $4H_2O$, 1.25 mM KNO_3, 0.25 mM KH_2PO_4, 0.5 mM Fe-EDTA, 0.5 mM NH_4NO_3, and a 1/1000th addition of a micronutrients mixture (H_3BO_3, 2.86 g L^{-1}; $MnCl_2$ $4H_2O$, 1.81 g L^{-1}; $ZnCl_2$, 0.095 g L^{-1}; $Cu(NO_3)_2$ xH_2O, 0.047 g L^{-1}; H_2MoO $4H_2O$, 0.09 g L^{-1}). The sulfate control solution contained 0.5 mM $MgSO_4$ $7H_2O$, and the sulfur-rich solution contained 2.0 mM $MgSO_4$ $7H_2O$. The nutrient solution was replaced every 3 days to maintain a consistently fresh nutrient supply. Seeds were harvested approximately 90 days after treatments began.

AMINO ACID ANALYSIS

Amino acid analysis was performed at the Donald Danforth Plant Science Center Proteomics and Mass Spectrometry Facility. Free amino acid content was determined following the protocol of Hacham et al. (2002). For the determination of the total amino acid content the seed powder was first subjected to hydrolysis with 6N HCl. For the quantification of methionine and cysteine, duplicate samples were first subjected to an initial oxidation step using performic acid prior to acid hydrolysis. Amino acids were quantified using the manufacturer's instructions of the Waters AccQ-Tag Ultra Kit on an Acquity UPLC system. Samples were run in quadruplicate and subjected to appropriate statistical analysis.

IMMUNOSTAINING OF PARAFFIN SECTIONS

Greenhouse-grown soybean seeds at R6 stage (Fehr et al., 1971) were cut into several pieces and fixed in FAA (10% formaldehyde, 50% ethyl alcohol, and 5% glacial acetic acid) for 8 h at room temperature. The tissue was dehydrated in a graded ethanol/xylene series and infiltrated with paraffin. Sections were cut with a microtome and processed for immunostaining following the protocol described earlier (Bilyeu et al., 2008). Sections were separately incubated with 1:1000 diluted antibodies raised against the β-subunit of β-conglycinin or Kunitz trypsin inhibitor. Following this step, the sections were treated sequentially with biotinylated linker, streptavidin conjugated to horseradish peroxidase, and substrate-chromogen solution (DAKO). The sections were examined under bright field optics.

ELECTRON MICROSCOPY

Mature dry seeds were surface sterilized by first sequentially incubating with 95% ethanol for 5 min and with 50% commercial bleach for 5 min. Following extensive washes in distilled water the seeds were transferred to 1% water agar plates and germinated in a 30°C incubator for 12 h. Seeds were sliced into several 2–4 mm cubes and fixed for 4 h in 2.5% glutaraldehyde buffered at pH 7.2 with 50 mM sodium phosphate. After several rinses in sodium phosphate buffer the seeds were post-fixed for 1 h with 1% aqueous osmium tetroxide. The seed tissue was dehydrated in a graded acetone series and infiltrated with Spurr's resin. Thin sections of the seed tissue were cut with a diamond knife and collected on 200 mesh copper grids. The grids were stained with 0.5% uranyl acetate and 0.4% lead citrate and examined at 80 kV under JEOL 1200 EX (Tokyo, Japan) transmission electron microscope.

RESULTS

CREATION OF B-CONGLYCININ KNOCKDOWN SOYBEAN SEEDS

The suppression of the 7S globulins accumulation in soybean seeds was accomplished by RNAi utilizing *Agrobacterium*-mediated cotyledonary node transformation protocol (Hinchee et al., 1988). Seeds from several independent transgenic lines were screened for their protein composition by SDS-PAGE and immunoblot analysis using antibodies raised against the β-subunit of β-conglycinin (**Figure 2**). An examination of the Coomassie stained gel revealed that in 50% of the transgenic plants no significant reduction in the accumulation of β-conglycinin had occurred when compared to the wild-type plant (**Figure 2A**). However, four transgenic RNAi lines showed a complete absence of the β-conglycinin accumulation in their seeds (**Figure 2A**). The lack of β-conglycinin in these RNAi lines was confirmed by immunoblot analysis using β-conglycinin antibodies (**Figure 2B**). These four lines were grown for five generations. Twenty individual T5 seeds from each transgenic line were subjected to Western blot analysis. Transgenic plants where all tested seeds failed to accumulate the β-conglycinin were considered to be homozygous plants. These homozygous transgenic RNAi plants are referred as SAM from this point onwards.

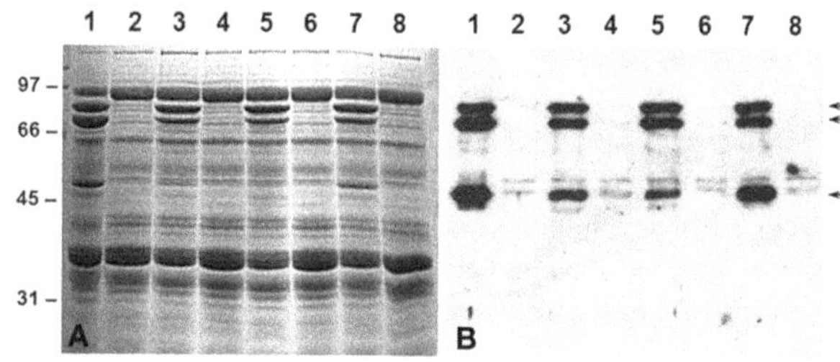

FIGURE 2 | Suppression of β-conglycinin accumulation in transgenic soybean lines. Total seed proteins from wild-type (lane 1) and from seven independent RNAi transgenic lines (lanes 2–8) were fractionated in duplicate gels by 10% SDS-PAGE and stained with Coomassie Blue **(A)** or transferred to nitrocellulose membranes. The membrane was probed with the β-conglycinin antibodies **(B)**. Immunoreactive proteins were identified using anti rabbit IgG-horseradish peroxidase conjugate followed by chemiluminescent detection. Note the absence of α′, α, and β-subunits of β-conglycinin in some of the RNAi lines.

We investigated the mRNA levels of the major seed storage proteins between the RNAi line SAM and the wild-type control by RT-PCR. When primers specific for *gy1*, *gy2*, *gy3*, and *gy4* were used in RT-PCR reactions, gene-specific products were amplified in both plants (**Figure 3A**). However, when primers specific for each of the three subunits of β-conglycinin were employed, RT-PCR products were seen only in the wild-type control and not in SAM (**Figure 3B**). The absence of the β-conglycinin accumulation in SAM seeds was further examined by immunocytochemical localization. Sections of soybean seed were incubated with either β-conglycinin or Kunitz trypsin inhibitor specific antibodies followed by incubation with streptavidin conjugated to horseradish peroxidase. Antibodies raised against the β-conglycinin did not detect their accumulation in the paraffin-embedded sections as evidenced by the absence of brown localization signal (Supplemental Figure 1A) while it was readily detected in the wild-type control seeds (Supplemental Figure 1B). In contrast, when the paraffin sections were incubated with Kunitz trypsin inhibitor specific antibodies positive reactions were seen in both SAM and the wild-type control seeds (Supplemental Figures 1C–D).

PROTEIN, OIL AND AMINO ACID COMPOSITION OF B-CONGLYCININ KNOCKDOWN SOYBEAN SEEDS

To investigate if knockdown of β-conglycinin resulted in any alterations of the seed components we first measured the total protein and oil content of these seeds. Both the wild-type control and SAM contain very similar concentrations of total protein and

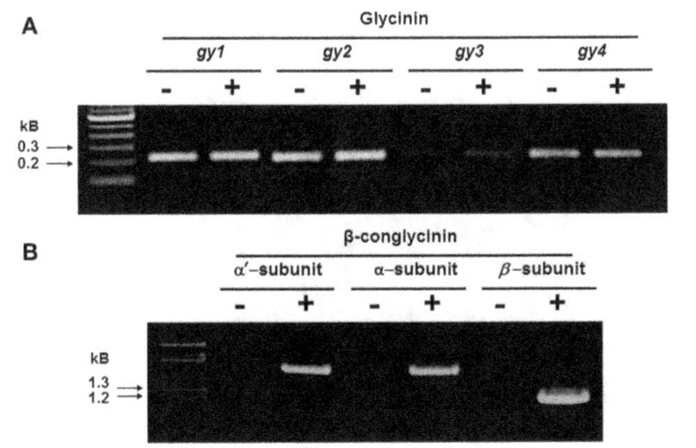

FIGURE 3 | RT-PCR analysis of 7S and 11S globulin gene expression in developing soybean seeds. Total RNA isolated from wild-type control (+) and β-conglycinin knockdown transgenic soybean line SAM (−) was used in RT-PCR analysis. The upper panel **(A)** shows the RT-PCR products for the *gy1, gy2, gy3,* and *gy4* genes (11S globulin) and the lower panel **(B)** shows the RT-PCR products for the α′, α, and β-subunits of β-conglycinin (7S globulin).

Table 1 | Protein, amino acid and oil content in wild-type and β-conglycinin knockdown line (SAM).

Seed Component	Wild-type	SAM	Significance level	*p*-value
Protein[a]	35.9 ± 0.4	36.0 ± 0.4	NS	0.84
Oil[a]	19.4 ± 0.1	19.7 ± 0.3	NS	0.05
Total Amino Acid (Free + Protein bound)[b]	2410.5 ± 113.4	2794.5 ± 130.1	**	4.53E-03
Fatty acids[a]				
Palmitic Acid (16:0)	11.6 ± 0.2	10.6 ± 0.1	***	3.76E-04
Stearic Acid (18:0)	4.2 ± 0.1	4.5 ± 0.3	NS	0.08
Oleic Acid (18:1)	20.5 ± 0.9	23.2 ± 0.7	**	3.38E-03
Linoleic Acid (18:2)	55.9 ± 0.6	55.0 ± 0.4	*	0.02
Linolenic Acid (18:3)	7.9 ± 0.2	6.7 ± 0.1	***	6.25E-04

*Significantly different means are indicated by *, **, or *** ($p \leq 0.05$, $p \leq 0.01$, $p \leq 0.001$, respectively) and insignificant differences are indicated by "NS."*

[a]*expressed as percentage.*

[b]*nanomoles/mg seed.*

oil (**Table 1**). Total amino acid content was significantly higher in β-conglycinin knockdown line (**Table 1**). Analysis of the five major fatty acids in these seeds by gas chromatograph also showed significant differences. The concentration of palmitic, linoleic and linolenic acids was higher in the wild-type while oleic acid content was higher in SAM (**Table 1**).

Previous studies have shown that soybean mutants lacking the seed storage proteins accumulate high levels of free amino acids (Takahashi et al., 2003; Schmidt et al., 2011). To examine if similar situation also occurred in the β-conglycinin knockdown soybean line we determined the total and free amino acid content (**Figure 4**). An examination of the total amino acid composition revealed a slight increase in several of the amino acids in SAM when compared to that of wild-type control (**Figure 4A**). In contrast, a comparison of the free amino acid composition revealed

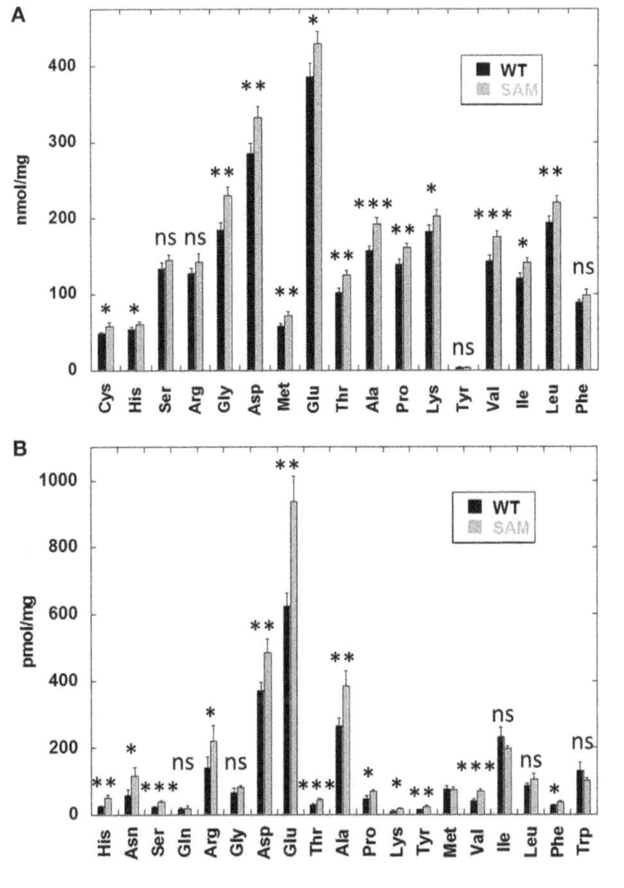

FIGURE 4 | Free and total amino acid composition of soybean seeds. Dry seed powder from quadruplicate samples was analyzed by HPLC to measure the total **(A)** and free amino acid **(B)** from wild-type (WT) and β-conglycinin knockdown soybean line (SAM).

significant increases in the concentration of several amino acids in SAM seeds especially glutamic acid, aspartic acid, asparagine, arginine, and alanine (**Figure 4B**). In previous studies arginine represented the main amino acid that contributed to the accumulation of high levels of free amino acids (Takahashi et al., 2003; Schmidt et al., 2011). However, in our studies we observed that arginine was not the most abundant amino acid in the overall free amino acid pool in SAM. Few other amino acids (glutamic acid, aspartic acid, asparagine, and alanine) also contributed significantly to the elevated levels of free amino acid content in SAM seeds (**Figure 4B**).

2D-GEL ANALYSIS OF SOYBEAN SEED PROTEIN COMPOSITION

Two dimensional-gel analysis was performed to evaluate the changes in the protein composition of RNAi soybean line with that of the wild-type control (**Figure 5**). This analysis clearly demonstrated that β-conglycinin knockdown RNAi line lacked all the three subunits (α′, α, and β-subunits) of β-conglycinin. Previous studies have shown that suppression of the seed 7S and 11S seed storage proteins resulted an increase in the accumulation of lipoxygenase, sucrose binding protein, basic 7S globulin, Kunitz trypsin inhibitor, soybean lectin, Gly m Bd 30k,

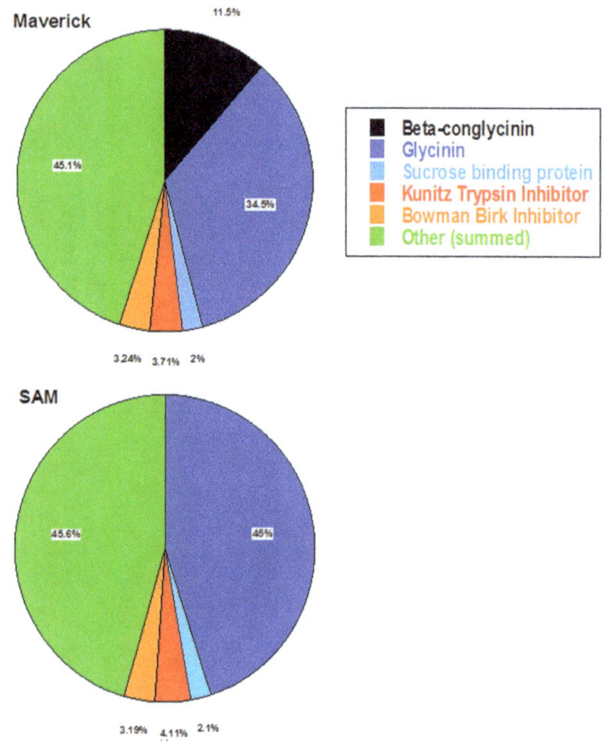

FIGURE 5 | Pie chart representation of the abundance of proteins in the proteome of Maverick (wild-type) and β-conglycinin knockdown soybean line (SAM). Equal amounts of proteins from soybean seeds were analyzed by high-resolution 2D gel electrophoresis. Spot volume measurements of the seed proteins averaged from triplicate 2-D gels were determined by using Delta2D image analysis software. The pie chart demonstrates that the suppression of the β-conglycinin in SAM results in increased accumulation of the glycinin.

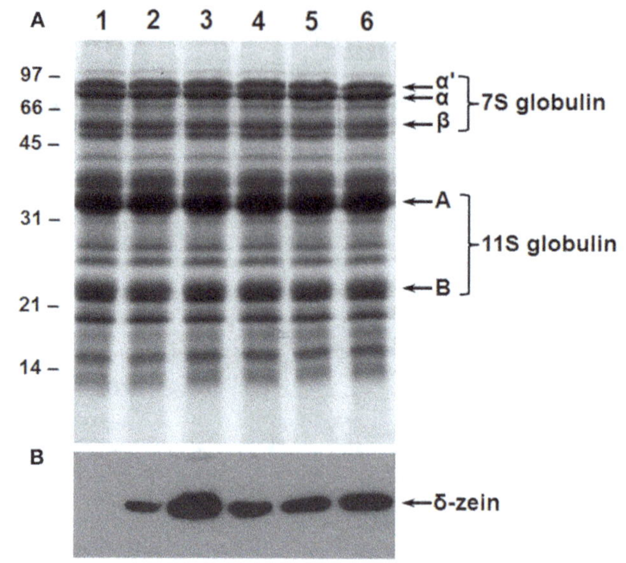

FIGURE 6 | Accumulation of 11 kDa delta zein in transgenic soybean seeds. Total seed proteins from dry soybean seeds from wild type (lane 1) and five independent transgenic lines (lanes 2–6) expressing the 11 kDa δ-zein under the control of glycinin promoter were resolved on a 15% SDS-PAGE. The gels were either stained with Coomassie Brilliant Blue **(A)** or subjected to western blot analysis using maize 11 kDa delta zein antibodies **(B)**. Immunoreactive proteins were detected by chemiluminescent method. Sizes of protein standards are shown in kilodaltons.

Glc-binding protein and seed maturation-associated protein in these seeds (Takahashi et al., 2003; Schmidt et al., 2011). To see if similar changes are also occurred in SAM 2D fractionated soybean seed proteins from SAM and wild-type control (run in triplicate) were analyzed with high-resolution image analysis software (Delta 2D). This analysis indicated that absence of the three subunits of β-conglycinin was associated with an increase in the glycinin content (**Figure 5**). A comparison of seed protein profiles between SAM and wild-type control did not reveal any substantial differences between other seed proteins (**Figure 5**).

PRODUCTION OF TRANSGENIC SOYBEAN LINES EXPRESSING THE 11 KDA δ-ZEIN AND IT'S INTROGRESSION INTO SAM

We had earlier generated transgenic soybean plants expressing 11 kDa δ-zein under the control of the β-conglycinin promoter (Kim and Krishnan, 2004). Since the β-conglycinin promoter was also used for suppressing the expression of β-conglycinin we wanted to use another seed-specific promoter to express the 11 kDa δ-zein. For this purpose a plasmid consisting of soybean glycinin promoter, the coding region of the 11 kDa δ-zein, the 3′ region of the potato proteinase inhibitor gene, together with the cassette containing *bar* herbicide resistance gene was constructed (**Figure 1**) and introduced into soybean cv. Williams 82 as described earlier (Kim and Krishnan, 2004). The accumulation of the 11 kDa δ zein in five independent transgenic events was verified by western blot analysis using antibodies raised against the purified 11 kDa δ-zein (**Figure 6**). The 11 kDa δ-zein expressing soybean lines were grown in the greenhouse for another three generations to produce T4 plants.

Earlier studies have shown proteome rebalancing in seeds can be exploited for elevated expression of foreign proteins (Goossens et al., 1999; Tada et al., 2003; Schmidt and Herman, 2008). We wanted to test if the δ-zein introgression into the β-conglycinin knockdown RNAi line would enhance the accumulation of the δ-zein. For this purpose, crosses were made between homozygous 11 kDa δ-zein expressing soybean plants and β-conglycinin knockdown RNAi soybean plants. Seeds from successful crosses were germinated and a small segment of the seed was assayed by immunoblot analysis for the absence of β-conglycinin and the accumulation of 11 kDa δ-zein. Seeds with the desired combination were grown in the green house and subjected to recurrent selection until homozygous plants with the desired traits were obtained.

The accumulation of the 11 kDa δ-zein in the β-conglycinin knockdown RNAi soybean plant was examined by immunoblot analysis. The 11 kDa δ-zein was readily detected in both original transgenic soybean plants and in β-conglycinin knockdown RNAi background. However, the 11 kDa δ-zein did not accumulate at higher amounts in soybean seeds lacking β-conglycinin, indicating that proteome rebalancing did not enhance the accumulation of the 11 kDa δ-zein.

It has been shown that availability of methionine and cysteine in legumes is the major limiting factor in enhancing the sulfur content of legumes (Tabe et al., 2002). We therefore, examined if the accumulation of the 11 kDa δ-zein can be promoted by providing the developing plants with sulfate-rich solution, which can be assimilated for sulfur metabolism. For this purpose, we grew the soybean plants in hydroponics under two levels of sulfate. In seeds from plants grown in presence 0.5 mM magnesium sulfate, the accumulation of the 11 kDa δ-zein was higher in wild-type background than in β-conglycinin knockdown background (**Figures 7A–B**). Interestingly, in seeds from soybean plants grown in presence of 2 mM magnesium sulfate there was a drastic increase in the accumulation of the 11 kDa protein. This difference in the accumulation of the δ-zein was much more striking in β-conglycinin knockdown background (**Figure 7B**). Densitometer scans of the immunoblots revealed that the seeds from plants grown in presence of 2 mM magnesium sulfate had about 3- to 16-fold increases in the accumulation of the 11 kDa δ-zein in the original and introgressed lines, respectively (**Figure 7B**). In contrast, the overall protein content of soybean seeds was not affected by sulfate treatment. Soybean plants grown in presence of 2 mM magnesium sulfate had 34.9 ± 0.6% while those grown in sulfate-rich solution contained 35.0 ± 0.4% protein. Our preliminary analysis also revealed that the transgenic soybean plants grown in sulfate-rich solution exhibited 1.2 fold-increase in the total sulfur amino acid content when compared with plants grown in presence of 0.5 mM magnesium sulfate.

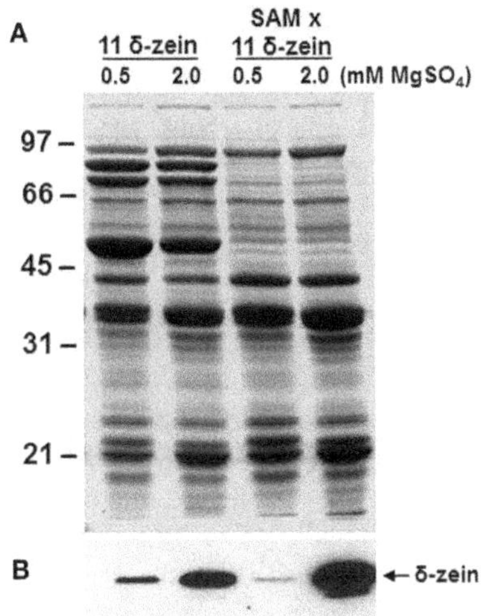

FIGURE 7 | Effect of sulfur nutrition on the accumulation of the 11 kDa δ-zein in transgenic soybean seeds. Total seed proteins from 11 kDa δ-zein expressing line and the 11 kDa δ-zein introgressed into the β-conglycinin knockdown soybean line (SAM) that were grown hydroponically in presence of 0.5 or 2 mM magnesium sulfate were fractionated in duplicate gels by 15% SDS-PAGE and stained with Coomassie Blue **(A)** or transferred to nitrocellulose membranes. The membrane was probed with the 11 kDa δ-zein specific antibodies **(B)**. Immunoreactive proteins were identified using anti rabbit IgG-horseradish peroxidase conjugate followed by chemiluminescent detection.

SULFUR SUPPLEMENTATION PROMOTES THE FORMATION OF PROTEIN BODIES

Previously we showed that expression of 11 kDa δ-zein results in the formation of endoplasmic reticulum derived protein bodies (Kim and Krishnan, 2004). Immunocytochemical localization studies also confirmed that δ-zein is localized within these protein bodies (Kim and Krishnan, 2004). Since sulfur supplementation resulted in a drastic increase in the accumulation of the 11 kDa δ-zein in transgenic soybean plants, we examined if this change was accompanied by an increase in the number of protein bodies. Thin-sections of soybean seeds expressing the 11 kDa δ-zein in β-conglycinin knockdown background grown in 0.5 and 2 mM of magnesium sulfate were examined by transmission electron microscopy (**Figure 8**). Electron microscopy observation of thin sections of soybean seeds grown in presence of 0.5 mM magnesium sulfate revealed prominent oil bodies and large protein storage vacuoles, the storage compartment for the native glycinin and β-conglycinin (**Figure 8A**). In addition few spherical electron-dense spherical protein bodies were also observed in the cotyledonary cells (**Figure 8A**). An examination of seeds grown in 2 mM of magnesium sulfate revealed the presence of numerous protein bodies (**Figure 8B**). Interestingly soybean seeds from plants grown in presence of 2 mM magnesium sulfate contained numerous small electron-dense protein bodies within vacuoles (**Figure 8C**).

DISCUSSION

Soybeans provide several advantages for the production of recombinant proteins for diagnostics, therapeutics, and industrial applications (Bost and Piller, 2011). Chief among them is the low cost of recombinant protein production and the ability to produce and store large amounts of transgenic proteins in seeds (Bost and Piller, 2011). Several laboratories have generated transgenic soybean plants that express 1 to 4% of the total soluble protein as recombinant protein (Piller et al., 2005; Ding et al., 2006; Garg et al., 2007; Moravec et al., 2007; Rao and Hildebrand, 2009; Cunha et al., 2010). Other groups have identified seed fill sensor proteins, such as cruiferin, that may be usable to alter seed composition (Lin et al., 2013). In spite of these achievements, problems remain in obtaining and maintaining high-level heterologous expression in seeds. In cases where high-level expression of transgene is achieved, invariably it is accomplished by down regulation of endogenous reserve proteins (Tabe and Droux, 2001; Hagan et al., 2003; Scossa et al., 2008).

Several naturally occurring, as well as induced, mutations that affect the accumulation of soybean seed storage proteins have been previously described (Kitamura and Kaizuma, 1981; Odanaka and Kaizuma, 1989; Takahashi et al., 1994; Yagasaki et al., 1996; Hayashi et al., 1998). By integrating these mutations by crossbreeding, a soybean line that lacks all the glycinin and β-conglycinin subunits has been developed (Takahashi et al., 2003). This soybean line, in spite of lacking all the major seed storage proteins, was able to grow and reproduce normally. Interestingly, the nitrogen content of the seeds was found to be similar to that

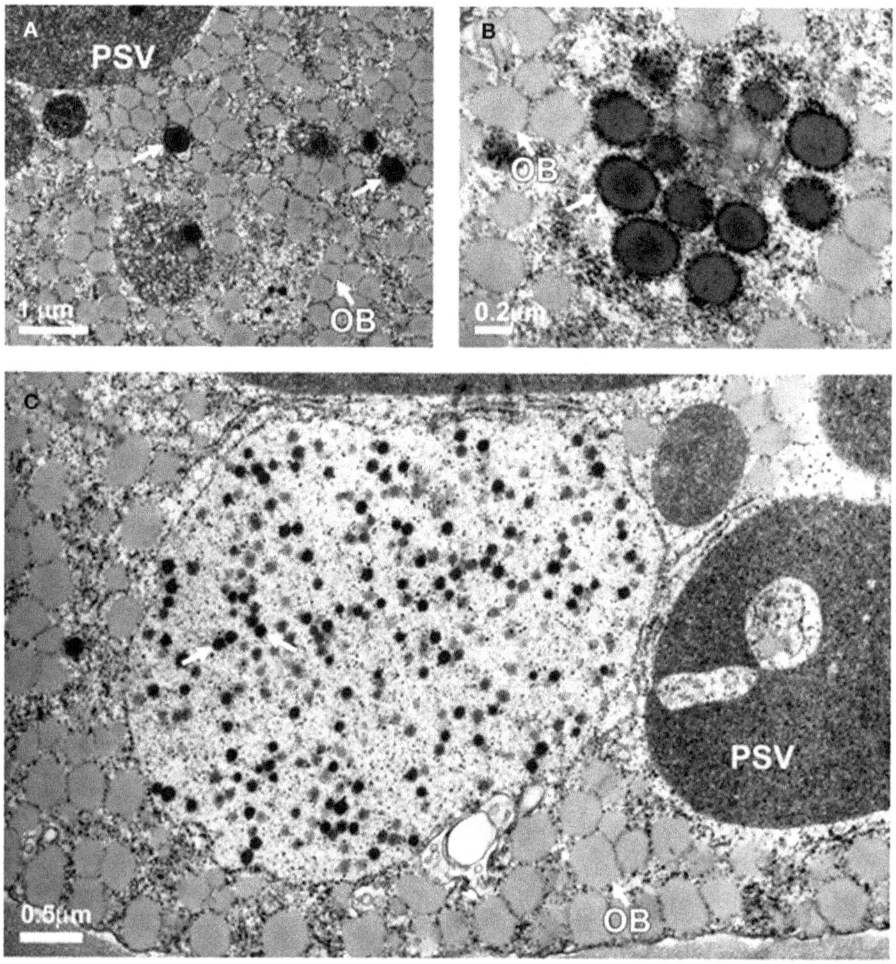

FIGURE 8 | Transmission electron microscopy observation of protein bodies in transgenic soybean seeds. Seeds grown in presence of 0.5 mM magnesium sulfate contain a few endoplasmic reticulum-derived spherical protein bodies (**A**, arrows) while seeds grown in presence of 2 mM magnesium sulfate reveal numerous dark staining spherical protein bodies **(B,C)**. PSV, protein storage vacuole; OB, oil bodies; PB, protein body.

of the wild-type cultivars (Takahashi et al., 2003). Additionally, the absence of the abundant seed proteins resulted in preferential increase in the accumulation of lipoxygenase, sucrose binding protein, agglutinin and the basic 7S globulin. Similarly, a soybean line lacking both glycinin and beta-conglycinin (SP-) was developed by RNA interference (Schmidt et al., 2011). The absence of glycinin and β-conglycinin in the SP- line was accompanied by selective increase in the accumulation of a few proteins similar to the situation encountered through integration of mutations (Takahashi et al., 2003). In both cases, the absence of major seed storage proteins was compensated by the accumulation of free amino acids with arginine accounting for more than 50% of the free amino acid content (Takahashi et al., 2003; Schmidt et al., 2011).

Here we used RNAi to suppress expression of the α, α′, and β-subunits of β-conglycinin in soybean seed. The resulting SAM line showed clear knockdown of these subunits, as confirmed by RT-PCR, immunoblot, and 2D-gel analyses (**Figures 2, 3, 5**). Although the SAM line retained normal overall protein and oil seed content compared to wild-type seed (**Table 1**), the distribution of amino acids in SAM showed increases in glutamic acid, aspartic acid, asparagine, and alanine compared to wild-type (**Figure 4**). In contrast to Takahashi et al. (2003), only a modest increase in arginine content in the β-conglycinin knockdown soybean line was observed (**Figure 4**). Moreover, this comparison revealed slight increase in the methionine and cysteine content in the β-conglycinin knockdown soybean line compared to wild-type. Glycinin is relatively rich in sulfur, while the β-conglycinin is poor in sulfur containing amino acids (Krishnan, 2005). Some earlier reports suggest that elimination of β-conglycinin could led to increased levels of sulfur-containing amino acids in soybean seeds (Ogawa et al., 1989; Panthee et al., 2004). Our analysis of the SAM seeds indicates that elimination of β-conglycinin only marginally increases the sulfur amino acid content of soybean seeds. This observation is consistent with other studies indicating that a lack of either glycinin or β-conglycinin has little effect on total amino acid composition of soybean seeds (Takahashi et al., 2003; Schmidt et al., 2011).

RNA interference has been successfully employed to alter the nutritional quality of the seed (Segal et al., 2003; Frizzi et al., 2010). For example, the suppression of the barley C-hordeins resulted in elevated levels of several essential amino acids in the transgenic seeds (Lange et al., 2007). Similarly, high-level expression of human growth hormone polypeptide was achieved in a glutelin and prolamin knockdown rice line (Shigemitsu et al., 2012). In contrast, our approach to increase the accumulation of the methionine-rich maize δ-zein in the β-conglycinin knockdown soybean line showed no appreciable changes in expression levels of the maize protein.

Previous attempts to overexpress foreign proteins in seed protein knockdown soybean lines have also met with mixed results. The introgression of green fluorescent protein (GFP-kdel) in a β-conglycinin suppressed soybean line resulted in a four-fold increase in the accumulation of the GFP-kdel (Schmidt and Herman, 2008). Subsequently, researchers from the same group reported that introgression of GFP in a glycinin and β-conglycinin knockdown (SP-) line did not lead to accumulation of GFP (Schmidt et al., 2011). The results from these studies clearly indicates that the capacity to overexpress foreign proteins in soybeans will be influenced by several factors such as the nature of the endogenous seed protein targeted for suppression by RNAi, the amino acid composition, and the localization of the heterologous protein.

An obstacle in achieving high-level expression of methionine-rich δ-zein may be the paucity of sulfur-containing amino acids in developing soybean seeds to support its production. Legumes in general are poor in sulfur-containing amino acids (Shewry, 2000). Often the expression of foreign proteins rich in sulfur amino acids is accompanied by reduction in the endogenous sulfur-rich proteins, indicating that sulfur availability in seeds is the limiting factor. Sulfur nutrition or methionine supplementation has been shown to influence the accumulation of methionine-rich storage proteins in legumes (Chiaiese et al., 2004; Amira et al., 2005). Common bean (*Phaseolus vulgaris* L.) accumulates large amounts of a γ-glutamyl dipeptide of S-methyl-cysteine, a non-protein amino acid (Taylor et al., 2008). Interestingly, the removal of 7S globulin and phytohemagglutinin in *Phaseolus vulgaris* enhanced the accumulation of sulfur-rich proteins (Marsolais et al., 2010). Additionally, an increase in cysteine and methionine was reported which occurred at the expense of S-methylcysteine. This indicates a redirection of sulfur from γ-glutamyl-S-methyl-cysteine to the protein cysteine pool.

When the transgenic soybean seeds were grown in presence of excess sulfur there was a significant increase in the accumulation of the methionine-rich δ-zein (**Figure 7**). Transmission electron microscopy observation (**Figure 8**) also confirmed that sulfur supplementation to soybean plants drastically increased the number of endoplasmic reticulum derived protein bodies, the site of δ-zein accumulation. These observations indicate that the intrinsic capacity to synthesize sulfur containing amino acids is very high but limited by the level of sulfate in the nutrient solution. Previous studies indicate that metabolic engineering of the sulfur assimilatory pathway can be used as a tool to increase the sulfur amino acid content of seeds (Avraham et al., 2005; Tabe et al., 2010; Song et al., 2013). Recently, it was reported that soybean seeds expressing feed-back-insensitive cystathionine-γ-synthase exhibited 1.8 to 2.3-fold increases in the total methionine content of their seeds (Song et al., 2013).

In this study we have shown the importance of sulfur nutrition on the accumulation of heterologous methionine-rich protein in soybean. Our results suggest that sink strength is not limiting for accumulation of methionine-rich δ-zein. Instead the limiting factor appears to be sulfate availability. Sulfur assimilation is inter-connected with nitrogen (N) and carbon (C) metabolism (Kopriva et al., 2002; Hawkesford and De Kok, 2006). Because of this inter-connection sulfur deficiency manifests in the form of poor plant growth and lower yields (Zhao et al., 1999a,b; Hawkesford, 2000). Insufficient sulfur supply causes changes in amino acid pools and alters the seed protein composition (Gayler and Sykes, 1985; Spencer et al., 1990). Additionally, sulfur nutrition has a pronounced effect on legume-rhizobium symbiosis. Legumes grown in sulfur-rich media exhibit elevated nitrogen fixation due to an increase in nodule development and function (Scherer et al., 2008; Varin et al., 2010). Compared to other organs nodules have much greater thiol concentrations due to active thiol synthesis in nodule tissue (Matamoros et al., 1999). Higher yields observed in legumes grown in sulfur-rich environment may be attributed to remobilization of nutrients from the nodules and other source organs to seeds. In a recent study, it was demonstrated that the response of plants to sulfur deficiency is dependent on the developmental stage of the plant (Zuber et al., 2013). Sulfur deficiency imposed at the mid-developmental stage of a model legume, *Medicago truncatula*, decreased yield and altered the allocation of nitrogen and carbon to seeds (Zuber et al., 2013). Interestingly, sulfur deficiency imposed during the reproductive period had little influence on the yield and nutrient allocation (Zuber et al., 2013).

Elevating the accumulation of sulfur-rich proteins in legumes possesses unique challenges. Recently, we reported a successful strategy to increase the sulfur amino acid content of soybean seed proteins by overexpressing a cytosolic isoform of *O*-acetylserine sulfhydrylase (Kim et al., 2012). These transgenic soybean plants contained elevated levels of sulfur-containing amino acids that promoted the accumulation of Bowman-Birk protease inhibitor, a cysteine-rich protein (Kim et al., 2012). Thus, introgression of foreign proteins rich in sulfur-containing amino acids into the *O*-acetylserine sulfhydrylase overexpressing soybean line may offer a viable strategy to increase the sulfur amino acid content of soybean seeds. It is worthwhile to note that co-transformation of potato plants with methionine insensitive cystathionine γ-synthase ($CgS_{\Delta 90}$) and 15 kD β-zein resulted in elevation of protein-bound methionine content (Dancs et al., 2008).

AUTHOR CONTRIBUTIONS

Won-Seok kim, Joseph M. Jez and Hari B. Krishnan designed research; Won-Seok kim and Hari B. Krishnan performed research; Hari B. Krishnan and Joseph M. Jez analyzed the data and wrote the paper.

ACKNOWLEDGMENTS

The authors thank Nathan Oehrle for technical assistance and Jason Gillman for help with statistical analysis. Soybean transformation and electron microscopy were carried out at the University of Missouri Plant Transformation and Electron Microscopy Core Facilities, respectively. Amino acid analysis was performed at the Proteomics and Mass Spectrometry facility at the Danforth Center, St. Louis, Missouri. Names are necessary to report factually on available data; however, either the University of Missouri or USDA neither guarantees nor warrants the standard of product and the use of the name by the University of Missouri or USDA implies no approval of the product to the exclusion of others that may be suitable.

REFERENCES

Amir, R., Han, T., and Ma, F. (2012). Bioengineering approaches to improve the nutritional values of seeds by increasing their methionine content. *Mol. Breed.* 29, 915–924. doi: 10.1007/s11032-011-9690-7

Amira, G., Ifat, M., Tal, A., Hana, B., Shmuel, G., and Rachel, A. (2005). Soluble methionine enhances accumulation of a 15 kDa zein, a methionine-rich storage protein, in transgenic alfalfa but not in transgenic tobacco plants. *J. Exp. Bot.* 56, 2443–2452. doi: 10.1093/jxb/eri237

Avraham, T., Badani, H., Galili, S., and Amir, R. (2005). Enhanced levels of methionine and cysteine in transgenic alfalfa (*Medicago sativa* L.) plants over-expressing the *Arabidopsis* cystathionine ϒ-synthase gene. *Plant Biotech. J.* 3, 71–79. doi: 10.1111/j.1467-7652.2004.00102.x

Beilinson, V., Chen, Z., Shoemaker, C., Fischer, L., Goldberg, B., and Nielsen, C. (2002). Genomic organization of glycinin genes in soybean. *Theor. Appl. Genet.* 104, 1132–1140. doi: 10.1007/s00122-002-0884-6

Bilyeu, K. D., Zeng, P., Coello, P., Zhang, Z. J., Krishnan, H. B., Bailey, A., et al. (2008). Quantitative conversion of phytate to inorganic phosphorus in soybean seeds expressing a bacterial phytase. *Plant Physiol.* 146, 468–477. doi: 10.1104/pp.107.113480

Bost, K., and Piller, K. (2011). "Protein expression systems: why soybean seeds?" in *Soybean - Molecular Aspects of Breeding ISBN: 978-953-307-240-1*, ed A. Sudaric (Rijeka: InTech), 1–18. doi: 10.5772/15376

Chiaiese, P., Ohkama-Ohtsu, N., Molvig, L., Godfree, R., Dove, H., Hocart, C., et al. (2004). Sulphur and nitrogen nutrition influence the response of chickpea seeds to an added, transgenic sink for organic sulphur. *J. Exp. Bot.* 55, 1889–1901. doi: 10.1093/jxb/erh198

Coates, J. B., Medeiros, J. S., Thanh, V. H., and Nielsen, N. C. (1985). Characterization of the subunits of β-conglycinin. *Arch. Biochem. Biophys.* 243, 184–194. doi: 10.1016/0003-9861(85)90787-8

Cunha, N. B., Araujo, A. C. G., Leite, A., Murad, A. M., Vianna, G. R., and Rech, E. L. (2010). Correct targeting of proinsulin in protein storage vacuoles of transgenic soybean seeds. *Genet. Mol. Res.* 9, 1163–1170. doi: 10.4238/vol9-2gmr849

Dancs, G., Kondrák, M., and Bánfalvi, Z. (2008). The effects of enhanced methionine synthesis on amino acid and anthocyanin content of potato tubers. *BMC Plant Biol.* 8:65 doi: 10.1186/1471-2229-8-65

Ding, S. H., Huang, L. Y., Wang, Y. D., Sun, H. C., and Xiang, Z. H. (2006). High-level expression of basic fibroblast growth factor in transgenic soybean seeds and characterization of its biological activity. *Biotechnol. Lett.* 28, 869–875. doi: 10.1007/s10529-006-9018-6

Dinkins, R. D., Reddy, R., Meurer, C. A., Yan, B., Trick, H., Thibaud-Nissen, F., et al. (2001). Increased sulfur amino acids in soybean plants overexpressing the maize 15 kD zein protein. *In Vitro Cell. Dev. Biol. Plant* 37, 742–747. doi: 10.1007/s11627-001-0123-x

Fehr, W. R., Caviness, C. E., Burmood, D. T., and Pennington, J. S. (1971). Stage of development descriptions for soybeans, *Glycine max* (L.) Merrill. *Crop Sci.* 11, 929–931. doi: 10.2135/cropsci1971.0011183X001100060051x

Friedman, A. M., Long, S. R., Brown, S. E., Buikema, W. J., and Ausubel, F. M. (1982). Construction of broad host range cosmid cloning vector and its use in the genetic analysis of *Rhizobium* mutants. *Gene* 18, 289–296. doi: 10.1016/0378-1119(82)90167-6

Frizzi, A., Caldo, R. A., Morrell, J. A., Wang, M., Lutfiyya, L. L., Brown, W. E., et al. (2010). Compositional and transcriptional analyses of reduced zein kernels derived from the opaque2 mutation and RNAi suppression. *Plant Mol. Biol.* 73, 569–585. doi: 10.1007/s11103-010-9644-1

Galili, G., and Amir, R. (2013). Fortifying plants with the essential amino acids lysine and methionine to improve nutritional quality. *Plant Biotech. J.* 11, 211–222. doi: 10.1111/pbi.12025

Garg, R., Tolbert, M., Oakes, J. L., Clemente, T. E., Bost, K. L., and Piller, K. J., (2007). Chloroplast targeting of FanC, the major antigenic subunit of *Escherichia coli* K99 fimbriae, in transgenic soybean. *Plant Cell Rep.* 26, 1011–1023. doi: 10.1007/s00299-007-0322-y

Gayler, K. R., and Sykes, G. E. (1985). Effects of nutritional stress on the storage proteins of soybeans. *Plant Physiol.* 78, 582–585. doi: 10.1104/pp.78.3.582

Goossens, A., Van Montagu, M., and Angenon, G. (1999). Co-introduction of an antisense gene for an endogenous seed storage protein can increase expression of a transgene in *Arabidopsis thaliana* seeds. *FEBS Lett.* 456, 160–164. doi: 10.1016/S0014-5793(99)00943-6

Hacham, Y., Avraham, T., and Amir, R. (2002). The N-terminal region of Arabidopsis cystathionine ϒ-synthase plays an important regulatory role in methionine metabolism. *Plant Physiol.* 128, 454–462. doi: 10.1104/pp.010819

Hagan, N. D., Upadhyaya, N., Tabe, L. M., and Higgins, T. J. (2003). The redistribution of protein sulfur in transgenic rice expressing a gene for a foreign, sulfur-rich protein. *Plant J.* 34, 1–11. doi: 10.1046/j.1365-313X.2003.01699.x

Harada, J. J., Baker, S. J., and Goldberg, R. B. (1989). Soybean beta-conglycinin genes are clustered in several DNA regions and are regulated by transcriptional and posttranscriptional processes. *Plant Cell* 1, 415–425.

Hawkesford, M. J. (2000). Plant responses to sulphur deficiency and the genetic manipulation of sulphate transporters to improve S-utilization efficiency. *J. Exp. Bot.* 51, 131–138. doi: 10.1093/jexbot/51.342.131

Hawkesford, M. J., and De Kok, L. J. (2006). Managing sulphur metabolism in plants. *Plant Cell Environ.* 29, 382–395. doi: 10.1111/j.1365-3040.2005.01470.x

Hayashi, M., Harada, K., Fujiwara, T., and Kitamura, K. (1998). Characterization of a 7S globulin-deficient mutant of soybean (*Glycine max* (L.) Merrill). *Mol. Genet. Genomics* 258, 208–214. doi: 10.1007/s004380050724

Herman, E. M. (2014). Soybean seed proteome rebalancing. *Front. Plant Sci.* 5:437. doi: 10.3389/fpls.2014.00437

Hinchee, M. A. W., Connor-Ward, D. V., Newell, C. A., McDonnell, R. E., Sato, S. J., Gasser, C. S., et al. (1988). Production of transgenic soybean plants using *Agrobacterium*-mediated DNA transfer. *Biotechnology* 6, 915–922. doi: 10.1038/nbt0888-915

Kim, W. S., Chronis, D., Juergens, M., Schroeder, A. C., Hyun, S. W., Jez, J. M., et al. (2012). Transgenic soybean plants overexpressing *O*-acetylserine sulfhydrylase accumulate enhanced levels of cysteine and Bowman-Birk protease inhibitor in seeds. *Planta* 235, 13–23. doi: 10.1007/s00425-011-1487-8

Kim, W. S., and Krishnan, H. B. (2003). Allelic variation and differential expression of methionine-rich δ-zeins in maize inbred lines B73 and W23a1. *Planta* 217, 66–74. doi: 10.1007/s00425-002-0971

Kim, W. S., and Krishnan, H. B. (2004). Expression of an 11 kDa methionine rich delta-zein in transgenic soybean results in the formation of two types of novel protein bodies in transitional cells situated between the vascular tissue and storage parenchyma cells. *Plant Biotech. J.* 2, 199–210. doi: 10.1111/j.1467-7652.2004.00063.x

Kitamura, K., and Kaizuma, N. (1981). Mutant strains with low level of subunits of 7S globulin in soybean (*Glycine max* Merr.) seed. *Jpn. J. Breed.* 31, 353–359. doi: 10.1270/jsbbs1951.31.353

Kopriva, S., Suter, M., von Ballmoos, P., Hesse, H., Krähenbühl, U., Rennenberg, H., et al. (2002). Interaction of sulfate assimilation with carbon and nitrogen metabolism in *Lemna minor*. *Plant Physiol.* 130, 1406–1413. doi: 10.1104/pp.007773

Krishnan, H. B. (2000). Biochemistry and molecular biology of soybean seed storage proteins. *J. New Seeds* 2, 1–25. doi: 10.1300/J153v02n03_01

Krishnan, H. B. (2005). Engineering soybean for enhanced sulfur amino acid content. *Crop Sci.* 45, 454–461. doi: 10.2135/cropsci2005.0454

Krishnan, H. B., and Nelson, R. L. (2011). Proteomic analysis of high protein soybean (*Glycine max*) accessions demonstrates the contribution of novel glycinin subunits. *J. Agric. Food Chem.* 59, 2432–2439. doi: 10.1021/jf104330n

Krishnan, H. B., Oehrle, N. W., and Natarajan, S. S. (2009). A rapid and simple procedure for the depletion of abundant storage proteins from legume seeds to advance proteome analysis: a case study using *Glycine max. Proteomics* 9, 3174–3188. doi: 10.1002/pmic.200800875

Lange, M., Vincze, E., Wieser, H., Schjørring, J. K., and Holm, P. B. (2007). Effect of an antisense C-hordein gene on the storage protein composition in the barley seed. *J. Agric. Food Chem.* 55, 6074–6081. doi: 10.1021/jf0709505

Lee, J. D., Woolard, M., Sleper, D. A., Smith, J. R., Pantalone, V. R., Nyinyi, C. N., et al. (2009). Environmental effects on oleic acid in soybean seed oil of plant introductions with elevated oleic concentration. *Crop Sci.* 49, 1762–1768. doi: 10.2135/cropsci2008.11.0663

Lin, Y., Pajak, A., Marsolais, F., McCourt, P., and Riggs, C. D. (2013). Characterization of a cruciferin deficient mutant of Arabidopsis and its utility for overexpression of foreign proteins in plants. *PLoS ONE* 8:e64980. doi: 10.1371/journal.pone.0064980

Marsolais, F., Pajak, A., Yin, F., Taylor, M., Gabriel, M., Merino, D. M., et al. (2010). Proteomic analysis of common bean seed with storage protein deficiency reveals up-regulation of sulfur-rich proteins and starch and raffinose metabolic enzymes, and down-regulation of the secretory pathway. *J. Proteomics* 73, 1587–1600. doi: 10.1016/j.jprot.2010.03.013

Matamoros, M. A., Moran, J. F., Iturbe-Ormaetxe, I., Rubio, M. C., and Becana, M. (1999). Glutathione and homoglutathione synthesis in legume root nodules. *Plant Physiol.* 121, 879–888. doi: 10.1104/pp.121.3.879

Moravec, T., Schmidt, M. A., Herman, E. M., and Woodford-Thomas, T. (2007). Production of *Escherichia coli* heat labile toxin (LT) B subunit in soybean seed and analysis of its immunogenicity as an oral vaccine. *Vaccine* 25, 1647–1657. doi: 10.1016/j.vaccine.2006.11.010

Nielsen, N. C. (1996). "Soybean seed composition," in *Soybean: Genetics, Molecular Biology and Biotechnology,* eds D. P. S. Verma and R. C. Shoemaker (Wallingford: CABI), 127–163.

Nielsen, N. C., Dickinson, C. D., Cho, T. J., Thanh, V. H., Scallon, B. J., Fischer, R. L., et al. (1989). Characterization of the glycinin gene family in soybean. *Plant Cell* 1, 313–328. doi: 10.1105/tpc.1.3.313

Nielsen, N. C., and Nam, Y. W. (1999). "Soybean globulins," in *Seed Proteins,* eds P. R. Shewry and R. Casey (Dordrecht: Kluwer Academic Publishers), 285–313. doi: 10.1007/978-94-011-4431-5_13

Odanaka, H., and Kaizuma, N. (1989). Mutants on soybean storage proteins induced with γ-ray irradiation. *Jpn. J. Breed.* 39, 430–431.

Ogawa, T., Tayama, E., Kitamura, K., and Kaizuma, N. (1989). Genetic improvement of seed storage protein using three variant alleles of 7S globulin subunits in soyabean. (*Glycine max* L.). *Jpn. J. Breed.* 39, 137–147. doi: 10.1270/jsbbs1951.39.137

Panthee, D. R., Kwanyuen, P., Sams, C. E., West, D. R., Saxton, A. M., and Pantalone, V. R. (2004). Quantitative trait loci for β-conglycinin (7S) and glycinin (11S) fractions of soybean storage protein. *J. Am. Oil Chem. Soc.* 81, 1005–1012. doi: 10.1007/s11746-004-1014-4

Piller, K. J., Clemente, T. E., Jun, S. M., Petty, C. C., Sato, S., Pascual, D. W., et al. (2005). Expression and immunogenicity of an *Escherichia coli* K99 fimbriae subunit antigen in soybean. *Planta* 222, 6–18. doi: 10.1007/s00425-004-1445-9

Rao, S. S., and Hildebrand, D. (2009). Changes in oil content of transgenic soybeans expressing the yeast SLC1 gene. *Lipids* 44, 945–951. doi: 10.1007/s11745-009-3337-z

Scherer, H. W., Pacyna, S., Spoth, K. R., and Schulz, M. (2008). Low levels of ferredoxin, ATP and leghemoglobin contribute to limited N_2 fixation of Peas (*Pisum sativum* L.) and alfalfa (*Medicago sativa* L.) under S deficiency conditions. *Biol. Fertil. Soils* 44, 909–916. doi: 10.1007/s00374-008-0273-7

Schmidt, M. A., Barbazuk, W. B., Sandford, M., May, G., Song, Z., Zhou, W., et al. (2011). Silencing of soybean seed storage proteins results in a rebalanced protein composition preserving seed protein content without major collateral changes in the metabolome and transcriptome. *Plant Physiol.* 156, 330–345. doi: 10.1104/pp.111.173807

Schmidt, M. A., and Herman, E. M. (2008). Proteome rebalancing in soybean seeds can be exploited to enhance foreign protein accumulation. *Plant Biotech. J.* 6, 832–842. doi: 10.1111/j.1467-7652.2008.00364.x

Scossa, F., Laudencia-Chingcuanco, D., Anderson, O. D., Vensel, W. H., Lafiandra, D., D'Ovidio, R., et al. (2008). Comparative proteomic and transcriptional profiling of a bread wheat cultivar and its derived transgenic line overexpressing a low molecular weight glutenin subunit gene in the endosperm. *Proteomics* 8, 2948–2966. doi: 10.1002/pmic.200700861

Segal, G., Song, R. T., and Messing, J. (2003). A new opaque variant of maize by a single dominant RNA-interference-inducing transgene. *Genetics* 165, 387–397.

Shewry, P. R. (2000). "Seed proteins," in *Seed Technology and its Biological Basis,* eds M. Black and J. D. Bewley (Sheffield: Sheffield Academic Press), 42–84.

Shigemitsu, T., Ozaki, S., Saito, Y., Kuroda, M., Morita, S., Satoh, S., et al. (2012). Production of human growth hormone in transgenic rice seeds: co-introduction of RNA interference cassette for suppressing the gene expression of endogenous storage proteins. *Plant Cell Rep.* 31, 539–549. doi: 10.1007/s00299-011-1191-y

Song, S., Hou, W., Godo, I., Wu, C., Yu, Y., Matityahu, I., et al. (2013). Soybean seeds expressing feedback-insensitive cystathionine-γ-synthase exhibit a higher content of methionine. *J. Exp. Bot.* 64, 1917–1926. doi: 10.1093/jxb/ert053

Spencer, D., Rerie, W., Randall, P., and Higgins, T. (1990). The regulation of pea seed storage protein genes by sulfur stress. *Funct. Plant Biol.* 17, 355–363.

Staswick, P., Hermodson, M. A., and Nielsen, N. C. (1984). Identification of the cystines which link the acidic and basic components of the glycinin subunits. *J. Biol. Chem.* 259, 13431–13435.

Streatfield, S. J. (2007). Approaches to achieve high-level heterologous protein production in plants. *Plant Biotech. J.* 5, 2–15. doi: 10.1111/j.1467-7652.2006.00216.x

Tabe, L., Hagan, N., and Higgins, T. J. (2002). Plasticity of seed protein composition in response to nitrogen and sulfur availability. *Curr. Opin. Plant Biol.* 5, 212–217. doi: 10.1016/S1369-5266(02)00252-2

Tabe, L. M., and Droux, M. (2001). Sulfur assimilation in developing lupin cotyledons could contribute significantly to the accumulation of organic sulfur reserves in the seed. *Plant Physiol.* 126, 176–187. doi: 10.1104/pp.126.1.176

Tabe, L. M., and Higgins, T. J. (1998). Engineering plant protein composition for improved nutrition. *Trends Plant Sci.* 3, 282–286. doi: 10.1016/S1360-1385(98)01267-9

Tabe, L. M., Wirtz, M., Molvig, L., Droux, M., and Hell, R. (2010). Overexpression of serine acetlytransferase produced large increases in *O*-acetylserine and free cysteine in developing seeds of a grain legume. *J. Exp. Bot.* 61, 721–733. doi: 10.1093/jxb/erp338

Tada, Y., Utsumi, S., and Takaiwa, F. (2003). Foreign gene products can be enhanced by introduction into low storage protein mutants. *Plant Biotech. J.* 1, 411–422. doi: 10.1046/j.1467-7652.2003.00038.x

Takahashi, K., Banba, H., Kikuchi, A., Ito, M., and Nakamura, S. (1994). An induced mutant line lacking the α-subunit of β-conglycinin in soybean (*Glycine max* (L.) Merrill). *Jpn. J. Breed.* 44, 65–66. doi: 10.1270/jsbbs1951.44.65

Takahashi, M., Uematsu, Y., Kashiwaba, K., Yagasaki, K., Hajika, M., Matsunaga, R., et al. (2003). Accumulation of high levels of free amino acids in soybean seeds through integration of mutations conferring seed protein deficiency. *Planta* 217, 577–586. doi: 10.1007/s00425-003-1026-3

Taylor, M., Chapman, R., Beyaert, R., Hernández-Sebastià, C., and Marsolais, F. (2008). Seed storage protein deficiency improves sulfur amino acid content in common bean (*Phaseolus vulgaris* L.): redirection of sulfur from gamma-glutamyl-S-methyl-cysteine. *J. Agric. Food Chem.* 56, 5647–5654. doi: 10.1021/jf800787y

Thanh, V. H., and Shibasaki, K. (1978). Major proteins of soybean seeds. subunit structure of β-conglycinin. *J. Agric. Food Chem.* 26, 692–695. doi: 10.1021/jf60217a026

Townsend, J. A., and Thomas, L. A. (1994). Factors which influence the *Agrobacterium*-mediated transformation of soybean. *J. Cell Biochem. Supp.* 18A, 78–86.

Varin, S., Cliquet, J. B., Personeni, E., Avice, J. C., and Lemauviel-Lavenant, S. (2010). How does sulphur availability modify N acquisition of white clover (*Trifolium repens* L.)?A *J. Exp. Bot.* 61, 225–234. doi: 10.1093/jxb/erp303

Wesley, S. V., Helliwell, C. A., Smith, N. A., Wang, M., Rouse, D. T., Liu, Q., et al. (2001). Construct design for efficient, effective and high-throughput gene silencing in plants. *Plant J.* 27, 581–590. doi: 10.1046/j.1365-313X.2001.01105.x

Yagasaki, K., Kaizuma, N., and Kitamura, K. (1996). Inheritance of glycinin subunits and characterization of glycinin molecules lacking the subunits in soybean (*Glycine max* (L.) Merr.). *Breeding Sci.* 46, 11–15.

Zhang, Z., Xing, A., Staswick, P. E., and Clemente, T. (1999). The use of glufosinate as a selective agent in *Agrobacterium*-mediated transformation of soybean. *Plant Cell Tiss. Org.* 37, 37–46. doi: 10.1023/A:1006298622969

Zhao, F. J., Hawkesford, M. J., and McGrath, S. P. (1999a). Sulphur assimilation and effects on yield and quality of wheat. *J. Cereal Sci.* 30, 1–17. doi: 10.1006/jcrs.1998.0241

Zhao, F. J., Wood, A. P., and McGrath, S. P. (1999b). Effects of sulphur nutrition on growth and nitrogen fixation of pea (*Pisum sativum* L.). *Plant Soil* 212, 207–217. doi: 10.1023/A:1004618303445

Zuber, H., Poignavent, G., Le Signor, C., Aimé, D., Vieren, E., Tadla, C., et al. (2013). Legume adaptation to sulfur deficiency revealed by comparing nutrient allocation and seed traits in *Medicago truncatula*. *Plant J.* 76, 982–996. doi: 10.1111/tpj.12350

Conflict of Interest Statement: The authors declare that the research was conducted in the absence of any commercial or financial relationships that could be construed as a potential conflict of interest.

16

Diversity and regulation of ATP sulfurylase in photosynthetic organisms

Laura Prioretti[1], Brigitte Gontero[2], Ruediger Hell[3] and Mario Giordano[1,4]*

[1] Laboratory of Algal and Plant Physiology, Dipartimento di Scienze della Vita e dell'Ambiente, Università Politecnica delle Marche, Ancona, Italy
[2] Aix-Marseille Université Centre National de la Recherche Scientifique, BL' Unité de Bioénergétique et Ingénierie des Protéines UMR 7281, Marseille, France
[3] Centre for Organismal Studies, University of Heidelberg, Heidelberg, Germany
[4] Institute of Microbiology Academy of Sciences of the Czech Republic, Trebon, Czech Republic

Edited by:
Stanislav Kopriva, University of Cologne, Germany

Reviewed by:
Ján A. Miernyk, University of Missouri, USA
Stanislav Kopriva, University of Cologne, Germany

Correspondence:
Mario Giordano, Dipartimento di Scienze della Vita e dell'Ambiente, Università Politecnica delle Marche, via Brecce Bianche, 60131 Ancona, Italy
e-mail: m.giordano@univpm.it

ATP sulfurylase (ATPS) catalyzes the first committed step in the sulfate assimilation pathway, the activation of sulfate prior to its reduction. ATPS has been studied in only a few model organisms and even in these cases to a much smaller extent than the sulfate reduction and cysteine synthesis enzymes. This is possibly because the latter were considered of greater regulatory importance for sulfate assimilation. Recent evidences (reported in this paper) challenge this view and suggest that ATPS may have a crucial regulatory role in sulfate assimilation, at least in algae. In the ensuing text, we summarize the current knowledge on ATPS, with special attention to the processes that control its activity and gene(s) expression in algae. Special attention is given to algae ATPS proteins. The focus on algae is the consequence of the fact that a comprehensive investigation of ATPS revealed that the algal enzymes, especially those that are most likely involved in the pathway of sulfate reduction to cysteine, possess features that are not present in other organisms. Remarkably, algal ATPS proteins show a great diversity of isoforms and a high content of cysteine residues, whose positions are often conserved. According to the occurrence of cysteine residues, the ATPS of eukaryotic algae is closer to that of marine cyanobacteria of the genera *Synechococcus* and *Prochlorococcus* and is more distant from that of freshwater cyanobacteria. These characteristics might have evolved in parallel with the radiation of algae in the oceans and the increase of sulfate concentration in seawater.

Keywords: algae, algal evolution, ATPS, cysteine, redox regulation, sulfur metabolism

INTRODUCTION

Sulfur is an element of primary importance for all living organisms because it is a component of a very large number of compounds with essential biological functions (Giordano and Prioretti, 2014; Giordano and Raven, 2014; Glaeser et al., 2014). Photosynthetic organisms acquire S at its highest oxidation number (as sulfate), but assimilate it in its most reduced form of sulfide. The assimilation into organic molecules occurs primarily through cysteine from which S is then redistributed to the other sulfur amino acid, methionine, and other S-containing compounds (Takahashi et al., 2011; Giordano and Prioretti, 2014; Giordano and Raven, 2014). Due to the high reactivity of thiol (–SH) groups, S-compounds such as cysteine and glutathione are pivotal for metabolic redox regulation (Couturier et al., 2013). Sulfur acquisition and assimilation have been thoroughly described for vascular plants (Takahashi et al., 2011) and appear to be mostly conserved in the other photoautotrophic organisms, although differences exist in parts of the pathway (Giordano and Prioretti, 2014; Giordano and Raven, 2014).

ATP sulfurylase (ATPS; ATP:sulfate adenylyltransferase, EC 2.7.7.4) is the first enzyme of the sulfate assimilation pathway (Takahashi et al., 2011). Although a fair amount of information is available for fungal and bacterial ATPS (Mueller and Shafqat, 2013), the ATPS of phototrophic organisms (especially of algae) has not been studied as much as the enzymes that catalyze sulfate reduction (adenosine 5′-phosphosulfate reductase, APR) and cysteine synthesis (cysteine synthase or O-acetylserine (thiol)lyase, OAS-TL). This is probably the consequence of the fact that sulfate assimilation, in vascular plants, is mostly regulated through these enzymes (Takahashi et al., 2011). New evidence (see below) however challenges this view and attributes a more important regulatory role to ATPS than previously believed. We attempted to put the structural and catalytic information on ATPS in an evolutionary context, in order to provide clues, although not yet definitive answers, on what selective processes led to the ATPS proteins present in the extant organisms.

ATP SULFURYLASE REACTION AND CATALYTIC MECHANISM

ATPS is a nucleotidyl transferase that belongs to the superfamily of α/β phosphodiesterases. It catalyzes the non-reductive adenylation of sulfate to adenosine 5′-phosphosulfate (APS) and pyrophosphate (PPi) (Bicknell et al., 1982; **Figure 1**). Early studies on fungi (Farley et al., 1976, 1978) and plant (Shaw and Anderson, 1974) suggested that ATPS catalyzes an ordered reaction, in which MgATP is the first substrate to bind the enzyme active site (E); sulfate (SO_4^{2-}) would then interact with the

E-MgATP complex and ATP is cleaved. Finally AMP is bound to sulfate, with a concomitant release of MgPPi; as the final step, APS is released from the active site (**Figure 2A**).

Later studies, however, disputed this mechanism and proposed that, at least in *Penicillium*, the entrance of ATP and SO_4^{2-} in the active site can occur in a random order (Seubert et al., 1983). The main evidence supporting the random catalytic model is that APS, the final product of ATPS catalysis, strongly competes with the binding of both MgATP and SO_4^{2-} to the enzyme (Seubert et al., 1983). Seubert et al. (1985) also showed that all kinetic data are consistent with a random substrate binding mechanism for the forward reaction of ATPS, whereas the reverse reaction follows an ordered mechanism in which MgPPi is the first substrate to bind and leave the ATPS active site (**Figure 2B**). A recent report suggested an ordered mechanism, with sulfate binding after ATP, in the ATPS from *Glycine max*. For the reverse reaction, kinetic analysis and isothermal titration calorimetry-binding studies for an *Arabidopsis thaliana* ATPS isoform indicate that APS binding occurs first, followed by addition of PPi (Ravilious et al., 2013), in contrast to what was reported for the fungal enzyme (Seubert et al., 1985). The investigation of the catalytic mechanisms and the resolution of three-dimensional structure of more ATPS proteins may lead to the clarification of these open questions.

SULFATE ACTIVATION

The incorporation of sulfate (SO_4^{2-}) into organic molecules requires the reduction of sulfate to sulfite (SO_3^-). The standard potential, E_0', of this redox pair, however, is extremely negative (sulfate/sulfite E_0' = −454 mV, pH 7 and 25°C), and this reaction is thermodynamically impossible in a biological context because there are no reductant carriers in the cell that have a sufficiently negative redox potential. Sulfate, therefore, needs to be activated to APS by ATPS. APS can in fact be reduced to sulfite at a much lower redox potential, since the E_0' for the pair APS/sulfite is of only −60 mV, well within the range of redox potential of the biological carriers of reducing power (Rauen, 1964; Segel, 1976).

Sulfate activation is energetically aided by the subsequent cleavage of the phosphoric-sulfuric acid anhydride bond of APS for the production of sulfite, with the catalysis of APS reductase (APR). Such reaction has a ΔG^0 of about −19 kcal mol^{-1} (Segel, 1976), a rather high value compared, for example, to the energy generated by the hydrolysis of the phosphate-phosphate bond of an ATP (ΔG^0 approximately −8 kcal mol^{-1}; Schiff and

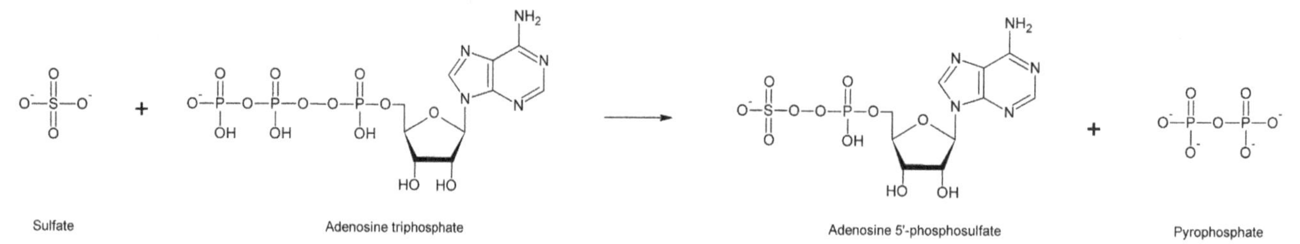

FIGURE 1 | Reaction catalyzed by ATP sulfurylase. The enzyme performs the hydrolysis of the bond between the α- and β-phosphate of ATP and, consequently, the attachment of AMP to SO_4^{2-}. MgPPi is finally released.

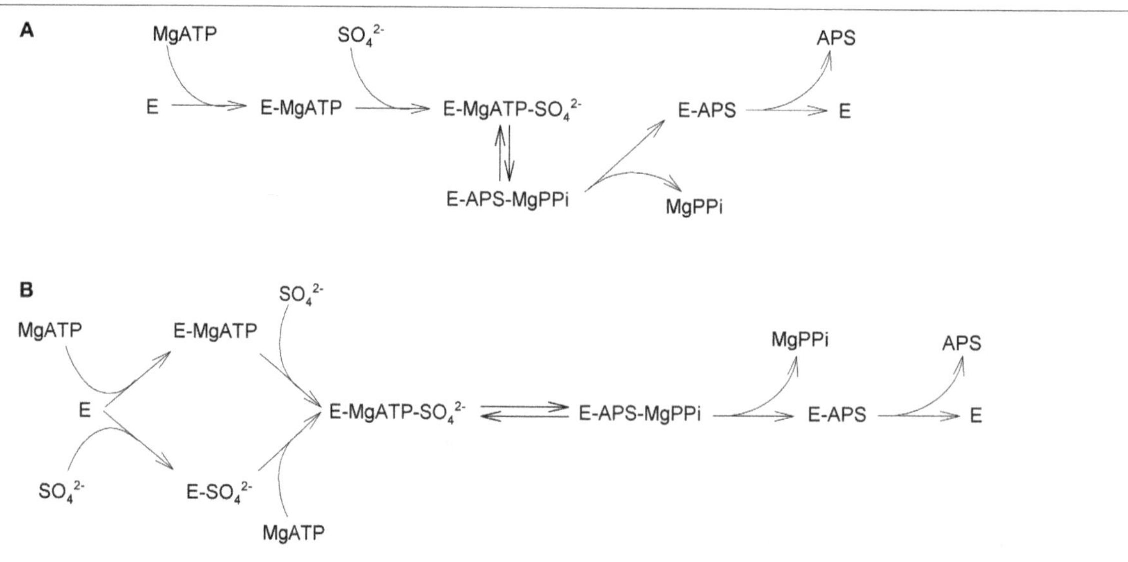

FIGURE 2 | ATP sulfurylase reaction mechanism. Panel **(A)** depicts the ordered reaction mechanism described by Farley et al. (1976) and Ravilious et al. (2013); panel **(B)** depicts the random reaction sequence as described by Seubert et al. (1983, 1985).

Hodson, 1973). The degradation of APS is also important because of the strong product inhibition of ATPS (Farley et al., 1976). It is thus not surprising that the enzymes that operate downstream of ATPS, i.e., APR and APS kinase (APK) (see Giordano and Prioretti, 2014 for details on the metabolic pathways downstream of ATPS) have a high affinity for APS [$K_{m(\text{APS})} = 2.1\,\mu\text{mol L}^{-1}$ for the red macroalga *Porphyra yezoensis* APR; (Bick and Leustek, 1998); $K_{m(\text{APS})} = 1\text{–}10\,\mu\text{mol L}^{-1}$ for *A. thaliana* APK; Lee and Leustek, 1998; Lillig et al., 2001]. *In vivo*, the activity of pyrophosphatases is believed to contribute to pull the reaction toward the production of APS, then facilitating the overall process of sulfate activation (ΔG^0 pyrophosphatase $= -5$ kcal mol^{-1}; Bicknell et al., 1982).

The fate of APS has been the subject of scholarly disputes for a rather long time. It was proposed that ATPS and APK worked together in a channeling mechanism that would allow the direct phosphorylation of APS to PAPS (Seubert et al., 1983; Sun and Leyh, 2006); this hypothesis was especially attractive because it could explain the existence of coupled ATPS-APK (PAPS synthetase) in metazoa (Mueller and Shafqat, 2013), fungi (MacRae et al., 2002), oomycetes and some algae (Patron et al., 2008). Experiments were carried out on enzymes from a variety of organisms to verify this hypothesis, but, so far, the channeling mechanism was demonstrated only for the purple bacterium *Rhodobacter sphaeroides* (Sun and Leyh, 2006). The existence of a channeling mechanism between ATPS and APR was also hypothesized for the *in vitro* formation of an ATPS-APR complex in *Allium cepa* (Cumming et al., 2007), but the occurrence of this complex *in vivo* and its functional significance are still unclear.

In plants and, as far as we know, in algae, the APS produced by ATPS is used (i) by APR for the production of sulfite in the chloroplast (only the euglenozoan *Euglena gracilis* seems to reduce sulfate in the mitochondrion; Brunold and Schiff, 1976) and (ii) by APK in plastid and cytosol for the production of PAPS, which is the main substrate for sulfation reactions (Takahashi et al., 2011; Giordano and Prioretti, 2014).

ATPS ISOFORMS

ATPS is present in both photosynthetic and non-photosynthetic organisms, in both prokaryotes and eukaryotes, and is involved in a variety of S-dependent processes. In proteobacteria, the genes for sulfate uptake and assimilation are organized in the *cys* operon, which in *Escherichia coli* and *Salmonella typhimurium* is composed of 18 genes (Leyh, 1993). Two of these genes, *cysD* and *cysN*, encode, respectively, the small (27 kDa) and the large (62 kDa) subunit of the heterodimeric ATPS (Leyh et al., 1988). The main peculiarity of bacterial ATPS is that the cysN subunit possesses a GTP-binding site and acts as a GTPase (Liu et al., 1994; Mougous et al., 2006). The energy derived from the hydrolysis of GTP fuels the ATPS activity of cysD (Leyh, 1993; Liu et al., 1994). Surprisingly, in the symbiotic diazotrophic bacterium *Rhizobium meliloti*, genes analogous to *cysD* and *cysN*, named *nodP* and *nodQ*, were found in the cluster of the nodulin genes where they participate to the biosynthesis of a sulfated nodulation factor (Schwedock and Long, 1990).

The crystalline structure of the thermophile *Thermus thermophilus* (Taguchi et al., 2004), the purple sulfur bacterium *Allochromatium vinosum* (Parey et al., 2013) and a symbiont of the hydrothermal vent tubeworm *Riftia pachyptila* (Beynon et al., 2001) revealed that, in these organisms, ATPS is a homodimeric enzyme. Also the ATPS crystal structure of the hyperthermophile bacterium *Aquifex aeolicus* shows that the enzyme is homodimeric and, differently from the above mentioned species, contains a functional APK domain at the ATPS C-terminus (Yu et al., 2007). Similarly to *A. aeolicus*, fungi possess an ATPS fused to an APK-like motif at the C-terminus (**Figure 3D**). In *Penicillium chrysogenum*, the two domains are organized in a homohexamer of 63.7 kDa subunits (see crystal structure in MacRae et al., 2001) and the APK domain carries out a PAPS-mediated allosteric regulation on the ATPS domain (MacRae et al., 2002). Also the ATPS of *Saccharomyces cerevisiae* is a homohexamer with 58 kDa monomers, as it emerges from the enzyme crystal structure (Ullrich et al., 2001); in this enzyme, however, the APK domain does not appear to have a catalytic function and probably has a structural role in the association of the ATPS monomers (Lalor et al., 2003).

Photosynthetic organisms contain a variable number of ATPS isoforms, with various degrees of sequence similarities. All vascular plants possess at least two ATPS isoforms, with the exception of the lycophyte *Selaginella moellendorffii* that only has one (Kopriva et al., 2009). In *A. thaliana*, an ATPS gene family of four members exists (*ATPS1-4*; Leustek et al., 1994; Murillo and Leustek, 1995; Hatzfeld et al., 2000). The four genes are located on different chromosomes. Although all four ATPS isoforms possess a plastidial transit peptide, there are indications that one of them is also expressed in the cytosol (Rotte and Leustek, 2000). Recently, the first plant ATPS crystal structure was obtained for *Glycine max* (Herrmann et al., 2014). The *Glycine*

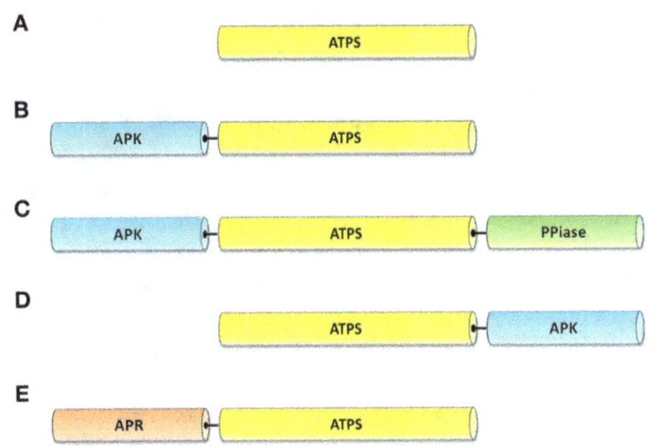

FIGURE 3 | Domain models of ATP sulfurylase (ATPS). (A) ATPS proteins of cyanobacteria, green algae, red algae, plastidial isoform of "red lineage" algae (with the exception of *T. pseudonana*). **(B)** metazoan PAPS synthetase with an APS kinase (APK) domain at the N-terminus of the ATPS domain. **(C)** cytosolic ATPS proteins of diatoms and haptophytes with an APK domain at the N-terminus and a pyrophosphatase (PPiase) domain at the C-terminus. **(D)** ATPS proteins of fungi and of the thermophylic bacterium *Aquifex aeolicus* fused with an APK domain at the C-terminus. **(E)** ATPS proteins of dinoflagellates, the apicomplexan *Chromera velia* and the filasterian *Capsaspora owczarzaki*, fused with an APS reductase (APR) at the N-terminus.

max ATPS is a homodimer of approximately 100 kDa, formed by two ~48 kDa monomers (Phartiyal et al., 2006; Ravilious et al., 2013). Surprisingly, sequence comparison and biochemical analyses revealed that plant ATPS is rather similar to the human enzyme (Patron et al., 2008), which, like soybean ATPS, is a homodimer (Harjes et al., 2005). Yet, the plant ATPS is a monofunctional enzyme (**Figure 3A**), whereas the human one is a bifunctional PAPS synthetase, with an APK domain fused through a linker at the ATPS N-terminus (Harjes et al., 2005; **Figure 3B**).

A very large degree of diversity has been found among algal ATPS proteins. The "green lineage" algae (those derived from the primary endosymbiotic event leading to Chl a+b algae; e.g., green algae) possess one plastidial ATPS isoform encoded by a single gene (Giordano and Prioretti, 2014; **Figure 3A**). The sole known exception is the freshwater green microalga *Chlamydomonas reinhardtii,* which possesses two ATPS proteins located in the plastid, which are encoded by two distinct nuclear genes, termed *ATS1* and *ATS2* (Patron et al., 2008). A larger degree of heterogeneity exists among the algae of the "red lineage" (those derived from the endosymbiosis leading to Chl a+c algae, e.g., red algae, diatoms, dinoflagellates and haptophytes). The few red algae for which an ATPS sequence is known possess one (*Galdieria sulphuraria, Pyropia yezoensis, Porphyra purpurea*) or two (*Cyanidioschyzon merolae* and *Chondrus crispus*) ATPS isoforms: *P. yezoensis* and *P. purpurea* ATPS enzymes are cytosolic; in *C. merolae* one isoform is plastidial and one is cytosolic (Patron et al., 2008); for the other species the location of ATPS is still unclear.

Diatoms, dinoflagellates and haptophytes possess two ATPS isoforms, one in the chloroplast, the other most likely in the cytosol (Patron et al., 2008). The plastidial enzyme is involved in primary sulfate assimilation. The cytosolic isoform of diatoms and haptophytes is presumably part of the sulfation pathway (Giordano and Prioretti, 2014) and possesses a functional APK domain fused at the ATPS N-terminus (Patron et al., 2008; **Figure 3C**). The cytosolic isoform of the diatoms and haptophytes is also characterized by a fusion with a pyrophosphatase at the C-terminus of the ATPS domain (Bradley et al., 2009), a similar domain configuration to that observed in oomycetes of the genus *Phytophthora* (Bradley et al., 2009). Although biochemical and kinetic data are not yet available for this enzyme, it has been hypothesized that the pyrophosphatase removes the PPi produced as a by-product of APS synthesis, thereby making the forward ATPS reaction irreversible. For what it is known, the only diatom that appears to constitute an exception to this pattern of isoform localization is *Thalassiosira pseudonana*: in this species, based on the presence/absence of a plastid transit peptide and sequence analysis, it was suggested that the isoform with the sole ATPS domain is located in the cytosol, whereas the APK-ATPS-pyrophosphatase isoform is in the chloroplast (Patron et al., 2008; Bromke et al., 2013). This localization is however to be taken with care, since the sequence was inferred from a raw contig at a time when few protein models were available for *T. pseudonana* genome (N. J. Patron, personal communication); even now, currently available softwares for the prediction of signal/transit peptides are not optimized for secondary endosymbiotic organisms and we were unable to unambiguously determine the nature of this transit peptide. We therefore believe that further and more thorough analyses are required to verify the location of ATPS in *T. pseudonana* and other related organisms.

A peculiar ATPS isoform was found in the dinoflagellates *Heterocapsa triquetra* (Patron et al., 2008), *Amphidinium klebsii* (Giordano and Prioretti, 2014) and *Amphidinium carterae,* in the photosynthetic apicomplexan *Chromera velia* (notice that Apicomplexa group into the superphylum Alveolata together with dinoflagellates) and in the filasterean snail symbiont *Capsaspora owczarzaki* (Suga et al., 2013; it is worthwhile mentioning that the filasterians are among the closest unicellular relative of metazoan). These organisms possess an APR domain fused to the ATPS domain at the N-terminus (**Figure 3E**). Although it is tempting to conclude that such an arrangement facilitates APS reduction during sulfate assimilation, no data are available on the activity and kinetic of these enzymes and the attribution of any function to this isoform is premature. The similarity between the enzyme of the Filasterea, a group rather closely related to metazoa, and that of alveolates makes the explanation of the origin of ATPS in these organisms difficult. The species of the genus *Amphidinium* also possess an ATPS sequence with the sole ATPS domain (Giordano and Prioretti, 2014). The sequences of *A. klebsii, A. carterae, C. velia,* and that of the green alga *Tetraselmis suecica,* which are not yet available in public databases, are shown in Table S1 (supplementary materials).

The high diversity of ATPS isoforms may reflect long and independent evolution of lineages as well as adaptation to specific habitats.

ATPS PHYLOGENY

The distribution of the different taxa in the ATPS tree (Patron et al., 2008; **Figure 4**) complicates the reconstruction of the phylogeny of this gene. It is noteworthy that the base of the tree is occupied by eukaryotic sequences. Patron et al. (2008) suggest that this may, at least to some extent, be the consequence of extensive lateral gene transfer. Interestingly, the ATPS of green algae cluster with that of the cyanobacteria and with the "red lineage" plastidial isoform. It is instead rather distant from the plant enzyme, making a plastidial origin of all extant assimilatory ATPS unlikely (Patron et al., 2008). Also the fungal ATPS-APK clusters with the bacterial enzyme and with the plastidial isoform of eukaryotic algae. On the other hand, the ATPS domains of the APK-ATPS of metazoa and algae with secondary red plastids have high sequence similarity (about 40–45%) with the ATPS of vascular plants, which does not possess the APK domain, but not with the ATPS-APK of fungi. The above ensemble of information is suggestive of the fact that the fused enzymes are derived from evolutionary events that occurred after the appearance of ATPS, whose phylogeny and evolutionary trajectories remain to be elucidated.

REGULATION OF ATPS EXPRESSION AND ACTIVITY

Although sulfate reduction catalyzed by APR is usually considered the most regulated step of S assimilation (Vauclare et al., 2002), there are hints of regulatory processes operating on ATPS activity as well (Giordano and Raven, 2014). The information of ATPS regulation is mostly limited to vascular plants (Leustek and

FIGURE 4 | Phylogenetic tree of ATP sulfurylase. All protein sequences, except those of *Tetraselmis suecica, Amphidinium klebsii, Amphidinium carterae, Heterocapsa triquetra,* and *Chromera velia* were obtained from either the NCBI protein database (http://www.ncbi.nlm.nih.gov/protein/), using the BLASTp (protein—Basic Local Alignment Search Tool) algorithm (http://blast.ncbi.nlm.nih.gov/Blast.cgi?PROGRAM=blastp&PAGE_TYPE=BlastSearch&LINK_LOC=blasthome), or the JGI (Joint Genome Institute) Genome Portal (http://genome.jgi.doe.gov/); *H. triquetra, A. carterae,* and *C. velia* ATPS sequences were kindly provided, respectively, by Stanislav Kopriva (University of Cologne), Charles F. Delwiche (University of Maryland) and Miroslav Obornik (Institute of Microbiology, Czech Academy of

(Continued)

FIGURE 4 | Continued
Sciences). *T. suecica* and *A. klebsii* sequences were determined by the authors (M.G.) in collaboration with Charles F. Delwiche. The sequences were then aligned using the software MUSCLE (MUltiple Sequence Comparison by Log-Expectation, http://www.ebi.ac.uk/Tools/msa/muscle/). The phylogenetic tree was finally constructed using the software SeaView (version 4, http://pbil.univ-lyon1.fr/software/seaview3.html). The aligned sequences were first modified using the Gblocks function to eliminate all the gaps and N- and C-termini in order to make the sequences comparable. A 10 bootstraps maximum-likelihood phylogenetic tree was then created using the PhyML program. The tree was finally edited with the software FigTree 1.4.0 (http://tree.bio.ed.ac.uk/software/figtree/).

Saito, 1999; Brunold, 2000; Koprivova et al., 2013), which, unfortunately, does not represent the full diversity of photosynthetic organisms.

In vascular plants and algae, ATPS activity is modulated in response to oxidative stress. For instance, an increase of ATPS activity was observed in roots of *Brassica napus* (Lappartient and Touraine, 1997) and *A. thaliana* subject to S deprivation (Lappartient et al., 1999). In *B. napus*, ATPS activity was inhibited by glutathione (Lappartient and Touraine, 1996) and H_2O_2 (Lappartient and Touraine, 1997). Also in the diatom *Phaeodactylum tricornutum* ATPS activity was very sensitive to H_2O_2 (Rosenwasser et al., 2014). The ATPS activity of the aquatic plants *Lemna gibba* and *Salvinia minima* exposed to arsenic greatly increased, together with the activities of γ-glutamylcysteine synthetase, glutathione S-transferase and glutathione reductase (Leao et al., 2014). An increase in the abundance of ATPS mRNA was observed in *Brassica juncea* roots exposed to Cd, but not in the presence of other heavy metals (Lee and Leustek, 1999). These data suggest that sulfur assimilation, on which ultimately glutathione and phytochelatins synthesis depends, is modulated at its beginning (i.e., at the sulfate activation step catalyzed by ATPS) when cells/plants are confronted with heavy metal and oxidative stress (typically the latter leading to the former type of stress).

In algae, ATPS gene expression (Yildiz et al., 1996; Zhang et al., 2004) and enzyme activity (Giordano et al., 2000; Prioretti and Giordano, unpublished) can be either down-regulated or up-regulated in response to sulfate availability. The transcription of both the genes *ATS1* and *-2* of *C. reinhardtii* is strongly up-regulated by S deprivation and their expression is under the control of the *SAC1* gene, which is responsible for the acclimation to S-limited conditions (Yildiz et al., 1996; Zhang et al., 2004). On the contrary, in the haptophyte *Emiliania huxleyi*, *ATPS1* expression is not affected by S limitation (Bochenek et al., 2013). In *Dunaliella salina*, ATPS activity is up-regulated in response to severe S-limitation leading to major metabolic adjustment concerning also N and C metabolism (Giordano et al., 2000). In the prasinophyte *T. suecica*, the diatom *T. pseudonana* and the marine cyanobacterium *Synechococcus* sp. strain WH7803 ATPS activity increases in response to S limitation. In the case of the dinoflagellate *A. klebsii*, ATPS activity is strongly down-regulated in response to S limitation (Prioretti and Giordano, unpublished).

Changes in the availability of other nutrients or possibly in elemental stoichiometry also appear to influence the expression and activity of ATPS. For example, a decrease of ATPS activity was detected in response to nitrate starvation (causing an increase in the S:N ratio) in *Lemna minor* (Brunold and Suter, 1984). In the haptophyte *Isochrysis galbana*, on the contrary, up-regulation of the ATPS gene expression was detected after nitrogen depletion (Song et al., 2013). Recently, an experiment carried out in order to test the effect of acid rains on plant sulfur metabolism revealed that ATPS gene expression, as the expression of most of the other genes of sulfur metabolism, was up-regulated after a treatment with a solution of 5:1 sulfate and nitrate (Liu et al., 2014). ATPS gene expression also responds to temperature: in *Glycine max* seeds, it was up-regulated in response to low temperature. This response appeared to be mediated by glutathione (Phartiyal et al., 2006). All these findings, together, demonstrate that ATPS is involved in a variety of cellular functions and is central to a number of metabolic adjustments possibly associated with the requirement for compositional homeostasis (Montechiaro and Giordano, 2010; Giordano, 2013).

The mechanisms of ATPS regulation, however are not fully understood. Recent findings indicate that ATPS gene expression in *A. thaliana* is, at least in part, controlled post-transcriptionally by the miR395 family of micro-RNAs (miRNAs, non-coding short RNAs), whose production is induced by sulfate deprivation (Jones-Rhoades and Bartel, 2004; Allen et al., 2005; Kawashima et al., 2011). miR395 is regulated by SLIM1 (Kawashima et al., 2011), a transcription factor responsible of the regulation of sulfate transporters *SULTR1;1*, *−1;2*, *−3;4*, and *−4;2* and of *ATPS4* during S limitation (Maruyama-Nakashita et al., 2006). The target of miR395 are *ATPS1*, *−3* and *−4* and *SULTR2;1* expression. *ATPS1* and *SULTR2;1* mRNA are post-transcriptionally degraded in the phloem companion cells where miR395 is expressed. However, the way these micro-RNAs perform their functions and their effects on *ATPS3* and *ATPS4* expression is still unclear.

Recent findings showed that miR395 expression is also sensitive to various kinds of environmental stresses, such as salt and drought stress (Wang et al., 2013). Oxidative stress mediated by copper and arsenate ions also induces miR395, whereas the addition of GSH suppresses this induction (Jagadeeswaran et al., 2014). These findings suggest that miR395 expression might be mediated also by redox signaling. The lack of a conserved miR395 family in *C. reinhardtii* suggests that green algae miRNAs may have a different mode of response to sulfur-deprivation (if any) than in higher plants (Shu and Hu, 2012).

Evidence for transcriptional regulation of *ATPS* expression was found in *A. thaliana* by Yatusevich et al. (2010). These authors reported that *ATPS1* and *ATPS3* expression was directly regulated by members of the R2R3-MYB transcription factors family, which are responsible of the synthesis of glucosinolates (GSs). GSs are a group of sulfated metabolites the synthesis of which involves catalysis by sulfotransferases (SOTs). These enzymes use 3′-phoshoadenosine-5′-phosphosulfate (PAPS) as a source of sulfate for the sulfation of amino acid-derived thioglucosides (Sonderby et al., 2010). Their synthesis is also regulated by SLIM1 when S is limiting (Maruyama-Nakashita et al., 2006). A specific control of GSs synthesis and of the expression of *ATPS1* and *ATPS3* by the MYB transcription factors was found: *ATPS1* expression is mainly regulated by MYB transcription factors which control the synthesis of aliphatic GSs, whereas *ATPS3* expression is associated with MYB transcription factors which control the synthesis of indolic GSs (Yatusevich et al., 2010). No relationships between MYB transcription factors and *ATPS2* and *ATPS4* expression has been detected until now. The above information point to a rather complex regulation of ATPS, although the exact mode of action is not always clear. The case of *A. thaliana* suggests that different ATPS isoforms may be differently regulated and may play different metabolic roles (Kopriva et al., 2009).

DO NUMBER AND POSITION OF CYSTEINE RESIDUES IN ATPS SEQUENCES HAVE SIGNIFICANCE FOR REDOX REGULATION?

ATPS enzymatic activity might be sensitive to redox regulation in some photosynthetic organisms and it was reported to be a target for thioredoxins (TRXs; see Buchanan and Balmer, 2005; Balsera et al., 2014 for specific reviews on TRXs) in both plants (Dixon et al., 2005; Marchand et al., 2006) and cyanobacteria (Lindahl and Florencio, 2003). The response of ATPS activity to oxidative stress mentioned above (Lappartient and Touraine, 1996, 1997; Rosenwasser et al., 2014) point to that direction.

Cysteine residues are among the most likely targets for redox regulation mechanisms (Couturier et al., 2013). Consequently, we studied the ATPS sequences of photosynthetic organisms to verify if the response to reducing and oxidizing agents was correlated with the number and position of cysteine residues. This study revealed that algal ATPS proteins contain an unexpected high number of cysteine residues with respect to the ATPS from plant and other organisms, and many of the algal cysteine residues are conserved (**Table 1**). Most algal ATPS proteins (except those of red algae—see below) contain 5–10 cysteine residues in their sequences, in contrast to the ATPS of vascular plants and fungi, which at the most contains two cysteine residues. The additional residues in algal sequences are often in positions different from those of the cysteine residues of other organisms. Among algae, five main groups of ATPS proteins can be identified with respect to the number and location of the cysteine residues:

(A) The ATPS of freshwater cyanobacteria and of marine cyanobacteria that do not belong to the genera *Synechococcus* and *Prochlorococcus*; these enzymes contain 4 conserved cysteine residues.
(B) The plastidial ATPS of Chlorophyta, Cryptophyta, Haptophyta and Heterokontophyta (notice the exception of *T. pseudonana* cytosolic ATPS mentioned above) and the enzyme of marine cyanobacteria of the genera *Synechococcus* and *Prochlorococcus* (possibly the most abundant marine cyanobacteria; Flombaum et al., 2013). These ATPS proteins constitute a consistent phylogenetic group (Patron et al., 2008) and contain 7–10 cysteine residues, 5 of which are highly conserved (**Figure 5**), although at positions different from those of the ATPS of the group A. The ATPS sequence from *E. gracilis* can be included in this group since it has 3 conserved cysteine residues at the same position as some of those of the group B.
(C) ATPS from red algae; 1–7 cysteine residues are present in these sequences and, although some cysteine residues are conserved among some of the species inside this group, their position varies and no common pattern can be identified.
(D) The cytosolic bi-functional APK-ATPS enzymes of algae with red secondary plastid; in this case, 5–8 cysteine residues are present in the ATPS domain of the protein, two of which are conserved, but at positions different from those of the cysteine residues of the ATPS of the other groups.
(E) Dinoflagellate ATPS; these proteins are different from all those mentioned above. At this stage, sequence information is available only for three species *H. triquetra* (Patron et al., 2008), *A. klebsii* (Prioretti and Giordano, unpublished) and *A. carterae*. It may therefore be unwise to draw general conclusion based on these species only. All the three species possess an enzyme with fused APR and ATPS domains. These APR-ATPS proteins contain 7 cysteine residues in the ATPS domain, one of which is at the same position as one of the cysteine residues of the enzymes from group A and two have the same position as two of the cysteine residues of group B ATPS. In *A. klebsii*, also a protein with the sole ATPS domain is present. It contains 6 cysteine residues, two of which are at the same positions as the cysteine residues of the bi-functional APR-ATPS enzyme (**Figure 5**).

Preliminary experiments conducted in MG laboratory showed that the number and location of the cysteines residues appear to be related to the sensitivity of the enzyme activity to thiol reducing and oxidizing agents, with the enzyme of group B being redox regulated, as opposite to those of group A and E. Further experiments are being conducted to check if and to what extent this information has general significance.

EVOLUTIONARY TRAJECTORIES OF ALGAL ATPS

The above grouping of alga ATPS and their phylogenetic relationships show one rather surprising fact: unlike the majority of cyanbacterial genes, which are phylogenetically closer within cyanobacteria than with respect to all other phyla (Zhaxybayeva et al., 2006), ATPS proteins of the *Synechococcus* and *Prochlorococcus* genera are closer to those of the eukaryotic algae than to those of all other cyanobacteria. Speculatively, the difference between the two groups of cyanobacteria may reflect the chemistry of the environments in which aquatic photosynthetic organisms live: sulfate concentration usually ranges between 10 and 800 μmol L^{-1} in freshwaters, 700–800 μmol L^{-1} can be found in eutrophic lakes, a maximum of 1 mmol L^{-1} is reached in brackish waters (Holmer and Storkholm, 2001). In the ocean, sulfate concentrations underwent a monotonic increase over time, until they reached the present maximum of 28–30 mmol L^{-1} in the Mesozoic era (Canfield, 2004; Ratti et al., 2011). The larger number of cysteines in the dominating oceanic cyanobacteria (Flombaum et al., 2013) may thus be associated to the higher availability of sulfate in extant seawater. On the other hand, the fact that most of eukaryotic algae, regardless of whether they inhabit freshwater or marine waters, have the same

Table 1 | ATP sulfurylase (ATPS) of algae and their cysteine content.

Phylum	Species	Environ.	Number and type of genes	Localization	Total number of Cys	Number of conserved Cys
Cyanobacteria	*Acaryochloris* sp.	M	1 ATPS	–	4	4 A
	Anabaena sp.	F	1 ATPS	–	4	4 A
	Arthrospira platensis	F	1 ATPS	–	4	4 A
	Crocosphaera sp.	M	1 ATPS	–	4	4 A
	Cyanobacterium sp.	M	1 ATPS	–	4	4 A
	Cyanothece sp.	M	1 ATPS	–	4	4 A
	Cylindrospermopsis sp.	F	1 ATPS	–	5	4 A
	Fischerella sp.	F	1 ATPS	–	4	4 A
	Gloeobacter sp.	F	1 ATPS	–	4	4 A
	Lyngbia sp.	M	1 ATPS	–	4	4 A
	Microcoleus vaginatus	F	1 ATPS	–	4	4 A
	Microcystis aeruginosa	F	1 ATPS	–	4	4 A
	Nodularia sp.	M	1 ATPS	–	4	4 A
	Nostoc sp.	F	1 ATPS	–	4	4 A
	Oscillatoria sp.	F	1 ATPS	–	4	4 A
	Prochlorococcus marinus	M	1 ATPS	–	9-11	5 B
	Raphidiopsis sp.	F	1 ATPS	–	4	4 A
	Spirulina subsalsa	M	1 ATPS	–	4	4 A
	Synechococcus elongatus	F	1 ATPS	–	4	4 A
	Synechococcus sp. CB0101	M	1 ATPS	–	8	5 B
	Synechococcus sp. CC9902	M	1 ATPS	–	10	5 B
	Synechococcus sp. JA-3-3Ab	F	1 ATPS	–	5	4 A
	Synechococcus sp. JA-2-3B'a	F	1 ATPS	–	6	4 A
	Synechococcus sp. RCC307	M	1 ATPS	–	8	5 B
	Synechococcus sp. RS9916	M	1 ATPS	–	9	5 B
	Synechococcus sp. WH7803	M	1 ATPS	–	10	5 B
	Synechococcus sp. WH8102	M	1 ATPS	–	9	5 B
	Synechocystis sp. PCC6803	F	1 ATPS	–	4	4 A
	Thermosynechococcus sp.	F	1 ATPS	–	6	4 A
	Trichodesmium sp.	M	1 ATPS	–	4	4 A
Chlorophyta	*Chlamydomonas reinhardtii*	F	1 ATPS	P	9	5 B
	Chlamydomonas reinhardtii	F	1 ATPS	P	10	5 B
	Chlorella variabilis	F	1 ATPS	P	12	5 B
	Coccomyxa subellipsoidea	F	1 ATPS	P	10	5 B
	Micromonas pusilla	M	1 ATPS	P	8	5 B
	Ostreococcus lucimarinus	M	1 ATPS	P	9	5 B
	Ostreococcus tauri	M	1 ATPS	P	8	5 B
	Tetraselmis suecica	M	1 ATPS	U	8	5 B
	Volvox carteri	F	1 ATPS	P	9	5 B
Euglenozoa	*Euglena gracilis*	F	1 ATPS	M	6	3 B
Rhodophyta	*Chondrus crispus*	M	1 ATPS	U	1	3 (1 B; 2 E)
	Chondrus crispus	M	1 ATPS	U	5	0
	Cyanidioschyzon merolae	M	1 ATPS	P	7	3 (1 B; 2 E)
	Cyanidioschyzon merolae	M	1 ATPS	C	4	0
	Galdieria sulphuraria	M	1 ATPS	U	5	3 (1 B; 2 E)
	Porphyra purpurea	M	1 ATPS	C	2	1 C
	Pyropia yezoensis	M	1 ATPS	C	1	1 C
Chromerida	*Chromera velia*	M	1 ATPS	U	7	2 B
Cryptophyta	*Guillardia theta*	M	1 ATPS	P	9	5 B
	Guillardia theta	M	1 ATPS	P	7	2 B
Dinophyta	*Amphidinium carterae*	M	1 ATPS	U	6	2 B
	Amphidinium carterae	M	1 APR-ATPS	U	7	3 (1 A; 2 B)
	Amphidinium klebsii	M	1 ATPS	U	6	2 B
	Amphidinium klebsii	M	1 APR-ATPS	U	7	3 (1 A; 2 B)
	Heterocapsa triquetra	M	1 APR-ATPS	P	7	3 (1 A; 2 B)
Haptophyta	*Emiliania huxleyi*	M	1 ATPS	P	9	5 B
	Emiliania huxleyi	M	1 APK-ATPS	C	5	2 D

(Continued)

Table 1 | Continued

Phylum	Species	Environ.	Number and type of genes	Localization	Total number of Cys	Number of conserved Cys
Heterokontophyta	*Aureococcus anophagefferens*	M	1 ATPS	U	17	5 B
	Aureococcus anophagefferens	M	1 APK-ATPS	U	7	2 D
	Ectocarpus siliculosus	M	1 ATPS	P	10	5 B
	Fragilariopsis cylindrus	M	1 ATPS	U	8	5 B
	Fragilariopsis cylindrus	M	1 APK-ATPS	U	6	2 D
	Phaeodactylum tricornutum	M	1 ATPS	P	9	5 B
	Phaeodactylum tricornutum	M	1 APK-ATPS	C	8	2 D
	Pseudo-nitzschia multiseries	M	1 ATPS	U	7	5 B
	Pseudo-nitzschia multiseries	M	1 APK-ATPS	U	5	2 D
	Thalassiosira pseudonana	M	1 ATPS	C	8	5 B
	Thalassiosira pseudonana	M	1 APK-ATPS	P	7	2 D

All species for which the ATPS sequence is known are included in the table, except for Cyanobacteria, for which only representative species or strains of the oceanic genera Prochlorococcus and Synechococcus are shown. The column "Environ." indicates whether the species is marine (M) or freshwater (F). The column "Number and type of genes" indicates how many ATPS are present and of what type (i.e., if the protein contains the sole ATPS domain, if the protein contains an APR domain, etc.). The column "Localization" shows whether the protein is plastidial (P), cytosolic (C), mitochondrial (M) or unknown (U); in the case of prokaryotes, no compartmentation is indicated. The column "Total number of Cys" indicates how many cysteine residues are contained in the ATPS domain of the enzyme. The column "Number of conserved Cys" refers to the number of cysteine residues that are retained at the same position in the sequences; residues in the same position are identified by the same letter (A, B, C, D or E). All sequences, except those of Tetraselmis suecica, Amphidinium klebsii, Amphidinium carterae, Heterocapsa triquetra and Chromera velia were obtained from either the NCBI protein database or the JGI genome database; H. triquetra, A. carterae, and C. velia ATPS sequences were kindly provided, respectively, by Stanislav Kopriva (University of Cologne), Charles F. Delwiche (University of Maryland) and Miroslav Obornik (Institute of Microbiology, Czech Academy of Sciences). T. suecica and A. klebsii sequences were produced by the authors in collaboration with Charles F. Delwiche.

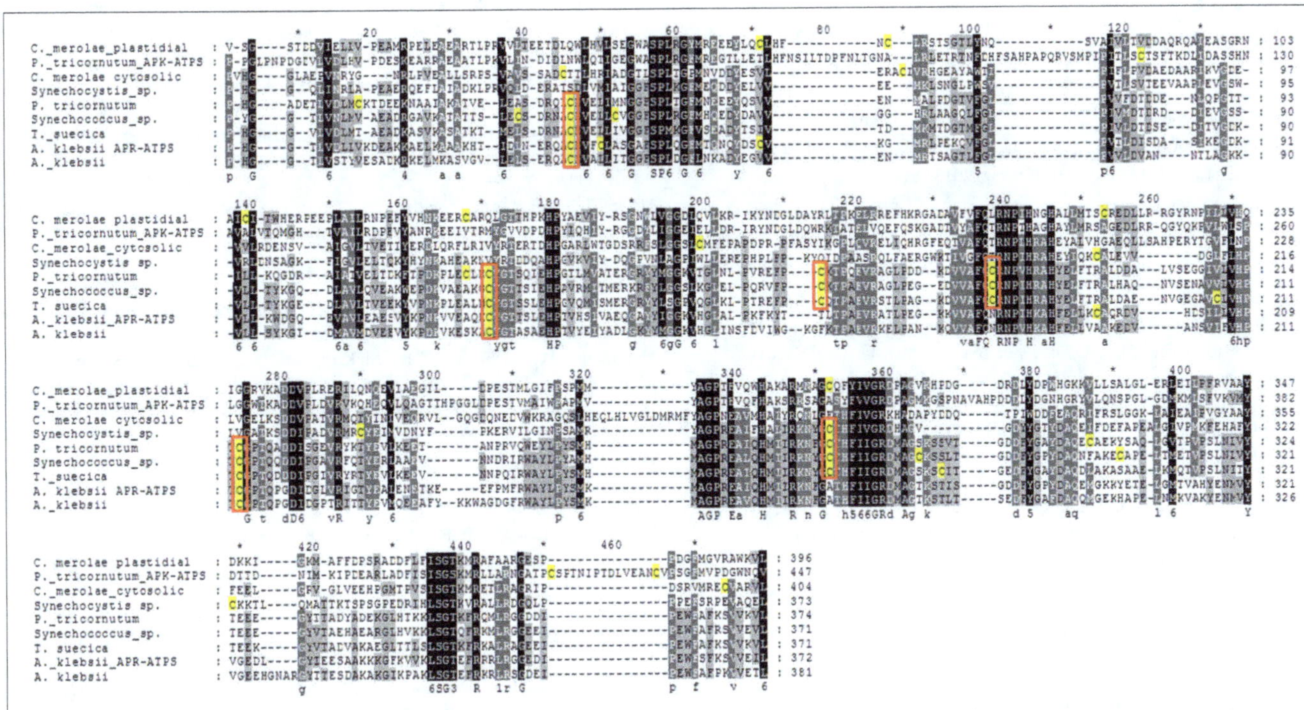

FIGURE 5 | Multiple sequence alignment of the sole ATP sulfurylase domain from different algal species showing cysteine residues (highlighted in yellow) and their conservation among species. A representative for each group of algal ATPS (see **Figure 3**) was chosen: *Tetraselmis suecica* was selected for green algae, *Phaeodactylum tricornutum* for diatoms and haptophytes (both the enzyme with the sole ATPS function and the ATPS domain in the APK-ATPS enzymes are shown), *Cyanidioschyzon merolae* for red algae (one plastidial and one cytosolic ATPS are present in this species, both constituted by the sole ATPS domain), *Amphidinium klebsii* for dinoflagellates (both the monofunctional ATPS and the ATPS domain of the APR-ATPS enzyme are shown), *Synechococcus* sp. strain WH7803 for marine cyanobacteria and *Synechocystis* sp. strain PCC6803 for freshwater cyanobacteria. The cysteine residues included in a red rectangle are present, and their position is conserved, in all the ATPS sequences belonging to the same group (as described in **Figure 3**). Alignments were performed using MUSCLE (http://www.ebi.ac.uk/Tools/msa/muscle/) and results exported and edited with GenDoc (http://www.nrbsc.org/gfx/genedoc/).

type of ATPS indicates that the number of cysteine residues is not a decisive feature for life in the oceans or it is a frozen accident of ATPS evolutionary trajectories in eukaryotes. Whether eukaryotic algae inherited their ATPS from marine cyanobacteria (or from a common ancestor which also gave rise to the ATPS of marine cyanobacteria of the *Synechococcus* and *Prochlorococcus* genera), or the extant distribution of ATPS isoforms is the consequence of lateral gene transfers between eukaryotic algae and marine cyanobacteria cannot be determined with the information at hand. Certainly, the existence of multiple and different ATPS isoforms in some algae (rhodophytes, diatoms and dinoflagellates, just to mention a few) is suggestive of a multiple and complex origin for this enzyme.

ACKNOWLEDGMENTS

Mario Giordano research on sulfur metabolism in algae was funded by the Italian Ministry for Agriculture (MIPAF, Bioforme project) and by the Assemble program of the European Union. Work by Ruediger Hell was supported by the Deutsche Forschungsgemeinschaft grants SPP1710 and He1848/13-1.

REFERENCES

Allen, E., Xie, Z. X., Gustafson, A. M., and Carrington, J. C. (2005). microRNA-directed phasing during trans-acting siRNA biogenesis in plants. *Cell* 121, 207–221. doi: 10.1016/j.cell.2005.04.004

Balsera, M., Uberegui, E., Schuermann, P., and Buchanan, B. B. (2014). Evolutionary development of redox regulation in chloroplasts. *Antioxid. Redox Signaling.* 21, 1327–1355. doi: 10.1089/ars.2013.5817

Beynon, J. D., Macrae, I. J., Huston, S. L., Nelson, D. C., Segel, I. H., and Fisher, A. J. (2001). Crystal structure of ATP sulfurylase from the bacterial symbiont of the hydrothermal vent tubeworm *Riftia pachyptila. Biochemistry* 40, 14509–14517. doi: 10.1021/bi015643l

Bick, J. A., and Leustek, T. (1998). Plant sulfur metabolism - the reduction of sulfate to sulfite. *Curr. Opin. Plant Biol.* 1, 240–244. doi: 10.1016/S1369-5266(98)80111-8

Bicknell, R., Cullis, P. M., Jarvest, R. L., and Lowe, G. (1982). The stereochemical course of nucleotidyl transfer catalyzed by ATP sulfurylase. *J. Biol. Chem.* 257, 8922–8927.

Bochenek, M., Etherington, G. J., Koprivova, A., Mugford, S. T., Malin, G., and Kopriva, S. (2013). Transcriptome analysis of the sulfate deficiency response in the marine microalga *Emiliania huxleyi. New Phytol.* 199, 650–662. doi: 10.1111/nph.12303

Bradley, M. E., Rest, J. S., Li, W.-H., and Schwartz, N. B. (2009). Sulfate activation enzymes: phylogeny and association with pyrophosphatase. *J. Mol. Evol.* 68, 1–13. doi: 10.1007/s00239-008-9181-6

Bromke, M. A., Hoefgen, R., and Hesse, H. (2013). Phylogenetic aspects of the sulfate assimilation genes from *Thalassiosira pseudonana. Amino Acids* 44, 1253–1265. doi: 10.1007/s00726-013-1462-8

Brunold, C. (2000). "Regulatory interactions between sulfate and nitrate assimilation," in *Sulfur Nutrition and Sulfur Assimilation in Higher Plants*, eds L. J. De Kok, I. Stulen, H. Rennenberg, C. Brunold, and W. E. Rauser (Hague: SPB Academic Publishing), 6–75.

Brunold, C., and Schiff, J. A. (1976). Studies of sulfate utilization of algae: 15. Enzymes of assimilatory sulfate reduction in euglena and their cellular localization. *Plant Physiol.* 57, 430–436. doi: 10.1104/pp.57.3.430

Brunold, C., and Suter, M. (1984). Regulation of sulfate assimilation by nitrogen nutrition in the duckweed *Lemna minor* L. *Plant Physiol.* 76, 579–583. doi: 10.1104/pp.76.3.579

Buchanan, B. B., and Balmer, Y. (2005). Redox regulation: a broadening horizon. *Annu. Rev. Plant Biol.* 56, 187–220. doi: 10.1146/annurev.arplant.56.032604.144246

Canfield, D. E. (2004). The evolution of the Earth surface sulfur reservoir. *Am. J. Sci.* 304, 839–861. doi: 10.2475/ajs.304.10.839

Couturier, J., Chibani, K., Jacquot, J.-P., and Rouhier, N. (2013). Cysteine-based redox regulation and signaling in plants. *Front. Plant Sci.* 4:105. doi: 10.3389/fpls.2013.00105

Cumming, M., Leung, S., McCallum, J., and McManus, M. T. (2007). Complex formation between recombinant ATP sulfurylase and APS reductase of *Allium cepa* (L.). *FEBS Lett.* 581, 4139–4147. doi: 10.1016/j.febslet.2007.07.062

Dixon, D. P., Skipsey, M., Grundy, N. M., and Edwards, R. (2005). Stress-induced protein S-glutathionylation in *Arabidopsis. Plant Physiol.* 138, 2233–2244. doi: 10.1104/pp.104.058917

Farley, J. R., Cryns, D. F., Yang, Y. H., and Segel, I. H. (1976). Adenosine triphosphate sulfurylase from *Penicillium chrysogenum.* Steady state kinetics of the forward and reverse reactions. *J. Biol. Chem.* 251, 4389–4397.

Farley, J. R., Nakayama, G., Cryns, D., and Segel, I. H. (1978). Adenosine triphosphate sulfurylase from *Penicillium chrysogenum* equilibrium binding, substrate hydrolysis, and isotope exchange studies. *Arch. Biochem. Biophys.* 185, 376–390. doi: 10.1016/0003-9861(78)90180-7

Flombaum, P., Gallegos, J. L., Gordillo, R. A., Rincon, J., Zabala, L. L., Jiao, N., et al. (2013). Present and future global distributions of the marine Cyanobacteria *Prochlorococcus* and *Synechococcus. Proc. Natl. Acad. Sci. U.S.A.* 110, 9824–9829. doi: 10.1073/pnas.1307701110

Giordano, M. (2013). Homeostasis: an underestimated focal point of ecology and evolution. *Plant Sci.* 211, 92–101. doi: 10.1016/j.plantsci.2013.07.008

Giordano, M., Pezzoni, V., and Hell, R. (2000). Strategies for the allocation of resources under sulfur limitation in the green alga *Dunaliella salina. Plant Physiol.* 124, 857–864. doi: 10.1104/pp.124.2.857

Giordano, M., and Prioretti, L. (2014). "Sulphur and algae: metabolism, ecology and evolution," in *The physiology of Microalgae*, eds M. A. Borowitzka, J. Beardall, and J. A. Raven (Dordrecht: Springer).

Giordano, M., and Raven, J. A. (2014). Nitrogen and sulfur assimilation in plants and algae. *Aquat. Bot.* 118, 45–61. doi: 10.1016/j.aquabot.2014.06.012

Glaeser, K., Kanawati, B., Kubo, T., Schmitt-Kopplin, P., and Grill, E. (2014). Exploring the *Arabidopsis* sulfur metabolome. *Plant J.* 77, 31–45. doi: 10.1111/tpj.12359

Harjes, S., Bayer, P., and Scheidig, A. J. (2005). The crystal structure of human PAPS synthetase 1 reveals asymmetry in substrate binding. *J. Mol. Biol.* 347, 623–635. doi: 10.1016/j.jmb.2005.01.005

Hatzfeld, Y., Lee, S., Lee, M., Leustek, T., and Saito, K. (2000). Functional characterization of a gene encoding a fourth ATP sulfurylase isoform from *Arabidopsis thaliana. Gene* 248, 51–58. doi: 10.1016/S0378-1119(00)00132-3

Herrmann, J., Ravilious, G. E., McKinney, S. E., Westfall, C. S., Lee, S. G., Baraniecka, P., et al. (2014). Structure and mechanism of soybean ATP sulfurylase and the committed step in plant sulfur assimilation. *J. Biol. Chem.* 289, 10919–10929. doi: 10.1074/jbc.M113.540401

Holmer, M., and Storkholm, P. (2001). Sulphate reduction and sulphur cycling in lake sediments: a review. *Freshw. Biol.* 46, 431–451. doi: 10.1046/j.1365-2427.2001.00687.x

Sonderby, I. E., Geu, F. F., and Halkier, B. A. (2010). Biosynthesis of glucosinolates - gene discovery and beyond. *Trends Plant Sci.* 15, 283–290. doi: 10.1016/j.tplants.2010.02.005

Jagadeeswaran, G., Li, Y.-F., and Sunkar, R. (2014). Redox signaling mediates the expression of a sulfate-deprivation-inducible microRNA395 in *Arabidopsis. Plant J.* 77, 85–96. doi: 10.1111/tpj.12364

Jones-Rhoades, M. W., and Bartel, D. P. (2004). Computational identification of plant MicroRNAs and their targets, including a stress-induced miRNA. *Mol. Cell* 14, 787–799. doi: 10.1016/j.molcel.2004.05.027

Kawashima, C. G., Matthewman, C. A., Huang, S., Lee, B.-R., Yoshimoto, N., Koprivova, A., et al. (2011). Interplay of SLIM1 and miR395 in the regulation of sulfate assimilation in *Arabidopsis. Plant J.* 66, 863–876. doi: 10.1111/j.1365-313X.2011.04547.x

Kopriva, S., Mugford, S. G., Matthewman, C., and Koprivova, A. (2009). Plant sulfate assimilation genes: redundancy versus specialization. *Plant Cell Rep.* 28, 1769–1780. doi: 10.1007/s00299-009-0793-0

Koprivova, A., Giovannetti, M., Baraniecka, P., Lee, B.-R., Grondin, C., Loudet, O., et al. (2013). Natural variation in the ATPS1 isoform of ATP sulfurylase contributes to the control of sulfate levels in *Arabidopsis*. *Plant Physiol.* 163, 1133–1141. doi: 10.1104/pp.113.225748

Lalor, D. J., Schnyder, T., Saridakis, V., Pilloff, D. E., Dong, A., Tang, H., et al. (2003). Structural and functional analysis of a truncated form of *Saccharomyces cerevisiae* ATP sulfurylase: C-terminal domain essential for oligomer formation but not for activity. *Protein Eng.* 16, 1071–1079. doi: 10.1093/protein/gzg133

Lappartient, A. G., and Touraine, B. (1996). Demand-driven control of root ATP sulfurylase activity and SO42- uptake in intact canola - The role of phloem-translocated glutathione. *Plant Physiol.* 111, 147–157.

Lappartient, A. G., and Touraine, B. (1997). Glutathione-mediated regulation of ATP sulfurylase activity, SO4(2-) uptake, and oxidative stress response in intact canola roots. *Plant Physiol.* 114, 177–183.

Lappartient, A. G., Vidmar, J. J., Leustek, T., Glass, A. D. M., and Touraine, B. (1999). Inter-organ signaling in plants: regulation of ATP sulfurylase and sulfate transporter genes expression in roots mediated by phloem-translocated compound. *Plant J.* 18, 89–95. doi: 10.1046/j.1365-313X.1999.00416.x

Leao, G. A., Oliveira, J. A., Farnese, F. S., Gusman, G. S., and Felipe, R. T. A. (2014). Sulfur metabolism: different tolerances of two aquatic macrophytes exposed to arsenic. *Ecotoxicol. Environ. Saf.* 105, 36–42. doi: 10.1016/j.ecoenv.2014.03.011

Lee, S., and Leustek, T. (1998). APS kinase from *Arabidopsis thaliana*: genomic organization, expression, and kinetic analysis of the recombinant enzyme. *Biochem. Biophys. Res. Commun.* 247, 171–175. doi: 10.1006/bbrc.1998.8751

Lee, S. M., and Leustek, T. (1999). The affect of cadmium on sulfate assimilation enzymes in Brassica juncea. *Plant Sci.* 141, 201–207. doi: 10.1016/S0168-9452(98)00231-3

Leustek, T., Murillo, M., and Cervantes, M. (1994). Cloning of a cDNA-encoding ATP sulfurylase from *Arabidopsis thaliana* by functional expression in *Saccharomyces cerevisiae*. *Plant Physiol.* 105, 897–902. doi: 10.1104/pp.105.3.897

Leustek, T., and Saito, K. (1999). Sulfate transport and assimilation in plants. *Plant Physiol.* 120, 637–643. doi: 10.1104/pp.120.3.637

Leyh, T. S. (1993). The physical biochemistry and molecular genetics of sulfate activation. *Crit. Rev. Biochem. Mol. Biol.* 28, 515–542. doi: 10.3109/10409239309085137

Leyh, T. S., Taylor, J. C., and Markham, G. D. (1988). The sulfate activation locus of *Escherichia coli* K12: cloning, genetic, and enzymatic characterization. *J. Biol. Chem.* 263, 2409–2416.

Lillig, C. H., Schiffmann, S., Berndt, C., Berken, A., Tischka, R., and Schwenn, J. D. (2001). Molecular and catalytic properties of *Arabidopsis thaliana* adenylyl sulfate (APS)-kinase. *Arch. Biochem. Biophys.* 392, 303–310. doi: 10.1006/abbi.2001.2453

Lindahl, M., and Florencio, F. J. (2003). Thioredoxin-linked processes in cyanobacteria are as numerous as in chloroplasts, but targets are different. *Proc. Natl. Acad. Sci. U.S.A.* 100, 16107–16112. doi: 10.1073/pnas.2534397100

Liu, C. X., Suo, Y., and Leyh, T. S. (1994). The energetic linkage of GTP hydrolysis and the synthesis of activated sulfate. *Biochemistry* 33, 7309–7314. doi: 10.1021/bi00189a036

Liu, T., Chen, J. A., Wang, W., Simon, M., Wu, F., Hu, W., et al. (2014). A combined proteomic and transcriptomic analysis on sulfur metabolism pathways of *Arabidopsis thaliana* under simulated acid rain. *PLoS ONE* 9:e90120. doi: 10.1371/journal.pone.0090120

MacRae, I. J., Segel, I. H., and Fisher, A. J. (2001). Crystal structure of ATP sulfurylase from *Penicillium chrysogenum*: insights into the allosteric regulation of sulfate assimilation. *Biochemistry* 40, 6795–6804. doi: 10.1021/bi010367w

MacRae, I. J., Segel, I. H., and Fisher, A. J. (2002). Allosteric inhibition via R-state destabilization in ATP sulfurylase from *Penicillium chrysogenum*. *Nat. Struct. Mol. Biol.* 9, 945–949. doi: 10.1038/nsb868

Marchand, C., Le Marechal, P., Meyer, Y., and Decottignies, P. (2006). Comparative proteomic approaches for the isolation of proteins interacting with thioredoxin. *Proteomics* 6, 6528–6537. doi: 10.1002/pmic.200600443

Maruyama-Nakashita, A., Nakamura, Y., Tohge, T., Saito, K., and Takahashi, H. (2006). Arabidopsis SLIM1 is a central transcriptional regulator of plant sulfur response and metabolism. *Plant Cell* 18, 3235–3251. doi: 10.1105/tpc.106.046458

Montechiaro, F., and Giordano, M. (2010). Compositional homeostasis of the dinoflagellate *Protoceratium reticulatum* grown at three different pCO_2. *J. Plant Physiol.* 167, 110–113. doi: 10.1016/j.jplph.2009.07.013

Mougous, J. D., Lee, D. H., Hubbard, S. C., Schelle, M. W., Vocadlo, D. J., Berger, J. M., et al. (2006). Molecular basis for G protein control of the prokaryotic ATP sulfurylase. *Mol. Cell* 21, 109–122. doi: 10.1016/j.molcel.2005.10.034

Mueller, J. W., and Shafqat, N. (2013). Adenosine-5-phosphosulfate - a multifaceted modulator of bifunctional 3-phospho-adenosine-5-phosphosulfate synthases and related enzymes. *FEBS J.* 280, 3050–3057. doi: 10.1111/febs.12252

Murillo, M., and Leustek, T. (1995). Adenosine-5′-triphosphate-sulfurylase from *Arabidopsis thaliana* and *Escherichia coli* are functionally equivalent but structurally and kinetically divergent - nucleotide sequence of two adenosine-5′-triphosphate-sulfurylase cDNAs from *Arabidopsis thaliana* and analysis of a recombinant enzyme. *Arch. Biochem. Biophys.* 323, 195–204. doi: 10.1006/abbi.1995.0026

Parey, K., Demmer, U., Warkentin, E., Wynen, A., Ermler, U., and Dahl, C. (2013). Structural, biochemical and genetic characterization of dissimilatory ATP sulfurylase from *Allochromatium vinosum*. *PLoS ONE* 8:e74707. doi: 10.1371/journal.pone.0074707

Patron, N. J., Durnford, D. G., and Kopriva, S. (2008). Sulfate assimilation in eukaryotes: fusions, relocations and lateral transfers. *BMC Evol. Biol.* 8:39. doi: 10.1186/1471-2148-8-39

Phartiyal, P., Kim, W.-S., Cahoon, R. E., Jez, J. M., and Krishnan, H. B. (2006). Soybean ATP sulfurylase, a homodimeric enzyme involved in sulfur assimilation, is abundantly expressed in roots and induced by cold treatment. *Arch. Biochem. Biophys.* 450, 20–29. doi: 10.1016/j.abb.2006.03.033

Ratti, S., Knoll, A. H., and Giordano, M. (2011). Did sulfate availability facilitate the evolutionary expansion of chlorophyll a plus c phytoplankton in the oceans? *Geobiology* 9, 301–312. doi: 10.1111/j.1472-4669.2011.00284.x

Rauen, H. M. (1964). *Biochemisches Taschenbuch*. Berlin; Heidelberg: Springer.

Ravilious, G. E., Herrmann, J., Lee, S. G., Westfall, C. S., and Jez, J. M. (2013). Kinetic mechanism of the dimeric ATP sulfurylase from plants. *Biosci. Rep.* 33, 585–591. doi: 10.1042/BSR20130073

Rosenwasser, S., Van Creveld, S. G., Schatz, D., Malitsky, S., Tzfadia, O., Aharoni, A., et al. (2014). Mapping the diatom redox-sensitive proteome provides insight into response to nitrogen stress in the marine environment. *Proc. Natl. Acad. Sci. U.S.A.* 111, 2740–2745. doi: 10.1073/pnas.1319773111

Rotte, C., and Leustek, T. (2000). Differential subcellular localization and expression of ATP sulfurylase and 5′-adenylylsulfate reductase during ontogenesis of *Arabidopsis* leaves indicates that cytosolic and plastid forms of ATP sulfurylase may have specialized functions. *Plant Physiol.* 124, 715–724. doi: 10.1104/pp.124.2.715

Schiff, J. A., and Hodson, R. C. (1973). The metabolism of sulfate. *Annu. Rev. Plant Physiol.* 24, 381–414. doi: 10.1146/annurev.pp.24.060173.002121

Schwedock, J., and Long, S. R. (1990). ATP sulphurylase activity of the nodP and nodQ gene products of *Rhizobium meliloti*. *Nature* 348, 644–647. doi: 10.1038/348644a0

Segel, I. H. (1976). *Biochemical Calculations: How to Solve Mathematical Problems in General Biochemistry*. New York, NY: John Wiley & Sons Inc.

Seubert, P. A., Hoang, L., Renosto, F., and Segel, I. H. (1983). ATP sulfurylase from *Penicillium chrysogenum*: measurements of the true specific activity of an enzyme subject to potent product inhibition and a reassessment of the kinetic mechanism. *Arch. Biochem. Biophys.* 225, 679–691. doi: 10.1016/0003-9861(83)90079-6

Seubert, P. A., Renosto, F., Knudson, P., and Segel, I. H. (1985). Adenosinetriphosphate sulfurylase from *Penicillium chrysogenum*: steady-state kinetics of the forward and reverse reactions, alternative substrate kinetics, and equilibrium binding studies. *Arch. Biochem. Biophys.* 240, 509–523. doi: 10.1016/0003-9861(85)90057-8

Shaw, W. H., and Anderson, J. W. (1974). The enzymology of adenosine triphosphate sulphurylase from spinach leaf tissue. *Biochem. J.* 139, 27–35.

Shu, L., and Hu, Z. (2012). Characterization and differential expression of microRNAs elicited by sulfur deprivation in *Chlamydomonas reinhardtii*. *BMC Genomics* 13:108. doi: 10.1186/1471-2164-13-108

Song, P., Li, L., and Liu, J. (2013). Proteomic analysis in nitrogen-deprived *Isochrysis galbana* during lipid accumulation. *PLoS ONE* 8:e82188. doi: 10.1371/journal.pone.0082188

Suga, H., Chen, Z., de Mendoza, A., Sebe-Pedros, A., Brown, M. W., Kramer, E., et al. (2013). The *Capsaspora* genome reveals a complex unicellular prehistory of animals. *Nat. Commun.* 4, 2325. doi: 10.1038/ncomms3325

Sun, M., and Leyh, T. S. (2006). Channeling in sulfate activating complexes. *Biochemistry* 45, 11304–11311. doi: 10.1021/bi060421e

Taguchi, Y., Sugishima, M., and Fukuyama, K. (2004). Crystal structure of a novel zinc-binding ATP sulfurylase from *Thermus thermophilus* HB8. *Biochemistry* 43, 4111–4118. doi: 10.1021/bi036052t

Takahashi, H., Kopriva, S., Giordano, M., Saito, K., and Hell, R. (2011). Sulfur assimilation in photosynthetic organisms: molecular functions and regulations of transporters and assimilatory enzymes. *Annu. Rev. Plant Biol.* 62, 157–184. doi: 10.1146/annurev-arplant-042110-103921

Ullrich, T. C., Blaesse, M., and Huber, R. (2001). Crystal structure of ATP sulfurylase from *Saccharomyces cerevisiae*, a key enzyme in sulfate activation. *EMBO J.* 20, 316–329. doi: 10.1093/emboj/20.3.316

Vauclare, P., Kopriva, S., Fell, D., Suter, M., Sticher, L., Von Ballmoos, P., et al. (2002). Flux control of sulphate assimilation in *Arabidopsis thaliana*: adenosine 5′-phosphosulphate reductase is more susceptible than ATP sulphurylase to negative control by thiols. *Plant J.* 31, 729–740. doi: 10.1046/j.1365-313X.2002.01391.x

Wang, M., Wang, Q., and Zhang, B. (2013). Response of miRNAs and their targets to salt and drought stresses in cotton (*Gossypium hirsutum* L.). *Gene* 530, 26–32. doi: 10.1016/j.gene.2013.08.009

Yatusevich, R., Mugford, S. G., Matthewman, C., Gigolashvili, T., Frerigmann, H., Delaney, S., et al. (2010). Genes of primary sulfate assimilation are part of the glucosinolate biosynthetic network in *Arabidopsis thaliana*. *Plant J.* 62, 1–11. doi: 10.1111/j.1365-313X.2009.04118.x

Yildiz, F. H., Davies, J. P., and Grossman, A. R. (1996). Sulfur availability and the *SAC1* gene control adenosine triphosphate sulfurylase gene expression in *Chlamydomonas reinhardtii*. *Plant Physiol.* 112, 669–675. doi: 10.1104/pp.112.2.669

Yu, Z., Lansdon, E. B., Segel, I. H., and Fisher, A. J. (2007). Crystal structure of the bifunctional ATP sulfurylase - APS kinase from the chemolithotrophic thermophile *Aquifex aeolicus*. *J. Mol. Biol.* 365, 732–743. doi: 10.1016/j.jmb.2006.10.035

Zhang, Z., Shrager, J., Jain, M., Chang, C. W., Vallon, O., and Grossman, A. R. (2004). Insights into the survival of *Chlamydomonas reinhardtii* during sulfur starvation based on microarray analysis of gene expression. *Eukaryot. Cell* 3, 1331–1348. doi: 10.1128/EC.3.5.1331-1348.2004

Zhaxybayeva, O., Gogarten, J. P., Charlebois, R. L., Doolittle, W. F., and Papke, R. T. (2006). Phylogenetic analyses of cyanobacterial genomes: quantification of horizontal gene transfer events. *Genome Res.* 16, 1099–1108. doi: 10.1101/gr.5322306

Conflict of Interest Statement: The authors declare that the research was conducted in the absence of any commercial or financial relationships that could be construed as a potential conflict of interest.

17

Alternative translational initiation of ATP sulfurylase underlying dual localization of sulfate assimilation pathways in plastids and cytosol in *Arabidopsis thaliana*

Anne-Sophie Bohrer[1†], Naoko Yoshimoto[2†], Ai Sekiguchi[2], Nicholas Rykulski[1], Kazuki Saito[2,3] and Hideki Takahashi[1]*

[1] *Department of Biochemistry and Molecular Biology, Michigan State University, East Lansing, MI, USA*
[2] *Graduate School of Pharmaceutical Sciences, Chiba University, Chiba, Japan*
[3] *RIKEN Center for Sustainable Resource Science, Yokohama, Japan*

Edited by:
Ruediger Hell, University of Heidelberg, Germany

Reviewed by:
Thomas Leustek, Rutgers, The State University of New Jersey, USA
Andreas Wachter, University Tuebingen, Germany

****Correspondence:***
Hideki Takahashi, Department of Biochemistry and Molecular Biology, Michigan State University, 603 Wilson Road, 209 Biochemistry Building, East Lansing, MI 48824, USA
e-mail: htakaha@msu.edu

†These authors have contributed equally to this work.

Plants assimilate inorganic sulfate into sulfur-containing vital metabolites. ATP sulfurylase (ATPS) is the enzyme catalyzing the key entry step of the sulfate assimilation pathway in both plastids and cytosol in plants. *Arabidopsis thaliana* has four *ATPS* genes (*ATPS1, –2, –3,* and *–4*) encoding ATPS pre-proteins containing N-terminal transit peptide sequences for plastid targeting, however, the genetic identity of the cytosolic ATPS has remained unverified. Here we show that *Arabidopsis ATPS2* dually encodes plastidic and cytosolic ATPS isoforms, differentiating their subcellular localizations by initiating translation at AUG^{Met1} to produce plastid-targeted ATPS2 pre-proteins or at AUG^{Met52} or AUG^{Met58} within the transit peptide to have ATPS2 stay in cytosol. Translational initiation of ATPS2 at AUG^{Met52} or AUG^{Met58} was verified by expressing a tandem-fused synthetic gene, $ATPS2_{(5'UTR-His12)}$*:Renilla luciferase:*$ATPS2_{(Ile13-Val77)}$*:firefly luciferase*, under a single constitutively active CaMV 35S promoter in *Arabidopsis* protoplasts and examining the activities of two different luciferases translated in-frame with split N-terminal portions of ATPS2. Introducing missense mutations at AUG^{Met52} and AUG^{Met58} significantly reduced the firefly luciferase activity, while AUG^{Met52} was a relatively preferred site for the alternative translational initiation. The activity of luciferase fusion protein starting at AUG^{Met52} or AUG^{Met58} was not modulated by changes in sulfate conditions. The dual localizations of ATPS2 in plastids and cytosol were further evidenced by expression of ATPS2-GFP fusion proteins in *Arabidopsis* protoplasts and transgenic lines, while they were also under control of tissue-specific *ATPS2* promoter activity found predominantly in leaf epidermal cells, guard cells, vascular tissues and roots.

Keywords: sulfur metabolism, ATP sulfurylase, alternative translational initiation, dual localization, *Arabidopsis*

INTRODUCTION

Sulfur is an essential macronutrient for plant growth, and it can be found in a wide variety of cellular components such as Cys, Met, GSH, sulfolipids, redox centers, and specialized metabolites involved in biotic and abiotic responses (Halkier and Gershenzon, 2006; Shimojima, 2011; Takahashi et al., 2011; Noctor et al., 2012). Sulfate is the main sulfur source available for plants in the environment. Therefore, reducing cofactors and carbon skeletons generated through photosynthesis are required in order to assimilate sulfate into organic sulfur metabolites. Following uptake of sulfate across the plasma membranes, several metabolic steps serve for reduction of sulfate. The first step of the sulfate assimilation pathway is catalyzed by ATP sulfurylase (ATPS) (ATP: sulfate adenylyltransferase, EC: 2.7.7.4) which uses ATP and sulfate to yield adenosine 5′-phosphosulfate (APS) and pyrophosphate. Following this step, APS is subsequently phosphorylated by APS kinase (APK, EC: 2.7.1.25) to form 3′-phosphoadenosine 5′-phosphosulfate (PAPS). PAPS is then used as a donor for sulfation reactions. Besides being utilized in the phosphorylation pathway, APS is supplied to the reductive assimilation pathway where it is converted to sulfite by APS reductase (APR, EC: 1.8.99.2). Sulfite is then reduced to sulfide by sulfite reductase (SiR, EC: 1.8.7.1). Cys biosynthesis from sulfide and *O*-acetylserine occurs following this reductive assimilation pathway.

ATPS is encoded by a multigenic family and its activity can be detected in cytosol and chloroplasts in plants (Lunn et al., 1990; Renosto et al., 1993; Klonus et al., 1994; Leustek et al., 1994; Murillo and Leustek, 1995; Logan et al., 1996; Hatzfeld et al., 2000; Rotte and Leustek, 2000; Phartiyal et al., 2006). APK is also found in both cytosol and plastids (Lee and Leustek, 1998; Lillig et al., 2001; Mugford et al., 2009). In contrast, reduction of APS catalyzed by APR occurs only in the plastids (Gutierrez-Marcos et al., 1996; Setya et al., 1996; Rotte and Leustek, 2000; Suter et al., 2000). APK and APR therefore compete for their common substrate APS in plastids, while no such competition happens in cytosol. PAPS biosynthesis in cytosol appears simple with ATPS and APK being direct enzymes involved in the pathway. However, it can eventually be affected by metabolic fluxes of APS phosphorylation and reduction in plastids, because PAPS is transported from plastids to cytosol (Gigolashvili et al., 2012). Thus, ATPS activities in plastids and cytosol contribute to provision of APS for downstream pathways in different ways, and their roles may vary depending on subcellular localizations.

Four *ATPS* genes (*ATPS1*, *-2*, *-3*, and *-4*) are present in the *Arabidopsis* genome (Leustek et al., 1994; Klonus et al., 1995; Murillo and Leustek, 1995; Logan et al., 1996; Hatzfeld et al., 2000). The protein coding regions of all four *ATPS* have the N-terminal leader sequences with characteristics for plastid-targeting transit peptides followed by the ATPS catalytic domains. Despite the presence of transit peptides in all four ATPS, the ATPS activity is detected in both chloroplasts and cytosol in *Arabidopsis* leaves (Rotte and Leustek, 2000). Thus, the identity of cytosolic ATPS has remained arguable, although *ATPS2* (Logan et al., 1996) has been proposed as a candidate gene to encode two isoforms (i.e., plastid- and cytosol-localizing ATPS) based on prediction of alternative translational initiation sites within the N-terminal transit peptide region (Hatzfeld et al., 2000).

In this study, we demonstrate experimental evidence that ATPS2 is alternatively translated into two different isoforms that dually localize in plastids and cytosol in *Arabidopsis*. The present study provides new insights into molecular mechanisms differentiating sulfate assimilation pathways in plastids and cytosol in plants.

MATERIALS AND METHODS

CHIMERIC GENE CONSTRUCTS FOR PROTOPLAST TRANSFECTION

Chimeric genes were generated using overlap-extension PCR methods. All the independent gene fragments were first amplified by PCRs using overlapping primers (Supplemental Table S1). The full-length chimeric genes were subsequently amplified by PCRs using 50 ng of each independent gene fragment, obtained from the initial PCRs, as templates and the primer pairs 1F/4R for *ATPS2-dual-Luc*, 1F/6R for $ATPS2_{(5'UTR-Val77)}$-*GFP* and $ATPS2_{FL}$-*GFP*, and 9F/6R for $ATPS1_{(5'UTR-Val63)}$-*GFP* genes (Supplemental Table S1). All PCRs were performed using Platinum Pfx DNA Polymerase (Thermo Fisher Scientific). The resultant PCR-amplified chimeric genes were cloned into pCRBlunt II-TOPO vector (Thermo Fisher Scientific) and fully sequenced. Each *Bam*HI-*Not*I-ended chimeric gene was ligated with the *Bam*HI-*Not*I fragment of *p35S:GFP* vector using a ligation kit Ligation Mighty Mix (Takara Bio) to generate *p35S:ATPS2-dual-Luc*, *p35S:*$ATPS2_{(5'UTR-Val77)}$-*GFP*, *p35S:*$ATPS2_{FL}$-*GFP*, and *p35S:*$ATPS1_{(5'UTR-Val63)}$-*GFP*.

The *p35S:GFP* vector [CaMV 35S:sGFP(S65T)] used in this study is a modified version of the *35Ω-sGFP(S65T)* vector (Chiu et al., 1996) from which the 35Ω promoter sequence was removed and replaced by the CaMV 35S promoter sequence of pBI221 (Clontech). The *Hind*III-*Bam*HI fragment (vector backbone, 3.65 kb) of *35Ω-sGFP(S65T)* and the *Hind*III-*Bam*HI fragment (CaMV 35S promoter, 0.8 kb) of pBI221 were ligated to obtain *p35S:GFP*.

Mutated versions of *p35S:ATPS2-dual-Luc* and *p35S:*$ATPS2_{(5'UTR-Val77)}$-*GFP* were generated by site-directed mutagenesis using the QuickChange Lightning Site-Directed Mutagenesis Kit (Agilent) according to the manufacturer's instructions. Oligonucleotide primers and DNA templates that were used to introduce various point mutations are listed in Supplemental Tables S2, S3.

PROTOPLAST ISOLATION AND TRANSFECTION

The protoplast isolation and transfections were performed as described by Yoo et al. (2007). *Arabidopsis* wild-type (Col-0) plants were grown on soil in an environment-controlled chamber under a 12-h-light / 12-h-dark cycle at 22°C with a light intensity of 80 μE m^{-2} s^{-1} and a relative humidity of 50%. For each protoplast transfection, 15–20 μg of plasmid DNAs were used, and the protoplasts were incubated in the transfection mixture for 10 min. Protoplasts were then re-suspended in WI solution (Yoo et al., 2007) supplemented with 1 mM $MgSO_4$ (+S condition) or 1 mM $MgCl_2$ (–S condition), and incubated in the dark for 16 h. Protoplasts were then harvested by centrifugation.

DUAL-LUCIFERASE ASSAYS

Dual luciferase assays were performed with the dual luciferase reporter assay system using the firefly luciferase reagent (LARII) and the *Renilla* luciferase reagent with firefly quenching (Stop & Glo) (Promega). All reagents were prepared as described by the manufacturer. Protoplasts were re-suspended in 50 μl 1X passive lysis buffer and incubated on ice for 15 min. The lysates were then centrifuged for 15 min at maximum speed at 4°C. Ten μl of undiluted supernatants were used to monitor the bioluminescence using a Centro SX3 luminometer (Berthold Technologies). Statistical significance was examined by One-Way analysis of the variance (ANOVA) and the Tukey's HSD *post-hoc* test with the level of significance set at 5%.

CREATION OF TRANSGENIC PLANTS EXPRESSING ATPS2 FUSED WITH GFP

For the creation of *ATPS2pro:ATPS2-GFP* fusion gene construct, oligonucleotide primers ATPS2-prom-FSal and ATPS2-CDSnstop-RNco (Supplemental Table S4) were used to amplify a genomic DNA fragment of *ATPS2* gene starting from 5′-region 3009-bp upstream of the plausible first translational initiation site and terminating just before the translational stop site. PCR was performed on genomic DNA prepared from *Arabidopsis thaliana* ecotype Col-0 using KOD plus DNA polymerase (Toyobo, Japan). The resultant PCR-amplified fragment of *ATPS2* was cloned into pCRBlunt II-TOPO (Thermo Fisher Scientific) and fully sequenced. The *Sal*I-*Nco*I-ended *ATPS2* gene fragment was inserted in the place of 35 Ω in the 35 Ω-sGFP(S65T) vector (Chiu et al., 1996) to obtain the *ATPS2pro:ATPS2-sGFP(S65T):NOSter* fusion gene. This fusion gene fragment was placed between the *Sal*I and *Eco*RI sites in the binary plasmid, pBI101 (Clontech), replacing the β-glucuronidase gene and the NOSter region (Figure S1).

The *ATPS2pro:ATPS2-GFP* chimeric gene constructs with mutated versions of *ATPS2* were created as follows. The *Bam*HI and *Xba*I sites, respectively located 748-bp upstream and 1127-bp downstream of the first translational initiation site of *ATPS2*, were used to cut out a DNA fragment from the binary plasmid harboring the *ATPS2pro:ATPS2-sGFP(S65T):NOSter* fusion gene (Figure S1), and this *Bam*HI-*Xba*I fragment was used as a template for overlap-extension PCRs. The nucleotide sequences of primer pairs used for the first PCRs are shown in Supplemental Table S4. The fragments obtained from the first PCR were mixed and amplified by PCR using primers ATPS2(–753)-FBam and ATPS2(+1136)-RXba. The DNA templates and the pairs of primers used for the construction of the mutated versions of *ATPS2pro:ATPS2-sGFP(S65T):NOSter* fusion gene are detailed in the Supplemental Table S5. The resultant PCR fragments containing mutations in *ATPS2* were cloned into pCR-Blunt II-TOPO and fully sequenced. The mutated *ATPS2* fragments were cut out as *Bam*HI-*Xba*I fragments, and used to replace the corresponding region of wild-type *ATPS2* gene in the binary plasmid harboring the *ATPS2pro:ATPS2-sGFP(S65T):NOSter* fusion gene construct (Figure S1).

The binary plasmids were transferred to *Agrobacterium tumefaciens* C58C1 GV3101 (pMP90) (Koncz and Schell, 1986) by a freeze-thaw method (Höfgen and Willmitzer, 1988). *Arabidopsis thaliana* ecotype Col-0 plants were transformed by a floral dip method (Clough and Bent, 1998). Transgenic plants were selected on GM agar medium (Valvekens et al., 1988) containing 50 mg L^{-1} kanamycin sulfate. Kanamycin-resistant T_2 progenies were used for the analyses.

MICROSCOPY AND IMAGING OF GFP

Fluorescence of ATPS2-GFP fusion proteins in protoplasts and transgenic plants was observed using confocal laser-scanning microscopes, Fluoview FV10i (Olympus) and LSM510 (Zeiss).

PROTEIN EXTRACTION AND IMMUNOBLOT ANALYSIS

Total protein was prepared from leaves and roots of plants grown for 2 weeks on GM agar medium (Valvekens et al., 1988). Tissues were ground under liquid nitrogen and homogenized in the extraction buffer [50 mM Tris-MES (pH 7.5), 300 mM sucrose, 150 mM NaCl, 10 mM CH_3COOK, 5 mM EDTA, 20 μM leupeptine, 100 μM 4-(2-aminoethyl)benzenesulfonyl fluoride, 1 mM phenylmethylsulfonyl fluoride]. The lysate was centrifuged at 10,000 g for 15 min, and the supernatant was collected. Protein concentrations were determined using a Bio-Rad protein assaying kit (Bio-Rad) based on the Bradford method (Bradford, 1976), using bovine serum albumin as a standard. Proteins were separated in 10% (w/v) polyacrylamide gel, and transferred to Immobilon-P membrane (Millipore) by electroblotting. Ten micrograms of crude proteins were loaded to each lane of the gels. The blot was incubated with anti-GFP mouse monoclonal antibody (Nacalai Tesque, Japan), followed by incubation with goat anti-mouse IgG conjugated to alkaline phosphatase (Promega). The presence of immuno-reactive protein was detected through the use of 5-bromo-4-chloro-3-indolyl-phosphate and nitro blue tetrazolium (Promega).

ACCESSION NUMBERS

The reference sequence information on *Arabidopsis ATPS* gene family members is available at The Arabidopsis Information Resource (TAIR, http://www.arabidopsis.org/) under the following accession numbers: *ATPS1* (At3g22890); *ATPS2* (At1g19920); *ATPS3* (At4g14680); *ATPS4* (At5g43780).

RESULTS

TRANSLATION OF ATPS2 CAN BE INITIATED FROM INTERNAL START SITES

The alignment of *Arabidopsis* ATPS protein sequences indicated that all four pre-proteins contained N-terminal extensions suggested to function as transit peptides for plastid targeting of polypeptides (**Figure 1**). Among them, only ATPS2 pre-protein contained four Met residues in its transit peptide, corresponding

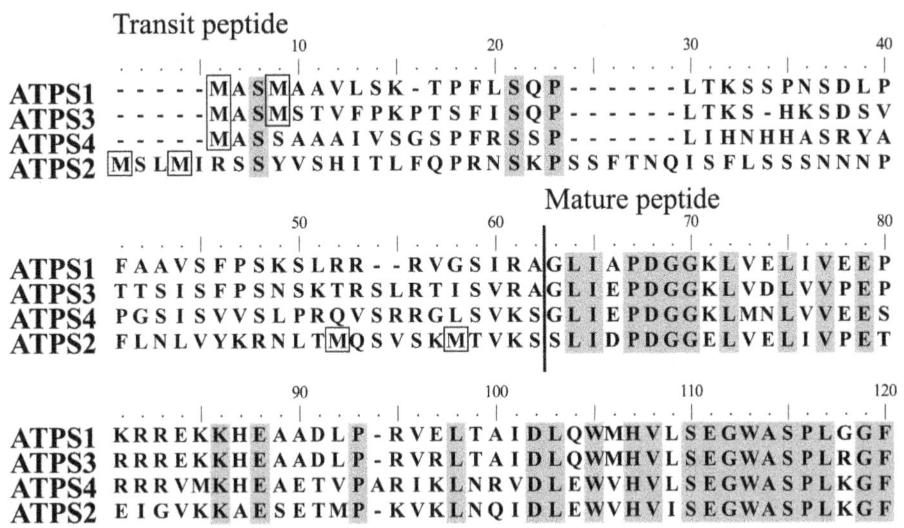

FIGURE 1 | Alignment of *Arabidopsis* ATP sulfurylase (ATPS) proteins. *Arabidopsis thaliana* ATPS complete protein sequences were aligned using Clustal W2. The predicted cleavage site of the transit peptide for plastid targeting is indicated. Identical residues in all four sequences are shaded and methionine residues in the transit peptides are boxed.

to Met1, Met4, Met52, and Met58 (**Figure 1**). The analysis of the nucleotide sequences surrounding AUG^{Met1}, AUG^{Met52}, and AUG^{Met58} codons indicated high similarities with the consensus sequence around translational initiation sites in dicot plants (Joshi et al., 1997; Hatzfeld et al., 2000). Translation of *ATPS2* mRNA was therefore predicted to start at multiple sites, AUG^{Met1} and either AUG^{Met52} or AUG^{Met58}, to produce the plastidic and the cytosolic ATPS2 isoforms, respectively, in *Arabidopsis*.

In order to test if the translation of ATPS2 can be alternatively initiated from either AUG^{Met52} or AUG^{Met58}, we constructed a tandem fusion gene, *p35S:ATPS2-dual-Luc*, that splits the portion from the 5′-untranslated region (5′UTR) through the N-terminal 77-amino-acid region of ATPS2 into two parts fused separately to two luciferase reporters (**Figure 2A**). A 184-bp fragment of DNA, from the 5′UTR to His12 of ATPS2, was cloned in frame with the coding sequence of *Renilla* luciferase (RLuc), and the remaining 195-bp fragment, from Ile13 to Val77 of ATPS2, was cloned in frame with the coding sequence of firefly luciferase (FLuc). This fusion construct was designed to express a tandem-fused single mRNA from the CaMV 35S promoter, and to observe subsequent translation of that transcription unit into two different luciferase-fusion proteins. The fusion construct was transfected into *Arabidopsis* protoplasts and both luciferase activities were monitored to determine the presence of translational products, M1M4-RLuc and M52M58-FLuc. A similar bicistronic gene construct containing GFP and FLuc open reading frames has been used to characterize the ability of a viral internal ribosome entry site (IRES) to mediate cap-independent internal translational initiation in plants (Urwin et al., 2000).

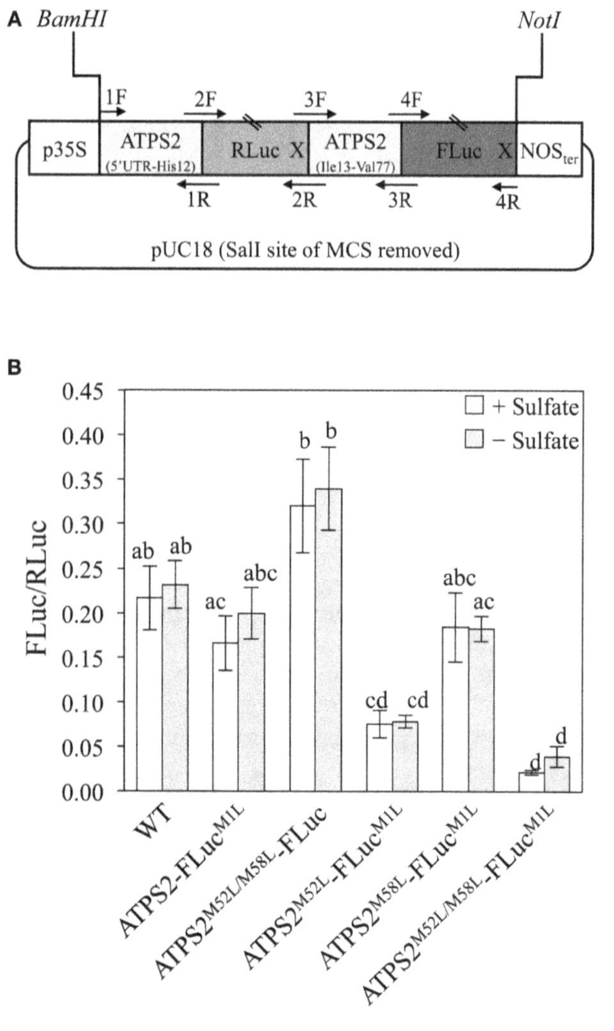

FIGURE 2 | Alternative translational initiation sites of ATPS2 determined by expression of dual-luciferase-tagged fusion constructs in *Arabidopsis* protoplasts. (A) Schematic representation of *p35S:ATPS2-dual-Luc* used for expression of tandem fusion gene $ATPS2_{(5'UTR-His12)}$:*Renilla luciferase*:$ATPS2_{(Ile13-Val77)}$:*firefly luciferase*. Positions of annealing sites of primers (1F-4R) used for chimeric gene construction and stop codons (X) are indicated. **(B)** FLuc activity of protein extracts from *Arabidopsis* protoplasts transfected with either wild-type (WT) or mutated versions of *p35:ATPS2-dual-Luc* chimeric genes. Results are presented as FLuc/Rluc activity ratios in the same transfection. Values indicate mean ± SD for the results of 8 independent transfections. Values marked with different letters indicate statistically significant differences (ANOVA followed by Tukey's HSD *post-hoc* test; $p < 0.05$).

The FLuc activity detected in the protoplasts indicated that M52M58-FLuc protein is produced (**Figure 2B**) and that an alternative translational initiation site is present and functional. To determine which of the potential start sites, AUG^{Met52} or AUG^{Met58}, can be used for producing the cytosolic isoform of ATPS2 (cyt-ATPS2), missense point mutations were introduced to the AUG codons. When the AUG^{Met1} of FLuc was mutagenized to CUG^{Leu} (ATPS2-$FLuc^{M1L}$), the FLuc/RLuc ratios were similar to or slightly lower than those in the wild type (WT). In contrast, when both AUG^{Met52} and AUG^{Met58} were mutagenized to CUG^{Leu} ($ATPS2^{M52L/M58L}$-FLuc), the FLuc/RLuc ratios were higher than in the WT. These results suggest that the endogenous AUG^{Met1} of FLuc is preferentially used as a translation initiation site in this fusion construct. Therefore, AUG^{Met1} of FLuc was included in mutant constructs to make cross comparisons with or among the double and triple mutants. The comparisons among the three experimental groups, ATPS2-$FLuc^{M1L}$, $ATPS2^{M52L}$-$FLuc^{M1L}$, and $ATPS^{M58L}$-$FLuc^{M1L}$, indicated that the mutation of AUG^{Met52} had the stronger impact showing approximately a 50% decrease in FLuc/RLuc ratios (**Figure 2B**). Moreover, the comparisons among $ATPS2^{M52L/M58L}$-$FLuc^{M1L}$, $ATPS2^{M52L}$-$FLuc^{M1L}$, and $ATPS^{M58L}$-$FLuc^{M1L}$ indicated that AUG^{Met52} was significant for translational initiation compared to AUG^{Met58} that only had a marginal effect (**Figure 2B**). These results indicated that translation of mRNA to FLuc fusion protein occurred more efficiently at AUG^{Met52} than AUG^{Met58}. The results obtained from this experimental system suggest that *ATPS2* mRNA can be translated preferentially at AUG^{Met52} to form the cytosolic ATPS2 isoform, M52-ATPS2, rather than starting at AUG^{Met58} producing M58-ATPS2. In protoplasts expressing $ATPS2^{M52L/M58L}$-$FLuc^{M1L}$ fusion construct, FLuc activity equivalent to approximately 10% of the wild-type level was still detectable. Therefore, it cannot be ruled out that translation might have initiated at another AUG downstream of AUG^{Met1} to produce a functional FLuc protein, resulting in a low residual FLuc activity. In this experiment, the protoplasts were divided into two fractions following the transfection, and incubated under sulfate-sufficient (+Sulfate) or sulfate-deficient conditions (–Sulfate). However, the FLuc activity was not modulated by changes in sulfate conditions.

ATPS2-GFP IS DUALLY LOCALIZED IN CHLOROPLASTS AND CYTOSOL

To determine the subcellular localizations of the alternatively translated products of ATPS2 (M1-ATPS2 and M52-ATPS2 or M58-ATPS2), the 5′UTR and the N-terminal 77-amino-acid region of ATPS2 ($ATPS2_{(5'UTR-Val77)}$) or the ATPS2 full-length ($ATPS2_{FL}$) sequences were fused to GFP, and resultant *p35S:ATPS2-GFP* constructs were transiently expressed in *Arabidopsis* protoplasts (**Figure 3**). The ATPS1-GFP fusion construct ($ATPS1_{(5'UTR-Val63)}$-GFP), containing the 5′UTR and the N-terminal 63-amino-acid region of ATPS1 fused to GFP following a Val residue conserved with Val77 of ATPS2, was prepared for comparison.

In protoplasts expressing either form of wild-type ATPS2-GFP fusion proteins ($ATPS2_{(5'UTR-Val77)}$-GFP or $ATPS2_{FL}$-GFP; **Figures 3A**), GFP signals were observed in both cytosol and chloroplasts ("GFP" column), where they were shown as GFP fluorescence excluded from or overlapping with chlorophyll fluorescence ("Merged" column). However, in the same transfection with the ATPS2-GFP fusion constructs, the GFP signals were found to be localized only in the cytosol in mesophyll protoplasts that contained a large number of fully developed chloroplasts (**Figure 3B**), unlike the dual localizations observed in protoplasts containing only a few small chloroplasts (**Figure 3A**). In contrast, the transfection of $ATPS1_{(5'UTR-Val63)}$-GFP showed exclusive localization of GFP in the chloroplasts (**Figure 3C**).

To investigate the roles of potential start codons within the transit peptide region of ATPS2 in differentiating its localization to chloroplasts and/or cytosol, missense point mutations changing the AUG^{Met} start codons to CUG^{Leu} were introduced to $ATPS2_{(5'UTR-V77)}$-GFP and these mutant forms were expressed in protoplasts. When CUG^{Leu} were introduced to both AUG^{Met1} and AUG^{Met4} in addition to AUG^{Met1} of GFP ($ATPS2^{M1L/M4L}$-GFP^{M1L}), GFP was exclusively localized in the cytosol of both cell-types with fully developed chloroplasts and small chloroplasts (**Figure 3D**). Furthermore, the protoplasts expressing the fusion constructs with either of these start codons intact ($ATPS2^{M1L/M4L/M52L}$-GFP^{M1L} or $ATPS2^{M1L/M4L/M58L}$-GFP^{M1L}) showed GFP fluorescence localized only in the cytosol (**Figure 3D**). These results indicate that cytosolic ATPS2 isoform can be translated from either AUG^{Met52} or AUG^{Met58}.

In contrast, the expression of a fusion protein $ATPS2^{M52L/M58L}$-GFP^{M1L} in protoplasts indicated chloroplastic localization of GFP (**Figure 3E**). The GFP signals were found only in protoplasts containing a few small chloroplasts, similar to the results showing dual localizations driven by the native forms of ATPS2-GFP (**Figure 3A**). Subcellular localizations of GFP fusion proteins were further tested using mutants, $ATPS2^{M1L/M52L/M58L}$-GFP^{M1L} and $ATPS2^{M4L/M52L/M58L}$-GFP^{M1L}. When AUG^{Met4} was mutated, GFP was still expressed and exclusively localized in these small chloroplasts ($ATPS2^{M4L/M52L/M58L}$-GFP^{M1L}; **Figure 3E**). However, when AUG^{Met1} was mutated, no GFP fluorescence could be detected (data not shown since no GFP signals were found as in non-transfected protoplasts). This indicates that the translation of chloroplastic ATPS2 pre-protein can only be initiated from the AUG^{Met1} start codon.

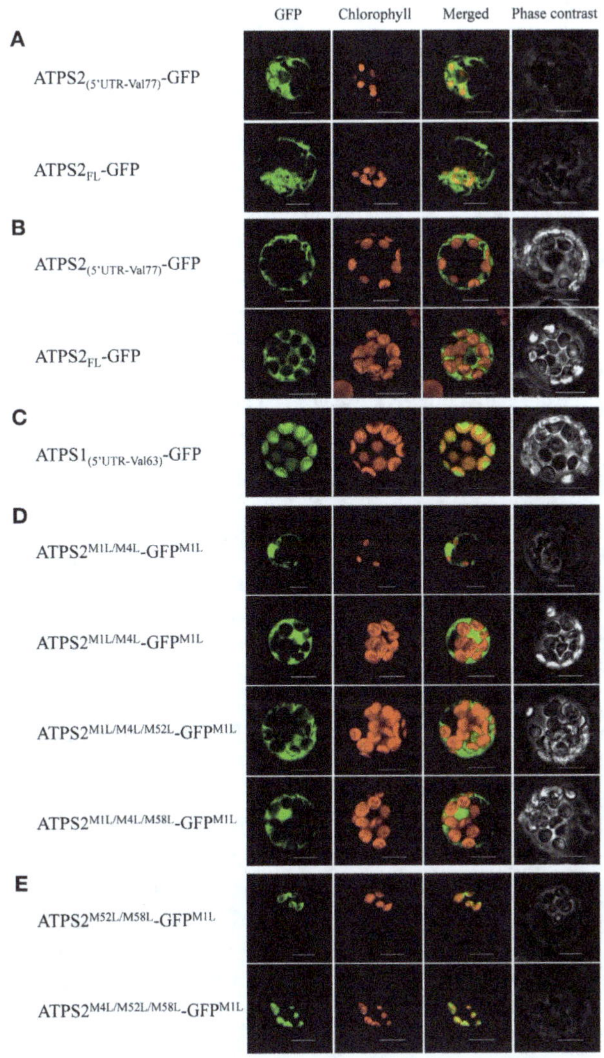

FIGURE 3 | Subcellular localization of ATPS2-GFP fusion proteins in *Arabidopsis* protoplasts. (A) Chloroplast-cytosol dual-localization of ATPS2-GFP ($ATPS2_{(5'UTR-Val77)}$-GFP and $ATPS2_{FL}$-GFP). **(B)** Cytosolic localization of ATPS2-GFP ($ATPS2_{(5'UTR-Val77)}$-GFP and $ATPS2_{FL}$-GFP) in mesophyll protoplasts. **(C)** Chloroplastic localization of ATPS1-GFP ($ATPS1_{(5'UTR-Val63)}$-GFP) in mesophyll protoplasts. **(D)** Cytosolic localization of $ATPS2^{M1L/M4L}$-GFP^{M1L}, $ATPS2^{M1L/M4L/M52L}$-GFP^{M1L}, and $ATPS2^{M1L/M4L/M58L}$-GFP^{M1L}. **(E)** Chloroplastic localization of $ATPS2^{M52L/M58L}$-GFP^{M1L} and $ATPS2^{M4L/M52L/M58L}$-GFP^{M1L}. Fluorescence was detected using a confocal laser-scanning microscope. GFP fluorescence (green), chlorophyll fluorescence (red), merged images (green and red) and bright-field phase contrast images are shown. Scale bars = 10 μm.

TISSUE AND SUBCELLULAR LOCALIZATIONS OF ATPS2-GFP IN PLANTS

To further study the spatial and subcellular localization of ATPS2 in plants, *ATPS2pro:ATPS2-GFP* fusion constructs with or without the mutations of translational start sites were prepared (Figure S1), and stable *Arabidopsis* transgenic lines were obtained for the microscopic analysis (**Figure 4**). The transgenic lines expressing the *ATPS1pro:ATPS1-GFP* fusion construct (Kawashima et al., 2011) were used for comparison. These chimeric constructs are designed to express full-length ATPS-GFP fusion proteins under control of native *ATPS* promoters in *Arabidopsis*.

The accumulations of ATPS2-GFP fusion proteins in transgenic lines were monitored by western blotting using anti-GFP antibody (**Figure 4A**). The fusion proteins were detected in leaves and roots of all lines except $ATPS2^{M1I/M52I/M58I}$-GFP, in which the fusion proteins were barely detectable in roots and only slightly produced in leaves (**Figure 4A**), despite the mRNAs being almost equally accumulated as those in transgenic lines made with other fusion constructs (Figure S2).

In transgenic lines expressing ATPS2-GFP fusion proteins, strong GFP fluorescence was observed in epidermal cells and guard cells as well as in vascular tissues and parenchyma cells on their abaxial side (**Figure 4B**). The same patterns of expression were observed in plants expressing the mutant constructs, $ATPS2^{M1I/M4I}$-GFP or $ATPS2^{M52I/M58I}$-GFP, prepared to determine their cytosolic or chloroplastic localizations of GFP fusion proteins (**Figure 4B**). The transgenic lines expressing the *ATPS1pro:ATPS1-GFP* fusion construct (Kawashima et al., 2011) also showed similar tissue localization patterns of GFP (**Figure 4B**). In all transgenic lines, the signals of GFP were much weaker in mesophyll cells, although they were detectable.

At subcellular levels, signals of GFP were dually localized in plastids and cytosol in both leaves and roots of ATPS2-GFP lines (**Figure 4C**), while they were present only in plastids in ATPS1-GFP lines (**Figure 4D**). When missense mutations (AUC^{Ile}) were introduced to both AUG^{Met1} and AUG^{Met4} ($ATPS2^{M1I/M4I}$-GFP), GFP was localized exclusively in the cytosol in both leaves and roots (**Figure 4E**). Moreover, in transgenic lines expressing $ATPS2^{M1I/M4I/M58I}$-GFP or $ATPS2^{M1I/M4I/M52I}$-GFP fusion proteins, the GFP fluorescence was similarly detected only in the cytosol (**Figure 4E**). In contrast, in transgenic lines expressing the mutant construct, $ATPS2^{M52I/M58I}$-GFP, the fluorescence of GFP was localized exclusively in chloroplasts (**Figure 4F**). The same pattern of GFP localization was observed with $ATPS2^{M4I/M52I/M58I}$-GFP (**Figure 4F**). However, in $ATPS2^{M1I/M52I/M58I}$-GFP transgenic lines, GFP fluorescence could not be detected in roots, and the GFP signals in the chloroplasts in leaves were very weak (**Figure 4F**). The results were consistent with the low levels of $ATPS2^{M1I/M52I/M58I}$-GFP fusion protein accumulation shown in the western blots (**Figure 4A**). The significance of AUG^{Met1} for the translational initiation of chloroplast-targeted isoform was also indicated by localization of GFP signals found only in the cytosol in leaves and roots of $ATPS2^{M1I}$-GFP lines (**Figure 4G**). In contrast, the point mutation at AUG^{Met4} resulted in dual localization maintained as in the wild-type ATPS2-GFP, suggesting that AUG^{Met1}, AUG^{Met52}, and AUG^{Met58} are viable as start codons (**Figure 4G**). The single point mutations at AUG^{Met52} or AUG^{Met58} showed similar dual localizations (**Figure 4H**), supporting the results obtained from triple mutants showing functional redundancies of these two start sites to produce the cytosolic isoform (**Figure 4E**).

DISCUSSION

The sequence analysis of four ATPS pre-proteins of *Arabidopsis* points to a uniqueness of ATPS2 containing four in-frame AUG codons within its transit peptide region (Met1, Met4, Met52, and Met58) (**Figure 1**). The present study demonstrates that translation of *ATPS2* mRNA starts at multiple AUG^{Met} translational initiation sites and produces plastidic and cytosolic ATPS2 isoforms in *Arabidopsis* (**Figures 2–4**). This appears to happen when the nucleotide contexts surrounding the AUG start codons are favorable for initiating translations. Several studies report that alternative translations of a single mRNA can produce protein isoforms located in different subcellular compartments in plants. Polγ2 organellar DNA polymerases are localized in chloroplast and mitochondria following alternative translational initiation (Wamboldt et al., 2009). AtMBP-1 is alternatively translated from *LOS2* transcript at the internal start codon and localizes to nucleus to modulate expression of transcriptional repressor for cold acclimation, whereas the full-length protein LOS2 functions as enolase in the glycolytic pathway in the cytosol (Lee et al., 2002; Kang et al., 2013). With regard to metabolic enzymes, NAD(P)HX dehydratase and epimerase are shown to localize in mitochondria, plastids, and cytosol by using alternative translational initiation sites (Niehaus et al., 2014).

Multiple lines of experimental evidence indicate alternative translation of ATPS2 and its relevance to plastid-cytosol dual subcellular localizations. Expression of a tandem luciferase fusion gene $ATPS2_{(5'UTR-His12)}$*:Renilla luciferase:*$ATPS2_{(Ile13-Val77)}$*:firefly luciferase* in *Arabidopsis* protoplasts suggests that M52M58-FLuc fusion protein is produced by translation initiated at internal start sites, AUG^{Met52} or AUG^{Met58} (**Figure 2**). Point mutations of these alternative start sites further suggest AUG^{Met52} being the preferred site over AUG^{Met58}. Furthermore, subcellular localizations of ATPS2-GFP proteins show alternative translational initiation underlying plastid-cytosol dual localizations (**Figures 3, 4**). The point mutation of AUG^{Met1} suggests the presence of this start site being essential for initiating the translation of ATPS2 pre-protein targeted to plastids. In contrast, both AUG^{Met52} and AUG^{Met58} are capable of initiating the translation of cytosolic ATPS2 isoform. These results clearly suggest *bona fide* relationships between the translational start sites and the duality of subcellular localizations of ATPS2 in *Arabidopsis*.

The exact mechanism that explains the alternative translation of *ATPS2* mRNA yet remains to be verified. Leaky ribosome scanning may be one of the possible scenarios where the same ribosome reads through the entire *ATPS2* mRNA to generate two ATPS2 protein isoforms with distinct subcellular localizations. Wamboldt et al. (2009) describes that such a mechanism of alternative translation of organellar DNA polymerase *Polγ2* mRNA allows dual localization of its translated products to mitochondria and chloroplasts in *Arabidopsis*. The other possible mechanism would be the presence of an internal ribosome entry site (IRES) around AUG^{Met52} and AUG^{Met58}, allowing *ATPS2* mRNA to produce the cytosolic ATPS2 independent of the plastid-targeted ATPS2 pre-protein. It is suggested that an IRES element in the 5′UTR mediates cap-independent selective translation of a maize heat shock protein *Hsp101* mRNA during heat stress (Dinkova

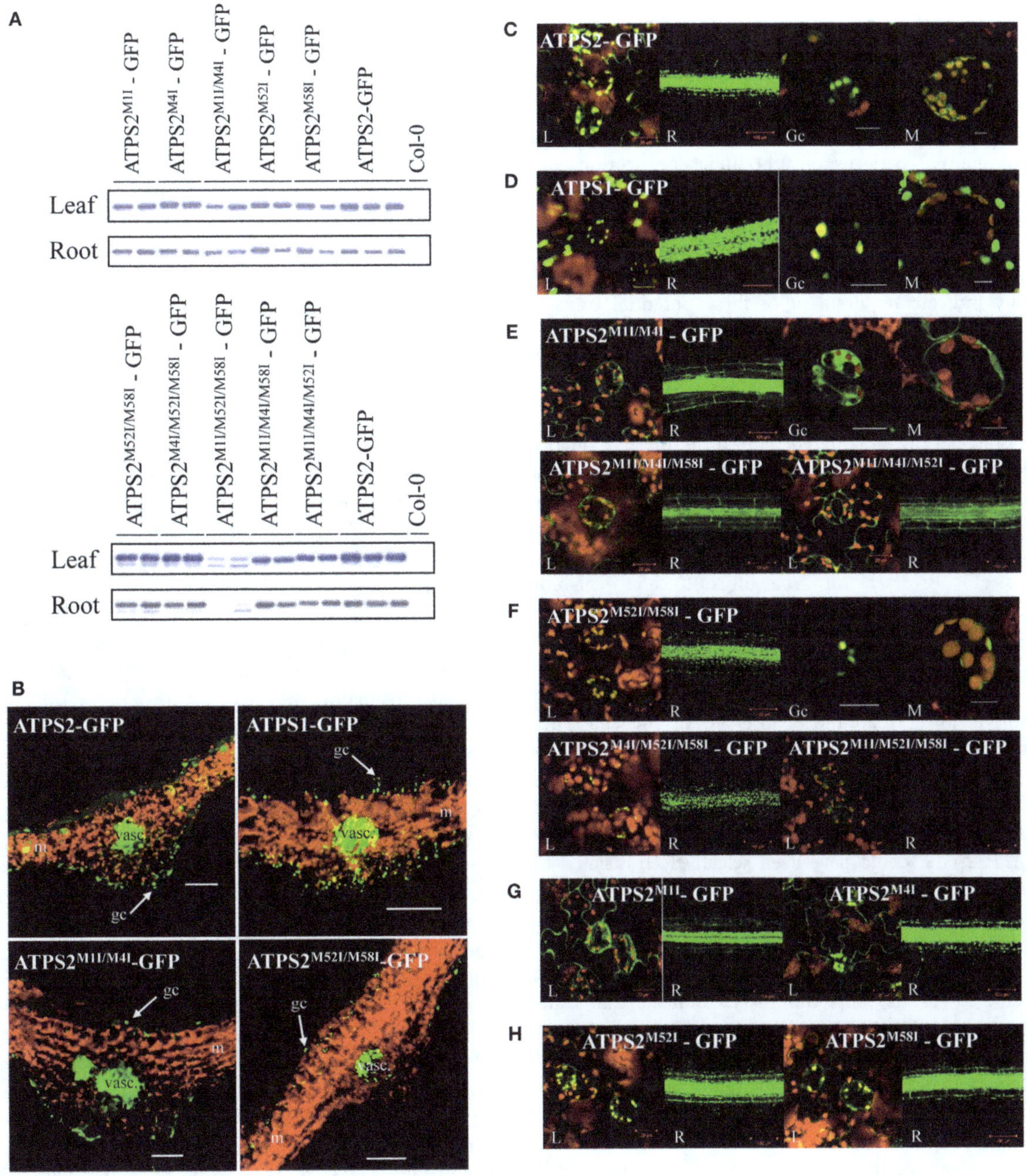

FIGURE 4 | Tissue and subcellular localization of ATPS2-GFP fusion proteins in *Arabidopsis* transgenic plants. (A) Detection of ATPS2-GFP fusion proteins in each transgenic line by western blotting using anti-GFP antibody. **(B)** GFP fluorescence (green) observed in transverse sections of leaves from the transgenic lines expressing *ATPS1pro:ATPS1:GFP*, *ATPS2pro:ATPS2:GFP* and mutated versions ($ATPS2^{M1I/M4I}$-GFP and $ATPS2^{M52I/M58I}$-GFP). Red indicates chlorophyll autofluorescence. Guard cells (gc), mesophyll cells (m) and vascular tissues (vasc) are indicated. Scale bars = 100 μm. **(C–H)** Subcellular localizations of GFP fluorescence (green) in leaves (L), roots (R), guard cells (Gc), and mesophyll cells (M). Red indicates chlorophyll autofluorescence. Yellow or orange indicates overlap between green and red signals. Scales bars = 20 μm (for leaves), 100 μm (for roots), and 10 μm (for guard cells and mesophyll cells).

et al., 2005). In addition to these mechanisms, translation efficiency of a bicistronic mRNA may be affected by alteration of sequence contexts having RLuc between the first (AUG^{Met1}) and alternative (AUG^{Met52} and AUG^{Met58}) start codons (**Figure 2A**). The overall stability of a long bicistronic mRNA with an internal stop codon can also be reduced by nonsense-mediated decay. Molecular mechanisms of alternative translational initiations need to be investigated with precautions of considering these additional possibilities.

The transient expression of *p35S:ATPS2-GFP* fusion gene in *Arabidopsis* protoplasts indicates two patterns of subcellular localizations of GFP fluorescence depending on cell types:

the dual localization is observed in protoplasts that contained a limited number of small chloroplasts, most likely derived from tissues that are less active in photosynthesis (**Figure 3A**), whereas the signal of GFP is found only in the cytosol of protoplasts containing a large number of fully developed chloroplasts, i.e., mesophyll cell protoplasts (**Figure 3B**). It is notable that such differential patterns of subcellular localizations are not seen with ATPS1-GFP (**Figure 3C**). The *Arabidopsis* transgenic lines expressing *ATPS2pro:ATPS2-GFP* fusion gene further demonstrate plastid-cytosol dual localization of ATPS2-GFP *in planta*. Mutations of AUG^{Met1}, AUG^{Met52}, and/or AUG^{Met58} unequivocally indicate requirement of these potential translational start sites for producing the isoforms localized to plastids and cytosol. The analysis of transgenic lines reveal that ATPS2-GFP is dually localized with its expression being the highest in epidermal cells, guard cells, vascular tissues including bundle sheath cells, and parenchyma cells present on the abaxial side of vasculature, which are considered less active in photosynthesis (**Figure 4B**). This pattern of expression is partly consistent with transcriptome data indicating relatively higher expression of *ATPS2* in guard cells than in mesophyll cells (Arabidopsis eFP Browser, http://bar.utoronto.ca/efp/cgi-bin/efpWeb.cgi). These results are likely consistent with dual subcellular localizations observed in protoplasts with small chloroplasts (**Figure 3A**). ATPS2-GFP is still found dually localized in mesophyll cells of transgenic lines (**Figure 4C**), although the level of expression is lower than those observed in the cell-types or tissues described above (**Figure 4B**). The subcellular localization of ATPS2-GFP in mesophyll protoplasts, which appears exclusive to the cytosol (**Figure 3B**), must therefore be speculated to have happened under unique mechanisms. Ectopically expressed *ATPS2-GFP* could have been selectively translated to the cytosolic isoform or post-translationally regulated to localize the polypeptides in cytosol but not in chloroplasts under specific conditions in protoplasts. Such mechanisms that may partially differentiate subcellular localizations in different cells appear specific to ATPS2, since our results indicate that transient expression of *p35S-ATPS1-GFP* in mesophyll protoplasts demonstrates clear localization of GFP signals in chloroplasts (**Figure 3C**) nevertheless cell-type specificities of gene expressions are similar between *ATPS1* and *ATPS2* in stable transformants (**Figure 4B**).

As mentioned above, both ATPS1 and ATPS2 are mainly expressed in epidermal cells, guard cells, vascular tissues, and cells in their vicinity (i.e., parenchyma cells on the abaxial side of the vasculature), but to a lower extent in mesophyll cells (**Figure 4B**). These patterns of expression of ATPS are similar to those of APK in leaves and roots of *Arabidopsis* (Mugford et al., 2009). Spatial co-localization of ATPS and APK is likely in accordance with their roles in providing PAPS for secondary metabolism producing sulfated compounds. It is known that, upon herbivore attack, one of the major sulfated compounds produced in Brassicaceae are glucosinolates (GLs). GLs are hydrolyzed by myrosinase and the by-products generated upon their hydrolysis serve as defense molecules against herbivores. Although GLs are found in the entire leaf, their abundance is higher in tissues surrounding mid-veins and in the periphery of leaf (Shroff et al., 2008). The specificity of GLs distribution seems likely a mechanism to limit herbivore feeding, as they tend to feed from the edges of plant leaves. Moreover, myrosinase is localized in myrosin cells in the phloem parenchyma (Andréasson et al., 2001). The spatial separation of the myrosinase-GLs system prevents the unnecessary hydrolysis of GLs. The co-localization of ATPS and APK in the vasculature or in the vicinity of the cells expressing myrosinase is therefore indicative of its potential contribution to providing substrates for GLs biosynthesis. Furthermore, a recent study of the translatome of vascular bundle sheath cells highlights the key role of these cells in sulfur metabolism in *Arabidopsis* (Aubry et al., 2014).

With regard to the control of sulfate assimilation, ATPS and APK in plastids appear to be expressed in favor of synthesizing PAPS for GLs biosynthesis under sulfur-sufficient conditions. MiR395s involved in post-transcriptional gene silencing of the chloroplastic ATPS1, –3, and –4, are repressed under sulfur-sufficient conditions (Kawashima et al., 2009). Chloroplastic APK (APK1 and APK2) play significant roles in providing PAPS for GLs biosynthesis (Mugford et al., 2009) while APR is repressed under sulfur-sufficient conditions. In contrast, the physiological relevance of the presence of cytosolic ATPS and APK and their molecular regulatory mechanisms are not well-documented to date. The present study unravels the molecular identity of the cytosolic ATPS2 and proposes alternative translational initiation as an underlying mechanism for its emergence. This translational mechanism specifically allows plastid-cytosol dual localization of ATPS2, a unique non-miR395 target among the *ATPS* gene family members. It is noteworthy that such distinction of molecular control mechanisms is apparent among the *ATPS* family members nevertheless the cell-type specificity resembles each other. The results shown in this study suggest that alternative translational initiation of ATPS2 is not significantly modulated by changes in sulfur conditions. Control of chloroplastic ATPS (ATPS1, –3, and –4) thus seems important for regulation of PAPS biosynthesis in response to sulfate supply, although the potential of the miR395-mediated post-transcriptional regulation may be limited for fine-tuning the ATPS1, –3, and –4 transcript levels (Kawashima et al., 2011). The physiological role of the cytosolic ATPS2 remains to be elucidated with relevance to its function in balancing PAPS biosynthesis between plastids and cytosol.

ACKNOWLEDGMENTS

The authors would like to thank Dr. Eva Farre (Michigan State University) for providing protocols and equipment for the bioluminescence experiments. Hideki Takahashi is supported by the National Science Foundation (MCB 1244300) and the AgBioResearch. Naoko Yoshimoto was supported by Grants-in-Aid for Scientific Research of Japan Society for the Promotion of Science (23770036).

REFERENCES

Andréasson, E., Bolt Jørgensen, L., Höglund, A.-S., Rask, L., and Meijer, J. (2001). Different myrosinase and idioblast distribution in Arabidopsis and *Brassica napus*. *Plant Physiol.* 127, 1750–1763. doi: 10.1104/pp.010334

Aubry, S., Smith-Unna, R. D., Boursnell, C. M., Kopriva, S., and Hibberd, J. M. (2014). Transcript residency on ribosomes reveals a key role for the *Arabidopsis thaliana* bundle sheath in sulfur and glucosinolate metabolism. *Plant J.* 78, 659–673. doi: 10.1111/tpj.12502

Bradford, M. M. (1976). A rapid and sensitive method for the quantitation of microgram quantities of protein utilizing the principle of protein-dye binding. *Anal. Biochem.* 72, 248–254. doi: 10.1016/0003-2697(76)90527-3

Chiu, W., Niwa, Y., Zeng, W., Hirano, T., Kobayashi, H., and Sheen, J. (1996). Engineered GFP as a vital reporter in plants. *Curr. Biol.* 6, 325–330. doi: 10.1016/S0960-9822(02)00483-9

Clough, S. J., and Bent, A. F. (1998). Floral dip: a simplified method for *Agrobacterium*-mediated transformation of *Arabidopsis thaliana*. *Plant J.* 16, 735–743. doi: 10.1046/j.1365-313x.1998.00343.x

Dinkova, T. D., Zepeda, H., Martínez-Salas, E., Martínez, L. M., Nieto-Sotelo, J., and de Jiménez, E. S. (2005). Cap-independent translation of maize Hsp101. *Plant J.* 41, 722–731. doi: 10.1111/j.1365-313X.2005.02333.x

Gigolashvili, T., Geier, M., Ashykhmina, N., Frerigmann, H., Wulfert, S., Krueger, S., et al. (2012). The *Arabidopsis* thylakoid ADP/ATP carrier TAAC has an additional role in supplying plastidic phosphoadenosine 5'-phosphosulfate to the cytosol. *Plant Cell* 24, 4187–4204. doi: 10.1105/tpc.112.101964

Gutierrez-Marcos, J. F., Roberts, M. A, Campbell, E. I., and Wray, J. L. (1996). Three members of a novel small gene-family from *Arabidopsis thaliana* able to complement functionally an *Escherichia coli* mutant defective in PAPS reductase activity encode proteins with a thioredoxin-like domain and "APS reductase" activity. *Proc. Natl. Acad. Sci. U.S.A.* 93, 13377–13382. doi: 10.1073/pnas.93.23.13377

Halkier, B. A., and Gershenzon, J. (2006). Biology and biochemistry of glucosinolates. *Annu. Rev. Plant Biol.* 57, 303–333. doi: 10.1146/annurev.arplant.57.032905.105228

Hatzfeld, Y., Lee, S., Lee, M., Leustek, T., and Saito, K. (2000). Functional characterization of a gene encoding a fourth ATP sulfurylase isoform from *Arabidopsis thaliana*. *Gene* 248, 51–58. doi: 10.1016/S0378-1119(00)00132-3

Höfgen, R., and Willmitzer, L. (1988). Storage of competent cells for *Agrobacterium* transformation. *Nucleic Acids Res.* 16:9877. doi: 10.1093/nar/16.20.9877

Joshi, C. P., Zhou, H., Huang, X., and Chiang, V. L. (1997). Context sequences of translation initiation codon in plants. *Plant Mol. Biol.* 35, 993–1001. doi: 10.1023/A:1005816823636

Kang, M., Abdelmageed, H., Lee, S., Reichert, A., Mysore, K. S., and Allen, R. D. (2013). AtMBP-1, an alternative translation product of LOS2, affects abscisic acid responses and is modulated by the E3 ubiquitin ligase AtSAP5. *Plant J.* 76, 481–493. doi: 10.1111/tpj.12312

Kawashima, C. G., Matthewman, C. A., Huang, S., Lee, B.-R., Yoshimoto, N., Koprivova, A., et al. (2011). Interplay of SLIM1 and miR395 in the regulation of sulfate assimilation in Arabidopsis. *Plant J.* 66, 863–876. doi: 10.1111/j.1365-313X.2011.04547.x

Kawashima, C. G., Yoshimoto, N., Maruyama-Nakashita, A., Tsuchiya, Y. N., Saito, K., Takahashi, H., et al. (2009). Sulphur starvation induces the expression of microRNA-395 and one of its target genes but in different cell types. *Plant J.* 57, 313–321. doi: 10.1111/j.1365-313X.2008.03690.x

Klonus, D., Höfgen, R., Willmitzer, L., and Riesmeier, J. W. (1994). Isolation and characterization of two cDNA clones encoding ATP-sulfurylases from potato by complementation of a yeast mutant. *Plant J.* 6, 105–112. doi: 10.1046/j.1365-313X.1994.6010105.x

Klonus, D., Riesmeier, J. W., and Willmitzer, L. (1995). A cDNA clone for an ATP-sulfurylase from *Arabidopsis thaliana*. *Plant Physiol.* 107, 653–654. doi: 10.1104/pp.107.2.653

Koncz, C., and Schell, J. (1986). The promoter of T_L-DNA gene 5 controls the tissue-specific expression of chimaeric genes carried by a novel type of *Agrobacterium* binary vector. *Mol. Genet. Genomics* 204, 383–396. doi: 10.1007/BF00331014

Lee, H., Guo, Y., Ohta, M., Xiong, L., Stevenson, B., and Zhu, J.-K. (2002). *LOS2*, a genetic locus required for cold-responsive gene transcription encodes a bi-functional enolase. *EMBO J.* 21, 2692–2702. doi: 10.1093/emboj/21.11.2692

Lee, S., and Leustek, T. (1998). APS kinase from *Arabidopsis thaliana*: genomic organization, expression, and kinetic analysis of the recombinant enzyme. *Biochem. Biophys. Res. Commun.* 247, 171–175. doi: 10.1006/bbrc.1998.8751

Leustek, T., Murillo, M., Cervantes, M., and Biology, A. M. (1994). Cloning of a cDNA encoding ATP sulfurylase from *Arabidopsis thaliana* by functional expression in *Saccharomyces cerevisiae*. *Plant Physiol.* 105, 897–902. doi: 10.1104/pp.105.3.897

Lillig, C. H., Schiffmann, S., Berndt, C., Berken, A., Tischka, R., and Schwenn, J. D. (2001). Molecular and catalytic properties of *Arabidopsis thaliana* adenylyl sulfate (APS)-kinase. *Arch. Biochem. Biophys.* 392, 303–310. doi: 10.1006/abbi.2001.2453

Logan, H. M., Cathala, N., Grignon, C., and Davidian, J-C. (1996). Cloning of a cDNA encoded by a member of the *Arabidopsis thaliana* ATP Sulfurylase multigene family: expression studies in yeast and in relation to plant sulfur nutrition. *J. Biol. Chem.* 271, 12227–12233. doi: 10.1074/jbc.271.21.12227

Lunn, J. E., Droux, M., Martin, J., and Douce, R. (1990). Localization of ATP sulfurylase and *O*-acetylserine(thiol)lyase in spinach leaves. *Plant Physiol.* 94, 1345–1352. doi: 10.1104/pp.94.3.1345

Mugford, S. G., Yoshimoto, N., Reichelt, M., Wirtz, M., Hill, L., Mugford, S. T., et al. (2009). Disruption of adenosine-5'-phosphosulfate kinase in *Arabidopsis* reduces levels of sulfated secondary metabolites. *Plant Cell* 21, 910–927. doi: 10.1105/tpc.109.065581

Murillo, M., and Leustek, T. (1995). Adenosine-5'-triphosphate-sulfurylase from *Arabidopsis thaliana* and *Escherichia coli* are functionally equivalent but structurally and kinetically divergent: nucleotide sequence of two adenosine-5'-triphosphate-sulfurylase cDNAs from *Arabidopsis thaliana* and anaysis of a recombinant enzyme. *Arch. Biochem. Biophys.* 323, 195–204. doi: 10.1006/abbi.1995.0026

Niehaus, T. D., Richardson, L. G. L., Gidda, S. K., ElBadawi-Sidhu, M., Meissen, J. K., Mullen, R. T., et al. (2014). Plants utilize a highly conserved system for repair of NADH and NADPH hydrates. *Plant Physiol.* 165, 52–61. doi: 10.1104/pp.114.236539

Noctor, G., Mhamdi, A., Chaouch, S., Han, Y., Neukermans, J., Marquez-Garcia, B., et al. (2012). Glutathione in plants: an integrated overview. *Plant Cell Environ.* 35, 454–484. doi: 10.1111/j.1365-3040.2011.02400.x

Phartiyal, P., Kim, W.-S., Cahoon, R. E., Jez, J. M., and Krishnan, H. B. (2006). Soybean ATP sulfurylase, a homodimeric enzyme involved in sulfur assimilation, is abundantly expressed in roots and induced by cold treatment. *Arch. Biochem. Biophys.* 450, 20–29. doi: 10.1016/j.abb.2006.03.033

Renosto, F., Patel, H. C., Martin, R. L., Christopher, T., Zimmerman, G., and Segel, I. H. (1993). ATP sulfurylase from higher plants: kinetic and structural characterization of the chloroplast and cytosol enzymes from spinach leaf. *Arch. Biochem. Biophys.* 307, 272–285. doi: 10.1006/abbi.1993.1590

Rotte, C., and Leustek, T. (2000). Differential subcellular localization and expression of ATP sulfurylase and 5'-adenylylsulfate reductase during ontogenesis of Arabidopsis leaves indicates that cytosolic and plastid forms of ATP sulfurylase may have specialized functions. *Plant Physiol.* 124, 715–724. doi: 10.1104/pp.124.2.715

Setya, A., Murillo, M., and Leustek, T. (1996). Sulfate reduction in higher plants: molecular evidence for a novel 5'-adenylylsulfate reductase. *Proc. Natl. Acad. Sci. U.S.A.* 93, 13383–13388. doi: 10.1073/pnas.93.23.13383

Shimojima, M. (2011). Biosynthesis and functions of the plant sulfolipid. *Prog. Lipid Res.* 50, 234–239. doi: 10.1016/j.plipres.2011.02.003

Shroff, R., Vergara, F., Muck, A., Svatos, A., and Gershenzon, J. (2008). Nonuniform distribution of glucosinolates in *Arabidopsis thaliana* leaves has important consequences for plant defense. *Proc. Natl. Acad. Sci. U.S.A.* 105, 6196–6201. doi: 10.1073/pnas.0711730105

Suter, M., von Ballmoos, P., Kopriva, S., den Camp, R. O., Schaller, J., Kuhlemeier, C., et al. (2000). Adenosine 5'-phosphosulfate sulfotransferase and adenosine 5'-phosphosulfate reductase are identical enzymes. *J. Biol. Chem.* 275, 930–936. doi: 10.1074/jbc.275.2.930

Takahashi, H., Kopriva, S., Giordano, M., Saito, K., and Hell, R. (2011). Sulfur assimilation in photosynthetic organisms: molecular functions and regulations of transporters and assimilatory enzymes. *Annu. Rev. Plant Biol.* 62, 157–184. doi: 10.1146/annurev-arplant-042110-103921

Urwin, P., Yi, L., Martin, H., Atkinson, H., and Gilmartin, P. M. (2000). Functional characterization of the EMCV IRES in plants. *Plant J.* 24, 583–589. doi: 10.1046/j.1365-313x.2000.00904.x

Valvekens, D., Van Montagu, M., and Van Lijsebettens, M. (1988). *Agrobacterium tumefaciens*-mediated transformation of *Arabidopsis thaliana* root explants by using kanamycin selection. *Proc. Natl. Acad. Sci. U.S.A.* 85, 5536–5540. doi: 10.1073/pnas.85.15.5536

Wamboldt, Y., Mohammed, S., Elowsky, C., Wittgren, C., de Paula, W. B. M., and Mackenzie, S. A. (2009). Participation of leaky ribosome scanning in protein dual targeting by alternative translation initiation in higher plants. *Plant Cell* 21, 157–167. doi: 10.1105/tpc.108.063644

Yoo, S.-D., Cho, Y.-H., and Sheen, J. (2007). *Arabidopsis* mesophyll protoplasts: a versatile cell system for transient gene expression analysis. *Nat. Protoc.* 2, 1565–1572. doi: 10.1038/nprot.2007.199

Conflict of Interest Statement: The authors declare that the research was conducted in the absence of any commercial or financial relationships that could be construed as a potential conflict of interest.

18

Update on the role of R2R3-MYBs in the regulation of glucosinolates upon sulfur deficiency

Henning Frerigmann and Tamara Gigolashvili*

Department of Molecular Plant Physiology, BioCenter, Botanical Institute and Cluster of Excellence on Plant Sciences (CEPLAS), University of Cologne, Cologne, Germany

Edited by:
Rüdiger Hell, University of Heidelberg, Germany

Reviewed by:
Masami Yokota Hirai, RIKEN Plant Science Center, Japan
Meike Burow, University of Copenhagen, Denmark

Correspondence:
Tamara Gigolashvili, Biozentrum Köln, Botanisches Institut, Universität zu Köln, Zülpicher Str. 47 B, 50674 Köln, Germany
e-mail: t.gigolashvili@uni-koeln.de

To balance the flux of sulfur (S) into glucosinolates (GSL) and primary metabolites plants exploit various regulatory mechanisms particularly important upon S deficiency (−S). The role of MYB34, MYB51 and MYB122 controlling the production of indolic glucosinolates (IGs) and MYB28, MYB29, and MYB76 regulating the biosynthesis of aliphatic glucosinolates (AGs) in *Arabidopsis thaliana* has not been fully addressed at −S conditions yet. We show that the decline in the concentrations of GSL during S depletion does not coincide with the globally decreased transcription of *R2R3-MYBs*. Whereas the levels of GSL are diminished, the expression of *MYB34, MYB51, MYB122*, and *MYB28* is hardly changed in early phase of S limitation. Furthermore, the mRNA levels of these *MYBs* start to raise under prolonged S starvation. In parallel, we found that SLIM1 can downregulate the *MYBs in vitro* as demonstrated in *trans*-activation assays in cultured *Arabidopsis* cells with *SLIM1* as effector and *ProMYB51:uidA* as a reporter construct. However, *in vivo*, only the mRNA of *MYB29* and *MYB76* correlated with the levels of GSL at −S. We propose that the negative effect of SLIM1 on GSL regulatory genes can be overridden by a "low GSL signal" inducing the transcription of *MYBs* in a feedback regulatory loop. In accordance with this hypothesis, the expression of *MYB34, MYB51, MYB122*, and *CYP83B1* was further induced in *cyp79b2 cyp79b3* mutant exposed to −S conditions vs. *cyp79b2 cyp79b3* plants grown on control medium. In addition, the possible role of *MYBs* in the regulation of essential S assimilation enzymes, in the regulation of GSL biosynthesis upon accelerated termination of life cycles, in the mobilization of auxin and lateral root formation at S deficiency is discussed.

Keywords: R2R3-MYB, *Arabidopsis thaliana*, glucosinolates (GSLs), regulation of glucosinolates, sulfur deficiency, SLIM1, MYB51, MYB28

INTRODUCTION

Sulfur (S) depletion leads to the decrease of the internal S levels, followed by a fast decrease in primary S-containing metabolites like glutathione as well as reduction in the levels of glucosinolates (GSLs) (Nikiforova et al., 2003, 2005; Hirai et al., 2004, 2005). Notably, the effects of S nutrition on GSL biosynthesis have been observed for years but the exact molecular mechanism by which changes in S supply modulate GSL metabolism are just starting to be understood. The backbone of GSLs contains from two to three S atoms, with one originating from 3′-phosphoadenosine 5′-phosphosulfate, the second one from glutathione, and the third being present in methionine derived aliphatic GSLs. This is the reason why the S status needs importantly to be regulated with GSL biosynthesis.

The analysis of transcript profile of *Arabidopsis thaliana* plants grown under S deficient conditions revealed the genes of the S assimilation pathway (sulfate transporters, cysteine biosynthesis, methionine biosynthesis and the glutathione cycle) upregulated in these plants after 48 h of S limitation (Hirai et al., 2003). Conversely, many genes of GSL biosynthesis were shown to be downregulated (Hirai et al., 2003; Maruyama-Nakashita et al., 2003; Nikiforova et al., 2003). Combining the metabolomic and transcriptomic studies demonstrates that S deficiency leads to reduced expression of all major GSL biosynthetic genes and, consequently, a reduction in GSL levels in plants. While decreasing the production of some S containing compounds, the plant maximizes uptake and utilization of S by increasing the expression of primary S assimilation genes.

In addition to changes in GSL biosynthesis rate, plants may also catabolize these secondary compounds. GSL catabolism has been postulated since the expression levels of genes coding for myrosinase-like proteins and thioglucosidases were upregulated in −S (Nikiforova et al., 2003, 2005; Hirai et al., 2004, 2005). During the myrosinase-catalyzed hydrolysis reaction (Bones and Rossiter, 1996; Rask et al., 2000), the GSL, which are normally stored in the vacuoles, need to come into contact (e.g., due to tissue disruption) with cytosolic myrosinases to be hydrolyzed. However, under conditions of −S, GSLs might be also degraded in intact plants by myrosinase-like proteins (Schnug and Haneklaus, 1993; Schnug et al., 1995), which do not require tissue damage. Similar mechanism of GSL degradation has been reported to be important for the plant innate immunity (Bednarek et al.,

2009). S released after *in vivo* GSL hydrolysis can be further incorporated into essential S-containing compounds and therefore maintain the vital processes in plant metabolism. Activation of GSL catabolism at −S is among the processes of −S response, which are least understood and cannot be explained directly by flux alterations because of changed concentrations of S-containing compounds (Hoefgen and Nikiforova, 2008). Along with the release of S, the hydrolysis of indolic GSLs (IGs) allows an increased synthesis of auxin which promotes lateral root formation and facilitates in this way the uptake of sulfate. Although the accumulation of auxin has not been shown to be induced in S-depletion experiments, several observations suggest the hydrolysis of IG upon this condition. These include an activation of genes involved in synthesis of tryptophan (Nikiforova et al., 2003), an activated GSL catabolism (Nikiforova et al., 2003, 2005; Hirai et al., 2004, 2005) and strong overexpression of nitrilases (Kutz et al., 2002). Additionally to that Nikiforova et al. (2003) reported transcriptional activation of genes involved in synthesis of indolic glucosinolates (IGs), downstream genes leading to auxin and its derivatives pointing to a possible flux to IAA biosynthesis under −S conditions.

Figure 1 summarizes finding of Hirai et al. (2003), Maruyama-Nakashita et al. (2003), Nikiforova et al. (2003), Li et al. (2013) on the effect of S deficiency of the transcription of GSL genes. The duration of S deficiency appears to determine the outcome of the gene expression, as S deficiency of 24–48 h duration was shown to inhibit gene expression, whereas under long-term depletion of S (lasting for 7 days or more), the activation of some GSL biosynthetic genes is registered. Remarkably, Li et al. (2013) has recently reported the activation of *MYB28* as measured after 3 weeks of mild −S conditions.

The primary and secondary S assimilation is positively controlled by the group of R2R3-MYB transcription factors, which are also known to regulate GSL biosynthesis (Hirai et al., 2007; Sønderby et al., 2007; Gigolashvili et al., 2007a; Yatusevich et al., 2009) (**Figure 2**). There are 6 different MYBs involved in GSL regulation, with MYB34, MYB51, and MYB122 controlling the production of IGs and MYB28, MYB29, and MYB76 controlling the production of aliphatic glucosinolates (AGs). Remarkably, these MYBs can also stimulate expression of primary S assimilation enzymes, enhancing substrate supply for GSL biosynthesis. Although all six MYB factors regulate adenosine-5′-phosphosulfate reductase (APR) and adenosine-5′-phosphosulfate kinase (APK), the trans-activation of ATP sulfurylase (ATPS) was isoform specific in relation to the aliphatic and indolic group.

S-deficiency response is largely controlled by Sulfur Limitation 1 (SLIM1). SLIM1 is a key transcriptional regulator of sulfate uptake identified from a genetic screen for *Arabidopsis* mutants disrupted in the S-limitation response. SLIM1 is the first transcription factor suggested to regulate plant metabolism upon −S, by, e.g., activating the sulfate acquisition (Maruyama-Nakashita et al., 2006). In addition, SLIM1 can probably activate the above described GSL catabolism process (like a putative thioglucosidase) and has been suggested as a negative regulator of *R2R3-MYB* genes controlling production of GSL in plants (**Figure 2**). In the support of this hypothesis, it was shown that the −S dependent decline in the expression of *MYB34* was not present any more in *slim1* knockout, pointing to the role of SLIM1 as a negative regulator of *MYB34* upon S limitation. The effect of SLIM1 on regulation of other *MYBs* was unclear (Maruyama-Nakashita et al., 2006). Still, a recent review on S assimilation in plants has suggested that *R2R3-MYBs* can be regulated by SLIM1 to repress the GSL biosynthesis (Takahashi et al., 2011). Despite the link that seems to exist between S and the biosynthesis of GSL (Mugford et al., 2009, 2010, 2011; Yatusevich et al., 2010; Kopriva et al., 2012; Huseby et al., 2013), the molecular mechanisms remain less clear. Particularly little is known about the role of R2R3-MYBs in the S-deficiency mediated regulation of GSL biosynthesis in *Arabidopsis*.

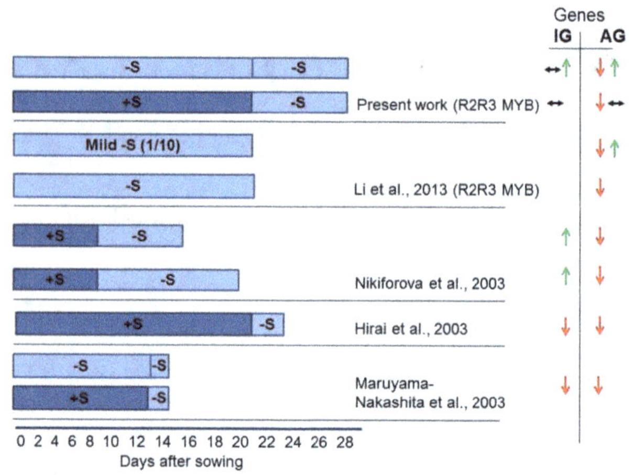

FIGURE 1 | Experimental scheme of sulfur (S) starvation showing number of days at S deficient conditions for four independent experiments conducted in different groups (Hirai et al., 2003; Maruyama-Nakashita et al., 2003; Nikiforova et al., 2003; Li et al., 2013; and present work). The seeds were either sown directly in −S medium (bright blue), or were grown in S sufficient (dark blue) medium followed by transfer to −S medium. This figure indicates that the duration of S deficiency determines the possible outcome on GSL gene expression. The S deficiency of 24–48 h (Hirai et al., 2003; Maruyama-Nakashita et al., 2003) was shown to inhibit GSL gene expression. The S deficiency applied for 7 days (present work) revealed absence of change in the expression of MYBs regulating IGs or downregulation of MYBs regulating AGs. The mild S shortage (1/10 of S levels) for 21 days revealed the activation of *MYB28* along with the downregulation of *MYB29* and *MYB76* (Li et al., 2013). The long-term depletion of S (Nikiforova et al., 2003—13 days; Present work—growth at −S for 28 days) revealed the activation of *MYB34, MYB51, MYB122* and *MYB28* and downregulation of *MYB29* and *MYB76*. Green arrows show increased gene expression. Red arrows show decreased gene expression. Black arrows indicate no significant change in the expression of gene. Two types of arrows shown simultaneously are indicative for the downregulation (↓), upregulation (↑), or no changes (↔) in the expression of different genes in GSL biosynthesis or regulation. As we do not possess original expression profiling data (Hirai et al., 2003; Maruyama-Nakashita et al., 2003; Nikiforova et al., 2003), the changes shown in this figure can be applied only to some selected genes discussed by the authors of original manuscripts. Hirai et al. (2003), Maruyama-Nakashita et al. (2003), Nikiforova et al. (2003) discuss only the expression of some genes in GSL biosynthesis, whereas Li et al. (2013) discuss the expression of R2R3-MYB regulators.

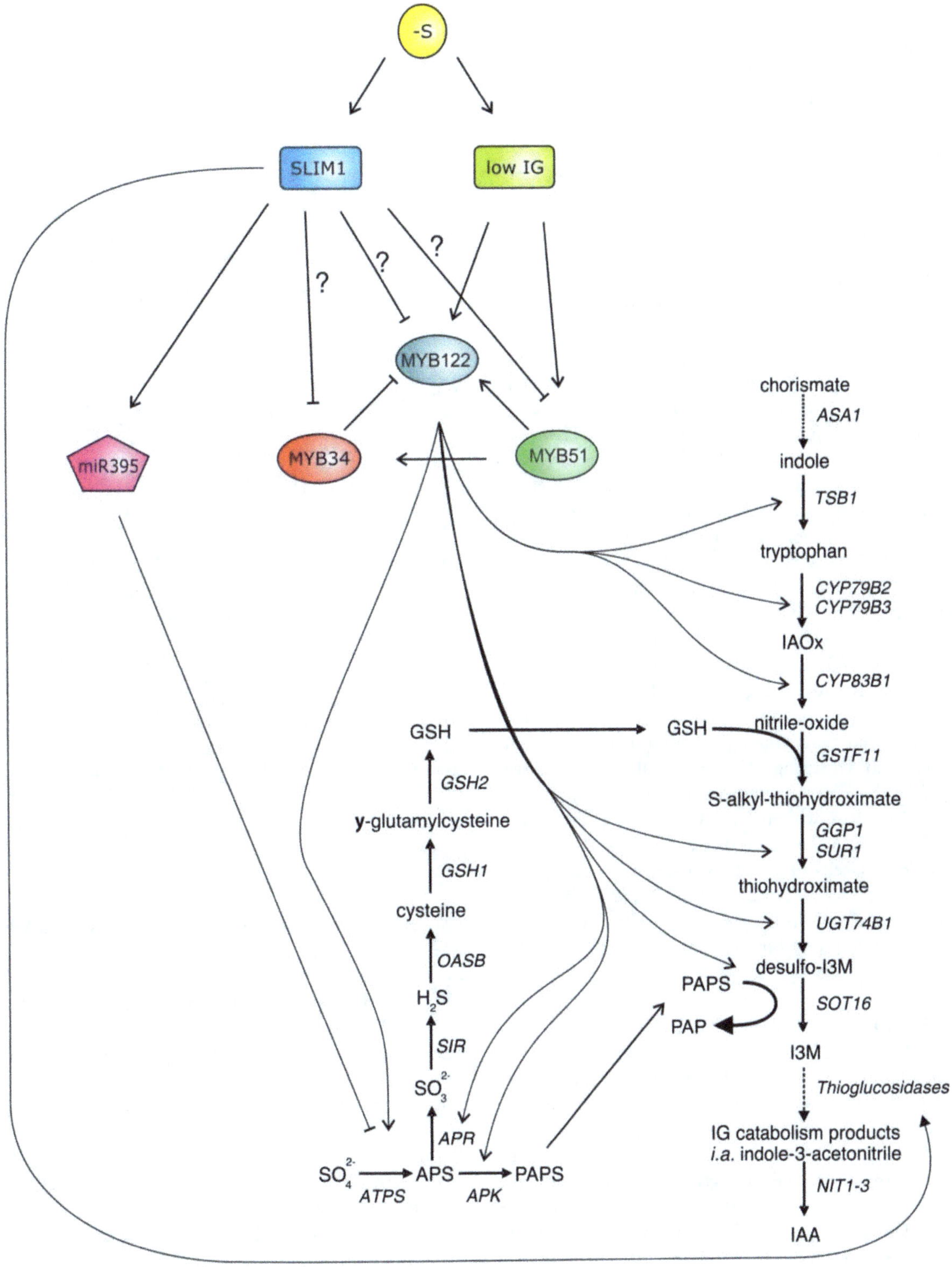

FIGURE 2 | MYB34, MYB51, and MYB122 are the central regulators of IG biosynthesis in *Arabidopsis* controlling transcription of the genes of core GSL biosynthesis and primary S assimilation. As accumulation of GSL is strongly diminished upon –S, it has been thought that these three *MYBs* along with the MYBs regulating aliphatic GSLs (*MYB28*, *MYB29*, and *MYB76*—not shown) need to be negatively regulated upon S deficiency. The SLIM1—a well-known regulator in S deficiency response has been suggested as an upstream negative regulator of *R2R3-MYBs* (Takahashi et al., 2011). In addition, SLIM1 seems to be able to stimulate GSL degradation by activating myrosinase-like proteins of thioglucosidases, which are able to degrade GSL and release the S. However, during this enzymatic hydrolysis of GSL, plants additionally release auxin, which is an important hormone to accelerate the lateral root formation, which can allow plant to acquire more sulfate.

We show that SLIM1 has a potential to downregulate the expression of *R2R3-MYBs* regulating GSL biosynthesis *in vitro*. However upon sulfate deficiency, the mRNA levels of main aliphatic and indolic GSL regulators *MYB28, MYB34, MYB51*, and *MYB122* are either not changed or increased. To explain this observation we suggested that the negative effect of SLIM1 on GSL regulatory genes can be overridden by a "low GSL signal" inducing the transcription of *MYBs* in a feedback regulatory loop.

RESULTS

SLIM1 IS CAPABLE OF REPRESSING THE TRANSCRIPTION OF *R2R3-MYBs* IN CULTURED *ARABIDOPSIS* CELLS *IN VITRO*

The S-deficiency regulator SLIM1 is an important regulator of −S response, which activates the sulfate acquisition and probably GSL catabolism with latter releasing S from these S-rich compounds. In addition, SLIM1 has been suggested to repress the GSL biosynthesis, probably by repressing the *R2R3-MYBs* which control their biosynthesis. To study how SLIM1 affects expression of *MYBs*, qRT-PCR analysis of *MYBs* in cultured *Arabidopsis* cells transiently over-expressing *SLIM1* was conducted. **Figure 3A** shows that *SLIM1* is capable of repressing the expression of *MYB34, MYB51, MYB28, MYB29*, and *MYB76 in vitro.*

Further SLIM-MYB interactions were performed with the help of *trans*-activation assay (Berger et al., 2007). In brief, the co-transformation assay with *Pro35S:SLIM1* as an effector construct and promoters of *R2R3-MYBs* as reporter was conducted. To be able to observe the repressing activity of SLIM1 on *MYBs in trans*, the promoter used in the assay should be strongly expressed in cultured cells. Among all promoters tested only *ProMYB51:GUS* showed strong GUS staining allowing the usage of *ProMYB51* in this assay. As shown on **Figure 3B**, the interaction of *Pro35S:SLIM1* with the *ProMYB51:GUS* revealed an inhibitory effect of *SLIM1* on the expression of *ProMYB51:GUS* cultured *Arabidopsis* cells (**Figure 3B**), confirming the potential of SLIM1 to repress transcription of *MYBs*. Despite the insights on the inhibitory role of SLIM *in vitro*, it's role on the expression of *R2R3-MYBs in vivo* remains to be studied in future.

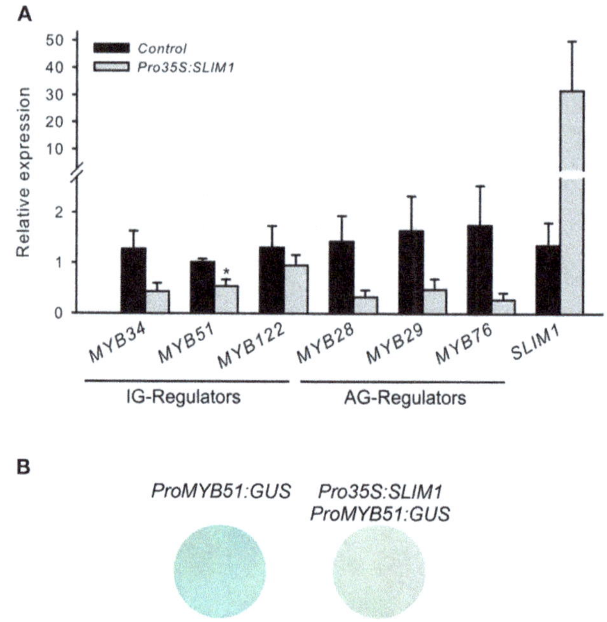

FIGURE 3 | SLIM1 is able to suppress the expression of *R2R3-MYBs*. **(A)** Expression of *R2R3-MYBs* regulating the biosynthesis of IG (*MYB34, MYB51, MYB122*) and AG (*MYB28, MYB29, MYB76*) in cultured *Arabidopsis* cells overproducing *SLIM1*; Data for *MYB28, MYB29* and *MYB76* are presented as means ± *SE* from five independent biological replicates ($n = 5$). Data for *MYB51, MYB34*, and *MYB122* are presented as means ± *SE* from 7 independent biological replicates ($n = 7$). Values marked with asterisks are significantly different from controls (+S to +S) (Student's *t*-test; $p < 0.05$). **(B)** Expression of promoter *ProMYB51:GUS* is suppressed by *Pro35S:SLIM1* in cultured *Arabidopsis* cells. This experiment was replicated twice with three independent biological replicates. Promoter of *MYB51* is the only MYB showing strong activity in cultured cells. Other *MYBs* regulators controlling the production of IG and AG show no staining in cultured cells, which hampered the potential visualization of inhibitory effects of SLIM1 on *MYBs*.

R2R3-MYBs REGULATING GLUCOSINOLATE BIOSYNTHESIS ARE DIFFERENTLY RESPONDING TO −S

To study the adaptive changes in GSL accumulation and MYB regulation upon S deficiency, wild-type seedlings of *Arabidopsis* were seeded out on Hoagland's media (+S) on agar plates and for the analyses of S deficiency were further cultivated on −S plates. Three-week-old *Arabidopsis* seedlings grown on +S or −S plates were transferred either to plates with the −S, or to new plates that maintained the existing S growth conditions and used for the analysis of expression levels of *MYBs* and GSL accumulation after 7 days of exposure to −S. This approach enabled us to describe the changes in the *MYB* expression after 7 and 28 days of S depletion conditions. Plants grown at −S for 7 days ("+S to −S") do not display any obvious symptoms, whereas plants grown for 28 days ("−S to −S") were retarded in growth. Plants transferred from +S to +S served as a control for the possible induction of genes by mechanical stress and were used as a calibrator for the relative expression analysis of *MYBs*. **Figure 4** shows that the expression levels of *MYBs* regulating IGs is not changed in seedlings transferred from +S to −S condition, (**Figure 4A**), although the amounts of IGs are significantly reduced (**Figure 4C**). Moreover, when Arabidopsis seedlings were transferred from −S to −S, these plants showed significantly increased expression of *MYB34*, *MYB51* and *MYB122* along with the further decreased levels of IGs. Notably, changes in the expression of *MYBs* went along with the increased expression of *CYP83B1* in "−S to −S" plants (**Figure 4A**). Thus, in "+S to −S" and "−S to −S" plants we have uncoupling of *MYB* and *SLIM1* (**Figure 4D**) transcripts with and GSL (**Figure 4C**) accumulation levels pointing on additional regulatory signal interfering with negative regulation of GSL regulation.

Not only the mRNA levels of *MYBs* regulating IG biosynthesis but also the mRNA of *MYBs* regulating AGs were different at "+S to −S" and at "−S to −S" conditions. The accumulation of AGs was significantly reduced in seedlings transferred from +S to −S and even further reduced in "−S to −S" plants (**Figure 4C**), which went along with the significantly declined mRNA levels of *MYB29, MYB76* and AG biosynthesis gene *CYP79F1* in both

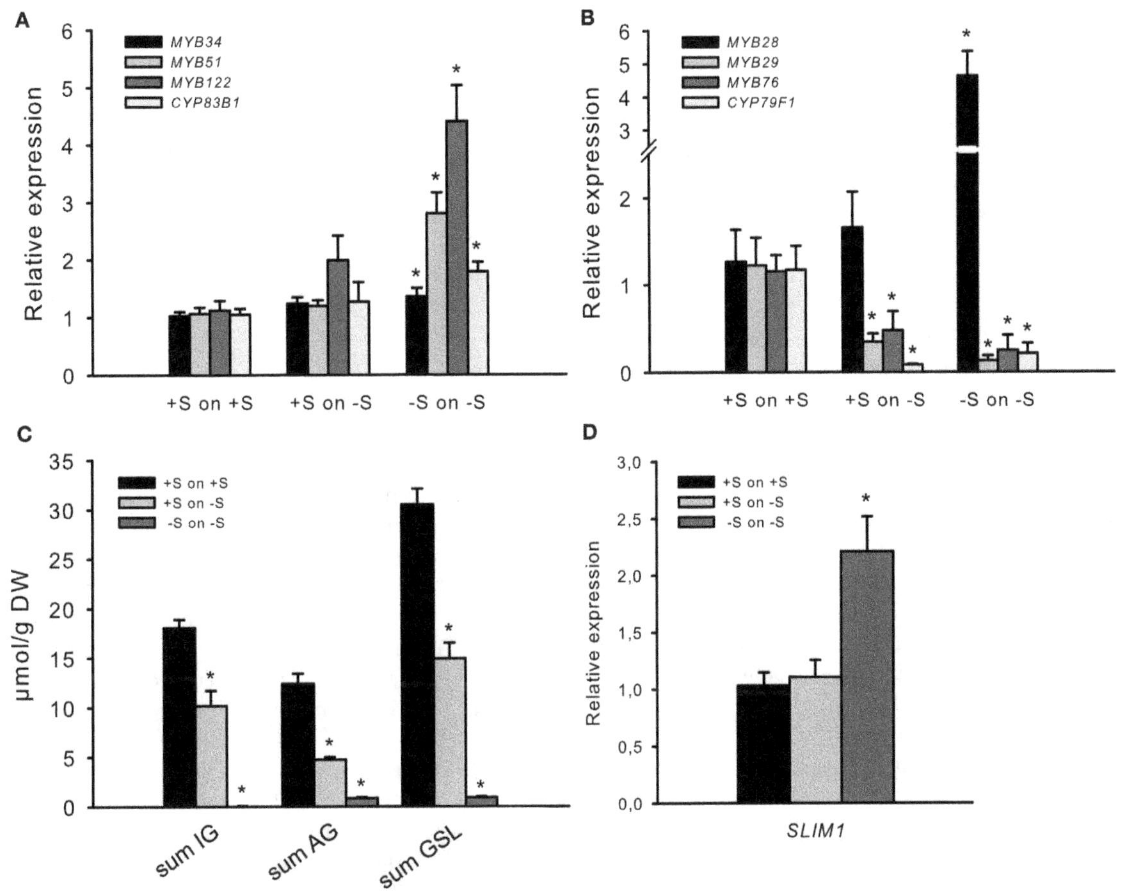

FIGURE 4 | Sulfur deficiency differently affects the expression of *R2R3-MYBs*. **(A)** Relative transcript levels of *MYB34, MYB51, MYB122*, and *CYP83B1*; **(B)** Relative transcript levels of *MYB28, MYB29, MYB76*, and *CYP79F1*; **(C)** Accumulation of indolic (IG) and aliphatic (AG) glucosinolates; **(D)** Relative transcript levels of *SLIM1*. Transcript levels of *R2R3-MYBs*, GSL biosynthesis genes (*CYP83B1* and *CYP79F1*) and *SLIM1* and GSL contents were determined by qPCR or UPLC analysis. Three-week-old seedlings grown on +S or −S plates were transferred to plates with +S and −S, respectively, grown for 7 more days followed by the analysis of gene transcript levels **(A,B,D)** and the sum of IG, AG, and GSL levels **(C)**. Relative gene expression values are given compared to plants transferred from +S to +S (+S to +S = 1). Data are presented as means ± *SE* from four independent cultivations with three biological replicates ($n = 12$). **(C)** For GSL analysis, two independent cultivations with four biological replicates were done ($n = 8$). GSLs were totalled either as the sum of IG (I3M, 4MO-I3M, 1MO-I3M), sum of AG (3MSOP, 4MSOB, 5MSOP, 8MSOO) or sum of all GSL (IG plus AG). Values marked with asterisks are significantly different from controls (+S to +S) (Student's *t*-test; $p < 0.05$).

conditions (**Figure 4B**). Furthermore, the transfer of seedlings from +S to −S did not negatively affect the expression of *MYB28*, whereas "−S to −S" plans manifested significantly increased level of *MYB28*. This observation also indicates uncoupling of *MYB28* transcript with the level of GSL, pointing to additional regulatory signal taking over the "−S signal" aiming the negative regulation of GSL biosynthesis.

Altogether our observation on the expression of *MYBs* suggest an existence of more complex SLIM1-independent signaling, which can positively regulate transcription of *MYB34, MYB51, MYB122*, and *MYB28* along with the negative regulation of *MYB29* and *MYB76* at −S. It's probably not a coincidence that the expression of two latter regulators of AG biosynthesis was significantly decreased at −S. One might suggest that AGs containing one more molecule of S need to be more tightly regulated at −S. This mechanism will probably allow to channel S into essential for the plant survival metabolites. In accordance with this suggestion the levels of AGs were strongly diminished after 7 days of −S than the levels of IGs. Still, this logic does not explain an increased expression of *MYB28* under the same condition and point to a specific role of *MYB28* at −S, differing from the role of MYB29 and MYB76.

Along with the interesting insights on the different regulation of MYBs regulating IG and AG biosynthesis we suggested that an additional regulatory signal positively controlling the transcription of *MYB34, MYB51*, and *MYB122* might be a "low IG level" in plants.

INCREASED EXPRESSION OF *MYB34, MYB51*, AND *MYB122* UPON −S CAN RESULT FROM THE LOW LEVELS OF IGs TRIGGERING THE TRANSCRIPTION OF THESE *MYBs* IN NEGATIVE FEEDBACK LOOP

To address whether low IGs may have signaling function we analyzed the transcript levels of *MYB34, MYB51* and *MYB122* and of IG biosynthetic gene *CYP83B1* in the *cyp79b2 cyp79b3* mutant

(Zhao et al., 2002). This mutant is known to be devoid of IG and therefore an increased level of *MYBs* especially at "−S to −S" was expected. **Figure 5** shows elevated expression of *CYP83B1* in *cyp79b2 cyp79b3* mutant already on +S to +S medium, together with an induction of *MYB51*. This observation supports previous findings on the negative feedback regulation of IG biosynthesis driven by low IG levels in IG deficient mutants like *cyp83b1, atr1* and *cyp83b1 atr1* (Celenza et al., 2005). Furthermore, the *cyp79b2 cyp79b3* mutant plants revealed a further increase in *MYB51, MYB122* and also of *CYP83B1* expression levels at "−S to −S" conditions. The mRNA levels of *MYB34* were also increased at "−S to −S" but only moderately. In sum, analysis of *MYB34, MYB51* and *MYB122* expression in *cyp79b2 cyp79b3* mutant plants grown at "−S to −S" indicated the possible role of "low IGs signal" in triggering the transcription of these transcription factors (TFs).

DISCUSSION

The biosynthesis of S-containing GSLs competes with primary S metabolism. SLIM1 is a S-deficiency induced TF, which was correlated with the induction of transcriptional changes leading to a downregulation of GSL biosynthetic genes and with the induction of genes involved in GSL catabolism (Maruyama-Nakashita et al., 2006). A decrease in the steady-state levels of *MYB34* in microarray experiments of wild-type vs. *slim1* mutants exposed to S deficiency (Maruyama-Nakashita et al., 2006) suggests that *MYBs* could be negatively regulated by SLIM1. Based on this observation and on the fact that GSL biosynthesis is negatively regulated at −S, a recent review work has summarized that MYB transcription factors should be negatively regulated by S limitation (Takahashi et al., 2011).

Still, the regulation of R2R3-MYBs at −S has not been specifically addressed. Furthermore, the recent finding of Li et al. (2013) reported counterintuitive results showing significantly increased expression of *MYB28* upon mild S deficiency. To find out how *R2R3-MYBs* controlling IG and AG biosynthesis are regulated upon S deficiency conditions, the expression of the *MYB34, MYB51*, and *MYB122* on one side and of the *MYB28, MYB29*, and *MYB76* on other side was analyzed at two different −S conditions.

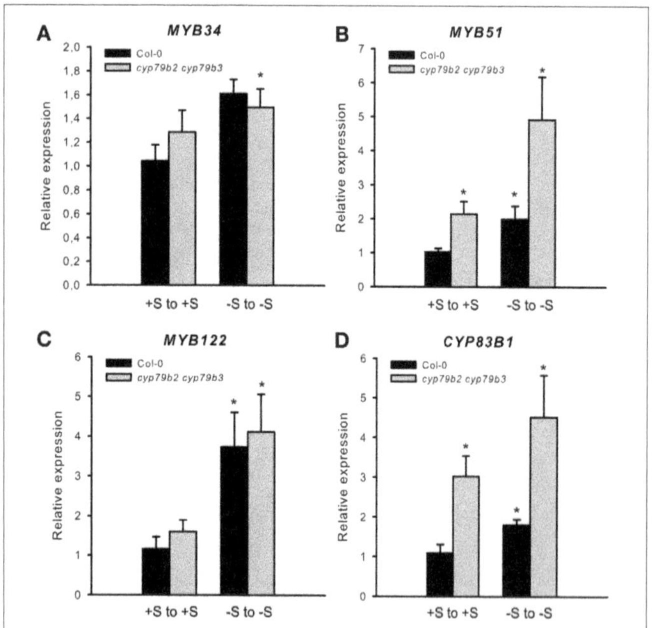

FIGURE 5 | Expression of *MYB34, MYB51*, and *MYB122* in *cyp79b3 cyp79b3* mutant deficient in IG biosynthesis. The relative expression of *MYB34* **(A)**, *MYB51* **(B)**, *MYB122* **(C)**, and *CYP83B1* **(D)** was measured in wild-type plants and *cyp79b2 cyp79b3* knockout mutant (Col-0 from +S on +S = 1). Data are presented as means ± *SE* from two independent cultivations with three biological replicates ($n = 6$). Values marked with asterisks are significantly different from the control (Col-0 from +S on +S) (Student's *t*-test; $p < 0.05$).

REGULATION OF *R2R3 MYBs* EXPRESSION AT −S

Whereas both 7 days of "+S to −S" and 28 days of "−S to −S" caused drastic decrease in GSL accumulation, the *R2R3 MYBs* responded differently to these −S conditions. The *MYB29* and *MYB76* were repressed in both −S conditions and correlated with the levels of GSL indicating the positive "feed-forward regulation" of these two genes at −S. Conversely, the expression of *MYB28, MYB51, MYB34*, and *MYB122* was not affected after 7 days of −S but was significantly increased after 28 days. The increased levels of mRNA of *MYBs* were counterintuitive, as we expected to find the correlation of the levels of GSL with the expression levels of *MYBs*. To explain this finding we suggested that a specific signal (activated by −S but SLIM1 independent) is interfering in negative GSL regulation and activates expression of *MYBs* (**Figure 6A**) for the reasons discussed below.

Several possible scenarios explaining high expression of *MYBs* under −S conditions could be suggested. Firstly, we propose that the downregulation of IG biosynthesis at −S could be overridden by "a low IG signal" inducing the transcription of *MYB51* and *MYB122* in a feedback regulatory loop. In analogy, "a low AG signal" can induce the transcription of *MYB28* in a feedback regulatory loop. The transcription of *MYB28, MYB51*, and *MYB122* is, therefore, increased to push the production of these compounds when GSL go below a certain threshold limit (summarized in **Figure 6A**). Remarkably, and in accordance with this hypothesis, the expression of *MYB34, MYB51*, and *MYB122* and of IG biosynthesis gene *CYP83B1* was further induced in *cyp79b2 cyp79b3* mutant exposed to −S conditions vs. *cyp79b2 cyp79b3* plants grown on Hoagland's medium (**Figure 5**). Secondly, the activation of these *MYBs* at −S could be important for plants because they regulate essential S assimilation enzymes like APR and APK. The role of MYBs in the specific regulation of these enzymes (e.g., APR) could be of special importance upon −S. Especially because the *APR* genes are known to be independent from SLIM1 and can be therefore controlled by MYBs. To further address this hypothesis, the primary S assimilation of plants need to be studied at −S in plants devoid of major *MYBs* (e.g., *myb28 myb51* and/or *myb28 myb29 myb34 myb51*). Thirdly, it is also possible that at continuous −S when the metabolism of plants exposed to stress is reprogrammed to the shortening of the life cycle and speeding up the seed production (Hoefgen and Nikiforova, 2008). Among the numerous changes happening in the plant metabolism are positive changes in the expression of

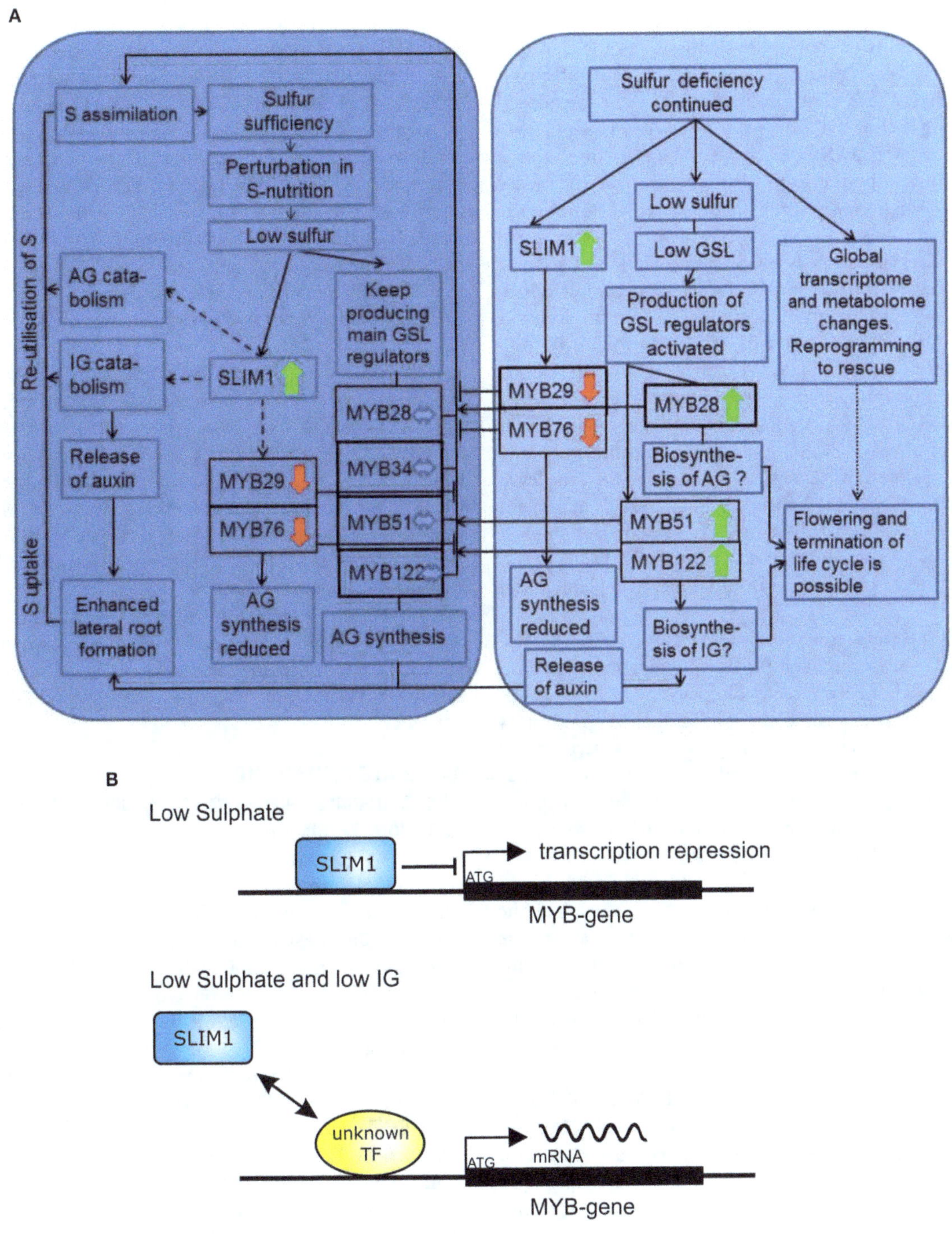

FIGURE 6 | The final model for the R2R3-MYB-mediated regulation of glucosinolate biosynthesis upon sulfur stress. (A) The early (7 days of −S; the dark blue chart) and late (28 days of −S; the bright blue chart) sulfur stress responses including changes in plant metabolism and expression of *R2R3-MYBs*. Only mRNA levels of *MYB29* and *MYB76* are downregulated in both early and late S deficiency responses. The expression of *MYB28, MYB34, MYB51*, and *MYB122* is not changed after 7 days of −S. The expression of *MYB28, MYB51*, and *MYB122* is significantly increased after 28 days of −S deficiency indicating that the inhibitory signal of SLIM on *MYBs* is overridden by "low GSL signal" (shown in detail in **Figure 6B**). Solid lines/arrows indicate positive (inducing) effects; Solid or dashed lines with an aslant dash indicate negative (inhibiting) effects; Dashed lines/arrows indicate postulated pathways; Dotted lines/arrows indicate complex changes with many and not highlight pathways affected. Bold green arrows indicate increased expression of *R2R3-MYB* gene. Bold red arrows indicate decreased expression of *R2R3-MYB* gene. **(B)** Model explaining the increased expression of *MYBs* under continuous sulfur deficiency. The SLIM inhibitory effect can be overridden by "low GSL signal," which positively regulates GSL biosynthesis with a so far unknown TF or regulatory switch by a negative feedback mechanism. This work demonstrated, that at −S low IGs can activate the expression of MYBs regulating their production.

MYBs required to produce GSL for the seeds (**Figure 6A**). In this case the MYB28 and MYB51 could be the important TFs taking over the responsibility to control the synthesis and the transport of GSL into the seeds ensuring the survival of plant offspring. Notably, this hypothesis is not in line with the previous observations showing decline in the expression of *MYB28* (Gigolashvili et al., 2007b) and *MYB51* (Gigolashvili et al., 2007a) on the onset of bolting. Conversely, the expression of genes closely related to MYB28 is strongly increased with the onset of flowering in *Brassica juncea* (Augustine et al., 2013). To verify this hypothesis it will be necessary to analyse GSL accumulation in seeds of triple *myb28 myb51 myb122* mutant "forced" to complete their life cycle at −S. Finally and according to the fourth scenario, which can explain the activation of *MYBs* controlling IG production, the positive regulation of *MYB34, MYB51*, and *MYB122* could be an important mechanism allowing the plant to produce auxin either via IAOx or via catabolism of IG (**Figure 6A**) with the involvement of nitrilases (Kutz et al., 2002). In conformity with this scenario, the levels of IG are stronger declined than the levels of IG in "−S to −S" plants (**Figure 4C**). To prove this hypothesis the catabolism of IG *in vivo* at −S needs to be addressed in more detail. Alternatively, accumulation of auxin at −S in triple *myb34 myb51 myb122* mutant in comparison to wild-type plants needs to be analyzed.

One more possible scenario, which can explain the up-regulation of *MYB28* along with the down regulation of *MYB29* and *MYB76* at −S has been recently discussed by Li et al. (2013). This hypothesis is based on the observation that MYB mutually regulate each other and therefore an increased transcription of *MYB28* may result from the decreased transcription of *MYB29* and *MYB76* (Sønderby et al., 2010). However, even if this hypothesis is applicable to AG pathway, it cannot be applied to explain the upregulation of *MYB34, MYB51*, and *MYB122*, at "−S to −S" conditions.

Even if each of the considered hypothesis can explain the observed positive regulation of *R2R3-MYBs* at −S alone, the GSL-S balance in plants is probably controlled in a complex combinatorial network integrating many signals (**Figure 6A**). It can be therefore assumed that several of suggested scenarios can be happening simultaneously.

FEEDBACK REGULATORY LOOP IN GSL BIOSYNTHESIS AS A TRIGGER ACTIVATING TRANSCRIPTION OF *R2R3-MYBs* UNDER S DEFICIENCY

Our first hypothesis suggested that negative feedback regulation of GSL is switched upon low GSL status in the cell, resulting in the activation *MYB28, MYB51*, and *MYB122* at −S. Low-GSL signal activating the transcription of *MYBs* seem to act together but against the SLIM1 to regulate the glucosinolate-sulfur balance in the cell (**Figure 6B**). In agreement with this hypothesis, the expression of *MYB34, MYB51*, and *MYB122* and of IG biosynthesis gene *CYP83B1* is induced in *cyp79b2 cyp79b3* mutant vs. wild-type plants and is further stimulated in *cyp79b2 cyp79b3* plants exposed to −S conditions (**Figure 5**).

It was also previously reported that *Arabidopsis* plants possess a mechanism reacting to low levels of GSLs as a signal for induction of their synthesis (Smolen and Bender, 2002; Mugford et al., 2009). For example, the negative feedback regulation of the IGs was shown in *cyp83b1* mutant having increased levels of *ASA1, TSB1, CYP79B2, CYP79B3*, and *MYB34* transcripts, suggesting feedback inhibition of expression of *MYB* regulators by IGs (or their intermediates) (Smolen and Bender, 2002; Celenza et al., 2005). Furthermore, expression of *CYP83B1* and *MYB34* was significantly enhanced in IG deficient *cyp79b2 cyp79b3* double mutant, substantiating the existence of a feedback regulatory loop (Celenza et al., 2005). Finally, the *CYP79B2* and *CYP79B3* transcripts were shown to be suppressed in the *myb34 cyp83b1 (atr1-2 cyp83b1)* double mutant indicating the MYBs to be an important element of the negative feedback loop. It has been recently suggested that GSL may bind MYB transcription factors and thereby modulate their activity toward to promoters of biosynthetic genes (Kopriva et al., 2012).

Despite this counterintuitive finding on the positive regulation of *R2R3-MYBs* at S deficiency, the question on the role of such regulation remains open. We suggest that the specificity of the response of different *MYBs* can be achieved by the combinatorial activation of various signaling components acting upstream of *MYBs*. These signaling components include the SLIM1 negatively regulating GSL accumulation, "low IG signal" activating negative feedback loop of IG biosynthesis as well as other known −S-responsive pathways suggested in several studies (Maruyama-Nakashita et al., 2003; Nikiforova et al., 2003) (e.g., the activation of GSL catabolism, production of auxin via IG biosynthesis pathway etc.) and shown in **Figure 6A**, which build up together a complex regulatory unit for the response of *R2R3-MYB* to S deficiency.

MATERIALS AND METHODS

PLANT MATERIALS AND GROWTH CONDITIONS IN SULFUR LIMITATION EXPERIMENT

Seeds of wild-type *A. thaliana* (Col-0) were grown in a temperature-controlled greenhouse or in a growth chamber in a light/dark cycle of 8 h/16 h at a day/night temperature of 21°C/18°C and 40% humidity.

To analyse the expression of different *MYBs* in response so −S, surface-sterilized seeds of wild type or *cyp79b2 cyp79b3* plants were plated on Hoagland's Media (+S) or corresponding sulfur limiting media (−S) [2 mM $Ca(NO_3)_2$, 0.5 mM KH_2PO_4, 0.75 mM $MgCl_2$, 10 mM KNO_3, 1.5 μM $CuCl_2$, 2 μM $ZnCl_2$, 10 μM $MnCl_2$, 50 μM H_3BO_3, 0.1 μM MoO_3, 50 μM KCl, 50 μM Fe-Na-EDTA] or Hoagland's media (Hoagland and Martin, 1950) (+S) [2 mM $Ca(NO_3)_2$, 0.5 mM KH_2PO_4, 0.75 mM $MgSO_4$, 10 mM KNO_3, 1.5 μM $CuSO_4$, 2 μM $ZnSO_4$, 10 μM $MnSO_4$, 50 μM H_3BO_3, 0.1 μM MoO_3, 50 μM KCl, 50 μM Fe-Na-EDTA] After 7 and 28 days of growth, the plants were harvested for the analysis of gene expression and GSL content.

RNA EXTRACTION AND EXPRESSION ANALYSIS

The isolation of RNA, first strand synthesis and qRT-PCR was performed as described recently (Dean and Annilo, 2005). Relative quantification of expression levels was performed using the comparative ΔΔCt method and the calculated relative expression values were normalized to *Actin2* and to wild-type expression levels (wild type = 1). Primers used for the qRT-PCR analysis are shown in Supplemental Table 1.

HPLC ANALYSIS OF DESULFO-GS

The isolation and analysis of GSL content was performed by UPLC (Waters, Eschborn) as described recently (Gigolashvili et al., 2012).

GROWTH OF *ARABIDOPSIS THALIANA* CELL SUSPENSION AND OVEREXPRESSION OF *SLIM1*

A. thaliana dark grown suspension-culture cell line was maintained in 50 mL of *A. thaliana* (AT) medium. The AT medium contained 4.3 g/L MS basal salts (Duchefa), 1 mg/L 2,4-dichlorophenoxyacetic acid (2,4-D), 4 mL vitamin B5 mixture (Sigma-Aldrich) and 30 g/L sucrose (pH 5.8). Cells were gently agitated at 160 rpm in the dark at 22°C.

To generate cells transiently overexpressing *SLIM1*, the full length coding sequence of *SLIM1* was amplified from the cDNA and cloned into the Gateway pDONR207 vector (Life Technologies) using primers containing attB1 and attB2 sequences (SLIM1_attB1:gggacaagtttgtacaaaaaagcaggcttcATGGGCGATCTT GCTATGTCCGTAGC and SLIM1_attB2: gggaccactttgtacaagaa agctgggtcAGCTCCAAACCATGAGAAATCATCAC). The obtained clone was recombined with the *pGWB2* to obtain *Pro35S-SLIM1-pGWB2*, which was used for transient expression assay.

Transformation of dark-grown cultured Arabidopsis cells was performed using the supervirulent *Agrobacteria* strains LBA4404. pBBR1MCS.virGN54D containing *Pro35S-SLIM1-pGWB2* as described by Koprivova et al. (2000).

PROMOTER *TRANS*-ACTIVATION ASSAY WITH *SLIM1* AND PROMOTER OF *MYB51* IN CULTURED *A. THALIANA* CELLS

Promoter of *MYB51* gene was generated as reported in Gigolashvili et al. (2007a). To assess the *trans*-activation potential of *SLIM1* against promoter of *MYB51*, the effector construct with *Pro35S-SLIM1-pGWB2* and the promoter reporter *uidA* construct driven by the promoters of *MYB51* gene were used. Thus, the effector construct in the supervirulent *Agrobacterium strain*, the anti-silencing *Agrobacteria* strain 19 K (Voinnet et al., 1999) and *ProMYB51-uidA-pGWB3i* constructs were taken from fresh YEB plates, grown overnight, resuspended in 1 mL AT medium and used for cotransfection (Berger et al., 2007). Three clones of *Agrobacterium* were mixed in 1:1:1 ratio and 75 μL of this suspension was added to 3 mL of cultured *A. thaliana* cells. After 4–5 days of co-culturing (in the dark, 22°C, 160 rpm), cells were treated with 100 μL 5-bromo-4-chloro-3- indolyl-β-D-glucuronid acid (X-Gluc) solution for 1 h to overnight at 37°C.

ACKNOWLEDGMENTS

This work was supported by the Deutsche Forschungsgemeinschaft and The Cluster of Excellence in Plant Sciences (Project Reference Number: GI 824/1-1 and EXC 1028). We would also like to thank Prof. Dr. Stanislav Kopriva for critical reading of the manuscript.

REFERENCES

Augustine, R., Majee, M., Gershenzon, J., and Bisht, N. C. (2013). Four genes encoding MYB28, a major transcriptional regulator of the aliphatic glucosinolate pathway, are differentially expressed in the allopolyploid Brassica juncea. *J. Exp. Bot.* 64, 4907–4921. doi: 10.1093/jxb/ert280

Bednarek, P., Pislewska-Bednarek, M., Svatos, A., Schneider, B., Doubsky, J., Mansurova, M., et al. (2009). A glucosinolate metabolism pathway in living plant cells mediates broad-spectrum antifungal defense. *Science* 323, 101–106. doi: 10.1126/science.1163732

Berger, B., Stracke, R., Yatusevich, R., Weisshaar, B., Fluegge, U.-I., and Gigolashvili, T. (2007). A simplified method for the analysis of transcription factor-promoter interactions that allows high-throughput data generation. *Plant J.* 50, 911–916. doi: 10.1111/j.1365-313X.2007.03088.x

Bones, A. M., and Rossiter, J. T. (1996). The myrosinase-glucosinolate system, its organisation and biochemistry. *Physiol. Plant.* 97, 194–208. doi: 10.1111/j.1399-3054.1996.tb00497.x

Celenza, J. L., Quiel, J. A., Smolen, G. A., Merrikh, H., Silvestro, A. R., Normanly, J., et al. (2005). The *Arabidopsis* ATR1 MYB transcription factor controls indolic glucosinolate homeostasis. *Plant Physiol.* 137, 253–262. doi: 10.1104/pp.104.054395

Dean, M., and Annilo, T. (2005). Evolution of the Atp-Binding Cassette (Abc) transporter superfamily in vertebrates*. *Annu. Rev. Genomics Hum. Genet.* 6, 123–142. doi: 10.1146/annurev.genom.6.080604.162122

Gigolashvili, T., Berger, B., Mock, H.-P., Mueller, C., Weisshaar, B., and Fluegge, U.-I. (2007a). The transcription factor HIG1/MYB51 regulates indolic glucosinolate biosynthesis in *Arabidopsis thaliana*. *Plant J.* 50, 886–901. doi: 10.1111/j.1365-313X.2007.03099.x

Gigolashvili, T., Geier, M., Ashykhmina, N., Frerigmann, H., Wulfert, S., Krueger, S., et al. (2012). The Arabidopsis thylakoid ADP/ATP carrier TAAC has an additional role in supplying plastidic phosphoadenosine 5′-phosphosulfate to the cytosol. *Plant Cell* 24, 4187–4204. doi: 10.1105/tpc.112.101964

Gigolashvili, T., Yatusevich, R., Berger, B., Mueller, C., and Fluegge, U.-I. (2007b). The R2R3-MYB transcription factor HAG1/MYB28 is a regulator of methionine-derived glucosinolate biosynthesis in *Arabidopsis thaliana*. *Plant J.* 51, 247–261. doi: 10.1111/j.1365-313X.2007.03133.x

Hirai, M. Y., Fujiwara, T., Awazuhara, M., Kimura, T., Noji, M., and Saito, K. (2003). Global expression profiling of sulfur-starved Arabidopsis by DNA macroarray reveals the role of O-acetyl-l-serine as a general regulator of gene expression in response to sulfur nutrition. *Plant J.* 33, 651–663. doi: 10.1046/j.1365-313X.2003.01658.x

Hirai, M. Y., Klein, M., Fujikawa, Y., Yano, M., Goodenowe, D. B., Yamazaki, Y., et al. (2005). Elucidation of gene-to-gene and metabolite-to-gene networks in Arabidopsis by integration of metabolomics and transcriptomics. *J. Biol. Chem.* 280, 25590–25595. doi: 10.1074/jbc.M502332200

Hirai, M. Y., Sugiyama, K., Sawada, Y., Tohge, T., Obayashi, T., Suzuki, A., et al. (2007). Omics-based identification of *Arabidopsis* MYB transcription factors regulating aliphatic glucosinolate biosynthesis. *Proc. Natl. Acad. Sci. U.S.A.* 104, 6478–6483. doi: 10.1073/pnas.0611629104

Hirai, M. Y., Yano, M., Goodenowe, D. B., Kanaya, S., Kimura, T., Awazuhara, M., et al. (2004). Integration of transcriptomics and metabolomics for understanding of global responses to nutritional stresses in *Arabidopsis thaliana*. *Proc. Natl. Acad. Sci. U.S.A.* 101, 10205–10210. doi: 10.1073/pnas.0403218101

Hoagland, D. R., and Martin, J. C. (1950). Availability of potassium to crops in relation to replaceable and nonreplaceable potassium and to effects of cropping and organic matter. *Soil Sci. Soc. Am. Proc.* 15, 272–278. doi: 10.2136/sssaj1951.036159950015000C0063x

Hoefgen, R., and Nikiforova, V. J. (2008). Metabolomics integrated with transcriptomics: assessing systems response to sulfur-deficiency stress. *Physiol. Plant.* 132, 190–198. doi: 10.1111/j.1399-3054.2007.01012.x

Huseby, S., Koprivova, A., Lee, B.-R., Saha, S., Mithen, R., Wold, A.-B., et al. (2013). Diurnal and light regulation of sulphur assimilation and glucosinolate biosynthesis in Arabidopsis. *J. Exp. Bot.* 64, 1039–1048. doi: 10.1093/jxb/ers378

Kopriva, S., Mugford, S. G., Baraniecka, P., Lee, B.-R., Matthewman, C. A., and Koprivova, A. (2012). Control of sulfur partitioning between primary and secondary metabolism in Arabidopsis. *Front. Plant Sci.* 3:163. doi: 10.3389/fpls.2012.00163

Koprivova, A., Suter, M., Den Camp, R. O., Brunold, C., and Kopriva, S. (2000). Regulation of sulfate assimilation by nitrogen in Arabidopsis. *Plant Physiol.* 122, 737–746. doi: 10.1104/pp.122.3.737

Kutz, A., Muller, A., Hennig, P., Kaiser, W. M., Piotrowski, M., and Weiler, E. W. (2002). A role for nitrilase 3 in the regulation of root morphology in sulphur-starving *Arabidopsis thaliana*. *Plant J.* 30, 95–106. doi: 10.1046/j.1365-313X.2002.01271.x

Li, Y., Sawada, Y., Hirai, A., Sato, M., Kuwahara, A., Yan, X., et al. (2013). Novel insights into the function of Arabidopsis R2R3-MYB transcription factors regulating aliphatic glucosinolate biosynthesis. *Plant Cell Physiol.* 54, 1335–1344. doi: 10.1093/pcp/pct085

Maruyama-Nakashita, A., Inoue, E., Watanabe-Takahashi, A., Yamaya, T., and Takahashi, H. (2003). Transcriptome profiling of sulfur-responsive genes in Arabidopsis reveals global effects of sulfur nutrition on multiple metabolic pathways. *Plant Physiol.* 132, 597–605. doi: 10.1104/pp.102.019802

Maruyama-Nakashita, A., Nakamura, Y., Tohge, T., Saito, K., and Takahashi, H. (2006). *Arabidopsis* SLIM1 is a central transcriptional regulator of plant sulfur response and metabolism. *Plant Cell* 18, 3235–3251. doi: 10.1105/tpc.106.046458

Mugford, S. G., Lee, B.-R., Koprivova, A., Matthewman, C., and Kopriva, S. (2011). Control of sulfur partitioning between primary and secondary metabolism. *Plant J.* 65, 96–105. doi: 10.1111/j.1365-313X.2010.04410.x

Mugford, S. G., Matthewman, C. A., Hill, L., Kopriva, S. (2010). Adenosine-5′-phosphosulfate kinase is essential for Arabidopsis viability. *FEBS lett.* 584, 119–123. doi: 10.1016/j.febslet.2009.11.014

Mugford, S. G., Yoshimoto, N., Reichelt, M., Wirtz, M., Hill, L., Mugford, S. T., et al. (2009). Disruption of Adenosine-5′-phosphosulfate kinase in arabidopsis reduces levels of sulfated secondary metabolites. *Plant Cell* 21, 910–927. doi: 10.1105/tpc.109.065581

Nikiforova, V., Freitag, J., Kempa, S., Adamik, M., Hesse, H., and Hoefgen, R. (2003). Transcriptome analysis of sulfur depletion in *Arabidopsis thaliana*: interlacing of biosynthetic pathways provides response specificity. *Plant J.* 33, 633–650. doi: 10.1046/j.1365-313X.2003.01657.x

Nikiforova, V. J., Kopka, J., Tolstikov, V., Fiehn, O., Hopkins, L., Hawkesford, M. J., et al. (2005). Systems rebalancing of metabolism in response to sulfur deprivation, as revealed by metabolome analysis of Arabidopsis plants. *Plant Physiol.* 138, 304–318. doi: 10.1104/pp.104.053793

Rask, L., Andréasson, E., Ekbom, B., Eriksson, S., Pontoppidan, B., and Meijer, J. (2000). Myrosinase: gene family evolution and herbivore defense in Brassicaceae. *Plant Mol. Biol.* 42, 93–114. doi: 10.1023/A:1006380021658

Schnug, E., and Haneklaus, S. (1993). Physiological backgrounds of different sulphur utilisation in *Brassica napus* varieties. *Aspects Appl. Biol.* 34, 235–242.

Schnug, E., Haneklaus, S., Borchers, A., and Polle, A. (1995). Relations between sulphur supply and glutathione and ascorbate concentrations in *Brassica napus*. *Z. Pflanz. Bodenkunde* 158, 67–69. doi: 10.1002/jpln.19951580113

Smolen, G., and Bender, J. (2002). *Arabidopsis* cytochrome P450 *cyp83B1* mutations activate the tryptophan biosynthetic pathway. *Genetics* 160, 323–332.

Sønderby, I. E., Burow, M., Rowe, H. C., Kliebenstein, D. J., and Halkier, B. A. (2010). A complex interplay of three R2R3 MYB transcription factors determines the profile of aliphatic glucosinolates in *Arabidopsis*. *Plant Physiol.* 153, 348–363. doi: 10.1104/pp.109.149286

Sønderby, I. E., Hansen, B. G., Bjarnholt, N., Ticconi, C., Halkier, B. A., and Kliebenstein, D. J. (2007). A systems biology approach identifies a R2R3 MYB gene subfamily with distinct and overlapping functions in regulation of aliphatic glucosinolates. *PLoS ONE* 2:e1322. doi: 10.1371/journal.pone.0001322

Takahashi, H., Kopriva, S., Giordano, M., Saito, K., and Hell, R. (2011). Sulfur assimilation in photosynthetic organisms: molecular functions and regulations of transporters and assimilatory enzymes. *Annu. Rev. Plant Biol.* 62, 157–184. doi: 10.1146/annurev-arplant-042110-103921

Voinnet, O., Pinto, Y. M., and Baulcombe, D. C. (1999). Suppression of gene silencing: a general strategy used by diverse DNA and RNA viruses of plants. *Proc. Natl. Acad. Sci. U.S.A.* 96, 14147–14152. doi: 10.1073/pnas.96.24.14147

Yatusevich, R., Mugford, S. G., Matthewman, C., Gigolashvili, T., Frerigmann, H., Delaney, S., et al. (2009). Genes of primary sulfate assimilation are part of the glucosinolate biosynthetic network in *Arabidopsis thaliana*. *Plant J.* 62, 1–11. doi: 10.1111/j.1365-313X.2009.04118.x

Yatusevich, R., Mugford, S. G., Matthewman, C., Gigolashvili, T., Frerigmann, H., Delaney, S., et al. (2010). Genes of primary sulfate assimilation are part of the glucosinolate biosynthetic network in *Arabidopsis thaliana*. *Plant J.* 62, 1–11. doi: 10.1111/j.1365-313X.2009.04118.x

Zhao, Y., Hull, A. K., Gupta, N. R., Goss, K. A., Alonso, J., Ecker, J. R., et al. (2002). Trp-dependent auxin biosynthesis in Arabidopsis: involvement of cytochrome P450s CYP79B2 and CYP79B3. *Genes Dev.* 16, 3100–3112. doi: 10.1101/gad.1035402

Conflict of Interest Statement: The authors declare that the research was conducted in the absence of any commercial or financial relationships that could be construed as a potential conflict of interest.

19

Metabolic responses to sulfur dioxide in grapevine (*Vitis vinifera* L.): photosynthetic tissues and berries

***Michael J. Considine*[1,2,3] *and Christine H. Foyer*[1,2] ***

[1] *School of Plant Biology, and The UWA Institute of Agriculture, University of Western Australia, Crawley, WA, Australia*
[2] *Centre for Plant Sciences, School of Biology, Faculty of Biological Sciences, University of Leeds, Leeds, Yorkshire, UK*
[3] *Irrigated Agriculture and Diversification, Department of Agriculture and Food Western Australia, South Perth, WA, Australia*

Edited by:
Ruediger Hell, University of Heidelberg, Germany

Reviewed by:
Heinz Rennenberg, University of Freiburg, Germany
Jochen Bogs, Fachhochschule Bingen, Germany

****Correspondence:***
Christine H. Foyer, Centre for Plant Sciences, School of Biology, Faculty of Biological Sciences, University of Leeds, Woodhouse Lane, Leeds, Yorkshire LS29JT, UK
e-mail: c.foyer@leeds.ac.uk

Research on sulfur metabolism in plants has historically been undertaken within the context of industrial pollution. Resolution of the problem of sulfur pollution has led to sulfur deficiency in many soils. Key questions remain concerning how different plant organs deal with reactive and potentially toxic sulfur metabolites. In this review, we discuss sulfur dioxide/sulfite assimilation in grape berries in relation to gene expression and quality traits, features that remain significant to the food industry. We consider the intrinsic metabolism of sulfite and its consequences for fruit biology and postharvest physiology, comparing the different responses in fruit and leaves. We also highlight inconsistencies in what is considered the "ambient" environmental or industrial exposures to SO_2. We discuss these findings in relation to the persistent threat to the table grape industry that intergovernmental agencies will revoke the industry's exemption to the worldwide ban on the use of SO_2 for preservation of fresh foods. Transcriptome profiling studies on fruit suggest that added value may accrue from effects of SO_2 fumigation on the expression of genes encoding components involved in processes that underpin traits related to customer satisfaction, particularly in table grapes, where SO_2 fumigation may extend for several months.

Keywords: fruit, oxidative stress, antioxidants, quality, glutathione, SO_2, elemental sulfur, wine

INTRODUCTION

Sulfur dioxide may be considered to be the "elephant in the room" of grape and wine industries and in agriculture more broadly, for both its health and environmental consequences. Or perhaps it is something of a "golden goose?" SO_2 is used in >99% of wine production. About 15% of more than 15,000 patents for biological study of SO_2 are related to wine (Chemical Abstracts Service, 2014). Therefore, grape berries and wine are an appropriate case for study and discussion of the metabolic responses of plant tissues and organs to SO_2exposure, particularly considering the responses of non-photosynthetic tissues to sulfur dioxide. Sulfur dioxide is considered here more as a food additive than an atmospheric pollutant and potentially phytotoxic agent. More specifically, we compare and contrast effects of SO_2 in grape berries, which are reproductive organs, with the more expansive knowledge of the consequences of SO_2 fumigation in leaves but we do not consider wine *per se*, which bears additional consequences beyond the life of the plant cell. In the following analysis, we do not distinguish between effects of SO_2 on table and wine grapes. However, it is important to acknowledge the difference in the SO_2 fumigation strategies that are applied in each case. Table grapes may be exposed to SO_2 fumigation for several weeks for conservation. In contrast, the application for wine grapes is often only a few hours, prior to fermentation.

The grape and wine industries place high value on the knowledge and control of reductive and oxidative (redox) processes, and of microbial populations. Sulfur is capable of a wide range of oxidation states (−2 to +6), and hence sulfur-derived compounds are a major feature of redox metabolism and post-translational modifications, as well as defense and detoxification of toxins or heavy metals. Thiols and sulfides are among the most important flavor and aroma compounds and precursors in many wine varieties and styles, both desirable and undesirable (Baumes, 2009). Sulfur-derived compounds also play a major role in the abundance and stability of other flavor, aroma and texture components, such as tannins, phenolic acids, anthocyanins, and aldehydes.

Sulfur is added to grapes and wine at several stages in several chemical forms, exploiting various chemical or toxicological properties (**Figure 1**). By mass, the greatest input is elemental sulfur (S^0), which has some fungicidal activity (Williams and Cooper, 2004) but is largely used as a slow-release substrate for SO_3^{2-}, or SO_2 when burned, which have fungicidal and fungistatic activities against most economically important pathogens, particularly powdery mildew (*Uncinula* spp.) and botrytis (*Botrytis cinerea* Pers. Fr.). It is also effective in control of mites, which can decimate bud vitality, damaging reproductive structures even before bud burst, and spiders for disinfestation of fruit postharvest. The abundance of elemental sulfur applied in

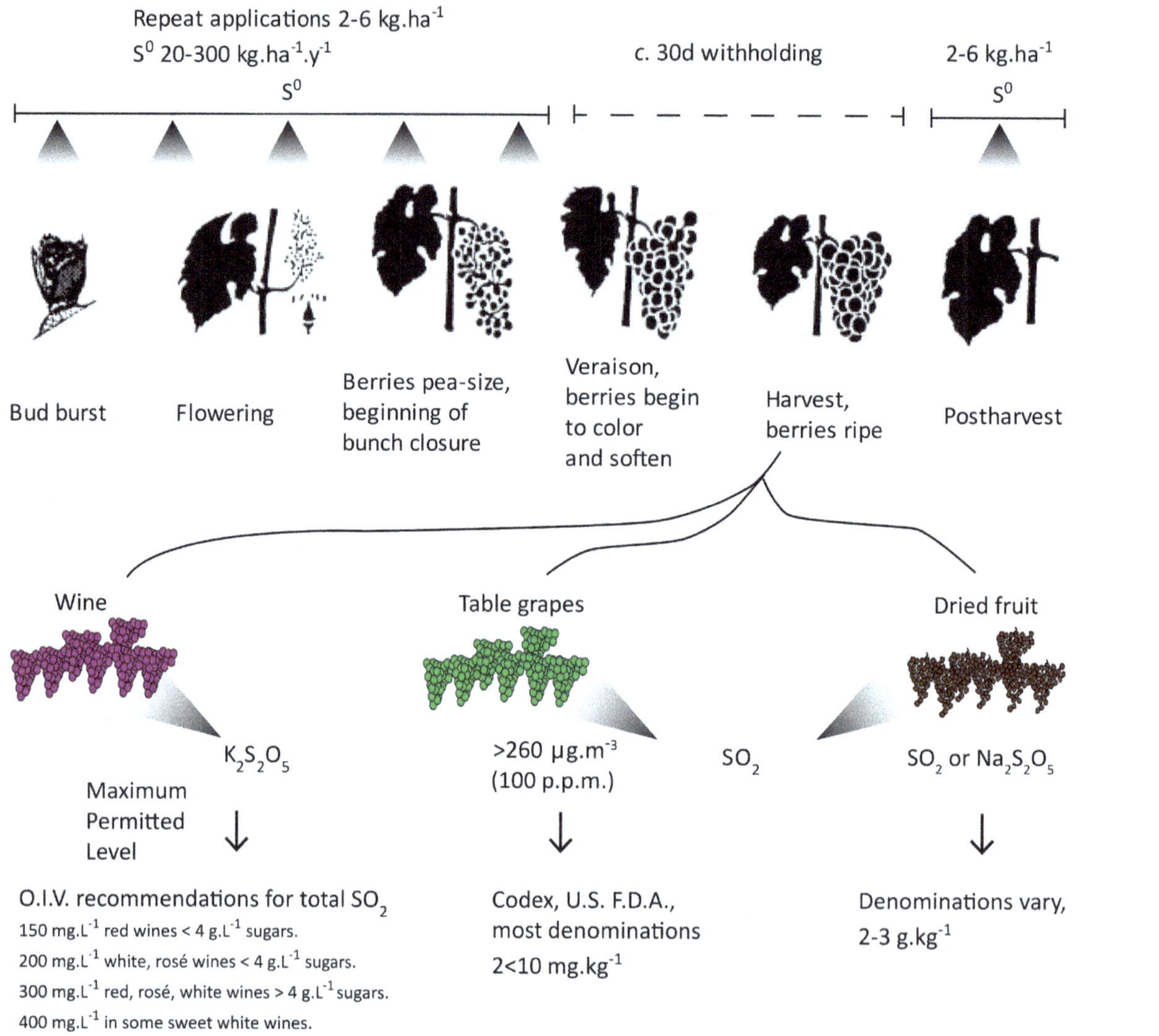

FIGURE 1 | Representative inputs of elemental sulfur (S^0) and SO_2, including sulfite salts, in grape and wine production. Elemental sulfur is typically sprayed on vines prior to bud burst at rates of 2–6 kg.ha^{-1}.y^{-1}, repeated throughout the season. Application rates vary greatly, according to climate, wine style, fungal risk, and commercial preference. Many denominations mandate a > 30d withholding period for S^0 application, largely due to adverse effects of residual sulfur on wine flavor and aroma qualities. For winemaking, sulfite salts (e.g., $K_2S_2O_5$) or SO_2 are added during the grape crush prior to fermentation for wine, as well as throughout fermentation and often at bottling to control microbial growth and for chemical antioxidant properties. For table grapes, SO_2 is commonly used in storage shed, whether by forced fumigation in the cold room or by controlled release SO_2-generating pads, particularly when fruit are intended for export or to be stored more than 1 week. Typically >260 μg.m^{-3} SO_2 (100 p.p.m.) is required to surface-sterilize effectively, and thereafter a maintenance of 5–8 μg.m^{-3} (2–3 p.p.m.) has a fungistatic effect. The use in dried fruit prevents oxidative browning but is shown here only for additional context. It should be noted that the stated maximum residue levels are often far in excess of the actual amounts used in industry and detected in food surveys, e.g., sultana 2 mg.kg^{-1}, 1000-fold below Maximum Permitted Level (MPL; F.S.A.N.Z., 2012). O.I.V., International Organization of Vine and Wine (www.oiv.it); Codex, Codex Alimentarius Commission (2014) (www.codexalimentarius.net/); U.S. F.D.A., U.S. Food and Drug Administration (2014) (www.fda.gov).

most vineyards precludes any risk of sulfur deficiency (Robinson, 1988). Yet there is a large knowledge-gap in the speciation, chemical and metabolic, of sulfur-derived compounds between application in the vineyard and fermentation in the winery. The same is true of SO_2 fumigation of fresh table grapes. To date there are no effective alternatives to S^0 or SO_2 application to control microbial infection, as grapes are highly susceptible to pathogens of different trophic habits; biotrophs, hemi-biotrophs and necrotrophs, and S^0/SO_2 are at least partly effective in controlling all, while being extremely cost-effective when used in combination with other agents such as copper. So while society and intergovernmental agencies maintain the threat of a complete ban, the grape industries would not survive, and typically act with great responsibility to minimize use within the Generally Recognized as Safe (G.R.A.S.) limits.

It is widely known that residual sulfur on berries promotes formation of off-flavors such as H_2S during fermentation (Rankine, 1963; Kwasniewski et al., 2014). SO_2 residues in fresh table grapes and wines are restricted by legislation to limit risks of ill-health effects in consumers. A few recent studies also reveal ecological pathways of applied sulfur in the vineyard (Hinckley et al., 2011; Hinckley and Matson, 2011). Each of these ignore the reality that atmospheric sulfur (e.g., SO_2, H_2S) is readily assimilated by plants (Rennenberg, 1984).

We recently demonstrated an expansive transcriptional reprogramming evoked by SO_2 application to table grape berries (Giraud et al., 2012). The exposure was non-phytotoxic and at levels far below those acceptable in production, even in the context of organic grapes and wine. Here we explore the metabolic and quality consequences of such a large transcriptional footprint in the context of other studies that have documented pathways of elemental sulfur applied to grapevine, including retention in the berries. Importantly, we draw contrast to the broader literature on the consequences of SO_2 exposure in leaves. While the impact of sulfur and sulfur-derived compounds in wine extends far beyond the strictly metabolic, or living activity of the grapes, this review is largely confined to that scope, but with hypotheses for the consequences for wine stability and sensory qualities. The chemistry of SO_2 and sulfites in wine is adequately described in text books (e.g., Boulton et al., 1999). The influence of sulfites on fermentation and microbial activities is also beyond this review, as are the consequences of the many additional forms of sulfur inputs in the vineyard, including polysulfides (lime sulfur) and various thiols. However, we do consider elemental sulfur, as its application assumes oxidation to the antimicrobial oxide anions.

SO_2 CONCENTRATIONS: ISSUES AND INCONSISTENCIES

There is a notable inconsistency concerning the units of SO_2 concentration used in the literature, particularly with regard to what constitutes a high or low concentration. This has led to difficulties in relating information in different studies. In fact, there are very few examples of the use of S.I. units of $kg.m^{-3}$, or whether the volumes used refer to either a liquid or gas. This is not helped by the fact that the agronomic preference is to express units as parts per million (p.p.m.). Other researchers and various food industries use either mass- or volume per volume, e.g., $\mu g.L^{-1}$ vs $\mu L.L^{-1}$, or per mass, however, these units are not equivalent in terms of the amount of SO_2 exposed to the plant or food. The information contained in this review largely concerns atmospheric concentration. We have therefore cited information in the S.I. units, referring to the density of 2.62 $kg.m^{-3}$ SO_2 at 25°C, 101.3 kPa (C.R.C., 2014), i.e., the volume-base unit is *c.* 382x the mass-base unit, hence if a particular study equates 1 p.p.m. to 1 $\mu L.L^{-1}$, i.e., v/v, that is actually 2.62 $\mu g.m^{-3}$ when expressed in S.I. units, i.e., 2.62 p.p.b.

An additional consideration when interpreting the results of different studies or contexts for atmospheric sulfur assimilation is the differences in the levels of flux of SO_2penetrating the tissues (Rennenberg and Herschbach, 2014). The numbers of stomata and their functional operation to control conductance is a major control point for SO_2 influx into metabolism. When comparing leaves and fruit, it is important to point out that stomatal density is comparatively low in fruit, typically < 10 stomata per berry in grapevine (Palliotti and Cartechini, 2001). Moreover, the stomata on the berries are at least partly blocked with epicuticular wax (Rogiers et al., 2004). Postharvest storage conditions also maintain very low vapor pressure deficits, with high levels of relative humidity and low temperatures. Hence the capacity for influx of atmospheric sulfur would be manifold lower for fruit than leaves, especially when the considerable differences in surface area to volume ratios are taken into account.

A further inconsistency exists among the data from the grape, food and wine industries, where SO_2 application is reported per unit volume, which may be gas or liquid, while residues are reported per unit mass or liquid. For example, postharvest application of SO_2 to table grapes is based on units per volume of gas, while residues are based on units per mass of extracted berry [e.g., maximum permitted level 30 $mg.kg^{-1}$ Codex Alimentarius Commission (2014), 10 $mg.kg^{-1}$ U.S. Food and Drug Administration (2014), refer **Figure 1**]. Hence, p.p.m. application of SO_2 does not directly relate to p.p.m. residues. We do not attempt to resolve this inconsistency here but are careful to distinguish the two.

It is pertinent to also provide a more environmental context to understand the range and trends in global SO_2 emissions and atmospheric concentrations in industrial and natural environments. Global SO_2 emissions have declined >15% since 1990, although only peaked in China *c.* 2006 and emissions in India were still increasing in 2010 (Klimont et al., 2013). From 1980 to 2013 atmospheric SO_2 surveys in the United States of America showed a mean atmospheric SO_2 declined from >400 $\mu g.m^{-3}$ to < 80 $\mu g.m^{-3}$ (E.P.A., 2014). In 2011 > 96% of Chinese cities were 20–150 $\mu g.m^{-3}$, which is within their grade I and II air quality guidelines, which refers to protected conservation environments, and rural and residential areas, respectively (State Environmental Protection Administration, 2013). The World Health Organization air quality guidelines are 20 $\mu g.m^{-3}$ (24 h mean) or 500 $\mu g.m^{-3}$ (10 min mean; W.H.O., 2006). Internationally, air quality standards vary, e.g., >350 $\mu g.m^{-3}$ (24 h mean) for Bangladesh, Indonesia, and Singapore (Clean Air Initiative, 2010). In addition, some denominations use 24- or 1-h means, while the WHO is committed to a 10-min mean (W.H.O., 2006).

SULFUR ASSIMILATION AND SEQUESTRATION IN LEAVES

The preservative effects of SO_2 have been exploited in winemaking since antiquity. Despite this, the post-industrial contexts of ecological and agricultural damage have attracted far more scientific enquiry on the mechanisms and consequences of SO_2 exposure to plants. In recent decades, pollution-prevention measures have decreased such threats, and acute SO_2 injury is now much less common. Sulfur deficiency can be experienced by plants in the natural environment, leading to changes in plant morphology, metabolism and gene expression (Honsel et al., 2012). In particular, levels of the antioxidant thiol metabolite, glutathione are decreased leading to a de-repression of sulfate uptake and assimilation (Hartmann et al., 2004). Similarly, it is not uncommon for field crops to suffer from low level sulfur deficiency, resulting in changes in nitrogen metabolism and leading to the accumulation of amino acids such as asparagine (Shewry et al., 1983). For example, asparagine accumulates to very high levels in wheat is grown under conditions of sulfur deficiency.

This is important because sulfur availability is the most important factor affecting the acrylamide-forming potential of wheat grain (Muttucumaru et al., 2006).

The highly regulated processes of sulfur uptake, assimilation and distribution throughout the plant have been extensively reviewed (refer to Takahashi et al., 2011; Koprivova and Kopriva, 2014, and references therein). Sulfur uptake is considered to be driven by the demand for core sulfur-containing compounds, such as cysteine and glutathione (Davidian and Kopriva, 2010). Sulfur depletion initially leads to an increase in sulfate uptake from the soil, while further limitation results in redistribution, driven by sink capacity. The multiple tiers of transcriptional to hormonal and metabolic regulation, including by sugars, triggered by sulfur depletion illustrate the vast importance of sulfur metabolism to plants.

Low sulfur-dependent restrictions on glutathione accumulation in plants (Nikiforova et al., 2003) are likely to limit the stress tolerance, because of the multiple roles of this abundant non-protein thiol, particularly in secondary metabolism and oxidative signaling (Noctor et al., 2012). *Arabidopsis* mutants lacking high affinity sulfate transporter, SULTR1:2, have decreased levels of glutathione (Maruyama-Nakashita et al., 2003). Moreover, overexpression of genes encoding sulfur-assimilation enzymes SERINE ACETYLTRANSFERASE (SAT) and *O*-ACETYLSERINE(THIOL)LYASE (OASTL) increased cysteine and glutathione contents in *Arabidopsis*, potato and tobacco leaves (Harms et al., 2000; Noji and Saito, 2002; Wirtz and Hell, 2007).

High atmospheric SO_2 concentrations can have both positive and negative effects on plant growth and development (Gayler and Sykes, 1985). Plants can rapidly assimilate SO_2 and H_2S into reduced sulfur pools and sulfates, leading to improved growth especially in soils with poor sulfur availability. An atmospheric level of $\geq$ 30 nL.L^{-1} SO_2 (79 ng.m^{-3}) can contribute 10-40% of leaf sulfur assimilation (De Kok et al., 2007; Zhao et al., 2008). Elevated SO_2 concentrations around natural CO_2 springs can lead to an enhanced accumulation of sulfur metabolites and proteins in surrounding vegetation (Rennenberg, 1984; Schulte et al., 2002). However, these effects vary greatly between species (Naito et al., 1994), as SO_2 exposures as low as 2-5 nL.L^{-1} (5-13 ng.m^{-3}) can cause reductions in growth (Heber and Huve, 1998), and even visible injury to leaves and other vegetative tissues. High SO_2 levels can lead to visible injury in young leaves is with chlorosis and necrotic inter-vein areas in broad-leaved species, and chlorotic spots and brown tips in pine conifers (Rennenberg, 1984). This is often caused by an accumulation of sulfite and sulfate and associated with very high leaf sulfur contents. SO_2 gas enters leaves via stomata and at apoplastic pH is hydrated and oxidized successively to sulfite and sulfate, both of which can inhibit photosynthesis and energy metabolism if they accumulate. The SO_2-induced inhibition of photosynthesis and associated increase in the oxidation state of leaf cells underpins the toxicity syndrome.

Within a normal physiological range (3–76 S g.kg^{-1} FW; Zhao et al., 2008), sulfate assimilation leads to the synthesis of L-cysteine, which is the precursor for the synthesis of a range of sulfur-containing metabolites such as methionine and glutathione. The two final reactions in this sequence are catalyzed by a cysteine synthase complex, which is comprised of two enzymes SAT and OASTL. SAT is responsible for the acetylation of L-serine by acetyl-CoA to produce *O*-acetylserine (OAS). OASTL catalyzes the formation of cysteine from H_2S and OAS. The SAT family consists of five members in *Arabidopsis*, three of which are localized to the cytosol, one in chloroplast stroma, and one in mitochondria (Kawashima et al., 2005). The mitochondrial and chloroplast SAT forms make the major contribution to cysteine synthesis under optimal and stress conditions (Haas et al., 2008; Watanabe et al., 2008). The SAT protein is unstable when not associated with OASTL, and hence SAT activity in the chloroplasts and mitochondria is regulated by the assembly and maintenance of the cysteine synthase complex. The chloroplast SAT form (SAT1), which is considered to be the rate-limiting enzyme in cysteine biosynthesis in leaves interacts with cyclophilin CYP20–3, which located in the chloroplast stroma. CYP20–3 foldase activity is influenced by thioredoxin-mediated reduction and is considered to link photosynthetic electron transport activity and oxidative regulation to the folding of SAT1, and hence SAT1 activity and cysteine biosynthesis (Dominguez-Solis et al., 2008). Thus, SO_2-mediated oxidation of the chloroplast stroma might directly influence the flux and capacity of cysteine synthesis.

In addition to the reductive pathway of sulfur assimilation described above, which is localized in chloroplasts, there is also an oxidative pathway for the removal of sulfite derived from SO_2 that is localized in the peroxisomes, in which SULFITE OXIDASE (SO) plays a predominate role. While the significance of the SO pathway relative to the reductive pathway in the chloroplasts remains to be established for example in terms of relative flux (Rennenberg and Herschbach, 2014), SO is considered to be important in the maintenance of intracellular sulfate pools, and to contribute to metabolic recycling and potentially act as a sink pathway for excessive sulfur (Hänsch et al., 2007; Brychkova et al., 2013).

Leaves exposed to non-phytotoxic levels of SO_2 (600 nL.L^{-1}; 1.6 ng.m^{-3}) show a wide range of transcriptome, metabolic and enzymatic changes in *Arabidopsis*, indicating a large scale reprogramming at both transcriptional and translational/post-translational levels. SO_2 (sulfite) enters the plastid sulfur assimilation pathway downstream of sulfate, immediately downstream of the major rate-limiting enzyme ADENOSINE 5′-PHOSPHOSULFATE REDUCTASE (APR), and upstream of SULFITE REDUCTASE (SIR) and OASTL/SAT. In general, enzyme activities upstream of sulfite were repressed, including APR, although its transcription was unaffected (Hamisch et al., 2012; Randewig et al., 2012). This indicated a repression of further sulfite synthesis, while sulfate accumulated. However, metabolism of sulfite was enhanced, via increased SIR and SAT activities, effecting increased cysteine and glutathione contents, although again, transcriptional regulation was more marginal. Transcripts encoding proteins involved in nitric oxide synthesis and antioxidant defenses as well as apoplastic peroxidases and defensins were also upregulated (Hamisch et al., 2012). These transcriptional signatures were very similar to those seen in *Arabidopsis* leaves exposed to much higher concentrations (30 μg.L^{-1}; 30 mg.m^{-3}; Zhao and Yi, 2014), which was phytotoxic and reduced the growth rate, but not lethal (Li et al., 2008). Insight can also be drawn from SO knock-out mutants in *Arabidopsis*, which showed even more marked transcript and activity reductions in APR when exposed to SO_2, indicating strong downregulation of sulfite synthesis, while cysteine, glutathione and thiols were markedly increased (Hamisch et al., 2012; Randewig et al., 2012).

There are a number of similarities between the responses of leaf metabolism to SO_2 and to the metabolic production of hydrogen peroxide (H_2O_2). For example, photorespiration-induced oxidative stress in *Arabidopsis* mutants deficient in CATALASE (*cat2*) led to extensive glutathione accumulation and triggered increases in transcripts encoding APR and SAT (Queval and Noctor, 2007). The chloroplast SAT is strongly induced by H_2O_2 and by glutathione (Queval and Noctor, 2007). Moreover, oxidation triggers post-translational activation of γ-GLUTAMYL CYSTEINE SYNTHETASE (γ-ECS) and APR, possibly by oxidation-triggered decreases in the reduced glutathione (GSH): glutathione disulfide (GSSG) ratio that may allow glutaredoxin (GRX)-mediated activation of both enzymes (Noctor et al., 2012). The H_2O_2-induced increases in glutathione accumulation in catalase-deficient barley mutants were accompanied by increased uptake of labeled sulfate (Smith et al., 1985). Similarly, the large increases in glutathione accumulation achieved in transgenic plants with ectopic expression of a bacterial enzyme having both γ-ECS and GLUTATHIONE SYNTHETASE activities were dependent on having a high sulfur supply (Liedschulte et al., 2010).

Taken together, these observations suggest that enhanced cellular oxidation is a hallmark of SO_2 action in leaves. However, SO_2-induced changes in cellular redox state are important in facilitating enhanced rates of sulfur assimilation, oxidative activation being a trigger for both cysteine and glutathione synthesis. Presumably, SO_2-induced damage only occurs when the oxidative activation of these pathways fails to restore the cellular redox balance. Major differences in the effects of SO_2 observed between vegetative and reproductive tissues may therefore be attributed to the presence or absence of photosynthesis, with its inherent sensitivity to oxidative inhibition and the relative metabolic activities of different types of plastids. In addition, variations in the barriers to gas exchange and the surface area to volume ratios may lead to differences in SO_2 sensitivity between vegetative tissues such as leaves and reproductive organs such as fruit.

FIGURE 2 | Application of elemental sulfur (S^0), as wettable sulfur (A) in the vineyard during the growing season, as well as postharvest application of SO_2 to table grapes from SO_2-generating pads (B). Applications of wettable sulfur, as well as other sulfurous pesticides vary greatly across industry and climatic zones (**Figure 1**). Unless destined for immediate sale, table grapes are treated with SO_2, typically with the use of SO_2-generating pads such as the one seen on top of berries in the right hand side box of **(B)**. The fruit **(B)** had been stored at 2–4°C for 4 weeks, with (right) or without (left) SO_2, showing no visible quality differences. Panel **(A)** rights purchased from ShutterStock (www.shutterstock.com).

SULFUR IN THE VINEYARD, WINERY AND PACKING SHED

Elemental sulfur (S^0) is widely and frequently applied during the growing season, typically in the form of wettable powders, sprayed directly on vines to provide a "protective" coating, or alternatively burned in the vineyard. Both assume oxidation to $SO_2/SO_3^{2-}/HSO_3^-$, which are effective in controlling the pathogen, albeit with differing efficacies. The reported range of wettable S^0 used in commercial vineyards, including certified organic vineyards, varies by several orders of magnitude (**Figures 1 and 2A**). For example, agrochemical companies in Australia typically recommend up to 100 $kg.ha^{-1}.yr^{-1}$, and while many wineries may use as little as 20 $kg.ha^{-1}.yr^{-1}$, reports internationally, where pathogen pressures are higher, vary up to 600 $kg.ha^{-1}.yr^{-1}$ (Hinckley and Matson, 2011). Recent studies have shown that much of the S^0 may be oxidized within minutes or hours and is ultimately lost from the vineyard via hydrological pathways (Hinckley et al., 2011; Hinckley and Matson, 2011). A significant pool of S^0 was retained in the soil surface, and likely the vegetative surface, until irrigation or rain events. Within 7–12 days, the initial surge in topsoil (0–0.5 m) sulfates had declined to pre-application levels. Using the dynamic changes in sulfur species in above- and below-ground fractions, and scaling to vineyard-scale, the authors concluded that any accumulation of sulfur in the soil and plant matter was lost during rain events in the dormant season. Yet, *c.* 2% (w/w) of applied sulfur was retained in the berries, which in the context of biomass represented 7-14 $kg.ha^{-1}.yr^{-1}$.

Although ecologically revealing, the above study isn't greatly informative for the biologist, grower, winemaker or consumer, as the study was dynamic, with no control *per se*, and the forms of sulfur on or within the berries could not be discriminated. Earlier studies in wheat using $^{35}S^0$ showed rapid assimilation of up to 2% (w/w) of applied sulfur into sulfate, glutathione and amino acids (Legris-Delaporte et al., 1987). The most prominent concern to winemakers is that S^0 residues may result in increased H_2S production during fermentation (Rankine, 1963; Kwasniewski et al., 2014). For this reason most practices require a >30 day withholding period (**Figure 1**), which typically results in S^0 residues of < 1 $mg.kg^{-1}$ (Kwasniewski et al., 2014). At this level, the effect on H_2S production is thought to be negligible, particularly for white wines, where settling and limited skin contact, result in near-complete removal of S^0 before fermentation. However, before and during the crushing process, SO_2 or $K_2S_2O_5$ is added at levels to give appropriate levels of free SO_2 (20-50 $g.m^{-3}$ in liquid). The metabolic impacts of those additions are difficult to dissect from chemical effects and beyond this review.

SULFUR DIOXIDE ASSIMILATION AND METABOLISM IN THE BERRY

The prolonged or repeated postharvest applications of SO_2 to fresh table grapes have been a mainstay of the table grape industry for decades. The maximum residue level for fresh table grapes is 10 p.p.m. (10 $mg.kg^{-1}$). Between the 1920's and 1980's, before the US Food and Drug Administration suspended sulfiting agents from the register of GRAS additives, an initial fumigation of 13–26 $mg.m^{-3}$ (5–10,000 p.p.m. in air) was common, followed by repeat fumigations of 6.5 $mg.m^{-3}$ at 7–10 day intervals (Nelson and Baker, 1963). Although this practice is still widely used in some regions, e.g., California, USA (Crisosto and Smilanick, 2014; Luvisi, 2014), international practice is far more conservative and increasingly sophisticated through the use of SO_2-generating pads, particularly where fruit are to be exported (**Figure 2B**). Such pads are impregnated with $Na_2S_2O_5$ in a polymeric matrix that, upon hydration enables a transient burst of >260 $\mu g.m^{-3}$ (100 p.p.m. in air), which is sufficient to surface-sterilize, followed by sustained release of 5-8 $\mu g.m^{-3}$ (2–3 p.p.m. in air) for several weeks to prevent re-infection (Clemes, 1986; Palou et al., 2002).

Only a limited number of studies have rigorously investigated the absorption and oxidation of SO_2 in the berry. Peiser and Yang (1985) used a combination of radiolabeled and unlabeled SO_2, and carefully managed extraction technique to control oxidation, to calculate that *c.* 10% (w/w) of the applied SO_2 rapidly accumulated in the berry as sulfites (4-5 $mg.kg^{-1}$ berry FW), in free and bound forms (hydroxyl sulfonic acids of aldehydes and methyl- and cyclic-ketones). The authors found *c.* 70% of the absorbed sulfites were rapidly oxidized to sulfate with a half-life of 4 h, with most of the remainder oxidizing with a half-life of 20 h. A more recent study with similar technique showed similar rate of uptake but more sustained retention of sulfites of $\geq$ 30% (Lagunas-Solar et al., 1992). Both studies concluded that inorganic sulfur formed the major pool of retained sulfur, with little or no evidence of assimilation to organic forms, such as thiol amino acids, proteins and sulfolipids. If so, this would contrast greatly with foliar assimilation, suggesting major differences in the metabolic activity of the plastids. To date in grape berries, only the ultrastructure of plastids have been presented (Fougere-Rifot et al., 1995), however, a recent study of the bioenergetics of tomato chromoplasts demonstrated significant functional rearrangement of electron transport (Renato et al., 2014), which may suggest that sulfur assimilation is also altered.

Previously, we have shown that substantial reprogramming of the grape berry transcriptome occurs after 21 days of fumigation with a commercial SO_2-generating pad, which generated up to 260 $\mu g.m^{-3}$ within 8 h of application, declining to 26 $\mu g.m^{-3}$ by 24 h and sustaining 3–8 $\mu g.m^{-3}$ for at least 8 weeks (**Figure 2B**; Giraud et al., 2012). The number of SO_2-responsive transcripts, both up- and down-regulated was several-fold larger and different from the sole or combined effects of salicylic acid or methyl jasmonate, which are both well-known elicitors of plant transcriptional response. The net transcriptome signature of sulfur assimilation suggested that oxidation to sulfate in the apoplast and peroxisome had reached a saturation point, and that sulfite was directed toward alternative paths, including conjugation, and sulfation. The data suggest that sulfur was also directed toward cysteine, methionine and particularly glutathione (**Figure 3**), as has been observed in *Arabidopsis* leaves, albeit to a limited extent (Van der Kooij et al., 1997; De Kok and Tausz, 2001). Glutathione and enzyme activities associated with glutathione metabolism, including GLUTATHIONE-S-TRANSFERASE (GST) and other thiols play important roles in plant responses and acclimation to a range of abiotic and biotic stresses. In SO_2-treated *Arabidopsis* leaves, water-soluble thiol accumulation comprised only 2% (w/w) of the assimilated sulfur (Van der Kooij et al., 1997), however, the berry differs in several ways, not least because sulfur cannot be mobilized to other organs. Previous studies have shown that "super-nutritional" levels of sulfur can enhance the innate defenses of plants and crops (Bloem et al., 2005; Kruse et al., 2007; Nakamura et al., 2009). Our transcript data showed up-regulation of several orthologs of *GST*, however, the microarray format was not completely representative of the sulfur metabolic pathways, for example lacking an ortholog of *GLUTATHIONE SYNTHETASE* (Giraud et al., 2012). That study was also limited in metabolic data, which we are currently investigating along lines of thiol metabolism.

The broader picture suggests that SO_2-fumigation may have a number of value-adding effects on the quality of the berry. In an earlier study, with comparable treatment, we'd shown that total phenolic acids and *in vitro* total antioxidant capacity were increased in SO_2-treated berries (Considine et al., 2009). Notwithstanding our reservations of *in vitro* assays of total antioxidants (Mubarak et al., 2012), the transcriptome signature suggested that anthocyanin synthesis was enhanced, as well as a number of other processes that may contribute to improved retention of berry quality postharvest, particularly preservation of texture and flavor qualities.

THE ADDED VALUES OF SO_2-FUMIGATION

Sugars, organic and amino acids, and soluble pectins are the major soluble solids in grapes. The fruit soluble solids concentration (SSC%, °Brix) and titratable acidity, together with texture are the major determinants of the fruit taste and quality. Postharvest practices implemented by the industry, however, focus on the weakest link, being infection, loss of turgor and cell wall degradation, rather than flavor, even though the taste and flavor of table grapes are key components of marketability. The transcriptome data suggest that SO_2-fumigation may have the potential to improve traits such as sugar profiles and soluble pectin content. For example, transcripts encoding grapevine orthologs of PECTIN METHYLESTERASE and PECTATE LYASE,

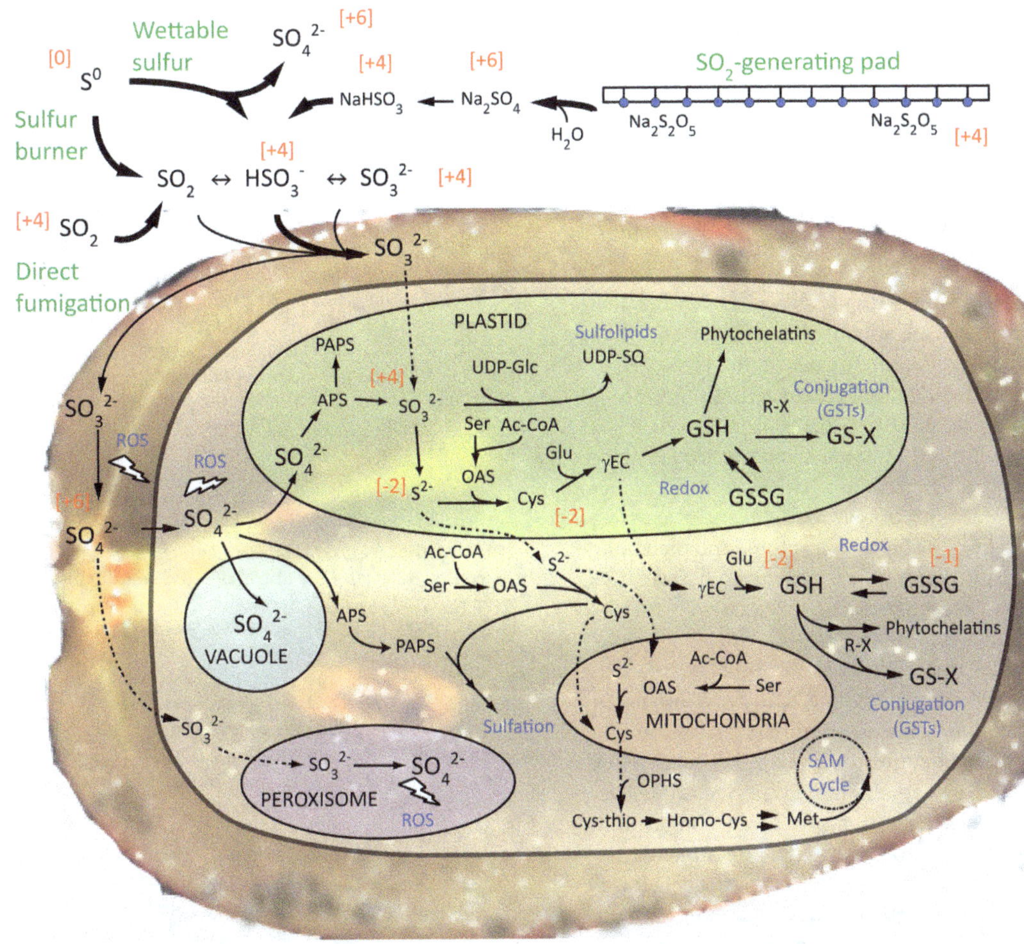

FIGURE 3 | Outline of the pathways of sulfur assimilation in the grape berry from major viticultural inputs of sulfur (S^0), SO_2 and sulfites. Atmospheric SO_2 and sulfites, shown here generated from elemental sulfur (wettable or burned), SO_2 fumigation or SO_2-generating pads, will hydrate predominantly to bisulfite (HSO_3^-) at apoplastic and subcellular pH ranges [apoplast pH5-6 (Grignon and Sentenac, 1991), cytosol, plastid stroma and nucleus *c.* pH7.2, peroxisome and mitochondria *c.* pH >8 (Shen et al., 2013) and vacuole *c.* pH3.5 (Fontes et al., 2012)]. However, for simplicity, we've only shown SO_3^{2-} within the cell. Arrow thickness and relative font size of metabolites represents hypothesized downregulation of sulfite synthesis and oxidation, and accumulation of cysteine, glutathione and related metabolites. Dotted lines represent unconfirmed transport steps. Numbers in square brackets represent oxidation states of the sulfur atom. ROS, reactive oxygen species; APS, adenosine 5′-phosphosulfate; PAPS 3′-phosphoadenosine 5′-phosphosulfate; OASi-acetyl serine; Ser, serine; Ac-CoA, acetyl co-enzyme A; UDP-Glc; UDP-glucose; UDP-SQ, UDP-sulfoquinovose; Cys, cysteine, γEC, γ-glutamylcysteine; GSH, glutathione (reduced); GSSG, glutathione disulfide (oxidized); R-X, substrate electrophile (e.g., xenobiotic, flavonoid); GS-X, glutathionylated substrate (by GSTs, GLUTATHIONE *S*-TRANSFERASES); OPHS, *O*-phosphohomoserine; Cys-thio, cystathionine; Homo-cys, homocysteine; Met, methionine; SAM, *S*-adenosylmethionine. Adapted from Giraud et al. (2012) with permission.

as well as GALACTINOL SYNTHASE, which is the first committed step in synthesis of raffinose family oligosaccharides, were increased in grape berries after SO_2-fumigation (Giraud et al., 2012). These transcripts have previously been shown to be highest late in the ripening stages of grape berries (Guillaumie et al., 2011). However, enhanced activities of these enzymes could also lead to softening of the berry and accumulation of raffinose oligosaccharides, which would tend to have a negative impact on grape quality. Nevertheless, it may be possible to maximize the effects of SO_2 fumigation to improve the outcome of current practices leading to long-term to enhanced postharvest soluble solids contents.

In relation to the wine industry, SO_2 serves several purposes postharvest, including limiting oxidation and controlling microbial populations at least until the inoculated yeast can dominate fermentation. As glutathione and thiols are widely known to be important determinants of wine sensory attributes, a major contribution of postharvest SO_2 is to maintain their stability. It is unknown to what extent the SO_2 may augment their levels through assimilation in the berry, whether from S^0 or SO_2, or through yeast assimilation.

While more in-depth studies are required to determine whether the observed changes in pectin and sugar metabolism-related transcripts are translated into effects on sugar composition, the possibility remains that such changes could result in alterations in soluble solids contents and hence improve berry quality. It would be worthwhile to explore this possibility, together with more comprehensive studies on the effects of SO_2 on the content and composition of secondary metabolites. For example, there is little evidence to date that anthocyanin synthesis is changed as a result of SO_2 fumigation (Giraud et al., 2012) although the lower abundance of flavan-3-ol transcripts after SO_2 fumigation suggest that anthocyanins are not degraded as rapidly. Further studies are required to explore such possibilities, together with the effects of the duration of SO_2 exposure on quality-linked traits such as tannin contents in wine and table grapes.

CONCLUSION

Evidence suggests that there are not only differences in the susceptibility of different plant species to SO_2, but variations in the effects of SO_2 on the different organs of the same plant. While gaps in current knowledge remain concerning the mechanisms that prevent SO_2-induced damage in some tissues but not others, the marked contrast in the metabolic consequences of SO_2-exposure in photosynthetic and non-photosynthetic tissues suggests that in the absence of photosynthesis plant organs are highly tolerant to SO_2. The available transcriptome and metabolic data from leaves and fruit demonstrate that in both vegetative and reproductive organs atmospheric SO_2 is preferentially metabolized to SO_4^{2-}. The high requirement of SO_2 metabolism for cellular reductants results in an increases in cellular oxidation. The resultant shift in cellular redox state that provokes much of the broader transcriptional reprogramming that is observed in leaves and berries, in an attempt to restore the cellular redox balance. At levels that are currently used in grapevine industries, SO_2 appears to have beneficial effects on quality, and more importantly to industry, does not appear to be damaging, or to compromise quality. In contrast, leaves have a much lower threshold of sensitivity that is orders of magnitude lower than the fruit, in terms of the potential to induce damage. This differential sensitivity is not just due to variations in the physical structure of the two tissue types in terms of the barriers to diffusion but also to functional organization of the plastid, particularly the operation of the photosynthetic electron transport chain in the thylakoid membrane together with the highly redox-sensitive enzymes of carbon assimilation and associated metabolism. However, such factors might not form the basis for a large distinction between table and wine grapes, in which there is photosynthetic activity at least during the first growth phase because of the limitations imposed by limited stomatal numbers and conduction on SO_2 penetration into the photosynthetic cells.

AUTHOR CONTRIBUTIONS

MC and CF co-wrote the manuscript.

ACKNOWLEDGMENTS

CF thanks BBSRC (UK) for financial support (BB/K010476/1). MC acknowledges the support of a grant funded by the Australian Research Council with co-investment from the Department of Agriculture and Food Western Australia and Gascoyne Table Grape Growers' Association (LP0990355). MC would also like to thank in-kind contributions of fruit from Peter Nuich (Swan Valley, Western Australia) and advice from, E/Prof John "jac" Considine (UWA), Colin Gordon (DAFWA, Western Australia), Carolyn Adams (GrapeTek Pty Ltd, South Africa) and Gill Ambler (FreshTec Pty Ltd, Australia), and Don Luvisi (University of California Cooperative Extension).

REFERENCES

Baumes, R. (2009). "Wine aroma precursors," in *Wine Chemistry and Biochemistry*, eds M. V. Moreno-Arribas and M. C. Polo (New York, NY: Springer), 251–274.

Bloem, E., Haneklaus, S., and Schnug, E. (2005). Significance of sulfur compounds in the protection of plants against pests and diseases. *J. Plant Nutr.* 28, 763–784. doi: 10.1081/pln-200055532

Boulton, R., Singleton, V., Bisson, L., and Kunkee, R. (1999). "The role of sulfur dioxide in wine," in *Principles and Practices of Winemaking* (New York, NY: Springer), 448–473.

Brychkova, G., Grishkevich, V., Fluhr, R., and Sagi, M. (2013). An essential role for tomato sulfite oxidase and enzymes of the sulfite network in maintaining leaf sulfite homeostasis. *Plant Physiol.* 161, 148–164. doi: 10.1104/pp.112.208660

C.R.C. (2014). "Physical constants of inorganic compounds," in *CRC Handbook of Chemistry and Physics, 95th Edition (Internet Version)*, ed. W. M. Haynes (Boca Raton: CRC Press/Taylor and Francis), 4–92.

Chemical Abstracts Service. (2014). *Sulfur Dioxide [Online]*. American Chemical Society. Available at: https://scifinder.cas.org (accessed July 29, 2014).

Clean Air Initiative. (2010). *Sulfur Dioxide (SO_2) Standards in Asia. CAI-Asia Factsheet No. 4 [Online]*. Available at: www.cleanairinitiative.org/ (accessed November 20, 2014).

Clemes, D. C. (1986). Packing material which releases sulfur dioxide in the presence of moisture. *Pat. Appl. ZA* 1985–4059.

Codex Alimentarius Commission. (2014). "Sulfites—04.1.1.2, surface-treated fresh fruit," in *Codex General Standard for Food Additives (GSFA) Online Database [Online]*. Available at: http://www.codexalimentarius.net/gsfaonline/groups/details.html?id=161 (accessed November 18, 2014).

Considine, M. J., Gordon, C., Ching, S., and Croft, K. D. (2009). Salicylic acid overrides the effect of methyl jasmonate on the total antioxidant capacity in table grapes. *Acta Hortic.* 841, 495–495.

Crisosto, C. H., and Smilanick, J. L. (2014). "Grape (Table)" in *The Commercial Storage of Fruits, Vegetables, and Florist and Nursery Stocks [Online], HB66*. Available at: http://www.ba.ars.usda.gov/hb66/grapeTable.pdf (accessed November 18, 2014).

Davidian, J.-C., and Kopriva, S. (2010). Regulation of sulfate uptake and assimilation—the same or not the same? *Mol. Plant* 3, 314–325. doi: 10.1093/mp/ssq001

De Kok, L. J., Durenkamp, M., Yang, L., and Stulen, I. (2007). "Atmospheric sulfur," in *Sulfur in Plants—and Ecological Perspective*, eds M. J. Hawkesford and L. J. De Kok (Dordrecht: Springer), 91–106.

De Kok, L. J., and Tausz, M. (2001). "The role of glutathione in plant reaction and adaptation to air pollutants," in *Significance of Glutathione in Plant Adaptation to the Environment*, eds D. Grill, M. Tausz, and L. J. De Kok (Netherlands: Kluwer Academic Publishers), 185–205.

Dominguez-Solis, J. R., He, Z., Lima, A., Ting, J., Buchanan, B. B., and Luan, S. (2008). A cyclophilin links redox and light signals to cysteine biosynthesis and stress responses in chloroplasts. *Proc. Natl. Acad. Sci. U.S.A.* 105, 16386–16391. doi: 10.1073/pnas.0808204105

E.P.A. (2014). *National Trends in Sulfur Dioxide Levels; SO_2 Air Qualitiy 1980–2013 (Annual 99th Percentile of Daily Max 1-Hour Average) National Trend based on 47 Sites*. Available at: www.epa.gov/airtrends/sulfur.html.(accessed November 22, 2014).

F.S.A.N.Z. (2012). *National Coordinated Survey of Sulphite Levels in Sausages, Cordial and Dried Fruit*. Canberra: Food Standards Australia New Zealand.

Fontes, N., Delrot, S., and Geros, H. (2012). "Grape cell vacuoles: structure-function and solute transport across the tonoplast," in *The Biochemistry of the Grape Berry*, eds H. Geros, M. M. Chaves, and S. Delrot (Dubai, UAE: Bentham Books), 160–171.

Fougere-Rifot, M., Park, H. S., Cherrad, M., Darne, G., and Bouard, J. (1995). Plastids, starch and chlorophyll in *Vitis vinifera* L. berry during its ontogenesis. *J. Int. Sci. Vigne Vin* 29, 213–225.

Gayler, K. R., and Sykes, G. E. (1985). Effects of nutritional stress on the storage proteins of soybean. *Plant Physiol.* 78, 582–585. doi: 10.1104/pp.78.3.582

Giraud, E., Ivanova, A., Gordon, C. S., Whelan, J., and Considine, M. J. (2012). Sulphur dioxide evokes a large scale reprogramming of the grape berry transcriptome associated with oxidative signalling and biotic defence responses. *Plant Cell Environ.* 35, 405–417. doi: 10.1111/j.1365-3040.2011.02379.x

Grignon, C., and Sentenac, H. (1991). pH and ionic conditions in the apoplast. *Annu. Rev. Plant Physiol. Plant Mol. Biol.* 42, 103–128. doi: 10.1146/annurev.pp.42.060191.000535

Guillaumie, S., Fouquet, R., Kappel, C., Camps, C., Terrier, N., Moncomble, D., et al. (2011). Transcriptional analysis of late ripening stages of grapevine berry. *BMC Plant Biol.* 11:165. doi: 10.1186/1471-2229-11-165

Haas, F. H., Heeg, C., Queiroz, R., Bauer, A., Wirtz, M., and Hell, R. (2008). Mitochondrial serine acetyltransferase functions as a pacemaker of cysteine synthesis in plant cells. *Plant Physiol.* 148, 1055–1067. doi: 10.1104/pp.108.125237

Hamisch, D., Randewig, D., Schliesky, S., Bräutigam, A., Weber, A. P. M., Geffers, R., et al. (2012). Impact of SO_2 on *Arabidopsis thaliana* transcriptome in wildtype and sulfite oxidase knockout plants analyzed by RNA deep sequencing. *New Phytol.* 196, 1074–1085. doi: 10.1111/j.1469-8137.2012.04331.x

Hänsch, R., Lang, A., Rennenberg, H., and Mendel, R. R. (2007). Significance of plant sulfite oxidase. *Plant Biol.* 9, 589–595. doi: 10.1055/s-2007-965433

Harms, K., Von Ballmoos, P., Brunold, C., Höfgen, R., and Hesse, H. (2000). Expression of a bacterial serine acetyltransferase in transgenic potato plants leads to increased levels of cysteine and glutathione. *Plant J.* 22, 335–343. doi: 10.1046/j.1365-313x.2000.00743.x

Hartmann, T., Hönicke, P., Wirtz, M., Hell, R., Rennenberg, H., and Kopriva, S. (2004). Regulation of sulphate assimilation by glutathione in poplars (*Populus tremula* × P. *alba*) of wild type and overexpressing γ-glutamylcysteine synthetase in the cytosol. *J. Exp. Bot.* 55, 837–845. doi: 10.1093/jxb/erh094

Heber, U., and Huve, K. (1998). Action of SO_2 on plants and metabolic detoxification of SO_2. *Int. Rev. Cytol.* 177, 255–286. doi: 10.1016/S0074-7696(08)62234-2

Hinckley, E.-L., Fendorf, S., and Matson, P. (2011). Short-term fates of high sulfur inputs in Northern California vineyard soils. *Nutr. Cycl. Agroecosyst.* 89, 135–142. doi: 10.1007/s10705-010-9383-3

Hinckley, E.-L., and Matson, P. A. (2011). Transformations, transport, and potential unintended consequences of high sulfur inputs to Napa Valley vineyards. *Proc. Natl. Acad. Sci. U.S.A.* 108, 14005–14010. doi: 10.1073/pnas.1110741108

Honsel, A., Kojima, M., Haas, R., Frank, W., Sakakibara, H., Herschbach, C., et al. (2012). Sulphur limitation and early sulphur deficiency responses in poplar: significance of gene expression, metabolites, and plant hormones. *J. Exp. Bot.* 63, 1873–1893. doi: 10.1093/jxb/err365

Kawashima, C. G., Berkowitz, O., Hell, R., Noji, M., and Saito, K. (2005). Characterization and expression analysis of a serine acetyltransferase gene family involved in a key step of the sulfur assimilation pathway in Arabidopsis. *Plant Physiol.* 137, 220–230. doi: 10.1104/pp.104.045377

Klimont, Z., Smith, S. J., and Cofala, J. (2013). The last decade of global anthropogenic sulfur dioxide: 2000–2011 emissions. *Environ. Res. Lett.* 8, 014003. doi: 10.1088/1748-9326/8/1/014003

Koprivova, A., and Kopriva, S. (2014). Molecular mechanisms of regulation of sulfate assimilation: first steps on a long road. *Front. Plant Sci.* 5:589. doi: 10.3389/fpls.2014.00589

Kruse, C., Jost, R., Lipschis, M., Kopp, B., Hartmann, M., and Hell, R. (2007). Sulfur-enhanced defence: effects of sulfur metabolism, nitrogen supply, and pathogen lifestyle. *Plant Biol.* 9, 608–619. doi: 10.1055/s-2007-965432

Kwasniewski, M. T., Sacks, G. L., and Wilcox, W. F. (2014). Persistence of elemental sulfur spray residue on grapes during ripening and vinification. *Am. J. Enol Vitic.* 65, 453–462. doi: 10.5344/ajev.2014.14027

Lagunas-Solar, M. C., Demateo, A., Fernandez, J. E., Oyarzun, J. I., Carvacho, O. F., Arancibia, R. A., et al. (1992). Radiotracer studies on the uptake and retention (conversion) of 35S-sulfur dioxide in table grapes. *Am. J. Enol. Vitic.* 43, 266–274.

Legris-Delaporte, S., Ferron, F., Landry, J., and Costes, C. (1987). Metabolization of elemental sulfur in wheat leaves consecutive to its foliar application. *Plant Physiol.* 85, 1026–1030. doi: 10.1104/pp.85.4.1026

Li, L., Yi, H., Wang, L., and Li, X. (2008). Effects of sulfur dioxide on the morphological and physiological biochemical parameters in *Arabidopsis thaliana* plants. *J. Agro Envrion. Sci.* 27, 525–529.

Liedschulte, V., Wachter, A., Zhigang, A., and Rausch, T. (2010). Exploiting plants for glutathione (GSH) production: uncoupling GSH synthesis from cellular controls results in unprecedented GSH accumulation. *Plant Biot. J.* 8, 807–820. doi: 10.1111/j.1467-7652.2010.00510.x

Luvisi, D. (2014). *Sulfur Dioxide Practices in California.* Personal Communication 14/11/2014. Hollister, CA: University of California Cooperative Extension.

Maruyama-Nakashita, A., Inoue, E., Watanabe-Takahashi, A., Yamaya, T., and Takahashi, H. (2003). Transcriptome profiling of sulfur-responsive genes in Arabidopsis reveals global effects of sulfur nutrition on multiple metabolic pathways. *Plant Physiol.* 132, 597–605. doi: 10.1104/pp.102.019802

Mubarak, A., Swinny, E. E., Ching, S. Y. L., Jacob, S. R., Lacey, K., Hodgson, J. M., et al. (2012). Polyphenol composition of plum selections in relation to total antioxidant capacity. *J. Agric. Food Chem.* 60, 10256–10262. doi: 10.1021/jf302903k

Muttucumaru, N., Halford, N. G., Elmore, J. S., Dodson, A. T., Parry, M., Shewry, P. R., et al. (2006). Formation of high levels of acrylamide during the processing of flour derived from sulfate-deprived wheat. *J. Agric. Food Chem.* 54, 8951–8955. doi: 10.1021/jf0623081

Naito, S., Hirai, M. Y., Chino, M., and Komeda, Y. (1994). Expression of a soybean (*Glycine max* [L.] Merr.) seed storage protein gene in transgenic *Arabidopsis thaliana* and its response to nutritional stress and to abscisic acid mutations. *Plant Physiol.* 104, 497–503. doi: 10.1104/pp.104.2.497

Nakamura, M., Kuramata, M., Kasugai, I., Abe, M., and Youssefian, S. (2009). Increased thiol biosynthesis of transgenic poplar expressing a wheat *O*-acetylserine(thiol) lyase enhances resistance to hydrogen sulfide and sulfur dioxide toxicity. *Plant Cell Rep.* 28, 313–323. doi: 10.1007/s00299-008-0635-5

Nelson, K. E., and Baker, G. A. (1963). Studies on the sulfur dioxide fumigation of table grapes. *Am. J. Enol. Vitic.* 14, 13–22.

Nikiforova, V., Freitag, J., Kempa, S., Adamik, M., Hesse, H., and Hoefgen, R. (2003). Transcriptome analysis of sulfur depletion in *Arabidopsis thaliana*: interlacing of biosynthetic pathways provides response specificity. *Plant J.* 33, 633–650. doi: 10.1046/j.1365-313X.2003.01657.x

Noctor, G., Mhamdi, A., Chaouch, S., Han, Y. I., Neukermans, J., Marquez-Garcia, B., et al. (2012). Glutathione in plants: an integrated overview. *Plant Cell Environ.* 35, 454–484. doi: 10.1111/j.1365-3040.2011.02400.x

Noji, M., and Saito, K. (2002). Molecular and biochemical analysis of serine acetyltransferase and cysteine synthase towards sulfur metabolic engineering in plants. *Amino Acids* 22, 231–243. doi: 10.1007/s007260200011

Palliotti, A., and Cartechini, A. (2001). Developmental changes in gas exchange activity in flowers, berries, and tendrils of field-grown Cabernet Sauvignon. *Am. J. Enol. Vitic.* 52, 317–323.

Palou, L., Crisosto, C., Garner, D., Basinal, L., Smilanick, J., and Zoffoli, J. (2002). Minimum constant sulfur dioxide emission rates to control gray mold of cold-stored table grapes. *Am. J. Enol. Vitic.* 53, 110–115.

Peiser, G. D., and Yang, S. F. (1985). "Biochemical and physiological effects of SO_2 on nonphotosynthetic processes in plants," in *Sulfur Dioxide and Vegetation*, eds W. E. Winner, H. A. Mooney, and R. A. Goldstein (Stanford, CA: Stanford University Press), 148–161.

Queval, G., and Noctor, G. (2007). A plate reader method for the measurement of NAD, NADP, glutathione, and ascorbate in tissue extracts: application to redox profiling during *Arabidopsis* rosette development. *Anal. Biochem.* 363, 58–69. doi: 10.1016/j.ab.2007.01.005

Randewig, D., Hamisch, D., Herschbach, C., Eiblmeier, M., Gehl, C., Jurgeleit, J., et al. (2012). Sulfite oxidase controls sulfur metabolism under SO_2 exposure in *Arabidopsis thaliana. Plant Cell Environ.* 35, 100–115. doi: 10.1111/j.1365-3040.2011.02420.x

Rankine, B. (1963). Nature, origin and prevention of hydrogen sulphide in wine fermentation. *J. Sci. Food Agric.* 14, 79–91. doi: 10.1002/jsfa.2740140204

Renato, M., Pateraki, I., Boronat, A., and Azcón-Bieto, J. (2014). Tomato fruit chromoplasts behave as respiratory bioenergetic organelles during ripening. *Plant Physiol.* 166, 920–933. doi: 10.1104/pp.114.243931

Rennenberg, H. (1984). The fate of excess sulfur in higher plants. *Annu. Rev. Plant Physiol.* 35, 121–153. doi: 10.1146/annurev.pp.35.060184.001005

Rennenberg, H., and Herschbach, C. (2014). A detailed view on sulphur metabolism at the cellular and whole-plant level illustrates challenges in metabolite flux analyses. *J. Exp. Bot.* 65, 5711–5724. doi: 10.1093/jxb/eru315

Robinson, J. B. (1988). "Grapevine nutrition," in *Viticulture* Vol. 2 *Practices*, eds B. G. Coombe and P. R. Dry (Adelaide: Winetitles), 178–208.

Rogiers, S. Y., Hatfield, J. M., Jaudzems, V. G., White, R. G., and Keller, M. (2004). Grape verry cv. Shiraz epicuticular wax and transpiration during ripening and preharvest weight loss. *Am. J. Enol. Vitic.* 55, 121–127.

Schulte, M., Von Ballmoos, P., Rennenberg, H., and Herschbach, C. (2002). Life-long growth of *Quercus ilex* L. at natural CO_2 springs acclimates sulphur, nitrogen and carbohydrate metabolism of the progeny to elevated pCO_2. *Plant Cell Environ.* 25, 1715–1727. doi: 10.1046/j.1365-3040.2002.00948.x

Shen, J., Zeng, Y., Zhuang, X., Sun, L., Yao, X., Pimpl, P., et al. (2013). Organelle pH in the *Arabidopsis* endomembrane system. *Mol. Plant* 6, 1419–1437. doi: 10.1093/mp/sst079

Shewry, P. R., Franklin, J., Parmar, S., Smith, S. J., and Miflin, B. J. (1983). The effects of sulphur starvation on the amino acid and protein compositions of barley grain. *J. Cereal Sci.* 1, 21–31. doi: 10.1016/S0733-5210(83)80005-8

Smith, I. K., Kendall, A. C., Keys, A. J., Turner, J. C., and Lea, P. J. (1985). The regulation of the biosynthesis of glutathione in leaves of barley (*Hordeum vulgare* L.). *Plant Sci.* 41, 11–17. doi: 10.1016/0168-9452(85)90059-7

State Environmental Protection Administration. (2013). "Atmospheric environment," in *SOE 2011 [Online]*. Available at: http://english.mep.gov.cn/standards_reports/soe/soe2011/201307/t20130712_255427.htm (accessed November 19, 2014).

Takahashi, H., Kopriva, S., Giordano, M., Saito, K., and Hell, R. (2011). Sulfur assimilation in photosynthetic organisms: molecular functions and regulations of transporters and assimilatory enzymes. *Annu. Rev. Plant Biol.* 62, 157–184. doi: 10.1146/annurev-arplant-042110-103921

U.S. Food and Drug Administration. (2014). "Part 101—Food labeling, subpart G-exemptions from food labeling requirements, Sec. 101.100 Food; exemptions from labeling," in *Code of Federal Regulations [Online]*. Available at: http://www.accessdata.fda.gov/scripts/cdrh/cfdocs/cfcfr/CFRSearch.cfm?fr=101.100 (accessed November 18, 2014).

Van der Kooij, T. A. W., De Kok, L. J., Haneklaus, S., and Schnug, E. (1997). Uptake and metabolism of sulphur dioxide by *Arabidopsis thaliana*. *New Phytol.* 135, 101–107. doi: 10.1046/j.1469-8137.1997.00619.x

W.H.O. (2006). *W.H.O. Air Quality Guidelines for Particulate Matter, Ozone, Nitrogen Dioxide and Sulfur Dioxide. Global Update 2005 [Online]*. Geneva: World Health Organization. Available at: http://whqlibdoc.who.int/hq/2006/WHO_SDE_PHE_OEH_06.02_eng.pdf?ua=1 (accessed November 17, 2014).

Watanabe, M., Mochidab, K., Katoc, T., Tabatac, S., Yoshimotoa, N., Nojid, M., et al. (2008). Comparative genomics and reverse genetics analysis reveal indispensable functions of the serine acetyltransferase gene family in *Arabidopsis*. *Plant Cell* 20, 2484–2496. doi: 10.1105/tpc.108.060335

Williams, J. S., and Cooper, R. M. (2004). The oldest fungicide and newest phytoalexin—a reappraisal of the fungitoxicity of elemental sulphur. *Plant Pathol.* 53, 263–279. doi: 10.1111/j.0032-0862.2004.01010.x

Wirtz, M., and Hell, R. (2007). Dominant-negative modification reveals the regulatory function of the multimeric cysteine synthase protein complex in transgenic tobacco. *Plant Cell* 19, 625–639. doi: 10.1105/tpc.106.043125

Zhao, F.-J., Tausz, M., and De Kok, L. J. (2008). "Role of sulfur for plant production in agricultural and natural ecosystems," in *Sulfur Metabolism in Phototrophic Organisms*, eds R. Hell, C. Dahl, D. B. Knaff, and T. Leustek (Dordrecht: Springer), 417–435.

Zhao, J., and Yi, H. (2014). Genome-wide transcriptome analysis of *Arabidopsis* response to sulfur dioxide fumigation. *Mol. Genet Genomics* 289, 989–999. doi: 10.1007/s00438-014-0870-0

Conflict of Interest Statement: The authors declare that the research was conducted in the absence of any commercial or financial relationships that could be construed as a potential conflict of interest.

20

The significance of cysteine synthesis for acclimation to high light conditions

*Anna Speiser[1], Stefan Haberland[1], Mutsumi Watanabe[2], Markus Wirtz[1], Karl-Josef Dietz[3], Kazuki Saito[4,5] and Rüdiger Hell[1]**

[1] Plant Molecular Biology, Centre for Organismal Studies, University of Heidelberg, Heidelberg, Germany
[2] Molecular Plant Physiology, Max Planck Institute of Molecular Plant Physiology, Potsdam, Germany
[3] Plant Biochemistry and Physiology, University of Bielefeld, Bielefeld, Germany
[4] RIKEN Center for Sustainable Resource Science, Yokohama, Japan
[5] Graduate School of Pharmaceutical Sciences, Chiba University, Chiba, Japan

Edited by:
Stanislav Kopriva, University of Cologne, Germany

Reviewed by:
Uener Kolukisaoglu, University of Tuebingen, Germany
Luis C. Romero, Consejo Superior de Investigaciones Científicas, Spain

***Correspondence:**
Rüdiger Hell, Plant Molecular Biology, Centre for Organismal Studies, Im Neuenheimer Feld 360, 69115 Heidelberg, Germany
e-mail: ruediger.hell@cos.uni-heidelberg.de

Situations of excess light intensity are known to result in the emergence of reactive oxygen species that originate from the electron transport chain in chloroplasts. The redox state of glutathione and its biosynthesis contribute importantly to the plant's response to this stress. In this study we analyzed the significance of cysteine synthesis for long-term acclimation to high light conditions in *Arabidopsis thaliana*. Emphasis was put on the rate-limiting step of cysteine synthesis, the formation of the precursor O-acetylserine (OAS) that is catalyzed by serine acetyltransferase (SERAT). Wild type Arabidopsis plants responded to the high light condition (800 μmol m^{-2} s^{-1} for 10 days) with synthesis of photo-protective anthocyanins, induction of total SERAT activity and elevated glutathione levels when compared to the control condition (100 μmol m^{-2} s^{-1}). The role of cysteine synthesis in chloroplasts was probed in mutant plants lacking the chloroplast isoform SERAT2;1 (*serat2;1*) and two knock-out alleles of CYP20-3, a positive interactor of SERAT in the chloroplast. Acclimation to high light resulted in a smaller growth enhancement than wild type in the *serat2;1* and *cyp20-3* mutants, less induction of total SERAT activity and OAS levels but similar cysteine and glutathione concentrations. Expression analysis revealed no increase in mRNA of the chloroplast SERAT2;1 encoding *SERAT2;1* gene but up to 4.4-fold elevated *SERAT2;2* mRNA levels for the mitochondrial SERAT isoform. Thus, lack of chloroplast SERAT2;1 activity or its activation by CYP20-3 prevents the full growth response to high light conditions, but the enhanced demand for glutathione is likely mediated by synthesis of OAS in the mitochondria. In conclusion, cysteine synthesis in the chloroplast is important for performance but is dispensable for survival under long-term exposure to high light and can be partially complemented by cysteine synthesis in mitochondria.

Keywords: high light stress, SERAT, CYP20-3, chloroplasts, mitochondria, glutathione

INTRODUCTION

Reactive oxygen species (ROS) play a dual role in plants since they function as regulators and if excessively produced as harmful reactive metabolites. Protein thiols are particularly sensitive to redox regulation, but also to oxidative damage by ROS (Meyer and Hell, 2005). Formation of ROS continuously occurs at low rates during photosynthetic electron transport. However, exposure of plants to high light stimulates release of ROS such as singlet oxygen, superoxide and hydrogen peroxide from the electron transport chain and may cause oxidative stress (Barber and Andersson, 1992; Fryer et al., 2002). The effects on growth of high light and the associated excess excitation energy in chloroplasts have been discussed in literature extensively (Foyer et al., 2009; Suzuki et al., 2012; Szechyńska-Hebda and Karpiński, 2013). Among these responses are morphological changes of the leaves (Eckardt et al., 1997), accumulation of anthocyanins as scavengers of evolving ROS (Chalker-Scott, 1999; Gould et al., 2002; Vanderauwera et al., 2005; Zeng et al., 2010) and the expression of high light-induced genes (Alsharafa et al., 2014). A well-characterized excess light-induced gene is *APX2* that encodes for ascorbate-peroxidase 2, as part of the ascorbate-glutathione ROS detoxification cycle (Rossel et al., 2006; Foyer and Noctor, 2011; Noctor et al., 2011).

Reduced glutathione and its oxidized form represent the most abundant low-molecular weight thiol redox couple found in eukaryotes and prokaryotes and plays a crucial role in adjustment of cellular redox potential and signaling of ROS stress (May et al., 1998; Noctor and Foyer, 1998; Rouhier et al., 2008). Changes of the ratio of reduced to oxidized glutathione and gene expression and activity of glutathione reductase (GR) in response to oxidative stress have frequently been reported (Noctor et al., 2011; Chan et al., 2013). Reduction of oxidized glutathione takes place in the plastids, mitochondria, cytosol and peroxisomes and is so essential for survival of plants that even back-up systems have evolved by NADPH-dependent thioredoxin reductases (Marty et al., 2009). Increases of the steady-state level of the glutathione pool in response to different light conditions are reportedly relatively small (Noctor et al., 2012), but excess oxidized glutathione has been hypothesized to be removed from the cytosolic pool and directed to the vacuole for degradation by γ-glutamyltransferase to recycle cysteine (Grzam et al., 2007; Noctor et al., 2012).

The functional component of glutathione is cysteine that is synthesized from sulfide, the endproduct of the reductive sulfate assimilation pathway in plastids, and from *O*-acetylserine (OAS) by the enzyme *O*-acetylserine (thiol) lyase (OAS-TL). OAS is synthesized from serine and acetyl coenzyme A by serine acetyltransferase (SERAT) whose activity limits the overall rate of cysteine synthesis (Hell and Wirtz, 2011). In *Arabidopsis thaliana* the SERAT protein family is encoded by five isoforms that localize to the cytosol, the mitochondria and the plastids. In unstressed leaves of Arabidopsis the major source of OAS synthesis (79%) are the mitochondria (SERAT2;2) (Haas et al., 2008; Watanabe et al., 2008b), whereas cytosolic SERAT1;1 and chloroplastidic SERAT2;1 contribute 9 and 13%, respectively, of total SERAT activity (Watanabe et al., 2008b). Transcriptional changes in *SERAT* expression levels have several times been observed. Howarth et al. (2003) reported kinetic changes in mRNA abundance of all three major SERAT genes in response to cadmium exposure in leaf and root. *SERAT2;2* and *SERAT2;1* transcripts were increased in roots by treatment with the oxidizing reagent menadione (Lehmann et al., 2009). Transfer of a catalase-deficient mutant (*cat2*) from high CO_2 concentration to a normal environment resulted in a strong (up to 10-fold) induction of plastidic *SERAT2;1* transcript levels (Queval et al., 2009). Correspondingly, public gene expression databases show up to 15-fold elevations of *SERAT2;1* expression upon H_2O_2 treatments (Genevestigator® V3; Hruz et al., 2008; https://genevestigator.com/gv/plant.jsp). With respect to transcription levels and specific activities the cytosolic isoforms SERAT3;1 and SERAT3;2 are considered to be the minor isoforms (reviewed in Hell and Wirtz, 2011).

SERAT activity in the three compartments of cysteine synthesis is modulated *in vivo* by two post-transcriptional mechanisms. In the first mechanism, SERAT and OAS-TL proteins reversibly interact with each other and form the hetero-oligomeric cysteine synthase complex (CSC). Formation and dissociation of the complex constitute a metabolic regulatory model, positioning the CSC as a sensor for cellular sulfur homeostasis (Hell and Wirtz, 2011). Sulfide stabilizes the complex and prevents SERAT from feedback inhibition by cysteine (Wirtz et al., 2012) whereas in the absence of sulfide high OAS concentrations dissociate the complex and thereby adjusting SERAT activity and thus OAS production depending on the actual sulfide supply (Hell and Wirtz, 2011). In the second mechanism SERAT2;1 is postulated to be activated by CYP20-3 mediated CSC association in the stroma (Dominguez-Solis et al., 2008; Park et al., 2013).

Cyclophilins (CYPs) belong to the superfamily of immunophilins, also including FK506- and rapamycin-binding proteins (FKBPs). Both groups of proteins harbor a peptidyl-prolyl *cis/trans* isomerase activity, favoring and accelerating the *cis/trans* transition of peptidyl-prolyl bonds during folding and multimerization of proteins and thus comprising foldase- and chaperone-like functions (Wang and Heitman, 2005). In the Arabidopsis genome 52 immunophilin-encoding genes were identified of which 23 are putative FKBPs and 29 are putative CYP proteins (He et al., 2004; Romano et al., 2004). Subcellular localization studies reveal that 13 CYPs are in the cytosol (Chou and Gasser, 1997), three CYPs are present in the nucleus (Romano et al., 2004), and two CYPs harbor a mitochondrial localization motif. Arabidopsis contains further six genes encoding plastidic CYPs of which only one isoform localizes to the stroma (CYP20-3) (Lippuner et al., 1994), whereas the remaining five CYPs are targeted to the thylakoid lumen. CYP20-3 is the sixth most abundant protein in the stroma (Lippuner et al., 1994; Peltier et al., 2006), suggesting multiple targets for its foldase and chaperone activities. It contains two internal disulfide bonds and its activity was shown to be dependent on thioredoxin-triggered and thus redox-related conformational changes (Motohashi et al., 2003; Laxa et al., 2007).

Investigations of the Arabidopsis *cyp20-3* (= *roc4*; rotamase CYP 4) mutant found a growth retardation under continuous elevated light exposure (up to 1000 μmol m^{-2} s^{-1}) and attributed one CYP20-3 function to the repair of photodamaged photosystem II (Cai et al., 2008). Enhanced light intensity (up to 300 μmol m^{-2} s^{-1}) and other ROS-inducing conditions (Dominguez-Solis et al., 2008) caused significant growth reduction of the same *cyp20-3* mutant line (Dominguez-Solis et al., 2008). CYP20-3 was shown to interact with SERAT2;1, suggesting some sort of activation or stabilization of the enzyme or the cysteine synthase complex. A mechanism was proposed in which the chloroplast 2-Cys peroxiredoxin may oxidize CYP20-3 and photoreduced thioredoxin then reduces and activates CYP20-3, which in turn would promote cysteine synthesis by its protein-folding capabilities (Dominguez-Solis et al., 2008). This model was extended based on the binding of 12-oxo-phytodienoic acid (OPDA), the precursor of jasmonic acid with independent hormone function, to CYP20-3. Facilitated interaction of CYP20-3 to SERAT2;1 by OPDA was concluded as a key step in hormonal signaling toward cellular redox homeostasis in stress responses (Park et al., 2013).

In this study we investigated whether cysteine synthesis in the chloroplast has a specific role for high light acclimation in Arabidopsis, since it provides the thiol-harboring building block for glutathione. Reasoning that the lack of SERAT2;1 protein in plastids of a null mutant should result in similar or even more severe stress phenotypes compared to CYP20-3 loss of function mutants, we analyzed growth patterns, stress symptoms, cysteine-related metabolites and *SERAT* gene expression. Thus, this study addresses the important question of stress-related redox homeostasis in context of site-specific activity of cysteine synthase complexes.

RESULTS

GENOMIC CHARACTERIZATION OF *cyp20-3* T-DNA INSERTION LINES

Two different T-DNA insertion lines of the *CYP20-3* gene as well as one of the *serat2;1* mutant, lacking plastidic SERAT (Watanabe et al., 2008b) were chosen to examine the effects of high light

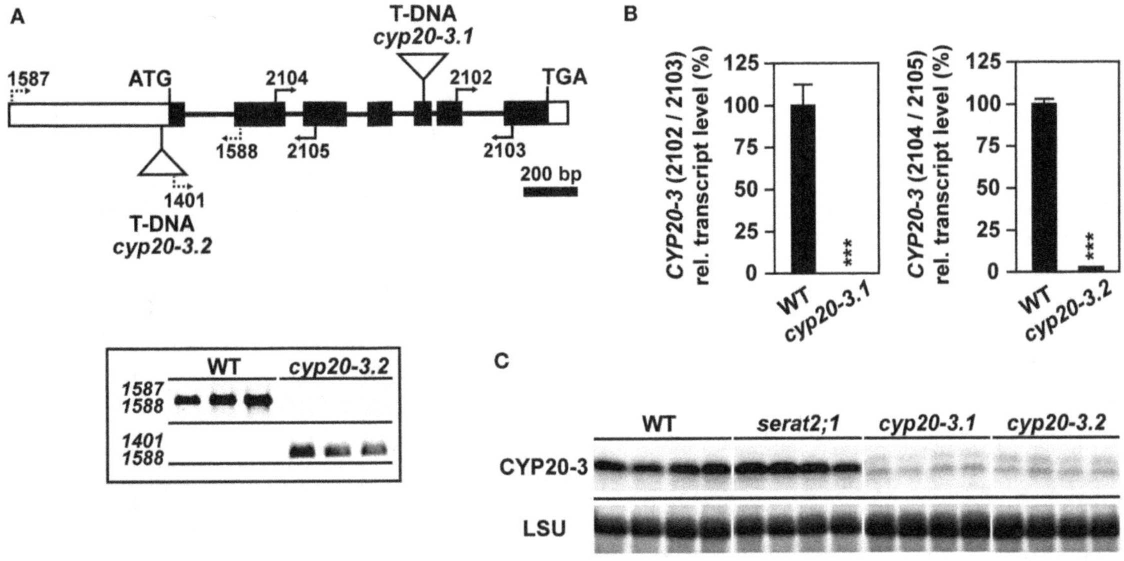

FIGURE 1 | Identification, molecular characterization and gene expression analysis of *cyp20-3* mutant lines. (A) Gene model of the *Arabidopsis thaliana* gene *CYP20-3* (AT3G62030). Exons and introns are illustrated as black boxes and thin lines, respectively. White boxes represent 5′- and 3′-UTRs. Triangles define the two T-DNA insertions of *cyp20-3.1* and *cyp20-3.2*, respectively. Black arrows and numbering indicate the positions of used primers. Molecular verification of *cyp20-3.2* by PCR using specific primer combinations for the wild type (WT) allele (upper panel) and mutant allele (lower panel). **(B)** *CYP20-3* gene expression levels by qRT-PCR with *cyp20-3.1* and *cyp20-3.2* cDNA and indicated primer combinations. WT transcript levels were set to 100%. Statistically significant differences are indicated as asterisks (*** $p \leq 0.001$; Student's *t*-test). Results represent means ± *SE* ($n = 3$). **(C)** Immunological detection of CYP20-3 using a polyclonal antibody against Arabidopsis CYP20-3 in leaf protein extracts of WT, *serat2;1*, *cyp20-3.1* and *cyp20-3.2* plants (four lanes each) grown under control conditions. Staining intensities of the large subunit of ribulose-1,5-bisphosphate carboxylase/oxygenase (LSU) protein in the same samples confirm equal loading in the individual lanes.

on growth and cysteine synthesis. T-DNA positions and primer binding sites which were used for the molecular characterization of both *cyp20-3* mutants are depicted in **Figure 1A**. One mutant line (SALK_001615) harbors a T-DNA insertion in the fifth exon and had been previously described as a full knock-out of the expression of the *CYP20-3* gene (Cai et al., 2008; Dominguez-Solis et al., 2008). In these studies this mutant allele was named *cyp20-3* or *roc4* and will herein be referred to as *cyp20-3.1*. Quantitative expression analysis of the *CYP20-3* transcript in *cyp20-3.1* was performed with qRT-PCR (**Figure 1B**) using 3′ nested primers and demonstrated a minimal expression level of 0.02% of the *CYP20-3* transcript compared to wild type. A so far uncharacterized second T-DNA insertion mutant allele of *CYP20-3*, *cyp20-3.2* was selected from the SALK-collection (SALK_054125) with the T-DNA residing upstream of the first exon in the 5′-UTR (**Figure 1A**). Homozygous plants were isolated based on PCR with specified primer combinations (**Figure 1A**). Transcript analysis with qRT-PCR in homozygous *cyp20-3.2* mutant lines revealed a residual *CYP20-3* expression level of 2.5% compared to wild type plants (**Figure 1B**). A polyclonal antibody against Arabidopsis CYP20-3 detected residual amounts of CYP20-3 in *cyp20-3.1* and *cyp20-3.2* that corresponded to the detected remaining mRNA levels (**Figure 1C**). It should be noted that no CYP20-3 protein was detected in *cyp20-3.1* using a different antibody in an earlier study (Cai et al., 2008). Thus, both alleles are strong knock-down mutants of the *cyp20-3* gene. The CYP20-3 protein level was unchanged in the *serat2;1* mutant compared to wild type, excluding any compensatory upregulation in response to the lack of this interaction target.

ACCLIMATION TO LONG-TERM EXPOSURE TO HIGH LIGHT CONDITIONS

The effect of high light on Arabidopsis mutants with eliminated plastidic SERAT activity (*serat2;1*) or strongly depleted *CYP20-3* expression (*cyp20-3.1* and *cyp20-3.2*), was analyzed with plants that were grown under short-day conditions at a light intensity of 100 μmol m^{-2} s^{-1} until the age of 3 weeks and subsequently challenged with long-day light of 800 μmol m^{-2} s^{-1} irradiance (high light) for additional 10 days. Under control conditions with 100 μmol m^{-2} s^{-1} the growth phenotype of the mutant lines was similar to wild type plants (**Figure 2A**). Growth under high light conditions caused a decrease in rosette diameter and downward curling of leaves in wild type, *serat2;1*, *cyp20-3.1*, and *cyp20-3.2* plants when compared to the phenotype of control plants. The mutant genotypes were affected in the same manner as the wild type plants, i.e., also the two *cyp20-3* lines did not differ with respect to phenotypes unlike earlier reports (Cai et al., 2008; Dominguez-Solis et al., 2008). Protein levels of CYP20-3 were neither responding in wild type plants under high light nor in the *serat2;1* mutant. The very low levels of CYP20-3 in *cyp20-3.1* and *cyp20-3.2* plants remained also unchanged (Supplementary Figure 1). Fresh weight determination revealed a 2.2-fold increase in biomass of high light-treated wild type plants compared to control wild type plants (**Figure 2B**). The fresh weight biomass of *serat2;1*, *cyp20-3.1*, and *cyp20-3.2* increased only 1.7-, 1.6-, and 2.1-fold, respectively, when grown under high light (**Figure 2B**). Dry weight measurements of whole rosettes also showed a significant increase of biomass in all genotypes upon high light exposure (**Figure 2C**). Wild type plants exhibited a 3.5-fold increase, *serat2;1* and *cyp20-3.1* a 2.7-fold increase and *cyp20-3.2* a 4.7-fold increase of dry weight upon high light treatment (**Figure 2C**).

Along with the stunted growth phenotype red leaf coloring was observed for the high light-treated plants at the abaxial side of the leaves (Supplementary Figure 2). Therefore, anthocyanin levels were determined on the basis of fresh weight (**Figure 2D**). Significant 2.3-fold to 3.1-fold increases in anthocyanin content could be observed for the wild type, *serat2;1* and both *cyp20-3* mutant lines when compared to anthocyanin levels of unstressed rosette leaves (**Figure 2D**). Noteworthy, the anthocyanin levels were indistinguishable between wild type and mutant lines, indicating that this stress response is independent of the presence of CYP20-3 or SERAT2;1. Thus, the applied high light conditions had two components: significant stress but also promoted growth. However, in comparison to wild type the *serat2;1* and *cyp20-3.1* and *cyp20-3.2* mutants grew slower under high light conditions. Apparently activation of chloroplast SERAT2;1 by CYP20-3 as well as the presence of SERAT2;1 and thus the chloroplast synthase complex are dispensable, but both systems enhance the performance of Arabidopsis plants under highlight.

IMPACT OF HIGH LIGHT ON CYSTEINE METABOLISM

To test whether the observed stress susceptibility phenotype of the mutant lines was due to alterations of cysteine metabolism the total SERAT activity and steady-state OAS levels of leaves were determined. Both parameters are established biochemical markers for the state of cysteine synthesis (Haas et al., 2008; Heeg et al., 2008; Watanabe et al., 2008b; Khan et al., 2010). Compared to control conditions a 1.9-fold increase of SERAT activity was observed in high light-treated wild type plants (**Figure 3A**). A smaller, but still significant increase under high light was also observed for *serat2;1* (1.5-fold) and *cyp20-3.2* (1.5-fold) mutants. SERAT activity was also 1.3-fold elevated in *cyp20-3.1*, but was found at the limit of statistical significance. Higher SERAT activity may result in higher OAS contents. Indeed, changes in OAS concentrations were in agreement with the increased SERAT activity (**Figure 3B**). A 2.6-fold increase in OAS content was observed for wild type plants exposed to high light conditions. In *serat2;1*, *cyp20.3.1* and *cyp20-3.2* OAS steady state levels increased 1.7-, 1.6-, and 2-fold, respectively. In accordance with the SERAT activity the raise of OAS concentrations upon high light treatment was smaller in the mutant plants than in the wild type plants. Statistical examination of SERAT activity and OAS content under control conditions revealed no difference between wild type and all three mutant lines. However, the SERAT activities and OAS concentrations in the three mutant lines increased significantly less under high light than those in wild type plants. The weak statistical significance ($p = 0.106$) of the difference of SERAT activity in *cyp20-3.2* to wild type under high light might be owed to the leaky insertion allele and the higher residual CYP20-3 level of this mutant. The same consideration applies to the OAS concentration under high light in *cyp20-3.2* ($p = 0.120$). Thus,

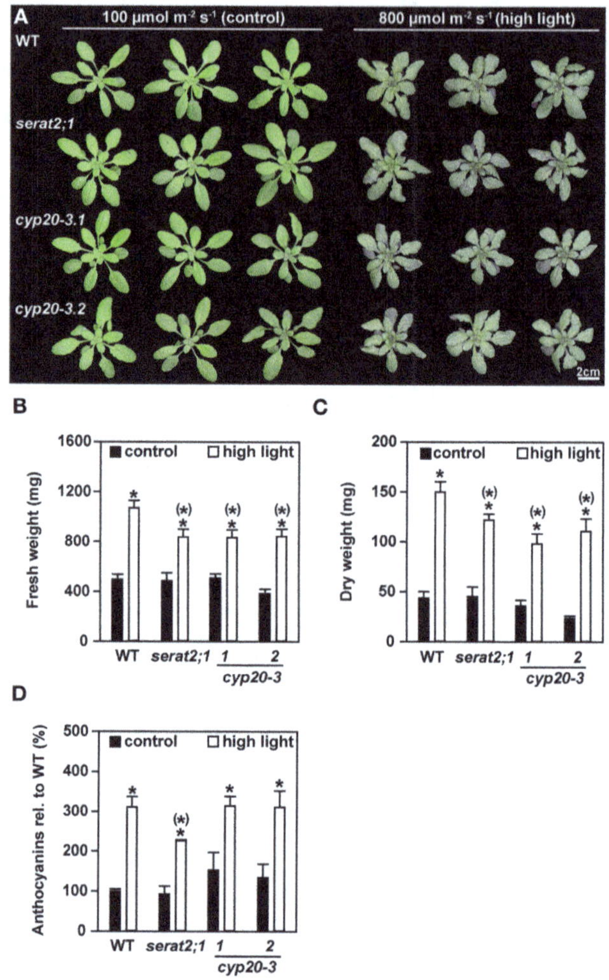

FIGURE 2 | Effect of high light (800 μmol m^{-2} s^{-1}, 10 days) on the phenotype, fresh weight and anthocyanin content in wild type (WT) and *serat2;1*, *cyp20-3.1*, and *cyp20-3.2* mutant lines. (A) Phenotype of three representative WT and mutant plants, respectively, that were subjected to control conditions of 100 μmol m^{-2}s^{-1} (long-day) and high light conditions of 800 μmol m^{-2}s^{-1} (long-day) for 10 days after 3 weeks of growth under short-day and low light conditions. **(B)** Fresh weight analysis of whole rosettes from control and high light-treated plants. Bars denote data means of 8–10 sampled individuals. **(C)** Dry weight determination of 3–5 sampled individuals from control and high light-treated plants. **(D)** Measurement of anthocyanin contents in control and high light-treated plant leaves. WT anthocyanin levels were set to 100% (0.2 μg mg^{-1} FW). Results represent means $\pm$ *SE* ($n = 3$). Black bars indicate control conditions, whereas white bars outline high light treatment. Asterisks indicate statistically significant differences between control and high light treatment ($p \leq 0.05$; Student's *t*-test). Asterisks with parentheses point to the comparison with high light treated WT ($p \leq 0.05$; Student's *t*-test). FW, fresh weight.

plastidic SERAT2;1 and its presumed activation by CYP20-3 is required for fully enhanced OAS synthesis, but about half of the observed increases of SERAT activity and OAS concentrations is likely due to events in mitochondria and cytosol.

Cysteine steady state levels were not significantly altered in high light-exposed wild type, *serat2;1*, *cyp20-3.1* and *cyp20-3.2*

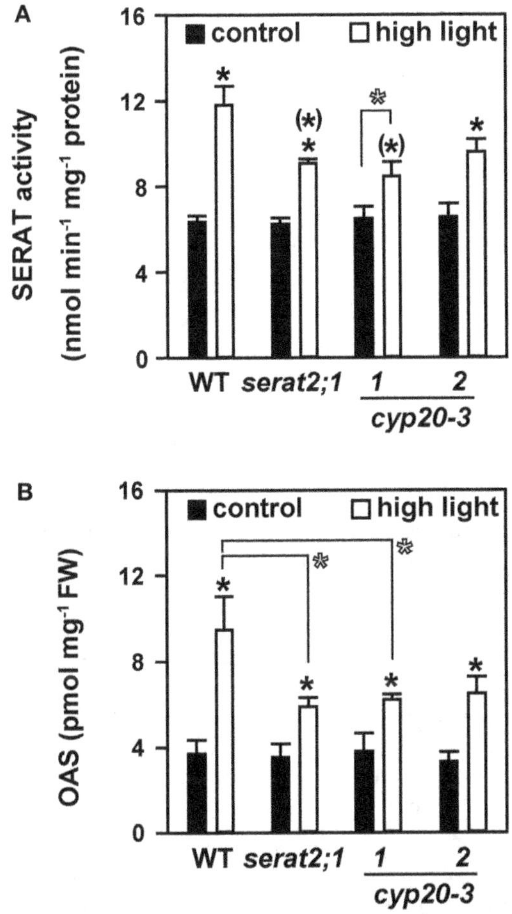

FIGURE 3 | Effect of 10 days high light treatment on SERAT activity and OAS steady state concentrations in wild type (WT) and *serat2;1*, *cyp20-3.1*, and *cyp20-3.2* mutant lines. (A) Total SERAT activity was measured in leaf protein extracts of control and high light-treated WT and mutant plants ($n = 3$). **(B)** OAS concentrations were determined from leaf metabolite extracts of control and light-stressed plants ($n = 5$). Black bars indicate control conditions, whereas white bars outline stress conditions. Results represent means ± *SE*. Statistical significances are based on Student's *t*-test. Differences between control and high light treatment are marked with asterisks ($p \leq 0.05$). Differences between high light treated WT and mutants are marked by asterisks in parentheses ($p \leq 0.05$). Open asterisks refer to indicated pairwise comparisons with ($p \leq 0.1$). FW, fresh weight.

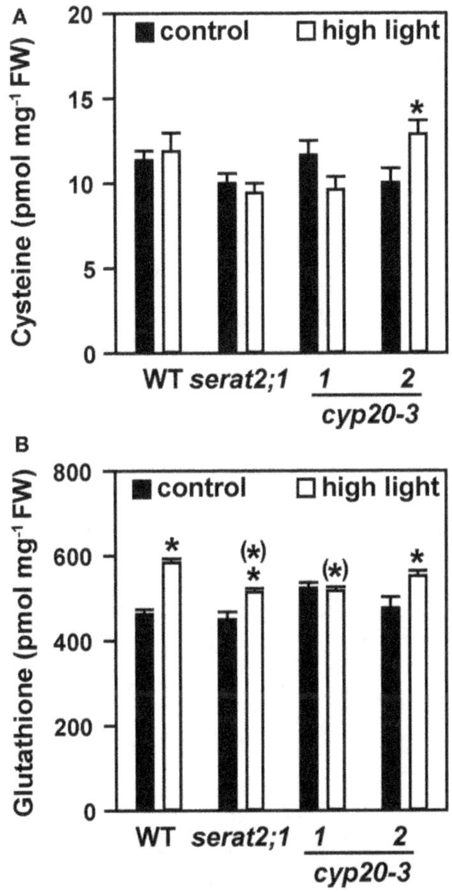

FIGURE 4 | Effect of 10 days high light exposure on leaf cysteine and glutathione steady state concentrations in wild type (WT) and *serat2;1*, *cyp20-3.1*, and *cyp20-3.2* mutant lines. (A) Cysteine concentrations were determined from metabolite extracts of control and high light-treated WT and mutant plants. **(B)** Glutathione concentrations. Black bars indicate control conditions, whereas white bars outline stress conditions. Results represent means ± *SE* ($n = 5$). Asterisks indicate statistically significant differences between control and high light treatment ($p \leq 0.05$; Student's *t*-test). Asterisks in parentheses point to the comparison with high light-treated WT ($p \leq 0.05$; Student's *t*-test). FW, fresh weight.

plants compared to unstressed plants (**Figure 4A**). Glutathione content was moderately but significantly increased in wild type (1.3-fold) upon high light exposure (**Figure 4B**) similar to previous reports (e.g., Alsharafa et al., 2014). In *serat2;1* (1.1-fold) and *cyp20-3.2* (1.2-fold) the increase was lower than in wild type and found to be unchanged for *cyp20-3.1*. Cysteine and glutathione levels from all three mutant lines were not different from wild type levels under both conditions (**Figure 4**). Taken together, glutathione steady state levels increase as part of the acclimation response but were similar between wild type and mutants. This steady-state level determination indicates that OAS synthesis is enhanced and, while keeping cysteine levels unchanged, allows for increased glutathione concentrations under high light conditions.

IMPACT OF HIGH LIGHT ON SERAT TRANSCRIPT LEVELS

The increase in total SERAT activity and OAS contents upon exposure to high light prompted us to determine the transcript levels of the three major SERAT isoforms. Transcript levels of the plastidic SERAT isoform (*SERAT2;1*) were indistinguishable between control and stressed plants in wild type, *cyp20-3.1* and *cyp20-3.2* lines (**Figure 5A**). *serat2;1* RNA was not tested in this respect, as this line was already shown to be a full gene knock-out mutant (Watanabe et al., 2008b). A significant increase of the mitochondrial *SERAT2;2* transcript was observed for high light-treated wild type plants (4.4-fold) when compared to wild type control plants (**Figure 5B**). *SERAT2;2* transcript levels in leaves of *serat2;1* (3.2-fold), *cyp20-3.1* (3.1-fold), and *cyp20-3.2* (1.9-fold) plants were also significantly increased (**Figure 5B**). Only wild type (1.8-fold) and *cyp20-3.1* (1.7-fold) plants but not *serat2;1*

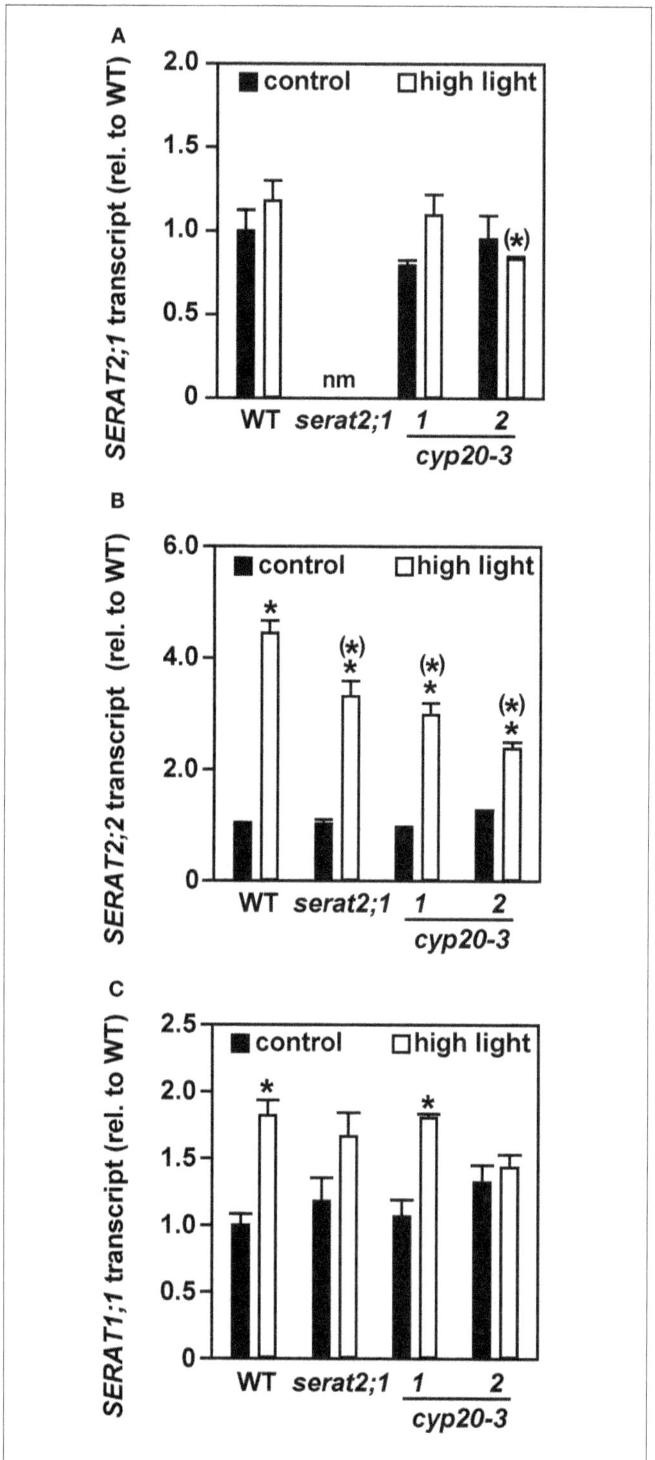

FIGURE 5 | Effect of high light (10 days) on SERAT transcript levels in leaves of wild type (WT) and *serat2;1, cyp20-3.1,* and *cyp20-3.2* mutant lines. *SERAT* gene expression levels were determined by qRT-PCR in WT and mutant plants grown under control and high light stress conditions. **(A)** *SERAT2;1* transcripts. **(B)** *SERAT2;2* transcripts. **(C)** *SERAT1;1* transcripts. Black bars indicate control conditions, whereas white bars outline stress conditions. Control WT transcript levels were set to 100%. Results represent means ± *SE* ($n = 3$). Asterisks indicate statistically significant differences between control and high light treatment ($p \leq 0.05$; Student's *t*-test). Asterisks in parentheses point to the comparison with high light treated WT ($p \leq 0.05$; Student's *t*-test).

and *cyp20-3.2* lines showed a significant increase in *SERAT1;1* transcript amount (**Figure 5C**). These expression analyses suggest that mitochondria and possibly to some extent the cytosol provide most of the SERAT activity at least on this intermediate time scale of a 10 day high light treatment for the enhanced synthesis of OAS and ultimately glutathione during high light acclimation to achieve stress mitigation and growth acceleration. The chloroplasts contribute to this process and improve it but are dispensable, at least under the conditions tested.

DISCUSSION

Earlier analyses of cysteine synthesis in Arabidopsis revealed that under regular growth conditions the chloroplasts provide the substrate sulfide, the mitochondria most of the substrate OAS and the cytosol synthesizes most of the cysteine in leaves, but that all compartments in principle are able to produce cysteine (Haas et al., 2008; Heeg et al., 2008; Watanabe et al., 2008a; Birke et al., 2013). This surprising finding raised the question if there is a specific function of cysteine synthesis in chloroplasts under oxidative stress, since significant upregulation of *SERAT2;1* gene expression was observed after transfer of the CATALASE2 deficient mutant *cat2* from high to ambient CO_2 concentrations and concomitant H_2O_2 stress (Queval et al., 2009) and directly upon H_2O_2 exposure (Genevestigator® V3; https://genevestigator.com/gv/plant.jsp). In line with this observation a post-transcriptional upregulation of SERAT2;1 activity by the peptidyl-prolyl *cis/trans*-isomerase CYP20-3 in the chloroplast stroma was linked to resistance against abiotic stress (Dominguez-Solis et al., 2008) and against biotic stress (Park et al., 2013). While a relatively low light intensity which hardly should be considered as stress-inducing for Arabidopsis (300 μmol m^{-2} s^{-1}) (Dominguez-Solis et al., 2008) or uncontrolled greenhouse conditions (Cai et al., 2008) had been applied earlier, the controlled acclimation to 800 μmol $m^{-2}s^{-1}$ of wild type and mutant plants, either deficient in chloroplast SERAT or strongly depleted of its activation partner CYP20-3 in two mutant alleles, was used as an improved approach to challenge chloroplast cysteine synthesis and indicate a role for this process.

When grown under high light the purple-colored rosette leaves of all genotypes indicated similar anthocyanin accumulation. This confirmed the stress-inducing conditions, since increased anthocyanin levels are considered to reflect stress response to enhanced accumulation of reductive power and carbohydrates (Chalker-Scott, 1999; Vanderauwera et al., 2005; Zeng et al., 2010) and to contribute to scavenging of ROS that are produced in high light-treated plants (Gould et al., 2002). In addition, the glutathione concentration increased in wild type leaves, confirming enhanced demand for cysteine synthesis. This observation was backed up by enriched OAS concentrations, whereas cysteine concentrations remained unchanged in control and high light-treated wild type and mutant plants. This unchanged steady-state might be explained by draining cysteine away to meet the increased demand for glutathione to function as a reductant. This would correspond to the metabolic concept of cysteine as a compound with high turnover but low and rarely changing cellular concentrations (see Hell and Wirtz, 2011, for review). Glutathione levels in turn are known to respond to increasing light intensity but to reach a plateau at 100–200 μmol m^{-2} s^{-1} (Ogawa et al., 2004; Alsharafa et al., 2014). Glutathione was furthermore shown to be essential for the vacuolar sequestration of anthocyanins in maize and petunia through glutathione *S*-transferases (Alfenito et al., 1998; Edwards et al., 2000; Xiang et al., 2001), suggesting that some of the glutathione was removed from the cytoplasmic pool and probably degraded.

However, under the high light conditions employed here (800 μmol m^{-2} s^{-1} for 10 days) the lack of CYP20-3 in the *cyp20-3.1* mutant or the absence of its activation target SERAT2;1 in the chloroplast had only a small effect on the stress phenotypes. Wild type and mutant plants acclimated after transfer to high light, but the increase in fresh and dry weight of mutants lacking SERAT2;1 or CYP20-3 was about 20–30% smaller than that of wild type. Interestingly, the increased total SERAT activities and the resulting OAS concentrations in the four genotypes followed very much the same pattern under high light stress. Thus, the fact that *serat2;1* and *cyp20-3* mutant plants achieve less biomass along with less SERAT activity under high light stress compared to wild type can be attributed to chloroplast SERAT activity. In contrast, Dominguez-Solis et al. (2008) reported an already lowered total SERAT activity in non-stressed *cyp20-3.1* plants and no increase under stress conditions. While it should be cautioned that the treatment conditions and age of plants were different (2 weeks as compared to 4.5 weeks here), the significant increase of total SERAT activity also in the *serat2;1* null mutant demonstrates that the SERAT activities in the other compartments must have responded to the high light treatment.

SERAT2;1 forms the cysteine synthase complex with the OAS-TL B isoform in the stroma of chloroplasts. However, in addition to OAS-TL B the related protein CS26 also consumes OAS: CS26 is a member of the same family of β-substituted alanine synthases like the OAS-TLs and catalyzes the formation of S-sulfocysteine from OAS and thiosulfate in the thylakoid lumen. Thus, CS26 and OAS-TL B potentially compete for the chloroplastidic OAS pool. Intriguingly, expression of the *CS26* gene increases upon transfer to high light according to public databases (Genevestigator® V3; https://genevestigator.com/gv/plant.jsp). Extended transfer of a *CS26* null mutant from short-day to long-day or permanent light (120–160 μmol m^{-2} s^{-1}) was reported to result in growth retardation as a result of disturbed redox processes in the thylakoid lumen (Bermúdez et al., 2010, 2012). The control treatment used here is similar to these experiments by shifting *serat2;1* and *cyp20-3* mutants from short-day to long-day. The *serat2;1* mutant was not affected while both *cyp20-3* mutants appeared to grow more slowly, suggesting a possible link to CS26 functions. However, a careful statistical analysis revealed only very weak significance for the growth retardation, at least under the conditions used here. An assessment of photosynthetic parameters of *cyp-20-3* mutants as carried out for the *cs26* mutant (Bermúdez et al., 2012) might be interesting to investigate redox functions of the system. Indeed, the recovery of photosystem II activity was slower in *cyp20-3.1* compared to wild type following photoinhibition by high light treatment (Cai et al., 2008). However, the assumed multiple targets of CYP20-3 reflected by its high abundance (Lippuner et al., 1994; Peltier et al., 2006) makes it difficult to dissect its various functions.

Indeed, expression analysis of the three major *SERAT* genes showed significant increases of *SERAT2;2* mRNA and much less of *SERAT1;1*. The expression of the *SERAT2;1* gene remained unchanged, although regulation could also occur on the transcriptional level: *SERAT2;1* transcript levels were reported to be up-regulated upon treatment of Arabidopsis roots with the oxidizing reagent menadione (Lehmann et al., 2009) and upon oxidative stress in the leaves of the catalase-deficient mutant (*cat2*; Queval et al., 2009). This suggests that the observed increases in total SERAT activity are at least only partially due to activation of SERAT2;1 by CYP20-3 but are complemented by the contributions of the other compartments. In fact, the increase of total SERAT activity in high light-treated wild type plants might be mainly attributed to the induction of *SERAT2;2* transcript amounts in these plants. Several lines of evidence support this assumption. The plastidic SERAT loss-of-function mutant demonstrates that SERAT2;1 makes a small contribution to total SERAT activity in leaves (Watanabe et al., 2008b; **Figure 3A**). In contrast mitochondrial SERAT2;2 represents approximately 80% of total SERAT activity in Arabidopsis leaves (Watanabe et al., 2008b) and consequently triggers the flux of cysteine synthesis, at least under normal growth conditions (Haas et al., 2008; Wirtz et al., 2012).

A specific contribution of CYP20-3 to the formation of the plastidic cysteine synthase complex by its chaperone function may be unnecessary, since formation of the Arabidopsis cysteine synthase complexes is known to occur spontaneously ($\Delta G = -33$ kcal mol^{-1}), without assistance of chaperones (Wirtz et al., 2010) and with an equilibrium dissociation constant of SERAT and OAS-TL subunits of $K_D = 25$ nM (Berkowitz et al., 2002). Given the several hundred-fold excess of OAS-TL over SERAT particularly in chloroplasts of many plant species (reviewed in Hell and Wirtz, 2011), essentially all SERAT molecules can be expected to be bound in the cysteine synthase complex. In addition, measurements of maximal activities of chloroplast SERATs in pea and Arabidopsis indicate only 10–13% of total SERAT activity in chloroplasts (Ruffet et al., 1995; Watanabe et al., 2008b). However, the function of CYP20-3 as a mediator of plastidic CSC formation was reinforced by the demonstration of strongly fostered CYP20-3 and SERAT2;1 interaction in the presence of OPDA (Park et al., 2013). In addition, a sequential pull-down assay showed that OPDA stimulates the interaction between SERAT2;1 and plastidic OAS-TL B with CYP20-3 as the mediator (Park et al., 2013). In line with this function the strong phenotypic and biochemical similarities between *serat2;1* and both *cyp20-3* mutants provide tentative evidence for a joint and specific mechanism. This indicates that, despite having multiple interaction partners, the activation of total SERAT activity upon high light stress is one of the major contributions of CYP20-3 to cope with this stress. The analysis of a mutant lacking chloroplast OAS-TL B protein (*oastlB*; Heeg et al., 2008; Birke et al., 2013) under high light conditions could be suitable to test this hypothesis.

Therefore, the following hypothesis is put forward: CYP20-3 and SERAT1 might contribute to retrograde-signaling of high light stress response instead of direct induction of cysteine synthesis in plastids for enhanced glutathione production. The advantage of this mechanism could be the speed of activation, since OPDA was shown to increase rapidly upon exposure to high light (Alsharafa et al., 2014) which could immediately activate chloroplastidic CSC. This scenario does not exclude that CYP20-3 has also additional functions, which are important under non-stress or other stress condition. Thus, it would not be the plastidic SERAT activity *per se* but the formation of the plastidic CSC itself. CYP20-3 promoted formation of the CSC upon high light might trigger through a yet not characterized signaling cascade the expression of *SERAT2;2* (and probably more) transcripts. The mitochondrial OAS might be transported from mitochondria to plastids and to the cytosol as evidenced (Wirtz et al., 2012; Birke et al., 2013; Lee et al., 2013), to generate cysteine to be used for glutathione biosynthesis. Kinetic analysis of OAS, cysteine and glutathione levels with subcellular resolution would have to be developed to validate this model.

MATERIALS AND METHODS

PLANT GENOTYPES AND GROWTH UNDER CONTROL AND STRESS CONDITIONS

All experiments were conducted using *Arabidopsis thaliana*, ecotype Columbia-0, as the wild type control and T-DNA insertion mutants, which derived from the same background. Seeds of *cyp20-3.1* (SALK_001615; AT3G62030) and *cyp20-3.2* (SALK_054125; AT3G62030) were obtained from the SALK collection (Salk Institute Genomic Analysis Laboratory) and the *serat2;1* knock-out mutant (SALK_099019; AT1G55920) from Watanabe et al. (2008b). *cyp20-3.2* plants were tested for homozygosity by PCR after Sambrook et al. (1989) with primers 1587_TTTGGCGAAAACTCTTAGCTG), 1588_TGGATTTAACACAAGCGGTTC and 1401_ATTTTGCC GATTTCGGAAC. Genomic leaf DNA was isolated according to Edwards et al. (1991).

Seeds were stratified on soil for 2 days at 4°C and subsequently transferred for germination to growth chambers. Plants were initially grown under short-day conditions with a day/night cycle of 8.5/15.5 h. Humidity was set to 60% and light intensity to 100 μmol m^{-2} s^{-1}. After 3 weeks the plants were transferred for 10 days to long-day conditions with a day/night cycle of 16/8 h and light intensities of either 800 μmol m^{-2} s^{-1} (high light) or 100 μmol m^{-2} s^{-1} (low light control). For fresh weight analysis whole rosettes were weighed, for dry weight determination the fresh material was kept for 3–5 days at 60°C beforehand.

ANTHOCYANIN DETERMINATION

Determination of anthocyanin content was performed based on the spectral characteristics of cyanidin-derived anthocyanins. The procedure was modified after Giusti and Wrolstad (2001) and Gou et al. (2011). Pigments were extracted in 1 ml 0.1% HCl (in ethanol) from 30 mg ground leaf tissue on a platform shaker at 40 rpm overnight at 4°C. Cell debris was sedimented at 20,000 × g for 10 min at 4°C. Absorbance of the anthocyanin was measured spectrophotometrically at 535 nm and 700 nm and anthocyanin concentration was calculated according to Giusti and Wrolstad (2001).

DETERMINATION OF METABOLITES

Hydrophilic metabolites were extracted from leaves of control and high light-treated Arabidopsis leaves as described in Wirtz and Hell (2003). Thiols and OAS contents were quantified according to Heeg et al. (2008) in cooperation with the Metabolomics Core Technology Platform Heidelberg funded by the DFG Excellence inititative.

DETERMINATION OF SERAT ENZYME ACTIVITY AND IMMUNOLOGICAL DETECTION OF PROTEINS

SERAT activity, coupled to the OAS-TL reaction, was determined based on spectrophotometrical cysteine detection described by Gaitonde (1967). Total soluble protein extracts were prepared from 200 mg ground leaf tissue according to Birke et al. (2013) using Spintrap G-25 columns (GE Healthcare, München). Subsequently protein concentrations as well as enzymatic activities were determined as described by Heeg et al. (2008). Equ a l amounts of the crude extract were separated by discontinuous SDS-PAGE and blotted on PVDF membrane using a Trans-Blot® Cell system (Bio-Rad, München). Immunological detection of CYP20-3 was done using α-CYP20-3 in combination with a horseradish peroxidase-conjugated secondary antibody and chemiluminescent detection. Loading of protein was tested by staining the gel with Coomassie Brilliant Blue G-250 (Merck, Darmstadt).

TRANSCRIPT QUANTIFICATION BY qRT-PCR

Gene expression levels were verified in leaves using the principles of the quantification of cDNA targets with quantitative real-time PCR (qRT-PCR). Total RNA was extracted from 100 mg of Arabidopsis leaf tissue using the peqGOLD Total RNA Kit (Peqlab, Erlangen). cDNA was synthesized using the M-MLV Reverse Transcriptase-Kit (Promega, Mannheim). Transcript amount of the respective genes were determined with the Rotor-Gene SYBRGreen PCR Kit (Qiagen, Hilden) and the Rotor-Gene Q system (Qiagen, Hilden).

Gene specific primers used for qRT-PCR: *CYP20-3* (2102_ CTGGACCTGGAATCTTGAGC; 2103_CTTGTCCAAACACGA CATGC; 2104_CCACCAAGCATCAGAGAACC; 2105_CCAGCA ACTTCACCTCCAAT); *SERAT2;1* (for_CACATGCCGAACCGG TAATAC; rev_GGTGAATCTTCCGGTTTACAGAGA); *SERAT2;2* (for_AATGGAACCCAGACCAAAACC; rev_GCCCAAACATCA TCGACTTCA); *SERAT1;1* (for_TGGACACAGATCAAGGCGG; rev_ATGAGAAAGAATCGTCGAATATAGATAGC); *PP2a_PDF2* (for_CTTCTCGCTCCAGTAATGGGACC; rev_GCTTGGTCGA CTATCGGAATGCGCG)

ACKNOWLEDGMENTS

The authors are grateful for financial support by the Deutsche Forschungsgemeinschaft to Rüdiger Hell and the Schmeil Foundation Heidelberg to Anna Speiser.

REFERENCES

Alfenito, M. R., Souer, E., Goodman, C. D., Buell, R., Mol, J., Koes, R., et al. (1998). Functional complementation of anthocyanin sequestration in the vacuole by widely divergent glutathione *S*-transferases. *Plant Cell* 10, 1135–1149. doi: 10.1105/tpc.10.7.1135

Alsharafa, K., Vogel, M. O., Oelze, M.-L., Moore, M., Stingl, N., König, K., et al. (2014). Kinetics of retrograde signalling initiation in the high light response of *Arabidopsis thaliana*. *Phil. Trans. R. Soc. B* 369, 20130424. doi: 10.1098/rstb.2013.0424

Barber, J., and Andersson, B. (1992). Too much of a good thing: light can be bad for photosynthesis. *Trends Biochem. Sci.* 17, 61–66. doi: 10.1016/0968-0004(92)90503-2

Berkowitz, O., Wirtz, M., Wolf, A., Kuhlmann, J., and Hell, R. (2002). Use of biomolecular interaction analysis to elucidate the regulatory mechanism of the cysteine synthase complex from *Arabidopsis thaliana*. *J. Biol. Chem.* 277, 30629–30634. doi: 10.1074/jbc.M111632200

Bermúdez, M. Á., Galmés, J., Moreno, I., Mullineaux, P. M., Gotor, C., and Romero, L. C. (2012). Photosynthetic adaptation to length of day is dependent on S-sulfocysteine synthase activity in the thylakoid lumen. *Plant Phys.* 160, 274–288. doi: 10.1104/pp.112.201491

Bermúdez, M. Á., Paez-Ochoa, M. A., Gotor, C., and Romero, L. C. (2010). Arabidopsis S-sulfocysteine synthase activity is essential for chloroplast function and long-day light-dependent redox control. *Plant Cell* 22, 403–416. doi: 10.1105/tpc.109.071985

Birke, H., Heeg, C., Wirtz, M., and Hell, R. (2013). Successful fertilization requires the presence of at least one major *O*-Acetylserine(thiol)lyase for cysteine synthesis in pollen of Arabidopsis. *Plant Physiol.* 163, 959–972. doi: 10.1104/pp.113.221200

Cai, W., Ma, J., Guo, J., and Zhang, L. (2008). Function of ROC4 in the efficient repair of photodamaged photosystem II in Arabidopsis. *Photochem. Photobiol.* 84, 1343–1348. doi: 10.1111/j.1751-1097.2008.00448.x

Chalker-Scott, L. (1999). Environmental significance of anthocyanins in plant stress responses. *Photochem. Photobiol.* 70, 1–9. doi: 10.1111/j.1751-1097.1999.tb01944.x

Chan, K. X., Wirtz, M., Phua, S. Y., Estavillo, G. M., and Pogson, B. J. (2013). Balancing metabolites in drought: the sulfur assimilation conundrum. *Trends Plant Sci.* 18, 18–29. doi: 10.1016/j.tplants.2012.07.005

Chou, I., and Gasser, C. (1997). Characterization of the cyclophilin gene family of *Arabidopsis thaliana* and phylogenetic analysis of known cyclophilin proteins. *Plant Mol. Biol.* 35, 873–892. doi: 10.1023/A:1005930024796

Dominguez-Solis, J. R., He, Z., Lima, A., Ting, J., Buchanan, B. B., and Luan, S. (2008). A cyclophilin links redox and light signals to cysteine biosynthesis and stress responses in chloroplasts. *Proc. Natl. Acad. Sci. U.S.A.* 105, 16386–16391. doi: 10.1073/pnas.0808204105

Eckardt, N. A., Snyder, C. W., Portis, A. R., and Ogren, W. L. (1997). Growth and photosynthesis under high and low irradiance of *Arabidopsis thaliana* antisense mutants with reduced ribulose-1,5-bisphosphate carboxylase/oxygenase activase content. *Plant Physiol.* 113, 575–586. doi: 10.1104/pp.113.2.575

Edwards, K., Johnstone, C., and Thompson, C. (1991). A simple and rapid method for the preparation of plant genomic DNA for PCR analysis. *Nucleic Acids Res.* 19:1349. doi: 10.1093/nar/19.6.1349

Edwards, R., Dixon, D. P., and Walbot, V. (2000). Plant glutathione *S*-transferases: enzymes with multiple functions in sickness and in health. *Trends Plant Sci.* 5, 193–198. doi: 10.1016/S1360-1385(00)01601-0

Foyer, C. H., Bloom, A. J., Queval, G., and Noctor, G. (2009). Photorespiratory metabolism: genes, mutants, energetics, and redox signaling. *Annu. Rev. Plant Biol.* 60, 455–484. doi: 10.1146/annurev.arplant.043008.091948

Foyer, C. H., and Noctor, G. (2011). Ascorbate and glutathione: the heart of the redox hub. *Plant Physiol.* 155, 2–18. doi: 10.1104/pp.110.167569

Fryer, M. J., Oxborough, K., Mullineaux, P. M., and Baker, N. R. (2002). Imaging of photo-oxidative stress responses in leaves. *J. Exp. Bot.* 53, 1249–1254. doi: 10.1093/jexbot/53.372.1249

Gaitonde, M. K. (1967). A spectrophotometric method for the direct determination of cysteine in the presence of other naturally occuring amino acids. *Biochem. J.* 104, 627–633.

Giusti, M. M., and Wrolstad, R. E. (2001). "Characterization and Measurement of Anthocyanins by UV-Visible Spectroscopy," in *Current Protocols in Food Analytical Chemistry* (Hoboken, NJ: John Wiley & Sons, Inc. F:F1:F1.2). doi: 10.1002/0471142913.faf0102s00

Gou, J.-Y., Felippes, F. F., Liu, C.-J., Weigel, D., and Wang, J.-W. (2011). Negative regulation of anthocyanin biosynthesis in arabidopsis by a miR156-targeted SPL transcription factor. *Plant Cell* 23, 1512–1522. doi: 10.1105/tpc.111.084525

Gould, K. S., McKelvie, J., and Markham, K. R. (2002). Do anthocyanins function as antioxidants in leaves? Imaging of H_2O_2 in red and green leaves after mechanical injury. *Plant Cell Environ.* 25, 1261–1269. doi: 10.1046/j.1365-3040.2002.00905.x

Grzam, A., Martin, M., Hell, R., and Meyer, A. (2007). γ-glutamyl transpeptidase GGT4 initiates vacuolar degradation of glutathione *S*-conjugates in Arabidopsis. *FEBS Lett.* 581, 3131–3138. doi: 10.1016/j.febslet.2007.05.071

Haas, F. H., Heeg, C., Queiroz, R., Bauer, A., Wirtz, M., and Hell, R. (2008). Mitochondrial serine acetyltransferase functions as a pacemaker of cysteine synthesis in plant cells. *Plant Physiol.* 148, 1055–1067. doi: 10.1104/pp.108.125237

He, Z., Li, L., and Luan, S. (2004). Immunophilins and parvulins. Superfamily of peptidyl prolyl isomerases in Arabidopsis. *Plant Physiol.* 134, 1248–1267. doi: 10.1104/pp.103.031005

Heeg, C., Kruse, C., Jost, R., Gutensohn, M., Ruppert, T., Wirtz, M., et al. (2008). Analysis of the Arabidopsis *O*-acetylserine(thiol)lyase gene family demonstrates compartment-specific differences in the regulation of cysteine synthesis. *Plant Cell* 20, 168–185. doi: 10.1105/tpc.107.056747

Hell, R., and Wirtz, M. (2011). Molecular Biology, Biochemistry and Cellular Physiology of Cysteine Metabolism in *Arabidopsis thaliana*. *Arabidopsis Book* 9:e0154. doi: 10.1199/tab.0154

Howarth, J. R., Dominguez-Solis, J. R., Gutierrez-Alcala, G., Wray, J. L., Romero, L. C., and Gotor, C. (2003). The serine acetyltransferase gene family in *Arabidopsis thaliana* and the regulation of its expression by cadmium. *Plant Mol. Biol.* 51, 589–598. doi: 10.1023/A:1022349623951

Hruz, T., Laule, O., Szabo, G., Wessendorp, F., Bleuler, S., Oertle, L., et al. (2008). Genevestigator V3: a reference expression database for the meta-analysis of transcriptomes. *Adv. Bioinformatics* 2008:420747. doi: 10.1155/2008/420747

Khan, M. S., Haas, F. H., Allboje Samami, A., Moghaddas Gholami, A., Bauer, A., Fellenberg, K., et al. (2010). Sulfite reductase defines a newly discovered bottleneck for assimilatory sulfate reduction and is essential for growth andd in *Arabidopsis thaliana*. *Plant Cell* 22, 1216–1231. doi: 10.1105/tpc.110.074088

Laxa, M., Konig, J., Dietz, K. J., and Kandlbinder, A. (2007). Role of the cysteine residues in *Arabidopsis thaliana* cyclophilin CYP20-3 in peptidyl-prolyl cis-trans isomerase and redox-related functions. *Biochem. J.* 401, 287–297. doi: 10.1042/BJ20061092

Lee, C. P., Wirtz, M., and Hell, R. (2013). Evidence for several cysteine transport mechanisms in the mitochondrial membranes of *Arabidopsis thaliana*. *Plant Cell Physiol.* 55, 64–73. doi: 10.1093/pcp/pct155

Lehmann, M., Schwarzlander, M., Obata, T., Sirikantaramas, S., Burow, M., Olsen, C. E., et al. (2009). The metabolic response of Arabidopsis roots to oxidative stress is distinct from that of heterotrophic cells in culture and highlights a complex relationship between the levels of transcripts, metabolites, and flux. *Mol. Plant* 2, 390–406. doi: 10.1093/mp/ssn080

Lippuner, V., Chou, I. T., Scott, S. V., Ettinger, W. F., Theg, S. M., and Gasser, C. S. (1994). Cloning and characterization of chloroplast and cytosolic forms of cyclophilin from *Arabidopsis thaliana*. *J. Biol. Chem.* 269, 7863–7868.

Marty, L., Siala, W., Schwarzlander, M., Fricker, M. D., Wirtz, M., Sweetlove, L. J., et al. (2009). The NADPH-dependent thioredoxin system constitutes a functional backup for cytosolic glutathione reductase in Arabidopsis. *Proc. Natl. Acad. Sci. U.S.A.* 106, 9109–9114. doi: 10.1073/pnas.0900206106

May, M. J., Vernoux, T., Sanchez-Fernandez, R., Van Montagu, M., and Inze, D. (1998). Evidence for posttranscriptional activation of γ-glutamylcysteine synthetase during plant stress responses. *Proc. Natl. Acad. Sci. U.S.A.* 95, 12049–12054. doi: 10.1073/pnas.95.20.12049

Meyer, A. J., and Hell, R. (2005). Glutathione homeostasis and redox-regulation by sulfhydryl groups. *Photosynth. Res.* 86, 435–457. doi: 10.1007/s11120-005-8425-1

Motohashi, K., Koyama, F., Nakanishi, Y., Ueoka-Nakanishi, H., and Hisabori, T. (2003). Chloroplast cyclophilin is a target protein of thioredoxin: thiol modulation of peptidyl-prolyl cis-trans isomerase activity. *J. Biol. Chem.* 278, 31848–31852. doi: 10.1074/jbc.M304258200

Noctor, G., and Foyer, C. H. (1998). Ascorbate and glutathione: keeping active oxygen under control. *Annu. Rev. Plant Physiol. Plant Mol. Biol.* 49, 249–279. doi: 10.1146/annurev.arplant.49.1.249

Noctor, G., Mhamdi, A., Chaouch, S., Han, Y. I., Neukermans, J., Marquez-Garcia, B., et al. (2012). Glutathione in plants an integrated overview. *Plant, Cell Environ.* 35, 454–484. doi: 10.1111/j.1365-3040.2011.02400.x

Noctor, G., Queval, G., Mhamdi, A., Chaouch, S., and Foyer, C. H. (2011). Glutathione. *Arabidopsis Book* 9, e0142. doi: 10.1199/tab.0142

Ogawa, K., Hatano-Iwasaki, A., Yanagida, M., and Iwabuchi, M. (2004). Level of glutathione is regulated by ATP-dependent ligation of glutamate and cysteine through photosynthesis in *Arabidopsis thaliana*: mechanism of strong interaction of light intensity with flowering. *Plant Cell Physiol.* 45, 1–8. doi: 10.1093/pcp/pch008

Park, S.-W., Li, W., Viehhauser, A., He, B., Kim, S., Nilsson, A. K., et al. (2013). Cyclophilin 20-3 relays a 12-oxo-phytodienoic acid signal during stress responsive regulation of cellular redox homeostasis. *Proc. Natl. Acad. Sci. U.S.A.* 110, 9559–9564. doi: 10.1073/pnas.1218872110

Peltier, J.-B., Cai, Y., Sun, Q., Zabrouskov, V., Giacomelli, L., Rudella, A., et al. (2006). The oligomeric stromal proteome of arabidopsis thaliana chloroplasts. *Mol. Cell. Proteomics* 5, 114–133. doi: 10.1074/mcp.M500180-MCP200

Queval, G., Thominet, D., Vanacker, H., Miginiac-Maslow, M., Gakiere, B., and Noctor, G. (2009). H_2O_2-activated up-regulation of glutathione in Arabidopsis involves induction of genes encoding enzymes involved in cysteine synthesis in the chloroplast. *Mol. Plant* 2, 344–356. doi: 10.1093/mp/ssp002

Romano, P. G. N., Horton, P., and Gray, J. E. (2004). The Arabidopsis cyclophilin gene family. *Plant Physiol.* 134, 1268–1282. doi: 10.1104/pp.103.022160

Rossel, J. B., Walter, P. B., Hendrickson, L., Chow, W. S., Poole, A., Mullineaux, P. M., et al. (2006). A mutation affecting ASCORBATE PEROXIDASE 2 gene expression reveals a link between responses to high light and drought tolerance. *Plant, Cell Environ.* 29, 269–281. doi: 10.1111/j.1365-3040.2005.01419.x

Rouhier, N., Lemaire, S. D., and Jacquot, J.-P. (2008). The role of glutathione in photosynthetic organisms: emerging functions for glutaredoxins and glutathionylation. *Annu. Rev. Plant Biol.* 59, 143. doi: 10.1146/annurev.arplant.59.032607.092811

Ruffet, M. L., Lebrun, M., Droux, M., and Douce, R. (1995). Subcellular distribution of serine acetyltransferase from *Pisum sativum* and characterization of an *Arabidopsis thaliana* putative cytosolic isoform. *Eur. J. Biochem.* 227, 500–509.

Sambrook, J., Fritsch, E. F., and Maniatis, T. (1989). *Molecular Cloning: A Laboratory Manual.* New York, NY: Cold Spring Harbour Laboratory Press.

Suzuki, N., Koussevitzky, S., Mittler, R. O. N., and Miller, G. A. D. (2012). ROS and redox signalling in the response of plants to abiotic stress. *Plant, Cell Environ.* 35, 259–270. doi: 10.1111/j.1365-3040.2011.02336.x

Szechyńska-Hebda, M., and Karpiński, S. (2013). Light intensity-dependent retrograde signalling in higher plants. *J. Plant Physiol.* 170, 1501–1516. doi: 10.1016/j.jplph.2013.06.005

Vanderauwera, S., Zimmermann, P., Rombauts, S., Vandenabeele, S., Langebartels, C., Gruissem, W., et al. (2005). Genome-wide analysis of hydrogen peroxide-regulated gene expression in Arabidopsis reveals a high light-induced transcriptional cluster involved in anthocyanin biosynthesis. *Plant Physiol.* 139, 806–821. doi: 10.1104/pp.105.065896

Wang, P., and Heitman, J. (2005). The cyclophilins. *Genome Biol.* 6:226. doi: 10.1186/gb-2005-6-7-226

Watanabe, M., Kusano, M., Oikawa, A., Fukushima, A., Noji, M., and Saito, K. (2008a). Physiological roles of the β -substituted alanine synthase gene family in Arabidopsis. *Plant Physiol.* 146, 310–320. doi: 10.1104/pp.107.106831

Watanabe, M., Mochida, K., Kato, T., Tabata, S., Yoshimoto, N., Noji, M., et al. (2008b). Comparative genomics and reverse genetics analysis reveal indispensable functions of the serine acetyltransferase gene family in Arabidopsis. *Plant Cell* 20, 2484–2496. doi: 10.1105/tpc.108.060335

Wirtz, M., Beard, K. F. M., Lee, C. P., Boltz, A., Schwarzländer, M., Fuchs, C., et al. (2012). Mitochondrial cysteine synthase Complex regulates *O*-acetylserine biosynthesis in plants. *J. Biol. Chem.* 287, 27941–27947. doi: 10.1074/jbc.M112.372656

Wirtz, M., Birke, H., Heeg, C., Mueller, C., Hosp, F., Throm, C., et al. (2010). Structure and function of the hetero-oligomeric cysteine synthase complex in plants. *J. Biol. Chem.* 285, 32810–32817. doi: 10.1074/jbc.M110.157446

Wirtz, M., and Hell, R. (2003). Production of cysteine for bacterial and plant biotechnology: application of cysteine feedback-insensitive isoforms of serine acetyltransferase. *Amino Acids* 24, 195–203. doi: 10.1007/s00726-002-0313-9

Xiang, C., Werner, B. L., Christensen, E. M., and Oliver, D. J. (2001). The biological functions of glutathione revisited in Arabidopsis transgenic plants with altered glutathione levels. *Plant Physiol.* 126, 564–574. doi: 10.1104/pp.126.2.564

Zeng, X.-Q., Chow, W. S., Su, L.-J., Peng, X.-X., and Peng, C.-L. (2010). Protective effect of supplemental anthocyanins on Arabidopsis leaves under high light. *Physiologia Plantarum* 138, 215–225. doi: 10.1111/j.1399-3054.2009.01316.x

Conflict of Interest Statement: The authors declare that the research was conducted in the absence of any commercial or financial relationships that could be construed as a potential conflict of interest.

21

Milestones in plant sulfur research on sulfur-induced-resistance (SIR) in Europe

Elke Bloem *, Silvia Haneklaus and Ewald Schnug

Federal Research Centre for Cultivated Plants, Julius Kühn-Institute, Institute for Crop and Soil Science, Braunschweig, Germany

Edited by:
Stanislav Kopriva, University of Cologne, Germany

Reviewed by:
Cecilia Gotor, Consejo Superior de Investigaciones Cientificas, Spain
Anja Riemenschneider, Leibniz Universität Hannover, Germany

****Correspondence:***
Elke Bloem, Federal Research Centre for Cultivated Plants – Julius Kühn-Institute (JKI), Institute for Crop and Soil Science, Bundesallee 50, 38116 Braunschweig, Germany
e-mail: elke.bloem@jki.bund.de

Until the 1970's of the last century sulfur (S) was mainly regarded as a pollutant being the main contributor of acid rain, causing forest dieback in central Europe. When Clean Air Acts came into force at the start of the 1980's SO_2 contaminations in the air were consequently reduced within the next years. S changed from an unwanted pollutant into a lacking plant nutrient in agriculture since agricultural fields were no longer "fertilized" indirectly by industrial pollution. S deficiency was first noticed in *Brassica* crops that display an especially high S demand because of its content of S-containing secondary metabolites, the glucosinolates. In Scotland, where S depositions decreased even faster than in continental Europe, an increasing disease incidence with *Pyrenopeziza brassicae* was observed in oilseed rape in the beginning 1990's and the concept of sulfur-induced-resistance (SIR) was developed after a relationship between the S status and the disease incidence was uncovered. Since then a lot of research was carried out to unravel the background of SIR in the metabolism of agricultural crops and to identify metabolites, enzymes and reactions, which are potentially activated by the S metabolism to combat fungal pathogens. The S status of the crop is affecting many different plant features such as color and scent of flowers, pigments in leaves, metabolite concentrations and the release of gaseous S compounds which are directly influencing the desirability of a crop for a variety of different organisms from microorganisms, over insects and slugs to the point of grazing animals. The present paper is an attempt to sum up the knowledge about the effect of the S nutritional status of agricultural crops on parameters that are directly related to their health status and by this to SIR. Milestones in SIR research are compiled, open questions are addressed and future projections were developed.

Keywords: nutrient induced resistance, S fertilization, plant S metabolism, fungal diseases, biotrophic and necrotrophic pathogens

NUTRIENT INDUCED RESISTANCE

Already Justus von Liebig identified in 1873 the nutritional status of a crop as crucial for its susceptibility against diseases. Interactions between mineral elements and plant diseases are established for several macro- and microelements. An overview of current knowledge on the effect of mineral nutrition on plant diseases was compiled by Datnoff et al. (2007).

A sufficient nutrient supply is the first agricultural measure against infection and determines the course of pathogenesis. In general, the greatest benefit can be expected when all essential nutrients are applied in sufficient amounts; however, the response to a particular nutrient may be different when going from deficiency to sufficiency than from sufficiency to excess (Huber and Haneklaus, 2007). For nitrogen it was shown that fertilizer application above recommended rates can lead to significantly greater disease incidences (Walters and Bingham, 2007). Strengthening the natural plant resistance is an important aspect of fertilization practice and modern fertilizers deliver the possibility to individually treat each kind of nutrient deficiency by tailored-made products. All essential plant nutrients have a direct impact on plants, pathogens, and microbial growth so that all of them as well as their proportions are important in disease control and will affect disease incidence or severity (Huber and Haneklaus, 2007). This illustrates an important problem in investigating the metabolic background of sulfur induced resistance (SIR): Plant pathogen response is determined by several interacting factors—different nutrients and their interactions, soil parameters, climatic conditions, pathogens, water supply and much more. Therefore, it is nearly impossible to investigate the response to a certain pathogen in relation to S under natural conditions without having interacting parameters.

PROGRESS IN RESEARCH ON SULFUR INDUCED RESISTANCE (SIR)

The fungicidal effect of foliar-applied elemental S (S^0) was already discovered by William Forsyth in 1802 and S^0 was used as the most important fungicide until the development of organic fungicides. The effects of foliar-applied elemental S have to be clearly distinguished from the health promoting effects of soil applied S on which SIR is based. The term SIR which denotes the reinforcement of the natural resistance of plants against fungal pathogens through triggering the stimulation of metabolic processes involving S by targeted soil-applied fertilizer strategies was first introduced by Schnug et al. (1995). In subsequent studies the term sulfur enhanced defense (SED) was used as synonym to prevent misinterpretation of the term resistance in a phytopathological context (Rausch and Wachter, 2005; Kruse et al., 2007).

Different research areas are of major relevance when investigating the background of SIR. The most important milestones in plant S research with respect to SIR are summarized in **Table 1**. Here important discoveries such as the detection of the *Foyer-Halliwell-Asada* pathway or the mustard oil bomb are listed as well as important technical developments.

The achievements made in gene transfer, by which the possibility to work with genetically modified plants was established, as well as the elucidation of the *Arabidopsis* genome promoted the progress in plant S research tremendously (Chilton et al., 1977). Experimentation with knock-out mutants delivered deep insight into plant metabolism and cross-talk between different pathways (Thomma et al., 1998; Kopriva, 2006; Parisy et al., 2006).

A lot of efforts were undertaken to understand the S assimilation pathway in plants, the transport of S into plants, and the storage and regulation of the S metabolism (**Table 1**). Since the completion of the *Arabidopsis* genome research has made considerable progress.

For example a range of S transporters carrying S containing metabolites within and between cells and over long-distance have been characterized, some of them just recently (Gigolashvili and Kopriva, 2014). In glucosinolate research the biosynthesis as well as its regulation was nearly explained in the last years (Halkier and Gershenzon, 2006).

Technical progress such as the development of macroarray hybridization can be seen as a further important milestone. Jost et al. (2005) recorded the reaction of more than 2000 selected genes of *Arabidopsis thaliana* to methyl jasmonate (JA) elicitation, a signaling compound in host-pathogen interactions. The authors could show that S-related genes were even more up-regulated due to methyl JA treatment than stress-related genes and that more than one pathway is involved in plant stress response. Gene expression of the ascorbate and glutathione metabolic pathways increased in response to JA as well as the synthesis of indole glucosinolates (Sasaki-Sekimoto et al., 2005). Moreover it was shown that imbalances in cytosolic cysteine alter the expression of groups of genes involved in pathogen response (Alvarez et al., 2012). Therefore, macroarray analysis delivers the opportunity to investigate and understand the network and cross-talk of metabolic pathways.

But despite of these great advances in scientific discoveries and technologies delivering several pieces of the puzzle of SIR, many questions remain open. It is still under discussion which reactions or compounds are responsible for the higher resistance of plants in relation to the S supply and how it is possible to induce a higher resistance and use this by advanced fertilizer application.

PHYSIOLOGICAL BACKGROUND OF SIR

Plants have developed several defense mechanisms in response to stress and react to a certain pathogen attack through a combination of constitutive and inducible defense with S-containing compounds being involved compiled by Bloem et al. (2005). In principle plants have three major strategies to combat pathogens: cell wall strengthening, apoplastic defense for inhibition of microbial enzymes and poisoning of the pathogen by toxic compounds like phytoalexins (Huckelhoven, 2007).

Initial pathogen recognition causes responses such as oxidative burst with the production of reactive oxygen species (ROS) and cell wall lignification (Swarupa et al., 2014). ROS serve as major signaling molecules in plant defense and are closely linked to the S metabolism via the *Foyer-Halliwell-Asada* pathway where glutathione is involved in the detoxification of ROS (Foyer and Halliwell, 1976). Via this link to ROS the S metabolism is linked to pathogen recognition and activation of the defense network.

The complexity of plant stress responses became obvious in several infection trials. S metabolites such as cysteine, glutathione, gaseous S emissions, phytoalexins, glucosinolates, and elemental S depositions have been investigated for their role in plant defense and how targeted S applications may prompt and enhance crop resistance to fungal pathogens (Bloem et al., 2007; Haneklaus et al., 2007, 2009). For most S containing metabolites a direct antifungal mode of action was proven (**Table 2**). Cysteine is the main precursor for all S containing compounds and is directly linked to stress response via its function related to systemic acquired resistance (Luckner, 1990). Cysteine displays a regulatory function in pathogen defense. It was shown that a specific cytosolic cysteine content is mandatory for the initiation of the plant immune response to pathogens and a link to the hypersensitive response (HR) was proven (Alvarez et al., 2012).

Glutathione displays a central function in plant defense as well: it is an important redox buffer in cells as it exists in a reduced form (GSH) which can react with another molecule of reduced glutathione (GSH) to form the oxidized disulfide form (GSSG) and which is restored by the enzyme glutathione reductase (Leustek et al., 2000). The ratio of reduced to oxidized glutathione delivers already an important information as it decreases under stress conditions that consume reducing equivalents. Moreover, glutathione is supposed to be involved in stress signaling, the detoxification of xenobiotics, it is the precursor of phytochelatines, which are important for heavy metal detoxification, acts as transport and storage form of reduced S and has a regulatory function in S assimilation (Leustek et al., 2000). These manifold functions illustrate the major importance of glutathione in plant S metabolism and stress response.

A direct antifungal mode of action was determined for S-rich proteins, phytoalexins such as camalexin, elemental S and the degradation products build from glucosinolates (Mithen, 1992; Kuć, 1994; Cooper et al., 1996; Wallsgrove et al., 1999; Hughes et al., 2000; Williams et al., 2002; Stec et al., 2004; Glawischnig, 2007). The toxicity of S-rich proteins such as thionins is explained by their ability to generate ion channels in cell membranes of pathogens and by this disturbing ion concentration gradients and cellular homeostasis (Shai, 1999; Hughes et al., 2000).

The antifungal mode of action of S^0 can be explained by its lipophilic character. S^0 may enter directly through the fungal cell wall disturbing redox reactions in the metabolism of the pathogen (Beffa, 1993). Beffa (1993) suggested that the fungicidal action of S^0 is mainly related to the oxidation of important sulfhydryl

Table 1 | Discoveries and progress in plant sulfur (S) research with respect to sulfur induced resistance (SIR) during the twentieth century.

Year	Scientific discoveries	References
1802	• William Forsyth discovered the fungicidal effect of elemental S	Forsyth, 1802
1860	• S was recognized as an essential plant nutrient, required for growth	Woodard, 1922
1872	• Robert Angus Smith coined the term "acid rain"	Seinfeld and Pandis, 1998
1956	• The common structure of glucosinolates was discovered	Ettlinger and Lundeen, 1956
1973	• Elucidation of the major steps in glucosinolate biosynthesis	Underhill et al., 1973
1976	• First description of the *Foyer-Halliwell-Asada*-cycle	Foyer and Halliwell, 1976
1977	• *Agrobacterium tumefaciens*- mediated gene transfer	Chilton et al., 1977
1979	• SO_2 exposure increase the glutathione content in sensitive trees	Grill et al., 1979
1982	• Description of the glutathione metabolism in higher plants and its function in transport, storage and detoxification of xenobiotics	Rennenberg, 1982
	• Detection of hydrogen sulfide (H_2S) emissions from leaf tissue in response to L-cysteine feeding	Sekiya et al., 1982
1984	• Description of the stimulating effect of abiotic stress and the restrictive impact of S deficiency on synthesis of S containing secondary plant metabolites	Gershenzon, 1984
	• Description of the "mustard oil bomb," a model of the subcellular organization of the glucosinolate-myrosinase system	Lüthy and Matile, 1984
1986	• Demonstration that leaf glucosinolates of *Brassica napus* can control fungal infection by *Leptospheria maculans*	Mithen et al., 1986
1989	• Plants can take up and use atmospheric H_2S as S source	De Kok et al., 1989
1990	• Localization of the γ-glutamylcysteine synthetase in higher plants	Hell and Bergmann, 1990
1994	• The term "sulfur induced resistance" (SIR) was introduced after field trials unraveled a relationship between S nutrition and plants susceptibility toward fungal diseases	Schnug et al., 1995
	• Significance of glutathione in plants under stress was demonstrated	Rennenberg and Brunold, 1994
	• Concept of "biofumigation" was developed	Angus et al., 1994
1995	• Isolation of three sulfate transporters for sulfate uptake by plant roots	Smith et al., 1995
1996	• Detection and cellular localization of elemental S in disease resistant genotypes of *Theobroma cacao*	Cooper et al., 1996
1999	• Overexpression of serineacetyltransferase (SAT) caused increased cysteine and glutathione contents accompanied by an increased resistance to oxidative stress	Blaszczyk et al., 1999
2000	• Interaction of sulfate reduction with N nutrition and major role of *O*-acetylserine in this regulation was shown at the transcriptional level	Koprivova et al., 2000
2001	• Identification and biochemical characterization of *Arabidopsis thaliana* sulfite oxidase	Eilers et al., 2001
2003	• Application of DNA macroarray technique to investigate the gene-to-metabolite networks regulating the S metabolism of *Arabidopsis*	Hirai et al., 2003
2004	• The regulatory function of the *O*-acetylserine(thiol)lyase (OAS-TL) in the S assimilation pathway was shown	Wirtz et al., 2004
2005	• Introduction of the term "sulfur enhanced defense" (SED)	Rausch and Wachter, 2005
	• Higher susceptibility of S deficient oilseed rape for different pathogens	Dubuis et al., 2005
	• The link between S assimilation and the stress hormone jasmonate (JA) was proven by macroarray technique	Jost et al., 2005
2006	• Identification of PAD2 as a γ-glutamylcysteine synthetase and the importance of glutathione in pathogen defense	Parisy et al., 2006
2009	• Indole glucosinolate biosynthesis and hydrolysis is required for callose accumulation in response to microbial pathogens	Clay et al., 2009
2012	• A shift from plant COS uptake to COS release with fungal infection	Bloem et al., 2012
	• Regulatory role of cytosolic cysteine/cytosolic OAS-TL in plant immune response	Alvarez et al., 2012; Tahir et al., 2013

groups. S^0 depositions in the vascular tissue of resistant varieties of *Theobroma cacao* in response to infection by *Verticillium dahliae* were considered as defense reaction causing the resistance of these varieties (Cooper et al., 1996; Resende et al., 1996).

Native glucosinolates display no fungal toxicity in contrast to their hydrolysis products, the isothiocyanates (ITC), which display a strong antimicrobial activity (Manici et al., 1997). Fungal inhibition is caused by irreversible reactions of ITC's with functional groups of proteins resulting in enzyme inactivation (Brown and Morra, 1997). In accordance with this not only the biosynthesis of indole glucosinolates was up-regulated by ethylene signaling after pathogen recognition in *Arabidopsis* but also the expression of myrosinase enzymes which catalyze their hydrolysis (Clay et al., 2009). Additionally the biosynthesis of indole glucosinolates was shown to be required for callose depositions in response to microbial pathogens (Clay et al., 2009). Therefore,

Table 2 | Possible mode of action of S-containing plant compounds in stress resistance and in response to fungal infection.

Compound	Mode of action in stress resistance and after fungal infection	References
Cysteine	–Precursor for all relevant S containing metabolites –Cytosolic cysteine has a regulatory function in the establishment and signaling of the plant response to pathogens –Increase with fungal infection –Link to salicylic acid and by this to systemic acquired resistance via CoASH and essential for the initiation of the hypersensitive response (HR)	Luckner, 1990; Bloem et al., 2007; Alvarez et al., 2012
Glutathione	–Participation in antioxidative defense –Detoxification of xenobiotics by targeting them into the vacuole –Involved in phytochelatine biosynthesis/ detoxification of heavy metals –Messenger in the hypersensitive response (HR) –Rapid accumulation after fungal attack	Edwards et al., 1991; Rea et al., 1998;Leustek and Saito, 1999; Cobbett, 2000; Foyer and Rennenberg, 2000; Vanacker et al., 2000
S-containing volatiles	–H_2S causes disturbances in redox reactions –Release of H_2S and COS increased with fungal infections	Bloem et al., 2007, 2012
S-rich proteins	–Pathogen-induced or constitutive expression (defensins) –Thionins are enhanced locally and systemically after infection –Toxic mode of thionins: disruption of the cell wall structure; generation of ion channels	Hughes et al., 2000; Stec et al., 2004; Kruse et al., 2007
Phytoalexins	–*De-novo* synthesis after pathogen attack	Kuć, 1994
S^0	–S^0 accumulates after fungal infection in vascular tissue –Disturbances of the respiratory chain –Oxidation of sulfhydryl groups	Beffa, 1993; Cooper et al., 1996; Williams et al., 2002
Glucosinolates	–Their degradation products (isothiocyanates) exhibit a toxic and repellent effect → reason for its use in biofumigation	Mithen, 1992; Wallsgrove et al., 1999

glucosinolate biosynthesis seems to be involved in pathogen defense in more than one way in glucosinolate containing plants.

The concentrations of all S containing metabolites, the amino acids cysteine and methionine as well as primary and secondary S compounds were reduced with S deficiency or can be increased by S fertilization (Salac et al., 2005; Bloem et al., 2007). It was observed that the gas exchange of H_2S and carbonyl sulfide (COS) between plants and atmosphere changed in relation to S supply and fungal infection. As long as enough S is available plants release H_2S into the atmosphere. This happens most likely to reduce excess S in their metabolism or as a signal molecule (Rennenberg, 1984; Bloem et al., 2012). Under conditions of S deficiency plants take up and use gaseous S compounds from the ambient air (De Kok et al., 1989).

Linear relationships were determined between the S supply and most of the mentioned S containing compounds. When next to the S supply a fungal infection was studied the results became less conclusive (Bloem et al., 2004; Salac et al., 2005). In many trials S fertilization decreased fungal infection (Wang et al., 2003). But in some trials no effect on disease severity could be determined despite of the fact that a stress response occurred, indicated by changes in the S metabolism (Salac et al., 2005). The S metabolism is only one branch of the overall plant stress response. Several other pathways and metabolites are involved (Bennett and Wallsgrove, 1994; Morrissey and Osbourn, 1999). Amongst others flavonoids and phenolics are shown to be major biochemical marker against fungal infections (Shanmugam et al., 2010; Datta and Lal, 2012). Cell wall strengthening is another important resistance response against fungi as it helps to inhibit pathogen entry. Accumulation of cell wall-bound phenolics, the monomers of lignin, is part of this process (Swarupa et al., 2014). It was shown that cell wall-bound phenolics increase together with soluble phenolics in plant tissue after fungal infection (Huckelhoven, 2007). Moreover several studies show a close link between the S metabolism, mineral deficiency or increased internal demand and hormonal signaling by methyl jasmonate and possibly other hormones (Hirai et al., 2003; Saito, 2004; Jost et al., 2005) (**Figure 1**).

Table 3 gives an example which changes occur in the S metabolism in response to S fertilization and fungal infection (Bloem et al., 2012). *Brassica napus* was artificially infected by *Sclerotinia sclerotiorum* and the plants displayed strong symptoms of S deficiency without S application.

With increasing S supply total S and SO_4-S increased in leaves as well as the cysteine and glutathione content (**Table 3**). Only γ–glutamylcysteine, which is an intermediate in the biosynthesis of glutathione, was lower with S fertilization indicating to a fast turn-over to build glutathione under these conditions. H_2S exchange shifted from uptake, indicated by the negative value in S deficient plants, to H_2S release in the fertilized ones. COS was taken up in S fertilized as well as in non-fertilized plants.

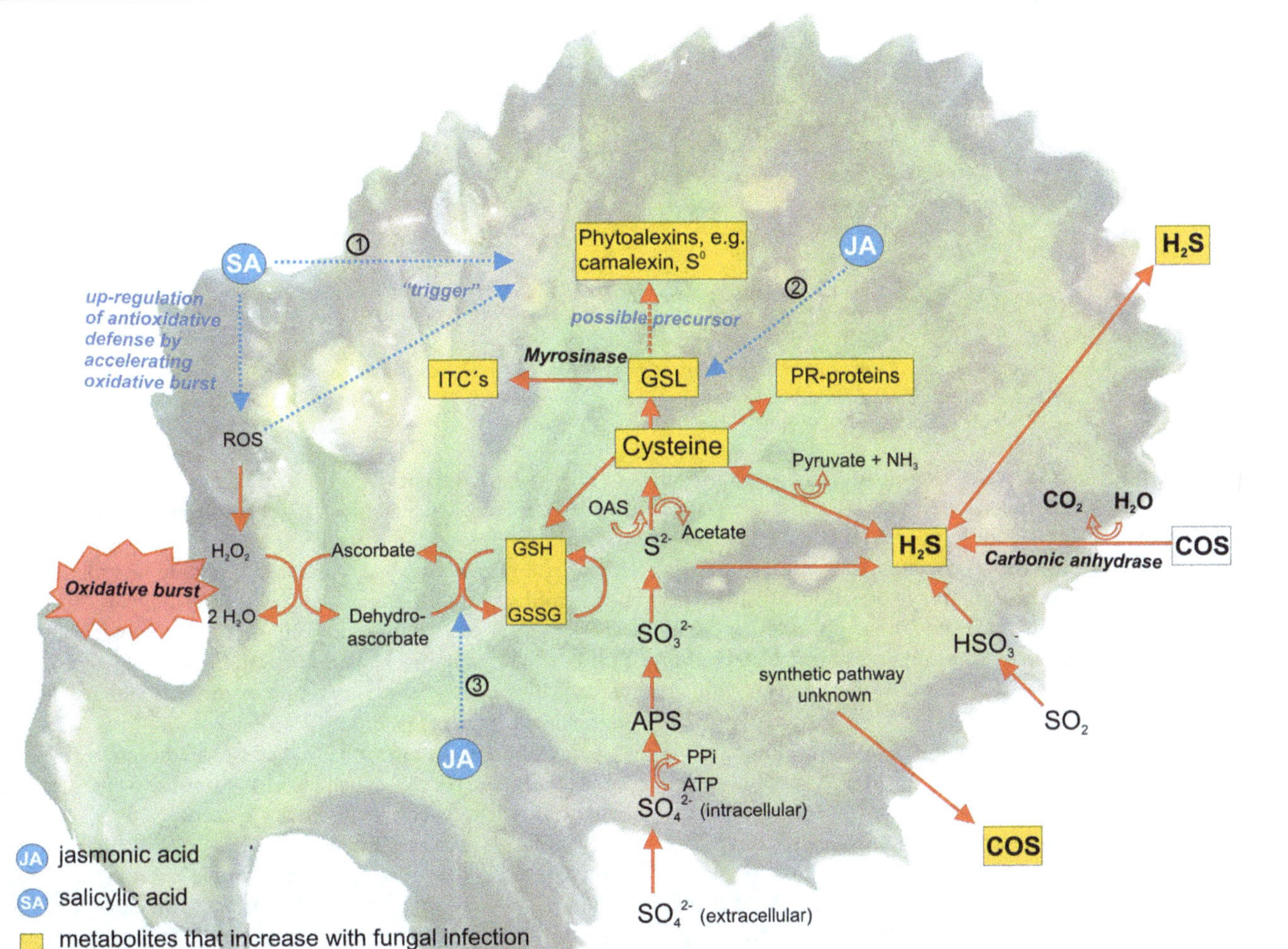

① SA is up-regulating the antioxidative defense by accelerating oxidative burst and by triggering camalexin biosynthesis.

② JA is inducing indol GSL biosynthesis and a set of S-related genes such as serineacetyltransferase (SAT) and adenosinephosphosulfate (APS) reductase (Jost et al., 2005).

③ JA can regulate the ascorbate and glutathione content under stress: ascorbate and GSH increase due to inhibition of ascorbate oxidase activity and enhancement of monodehydroascorbate reductase, dehydroascorbate reductase and glutathione reductase activities (Cai et al., 2011).

Abbreviations: ROS reactive oxygen species; ITC isothiocyanate; GSL glucosinolate; H_2S hydrogen sulfide; COS carbonyl sulfide; APS adenosinephosphosulfate ; GSH and GSSG reduced and oxidized glutathione; PR-proteins pathogenicity related proteins

FIGURE 1 | Sulfur metabolites and pathways involved in the defense against fungal pathogens in *Brassica* species. Metabolites in yellow boxes were found to increase after fungal attack (Williams and Cooper, 2004; Glazebrook, 2005; Jost et al., 2005; Kruse et al., 2007; Bloem et al., 2012).

Infection with *S. sclerotiorum* caused significant changes in the S metabolism. The total S content decreased as well as cysteine and γ–glutamylcysteine while glutathione significantly increased. Additionally plants were analyzed for their potential to take up or release H_2S and COS in the first days after infection (Bloem et al., 2012). In **Table 3** the data from 2 days after infection are shown when the strongest plant response was observed. H_2S release was significantly increased by infection. The change in COS was even more striking as COS was changed from uptake to release (Bloem et al., 2012). The data clearly revealed that plants responded to the infection by several changes in their S metabolism. Nevertheless, the visual scoring revealed that the infection rate was not reduced by the higher S supply at this stage of infection (see Bloem et al., 2012).

Likewise Raj and Srivastava (1977) showed that the total S content of infected tissue of *Brassica juncea* was inversely correlated with the pathogenicity of different isolates of *Macrophomina phaseolina* and suggested that the pathogens are able to

Table 3 | Impact of S nutrition and fungal infection with *Sclerotinia sclerotiorum* on the S status, S-containing metabolites and the release of gaseous S compounds from *Brassica napus* (variety *Heros*) at stem elongation (data derived from Bloem et al., 2012).

		Total S	SO_4-S	Cysteine	γ-GC	GSH^1_{tot}	H_2S^2	COS^2
		[mg g^{-1} dw]		[nmol g^{-1} dw]			[pg min^{-1} g^{-1} dw]	
S fertilization	0	0.74 b	0.11 b	37.4 b	87.6 a	276 b	−91 b	−63 a
[mg pot^{-1}]	250	5.63 a	1.34 a	232.0 a	39.8 b	2370 a	41 a	−174 a
Infection with	no	4.28 a	0.73 a	236.0 a	88.4 a	1383 b	41 a	−174 b
Sclerotinia sclerotiorum	yes	2.75 b	0.83 a	114.2 b	38.9 b	1851 a	123 b	382 a

[1] GSH_{tot}, total glutathione content.

[2] The gas measurement was performed on non-infected control plants to determine the effect of S fertilization and from S fertilized plants that were infected for 2 days for the impact of infection. Sulfur contents and metabolites were determined in leaf material of B. napus while the gas release was measured from whole intact plants.

metabolize S from the host plant. Losses in total S with fungal infection could be also explained by the release of gaseous S compounds like shown in **Table 3**. In some studies also a higher total S content was observed in response to infection indicating to an up-regulation of the S assimilation due to infection. Most likely it is the timing of sampling or the degree of infection which determine if an increase or decrease of a compound is determined in response to infection as a cascade of reactions takes place (Bloem et al., 2007).

A direct relationship between fungal infection and S metabolism as shown exemplary in **Figure 1** was also found for other host-pathogen interactions. Infection of oilseed rape with *Pyrenopeziza brassicae* increased the cysteine and glutathione content in leaves as well as the activity of the L-cysteine-desulfhydrase, an enzyme that releases H_2S during cysteine degradation (Bloem et al., 2004). A higher release of H_2S after fungal infection was determined in grapes (*Vitis vinifera* L.) infected by *Uncinula necator* (Bloem et al., 2007). Gaseous S compounds seem to be involved in stress response but to date their function is not fully understood. A possible role could be in stress signaling or as regulatory compounds comparable to the effect in mammalian cells where H_2S is involved in the regulation of the intracellular redox-homeostasis and glutathione generation (Ju et al., 2013).

Also Kruse et al. (2007) determined a steep and fast increase not only for H_2S, but also for cysteine, glutathione and phytoalexins during the initial phase of pathogenesis. The important role of cysteine in pathogenesis was proven by Alvarez et al. (2012) and Tahir et al. (2013). Alvarez et al. (2012) could show that mutants with increased cytosolic cysteine content are resistant to biotrophic as well as necrotrophic pathogens, while mutants with decreased cytosolic cysteine contents are more susceptible. Also Tahir et al. (2013) found that decreased cytosolic cysteine contents resulted in enhanced disease susceptibility against infection with virulent and non-virulent *Pseudomonas syringae* strains.

Though the sequence, magnitude and efficacy of all individual S metabolites involved in the activation and strengthening of plant defenses by S fertilization are not yet fully known, these could be released in a chain reaction triggered by the pathogen and mediated by the S status of the plant (Haneklaus et al., 2006). It seems possible that infection triggers the activation of all effective resistance mechanisms of the host.

Trials where the effect of S nutrition on fungal infection was studied are compiled in **Table 4**. Plant pathogens are often divided into biotrophs and necrotrophs despite of the fact that there are several transitions. Biotrophs feed on living host tissue while necrotrophs cause die-off and feed on the remains (Glazebrook, 2005). SIR was proven for biotrophic and necrotrophic pathogens (**Table 4**). Different mechanisms in pathogen response are important for biotrophic and necrotrophic pathogens and a schematic model is summarized in **Figure 2**.

Generally defense reactions that cause the die-off of cells such as oxidative burst and hypersensitive response (HR) are only beneficial when repelling the attack of a biotrophic pathogen. In contrast, it is not predicted that the cell death of a host plant will limit the growth of necrotrophic pathogens. It is the opposite way round; necrotrophic fungi can elicit a defense response such as oxidative burst in a susceptible host plant causing necrosis (Winterberg et al., 2014). SA dependent defense is more frequently observed against biotrophs and JA/ET dependent defense against necrotrophs but there are exceptions. The fact that pathogenesis-related proteins are not expressed and JA dependent signaling is not activated against a special biotrophic pathogen does not mean that they are not active in case they are triggered (Glazebrook, 2005). S containing compounds are involved in both defense lines (see also **Figure 1**). Glutathione is involved in detoxification of ROS, many pathogenesis-related proteins contain S (phytoalexins, thionins, defensins) and SA needs coenzyme A (CoASH) as a precursor. Therefore, it is hardly possible to predict the efficacy of S against special fungi based on the lifestyle of the pathogen.

It is hard to explain why in some trials a clear relationship between the S supply and the extent of fungal infection was found whilst in others with the same pathosystem no such response was observed. Probably it is the timing and extent of plants defense response which decides over resistance or susceptibility while the nutritional status of the crop determines the extent of defense. Moreover, the type of pathogen and its pathogenicity, infection severity and other environmental factors are important as well.

PRACTICAL RELEVANCE OF SIR

Optimizing the S nutritional status of a plant is equivalent of enhancing the capability of a plant to cope with stress. The identification of the mechanisms causing SIR is an important

Table 4 | Influence of soil S application on pathogen development of different host pathogen interactions.

Host	Pathogen	Pathogen classification[1]	Lifestyle	Trial[2]	S-fertilization	Change in pathogen development	References
CONFIRMED SIR							
Brassica napus L.	*Sclerotinia sclerotiorum*	A	necrotrophic (heterotrophic)	PT	120 mg S kg^{-1} soil	Disease index (DI) was reduced by 5% in comparison to a control without S application	Wang et al., 2003
Zea mays	*Bipolaris maydis* (Southern leaf blight)	A	necrotrophic	PT	120 mg S kg^{-1} soil	DI was reduced by 37% in comparison to a control without S application	Wang et al., 2003
Brassica napus L.	*Botrytis cinerea* (Gray mold)	A	necrotrophic	WC	0.5 mM $MgSO_4$	Lesions were 24-times larger in S-starved plants of cultivar *Express* and 3.7-fold larger in cultivar *Bienvenue*	Dubuis et al., 2005
Brassica napus L.	*Leptosphaeria maculans*	A	facultative necrotrophic, initially biotrophic	WC	0.5 mM $MgSO_4$	Lesions were 1.9-times larger in S-starved plants (cultivar *Bienvenue*)	Dubuis et al., 2005
Arabidopsis thaliana	*Alternaria brassicicola*	A	necrotrophic	WC	50 μM vs. 500 μM SO_4	DNA from *A. brassicicola* was 3-time more abundant on plants grown on 50μM SO_4 in comparison to plants grown on 500μM	Kruse et al., 2012
Brassica napus L.	*Pyrenopeziza brassicae*	A	hemi-biotrophic	FT	Plots with and without S fertilization	A non-resistant and a resistant oilseed rape variety were compared with and without S application and fungicide treatment: the non-resistant variety showed a much stronger response to fungicide under S deficiency	Schnug et al., 1995
Solanum lycopersicum L.	*Verticillium dahliae*	A	hemi-biotrophic	WC	0.016 mM vs. 25 mM K_2SO_4	Supra-optimal S supply significantly reduced the number of infected cells and the amount of *V. dahliae* gDNA in vascular tissue of the hypocotyl	Bollig et al., 2013
Vitis vinifera L.	*Uncinula necator*	A	obligate biotrophic	FT	250 or 500 kg S^0 ha^{-1} (soil applied)	Proportion of infected leaves and berries decreased by more than 80% with soil S application	Haneklaus et al., 2007
Solanum tuberosum L.	*Rhizoctonia solani*	B	necrotrophic	FT	50 kg S^0 ha^{-1} (soil applied)	Soil applied S^0 reduced infection rate by 41% in comparison to control without S application	Klikocka et al., 2005
Triticum aestivum L.	*Rhizoctonia cerealis*	B	necrotrophic	PT	120 mg S kg^{-1} soil	DI was reduced by 44% in comparison to a control without S application	Wang et al., 2003
Brassica napus L.	*Peronospora parasitica*	O	obligate biotrophic	FT	100 kg S ha^{-1}	Decrease in disease incidence and severity was found	Salac et al., 2005
Brassica napus L.	*Phytophthora brassicae*	O	hemi-biotrophic	WC	0.5 mM $MgSO_4$	Lesions were 3.3-times larger in S-starved plants of cultivar *Express*	Dubuis et al., 2005

(Continued)

Table 4 | Continued

Host	Pathogen	Pathogen classification[1]	Lifestyle	Trial[2]	S-fertilization	Change in pathogen development	References
CONFLICTING RESULTS OF S FERTILIZATION ON PATHOGEN DEVELOPMENT							
Gossypium L.	*Fusarium oxisporum* (Fusarium wilt)	A	necrotrophic	PT	160 mg S kg^{-1} soil	DI was reduced by 8% in comparison to a control; with all other S rates (40, 80, 120 mg S kg^{-1} soil) DI was higher than in the control	Wang et al., 2003
Brassica napus L.	*Leptosphaeria maculans*	A	facultative necrotrophic, initially biotrophic	FT	100 kg S ha^{-1}	Depending on season, year and site controversial effects on disease incidence and severity were found	Salac et al., 2005
Gossypium L.	*Verticillium dahliae*	A	hemi-biotrophic	PT	40 mg S kg^{-1} soil	DI was reduced by 47% in comparison to a control, but with higher S application rates DI increased again and was significantly higher than in the control when 160 mg S kg^{-1} soil were applied	Wang et al., 2003
Brassica napus L.	*Pyrenopeziza brassicae*	A	hemi-biotrophic	FT	100 kg S ha^{-1}	Depending on season, year and site controversial effects on disease incidence and severity were found	Salac et al., 2005

[1] *Pathogen classification: A, Ascomycete; B, Basidiomycete; O, Oomycete.*
[2] *Trial: PT, Pot trial; WC, Water culture; FT, Field trial; VWC, Vermiculite water culture.*

milestone for a sustainable agricultural production as the input of fungicides can be minimized by crop specific S fertilization and a higher resistance due to S will not be rapidly broken by new pathotypes. It is possible to optimize the S nutrition without understanding all mechanisms underlying SIR. For winter crops a first S application in autumn was shown to be advantage with respect to disease resistance followed by the regular S application in spring. An increasing S supply is associated with higher contents of cysteine, glutathione, H_2S, and glucosinolates in *Brassica* crops so that plants with a higher content of phytoanticipins might not only have *a priori* a better protection against pathogens, but also be able to activate resistance mechanisms more rapidly and intensely. In addition, an instantly high S supply satisfying the elevated S demand after a fungal attack may play a pivotal role in SIR, even when the nutrient demand of the crop is well exceeded by such an S application (Haneklaus et al., 2009).

OPEN QUESTIONS AND FUTURE PROJECTIONS

It is generally difficult to assign a change in plant metabolism to a specific stress factor, as usually a variety of abiotic and biotic stress factors occur at the same time and can induce antagonistic responses resulting in accumulation, degradation and consumption of primary and secondary metabolites. Therefore, standardized experimental conditions are important to improve comparability of results.

Moreover, more field studies and infection trials are necessary accompanied by molecular research to unravel the relationship between S supply and fungal infection and by this to enable researchers and farmers to adopt the results into new fertilizer concepts. There are many unknowns affecting plant response such as timing of application, kind of fungal pathogen, crop species, climatic conditions, and cross-talk with other macro- and micronutrients. Therefore, to date it is not possible to induce a stress response by S application that will certainly reduce or prevent a crop from fungal infection. It is necessary to further elucidate the cross-talk between different pathways to understand which other parameters need to be optimized in order to reach the full potential of plants own pathogen defense.

Much more work in the field of phytopathology is necessary to solve the questions why only some pathogens are affected by S nutrition and which is the exact mode of action by which the S supply is affecting fungal pathogens. Could be the lifestyle (biotrophic, necrotrophic, heterotrophic) of a pathogen important for plants defense in relation to S or is the timing of infection in relation to plant development most important for the course of infection? Which part of metabolite changes is caused by the host and which one by the pathogen? These are only some of the manifold topics and questions where further research is necessary.

Moreover, it is not possible to transfer all results obtained from research with model plants such as *Arabidopsis* to other species so that studies on major agricultural crops are important especially as most agricultural crops do not contain glucosinolates.

Nevertheless, the manifold results, which point to a relationship between S nutrition and crop resistance, indicate that in future the crucial factors will be identified. There are still some agricultural diseases where the efficiency of chemical fungicides is limited. For example currently, no fungicides are available

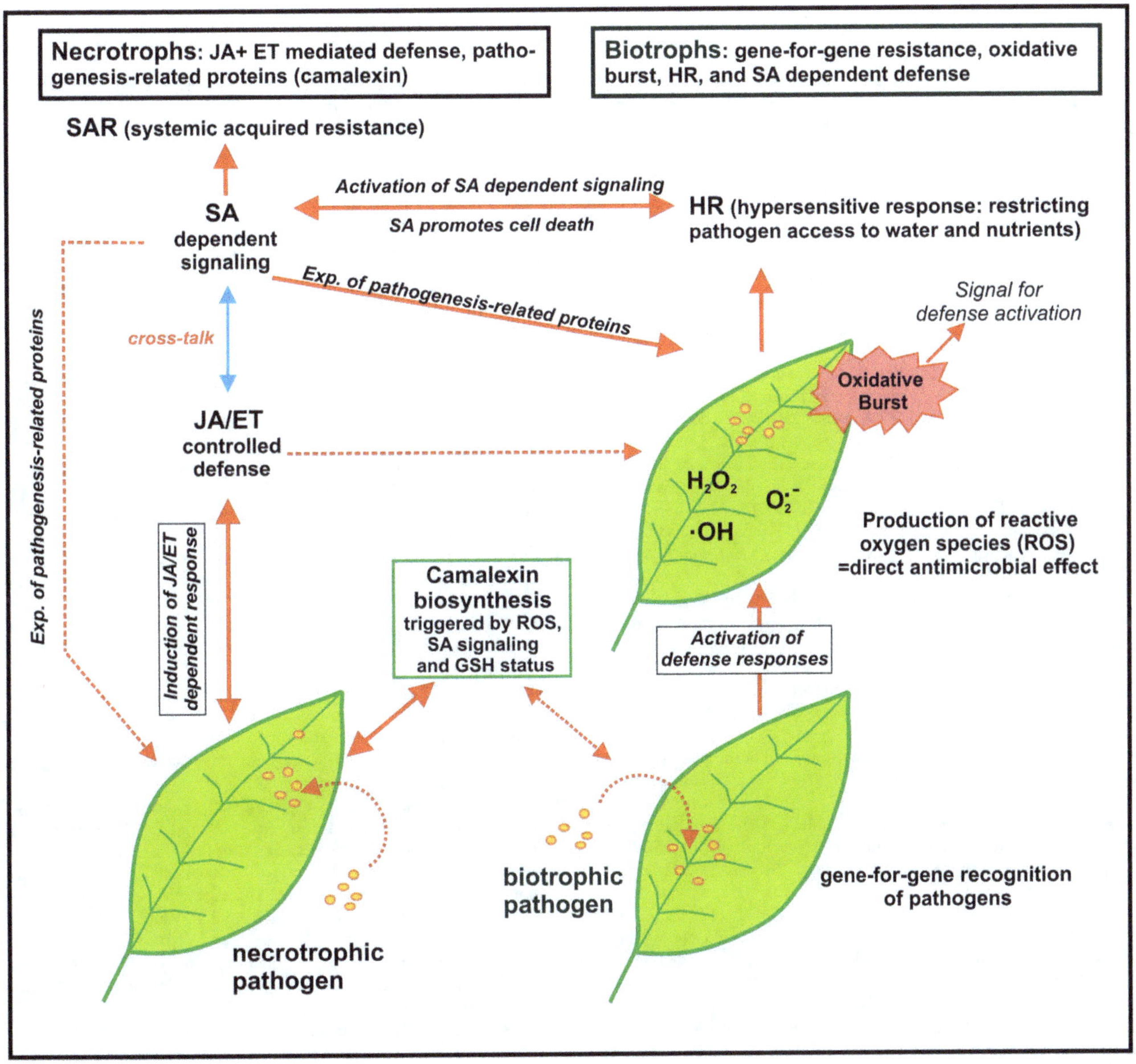

FIGURE 2 | Model of the response of plants to biotrophic and necrotrophic plant pathogens (adapted from Glazebrook (2005): displayed are the interactions of *Arabidopsis* with the biotrophs *Peronospora parasitica* and *Erysiphe* ssp. and with the necrotrophs *Alternaria brassicicola* and *Botrytis cinerea*; SA, salicylic acid, JA, jasmonic acid, ET, ethylene; broken line arrows indicate to a possible interaction but which was not found in the chosen experiments while the solid line arrows indicate to the observed plant-pathogen-response). The defense reaction of *Arabidopsis* against biotrophic pathogens start with gene-for-gene recognition of the pathogen followed by rapid activation of defense and the production of reactive oxygen species (ROS), the so-called "oxidative burst," which is by self a signal for defense activation. ROS production is connected with the hypersensitive response (HR), also called "programmed cell death," which limits the access of biotrophs that feed on living tissue to water and nutrients. HR is associated with the activation of the salicylic acid (SA) dependent signaling pathway that is connected with systemic acquired resistance (SAR) and the expression of pathogenesis-related proteins. For necrotrophic pathogens a different defense line takes place as they feed on dead plant tissue and host cell death is not predicted to limit their growth. Defense against necrotrophic pathogens is mainly mediated by JA and ET controlled defense as well as production of phytoalexins such as camalexin. The broken line arrows indicate that also mixed defense lines are possible for other biotrophic or nectrotrophic pathogens.

to control *Verticillium* wilt. Therefore, fertilizer strategies which improve the plants potential and resistance against fungal diseases are still of high importance not only in organic farming but in conventional agriculture as well.

ACKNOWLEDGMENTS

The authors thank Stan Kopriva and Kerstin Panten for their valuable comments on the manuscript. We apologize to the colleagues whose papers we could not cite due to space limitations.

REFERENCES

Alvarez, C., Bermudez, M. A., Romero, L. C., Gotor, C., and Garcia, I. (2012). Cysteine homeostasis plays an essential role in plant immunity. *New Phytol.* 193, 165–177. doi: 10.1111/j.1469-8137.2011.03889.x

Angus, J. F., Gardner, P. A., Kirkegaard, J. A., and Desmarchelier, J. M. (1994). Biofumigation: isothiocyanates released from Brassica roots inhibit growth of the take-all fungus. *Plant Soil* 162, 107–112. doi: 10.1007/BF01416095

Beffa, T. (1993). Inhibitory action of elemental sulphur (S^0) on fungal spores. *Can. J. Microbiol.* 39, 731–735. doi: 10.1139/m93-107

Bennett, R. N., and Wallsgrove, R. M. (1994). Secondary metabolites in plant defence mechanisms. *New Phytol.* 127, 617–633. doi: 10.1111/j.1469-8137.1994.tb02968.x

Blaszczyk, A., Brodzik, R., and Sirko, A. (1999). Increased resistance to oxidative stress in transgenic tobacco plants overexpressing bacterial serine acetyltransferase. *Plant J.* 20, 237–243. doi: 10.1046/j.1365-313x.1999.00596.x

Bloem, E., Haneklaus, S., Kesselmeier, J., and Schnug, E. (2012). Sulfur fertilization and fungal infections affect the exchange of H_2S and COS from agricultural crops. *J. Agric. Food Chem.* 60, 7588–7596. doi: 10.1021/jf301912h

Bloem, E., Haneklaus, S., Salac, I., Wickenhäuser, P., and Schnug, E. (2007). Facts and fiction about sulphur metabolism in relation to plant-pathogen interactions. *Plant Biol.* 9, 596–607. doi: 10.1055/s-2007-965420

Bloem, E., Haneklaus, S., and Schnug, E. (2005). Significance of sulfur compounds in the protection of plants against pests and diseases. *J. Plant Nutr.* 28, 763–784. doi: 10.1081/PLN-200055532

Bloem, E., Riemenschneider, A., Volker, J., Papenbrock, J., Schmidt, A., Salac, I., et al. (2004). Sulphur supply and infection with *Pyrenopeziza brassicae* influence L-cysteine desulphydrase activity in *Brassica napus* L. *J. Exp. Bot.* 55, 2305–2312. doi: 10.1093/jxb/erh236

Bollig, K., Specht, A., Myint, S. S., Zahn, M., and Horst, W. J. (2013). Sulphur supply impairs spread of *Verticillium dahliae* in tomato. *Eur. J. Plant Pathol.* 135, 81–96. doi: 10.1007/s10658-012-0067-5

Brown, P. D., and Morra, M. J. (1997). Control of soil-borne plant pests using glucosinolate-containing plants. *Adv. Agron.* 61, 167–231. doi: 10.1016/S0065-2113(08)60664-1

Cai, Y., Cao, S., Yang, Z., and Zheng, Y. (2011). MeJA regulates enzymes involved in ascorbic acid and glutathione metabolism and improves chilling tolerance in loquat fruit. *Postharvest Biol. Technol.* 59, 324–326. doi: 10.1016/j.postharvbio.2010.08.020

Chilton, M. D., Drummond, M. H., Merlo, D. J., Sciaky, D., Montoya, A. L., Gordon, M. P., et al. (1977). Stable incorporation of plasmid DNA into higher plant cells: molecular basis of crown gall tumorigenesis. *Cell* 11, 263–271. doi: 10.1016/0092-8674(77)90043-5

Clay, N. K., Adio, A. M., Denoux, C., Jander, G., and Ausubel, F. M. (2009). Glucosinolate metabolites required for an *Arabidopsis* innate immune response. *Science* 323, 95–101. doi: 10.1126/science.1164627

Cobbett, C. S. (2000). Phytochelatines and their roles in heavy metal detoxification. *Plant Physiol.* 123, 825–832. doi: 10.1104/pp.123.3.825

Cooper, R. M., Resende, M. L. V., Flood, J., Rowan, M. G., Beale, M. H., and Potter, U. (1996). Detection and cellular localization of elemental sulphur in disease resistant genotypes of *Theobroma cacao*. *Nature* 379, 159–162. doi: 10.1038/379159a0

Datnoff, L. E., Elmer, W. H., and Huber, D. M. (2007). *Mineral Nutrition and Plant Disease*. Minnesota, MN: APS Press, St. Paul.

Datta, J., and Lal, N. (2012). Temporal and special changes in phenolic compounds in response to *Fusarium* wilt in chickpea and pigeonpea. *Cell. Mol. Biol.* 58, 96–102.

De Kok, L. J., Stahl, K., and Rennenberg, H. (1989). Fluxes of atmospheric hydrogen sulphide to plant shoots. *New Phytol.* 112, 533–542. doi: 10.1111/j.1469-8137.1989.tb00348.x

Dubuis, P. H., Marazzi, C., Städler, E., and Mauch, F. (2005). Sulphur deficiency causes a reduction in antimicrobial potential and leads to increased disease susceptibility of oilseed rape. *J. Phytopathol.* 153, 27–36. doi: 10.1111/j.1439-0434.2004.00923.x

Edwards, R., Blount, J. W., and Dixon, R. A. (1991). Glutathione and elicitation of the phytoalexin response in legume cultures. *Planta* 184, 403–409. doi: 10.1007/BF00195343

Eilers, T., Schwarz, G., Brinkmann, H., Witt, C., Richter, T., Nieder, J., et al. (2001). Identification and biochemical characterization of *Arabidopsis thaliana* sulfite oxidase. *J. Biol. Chem.* 276, 46989–46994. doi: 10.1074/jbc.M108078200

Ettlinger, M. G., and Lundeen, A. J. (1956). The structures of sinigrin and sinalbin—an enzymatic rearrangement. *J. Am. Chem. Soc.* 78, 4172–4173. doi: 10.1021/ja01597a090

Forsyth, W. (1802). *A Treatise on the Culture and Management of Fruit Trees*. London: Nichols and Son.

Foyer, C. H., and Halliwell, B. (1976). The presence of glutathione and glutathione reductase in chloroplasts: a proposed role in ascorbic acid metabolism. *Planta* 133, 21–25. doi: 10.1007/BF00386001

Foyer, C. H., and Rennenberg, H. (2000). "Regulation of glutathione synthesis and its role in abiotic and biotic stress defence," in *Sulfur nutrition and sulfur assimilation in higher plants*, eds C. Brunold, H. Rennenberg, L. J. De Kok, I. Stulen, and J. C. Davidian (Berne: Paul Haupt Publishers), 127–153.

Gershenzon, J. (1984). "Changes in the levels of plant secondary metabolites under water and nutrient stress," in *Phytochemical Adaptations to Stress: Proceedings*, eds B.N. Timmermann, C. Steelink, and F.A. Loewus (New York, NY: Plenum Press), 273–320.

Gigolashvili, T., and Kopriva, S. (2014). Transporters in plant sulfur metabolism. *Front. Plant Sci.* 5:442. doi: 10.3389/fpls.2014.00442

Glawischnig, E. (2007). Camalexin. *Phytochemistry* 68, 401–406. doi: 10.1016/j.phytochem.2006.12.005

Glazebrook, J. (2005). Contrasting mechanisms of defense against biotrophic and necrotrophic pathogens. *Annu. Rev. Phytopathol.* 43, 205–227. doi: 10.1146/annurev.phyto.43.040204.135923

Grill, D., Esterbauer, H., and Klösch, U. (1979). Effect of sulphur dioxide on glutathione in leaves of plants. *Environ. Pollut.* 19, 187–194. doi: 10.1016/0013-9327(79)90039-9

Halkier, B. A., and Gershenzon, J. (2006). Biology and biochemistry of glucosinolates. *Annu. Rev. Plant Biol.* 57, 303–333. doi: 10.1146/annurev.arplant.57.032905.105228

Haneklaus, S., Bloem, E., and Schnug, E. (2006). Disease control by sulphur induced resistance. *Asp. Appl. Biol.* 79, 221–224.

Haneklaus, S., Bloem, E., and Schnug, E. (2007). "Sulfur and Plant Disease," in *Mineral Nutrition and Plant Disease*, eds L. E. Datnoff, W. H. Elmer, and D. M. Huber (Minnesota, MN: APS Press; St. Paul), 101–118.

Haneklaus, S., Bloem, E., and Schnug, E. (2009). "Plant disease control by nutrient management: sulphur," in *Disease Control in Crops – Biological and Environmentally Friendly Approaches*, ed D. Walters (Chichester: Wiley-Blackwell), 221–236.

Hell, R., and Bergmann, L. (1990). γ-Glutamylcysteine synthetase in higher plants: catalytic properties and subcellular localization. *Planta* 180, 603–612. doi: 10.1007/BF02411460

Hirai, M. Y., Fujiwara, T., Awazuhara, M., Kimura, T., Noji, M., and Saito, K. (2003). Global expression profiling of sulfur-starved *Arabidopsis* by DNA macroarray reveals the role of O-acetyl-L-serine as a general regulator of gene expression in response to sulfur nutrition. *Plant J.* 33, 651–663. doi: 10.1046/j.1365-313X.2003.01658.x

Huber, D. M., and Haneklaus, S. (2007). Managing nutrition to control plant disease. *FAL Agricul. Res.* 57, 313–322.

Huckelhoven, R. (2007). Cell wall-associated mechanisms of disease resistance and susceptibility. *Annu. Rev. Phytopathol.* 45, 101–127. doi: 10.1146/annurev.phyto.45.062806.094325

Hughes, P., Dennis, E., Whitecross, M., Llewellyn, D., and Gage, P. (2000). The cytotoxic plant protein, beta-purothionin, forms ion channels in liquid membranes. *J. Biol. Chem.* 275, 823–827. doi: 10.1074/jbc.275.2.823

Jost, R., Altschmied, L., Bloem, E., Bogs, J., Gershenzon, J., Hähnel, U., et al. (2005). Expression profiling of metabolic genes in response to methyl jasmonate reveals regulation of genes of primary and secondary sulfur-related pathways in *Arabidopsis thaliana*. *Photosynth. Res.* 86, 491–508. doi: 10.1007/s11120-005-7386-8

Ju, Y., Zhang, W., Pei, Y., and Yang, G. (2013). H_2S signaling in redox regulation of cellular functions. *Can. J. Physiol. Pharmacol.* 91, 8–14. doi: 10.1139/cjpp-2012-0293

Klikocka, H., Haneklaus, S., Bloem, E., and Schnug, E. (2005). Influence of sulfur fertilization on infection of potato tubers with *Rhizoctonia solani* and *Streptomyces scabies*. *J. Plant Nutr.* 28, 819–833. doi: 10.1081/PLN-200055547

Kopriva, S. (2006). Regulation of sulfate assimilation in *Arabidopsis* and beyond. *Ann. Bot.* 97, 479–495. doi: 10.1093/aob/mcl006

Koprivova, A., Suter, M., Op den Camp, R., Brunold, C., and Kopriva, S. (2000). Regulation of sulfate assimilation by nitrogen in *Arabidopsis*. *Plant Physiol.* 122, 737–746. doi: 10.1104/pp.122.3.737

Kruse, C., Haas, F. H., Jost, R., Reiser, B., Reichelt, M., Wirtz, M., et al. (2012). Improved sulfur nutrition provides the basis for enhanced production of sulfur-containing defense compounds in *Arabidopsis thaliana* upon inoculation with *Alternaria brassicicola*. *J. Plant Physiol.* 169, 740–743. doi: 10.1016/j.jplph.2011.12.017

Kruse, C., Jost, R., Lipschis, M., Kopp, B., Hartmann, M., and Hell, R. (2007). Sulfur-enhanced defence: effects of sulfur metabolism, nitrogen supply, and pathogen lifestyle. *Plant Biol.* 9, 608–619. doi: 10.1055/s-2007-965432

Kuć, J. (1994). Relevance of phytoalexins – a critical review. *Acta Horticulturae* 381, 526–539.

Leustek, T., Martin, M. N., Bick, J. A., and Davies, J. P. (2000). Pathways and regulation of sulfur metabolism revealed through molecular and genetic studies. *Annu. Rev. Plant Physiol. Plant Mol. Biol.* 51, 141–165. doi: 10.1146/annurev.arplant.51.1.141

Leustek, T., and Saito, K. (1999). Sulfate transport and assimilation in plants. *Plant Physiol.* 120, 637–643. doi: 10.1104/pp.120.3.637

Luckner, M. (1990). *Secondary Metabolism in Microorganisms, Plants, and Animals*. Jena: VEB Gustav Fischer Verlag.

Lüthy, B., and Matile, P. (1984). The mustard oil bomb: rectified analysis of the subcellular organization of the myrosinase system. *Biochem. Physiol. Pflanzen.* 179, 5–12. doi: 10.1016/S0015-3796(84)80059-1

Manici, L. M., Lazzeri, L., and Palmieri, S. (1997). *In vitro* fungitoxic activity of some glucosinolates and their enzyme-derived products toward plant pathogenic fungi. *J. Agric. Food Chem.* 45, 2768–2773. doi: 10.1021/jf9608635

Mithen, R. (1992). Leaf glucosinolate profiles and their relationship to pest and disease resistance in oilseed rape. *Euphytica* 63, 71–83. doi: 10.1007/BF00023913

Mithen, R. F., Lewis, B. G., and Fenwick, G. R. (1986). *In vitro* activity of glucosinolates and their products against *Leptosphaeria maculans*. *Trans. Br. Mycol. Soc.* 87, 433–440. doi: 10.1016/S0007-1536(86)80219-4

Morrissey, J. P., and Osbourn, A. E. (1999). Fungal resistance to plant antibiotics as a mechanism of pathogenesis. *Microbiol. Mol. Biol. Rev.* 63, 708–724.

Parisy, V., Poinssot, B., Owsianowski, L., Buchala, A., Glazebrook, J., and Mauch, F. (2006). Identification of PAD2 as a gamma-glutamylcysteine synthetase highlights the importance of glutathione in disease resistance of *Arabidopsis*. *Plant J.* 49, 159–172. doi: 10.1111/j.1365-313X.2006.02938.x

Raj, J. N., and Srivastava, S. K. (1977). Changes in the total sulphur content of *Brassica juncea* plants infected with isolates of *Macrophomina phaseolina* and its possible relationship with the virulence of isolates. *Ind. Phytopathol.* 30, 486–488.

Rausch, T., and Wachter, A. (2005). Sulfur metabolism: a versatile platform for launching defence operations. *Trends Plant Sci.* 10, 503–509. doi: 10.1016/j.tplants.2005.08.006

Rea, P. A., Li, Z. S., Lu, Y. P., Drozdowicz, Y. M., and Martinoia, E. (1998). From vacuolar GS-X pumps to multispecific ABC transporters. *Annu. Rev. Plant Physiol. Plant Mol. Biol.* 49, 727–760. doi: 10.1146/annurev.arplant.49.1.727

Rennenberg, H. (1982). Glutathione metabolism and possible biological roles in higher plants. *Phytochem* 21, 2771–2781. doi: 10.1016/0031-9422(80)85045-X

Rennenberg, H. (1984). The fate of excess sulfur in higher plants. *Annu. Rev. Plant Physiol. Plant Mol. Biol.* 35, 121–153. doi: 10.1146/annurev.pp.35.060184.001005

Rennenberg, H., and Brunold, C. (1994). Significance of glutathione metabolism in plants under stress. *Prog. Bot.* 55, 142–156.

Resende, M. L. V., Flood, J., Ramsden, J. D., Rowan, M. G., Beale, M. H., and Cooper, R. M. (1996). Novel phytoalexins including elemental sulphur in the resistance of cocoa (*Theobroma cacao* L.) to *Verticillium* wilt (*Verticillium dahliae* Kleb.). *Physiol. Mol. Plant P.* 48, 347–359. doi: 10.1006/pmpp.1996.0028

Saito, K. (2004). Sulfur assimilatory metabolism. The long and smelly road. *Plant Physiol.* 136, 2443–2450 doi: 10.1104/pp.104.046755

Salac, I., Haneklaus, S. H., Bloem, E., Booth, E. J., Sutherland, K. G., Walker, K. C., et al. (2005). Sulfur nutrition and its significance for crop resistance—a case study from Scotland. *Landbauforsch. Volk. Spec. Issue* 283, 111–119.

Sasaki-Sekimoto, Y., Taki, N., Obayashi, T., Aono, M., Matsumoto, F., Sakurai, N., et al. (2005). Coordinated activation of metabolic pathways for antioxidants and defence compounds by jasmonates and their roles in stress tolerance in *Arabidopsis*. *Plant J.* 44, 653–668. doi: 10.1111/j.1365-313X.2005.02560.x

Schnug, E., Booth, E., Haneklaus, S., and Walker, K. C. (1995). *Sulphur Supply and Stress Resistance in Oilseed Rape*. Cambridge: Proceedings of the 9th International Rapeseed Congress.

Seinfeld, J. H., and Pandis, S. N. (1998). *Atmospheric Chemistry and Physics—From Air Pollution to Climate Change*. New York, NY: John Wiley and Sons, Inc.

Sekiya, J., Schmidt, A., Wilson, L. G., and Filner, P. (1982). Emission of hydrogen sulfide by leaf tissue in response to *L*-cysteine. *Plant Physiol.* 70, 430–436. doi: 10.1104/pp.70.2.430

Shai, Y. (1999). Mechanism of the binding, insertion and destabilization of phospholipid bilayer membranes by alpha-helical antimicrobial and cell non-selective membrane-lytic peptides. *BBA-Rev. Biomembr.* 1462, 55–70. doi: 10.1016/S0005-2736(99)00200-X

Shanmugam, V., Ronen, M., Shalaby, S., Larkov, O., Rachamim, Y., Hadar, R., et al. (2010). The fungal pathogen *Cochliobolus heterostrophus* responds to maize phenolics: novel small molecule signals in a plant-fungal interaction. *Cell. Microbiol.* 12, 1421–1434. doi: 10.1111/j.1462-5822.2010.01479.x

Smith, F. W., Ealing, P. M., Hawkesford, M. J., and Clarkson, D. T. (1995). Plant members of a family of sulfate transporters reveal functional subtypes. *Proc. Natl Acad. Sci. U.S.A.* 92, 9373–9377. doi: 10.1073/pnas.92.20.9373

Stec, B., Markman, O., Rao, U., Heffron, G., Henderson, S., Vernon, L. P., et al. (2004). Proposal for molecular mechanism of thionins deduced from physico-chemical studies of plant toxins. *J. Pept. Res.* 64, 210–224. doi: 10.1111/j.1399-3011.2004.00187.x

Swarupa, V., Ravishankar, K. V., and Rekha, A. (2014). Plant defense response against *Fusarium oxysporum* and strategies to develop tolerant genotypes in banana. *Planta* 239, 735–751. doi: 10.1007/s00425-013-2024-8

Tahir, J., Watanabe, M., Jing, H. C., Hunter, D. A., Tohge, T., Nunes-Nesi, A., et al. (2013). Activation of *R*-mediated innate immunity and disease susceptibility is affected by mutations in a cytosolic *O*-acetylserine (thiol) lyase in *Arabidopsis*. *Plant J.* 73, 118–130. doi: 10.1111/tpj.12021

Thomma, B. P. H. J., Eggermont, K., Penninckx, I. A. M. A., Mauch-Mani, B., Vogelsang, R., Cammue, B. P. A., et al. (1998). Separate jasmonate-dependent and salicylate-dependent defense-response pathways in *Arabidopsis* are essential for resistance to distinct microbial pathogens. *Proc. Natl. Acad. Sci. U.S.A.* 95, 15107–15111. doi: 10.1073/pnas.95.25.15107

Underhill, E. W., Wetter, L. R., and Chisholm, M. D. (1973). Biosynthesis of glucosinolates. *Biochem. Soc. Symp.* 38, 303–326.

Vanacker, H., Carver, T. L. W., and Foyer, C. (2000). Early H_2O_2 accumlation in mesophyll cells leads to induction of glutathione during the hypersensitive response in the barley-powdery mildew interaction. *Plant Physiol.* 123, 1289–1300. doi: 10.1104/pp.123.4.1289

Wallsgrove, R., Benett, R., Kiddle, G., Bartlet, E., and Ludwig-Mueller, J. (1999). *Glucosinolate Biosynthesis and Pest Disease Interactions*. Canbarra,CBR: Proceedings of the 10th Intenational Rapeseed Congress.

Walters, D. R., and Bingham, I. J. (2007). Influence of nutrition on disease development caused by fungal pathogens: implications for plant disease control. *Ann. Appl. Biol.* 151, 307–324. doi: 10.1111/j.1744-7348.2007.00176.x

Wang, J., Zhang, J., Ma, Y., Wang, L., Yang, L., Shi, S., et al. (2003). "Crop resistance to diseases as influenced by sulphur application rates," in *Proceedings of the 12th International World Fertilizer Congress - Fertilization in the Third Millenium - Fertilizer, Food Security and Environmental Protection,* eds L. Ji, G. Chen, E. Schnug, C. Hera, and S. Haneklaus (Shengyang: Liaoning Science and Technology Publishing House), 1285-1296.

Williams, J. S., and Cooper, R. M. (2004). The oldest fungicide and newest phytoalexin—a reappraisal of the fungitoxicity of elemental sulphur. *Plant Pathol.* 53, 263–279. doi: 10.1111/j.0032-0862.2004.01010.x

Williams, J. S., Hall, S. A., Hawkesford, M. J., Beale, M. H., and Cooper, R. M. (2002). Elemental sulfur and thiol accumulation in tomato and defense against a fungal vascular pathogen. *Plant Physiol.* 128, 150–159. doi: 10.1104/pp.010687

Winterberg, B., Du Fall, L. A., Song, X., Pascovici, D., Care, N., Molloy, M., et al. (2014). The necrotrophic effector protein SnTox3 re-programs metabolism and elicits a strong defence response in susceptible wheat leaves. *BMC Plant Biol.* 14:215. doi: 10.1186/s12870-014-0215-5

Wirtz, M., Droux, M., and Hell, R. (2004). *O*-Acetylserine(thiol)lyase: an enigmatic enzyme of plant cysteine biosynthesis revisited in *Arabidopsis thaliana*. *J. Ex. Bot.* 55, 1785–1798. doi: 10.1093/jxb/erh201

Woodard, J. (1922). Sulphur as a factor in soil fertility. *Bot. Gaz.* 73, 81–109. doi: 10.1086/332963

Conflict of Interest Statement: The authors declare that the research was conducted in the absence of any commercial or financial relationships that could be construed as a potential conflict of interest.

22

The poplar Phi class glutathione transferase: expression, activity and structure of GSTF1

Henri Pégeot[1,2], Cha San Koh[3,4†], Benjamin Petre[1,2†], Sandrine Mathiot[3,4], Sébastien Duplessis[1,2], Arnaud Hecker[1,2], Claude Didierjean[3,4] and Nicolas Rouhier[1,2*]

[1] Interactions Arbres - Microorganismes, Université de Lorraine, UMR1136, Vandoeuvre-lès-Nancy, France
[2] INRA, Interactions Arbres - Microorganismes, UMR1136, Champenoux, France
[3] Faculté des Sciences et Technologies, Université de Lorraine, CRM[2], Equipe BioMod, UMR 7036, Vandoeuvre-lès-Nancy, France
[4] Faculté des Sciences et Technologies, CNRS, CRM[2], Equipe BioMod, UMR 7036, Vandoeuvre-lès-Nancy, France

Edited by:
Stanislav Kopriva, University of Cologne, Germany

Reviewed by:
Georg Groth, University of Duesseldorf, Germany
Wayne Snedden, Queen's University, Canada

***Correspondence:**
Nicolas Rouhier, Interactions Arbres - Microorganismes, Université de Lorraine, UMR1136, Boulevard des aiguilettes, Faculté des sciences et technologies, F-54500 Vandoeuvre-lès-Nancy, France
e-mail: nicolas.rouhier@univ-lorraine.fr

†Present address:
Cha San Koh, RNA Therapeutics Institute, University of Massachusetts Medical School, Worcester, MA, USA;
Benjamin Petre, The Sainsbury Laboratory, Norwich, UK

Glutathione transferases (GSTs) constitute a superfamily of enzymes with essential roles in cellular detoxification and secondary metabolism in plants as in other organisms. Several plant GSTs, including those of the Phi class (GSTFs), require a conserved catalytic serine residue to perform glutathione (GSH)-conjugation reactions. Genomic analyses revealed that terrestrial plants have around ten GSTFs, eight in the *Populus trichocarpa* genome, but their physiological functions and substrates are mostly unknown. Transcript expression analyses showed a predominant expression of all genes both in reproductive (female flowers, fruits, floral buds) and vegetative organs (leaves, petioles). Here, we show that the recombinant poplar GSTF1 (PttGSTF1) possesses peroxidase activity toward cumene hydroperoxide and GSH-conjugation activity toward model substrates such as 2,4-dinitrochlorobenzene, benzyl and phenetyl isothiocyanate, 4-nitrophenyl butyrate and 4-hydroxy-2-nonenal but interestingly not on previously identified GSTF-class substrates. In accordance with analytical gel filtration data, crystal structure of PttGSTF1 showed a canonical dimeric organization with bound GSH or 2-(N-morpholino)ethanesulfonic acid molecules. The structure of these protein-substrate complexes allowed delineating the residues contributing to both the G and H sites that form the active site cavity. In sum, the presence of GSTF1 transcripts and proteins in most poplar organs especially those rich in secondary metabolites such as flowers and fruits, together with its GSH-conjugation activity and its documented stress-responsive expression suggest that its function is associated with the catalytic transformation of metabolites and/or peroxide removal rather than with ligandin properties as previously reported for other GSTFs.

Keywords: glutathione transferase, protein structure, crystallography, *Populus*, enzyme characterization, transcript profiling

INTRODUCTION

Glutathione transferases (GSTs; EC 2.5.1.18) represent a ubiquitous multigenic family of enzymes that conjugate the reduced tripeptide glutathione (GSH, γ-Glu-Cys-Gly) on a wide range of endogenous and exogenous electrophilic molecules (Hayes et al., 2005). From the most recent genomic and phylogenetic analyses, the GST family is subdivided into 14 classes in photosynthetic organisms: Phi (F), Tau (U), Theta (T), Zeta (Z), Lambda (L), Hemerythrin (H), Iota (I), Ure2p, glutathionyl-hydroquinone reductase (GHR), elongation factor 1B Gamma (EF1Bγ), dehydroascorbate reductase (DHAR), tetrachlorohydroquinone dehalogenase (TCHQD), metaxin, microsomal prostaglandin E synthase type 2 (mpges-2) (Lallement et al., 2014a). Behind Tau GSTs, Phi GSTs (GSTFs) represent the second largest class in plants and this expansion probably results from several rounds of gene duplication (Lan et al., 2009). This class is often presented in the literature as plant-specific, however, basidiomycetes also possess GSTFs (Morel et al., 2013).

Along with GSTUs, plant GSTFs have been extensively studied for their involvement in herbicide detoxification and for this reason they could be considered as the counterparts of the mammalian drug metabolizing GSTs. By catalyzing GSH-conjugation reactions of electrophilic molecules that are subsequently recognized by vacuolar ABC transporters, GSTFs participate to the vacuolar sequestration and thus detoxification of exogenous compounds. However, other biochemical activities can account for the observed increased herbicide resistance. For instance, it was shown that the GSTF1 from the black grass *Alopecurus myosuroides*, a weed of cereals, possesses a glutathione peroxidase activity which lowers the levels of hydroperoxides produced in response to herbicides (Cummins et al., 1999). *Arabidopsis thaliana* transgenic plants expressing this GSTF1 acquire multiple herbicide resistance and accumulate protective flavonoids as initially observed in the black grass (Cummins et al., 2009, 2013). Another facet of GSTs is their involvement in secondary metabolism, in stress response and in their associated signaling. For instance, *A. thaliana* GSTF6 is required for the synthesis of the defense compound camalexin, by catalyzing the conjugation of glutathione onto indole-3-acetonitrile (Su et al., 2011) whereas *A. thaliana* GSTF2 binds tightly to camalexin and might be required for its transport (Dixon et al., 2011). On the other hand, *A. thaliana* GSTF8 catalyzes glutathione conjugation to prostaglandin 12-oxophytodienoic acids and A_1-phytoprostanes, two stress signaling molecules (Mueller et al., 2008). Consistent with these functions, the expression of *GST* gene belonging to all classes is often highly induced in response to biotic and abiotic stresses or to hormone treatments, and this often correlated with an increase in the protein amount. For instance, the expression of several *GSTF* genes is enhanced in response to plant hormones such as ethylene, methyl jasmonate, salicylic acid and auxin, to herbicides and to herbicide safeners, to pathogen infection, and more generally to treatments leading to oxidative stress (Deridder et al., 2002; Wagner et al., 2002; Lieberherr et al., 2003; Smith et al., 2003, 2004; Sappl et al., 2004, 2009).

Interestingly, previous biochemical analyses have shown that GSTFs can bind to metabolites for non-catalytic functions. The best characterized example of carrier/transport functions for a Phi GST concerns the requirement of *A. thaliana* transparent testa 19 (tt19)/AtGSTF12 and of the petunia ortholog AN9 for the correct vacuolar localization of anthocyanins and proanthocyanidins (Alfenito et al., 1998; Kitamura et al., 2004). While it was initially thought that these GSTFs could catalyze GSH-conjugation reactions, it was determined that they serve as flavonoid carrier proteins (Mueller et al., 2000). Moreover, photoaffinity-labeling experiments or competition activity assays pointed to the capacity of GSTFs to bind plant hormones such as gibberellic acid (Axarli et al., 2004), cytokinin and auxin (Bilang et al., 1993; Bilang and Sturm, 1995; Gonneau et al., 2001). A screen for metabolites able to bind to *A. thaliana* GSTF2, either from pure molecules or from plant or *Escherichia coli* extracts also identified other interacting molecules. Besides camalexin, flavonoids (quercetin, quercetin-3-O-rhamnoside and kaempferol) and other heterocyclic compounds structurally close to flavonoids (harmane, norharmane, indole-3-aldehyde, and lumichrome) have been shown to bind to AtGSTF2 (Smith et al., 2003; Dixon et al., 2011). The absence of GSH-conjugation activity with these compounds indicated that AtGSTF2 functions as a carrier protein. Moreover, competition binding experiments or activity assays in the presence of several of these binding molecules showed that they either did not alter AtGSTF2 conjugating activity or even increased it, hinting the existence of multiple ligand/substrate binding sites.

At the structural level, GSTFs exist as homodimers, the dimerization interface involving mainly hydrophobic surface patches (Armstrong, 1997). Each monomer comprises an active site region formed by a glutathione binding pocket (G-site) primarily involving residues from the conserved N-terminal thioredoxin domain and an hydrophobic pocket (H-site) primarily involving residues from the less conserved C-terminal domain (Prade et al., 1998). In their active sites, most GSTFs present a serine residue that is located in the N-terminal end of the $\alpha1$ helix which promotes the formation of the active thiolate anion on the sulphydryl group of the cysteine of GSH that is required for catalysis. However, the non-catalytic functions observed for some GSTFs suggested the existence of a ligandin site (L-site) but structural details of the latter are still lacking. From mutagenesis experiments performed on *Zea mays* GST-I, the L-site is likely overlapping with the G- and H-sites (Axarli et al., 2004).

In this study, the transcript levels of the eight poplar *GSTFs* have been analyzed in various organs. Then, the biochemical and structural properties of the stress-responsive GSTF1 have been further characterized, examining the enzymatic properties of recombinant proteins (WT protein and variants mutated for the catalytic serine) and solving the 3D structure of the protein in complex with substrates/ligands.

MATERIALS AND METHODS

GENOMIC AND PHYLOGENETIC ANALYSES

In order to identify all poplar *GSTF* genes, homology searches with the BLAST algorithm have been performed on the different versions of the *P. trichocarpa* genome including the version 3.0 available on the phytozome v10 portal (http://phytozome.jgi.doe.gov/pz/portal.html). Genome analyses for other terrestrial plants have been also performed on the phytozome v10 portal whereas cyanobacterial and algal genomes have been analyzed from cyanobase (http://genome.microbedb.jp/cyanobase) and the JGI genome portal (http://genome.jgi.doe.gov) respectively. The protein sequences and corresponding accession numbers can be found as Supplementary Table 1. When possible, GSTF sequences were corrected on the basis of available ESTs.

BIOLOGICAL MATERIAL, GROWTH CONDITIONS, AND INOCULATION PROCEDURES

Hybrid poplar cultivar 'Beaupré' (*Populus trichocarpa* × *Populus deltoides*) greenhouse cultivation, *Melampsora larici-populina* urediniospore multiplication and leaf inoculation procedures were done as previously described (Rinaldi et al., 2007). The *M. larici-populina* isolates used in this study are 98AG31 (pathotype 3-4-7) and 93ID6 (pathotype 3-4) respectively virulent and avirulent on "Beaupré." Poplar organs have been harvested from a naturally growing male and female *P. trichocarpa* adult trees found on the faculty of sciences campus located in Vandoeuvre-lès-Nancy (France).

RT-PCR EXPERIMENTS

Total RNAs were extracted from 150 mg of *P. trichocarpa* stamens, male flowers, female flowers, fruits, petioles, leaves, buds, and roots using the RNeasy Plant Mini Kit (Qiagen) according to the Manufacturer's instructions with minor modifications described before (Lallement et al., 2014b). Then, mRNAs were reverse-transcribed to obtain cDNAs by using the iScript cDNA Synthesis kit (Bio-Rad) following the manufacturer's instructions. PCR amplifications were performed for 25, 30, or 35 cycles using Go-Taq polymerase (Promega). Specific forward and reverse primers (Supplementary Table 2) have been designed to amplify *ca* 300 bp fragments of each *GSTF* gene. The ubiquitin gene (Potri.015G013600) was used as a control of the cDNA concentration used for PCR amplification and incidentally of cDNA integrity (Lallement et al., 2014b). The PCR products have been separated by electrophoresis on 1% agarose gel and visualized by ethidium bromide staining.

PROTEIN EXTRACTION AND WESTERN-BLOT ANALYSIS

Extraction of soluble proteins from leaves, petioles, stems, roots, fruits, stamens, and buds or from rust-infected leaves was performed as previously described (Vieira Dos Santos et al., 2005). The proteins were separated by 15% SDS–PAGE and electro-transferred onto nitrocellulose membranes (LI-COR Biosciences). After rinsing in 13.7 mM NaCl, 0.27 mM KCl, 10 mM Na_2HPO_4, and 0.2 mM KH_2PO_4 buffer (phosphate buffered saline: PBS), membranes were blocked during 45 min at room temperature using the Odyssey blocking buffer (LI-COR Biosciences). Then, membranes were incubated with rabbit polyclonal antibodies (diluted 1:1000, synthesis by Genecust) raised against PttGSTF1 for 30 min in the presence of 0.05% of tween 20. After several washing steps with a PBS buffer supplemented with 0.05% tween 20 (PBST), membranes were incubated for 30 min with IRDye 800 CW goat or donkey anti-rabbit secondary antibodies (LI-COR Biosciences) diluted 1:5000 in the Odyssey blocking buffer supplemented with 0.05% tween 20 and 0.01% SDS. After extensive washes with PBST and PBS, immunodetection of proteins on the membrane was performed by exciting the IRDye with an Odyssey Infrared Imager (LI-COR Biosciences).

PCR CLONING AND SITE-DIRECTED MUTAGENESIS

The sequence coding for GSTF1 was amplified by PCR from *Populus tremula* × *P. tremuloides* leaf cDNAs using specific forward and reverse primers (Supplementary Table 2) and cloned into pET-3d between *Nco*I and *Bam*HI restriction sites. Hence, the sequence is subsequently referred to as PttGSTF1. PttGSTF1 S13C and PttGSTF1 S13A variants where the serine found at position 13 is substituted into cysteine or alanine were generated by site-directed mutagenesis using two complementary mutagenic primers (Supplementary Table 2). Two overlapping mutated fragments were generated in a first PCR reaction and were subsequently used in a second PCR to generate the full-length mutated sequences which have been then cloned into pET-3d.

HETEROLOGOUS EXPRESSION IN *E. COLI* AND PURIFICATION

PttGSTF1 expression was performed in an *E. coli* BL21 (DE3) strain (Novagen) containing the pSBET plasmid upon transformation with the recombinant pET-3d plasmids. Bacteria were cultivated at 37°C in LB medium containing kanamycin (50 µg/ml) and ampicillin (50 µg/ml). When the cell culture reached an OD_{600nm} of 0.7, PttGSTF1 expression was induced by the addition of 0.1 mM isopropyl β-D-1-thiogalactopyranoside (IPTG) and cells were further grown for 4 h. Cells were harvested by centrifugation, resuspended in a 30 mM Tris-HCl pH 8.0, 1 mM EDTA, 200 mM NaCl buffer and stored at −80°C. Cell lysis was achieved by two rounds of 1 min sonication. The cell extract was then centrifuged at 40,000 g for 30 min at 4°C to remove cellular debris and aggregated proteins. The fraction precipitating between 40 and 80% of the saturation in ammonium sulfate was subjected to a size-exclusion chromatography by loading the protein extract on an Ultrogel® ACA44 (5 × 75 cm, Biosepra) column equilibrated with 30 mM Tris-HCl pH 8.0, 200 mM NaCl buffer. The fractions containing the recombinant protein were then pooled, dialyzed by ultrafiltration in Amicon cells using a YM10 membrane (Millipore) and loaded onto a DEAE-cellulose column (Sigma Aldrich) equilibrated in 30 mM Tris-HCl pH 8.0. The proteins were eluted using a 0–400 mM NaCl gradient, concentrated by ultrafiltration and stored in 30 mM Tris-HCl pH 8.0, 200 mM NaCl buffer. The protein purity was then analyzed by 15% SDS-PAGE and protein concentration was determined after measuring the absorbance at 280 nm using a theoretical molar absorption coefficient of 33,982 M^{-1} cm^{-1} for PttGSTF1, PttGSTF1 S13C, and PttGSTF1 S13A.

DETERMINATION OF THE MOLECULAR MASS AND OLIGOMERIZATION STATE OF PURIFIED RECOMBINANT PROTEINS

The molecular masses of purified recombinant proteins were analyzed using a Bruker microTOF-Q spectrometer (Bruker Daltonics, Bremen, Germany) equipped with an Apollo II electrospray ionization source as described previously (Couturier et al., 2011). The oligomerization state of purified recombinant proteins was analyzed on a Superdex 200 10/300 column equilibrated in 30 mM Tris-HCl pH 8.0, 200 mM NaCl and connected to an Akta purifier system (GE Healthcare) by injecting 100 µg of purified recombinant proteins at a flow rate of 0.5 ml/min. The column was calibrated using the molecular weight standards (6500–700,000 Da) from Sigma.

ENZYMATIC ACTIVITIES

The GSH-conjugation activity toward phenetyl isothiocyanate (PITC), benzyl isothiocyanate (BITC), 1-chloro-2,4-dinitrobenzene (CDNB), 4-hydroxy-2-nonenal (HNE), 4-nitrophenyl butyrate (PNP-butyrate) was assayed at 25°C by following absorbance at 274 nm for isothiocyanate derivatives, or at 224, 340, 412 nm for HNE, CDNB, and PNP-butyrate respectively. Reactions were carried out in 500 µL of 100 mM phosphate buffer pH 6.5 for both isothiocyanate derivatives and HNE; 100 mM sodium phosphate buffer pH 7.5 for PNP-butyrate and 30 mM Tris-HCl pH 8.0, 1 mM EDTA for CDNB. Various concentrations of PITC (50–500 µM), HNE (12.5–125 µM), CDNB (500–6000 µM), BITC (100–1000 µM) or PNP-butyrate (50–3000 µM) have been tested at a fixed GSH concentration of 1 mM. When using HNE as a substrate, the GSH concentration was fixed at 0.7 mM to limit interferences with the detection of HNE at 224 nm.

Thiol-transferase, dehydroascorbate (DHA) reductase and peroxidase activities have been measured toward 2-hydroxyethyl disulfide (HED), DHA, and cumene hydroperoxide (CuOOH) or tert-butyl hydroperoxide (t-BOOH) respectively using an NADPH-coupled spectrophotometric method. The reactions were carried out at 25°C in 500 µL of 30 mM Tris-HCl, pH 8.0, 1 mM EDTA buffer containing 150 µM NADPH, 0.5 units of yeast glutathione reductase and various concentrations of HED (25–1000 µM), DHA (250–5000 µM), CuOOH (500–6000 µM), t-BOOH (250–5000 µM) at a fixed GSH concentration of 2 mM.

For all these assays, reactions were started by the addition of the enzyme and protein concentrations used were within the linear response range. The measured velocities were corrected by subtracting the rate of spontaneous non-enzymatic reaction and three independent experiments were performed at each substrate concentration. Changes in absorbance were followed with a Cary 50 spectrophotometer (Agilent Technologies). The kinetic parameters (k_{cat} and apparent K_m) were obtained by fitting the data to the non-linear regression Michaelis–Menten model in GraphPad Prism 5 software. The k_{cat} values are expressed as μmol of substrate oxidized per second per μmol of enzyme (i.e., the turnover number in s^{-1}), using specific molar absorption coefficients of 6220 M^{-1} cm^{-1} at 340 nm for NADPH, 8890 M^{-1} cm^{-1} at 274 nm for PITC, 9250 M^{-1} cm^{-1} at 274 nm for BITC, 9600 M^{-1} cm^{-1} at 340 nm for CDNB, 17700 M^{-1} cm^{-1} at 412 nm for PNP-butyrate and 13750 M^{-1} cm^{-1} at 224 nm for HNE.

CRYSTALLIZATION AND STRUCTURE DETERMINATION OF PttGSTF1 AND PttGSTF1 S13C

Initial screening of crystallization conditions was carried out by the microbatch-under-oil method. Sitting drops were set up using 1 μl of a 1:1 mixture of protein and crystallization solutions (672 different commercially available conditions) in Terasaki microbatch multiwell plates (Molecular Dimensions). The crystallization plates were stored at 4°C. Single crystals of sufficient size were obtained using Jena Bioscience 2D1 condition (30% w/v PEG 4000, 100 mM 2-(N-morpholino)ethanesulfonic acid (MES) sodium salt, pH 6.5). Best crystals were obtained with a protein concentration of 14 mg/ml for PttGSTF1 and of 10 mg/ml for PttGSTF1 S13C. The single crystals were flash-cooled in liquid nitrogen using a mixture of the crystallization solution and 20% glycerol as cryoprotectant. For PttGSTF1, before crystallization, the protein (*ca* 1 mL at 600 μM) was treated with 10 mM GSH for 30 min, desalted on G25 columns and concentrated to the indicated concentration using Amicon Ultra centrifugal filters, Ultracel 10 K Membrane from Millipore.

PttGSTF1 X-ray diffraction data were collected on beamline EMBL-X11 at the DORIS storage ring (DESY, Hamburg, Germany) and PttGSTF1 S13C X-ray diffraction data were collected on beamline BM30A at synchrotron ESRF (Grenoble, France). PttGSTF1 and PttGSTF1 S13C diffraction images were integrated with the program HKL2000 (Otwinowski and Minor, 1997) and the program XDS (Kabsch, 2010), respectively. Crystallographic calculations were carried out with programs from the CCP4 program suite (Winn et al., 2011). The structure of PttGSTF1 was solved by the molecular replacement method with the program Molrep (Vagin and Teplyakov, 2010) using *A. thaliana* GSTF2 as a template (PDB code: 1GNW). PttGSTF1 and PttGSTF1 S13C structures were refined by alternate cycles of restrained maximum-likelihood refinement with the program Phenix (Adams et al., 2010) and manual adjustments were made to the models with Coot (Emsley et al., 2010). The crystal parameters, data statistics, and final refinement parameters are shown in **Table 1**. All structural figures were generated with PyMol Molecular Graphics System (Schrödinger, LLC). The atomic coordinates and structure factors (codes 4RI6 and 4RI7 for PttGSTF1 and PttGSTF1 S13C, respectively) have been deposited in the Protein Data Bank, Research Collaboratory for Structural Bioinformatics, Rutgers University, New Brunswick, NJ (http://www.rcsb.org/).

Table 1 | Statistics of X-ray diffraction data collection and model refinement.

	PttGSTF1	PttGSTF1 S13C
DATA COLLECTION		
Space group	$P2_12_12_1$	
Nb of monomers in the ASU[a]	2	
Cell dimensions a, b, c (Å)	58.84 65.94 109.84	55.30 60.65 119.90
Resolution (Å)	18.28-1.52 (1.55-1.52)[b]	40.86-1.80 (1.90-1.80)
Rmerge	0.034 (0.20)	0.124 (0.614)
Mean I/σ (I)	40.6 (8.4)	11.7 (2.1)
Completeness (%)	99.2 (95.2)	96.5 (79.0)
n observations	7,26,145 (29,691)	2,41,669 (17,902)
Average redundancy	11.0 (9.0)	6.6 (4.2)
Wilson B factor ($Å^2$)	19.4	14.8
REFINEMENT		
Resolution (Å)	18.28-1.52 (1.54-1.52)	40.86-1.80 (1.85-1.80)
n reflections	65,579 (2554)	36,817 (2118)
Cutoff	$F > 0\sigma(F)$	$F > 0\sigma(F)$
Rall (%)[c]	15.3	15.3
Rfree (%)[c]	18.3 (19.6)	19.5 (28.9)
Average B-factor ($Å^2$)		
Protein atoms	24.2	18.5
Ligand atoms	25.8	17.2
Solvent atoms	35.8	29.2
Ramachandran statistics (%)		
Residues in preferred regions	98.6	97.9
Residues in allowed regions	1.2	1.9
Outlier residues	0.2	0.2
R.m.s.[d] deviations		
Bond length (Å)	0.009	0.01
Bond angle (°)	1.26	1.3

[a] *ASU, Asymmetric unit.*
[b] *Values in parentheses are for highest resolution shell.*
[c] *Rall was determined from all the reflections (working set + test set) whereas Rfree corresponds to a subset of reflections (test set).*
[d] *R.m.s.: Root mean square.*

RESULTS

PHYLOGENETIC AND SEQUENCE ANALYSES OF *P. TRICHOCARPA* GSTFs

In silico analysis of the various versions of *P. trichocarpa* genome led to the identification of eight genes coding for GSTFs. All the *P. trichocarpa GSTF* genes encode predicted proteins with a size ranging from 213 to 218 amino acids (**Figure 1**). None of these sequences exhibits a targeting sequence, suggesting a cytosolic localization. Based on sequence similarities and phylogenetic analysis, four subgroups can be distinguished in poplar: PtGSTF1/2, PtGSTF3/7, PtGSTF4/5/6, and PtGSTF8 (**Figures 1, 2**). In terms of sequence similarity, the percentage identity within a subgroup ranges from 65 to 98% whereas it is comprised between 40 and 48% between subgroups. The protein similarity somehow reflects the gene arrangement since *PtGSTF1*

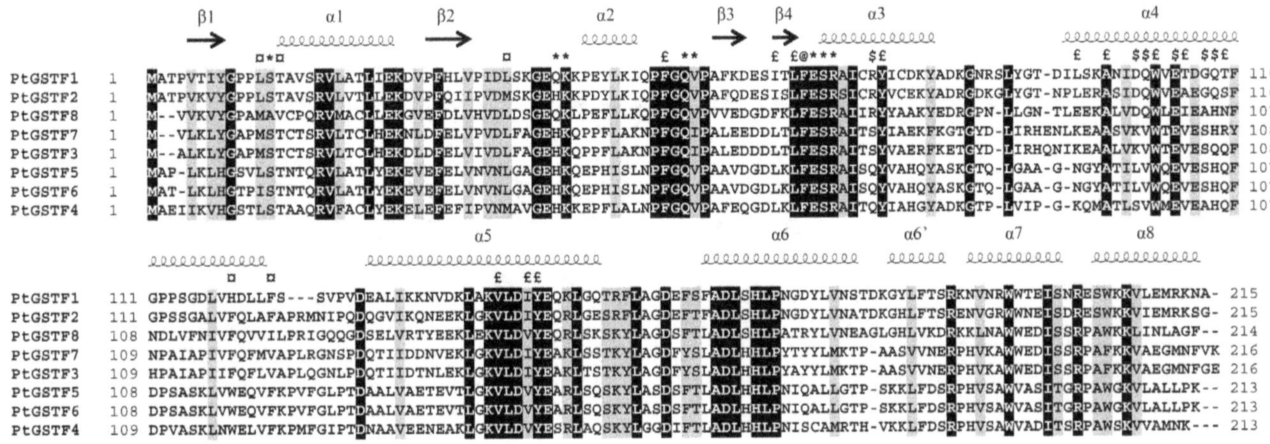

FIGURE 1 | Structure based sequence alignment of *Populus trichocarpa* GSTFs. The structural alignment was calculated with PROMALS3D using the structure of PttGSTF1 and rendered using ESpript 3.0. Accession numbers of GSTFs from the version 3.0 of *P. trichocarpa* genome are the following: PtGSTF1: Potri.002G015100, PtGSTF2: Potri.002G015200, PtGSTF3: Potri.014G132200, PtGSTF4: Potri.T035400, PtGSTF5: Potri.T035300, PtGSTF6: Potri.T035100, PtGSTF7: Potri.T035000, PtGSTF8: Potri.017G138800. From the structure of PttGSTF1, residues contributing to the dimerization and those involved in the G- or H-sites are highlighted as follows: *, glutathione-interacting residues; ◻, MES-interacting residues; $, dimer interface via hydrogen bond; £, dimer interface via van der Waals interactions; @, dimer interface via hydrogen bond and van der Waals interactions.

and *PtGSTF2* genes are present in tandem on chromosome 2, *PtGSTF4*, 5, 6, and 7 genes cluster on the scaffold 36, whereas *PtGSTF3* and *PtGSTF8* are found at isolated loci on the chromosomes 14 and 17, respectively. Hence, the only peculiarity is the genomic association of *PtGSTF7* with *PtGSTF4, 5, 6* whereas the sequence proximity to *PtGSTF3* suggested a common origin. The sequence differences between members of each group are also visible by looking to the four amino acid signature typical of proteins of the thioredoxin superfamily and containing the catalytic serine. Indeed, PtGSTF1 and PtGSTF2 display a STAV active site motif, PtGSTF3 and 7 a STCT motif and PtGSTF4, 5, and 6 display STAA or STNT motifs (**Figure 1**). PtGSTF8 is clearly particular since it has an alanine (AVCP motif) instead of the catalytic serine. This is not specific to the poplar isoform as this particularity is found in several plant orthologs, including the petunia AN9 protein for example.

An exhaustive search of GSTF homologs in available genomes from photosynthetic organisms indicated that *GSTF* genes are absent in cyanobacteria and green algae. On the other hand, there are considerable variations in the number of genes in terrestrial plants since there is only one gene in *Selaginella moellendorffii* but 27 predicted genes in *Aquilegia coerulea* (Supplementary Table 1). However, the average number of genes is close to 10. A phylogenetic tree constructed using the 400 retrieved sequences (**Figure 2**) shows several distinct clades that can be distinguished according to the protein active site signature even though some groups can be also differentiated on the basis of the presence of C-terminal or N-terminal extensions. The sequences identified in *P. patens* and *S. moellendorffii*, which are supposed to represent the ancestral versions, form an isolated clade and do not display a clear consensus active site motif. The four subgroups observed for poplar GSTFs are found again in the phylogenetic tree and fell within separate clades. It is worth noting that PtGSTF8 stands out within a clade containing proteins lacking the catalytic serine but displaying a conserved cysteine residue two residues away (AxC motif). Interestingly, this cysteine is also found in PtGSTF3 and PtGSTF7 and in all orthologs of the same clade whereas the catalytic serine is present.

Overall this indicates that numerous species-specific duplication events occurred during evolution and this raises the question of the appearance of the GSTF group in photosynthetic organisms since it appears to be an innovation specifically found in terrestrial plants. Moreover, the divergences observed in the active site signatures suggest that the proteins may have different properties.

TRANSCRIPT EXPRESSION OF GSTFs IN POPLAR ORGANS

In order to determine whether the expression territories could allow discriminating *GSTF* genes, RT-PCR experiments were performed from different tissues of an adult, naturally-growing *P. trichocarpa* individual. Experiments were performed with 25, 30, or 35 amplification cycles in order to examine gene expression in the linear range of PCR amplification. The best detection was obtained after 30 cycles as transcripts were barely detected at 25 cycles whereas the signal for some genes was saturated at 35 cycles. All GSTF transcripts were weakly detected in roots and in the male reproductive organ either as whole (male flower) or in stamen, whereas they were all detected in female flowers, fruits, petioles, leaves, and buds (**Figure 3**). Moreover, the *GSTF1, F2, F5, F6*, and *F7* genes are globally more expressed than *GSTF3, F4*, and *F8* genes. Comparing the expression of duplicated genes, we observed that they generally have the same expression profiles although variations in transcript abundance can sometimes be observed. A difference between *PtGSTF1* and *PtGSTF2* transcripts is the presence of *PtGSTF2* in male flowers. In the *PtGSTF4/F5/F6* subgroup and incidentally among all GSTFs tested, *PtGSTF5* is the most expressed in male flowers/stamen and in roots together with *PtGSTF7*.

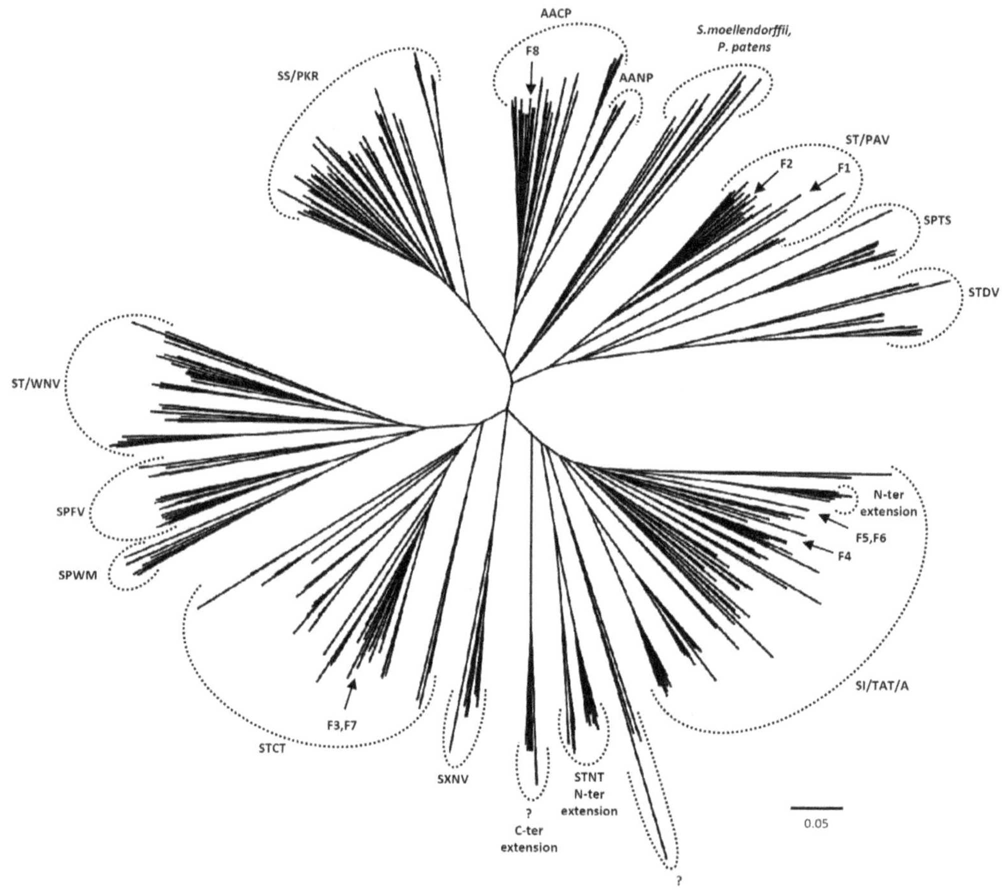

FIGURE 2 | Unrooted phylogenetic tree of GSTFs from terrestrial plants. The alignment was performed with PROMALS3D using 1BYE, 1AXD, 1AW9, 1GNW, and 1BX9 protein structure models as templates. The alignment was subsequently manually adjusted by using Seaview software. Phylogenetic tree was built with BioNJ and edited with Figtree software (http://tree.bio.ed.ac.uk/software/figtree/). Five hundred bootstrap replicates were performed in order to test the robustness of the tree. The scale marker represents 0.05 substitutions per residue. Sequence names have been removed for clarity but all sequences used are available as Supplementary Table S1. For each major branch, the consensus active site signature containing the catalytic residue is indicated, x is used when the variability is too high. *P. trichocarpa* isoforms have been indicated by an arrow on the tree (F1–F8).

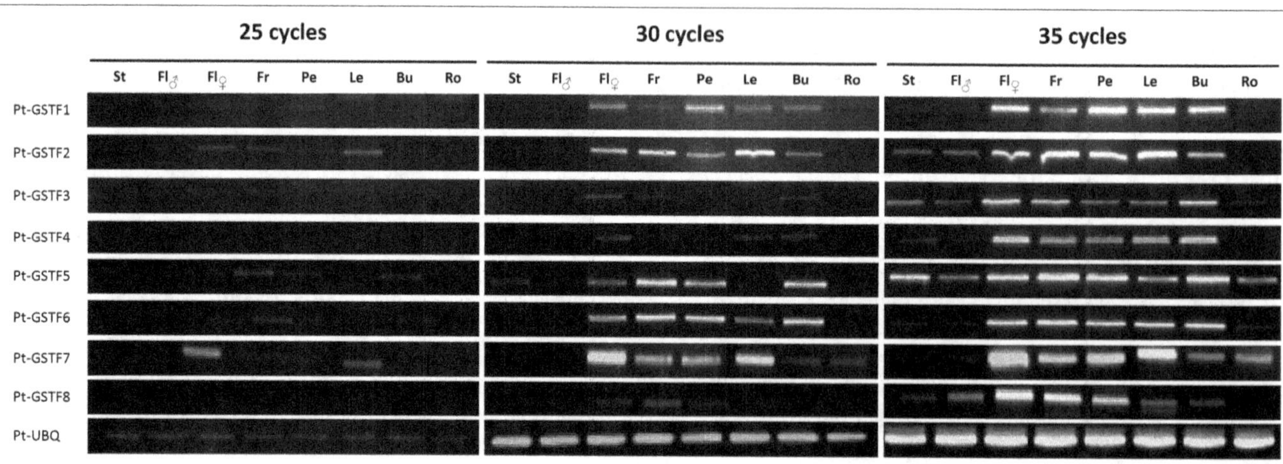

FIGURE 3 | Transcript accumulation of *GSTFs* in poplar organs. RT-PCR experiments were performed using cDNAs from stamens (St), male flowers (Fl♂), female flowers (Fl♀), fruits (Fr), petioles (Pe), leaves (Le), buds (Bu), and roots (Ro). Ubiquitin was used as a reference gene.

PtGSTF1 PROTEIN ACCUMULATES IN ALL ORGANS ANALYZED BUT ITS LEVEL IS NOT AFFECTED IN LEAVES INFECTED BY THE RUST FUNGAL PATHOGEN *MELAMPSORA LARICI-POPULINA*

In the subsequent parts, we focused our analysis on poplar GSTF1 since several studies showed that it is regulated in many stress conditions. For instance, it is up-regulated in poplar leaves exposed to the tent caterpillar *Malacosoma disstria* (Ralph et al., 2006), in root apices of drought-sensitive (Soligo) and tolerant (Carpaccio) poplar cultivars and in leaves of Carpaccio cultivar subjected to a water deficit (Cohen et al., 2010) and in leaves of 2 month-old *P. trichocarpa* cuttings treated with CDNB or H_2O_2 (Lan et al., 2009). Contrasting results have been obtained in the case of poplar infection by rust fungi, GSTF1 was found to be up-regulated in some (Miranda et al., 2007) but not all studies (Rinaldi et al., 2007; Azaiez et al., 2009). Besides, proteomic studies pointed to an increased GSTF1 protein level in roots of *Populus tremula* exposed to a cadmium stress (Kieffer et al., 2009) and in leaves of *Populus cathayana* male cuttings exposed to chilling or salt stresses (Chen et al., 2011; Zhang et al., 2012).

Hence, taking advantage of the production of the recombinant protein (see below), we have raised an antibody against GSTF1 first to investigate its protein level in several poplar organs, i.e., leaves, petioles, stems, roots, fruits, stamens, and buds by Western Blotting (**Figure 4A**). A major band around 25 kDa likely corresponding to GSTF1 was detected in protein extracts from various organs, indicating that the protein is present in many tissues, though a higher protein amount was found in leaves, petioles, stems, roots and stamens. Considering that GSTF2 is a close paralog, it is possible that the detected signal represents the sum of both GSTFs. Next, considering the discrepancy observed at the transcript level as detailed above, we sought to evaluate GSTF1 protein abundance in a poplar-rust pathosystem. The model used is *P. trichocarpa* × *P. deltoides* leaves either untreated or inoculated by two *M. larici-populina* isolates, virulent, or avirulent, leading to compatible and incompatible reactions respectively (**Figure 4B**). However, no significant variation in protein abundance was detected over a 7-day time-course infection which represents a whole asexual cycle from spore germination to urediniospore formation. This result suggests that GSTF1 protein levels are not affected by *M. larici-populina* infections.

POPLAR GSTF1 IS A HOMODIMERIC PROTEIN WITH GSH-CONJUGATING ACTIVITIES

In order to investigate the biochemical and structural properties of GSTF1, the mature form was expressed in *E. coli* as well as single mutated protein variants, the catalytic serine of which was replaced by a cysteine or an alanine residue. Having used a *P. tremula* × *P. tremuloides* leaf cDNA library, the amplified coding sequence, which is perfectly similar to the DN500362 EST sequence, is slightly different from the GSTF1 version found in the *P. trichocarpa* reference genome. Hence, the sequence will be referred to as PttGSTF1 in the following parts for *P. tremula* × *P. tremuloides* GSTF1. At the protein level, two very conservative changes are present, Ile33 is replaced by a Val and Lys86 by an Arg. After purification, around 30 mg of protein was obtained per liter of culture.

The purified proteins have been first analyzed by mass spectrometry. A single species was detected for each protein with molecular masses of 24192, 24511, and 24172 Da for PttGSTF1, PttGSTF1 S13C, and PttGSTF1 S13A respectively (Supplementary Table 3). Compared to theoretical masses, these values are compatible with proteins where the N-terminal methionine is cleaved, which was expected form the presence of an alanine as the second residue. A mass increment of 305 Da was specifically present in PttGSTF1 S13C, which suggested that a glutathione molecule is covalently bound to the newly introduced cysteine residue via a disulfide bridge. Accordingly, PttGSTF1 S13C is not retained on GSH Sepharose columns contrary to PttGSTF1 and PttGSTF1 S13A. Then, the oligomeric state of wild-type and mutated proteins was estimated using calibrated

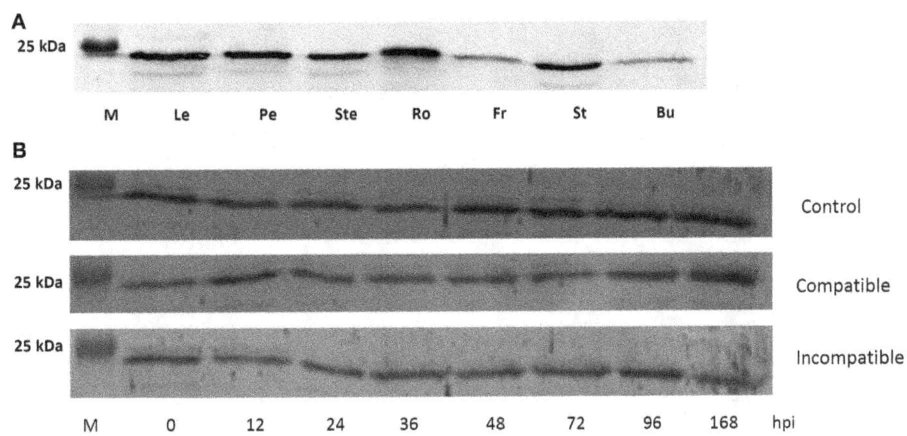

FIGURE 4 | GSTF1 protein abundance in poplar organs and in rust-infected leaves. (A) Western blot analysis was performed from 30 μg of soluble protein extracts from leaves (Le), petioles (Pe), stems (Ste), roots (Ro), fruits (Fr), stamens (St), and buds (Bu). **(B)** Western blot analysis from 30 μg of soluble protein extracts from leaves infected or not with virulent or avirulent isolates of *M. larici populina* leading respectively to compatible or incompatible interactions. Control refers to as mock inoculated treatment. Time-points correspond to key developmental stages, i.e., penetration through the stomata (12 hpi), formation of the first haustorial infection structures (24 hpi), arrest of avirulent isolate growth (48 hpi) and formation of uredinia symptoms by the virulent isolate and urediniospores release (168 hpi).

size exclusion chromatography. All purified proteins eluted as a single peak whose estimated mass (45–47 kDa) is consistent with a dimeric arrangement (Supplementary Table 3) as reported for example for Arabidopsis GSTF2 or maize GST-I proteins (Reinemer et al., 1996; Neuefeind et al., 1997a).

Next, in order to characterize the enzymatic properties of PttGSTF1, its activity was measured toward various model substrates (CDNB, BITC, PITC, PNP-butyrate, and HNE) usually employed to measure the activities of GSTs catalyzing GSH-conjugation reactions (**Table 2**). An activity was detected toward all these substrates with catalytic efficiencies (k_{cat}/K_m) ranging from $6.6 \times 10^2\ M^{-1}\ s^{-1}$ for PNP-butyrate to $3.1 \times 10^3\ M^{-1}\ s^{-1}$ for HNE. The slightly better catalytic efficiency obtained for HNE compared to other substrates is due to a better affinity of PttGSTF1 for this substrate. On the other hand, the lower efficiency observed with PNP-butyrate is due to a weak turnover number (k_{cat}). The kinetic parameters for the two tested isothiocyanate derivatives were in the same range. The difference by a factor around two of the apparent K_m value indicates that variations in the aromatic groups (benzyl vs phenetyl) do not affect much substrate recognition. Comparing all substrates, the highest K_m value was for CDNB but this is compensated by a better turnover number which is around 6–20 fold better than for the other substrates tested. Using PNP-butyrate as the second substrate, an apparent affinity of GSTF1 for GSH was determined. The K_m value is $97.6 \pm 6.0\ \mu M$.

Contrary to Tau GSTs, GSTFs often proved to have peroxidase activities. For this reason, we have also tested cumene hydroperoxide (CuOOH) and tert-butyl hydroperoxide (t-BOOH). Whereas no activity was detected with t-BOOH, the catalytic efficiency obtained in steady-state conditions for the reduction of CuOOH into the corresponding alcohol is $3.2 \times 10^3\ M^{-1}s^{-1}$. This is in fact quite close to the value obtained for example with a mitochondrial Prx IIF from poplar, the role of which is assumed to significantly contribute to peroxide detoxification or signaling (Gama et al., 2007).

With most substrates, the substitution of the catalytic serine into alanine (PttGSTF1 S13A variant) generally led to a completely inactive enzyme. However, a residual activity was still observed with CDNB and PNP-butyrate, the catalytic efficiency being decreased by a factor of 40 and 20 respectively compared to the results obtained with PttGSTF1. Whereas this suggested that one or several residues other than the serine contribute to the decrease of the pKa of the thiol group of GSH, the PttGSTF1 S13C variant had no or negligible activity toward all these substrates. According to the mass spectrometry results, the reason may be the formation of a covalent adduct. Hence, this prompted us to investigative whether PttGSTF1 S13C has acquired properties similar to GSTs naturally having a cysteine residue in their active site signature by testing the thioltransferase activity using DHA and HED, two substrates usually employed for characterizing Grxs and cysteine-containing GSTs. As expected, PttGSTF1 had no activity both with HED and DHA. Concerning PttGSTF1 S13C, whereas no activity has been detected with DHA, a reasonably good catalytic efficiency ($1.1 \times 10^3\ M^{-1}\ s^{-1}$) was obtained with HED, essentially because of a good apparent affinity (K_m value of $33.7\ \mu M$).

Besides these classical assays, we sought to examine more unusual substrates/ligands that have been isolated with orthologous GSTF members i.e., auxin/indole-3-acetic acid (IAA) or a synthetic analog, 2,4-dichlorophenoxyacetic (2,4-D) (Bilang et al., 1993; Bilang and Sturm, 1995) and norharmane, indole-3-aldehyde and quercetin (Smith et al., 2003; Dixon et al., 2011). Hence, we investigated whether these compounds could constitute poplar GSTF1 substrates first by simply analyzing changes in the UV-visible spectra of each compounds as a function of time upon successive addition of GSH and PttGSTF1. However, we did not detect any significant spectral shifts (data not shown). Thinking that the glutathionylation may eventually not modify the absorption spectra of these molecules, the product of a reaction of several hours was analyzed by reverse phase-HPLC on a

Table 2 | Kinetic parameters of PttGSTF1.

	BITC	PITC	CDNB	PNP-butyrate	HNE	CuOOH	HED
K_m (μM)							
PttGSTF1	380.6 ± 43.7	148.9 ± 5.9	3065.6 ± 286.5	360.3 ± 32.2	67.1 ± 6.6	592.1 ± 51.6	ND
PttGSTF1 S13A	ND	ND	3313.9 ± 344.9	1510.9 ± 119.9	ND	ND	ND
PttGSTF1 S13C	ND	ND	ND	ND	ND	ND	33.7 ± 3.5
k_{cat} (s^{-1})							
PttGSTF1	0.70 ± 0.03	0.21 ± 0.02	4.20 ± 0.18	0.23 ± 0.40	0.21 ± 0.01	1.92 ± 0.04	ND
PttGSTF1 S13A	ND	ND	0.11 ± 0.01	0.060 ± 0.002	ND	ND	ND
PttGSTF1 S13C	ND	ND	ND	ND	ND	ND	0.040 ± 0.001
k_{cat}/K_m ($M^{-1}s^{-1}$)							
PttGSTF1	1839.2 ± 87.2	1410.3 ± 19.0	1370.0 ± 3.6	661.9 ± 17.3	3141.6 ± 149.9	3245.7 ± 76.9	ND
PttGSTF1 S13A	ND	ND	35.4 ± 1.8	37.5 ± 1.0	ND	ND	ND
PttGSTF1 S13C	ND	ND	ND	ND	ND	ND	1124.7 ± 24.5

The apparent K_m values for all compounds were determined by varying substrate concentrations at a fixed saturating GSH concentration. The apparent K_m and k_{cat} values were calculated by non-linear regression using the Michaelis–Menten equation. Results are means ± S.D. (n = 3). ND means not detected. BITC, benzyl isothiocyanate; PITC, phenetyl isothiocyanate; CDNB, 1-chloro-2,4-dinitrobenzene; PNP-butyrate, 4-nitrophenyl butyrate; HNE, 4-hydroxy-2-nonenal; CuOOH, cumene hydroperoxide; HED, hydroxyethyl disulfide.

C18 column. However, no glutathionylated species can be separated and identified using this approach. Considering that some of these molecules may represent ligands and that the ligandin and catalytic sites in GSTs are generally overlapping at least partially, we have examined whether the addition of these molecules modulated PttGSTF1 activity. Despite using concentrations in the millimolar range, no effect was observed both using CDNB and PNP butyrate assays. We concluded that these compounds do not bind to PttGSTF1.

THE STRUCTURES OF PttGSTF1 AND PttGSTF1 S13C IN COMPLEX WITH GSH AND MES REVEAL THE RESIDUES PARTICIPATING TO SUBSTRATE BINDING

The crystallographic structures of PttGSTF1 and PttGSTF1 S13C, bound with ligands, have been obtained and refined to 1.5 and 1.8 Å resolutions (**Table 1**). The crystals belonged to the space group $P2_12_12_1$, and the asymmetric unit consisted of one biological dimer (residues Ala2-Ala215 in both monomers, Root Mean Square Deviation of 0.18 Å for 175 superimposed Cα atoms). The analysis of the Fourier difference maps of PttGSTF1 revealed the presence of two ligands in the active site in each monomer: a glutathione molecule originating from the pre-treatment performed with an excess of GSH and a MES molecule present in the crystallization buffer. They are located respectively in the G and H sites (**Figure 5A**). Unless covalently bound, both ligands cannot occupy the active site simultaneously. Currently, we do not have any evidence for a GSH-conjugation reaction with MES nor data for any non-catalytic binding. Both glutathione and MES molecules were refined with complementary occupancies. In monomer A, the refined occupancies of glutathione and MES molecules were 58 and 42%, respectively. In monomer B, the corresponding refined occupancies were 71 and 29%, respectively. Therefore, PttGSTF1 structure can be described as two structures: PttGSTF1 in complex with glutathione and PttGSTF1 in complex with a MES molecule. Concerning PttGSTF1 S13C, the structure refinement confirmed that Cys13 is glutathionylated but this modification did not induce significant conformational changes in comparison to PttGSTF1 (RMSD of 0.33Å based on alignments of 350 Cα positions).

In order to understand possible differences among GSTF isoforms, a detailed comparison was performed with the three other GSTFs (AtGSTF2, maize GST-I and GST-III) whose structures are known (Reinemer et al., 1996; Neuefeind et al., 1997a,b; Prade et al., 1998). The AtGSTF2 structure was solved in complex with S-hexylglutathione or with an acetamide herbicide like molecule-glutathione conjugate, ZmGST-I was in complex with lactoylglutathione or an atrazine-glutathione conjugate, and ZmGST-III was in an apoform. Interestingly, PttGSTF1 belongs to a distinct, uncharacterized GSTF subgroup (**Figure 2**). A PttGSTF1 monomer consists of an N-terminal domain (β1α1β2α2β3β4α3) and a C-terminal domain composed of α-helices (α4α5α6α6′α7α8) (**Figure 5A**) as classically observed in most GST classes. As expected, structures of plant GSTFs superimposed relatively well with a mean RMSD of 0.92 Å. Prominent differences are nevertheless observed in three regions (**Figure 5B**). In PttGSTF1, an additional α-helix is observed in the segment between the strands β2 and β3 while others exhibit 1–3 short 3_{10}-helices. This segment is involved in substrate binding and contains a conserved phenylalanine (Phe53 in PttGSTF1), which is assumed to be essential for dimerization (Prade et al., 1998). This phenylalanine represents the major inter-monomer contact, its side chain being buried in a hydrophobic pocket composed of Trp102, Thr105, Thr109, Val143, Ile146, and Tyr147 in PttGSTF1 and located between α4 and α5 of the other subunit.

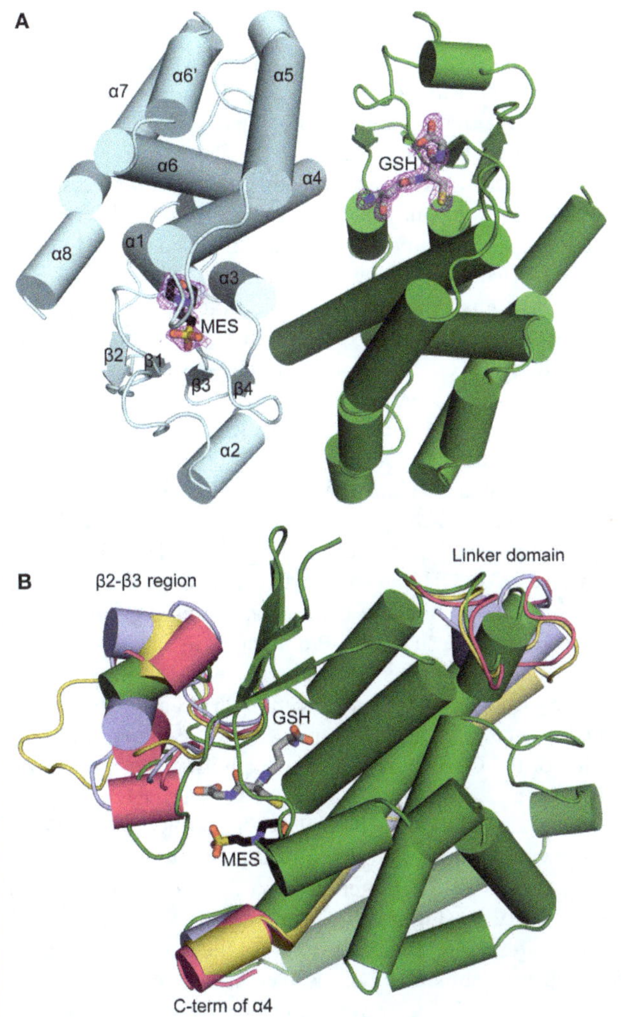

FIGURE 5 | Overall structure of PttGSTF1 and comparison with other plant GSTFs. (A) Architecture of PttGSTF1 dimer. The MES molecule (black) is located in the H site, as shown in monomer A whereas the GSH molecule (gray) bound to the G site is depicted in monomer B. Monomers A and B are colored in light cyan and dark green respectively. Secondary structures are labeled only for monomer A for clarity. The GSH and MES molecules are shown as sticks and colored according to atom type (nitrogen, blue; oxygen, red; sulfur, yellow; and carbon, gray/black). Omit map (colored in purple) of contour level of 0.8 σ is shown around each bound ligand. It was built using the Composite omit map command of the Phenix software suite. **(B)** Superimposition of monomers of PttGSTF1 (dark green), *A. thaliana* GSTF2 (pink) and *Z. mays* GST-I (blue) and GST-III (yellow). Only noticeable secondary structural differences between GSTFs are shown for clarity. The GSH (gray) and MES (black) molecules highlight the putative positions of the G and H sites respectively. GSH and MES molecules are colored according to atom type and shown as sticks.

Interestingly, among the residues involved in the dimer interface, the hydrophobic ones are those that are the most conserved in PtGSTFs (**Figure 1**). Another variation concerns the length and conformation of the linker found between α3 and α4 helices and that connects the N- and C-terminal domains. Considering the variable length of the linker, such conformational differences were expected. However, the central residue of this connecting region, Leu88 in PttGSTF1, is highly conserved and adopts a superimposable position in all plant GSTF structures. Its side chain, wedging between helices α3 and α6, connects the two domains. The last noticeable difference is likely to be a class-specific feature of PttGSTF1 in which the absence of 3 residues found in other poplar GSTFs (**Figure 1**) shortens the α4 helix.

In PttGSTF1, a glutathione molecule is positioned in the G site groove which is mainly populated by polar residues from the N-terminal domain (**Figure 6A**). The Glu68, Ser69, and Arg70 residues, situated in the β4-α3 loop and in α3, stabilize the glutamyl group of GSH through hydrogen bonds and Coulomb interactions. The NH and carbonyl groups of the cysteinyl moiety are hydrogen-bonded to the backbone amino group of Val56 that precedes the invariant cis-Pro57 found in all GSTs and in all Trx superfamily members. The carboxylate of the glycinyl residue interacts with the side chains of Gln42, Lys43 and Gln55 found in the loops connecting β2-α2 and α2-β3. The thiol group of the cysteine of the GSH moiety is quasi-equidistant to the hydroxyl groups of Ser13 and Thr14 (3.2 and 3.4 Å respectively). According to mass spectrometry data, in the PttGSTF1 S13C variant, GSH is covalently bound to the modified residue (Cys13). Apart this difference, the same GSH-protein interactions are observed in both crystal structures (**Figure 6B**). With regard to the electrophilic substrate site, a MES molecule occupies the position adopted by other substrates in known GSTF structures. Hence, the H site is delimitated by residues from three regions: residues 12–14 found at the end of the β1-α1 loop and in α1, residues 36–40 that are part of the β2-α2 loop and residues 119–123 which are located in the C-terminal end of α4 (**Figure 6C**). The MES molecule is surrounded by the hydrophobic residues Leu12, Leu37 and Phe123. Moreover, the oxygen atom of the morpholino ring is hydrogen-bonded to the NH and OH groups of Thr14 and the sulfonic group forms a salt bridge with His119. However, the latter two residues are less conserved as compared to the three others suggesting that they might confer substrate specificities to PttGSTF1.

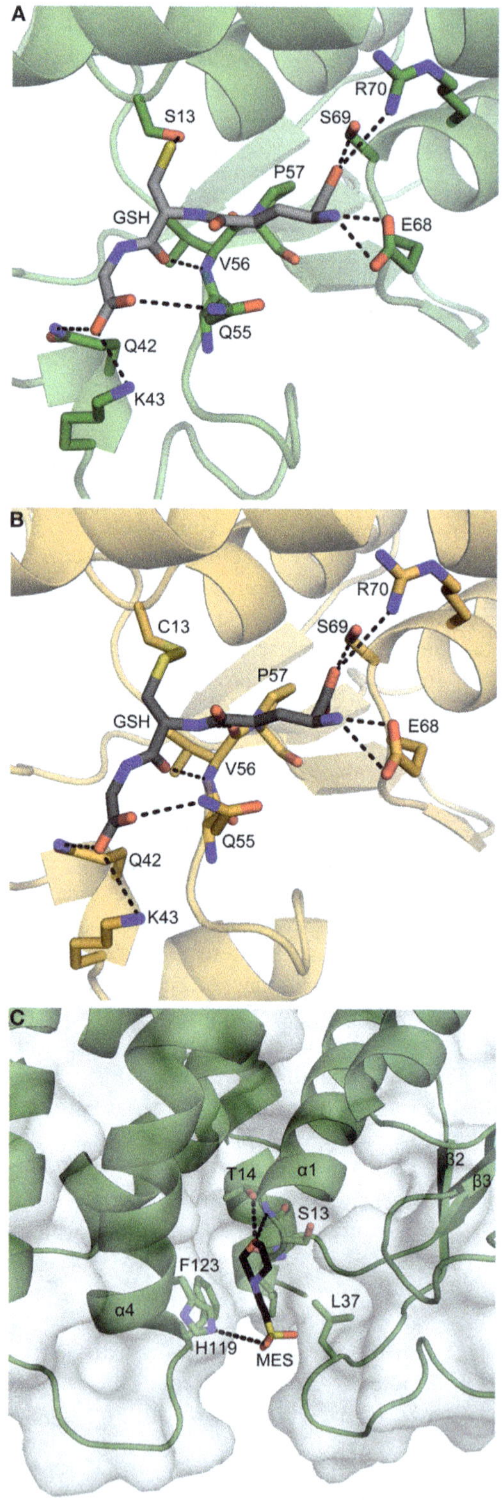

FIGURE 6 | Close up view of the G- and H-sites. (A,B) Glutathione binding site of PttGSTF1 and PttGSTF1 S13C. **(C)** Electrophilic substrate binding site of PttGSTF1. The PttGSTF1 (dark green) and PttGSTF1 S13C (orange) monomers are shown in cartoon with a transparent molecular surface in **(C)**. Residues involved in the binding of GSH (gray) and MES (black) molecules are shown as sticks, labeled and colored according to atom type. Putative hydrophilic interactions between the substrates and the enzyme are shown as black dashed lines.

DISCUSSION

The existence of multigenic families is frequently explained by the functional divergence i.e., the acquirement of new or specific functions, appearing following gene duplication. With the complete sequencing of several plant genomes, it appeared that many species-specific duplication events occurred, leading to the expansion of the *GSTF* gene family. The maintenance of so many GSTF genes in the genomes (eight in poplar but up to *ca* 27 in some terrestrial plants) might be attributed for example (i) to a specific cellular/tissular expression associated to certain developmental stages or stress conditions, (ii) to specific subcellular localizations or (iii) to specific biochemical and structural characteristics. An additional layer of complexity and possible redundancy is the existence of several tens of GSTUs in plants which have quite similar enzymatic and biochemical properties. Indeed, owing to the presence of the same conserved serine residue, GSTUs also possess glutathionylation activities toward herbicides, safeners and several other cyclic/aromatic compounds. An intriguing example illustrating the possible redundancy between GSTUs and GSTFs is the fact that petunia *AN9*, a GSTF gene, and maize *Bz2*, a GSTU gene, can complement mutants for the other gene (Alfenito et al., 1998). Some differences can however be sometimes noticed. For instance, contrary to most GSTFs, GSTUs usually do not have peroxidase activity. However, redundancy could exist with other GST classes, notably the Theta GSTs that do have such a peroxidase activity.

In this study, we provide the first elements exploring the question of the redundancy among GSTF members and functions in poplar. Focusing on the particularities among poplar GSTFs that could explain the presence of eight genes, their putative subcellular localizations were first examined from the bioinformatic analysis of primary sequences. According to the absence of clear N- or C -terminal targeting sequences, all poplar GSTFs are predicted to be cytosolic proteins. This is generally in accordance with data obtained in other organisms either from translational GFP fusion as for several GSTFs of *Physcomitrella patens* (Liu et al., 2013) or from the absence of GSTF detection in studies of organellar proteomes. A plasma membrane localization was suggested for AtGSTF2 (Murphy et al., 2002) and a dual targeting in the cytosol and chloroplast was demonstrated for AtGSTF8 owing to the presence of an alternative transcription start site (Thatcher et al., 2007). However, only a few AtGSTF8 orthologs in other plant species have a similar extension. With regard to expression profiles, all poplar *GSTF* genes are redundantly expressed in some organs as leaves or reproductive organs. Moreover, the transcript levels are not necessarily correlated with protein levels. For instance, we did not detect GSTF1 transcripts in roots whereas quite important protein amounts were detected by western blot. It certainly illustrates the variations inherent to the plant developmental stages or to the fluctuations of environmental constraints as we have harvested our samples from a naturally growing tree and at different periods. Considering that many GSTFs could have similar cellular and subcellular expression territories, the difference should come from specific biochemical and/or structural properties. This parameter has been examined by producing PttGSTF1 as a recombinant protein and assessing its activity toward model substrates representing various types of biochemical activities as well as by solving the 3D structure of the first GSTF representative from a tree. Indeed, structures for only three GSTFs have been solved in the late 90's and nothing since that time.

As other GSTF members bearing a conserved serine in the active site motif, enzymatic analysis showed that GSTF1 possesses glutathione-conjugating activity toward structurally diverse substrates and glutathione peroxidase activity. The kinetic parameters of GSTF1 activity toward the model substrate CDNB (k_{cat}/K_m of 1.3×10^3 $M^{-1}s^{-1}$) are within the range of reported values for some GSTFs as *P. patens* GSTF1 (k_{cat}/K_m of 1.5×10^3 $M^{-1}s^{-1}$) (Liu et al., 2013) although important variations can be sometimes detected. *Triticum aestivum* GSTF1 exhibits a 50 fold higher catalytic efficiency (k_{cat}/K_m of 7.2×10^4 $M^{-1}s^{-1}$) (Cummins et al., 2003). While CDNB is an artificial substrate that may somehow mimic the structure of some herbicides and that is usually modified by all GSTFs, the other substrates used may be more physiologically relevant. BITC and PITC are representatives of a family of natural compounds found in Brassicaceae and produced by the enzymatic degradation of glucosinolates. Surprisingly, whereas glucosinolates are found in Arabidopsis, only a few Arabidopsis GSTF members among the 13 isoforms are able to catalyze conjugation reactions on BITC (Wagner et al., 2002; Nutricati et al., 2006; Dixon et al., 2009). The quite important turnover number obtained for the GSH-conjugation reaction of BITC by PttGSTF1 (k_{cat} of $0.70\,s^{-1}$) indicates that poplar GSTF1 may have the particular ability to recognize related molecules. As a matter of comparison, higher turnover numbers, around $25\,s^{-1}$, have been reported for *Homo sapiens* GST M1-1 or P1-1 both using BITC and PITC (Kolm et al., 1995). CuOOH is used as a molecule representative of bulky peroxides such as peroxidized lipids whereas HNE is a toxic aldehyde formed as a major end product of lipid peroxidation (Esterbauer et al., 1991). Both types of molecules have a dual function, being deleterious by promoting DNA damages or membrane protein inactivation, but at the same time, they represent signaling molecules. Whereas peroxide activity is systematically tested for GSTFs, the GSH-conjugation of HNE has been rarely evaluated. One example is the demonstration that a *Sorghum bicolor* B1/B2 GSTF heterodimer purified from shoots of fluxofenim-treated plants exhibits a catalytic efficiency about 15 fold higher (calculated k_{cat}/K_m for this protein is around 2×10^4 $M^{-1}s^{-1}$) than for PttGSTF1 (k_{cat}/K_m of 1.3×10^3 $M^{-1}s^{-1}$) (Gronwald and Plaisance, 1998). With regard to peroxides, most GSTFs tested so far, whatever their origin, exhibit a glutathione peroxidase activity. Compared to other characterized GSTFs, poplar GSTF1 possesses quite elevated peroxidase activity toward cumene hydroperoxide with a turnover number of 1.92 s^{-1} (Dixon et al., 2009). It is for instance in the same range as those reported for *Lolium rigidum* and *Alopecurus myosuroides* GSTF1 which are considered as highly active peroxidases (k_{cat} of 2.64 and 1.3 s^{-1} respectively) (Cummins et al., 2013). From a physiological perspective, it is worth noting that pathogen attacks are often accompanied by an oxidative stress that triggers, among other symptoms, lipid peroxidation. Also, important amounts of HNE are accumulated in *Phaseolus vulgaris* upon fungal infection by *Botrytis cinerea* (Muckenschnabel et al., 2002). With the known induction of *GSTF* genes by defense hormones or biotic stresses (Wagner et al., 2002), their known involvement in the synthesis of defense compounds as camalexin (Su et al., 2011), the peroxidase and GSH-conjugating HNE activities, it is conceivable that GSTF1 is involved in oxidative stress tolerance and/or oxidative signaling occurring in particular during pathogen or insect attacks. In fact, whereas *GSTF1* expression is induced in poplar attacked by the tent caterpillar *Malacosoma disstria* (Ralph et al., 2006) contrasting results have been obtained for GSTF1 in the case of rust infected poplars. Indeed *GSTF1* gene was found to be up-regulated at six dpi in a former study investigating gene expression in *Populus trichocarpa* × *P. deltoides* leaves infected by *Melampsora medusae* which represent a compatible interaction (Miranda et al., 2007).

On the other hand, no regulation was detected when *Populus nigra* × *P. maximowiczii* leaves are infected by *M. medusae* or *M. larici-populina* (Azaiez et al., 2009) or when *P. trichocarpa* × *P. deltoides* leaves are infected by *M. larici-populina* either by a virulent (compatible) or an avirulent (incompatible) isolate (Rinaldi et al., 2007). According to the transcript measurements, no variation of GSTF1 protein level has been detected in *P. trichocarpa* × *P. deltoides* leaves during both compatible and incompatible reactions with *M. larici populina*. Here, the observed differences might simply be explained by differences in the poplar cultivars, rust isolates and time-points used in these independent studies, which altogether generate some specificity in these interactions. It would be informative to systematically analyze transcript and protein variations for all poplar GSTFs in different biotic interactions as done previously for example for the Arabidopsis or wheat GSTF families (Wagner et al., 2002; Cummins et al., 2003). To summarize this part, the biochemical and expression analyses demonstrated that, through its peroxidase and its GSH-conjugating activities, GSTF1 may have multiple roles notably related to xenobiotic detoxification or to oxidative stress tolerance both under biotic and abiotic constraints.

Contrary to other GSTFs, we have not observed an interaction or an activity with auxin and other heterocyclic compounds such as norharmane, indole-3-aldehyde and quercetin that were previously found to interact with other GSTFs and AtGSTF2 in particular (Bilang and Sturm, 1995; Smith et al., 2003; Dixon et al., 2011). This may indicate that PttGSTF1 has no ligandin function. In fact, several GSTFs for which ligandin function has been demonstrated, such as AN8 or Bz2, do not have the catalytic serine but an alanine instead in a AAxP motif. It does not mean however, that these GSTFs do not have catalytic functions. In fact, when the catalytic serine of PttGSTF1 was replaced by an alanine, the glutathionylation activity is not totally abolished as we would expect and a weak activity toward certain substrates was still measurable. This suggests that the catalytic serine is important but not mandatory for GSH-conjugating reactions and that residues other than the catalytic serine could be involved in glutathione activation. In support of this view, it has been reported that human GSTO1-1, a Cys-GST, loses deglutathionylation activity and acquires glutathionylation activity when the catalytic cysteine is replaced by an alanine (Whitbread et al., 2005). While Ser13 likely corresponds to the catalytic residue found in most GSTFs and is the primary candidate for GSH activation, the hydroxyl group of the adjacent Thr14 is found approximately at the same distance in the PttGSTF1 structure. Hence, it is tempting to conclude that it might substitute to Ser13, at least in its absence. It is worth noting that with the exception of some GSTF clades, the members of which have a proline, the catalytic serine is often followed by another Ser or Thr in most members of other clades (**Figure 2**). Interestingly, neither the serine nor the threonine is conserved in the clade containing poplar GSTF8 which harbors aliphatic residues at these positions (AACP signature). In this specific case, we could speculate that the cysteine found after the threonine position acts as the catalytic residue. Although this will have to be confirmed experimentally, it is interesting to note that poplar GSTF3 and F7 and their close orthologs also have a cysteine at this position, and that fungal Ure2p-like enzymes have an asparagine that was recently assumed to be important for catalysis (Thuillier et al., 2013). Overall, this suggests that all residues forming the active site signature and present around the N-terminal end of α1 could substitute to each other. Another proof of this assumption is that the PttGSTF1 S13C mutant lost its glutathione peroxidase and glutathionylating activity but acquired the capacity to perform deglutathionylation reaction toward HED, an activity typical of Cys-GSTs. Although the detected activity is weaker than the one obtained with naturally-existing Cys-GSTs (Lallement et al., 2014a), it shows that changing the nature of the catalytic residue is sufficient to determine the type of GST activity. Accordingly, when the catalytic cysteine of poplar Lambda GSTs is mutated into a serine, a shift from the original deglutathionylation to glutathionylation activity was observed (Lallement et al., 2014b).

Complementary to the biochemical and enzymatic analyses, the structural analysis should help understanding why GSTFs accept such diverse substrates but at the same time what are the fine differences that would generate substrate specificity. A comparison of poplar PttGSTF1 with AtGSTF2 structure does not point to dramatic structural changes. In fact, the glutathione binding site is in general not very different within a GST class but also among diverse GST classes. This is what we observed by superimposing GSTF structures. Rather, structural differences if any should come from variations in the H-site. However, owing to the lack of structures of GST in complex with their ligands, this H-site is often not very well defined. In the PttGSTF1 structure, the MES molecule, which likely mimics an H-site substrate (although it does not seem to be catalytically glutathionylated), is stabilized by five residues, Leu12, Thr14, Leu37, and His119 and Phe123. The Thr14 which is present in all poplar GSTFs except GSTF8, is in fact not found in other proteins whose structures are known, AtGSTF2 (SIAT signature), ZmGST-I (SWNL signature), and ZmGST-III (SPNV signature). Similarly, the His119 position is variable and it is occupied by an aromatic residue (Phe or Trp) in other poplar GSTFs as well as in AtGSTF2, ZmGST-I and ZmGST-III. Hence it is possible that these residues contribute to the recognition of specific substrates by PttGSTF1. In particular, the presence of His119 may be responsible for the binding of the MES molecule. On the contrary, the residues found at positions equivalent to Leu12, Leu37, and Phe123 in PttGSTF1 are also hydrophobic in most plant GSTFs and they are involved in the stabilization of the substrate in known structures of plant GSTFs in complex with herbicides. Thus, they seem to be critical for the electrophilic substrate recognition and they could constitute the core residues required for the general recognition of substrates. Supporting this view, it was shown that the Phe123 to Ile substitution in AtGSTF2 altered its ligand affinity and specificity (Dixon et al., 2011). Leu37 is found between β2 and β3, a region which is not well superimposable from one structure to another. For instance, five residues from this region are not visible in the electron density of the crystal structure of apo ZmGSTIII (Neuefeind et al., 1997b). A Phe35 modification in ZmGST-I (the residue equivalent to Leu37 in PttGSTF1) affects the enzyme affinity for its ligand (Axarli et al., 2004). Overall, this indicates that the β2-β3 region could be the protein area used by GSTFs to accommodate such a large spectrum of ligands/substrates. In GSTUs, the end of α4 helix and the C-terminal part are other regions that contribute to the correct positioning of the substrate in the H-site (Axarli et al., 2009). Similarly, the residues found at the end of α4 helix are also used by Lambda and Omega GSTs for substrate recognition which also involves the α4-α5 l **o**p and a C-terminal helix (α9) which is specific to these two classes (Lallement et al., 2014b). In contrast, in GHR/Xi GSTs, proteins specialized in the reduction of glutathionylated quinones, no ample conformational change occurs upon substrate binding (Lallement et al., 2014c).

To conclude on these biochemical and structural analyses, it is conceivable that most GSTFs display a common set of enzymatic activities on typical substrates that is linked to the conservation of core residues. The persistence of closely related genes in single species may be explained by subtle sequence changes that confer the ability to the enzymes to accommodate specific substrates and thus to acquire specific functions. Hence, to address this question of the enzyme divergence and substrate specificity, isolating and identifying physiological GSTF substrates should become a priority as well as accumulating more 3D structures of GSTFs from poplar and other plants, alone or more importantly in complex with their physiological substrates.

AUTHOR CONTRIBUTIONS

Henri Pégeot, Cha San Koh, Benjamin Petre, and Sandrine Mathiot performed the experiments under the supervision of Sébastien Duplessis, Arnaud Hecker, Claude Didierjean, and Nicolas Rouhier. All authors contributed to the writing of the manuscript, have read and approved the final manuscript.

ACKNOWLEDGMENTS

This work is supported by a grant overseen by the French National Research Agency (ANR) as part of the "Investissements d'Avenir" program (ANR-11-LABX-0002-01, Lab of Excellence ARBRE). The PhD grant of Henri Pégeot is funded by ANR-11-LABX-0002-01 and the région Lorraine. Access to the X-ray diffraction facilities of the Université de Lorraine was appreciated. We are grateful to the staff members of X11 beamline of the EMBL/DESY Synchrotron (Hamburg, Germany) and of BM30A beamline at ESRF (Grenoble, France).

REFERENCES

Adams, P. D., Afonine, P. V., Bunkoczi, G., Chen, V. B., Davis, I. W., Echols, N., et al. (2010). PHENIX: a comprehensive Python-based system for macromolecular structure solution. *Acta Crystallogr. D Biol. Crystallogr.* 66, 213–221. doi: 10.1107/S0907444909052925

Alfenito, M. R., Souer, E., Goodman, C. D., Buell, R., Mol, J., Koes, R., et al. (1998). Functional complementation of anthocyanin sequestration in the vacuole by widely divergent glutathione S-transferases. *Plant Cell* 10, 1135–1149. doi: 10.1105/tpc.10.7.1135

Armstrong, R. N. (1997). Structure, catalytic mechanism, and evolution of the glutathione transferases. *Chem. Res. Toxicol.* 10, 2–18. doi: 10.1021/tx960072x

Axarli, I., Dhavala, P., Papageorgiou, A. C., and Labrou, N. E. (2009). Crystallographic and functional characterization of the fluorodifen-inducible glutathione transferase from *Glycine max* reveals an active site topography suited for diphenylether herbicides and a novel L-site. *J. Mol. Biol.* 385, 984–1002. doi: 10.1016/j.jmb.2008.10.084

Axarli, I. A., Rigden, D. J., and Labrou, N. E. (2004). Characterization of the ligandin site of maize glutathione S-transferase I. *Biochem. J.* 382, 885–893. doi: 10.1042/BJ20040298

Azaiez, A., Boyle, B., Levee, V., and Seguin, A. (2009). Transcriptome profiling in hybrid poplar following interactions with Melampsora rust fungi. *Mol. Plant Microbe Interact.* 22, 190–200. doi: 10.1094/MPMI-22-2-0190

Bilang, J., Macdonald, H., King, P. J., and Sturm, A. (1993). A soluble auxin-binding protein from *Hyoscyamus muticus* is a glutathione S-transferase. *Plant Physiol.* 102, 29–34. doi: 10.1104/pp.102.1.29

Bilang, J., and Sturm, A. (1995). Cloning and characterization of a glutathione S-transferase that can be photolabeled with 5-azido-indole-3-acetic acid. *Plant Physiol.* 109, 253–260. doi: 10.1104/pp.109.1.253

Chen, F., Zhang, S., Jiang, H., Ma, W., Korpelainen, H., and Li, C. (2011). Comparative proteomics analysis of salt response reveals sex-related photosynthetic inhibition by salinity in *Populus cathayana* cuttings. *J. Proteome Res.* 10, 3944–3958. doi: 10.1021/pr200535r

Cohen, D., Bogeat-Triboulot, M. B., Tisserant, E., Balzergue, S., Martin-Magniette, M. L., Lelandais, G., et al. (2010). Comparative transcriptomics of drought responses in Populus: a meta-analysis of genome-wide expression profiling in mature leaves and root apices across two genotypes. *BMC Genomics* 11:630. doi: 10.1186/1471-2164-11-630

Couturier, J., Stroher, E., Albetel, A. N., Roret, T., Muthuramalingam, M., Tarrago, L., et al. (2011). Arabidopsis chloroplastic glutaredoxin C5 as a model to explore molecular determinants for iron-sulfur cluster binding into glutaredoxins. *J. Biol. Chem.* 286, 27515–27527. doi: 10.1074/jbc.M111.228726

Cummins, I., Bryant, D. N., and Edwards, R. (2009). Safener responsiveness and multiple herbicide resistance in the weed black-grass (*Alopecurus myosuroides*). *Plant Biotechnol. J.* 7, 807–820. doi: 10.1111/j.1467-7652.2009.00445.x

Cummins, I., Cole, D. J., and Edwards, R. (1999). A role for glutathione transferases functioning as glutathione peroxidases in resistance to multiple herbicides in black-grass. *Plant J.* 18, 285–292. doi: 10.1046/j.1365-313X.1999.00452.x

Cummins, I., O'Hagan, D., Jablonkai, I., Cole, D. J., Hehn, A., Werck-Reichhart, D., et al. (2003). Cloning, characterization and regulation of a family of phi class glutathione transferases from wheat. *Plant Mol. Biol.* 52, 591–603. doi: 10.1023/A:1024858218804

Cummins, I., Wortley, D. J., Sabbadin, F., He, Z., Coxon, C. R., Straker, H. E., et al. (2013). Key role for a glutathione transferase in multiple-herbicide resistance in grass weeds. *Proc. Natl. Acad. Sci. U.S.A.* 110, 5812–5817. doi: 10.1073/pnas.1221179110

Deridder, B. P., Dixon, D. P., Beussman, D. J., Edwards, R., and Goldsbrough, P. B. (2002). Induction of glutathione S-transferases in Arabidopsis by herbicide safeners. *Plant Physiol.* 130, 1497–1505. doi: 10.1104/pp.010066

Dixon, D. P., Hawkins, T., Hussey, P. J., and Edwards, R. (2009). Enzyme activities and subcellular localization of members of the Arabidopsis glutathione transferase superfamily. *J. Exp. Bot.* 60, 1207–1218. doi: 10.1093/jxb/ern365

Dixon, D. P., Sellars, J. D., and Edwards, R. (2011). The Arabidopsis phi class glutathione transferase AtGSTF2: binding and regulation by biologically active heterocyclic ligands. *Biochem. J.* 438, 63–70. doi: 10.1042/BJ20101884

Emsley, P., Lohkamp, B., Scott, W. G., and Cowtan, K. (2010). Features and development of Coot. *Acta Crystallogr. D Biol. Crystallogr.* 66, 486–501. doi: 10.1107/S0907444910007493

Esterbauer, H., Schaur, R. J., and Zollner, H. (1991). Chemistry and biochemistry of 4-hydroxynonenal, malonaldehyde and related aldehydes. *Free Radic. Biol. Med.* 11, 81–128. doi: 10.1016/0891-5849(91)90192-6

Gama, F., Keech, O., Eymery, F., Finkemeier, I., Gelhaye, E., Gardeström, P., et al. (2007). The mitochondrial type II peroxiredoxin from poplar. *Physiol. Plant.* 129, 196–206. doi: 10.1111/j.1399-3054.2006.00785.x

Gonneau, M., Pagant, S., Brun, F., and Laloue, M. (2001). Photoaffinity labelling with the cytokinin agonist azido-CPPU of a 34 kDa peptide of the intracellular pathogenesis-related protein family in the moss *Physcomitrella patens*. *Plant Mol. Biol.* 46, 539–548. doi: 10.1023/A:1010693213437

Gronwald, J. W., and Plaisance, K. L. (1998). Isolation and characterization of glutathione S-transferase isozymes from sorghum. *Plant Physiol.* 117, 877–892. doi: 10.1104/pp.117.3.877

Hayes, J. D., Flanagan, J. U., and Jowsey, I. R. (2005). Glutathione transferases. *Annu. Rev. Pharmacol. Toxicol.* 45, 51–88. doi: 10.1146/annurev.pharmtox.45.120403.095857

Kabsch, W. (2010). Xds. *Acta Crystallogr. D Biol. Crystallogr.* 66, 125–132. doi: 10.1107/S0907444909047337

Kieffer, P., Schroder, P., Dommes, J., Hoffmann, L., Renaut, J., and Hausman, J. F. (2009). Proteomic and enzymatic response of poplar to cadmium stress. *J. Proteomics* 72, 379–396. doi: 10.1016/j.jprot.2009.01.014

Kitamura, S., Shikazono, N., and Tanaka, A. (2004). TRANSPARENT TESTA 19 is involved in the accumulation of both anthocyanins and proanthocyanidins in Arabidopsis. *Plant J.* 37, 104–114. doi: 10.1046/j.1365-313X.2003.01943.x

Kolm, R. H., Danielson, U. H., Zhang, Y., Talalay, P., and Mannervik, B. (1995). Isothiocyanates as substrates for human glutathione transferases: structure-activity studies. *Biochem. J.* 311(Pt 2), 453–459.

Lallement, P. A., Brouwer, B., Keech, O., Hecker, A., and Rouhier, N. (2014a). The still mysterious roles of cysteine-containing glutathione transferases in plants. *Front. Pharmacol.* 5:192. doi: 10.3389/fphar.2014.00192

Lallement, P. A., Meux, E., Gualberto, J. M., Dumarcay, S., Favier, F., Didierjean, C., et al. (2014c). Glutathionyl-hydroquinone reductases from poplar are plastidial proteins that deglutathionylate both reduced and oxidized quinones. *FEBS Lett.* doi: 10.1016/j.febslet.2014.11.021. [Epub ahead of print].

Lallement, P. A., Meux, E., Gualberto, J. M., Prosper, P., Didierjean, C., Saul, F., et al. (2014b). Structural and enzymatic insights into Lambda glutathione transferases from *Populus trichocarpa*, monomeric enzymes constituting an early divergent class specific to terrestrial plants. *Biochem. J.* 462, 39–52. doi: 10.1042/BJ20140390

Lan, T., Yang, Z. L., Yang, X., Liu, Y. J., Wang, X. R., and Zeng, Q. Y. (2009). Extensive functional diversification of the Populus glutathione S-transferase supergene family. *Plant Cell* 21, 3749–3766. doi: 10.1105/tpc.109.070219

Lieberherr, D., Wagner, U., Dubuis, P. H., Metraux, J. P., and Mauch, F. (2003). The rapid induction of glutathione S-transferases AtGSTF2 and AtGSTF6 by avirulent *Pseudomonas syringae* is the result of combined salicylic acid and ethylene signaling. *Plant Cell Physiol.* 44, 750–757. doi: 10.1093/pcp/pcg093

Liu, Y. J., Han, X. M., Ren, L. L., Yang, H. L., and Zeng, Q. Y. (2013). Functional divergence of the glutathione S-transferase supergene family in *Physcomitrella patens* reveals complex patterns of large gene family evolution in land plants. *Plant Physiol.* 161, 773–786. doi: 10.1104/pp.112.205815

Miranda, M., Ralph, S. G., Mellway, R., White, R., Heath, M. C., Bohlmann, J., et al. (2007). The transcriptional response of hybrid poplar (*Populus trichocarpa* × *P. deltoides*) to infection by *Melampsora medusae* leaf rust involves induction of flavonoid pathway genes leading to the accumulation of proanthocyanidins. *Mol. Plant Microbe Interact.* 20, 816–831. doi: 10.1094/MPMI-20-7-0816

Morel, M., Meux, E., Mathieu, Y., Thuillier, A., Chibani, K., Harvengt, L., et al. (2013). Xenomic networks variability and adaptation traits in wood decaying fungi. *Microb. Biotechnol.* 6, 248–263. doi: 10.1111/1751-7915.12015

Muckenschnabel, I., Goodman, B. A., Williamson, B., Lyon, G. D., and Deighton, N. (2002). Infection of leaves of *Arabidopsis thaliana* by *Botrytis cinerea*: changes in ascorbic acid, free radicals and lipid peroxidation products. *J. Exp. Bot.* 53, 207–214. doi: 10.1093/jexbot/53.367.207

Mueller, L. A., Goodman, C. D., Silady, R. A., and Walbot, V. (2000). AN9, a petunia glutathione S-transferase required for anthocyanin sequestration, is a flavonoid-binding protein. *Plant Physiol.* 123, 1561–1570. doi: 10.1104/pp.123.4.1561

Mueller, S., Hilbert, B., Dueckershoff, K., Roitsch, T., Krischke, M., Mueller, M. J., et al. (2008). General detoxification and stress responses are mediated by oxidized lipids through TGA transcription factors in Arabidopsis. *Plant Cell* 20, 768–785. doi: 10.1105/tpc.107.054809

Murphy, A. S., Hoogner, K. R., Peer, W. A., and Taiz, L. (2002). Identification, purification, and molecular cloning of N-1-naphthylphthalmic acid-binding plasma membrane-associated aminopeptidases from Arabidopsis. *Plant Physiol.* 128, 935–950. doi: 10.1104/pp.010519

Neuefeind, T., Huber, R., Dasenbrock, H., Prade, L., and Bieseler, B. (1997a). Crystal structure of herbicide-detoxifying maize glutathione S-transferase-I in complex with lactoylglutathione: evidence for an induced-fit mechanism. *J. Mol. Biol.* 274, 446–453. doi: 10.1006/jmbi.1997.1402

Neuefeind, T., Huber, R., Reinemer, P., Knablein, J., Prade, L., Mann, K., et al. (1997b). Cloning, sequencing, crystallization and X-ray structure of glutathione S-transferase-III from Zea mays var. mutin: a leading enzyme in detoxification of maize herbicides. *J. Mol. Biol.* 274, 577–587. doi: 10.1006/jmbi.1997.1401

Nutricati, E., Miceli, A., Blando, F., and De Bellis, L. (2006). Characterization of two *Arabidopsis thaliana* glutathione S-transferases. *Plant Cell Rep.* 25, 997–1005. doi: 10.1007/s00299-006-0146-1

Otwinowski, Z., and Minor, W. (1997). Processing of X-ray diffraction data collected in oscillation mode. *Methods Enzymol.* 276, 307–326. doi: 10.1016/S0076-6879(97)76066-X

Prade, L., Huber, R., and Bieseler, B. (1998). Structures of herbicides in complex with their detoxifying enzyme glutathione S-transferase - explanations for the selectivity of the enzyme in plants. *Structure* 6, 1445–1452. doi: 10.1016/S0969-2126(98)00143-9

Ralph, S., Oddy, C., Cooper, D., Yueh, H., Jancsik, S., Kolosova, N., et al. (2006). Genomics of hybrid poplar (*Populus trichocarpa* × *deltoides*) interacting with forest tent caterpillars (*Malacosoma disstria*): normalized and full-length cDNA libraries, expressed sequence tags, and a cDNA microarray for the study of insect-induced defences in poplar. *Mol. Ecol.* 15, 1275–1297. doi: 10.1111/j.1365-294X.2006.02824.x

Reinemer, P., Prade, L., Hof, P., Neuefeind, T., Huber, R., Zettl, R., et al. (1996). Three-dimensional structure of glutathione S-transferase from *Arabidopsis thaliana* at 2.2 A resolution: structural characterization of herbicide-conjugating plant glutathione S-transferases and a novel active site architecture. *J. Mol. Biol.* 255, 289–309. doi: 10.1006/jmbi.1996.0024

Rinaldi, C., Kohler, A., Frey, P., Duchaussoy, F., Ningre, N., Couloux, A., et al. (2007). Transcript profiling of poplar leaves upon infection with compatible and incompatible strains of the foliar rust *Melampsora larici-populina*. *Plant Physiol.* 144, 347–366. doi: 10.1104/pp.106.094987

Sappl, P. G., Carroll, A. J., Clifton, R., Lister, R., Whelan, J., Harvey Millar, A., et al. (2009). The Arabidopsis glutathione transferase gene family displays complex stress regulation and co-silencing multiple genes results in altered metabolic sensitivity to oxidative stress. *Plant J.* 58, 53–68. doi: 10.1111/j.1365-313X.2008.03761.x

Sappl, P. G., Onate-Sanchez, L., Singh, K. B., and Millar, A. H. (2004). Proteomic analysis of glutathione S -transferases of *Arabidopsis thaliana* reveals differential salicylic acid-induced expression of the plant-specific phi and tau classes. *Plant Mol. Biol.* 54, 205–219. doi: 10.1023/B:PLAN.0000028786.57439.b3

Smith, A. P., Deridder, B. P., Guo, W. J., Seeley, E. H., Regnier, F. E., and Goldsbrough, P. B. (2004). Proteomic analysis of Arabidopsis glutathione S-transferases from benoxacor- and copper-treated seedlings. *J. Biol. Chem.* 279, 26098–26104. doi: 10.1074/jbc.M402807200

Smith, A. P., Nourizadeh, S. D., Peer, W. A., Xu, J., Bandyopadhyay, A., Murphy, A. S., et al. (2003). Arabidopsis AtGSTF2 is regulated by ethylene and auxin, and encodes a glutathione S-transferase that interacts with flavonoids. *Plant J.* 36, 433–442. doi: 10.1046/j.1365-313X.2003.01890.x

Su, T., Xu, J., Li, Y., Lei, L., Zhao, L., Yang, H., et al. (2011). Glutathione-indole-3-acetonitrile is required for camalexin biosynthesis in *Arabidopsis thaliana*. *Plant Cell* 23, 364–380. doi: 10.1105/tpc.110.079145

Thatcher, L. F., Carrie, C., Andersson, C. R., Sivasithamparam, K., Whelan, J., and Singh, K. B. (2007). Differential gene expression and subcellular targeting of Arabidopsis glutathione S-transferase F8 is achieved through alternative transcription start sites. *J. Biol. Chem.* 282, 28915–28928. doi: 10.1074/jbc.M702207200

Thuillier, A., Roret, T., Favier, F., Gelhaye, E., Jacquot, J. P., Didierjean, C., et al. (2013). Atypical features of a Ure2p glutathione transferase from *Phanerochaete chrysosporium*. *FEBS Lett.* 587, 2125–2130. doi: 10.1016/j.febslet.2013.05.031

Vagin, A., and Teplyakov, A. (2010). Molecular replacement with MOLREP. *Acta Crystallogr. D Biol. Crystallogr.* 66, 22–25. doi: 10.1107/S0907444909042589

Vieira Dos Santos, C., Cuine, S., Rouhier, N., and Rey, P. (2005). The Arabidopsis plastidic methionine sulfoxide reductase B proteins. Sequence and activity characteristics, comparison of the expression with plastidic methionine sulfoxide reductase A, and induction by photooxidative stress. *Plant Physiol.* 138, 909–922. doi: 10.1104/pp.105.062430

Wagner, U., Edwards, R., Dixon, D. P., and Mauch, F. (2002). Probing the diversity of the Arabidopsis glutathione S-transferase gene family. *Plant Mol. Biol.* 49, 515–532. doi: 10.1023/A:1015557300450

Whitbread, A. K., Masoumi, A., Tetlow, N., Schmuck, E., Coggan, M., and Board, P. G. (2005). Characterization of the omega class of glutathione transferases. *Meth. Enzymol.* 401, 78–99. doi: 10.1016/S0076-6879(05)01005-0

Winn, M. D., Ballard, C. C., Cowtan, K. D., Dodson, E. J., Emsley, P., Evans, P. R., et al. (2011). Overview of the CCP4 suite and current developments. *Acta Crystallogr. D Biol. Crystallogr.* 67, 235–242. doi: 10.1107/S0907444910045749

Zhang, S., Feng, L., Jiang, H., Ma, W., Korpelainen, H., and Li, C. (2012). Biochemical and proteomic analyses reveal that *Populus cathayana* males and females have different metabolic activities under chilling stress. *J. Proteome Res.* 11, 5815–5826. doi: 10.1021/pr3005953

Conflict of Interest Statement: The authors declare that the research was conducted in the absence of any commercial or financial relationships that could be construed as a potential conflict of interest.

Permissions

All chapters in this book were first published in FSMPGDSR, by Frontiers Media SA; hereby published with permission under the Creative Commons Attribution License or equivalent. Every chapter published in this book has been scrutinized by our experts. Their significance has been extensively debated. The topics covered herein carry significant findings which will fuel the growth of the discipline. They may even be implemented as practical applications or may be referred to as a beginning point for another development.

The contributors of this book come from diverse backgrounds, making this book a truly international effort. This book will bring forth new frontiers with its revolutionizing research information and detailed analysis of the nascent developments around the world.

We would like to thank all the contributing authors for lending their expertise to make the book truly unique. They have played a crucial role in the development of this book. Without their invaluable contributions this book wouldn't have been possible. They have made vital efforts to compile up to date information on the varied aspects of this subject to make this book a valuable addition to the collection of many professionals and students.

This book was conceptualized with the vision of imparting up-to-date information and advanced data in this field. To ensure the same, a matchless editorial board was set up. Every individual on the board went through rigorous rounds of assessment to prove their worth. After which they invested a large part of their time researching and compiling the most relevant data for our readers.

The editorial board has been involved in producing this book since its inception. They have spent rigorous hours researching and exploring the diverse topics which have resulted in the successful publishing of this book. They have passed on their knowledge of decades through this book. To expedite this challenging task, the publisher supported the team at every step. A small team of assistant editors was also appointed to further simplify the editing procedure and attain best results for the readers.

Apart from the editorial board, the designing team has also invested a significant amount of their time in understanding the subject and creating the most relevant covers. They scrutinized every image to scout for the most suitable representation of the subject and create an appropriate cover for the book.

The publishing team has been an ardent support to the editorial, designing and production team. Their endless efforts to recruit the best for this project, has resulted in the accomplishment of this book. They are a veteran in the field of academics and their pool of knowledge is as vast as their experience in printing. Their expertise and guidance has proved useful at every step. Their uncompromising quality standards have made this book an exceptional effort. Their encouragement from time to time has been an inspiration for everyone.

The publisher and the editorial board hope that this book will prove to be a valuable piece of knowledge for researchers, students, practitioners and scholars across the globe.

List of Contributors

Felix Hirschmann, Florian Krause and Jutta Papenbrock
Institute of Botany, Leibniz University Hannover, Hannover, Germany

Tahereh Aghajanzadeh and Luit J. De Kok
Laboratory of Plant Physiology, University of Groningen, Groningen, Netherlands

Malcolm J. Hawkesford
Plant Biology and Crop Science Department, Rothamsted Research, Harpenden, UK

Anna Wawrzyńska and Agnieszka Sirko
Institute of Biochemistry and Biophysics, Polish Academy of Sciences, Warsaw, Poland

Anna Koprivova and Stanislav Kopriva
Botanical Institute and Cluster of Excellence on Plant Sciences, Cologne Biocenter, University of Cologne, Cologne, Germany

Bernd Zechmann
Center for Microscopy and Imaging, Baylor University, Waco, TX, USA

Alexander Calderwood and Richard J. Morris
Department of Computational and Systems Biology, John Innes Centre, Norwich, UK

Stanislav Kopriva
Botanical Institute and Cluster of Excellence on Plant Sciences, University of Cologne, Cologne Biocenter, Cologne, Germany

Jacinta Gahan and Achim Schmalenberger
Department of Life Sciences, University of Limerick, Limerick, Ireland

Alexandra Girondé, Lucie Dubousset, Jacques Trouverie, Philippe Etienne and Jean-Christophe Avice
Normandie University, Caen, France
Université de Caen Basse Normandie, UMR INRA-UCBN 950 Ecophysiologie Végétale, Agronomie and Nutritions N.C.S., Caen, France
Institut National de la Recherche Agronomique, UMR INRA-UCBN 950 Ecophysiologie Végétale, Agronomie and Nutritions N.C.S., Caen, France

Silke C. Weckopp and Stanislav Kopriva
Botanical Institute and Cluster of Excellence on Plant Sciences, Cologne Biocenter, University of Cologne, Cologne, Germany

Sudhakar Pandurangan and Frédéric Marsolais
Department of Biology, University of Western Ontario, London, ON, Canada
Genomics and Biotechnology, Southern Crop Protection and Food Research Centre, Agriculture and Agri-Food Canada, London, ON, Canada

Ronald Beyaert and Kenneth L. Conn
Genomics and Biotechnology, Southern Crop Protection and Food Research Centre, Agriculture and Agri-Food Canada, London, ON, Canada

Anfu Hou and Mark Sandercock
Cereal Research Centre Morden, Agriculture and Agri-Food Canada, Morden, MB, Canada

Karine Gallardo, Christine Le Signor and Vanessa Vernoud
Institut National de la Recherche Agronomique, UMR1347 Agroécologie, Dijon, France

Pierre-Emmanuel Courty
Zurich-Basel Plant Science Center, Department of Environmental Sciences, Botany, University of Basel, Basel, Switzerland

Daniel Wipf
Université de Bourgogne, UMR1347 Agroécologie, Dijon, France

Naoko Yoshimoto, Ayami Yabe, Yuka Sugino, Soichiro Murakami, Niti Sai-ngam
Graduate School of Pharmaceutical Sciences, Chiba University, Chiba, Japan

Shin-ichiro Sumi
Research Planning Department, Wakunaga Pharmaceutical Company, Akitakata, Japan

Tadamitsu Tsuneyoshi
Central Research Institute, Wakunaga Pharmaceutical Company, Akitakata, Japan

Kazuki Saito
Graduate School of Pharmaceutical Sciences, Chiba University, Chiba, Japan
RIKEN Center for Sustainable Resource Science, Yokohama, Japan

Atsushi Sato, Mikio Tsuzuki and Norihiro Sato
School of Life Sciences, Tokyo University of Pharmacy and Life Sciences, Hachioji, Japan
Japan Science and Technology Agency, Core Research for Evolutionary Science and Technology, Chiyoda-ku, Japan

Rie Matsumura and Naomi Hoshino
School of Life Sciences, Tokyo University of Pharmacy and Life Sciences, Hachioji, Japan

Agnieszka Sirko, Anna Wawrzyńska, Milagros Collados Rodríguez and Paweł Sęktas
Institute of Biochemistry and Biophysics, Polish Academy of Sciences, Warsaw, Poland

Won-Seok Kim and Hari B. Krishnan
Plant Genetics Research Unit, Agricultural Research Service, U.S. Department of Agriculture, University of Missouri, Columbia, MO, USA

Joseph M. Jez
Department of Biology, Washington University, St. Louis, MO, USA

Laura Prioretti
Laboratory of Algal and Plant Physiology, Dipartimento di Scienze della Vita e dell'Ambiente, Università Politecnica delle Marche, Ancona, Italy

Brigitte Gontero
Aix-Marseille Université Centre National de la Recherche Scientifique, BL' Unité de Bioénergétique et Ingénierie des Protéines UMR 7281, Marseille, France

Ruediger Hell
Centre for Organismal Studies, University of Heidelberg, Heidelberg, Germany

Mario Giordano
Laboratory of Algal and Plant Physiology, Dipartimento di Scienze della Vita e dell'Ambiente, Università Politecnica delle Marche, Ancona, Italy
Institute of Microbiology Academy of Sciences of the Czech Republic, Trebon, Czech Republic

Anne-Sophie Bohrer, Nicholas Rykulski and Hideki Takahashi
Department of Biochemistry and Molecular Biology, Michigan State University, East Lansing, MI, USA

Naoko Yoshimoto and Ai Sekiguchi
Graduate School of Pharmaceutical Sciences, Chiba University, Chiba, Japan

Michael J. Considine
School of Plant Biology, and The UWA Institute of Agriculture, University of Western Australia, Crawley, WA, Australia
Centre for Plant Sciences, School of Biology, Faculty of Biological Sciences, University of Leeds, Leeds, Yorkshire, UK
Irrigated Agriculture and Diversification, Department of Agriculture and Food Western Australia, South Perth, WA, Australia

Christine H. Foyer
School of Plant Biology, and The UWA Institute of Agriculture, University of Western Australia, Crawley, WA, Australia
Centre for Plant Sciences, School of Biology, Faculty of Biological Sciences, University of Leeds, Leeds, Yorkshire, UK

Anna Speiser, Stefan Haberland, Markus Wirtz and Rüdiger Hell
Plant Molecular Biology, Centre for Organismal Studies, University of Heidelberg, Heidelberg, Germany

Mutsumi Watanabe
Molecular Plant Physiology, Max Planck Institute of Molecular Plant Physiology, Potsdam, Germany

Karl-Josef Dietz
Plant Biochemistry and Physiology, University of Bielefeld, Bielefeld, Germany

Kazuki Saito
RIKEN Center for Sustainable Resource Science, Yokohama, Japan
Graduate School of Pharmaceutical Sciences, Chiba University, Chiba, Japan

Elke Bloem, Silvia Haneklaus and Ewald Schnug
Federal Research Centre for Cultivated Plants, Julius Kühn-Institute, Institute for Crop and Soil Science, Braunschweig, Germany

Henri Pégeot, Benjamin Petre, Sébastien Duplessis, Arnaud Hecker and Nicolas Rouhier
Interactions Arbres - Microorganismes, Université de Lorraine, UMR1136, Vandoeuvre-lès-Nancy, France
INRA, Interactions Arbres - Microorganismes, UMR1136, Champenoux, France

Cha San Koh, Sandrine Mathiot and Claude Didierjean
Faculté des Sciences et Technologies, Université de Lorraine, CRM2, Equipe BioMod, UMR 7036, Vandoeuvre-lès-Nancy, France
Faculté des Sciences et Technologies, CNRS, CRM2, Equipe BioMod, UMR 7036, Vandoeuvre-lès-Nancy, France

Index

www.ingramcontent.com/pod-product-compliance
Lightning Source LLC
LaVergne TN
LVHW060940220926
848485LV00015B/1142
* 9 7 8 1 6 3 9 8 9 5 0 0 7 *